Tom Holthe
september 2013

Petroleum Production Systems

SECOND EDITION

Petroleum Production Systems

SECOND EDITION

Michael J. Economides
A. Daniel Hill
Christine Ehlig-Economides
Ding Zhu

Upper Saddle River, NJ • Boston • Indianapolis • San Francisco
New York • Toronto • Montreal • London • Munich • Paris • Madrid
Capetown • Sydney • Tokyo • Singapore • Mexico City

Many of the designations used by manufacturers and sellers to distinguish their products are claimed as trademarks. Where those designations appear in this book, and the publisher was aware of a trademark claim, the designations have been printed with initial capital letters or in all capitals.

The authors and publisher have taken care in the preparation of this book, but make no expressed or implied warranty of any kind and assume no responsibility for errors or omissions. No liability is assumed for incidental or consequential damages in connection with or arising out of the use of the information or programs contained herein.

The publisher offers excellent discounts on this book when ordered in quantity for bulk purchases or special sales, which may include electronic versions and/or custom covers and content particular to your business, training goals, marketing focus, and branding interests. For more information, please contact:

U.S. Corporate and Government Sales
(800) 382-3419
corpsales@pearsontechgroup.com

For sales outside the United States please contact:

International Sales
international@pearson.com

Visit us on the Web: informit.com/ph

Library of Congress Cataloging-in-Publication Data

Petroleum production systems / Michael J. Economides. — 2nd ed.
p. cm.
Includes bibliographical references and index.
ISBN 0-13-703158-0 (hardcover : alk. paper)
1. Oil fields—Production methods. 2. Petroleum engineering.
I. Economides, Michael J.
TN870.E29 2013
622'.338—dc23 2012022357

ISBN-13: 978-0-13-703158-0
ISBN-10: 0-13-703158-0

Text printed in the United States on recycled paper at Courier in Westford, Masscachusetts.
Second printing, January, 2013

Executive Editor:
Bernard Goodwin

Managing Editor:
John Fuller

Project Editor:
Elizabeth Ryan

Packager:
Laserwords

Copy Editor:
Laura Patchkofsky

Indexer:
Constance Angelo

Proofreader:
Susan Gall

Developmental Editor:
Michael Thurston

Cover Designer:
Chuti Prasertsith

Compositor:
Laserwords

Contents

Foreword

I have waited on this book for the last 10 years. It is a modernized version of the classic first edition, thousands of copies of which have been distributed to my former trainees, engineers, and associates. The authors of the book have worked with me in a number of capacities for 25 years and we have become kindred spirits both in how we think about oil and gas production enhancement and, especially, in knowing how bad production management can be, even in the most unexpected places and companies.

It is a comprehensive book that describes the "production system," or what I refer to as "nodal analysis," artificial lift, well diagnosis, matrix stimulation, hydraulic fracturing, and sand control.

There are some important points that are made in this book, which I have made repeatedly in the past:

1. To increase field production, well improvement can be more effective than infill drilling, especially when the new wells are just as suboptimum as existing wells. We demonstrated this while I was managing Yukos E&P in Russia. During that time appropriate production enhancement actions improved field production by more than 15% even after stopping all drilling for as long as a year.
2. In conventional reservoirs, optimized well completions do not sacrifice ultimate field recovery as long as they are achieved with adequate reservoir pressure support from either natural gas cap or water drive mechanisms or through injection wells.
3. Many, if not most, operators fail to address well performance, and few wells are produced at their maximum flow potential. This book takes great steps to show that proper production optimization is far more important to success than just simply executing blindly well completions and even stimulation practices. In particular, I consider the Unified Fracture Design (UFD) approach, the brainchild of the lead author, to be the only coherent approach to hydraulic fracture design. I have been using it exclusively and successfully in all my hydraulic fracture design work.

This book provides not only best practices but also the rationale for new activities. The strategies shown in this book explain why unconventional oil and gas reservoirs are successfully produced today.

The book fills a vacuum in the industry and has come not a moment too soon.

—Joe Mach
Inventor, Nodal Analysis
Former Executive VP, Yukos
Former VP, Schlumberger

Preface

Since the first edition of this book appeared in 1994, many advances in the practice of petroleum production engineering have occurred. The objective of this book is the same as for the first edition: to provide a comprehensive and relatively advanced textbook in petroleum production engineering, that suffices as a terminal exposure to senior undergraduates or an introduction to graduate students. This book is also intended to be used in industrial training to enable nonpetroleum engineers to understand the essential elements of petroleum production. Numerous technical advances in the years since the first edition have led to the extensive revisions that readers will notice in this second edition. In particular, widespread use of horizontal wells and much broader application of hydraulic fracturing have changed the face of production practices and justified critical updating of the text. The authors have benefited from wide experience in both university and industrial settings. Our areas of interest are complementary and ideally suited for this book, spanning classical production engineering, well testing, production logging, artificial lift, and matrix and hydraulic fracture stimulation. We have been contributors in these areas for many years. Among the four of us, we have taught petroleum production engineering to literally thousands of students and practicing engineers using the first edition of this book, both in university classes and in industry short courses, and this experience has been one of the key guiding factors in the creation of the second edition.

This book offers a structured approach toward the goal defined above. Chapters 2–4 present the inflow performance for oil, two-phase, and gas reservoirs. Chapter 5 deals with complex well architecture such as horizontal and multilateral wells, reflecting the enormous growth of this area of production engineering since the first edition of the book. Chapter 6 deals with the condition of the near-wellbore zone, such as damage, perforations, and gravel packing. Chapter 7 covers the flow of fluids to the surface. Chapter 8 describes the surface flow system, flow in horizontal pipes, and flow in horizontal wells. Combination of inflow performance and well performance versus time, taking into account single-well transient flow and material balance, is shown in Chapters 9 and 10. Therefore, Chapters 1–10 describe the workings of the reservoir and well systems.

Gas lift is outlined in Chapter 11, and mechanical lift in Chapter 12. For an appropriate production engineering remedy it is essential that well and reservoir diagnosis be done. Chapter 13 presents the state-of-the-art in modern diagnosis that includes well testing, production logging, and well monitoring with permanent downhole instruments.

From the well diagnosis it can be concluded whether the well is in need of matrix stimulation, hydraulic fracturing, artificial lift, combinations of the above, or none. Matrix stimulation for all major types of reservoirs is presented in Chapters 14, 15, and 16, while hydraulic fracturing is treated in Chapters 17 and 18. Chapter 19 is a new chapter dealing with advances in sand management.

To simplify the presentation of realistic examples, data for three characteristic reservoir types—an undersaturated oil reservoir, a saturated oil reservoir, and a gas reservoir—are presented in the Appendixes. These data sets are used throughout the book.

Revising this textbook to include the primary production engineering of the past 20 years has been a considerable task, requiring a long and concerted (and only occasionally contentious!) effort from the authors. We have also benefited from the efforts of many of our graduate students and support staff. Discussions with many of our colleagues in industry and academia have also been a key to the completion of the book. We would like to thank in particular the contributions of Dr. Paul Bommer, who provided some very useful material on artificial lift; Dr. Chen Yang, who assisted with some of the new material on carbonate acidizing; Dr. Tom Blasingame and Mr. Chih Chen, who shared well data used as pressure buildup and production data examples; Mr. Tony Rose, who created the graphics; and Ms. Katherine Brady and Mr. Imran Ali for their assistance in the production of this second edition.

As we did for the first edition, we acknowledge the many colleagues, students, and our own professors who contributed to our efforts. In particular, feedback from all of our students in petroleum production engineering courses has guided our revision of the first edition of this text, and we thank them for their suggestions, comments, and contributions.

We would like to gratefully acknowledge the following organizations and persons for permitting us to reprint some of the figures and tables in this text: for Figs. 3-2, 3-3, 5-2, 5-4, 5-7, 6-15, 6-16, 6-18, 6-19, 6-20, 6-21, 6-22, 624, 6-24, 6-26, 6-27, 6-28, 6-29, 7-1, 7-9, 7-12, 7-13, 7-13, 7-14, 8-1, 8-4, 8-6, 8-7, 8-17, 13-13, 13-19, 14-3, 15-1, 15-2, 15-4, 15-7, 15-10, 15-12, 16-1, 16-2, 16-4, 16-5, 16-6, 16-7, 16-8, 16-14, 16-16, 16-17, 16-20, 17-2, 17-3, 17-6, 17-11, 17-12, 17-13, 17-14, 17-15, 17-16, 17-17, 17-18, 17-19, 18-20, 18-21, 18-22, 18-23, 18-25, 18-26, 19-1, 19-6, 19-7, 19-8, 19-9, 19-10, 19-17, 19-18, 19-19, 19-20, 19-21a, 19-21b, and 19-22, the Society of Petroleum Engineers; for Figs. 6-13, 6-14, 13-2, 13-18, 18-13, 18-14, 18-19, 19-2, and 19-3, Schlumberger; for Figs. 6-23, 12-5, 12-6, 15-3, 15-6, 16-17, and 16-19, Prentice Hall; for Figs. 8-3, 8-14, 12-15, 12-16, and 16-13, Elsevier Science Publishers; for Figs. 4-3, 19-12, 19-13, 19-14, and 19-15, Gulf Publishing Co., Houston, TX; for Figs. 13-5, 13-6, 13-8, 13-9, 13-11, and 13-12, Hart Energy, Houston, TX; for Figs. 7-11 and 8-5, the American Institute of Chemical Engineers; for Figs. 7-6 and 7-7, the American Society of Mechanical Engineers; for Figs. 8-11 and Table 8-1, Crane Co., Stamford, CT; for Figs. 12-8, 12-9, and 12-10, Editions Technip, Paris, France; for Fig. 2-3, the American Institute of Mining, Metallurgical & Petroleum Engineers; for Fig. 3-4, McGraw-Hill; for Fig. 7-10, World Petroleum Council; for Fig. 12-11, Baker Hughes; for Fig. 13-1, PennWell Publishing Co., Tulsa, OK; for Fig. 13-3, the Society of Petrophysicists and Well Log Analysts; for Fig. 18-16, Carbo Ceramics, Inc.; for Figs. 12-1, 12-2, and 12-7, Dr. Michael Golan and Dr. Curtis Whitson; for Fig. 6-17, Dr. Kenji Furui; for Fig. 8-8, Dr. James P. Brill; for Fig. 15-8, Dr. Eduardo Ponce da Motta; for Figs. 18-11 and 18-15, Dr. Harold Brannon. Used with permission, all rights reserved.

About the Authors

MICHAEL J. ECONOMIDES

A chemical and petroleum engineer and an expert on energy geopolitics, Dr. Michael J. Economides is a professor at the University of Houston and managing partner of Economides Consultants, Inc., with a wide range of industrial consulting, including major retainers by Fortune 500 companies and national oil companies. He has written 15 textbooks and almost 300 journal papers and articles.

September 6, 1949

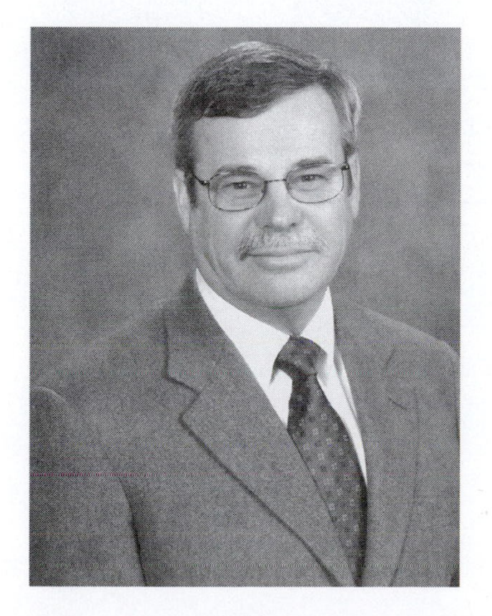

A. DANIEL HILL

Dr. A. Daniel Hill is professor and holder of the Noble Chair in Petroleum Engineering at Texas A&M University. The author of 150 papers, three books, and five patents, he teaches and conducts research in the areas of production engineering, well completions, well stimulation, production logging, and complex well performance.

CHRISTINE EHLIG-ECONOMIDES

Dr. Christine Ehlig-Economides holds the Albert B. Stevens Endowed Chair and is professor of petroleum engineering at Texas A&M University and Senior Partner of Economides Consultants, Inc. Dr. Ehlig-Economides provides industry consulting and training and supervises student research in well production and reservoir analysis. She has authored more than 70 papers and journal articles and is a member of the U. S. National Academy of Engineering.

DING ZHU

Dr. Ding Zhu is associate professor and holder of the W. D. Von Gonten Faculty Fellowship in Petroleum Engineering at Texas A&M University. Dr. Zhu's main research areas include general production engineering, well stimulation, and complex well performance. Dr. Zhu is a coauthor of more than 100 technical papers and one book.

CHAPTER 1

The Role of Petroleum Production Engineering

1.1 Introduction

Petroleum production involves two distinct but intimately connected general systems: the reservoir, which is a porous medium with unique storage and flow characteristics; and the artificial structures, which include the well, bottomhole, and wellhead assemblies, as well as the surface gathering, separation, and storage facilities.

Production engineering is that part of petroleum engineering that attempts to maximize production (or injection) in a cost-effective manner. In the 15 years that separated the first and second editions of this textbook worldwide production enhancement, headed by hydraulic fracturing, has increased tenfold in constant dollars, becoming the second largest budget item of the industry, right behind drilling. Complex well architecture, far more elaborate than vertical or single horizontal wells, has also evolved considerably since the first edition and has emerged as a critical tool in reservoir exploitation.

In practice one or more wells may be involved, but in distinguishing production engineering from, for example, reservoir engineering, the focus is often on specific wells and with a short-time intention, emphasizing production or injection optimization. In contrast, reservoir engineering takes a much longer view and is concerned primarily with recovery. As such, there may be occasional conflict in the industry, especially when international petroleum companies, whose focus is accelerating and maximizing production, have to work with national oil companies, whose main concerns are to manage reserves and long-term exploitation strategies.

Production engineering technologies and methods of application are related directly and interdependently with other major areas of petroleum engineering, such as formation evaluation, drilling, and reservoir engineering. Some of the most important connections are summarized below.

Modern formation evaluation provides a composite reservoir description through three-dimensional (3-D) seismic, interwell log correlation and well testing. Such description leads to the identification of geological flow units, each with specific characteristics. Connected flow units form a reservoir.

Drilling creates the all-important well, and with the advent of directional drilling technology it is possible to envision many controllable well configurations, including very long horizontal sections and multilateral, multilevel, and multibranched wells, targeting individual flow units. The drilling of these wells is never left to chance but, instead, is guided by very sophisticated measurements while drilling (MWD) and logging while drilling (LWD). Control of drilling-induced, near-wellbore damage is critical, especially in long horizontal wells.

Reservoir engineering in its widest sense overlaps production engineering to a degree. The distinction is frequently blurred both in the context of study (single well versus multiple well) and in the time duration of interest (long term versus short term). Single-well performance, undeniably the object of production engineering, may serve as a boundary condition in a fieldwide, long-term reservoir engineering study. Conversely, findings from the material balance calculations or reservoir simulation further define and refine the forecasts of well performance and allow for more appropriate production engineering decisions.

In developing a petroleum production engineering thinking process, it is first necessary to understand important parameters that control the performance and the character of the system. Below, several definitions are presented.

1.2 Components of the Petroleum Production System

1.2.1 Volume and Phase of Reservoir Hydrocarbons

1.2.1.1 Reservoir

The reservoir consists of one or several interconnected geological flow units. While the shape of a well and converging flow have created in the past the notion of radial flow configuration, modern techniques such as 3-D seismic and new logging and well testing measurements allow for a more precise description of the shape of a geological flow unit and the ensuing production character of the well. This is particularly true in identifying lateral and vertical boundaries and the inherent heterogeneities.

Appropriate reservoir description, including the extent of heterogeneities, discontinuities, and anisotropies, while always important, has become compelling after the emergence of horizontal wells and complex well architecture with total lengths of reservoir exposure of many thousands of feet.

Figure 1-1 is a schematic showing two wells, one vertical and the other horizontal, contained within a reservoir with potential lateral heterogeneities or discontinuities (sealing faults), vertical boundaries (shale lenses), and anisotropies (stress or permeability).

While appropriate reservoir description and identification of boundaries, heterogeneities, and anisotropies is important, it is somewhat forgiving in the presence of only vertical wells. These issues become critical when horizontal and complex wells are drilled.

The encountering of lateral discontinuities (including heterogeneous pressure depletion caused by existing wells) has a major impact on the expected complex well production. The well branch trajectories vis à vis the azimuth of directional properties also has a great effect on well production. Ordinarily, there would be only one set of optimum directions.

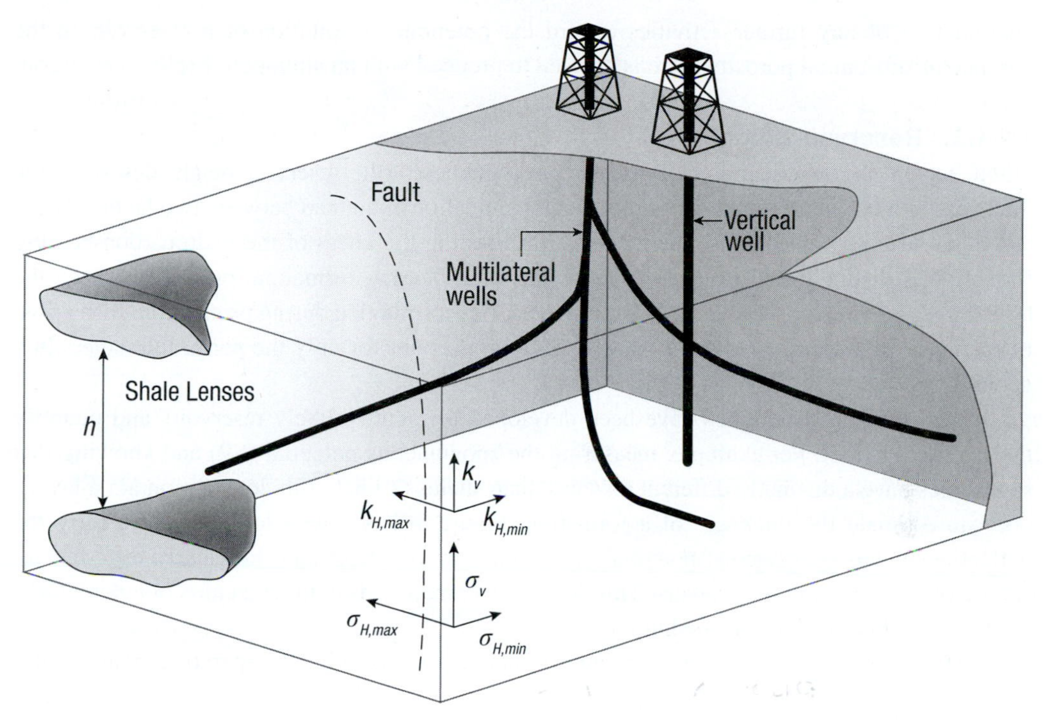

Figure 1-1 Common reservoir heterogeneities, anisotropies, discontinuities, and boundaries affecting the performance of vertical, horizontal, and complex-architecture wells.

Understanding the geological history that preceded the present hydrocarbon accumulation is essential. There is little doubt that the best petroleum engineers are those who understand the geological processes of deposition, fluid migration, and accumulation. Whether a reservoir is an anticline, a fault block, or a channel sand not only dictates the amount of hydrocarbon present but also greatly controls well performance.

1.2.1.2 Porosity

All of petroleum engineering deals with the exploitation of fluids residing within porous media. Porosity, simply defined as the ratio of the pore volume, V_p, to the bulk volume, V_b,

$$\phi = \frac{V_p}{V_b} \tag{1-1}$$

is an indicator of the amount of fluid in place. Porosity values vary from over 0.3 to less than 0.1. The porosity of the reservoir can be measured based on laboratory techniques using reservoir cores or with field measurements including logs and well tests. Porosity is one of the very first measurements obtained in any exploration scheme, and a desirable value is essential for the continuation of any further activities toward the potential exploitation of a reservoir. In the absence of substantial porosity there is no need to proceed with an attempt to exploit a reservoir.

1.2.1.3 Reservoir Height

Often known as "reservoir thickness" or "pay thickness," the reservoir height describes the thickness of a porous medium in hydraulic communication contained between two layers. These layers are usually considered impermeable. At times the thickness of the hydrocarbon-bearing formation is distinguished from an underlaying water-bearing formation, or aquifer. Often the term "gross height" is employed in a multilayered, but co-mingled during production, formation. In such cases the term "net height" may be used to account for only the permeable layers in a geologic sequence.

Well logging techniques have been developed to identify likely reservoirs and quantify their vertical extent. For example, measuring the spontaneous potential (SP) and knowing that sandstones have a distinctly different response than shales (a likely lithology to contain a layer), one can estimate the thickness of a formation. Figure 1-2 is a well log showing clearly the deflection of the spontaneous potential of a sandstone reservoir and the clearly different response of the adjoining shale layers. This deflection corresponds to the thickness of a *potentially* hydrocarbon-bearing, porous medium.

The presence of satisfactory net reservoir height is an additional imperative in any exploration activity.

1.2.1.4 Fluid Saturations

Oil and/or gas are never alone in "saturating" the available pore space. Water is always present. Certain rocks are "oil-wet," implying that oil molecules cling to the rock surface. More frequently, rocks are "water-wet." Electrostatic forces and surface tension act to create these wettabilities, which may change, usually with detrimental consequences, as a result of injection of fluids, drilling, stimulation, or other activity, and in the presence of surface-acting chemicals. If the water is present but does not flow, the corresponding water saturation is known as "connate" or "interstitial." Saturations larger than this value would result in free flow of water along with hydrocarbons.

Petroleum hydrocarbons, which are mixtures of many compounds, are divided into oil and gas. Any mixture depending on its composition and the conditions of pressure and temperature may appear as liquid (oil) or gas or a mixture of the two.

Frequently the use of the terms *oil* and *gas* is blurred. Produced oil and gas refer to those parts of the total mixture that would be in liquid and gaseous states, respectively, after surface separation. Usually the corresponding pressure and temperature are "standard conditions," that is, usually (but not always) 14.7 psi and 60° F.

Flowing oil and gas in the reservoir imply, of course, that either the initial reservoir pressure or the induced flowing bottomhole pressures are such as to allow the concurrent presence of two phases. Temperature, except in the case of high-rate gas wells, is for all practical purposes constant.

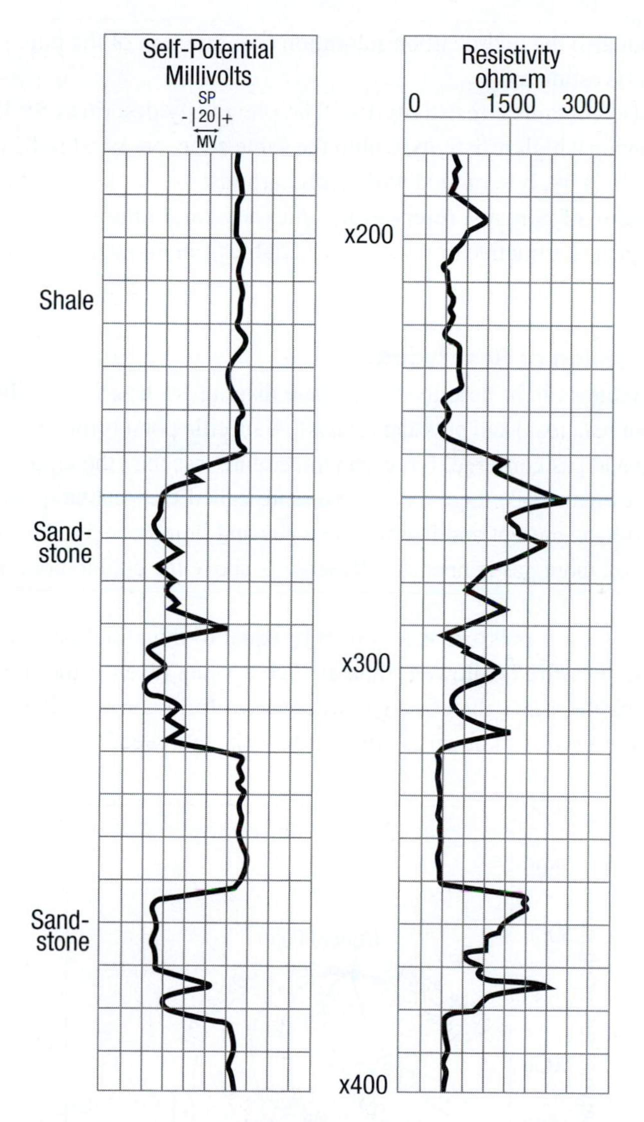

Figure 1-2 Spontaneous potential and electrical resistivity logs identifying sandstones versus shales, and water-bearing versus hydrocarbon-bearing formations.

An attractive hydrocarbon saturation is the third critical variable (along with porosity and reservoir height) to be determined before a well is tested or completed. A classic method, currently performed in a variety of ways, is the measurement of the formation electrical resistivity. Knowing that formation brines are good conductors of electricity (i.e., they have poor resistivity) and hydrocarbons are the opposite, a measurement of this electrical property in a porous formation of sufficient height can detect the presence of hydrocarbons. With proper calibration, not just the presence but also the hydrocarbon saturation (i.e., fraction of the pore space occupied by hydrocarbons) can be estimated.

Figure 1-2 also contains a resistivity log. The previously described SP log along with the resistivity log, showing a high resistivity within the same zone, are good indicators that the identified porous medium is likely saturated with hydrocarbons.

The combination of porosity, reservoir net thickness, and saturations is essential in deciding whether a prospect is attractive or not. These variables can allow the estimation of hydrocarbons near the well.

1.2.1.5 Classification of Reservoirs

All hydrocarbon mixtures can be described by a phase diagram such as the one shown in Figure 1-3. Plotted are temperature (x axis) and pressure (y axis). A specific point is the *critical point*, where the properties of liquid and gas converge. For each temperature less than the critical-point temperature (to the left of T_c in Figure 1-3) there exists a pressure called the "bubble-point" pressure, above which only liquid (oil) is present and below which gas and liquid coexist. For lower pressures (at constant temperature), more gas is liberated. Reservoirs above the bubble-point pressure are called "undersaturated."

If the initial reservoir pressure is less than or equal to the bubble-point pressure, or if the flowing bottomhole pressure is allowed to be at such a value (even if the initial reservoir pressure is above the bubble point), then free gas will at least form and will likely flow in the reservoir. This type of a reservoir is known as "two-phase" or "saturated."

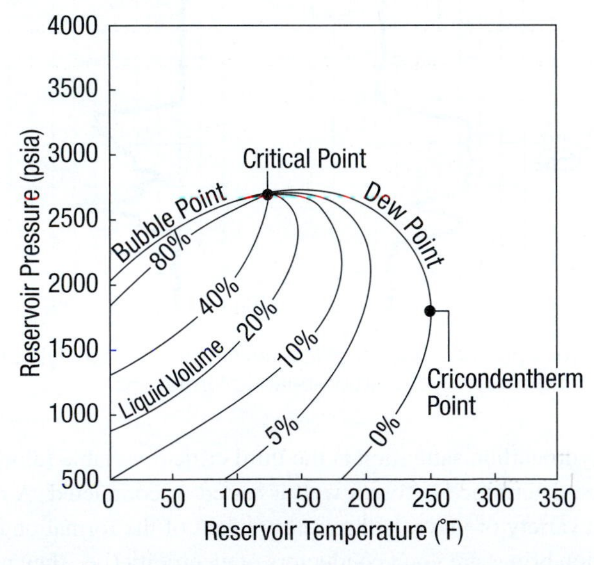

Figure 1-3 Oilfield hydrocarbon phase diagram showing bubble-point and dew-point curves, lines of constant-phase distribution, region of retrograde condensation, and the critical and cricondentherm points.

For temperatures larger than the critical point (to the right of T_c in Figure 1-3), the curve enclosing the two-phase envelop is known as the "dew-point" curve. Outside, the fluid is gas, and reservoirs with these conditions are "lean" gas reservoirs.

The maximum temperature of a two-phase envelop is known as the "cricondentherm." Between these two points there exists a region where, because of the shape of the gas saturation curves, as the pressure decreases, liquid or "condensate" is formed. This happens until a limited value of the pressure, after which further pressure reduction results in revaporization. The region in which this phenomenon takes place is known as the "retrograde condensation" region, and reservoirs with this type of behavior are known as "retrograde condensate reservoirs."

Each hydrocarbon reservoir has a characteristic phase diagram and resulting physical and thermodynamic properties. These are usually measured in the laboratory with tests performed on fluid samples obtained from the well in a highly specialized manner. Petroleum thermodynamic properties are known collectively as *PVT* (*pressure–volume–temperature*) properties.

1.2.1.6 Areal Extent

Favorable conclusions on the porosity, reservoir height, fluid saturations, and pressure (and implied phase distribution) of a petroleum reservoir, based on single well measurements, are insufficient for both the decision to develop the reservoir and for the establishment of an appropriate exploitation scheme.

Advances in 3-D and wellbore seismic techniques, in combination with well testing, can increase greatly the region where knowledge of the reservoir extent (with height, porosity, and saturations) is possible. Discontinuities and their locations can be detected. As more wells are drilled, additional information can enhance further the knowledge of the reservoir's peculiarities and limits.

The areal extent is essential in the estimation of the "original-oil (or gas)-in-place." The hydrocarbon volume, V_{HC}, in reservoir cubic ft is

$$V_{HC} = Ah\phi(1 - S_w) \quad \textbf{(1-2)}$$

where A is the areal extent in ft^2, h is the reservoir thickness in ft, ϕ is the porosity, and S_w is the water saturation. (Thus, $1 - S_w$ is the hydrocarbon saturation.) The porosity, height, and saturation can of course vary within the areal extent of the reservoir.

Equation (1-2) can lead to the estimation of the oil or gas volume under standard conditions after dividing by the oil formation volume factor, B_o, or the gas formation volume factor, B_g. This factor is simply a ratio of the volume of liquid or gas under reservoir conditions to the corresponding volumes under standard conditions. Thus, for oil,

$$N = \frac{7758Ah\phi(1 - S_w)}{B_o} \quad \textbf{(1-3)}$$

where N is in stock tank barrels (STB). In Equation (1-3) the area is in acres. For gas,

$$G = \frac{Ah\phi(1 - S_w)}{B_g} \tag{1-4}$$

where G is in standard cubic ft (SCF) and A is in ft^2.

The gas formation volume factor (traditionally, res ft^3/SCF), B_g, simply implies a volumetric relationship and can be calculated readily with an application of the real gas law. The gas formation volume factor is much smaller than 1.

The oil formation volume factor (res bbl/STB), B_o, is not a simple physical property. Instead, it is an empirical thermodynamic relationship allowing for the reintroduction into the liquid (at the elevated reservoir pressure) of all of the gas that would be liberated at standard conditions. Thus the oil formation volume factor is invariably larger than 1, reflecting the swelling of the oil volume because of the gas dissolution.

The reader is referred to the classic textbooks by Muskat (1949), Craft and Hawkins (revised by Terry, 1991), and Amyx, Bass, and Whiting (1960), and the newer book by Dake (1978) for further information. The present textbook assumes basic reservoir engineering knowledge as a prerequisite.

1.2.2 Permeability

The presence of a substantial porosity usually (but not always) implies that pores will be interconnected. Therefore the porous medium is also "permeable." The property that describes the ability of fluids to flow in the porous medium is permeability. In certain lithologies (e.g., sandstones), a larger porosity is associated with a larger permeability. In other lithologies (e.g., chalks), very large porosities, at times over 0.4, are not necessarily associated with proportionately large permeabilities.

Correlations of porosity versus permeability should be used with a considerable degree of caution, especially when going from one lithology to another. For production engineering calculations these correlations are rarely useful, except when considering matrix stimulation. In this instance, correlations of the *altered* permeability with the *altered* porosity after stimulation are useful.

The concept of permeability was introduced by Darcy (1856) in a classic experimental work from which both petroleum engineering and groundwater hydrology have benefited greatly.

Figure 1-4 is a schematic of Darcy's experiment. The flow rate (or fluid velocity) can be measured against pressure (head) for different porous media.

Darcy observed that the flow rate (or velocity) of a fluid through a specific porous medium is linearly proportional to the head or pressure difference between the inlet and the outlet and a characteristic property of the medium. Thus,

$$u \; \alpha \; k\Delta p \tag{1-5}$$

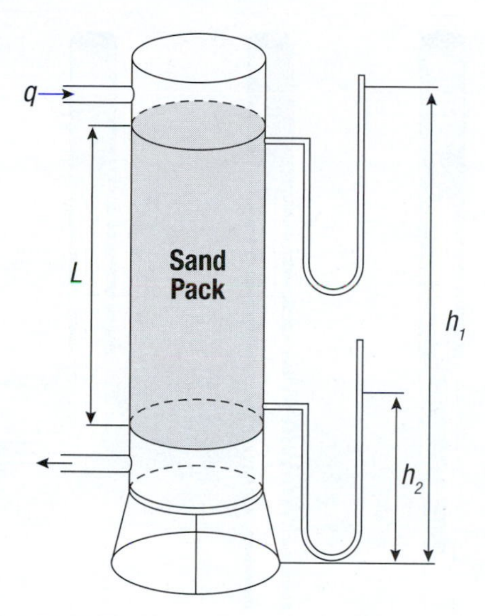

Figure 1-4 Darcy's experiment. Water flows through a sand pack and the pressure difference (head) is recorded.

where k is the permeability and is a characteristic property of the porous medium. Darcy's experiments were done with water. If fluids of other viscosities flow, the permeability must be divided by the viscosity and the ratio k/μ is known as the "mobility."

1.2.3 The Zone near the Well, the Sandface, and the Well Completion

The zone surrounding a well is important. First, even without any man-made disturbance, converging, radial flow results in a considerable pressure drop around the wellbore and, as will be demonstrated later in this book, the pressure drop away from the well varies logarithmically with the distance. This means that the pressure drop in the first foot away from the well is naturally equal to that 10 feet away and equal to that 100 feet away, and so on. Second, all intrusive activities such as drilling, cementing, and well completion are certain to alter the condition of the reservoir near the well. This is usually detrimental and it is not inconceivable that in some cases 90% of the total pressure drop in the reservoir may be consumed in a zone just a few feet away from the well.

Matrix stimulation is intended to recover or even improve the near-wellbore permeability. (There is damage associated even with stimulation. It is the net effect that is expected to be beneficial.) Hydraulic fracturing, today one of the most widely practiced well-completion techniques, alters the manner by which fluids flow to the well; one of the most profound effects is that near-well radial flow and the damage associated with it are eliminated.

Many wells are cemented and cased. One of the purposes of cementing is to support the casing, but at formation depths the most important reason is to provide zonal isolation. Contamination of the produced fluid from the other formations or the loss of fluid *into* other formations

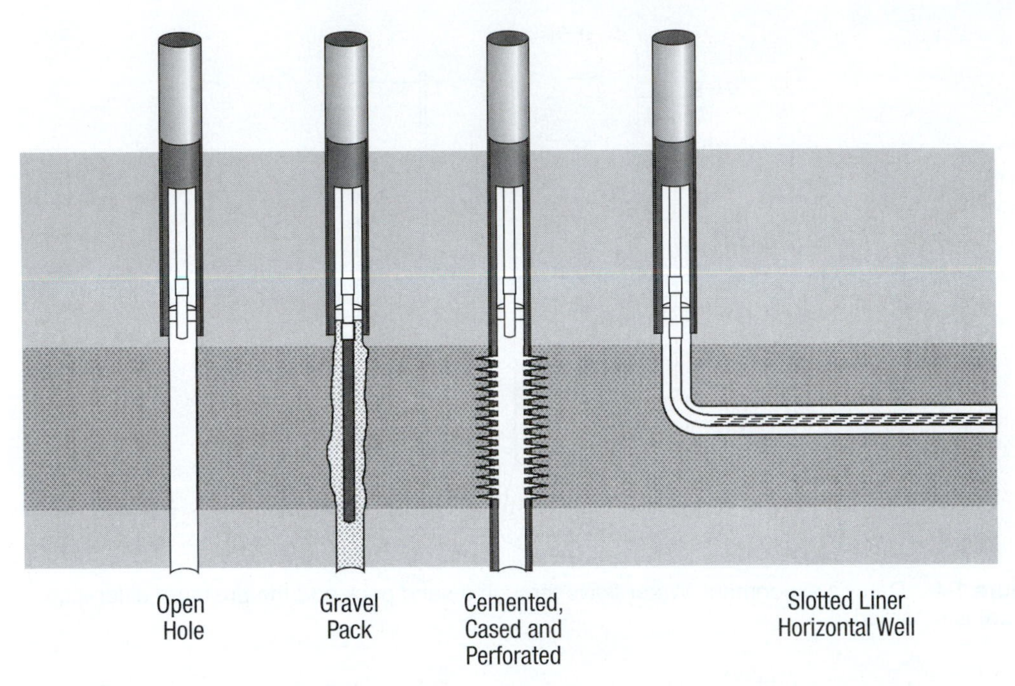

Figure 1-5 Options for well completions.

can be envisioned readily in an open-hole completion. If no zonal isolation or wellbore stability problems are present, the well can be open hole. A cemented and cased well must be perforated in order to reestablish communication with the reservoir. Slotted liners can be used if a cemented and cased well is not deemed necessary and are particularly common in horizontal wells where cementing is more difficult.

Finally, to combat the problems of sand or other fines production, screens can be placed between the well and the formation. Gravel packing can be used as an additional safeguard and as a means to keep permeability-reducing fines away from the well.

The various well completions and the resulting near-wellbore zones are shown in Figure 1-5.

The ability to direct the drilling of a well allows the creation of highly deviated, horizontal, and complex wells. In these cases, a longer to far longer exposure of the well with the reservoir is accomplished than would be the case for vertical wells.

1.2.4 The Well

Entrance of fluids into the well, following their flow through the porous medium, the near-well zone, and the completion assembly, requires that they are lifted through the well up to the surface.

There is a required flowing pressure gradient between the bottomhole and the well head. The pressure gradient consists of the potential energy difference (hydrostatic pressure) and the frictional pressure drop. The former depends on the reservoir depth and the latter depends on the well length.

If the bottomhole pressure is sufficient to lift the fluids to the top, then the well is "naturally flowing." Otherwise, artificial lift is indicated. Mechanical lift can be supplied by a pump. Another technique is to reduce the density of the fluid in the well and thus to reduce the hydrostatic pressure. This is accomplished by the injection of lean gas in a designated spot along the well. This is known as "gas lift."

1.2.5 The Surface Equipment

After the fluid reaches the top, it is likely to be directed toward a manifold connecting a number of wells. The reservoir fluid consists of oil, gas (even if the flowing bottomhole pressure is larger than the bubble-point pressure, gas is likely to come out of solution along the well), and water.

Traditionally, the oil, gas, and water are not transported long distances as a mixed stream, but instead are separated at a surface processing facility located in close proximity to the wells. An exception that is becoming more common is in some offshore fields, where production from subsea wells, or sometimes the commingled production from several wells, may be transported long distances before any phase separation takes place.

Finally, the separated fluids are transported or stored. In the case of formation water it is usually disposed in the ground through a reinjection well.

The reservoir, well, and surface facilities are sketched in Figure 1-6. The flow systems from the reservoir to the entrance to the separation facility are the production engineering systems that are the subjects of study in this book.

1.3 Well Productivity and Production Engineering

1.3.1 The Objectives of Production Engineering

Many of the components of the petroleum production system can be considered together by graphing the inflow performance relationship (IPR) and the vertical flow performance (VFP). Both the IPR and the VFP relate the wellbore flowing pressure to the surface production rate. The IPR represents what the reservoir can deliver, and the VFP represents what the well can deliver. Combined, as in Figure 1-7, the intersection of the IPR with the VFP yields the well deliverability, an expression of what a well will actually produce for a given operating condition. The role of a petroleum production engineer is to maximize the well deliverability in a cost-effective manner. Understanding and measuring the variables that control these relationships (well diagnosis) becomes imperative.

While these concepts will be dealt with extensively in subsequent chapters, it is useful here to present the productivity index, J, of an oil well (analogous expressions can be written for gas and two-phase wells):

$$J = \frac{q}{p - p_{wf}} = \frac{kh}{\alpha_r B\mu} J_D. \qquad \text{(1-6)}$$

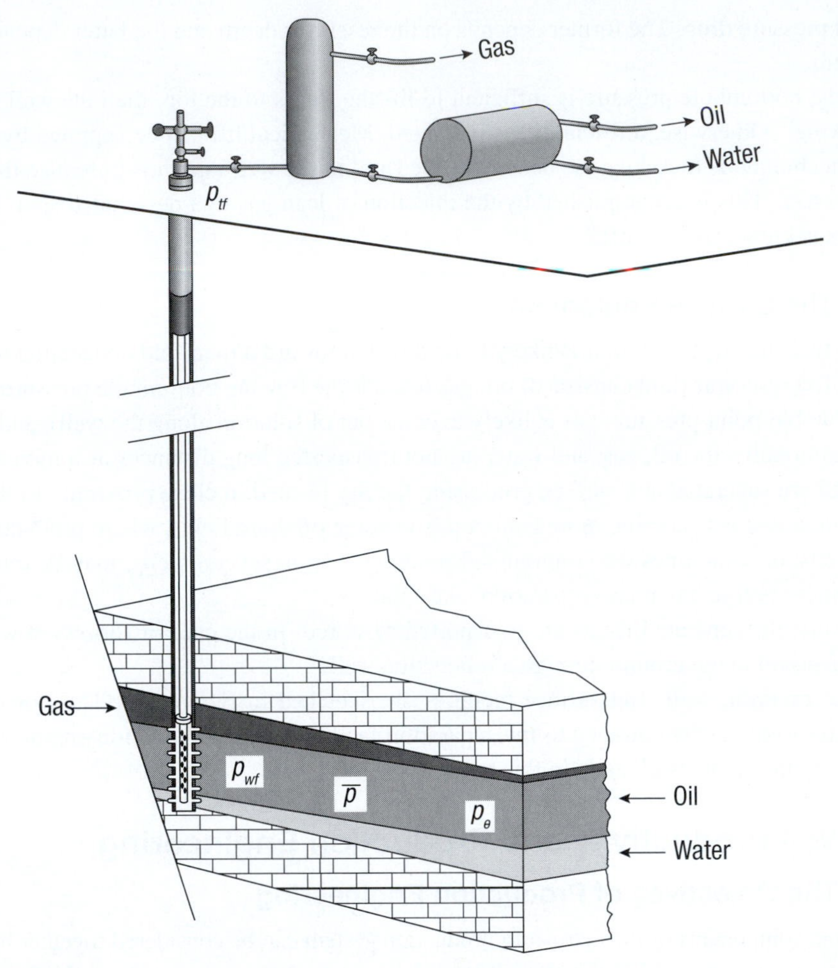

Figure 1-6 The petroleum production system, including the reservoir, underground well completion, the well, wellhead assembly, and surface facilities.

Equation (1-6) succinctly describes what is possible for a petroleum production engineer. First, the dimensioned productivity index with units of flow rate divided by pressure is proportional to the dimensionless (normalized) productivity index J_D. The latter, in turn, has very well-known representations. For steady-state flow to a vertical well,

$$J_D = \frac{1}{\ln\left(\frac{r_e}{r_w}\right) + s}. \tag{1-7}$$

For pseudosteady state flow,

$$J_D = \frac{1}{\ln\left(\frac{r_e}{r_w}\right) - 0.75 + s}, \tag{1-8}$$

and for transient flow,

$$J_D = \frac{1}{p_D + s} \tag{1-9}$$

where p_D is the dimensionless pressure. The terms steady state, pseudosteady state, and transient will be explained in Chapter 2. The concept of the dimensionless productivity index combines flow geometry and skin effects, and can be calculated for any well by measuring flow rate and pressure (reservoir and flowing bottomhole) and some other basic but important reservoir and fluid data.

For a specific reservoir with permeability k, thickness h, and with fluid formation volume factor B and viscosity μ, the only variable on the right-hand side of Equation (1-6) that can be engineered is the dimensionless productivity index. For example, the skin effect can be reduced or eliminated through matrix stimulation if it is caused by damage or can be otherwise remedied if it is caused by mechanical means. A negative skin effect can be imposed if a successful hydraulic fracture is created. Thus, stimulation can improve the productivity index. Finally, more favorable well geometry such as horizontal or complex wells can result in much higher values of J_D.

In reservoirs with pressure drawdown-related problems (fines production, water or gas coning), increasing the productivity can allow lower drawdown with economically attractive production rates, as can be easily surmised by Equation (1-6).

Increasing the drawdown $(p - p_{wf})$ by lowering p_{wf} is the other option available to the production engineer to increase well deliverability. While the IPR remains the same, reduction of the flowing bottomhole pressure would increase the pressure gradient $(p - p_{wf})$ and the flow rate, q, must increase accordingly. The VFP change in Figure 1-7 shows that the flowing bottomhole pressure may be lowered by minimizing the pressure losses between the bottomhole and the separation facility (by, for example, removing unnecessary restrictions, optimizing tubing size, etc.), or by implementing or improving artificial lift procedures. Improving well deliverability by optimizing the flow system from the bottomhole location to the surface production facility is a major role of the production engineer.

In summary, well performance *evaluation* and *enhancement* are the primary charges of the production engineer. The production engineer has three major tools for well performance evaluation: (1) the measurement of (or sometimes, simply the understanding of) the

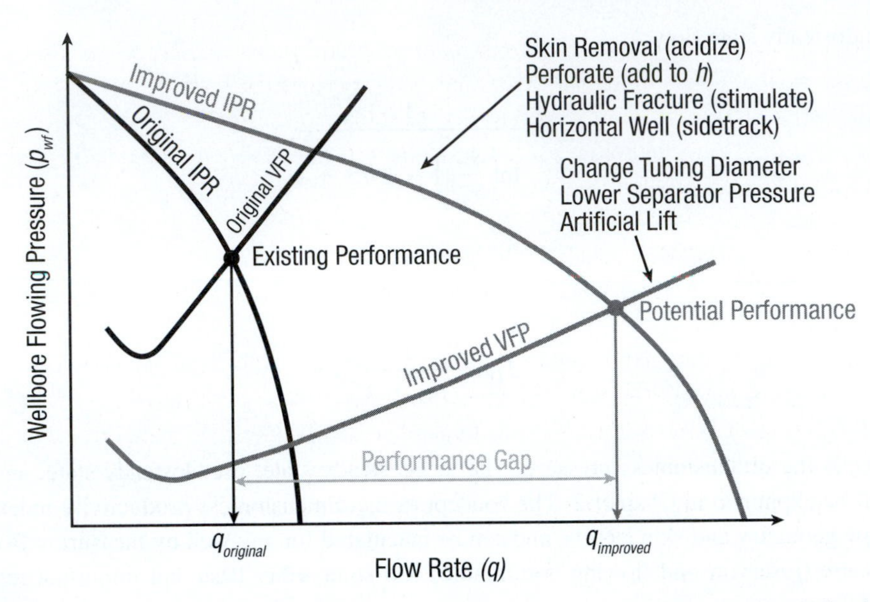

Figure 1-7 Well deliverability gap between the original well performance and optimized well performance.

rate-versus-pressure drop relationships for the flow paths from the reservoir to the separator; (2) well testing, which evaluates the reservoir potential for flow and, through measurement of the skin effect, provides information about flow restrictions in the near-wellbore environmental; and (3) production logging measurements or measurements of pressure, temperature, or other properties by permanently installed downhole instruments, which can describe the distribution of flow into the wellbore, as well as diagnose other completion-related problems.

With diagnostic information in hand, the production engineer can then focus on the part or parts of the flow system that may be optimized to enhance productivity. Remedial steps can range from well stimulation procedures such as hydraulic fracturing that enhance flow in the reservoir to the resizing of surface flow lines to increase productivity. This textbook is aimed at providing the information a production engineer needs to perform these tasks of well performance evaluation and enhancement.

1.3.2 Organization of the Book

This textbook offers a structured approach toward the goal defined above. Chapters 2–4 present the inflow performance for oil, two-phase, and gas reservoirs. Chapter 5 deals with complex well architecture such as horizontal and multilateral wells, reflecting the enormous growth of this area of production engineering since the first edition of the book. Chapter 6 deals with the condition of the near-wellbore zone, such as damage, perforations, and gravel packing. Chapter 7 covers the flow of fluids to the surface. Chapter 8 describes the surface flow system, flow in horizontal pipes, and flow in horizontal wells. Combination of inflow performance and well performance versus time, taking into account single-well transient flow and material balance, is shown in Chapters 9 and 10. Therefore, Chapters 1–10 describe the workings of the reservoir and well systems.

Gas lift is outlined in Chapter 11, and mechanical lift in Chapter 12.

For an appropriate product engineering remedy, it is essential that well and reservoir diagnosis be done.

Chapter 13 presents the state-of-the-art in modern diagnosis that includes well testing, production logging, and well monitoring with permanent downhole instruments.

From the well diagnosis it can be concluded whether the well is in need of matrix stimulation, hydraulic fracturing, artificial lift, combinations of the above, or none.

Matrix stimulation for all major types of reservoirs is presented in Chapters 14, 15, and 16. Hydraulic fracturing is discussed in Chapters 17 and 18.

Chapter 19 is a new chapter dealing with advances in sand management.

This textbook is designed for a two-semester, three-contact-hour-per-week sequence of petroleum engineering courses, or a similar training exposure.

To simplify the presentation of realistic examples, data for three characteristic reservoir types—an undersaturated oil reservoir, a saturated oil reservoir, and a gas reservoir—are presented in Appendixes. These data sets are used throughout the book. Examples and homework follow a more modern format than those used in the first edition. Less emphasis is given to hand-done calculations, although we still think it is essential for the reader to understand the salient fundamentals. Instead, exercises require application of modern software such as Excel spreadsheets and the PPS software included with this book, and trends of solutions and parametric studies are preferred in addition to single calculations with a given set of variables.

1.4 Units and Conversions

We have used “oilfield” units throughout the text, even though this system of units is inherently inconsistent. We chose this system because more petroleum engineers “think” in bbl/day and psi than in terms of m^3/s and Pa. All equations presented include the constant or constants needed with oilfield units. To employ these equations with SI units, it will be easiest to first convert the SI units to oilfield units, calculate the desired results in oilfield units, then convert the results to SI units. However, if an equation is to be used repeatedly with the input known in SI units, it will be more convenient to convert the constant or constants in the equation of interest. Conversion factors between oilfield and SI units are given in Table 1-1.

Table 1-1 Typical Units for Reservoir and Production Engineering Calculations

Variable	Oilfield Unit	SI Unit	Conversion (Multiply SI Unit)
Area	acre	m^2	2.475×10^{-4}
Compressibility	psi^{-1}	Pa^{-1}	6897
Length	ft	m	3.28
Permeability	md	m^2	1.01×10^{15}
Pressure	psi	Pa	1.45×10^{-4}
Rate (oil)	STB/d	m^3/s	5.434×10^5
Rate (gas)	Mscf/d	m^3/s	3049
Viscosity	cp	Pa-s	1000

Example 1-1 Conversion from Oilfield to SI Units

The steady-state, radial flow form of Darcy's law in oilfield units is given in Chapter 2 as

$$p_e - p_{wf} = \frac{141.2qB\mu}{kh}\left(\ln\frac{r_e}{r_w} + s\right) \tag{1-10}$$

for p in psi, q in STB/d, B in res bbl/STB, μ in cp, k in md, h in ft, and r_e and r_w in ft (s is dimensionless). Calculate the pressure drawdown $(p_e - p_{wf})$ in Pa for the following SI data, first by converting units to oilfield units and converting the result to SI units, then by deriving the constant in this equation for SI units.

Data

$q = 0.001\ m^3/s$, $B = 1.1$ res $m^3/ST\ m^3$, $\mu = 2 \times 10^{-3}$ Pa-s, $k = 10^{-14}\ m^2$, $h = 10$ m, $r_e = 575$ m, $r_w = 0.1$ m, and $s = 0$.

Solution

Using the first approach, we first convert all data to oilfield units. Using the conversion factors in Table 1-1,

$$q = \left(0.001\frac{\mathrm{m}^3}{\mathrm{s}}\right)(5.434 \times 10^5) = 543.4 \text{ STB/d} \tag{1-11}$$

$$B = 1.1 \text{ res bbl/STB} \tag{1-12}$$

$$\mu = (2 \times 10^{-3}\ \text{Pa-s})(10^3) = 2 \text{ cp} \tag{1-13}$$

$$k = (10^{-14}\ \mathrm{m}^2)(1.01 \times 10^{15}) = 10.1 \text{ md} \tag{1-14}$$

$$h = (10\ \mathrm{m})(3.28) = 32.8 \text{ ft.} \tag{1-15}$$

Since r_e is divided by r_w, the units for these radii do not have to be converted. Now, from Equation (1-10),

$$p_e - p_{wf} = \frac{(141.2)(543.4)(1.1)(2)}{(10.1)(32.8)}\left[\ln\left(\frac{575}{0.1}\right) + 0\right] = 4411 \text{ psi} \tag{1-16}$$

and converting this results to Pascals,

$$p_e - p_{wf} = (4411 \text{ psi})(6.9 \times 10^3) = 3.043 \times 10^7 \text{ Pa} \tag{1-17}$$

Alternatively, we can convert the constant 141.2 to the appropriate constant for SI units, as follows (including only-to-be-converted variables):

$$p_e - p_{wf}(\text{Pa}) = \frac{(141.2)[q(\mathrm{m}^3/s)(5.43 \times 10^5)][\mu(Pa - s)(10^3)]}{[k(\mathrm{m}^2)(1.01 \times 10^{15})][h(\mathrm{m})(3.28)]}(6.9 \times 10^3) \tag{1-18}$$

or

$$p_e - p_{wf} = \frac{0.159qB\mu}{kh}\left(\ln\frac{r_e}{r_w} + s\right) = \frac{qB\mu}{2\pi kh}\left(\ln\frac{r_e}{r_w} + s\right). \tag{1-19}$$

The constant derived, 0.159, is $1/2\pi$, as it should be for this consistent set of units. Substituting the parameters in SI units directly into Equation (1-19), we again calculate that $p_e - p_{wf} = 3.043 \times 10^7$ Pa.

Often, in regions where metric units are customary, a mix of SI and non-SI units is sometimes employed. For example, in using Darcy's law, the units for flow rate may be m^3/d; for viscosity, cp; for permeability, md; and so on. In this instance, units can be converted to oilfield units in the same manner demonstrated here for consistent SI units.

References

1. Amyx, J.W., Bass, D.M., Jr., and Whiting, R.L., *Petroleum Reservoir Engineering*, McGraw-Hill, New York, 1960.
2. Craft, B.C., and Hawkins, M. (revised by Terry, R.E.), *Applied Petroleum Engineering*, 2nd ed., Prentice Hall, Englewood Cliffs, NJ, 1991.
3. Dake, L.P., *Fundamentals of Reservoir Engineering*, Elsevier, Amsterdam, 1978.
4. Darcy, H., *Les Fontaines Publiques de la Ville de Dijon*, Victor Dalmont, Paris, 1856.
5. Earlougher, R.C., Jr., *Advances in Well Test Analysis*, SPE Monograph, Vol. 5, SPE, Richardson, TX, 1977.
6. Muskat, M., *Physical Principles of Oil Production*, McGraw-Hill, New York, 1949.

CHAPTER 2

Production from Undersaturated Oil Reservoirs

2.1 Introduction

Well deliverability analysis predicts the wellbore flowing pressure for a given surface flowrate. Chapters 2–5 deal with well inflow performance and describe the reservoir variables that control well productivity under different conditions. Wells drilled in oil reservoirs drain a porous medium of porosity ϕ, net thickness h, and permeability k. To understand the process of flow from the reservoir and into the well sandface, a simple expression of Darcy's (1856) law in radial coordinates can be used first:

$$q = \frac{kA}{\mu}\frac{dp}{dr} \tag{2-1}$$

where A is radial flow area at a distance r given by $A = 2\pi rh$.

Equation (2-1) is general and suggests a number of interesting observations. The flow rate is large if the pressure gradient dp/dr, the permeability k, and the reservoir height h are large or if the viscosity of the flowing fluid, μ, is small. Also, this expression assumes a single-phase fluid flowing and saturating the reservoir.

2.2 Steady-State Well Performance

Steady-state performance means that all parameters, including flow rate and all pressures, are invariant with time. For a vertical well draining a region with radius r_e, this requires that the pressure at the well boundary, p_e, and the bottomhole flowing pressure, p_{wf}, are constant with time. Practically, the boundary pressure, p_e, for a production well can remain constant only in the presence of pressure maintenance, either by natural water influx from an aquifer or by injection to maintain pressure in the reservoir. A reservoir being waterflooded is the most common situation for which steady-state behavior approximates the actual production well conditions.

The steady-state performance relationship is easily obtained from Darcy's law. In a well within a reservoir, as shown in Figure 2-1, the area of flow at any distance, r, is given by $2\pi rh$, and Equation (2-1) becomes

$$q = \frac{2\pi krh}{\mu}\frac{dp}{dr}. \tag{2-2}$$

Assuming that q is constant, separation of variables and integration leads to

$$\int_{p_{wf}}^{p} dp = \frac{q\mu}{2\pi kh}\int_{r_w}^{r}\frac{dr}{r} \tag{2-3}$$

and finally,

$$p - p_{wf} = \frac{q\mu}{2\pi kh}\ln\frac{r}{r_w}. \tag{2-4}$$

Equation (2-4) is semi-logarithmic, meaning that the pressure drop doubles or triples as the radial distance increases by one or two orders of magnitude. Thus, the near-wellbore region is extremely important in well production because that is where much of the pressure drop occurs.

Van Everdingen and Hurst (1949) quantified the condition of the near-wellbore region with the introduction of the concept of the skin effect. This is analogous to the film coefficient in heat transfer. This skin effect results in an additional steady-state pressure drop, given by

$$\Delta p_s = \frac{q\mu}{2\pi kh}s, \tag{2-5}$$

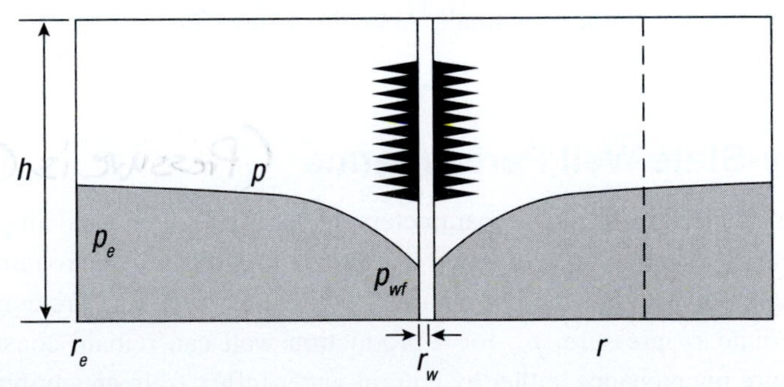

Figure 2-1 Reservoir schematic for steady-state flow into a well.

which can be added to the pressure drop in the reservoir. Thus Equation (2-4) becomes

$$p - p_{wf} = \frac{q\mu}{2\pi kh}\left(\ln\frac{r}{r_w} + s\right). \tag{2-6}$$

If the reservoir exhibits a constant-pressure outer boundary (at r_e), that is, the well operates under steady-state conditions, then if that pressure is p_e, the radial flow equation becomes

$$p_e - p_{wf} = \frac{q\mu}{2\pi kh}\left(\ln\frac{r_e}{r_w} + s\right). \tag{2-7}$$

In oilfield units, where p_e and p_{wf} are in psi, q is in STB/d, μ is in cp, k is in md, h is in ft, and B is the formation volume factor to convert STB into res bbl, Equation (2-7) becomes

$$p_e - p_{wf} = \frac{141.2qB\mu}{kh}\left(\ln\frac{r_e}{r_w} + s\right) \tag{2-8}$$

and with rearrangement for the rate, one of the best-known expressions in production engineering can be extracted:

IPR Curve:

$$q = \frac{kh(p_e - p_{wf})}{141.2B\mu\left(\ln\frac{r_e}{r_w} + s\right)}. \tag{2-9}$$

Two other important concepts are outlined below. They apply to all types of flow.

- The effective wellbore radius r'_w can be derived from a simple rearrangement of Equation (2-8),

$$p_e - p_{wf} = \frac{141.2qB\mu}{kh}\left(\ln\frac{r_e}{r_w} + \ln e^s\right) \tag{2-10}$$

and thus,

$$p_e - p_{wf} = \frac{141.2qB\mu}{kh}\left(\ln\frac{r_e}{r_w e^{-s}}\right). \tag{2-11}$$

The effective wellbore radius, r'_w, is defined by

$$r'_w = r_w e^{-s}. \quad \text{(2-12)}$$

This is an interesting finding. In a damaged well, with, for example, $s = 10$, the reservoir drains into a well with an effective radius equal to $4.5 \times 10^{-5}\, r_w$. Conversely, in a stimulated well with, for example, $s = -2$ (acidized well), the effective wellbore radius is $7.4r_w$ and if $s = -6$ (for a fractured well), then the effective wellbore radius is $402r_w$.

- Introduced already in Chapter 1, the *productivity index*, J, of a well is simply the production rate divided by the pressure difference (called drawdown). For steady-state production,

$$J = \frac{q}{p_e - p_{wf}} = \frac{kh}{141.2B\mu[\ln(r_e/r_w) + s]} = \frac{kh}{141.2B\mu} J_D \quad \text{(2-13)}$$

where

$$J_D = \frac{1}{\ln(r_e/r_w) + s} = \frac{1}{\ln(r_e/r'_w)}. \quad \text{(2-14)}$$

One of the main purposes of production engineering is to maximize the productivity index in a cost-effective manner, that is, to increase the flow rate for a given driving force (drawdown) or to minimize the drawdown for a given rate. Usually, this can be accomplished with a decrease of the skin effect (through matrix stimulation and removal of near-wellbore damage) or through the superposition of a negative skin effect from an induced hydraulic fracture.

In reservoirs where the viscosity is very large ($\mu > 100$ cp), thermal recovery may be indicated to reduce the viscosity.

Example 2-1 Steady-State Production Rate Calculation and Rate Improvement (Stimulation)

Assume that a well in the reservoir described in Appendix A has a drainage area equal to 640 acres ($r_e = 2980$ ft) and is producing at steady state with an outer boundary (constant) pressure equal to 5651 psi. Calculate the steady-state production rate if the flowing bottomhole pressure is equal to 4500 psi. Use a skin effect equal to +10.

Describe two mechanisms to increase the flow rate by 50%. Show calculations.

Solution

From Equation (2-9),

$$q = \frac{(8.2)(53)(5651 - 4500)}{(141.2)(1.117)(1.72)[\ln(2980/0.328) + 10]} = 92 \text{ STB/d}. \quad \text{(2-15)}$$

To increase the production rate by 50%, one possibility is to increase the drawdown, $p_e - p_{wf}$, by 50%. Therefore,

$$(5651 - p_{wf})_2 = 1.5(5651 - 4500), \tag{2-16}$$

leading to $p_{wf} = 3925$ psi.

A second possibility is to increase J_D by reducing the skin effect. In this case,

$$1.5J_D = 1.5\frac{1}{\ln\frac{2980}{0.328} + 10} = \frac{1}{\ln\frac{2980}{0.328} + s_2}, \tag{2-17}$$

leading to $s_2 = 3.6$.

Example 2-2 Effect of Drainage Area on Well Performance

Demonstrate the effect of drainage area on oilwell production rate by calculating the ratios of production rates from 80-, 160-, and 640-acre drainage areas to that obtained from a 40-acre drainage area. The well radius is 0.328 ft.

Solution

Assuming that the skin effect is zero (this would result in the most pronounced difference in the production rate), the ratios of the production rates (or productivity indices) can be given by an expression of the form

$$\frac{q}{q_{40}} = \frac{(\ln r_e/r_w)_{40}}{\ln r_e/r_w}. \tag{2-18}$$

The drainage radius for a given drainage area is calculated by assuming that the well is in the center of a circular drainage area. Thus,

$$r_e = \sqrt{\frac{(A)(43{,}560)}{\pi}}. \tag{2-19}$$

The results are shown in Table 2-1. These ratios indicate that the drainage area assigned to a well has a small impact on the production rate. For tight reservoirs, cumulative production differences are particularly immune to the drainage area because transient behavior is evident for much of the time.

Table 2-1 Production Rate Increases (over a 40-acre spacing)

A (acres)	r_e (ft)	$\ln(r_e/r_w)$	q/q_{40}
40	745	7.73	1
80	1053	8.07	0.96
160	1489	8.42	0.92
640	2980	9.11	0.85

2.3 Transient Flow of Undersaturated Oil

The diffusivity equation describes the pressure profile in an infinite-acting, radial reservoir, with a slightly compressible and constant viscosity fluid (undersaturated oil or water). This equation, with similar expressions in wide use in a number of engineering fields such as heat transfer (Carslaw and Jaeger, 1959), has the classic form

$$\frac{\partial^2 p}{\partial r^2} + \frac{1}{r}\frac{\partial p}{\partial r} = \frac{\phi \mu c_t}{k}\frac{\partial p}{\partial t}. \quad \text{(2-20)}$$

Its generalized solution is

$$p(r,t) = p_i + \frac{q\mu}{4\pi kh} E_i(-x) \quad \text{(2-21)}$$

where $E_i(x)$ is the exponential integral and x is given by

$$x = \frac{\phi \mu c_t r^2}{4kt}. \quad \text{(2-22)}$$

For $x < 0.01$ (i.e., for large values of time or for small distances, such as at the wellbore), the exponential integral $-E_i(-x)$ can be approximated by $-\ln(\gamma x)$, where γ is Euler's constant and is equal to 1.78.

Therefore, Equation (2-21), at the wellbore and shortly after production, can be approximated by $(p(r,t) \equiv p_{wf})$

$$p_{wf} = p_i - \frac{q\mu}{4\pi kh} \ln \frac{4kt}{\gamma \phi \mu c_t r_w^2}. \quad \text{(2-23)}$$

Finally, introducing variables in oilfield units as listed in Table 1-1 and converting the natural log to log base 10, Equation (2-23) becomes

$$p_{wf} = p_i - \frac{162.6 q B \mu}{kh}\left(\log t + \log \frac{k}{\phi \mu c_t r_w^2} - 3.23\right). \tag{2-24}$$

This expression is often known as the pressure drawdown equation describing the declining flowing bottomhole pressure, p_{wf}, while the well is flowing at a constant rate q.

Because a producing well is usually flowing for long times with the same wellhead pressure (which is imposed by the well hardware, such as chokes, etc.), the resulting bottomhole pressure is also largely constant. Therefore, Equation (2-24), which is for constant rate, must be adjusted. More commonly, the constant-bottomhole-pressure situation results in a similar expression, which, although it appears as a mere algebraic rearrangement, is an approximation of the analytical solution to Equation (2-20) with the appropriate inner boundary condition (Earlougher, 1977):

$$q = \frac{kh(p_i - p_{wf})}{162.6 B \mu}\left(\log t + \log \frac{k}{\phi \mu c_t r_w^2} - 3.23\right)^{-1} \tag{2-25}$$

where the time, t, must be in hours.

Equation (2-25), including the skin factor, becomes

Used to forecast rate as a function of time

$$q = \frac{kh(p_i - p_{wf})}{162.6 B \mu}\left(\log t + \log \frac{k}{\phi \mu c_t r_w^2} - 3.23 + 0.87 s\right)^{-1}. \tag{2-26}$$

Example 2-3 Prediction of Production Rate in an Infinite-Acting Oil Well

Using the well and reservoir variables in Appendix A, develop a production rate profile for 1 year assuming that no boundary effects emerge. Do this in increments of 2 months and use a flowing bottomhole pressure equal to 3500 psi.

Solution

From Equation (2-25) and substitution of the appropriate variables in Appendix A, the well production rate is given by

$$q = \frac{(8.2)(53)(5651 - 3500)}{(162.6)(1.2)(1.03)}\left[\log t + \log \frac{8.2}{(0.19)(1.03)(1.29 \cdot 10^{-5})(0.328)^2} - 3.23\right]^{-1}$$

$$= \frac{4651}{\log(t) + 4.25} \tag{2-27}$$

For $t = 2$ months, for Equation (2-27) the production rate $q = 627$ STB/d.

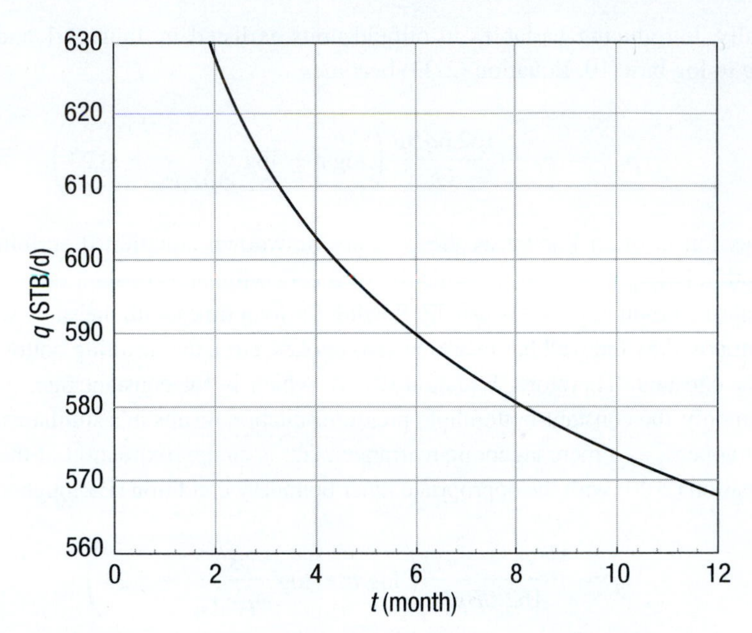

Figure 2-2 Rate decline for an infinite-acting oil reservoir (Example 2-3).

Figure 2-2 is a rate-decline curve for this oil well for the first year assuming infinite-acting behavior. The rate decline is from 627 STB/d (after 2 months) to 568 STB/d (after 1 year).

2.4 Pseudosteady-State Flow

Almost all wells eventually "feel" their boundaries. In Section 2-2 the steady-state condition implied a constant-pressure outer boundary. Naturally, this boundary can approximate the impact of a larger aquifer. Induced constant pressure may be the result of injector–producer configurations.

For no-flow boundaries, drainage areas can either be described by natural limits such as faults, pinchouts, and so on, or can be artificially induced by the production of adjoining wells.

The pressure at the outer boundary is no longer constant but instead declines at a constant rate with time, as does the pressure at every point in the well drainage volume. That is, $\partial p/\partial t$ = constant everywhere. Because the pressure profile is not changing, this condition is often referred to as "pseudosteady state."

Solving the radial diffusivity equation during pseudosteady state gives the pressure p at any point r in a reservoir of radius r_e as (Dake, 1978)

$$p = p_{wf} + \frac{141.2qB\mu}{kh}\left(\ln\frac{r}{r_w} - \frac{r^2}{2r_e^2}\right). \quad \textbf{(2-28)}$$

At $r = r_e$, Equation (2-28) can be converted to

$$p_e = p_{wf} + \frac{141.2qB\mu}{kh}\left(\ln\frac{r_e}{r_w} - \frac{1}{2}\right). \tag{2-29}$$

This equation is not particularly useful under pseudosteady state conditions, since p_e is not known at any given time. However, as shown in Chapter 13, the average reservoir pressure, $\overline{p}$, can be obtained from periodic pressure buildup tests. Therefore, a more useful expression for the pseudosteady state equation would be one using the average reservoir pressure. This is defined as a volumetrically weighted pressure,

$$\overline{p} = \frac{\int_{r_w}^{r_e} pdV}{\pi(r_e^2 - r_w^2)h\phi} \approx \frac{\int_{r_w}^{r_e} pdV}{\pi r_e^2 h\phi} \tag{2-30}$$

and since $dV = 2\pi rh\phi\, dr$, Equation (2-30) becomes

$$\overline{p} = \frac{2}{r_e^2}\int_{r_w}^{r_e} pr\, dr. \tag{2-31}$$

The expression for the pressure at any point r can be substituted from Equation (2-28), and therefore

$$\overline{p} - p_{wf} = \frac{2}{r_e^2}\frac{141.2qB\mu}{kh}\int_{r_w}^{r_e}\left(\ln\frac{r}{r_w} - \frac{r^2}{2r_e^2}\right)rdr. \tag{2-32}$$

Performing the integration results in

$$\overline{p} - p_{wf} = \frac{141.2qB\mu}{kh}\left(\ln\frac{r_e}{r_w} - \frac{3}{4}\right). \tag{2-33}$$

Introducing the skin effect and incorporating the term 3/4 into the logarithmic expression leads to the inflow relationship for a no-flow boundary oil reservoir:

$$\overline{p} - p_{wf} = \frac{141.2qB\mu}{kh}\left(\ln\frac{0.472r_e}{r_w} + s\right) \tag{2-34}$$

and rearranged for rate

$$q = \frac{kh(\overline{p} - p_{wf})}{141.2B\mu\left[\ln\left(\frac{0.472r_e}{r_w} + s\right)\right]}. \tag{2-35}$$

Equation (2-35) is particularly useful because it provides the relationship between the average reservoir pressure, $\overline{p}$, and the rate q. The average pressure, $\overline{p}$, is a variable that *can* be determined from a pressure buildup test. It depends on the drainage area and the properties of the fluid and rock. Material balance calculations presented in Chapter 10 combine depletion mechanisms with inflow relationships and lead to forecasts of well performance and cumulative production.

Finally, while Equation (2-35) appears similar to Equation (2-9) (for steady-state flow), the two should never be confused. They represent distinctly different reservoir production mechanisms. However, both lead to similar expressions for the dimensionless productivity index, J_D. For steady state:

$$J_D = \frac{1}{\ln\frac{r_e}{r_w} + s} \tag{2-36}$$

and for pseudosteady state:

$$J_D = \frac{1}{\ln\frac{0.472r_e}{r_w} + s}. \tag{2-37}$$

In both cases, for typical drainage and well radii, the logarithmic term ranges between 7 and 9, and thus for an undamaged or unstimulated well the J_D is of the order of 0.1; smaller values denote damage, and larger values denote stimulation, hydraulic fracturing, or more favorable well geometry such as horizontal or multilateral wells.

Example 2-4 Production from a No-Flow Boundary Reservoir

What would be the average reservoir pressure if the outer boundary pressure is 6000 psi, the flowing bottomhole pressure is 3000 psi, the drainage area is 640 acres, and the well radius is 0.328 ft? What would be the ratio of the flow rates before (q_1) and after (q_2) the average reservoir pressure drops by 1000 psi? Assume that $s = 0$.

Solution

A ratio of Equations (2-29) and (2-33) results in

$$\frac{p_e - p_{wf}}{\overline{p} - p_{wf}} = \frac{\ln(r_e/r_w) - 1/2}{\ln(r_e/r_w) - 3/4}. \tag{2-38}$$

The drainage area $A = 640$ and therefore $r_e = 2980$ ft. Substituting the given variables in Equation (2-38) results in

$$\overline{p} = \frac{(6000 - 3000)(8.36)}{8.61} + 3000 = 5913 \text{ psi}. \tag{2-39}$$

The flow-rate ratio after the 1000 psi average pressure decline would be

$$\frac{q_2}{q_1} = \frac{4913 - 3000}{5913 - 3000} = 0.66. \tag{2-40}$$

2.4.1 Transition to Pseudosteady State from Infinite Acting Behavior

Earlougher (1977) indicated that the time, t_{pss}, at which pseudosteady state begins is given by

$$t_{pss} = \frac{\phi \mu c_t A}{0.000264k} t_{DApss} \tag{2-41}$$

where A is the drainage area and t_{DA} has a characteristic value that depends on the drainage shape. For a regular shape such as a circle or a square, the dimensionless time at the onset of pseudosteady state, t_{DApss}, is equal to 0.1. For a well in a 1×2 rectangle, t_{DApss} is equal to 0.3; and for a 1×4 rectangle, it is equal to 0.8. Off-centered wells in irregular patterns have even larger values of t_{DApss}, implying that the well will "feel" the farther-off boundaries after a significantly longer time.

If the drainage area can be approximated by a circle with an equivalent drainage radius r_e, then Equation (2-41) (with $t_{DApss} = 0.1$) yields

$$t_{pss} \approx 1200 \frac{\phi \mu c_t r_e^2}{k} \tag{2-42}$$

at the onset of pseudosteady state. The time t_{pss} is in hours, and all other variables are in the customary oilfield units.

2.5 Wells Draining Irregular Patterns

Rarely do wells drain regular-shaped drainage areas. Even if they are assigned regular *geographic* drainage areas, these are distorted after production commences, either because of the presence of natural boundaries or because of lopsided production rates in adjoining wells. The drainage area is then *shaped* by the assigned production duty of a particular well.

To account for irregular drainage shapes or asymmetrical positioning of a well within its drainage area, a series of "shape" factors was developed by Dietz (1965).

Equation (2-33) is for a well at the center of a circle. The logarithmic expression can be modified (by multiplying and dividing the expressions within the logarithm by 4π and performing a simple property of logarithms) into

$$\ln\frac{r_e}{r_w} - \frac{3}{4} = \frac{1}{2}\ln\frac{4\pi r_e^2}{4\pi e^{3/2} r_w^2}. \tag{2-43}$$

The argument πr_e^2 is the drainage area of a circle of radius r_e. The product $4\pi e^{3/2}$ in the denominator is equal to 56.32 or (1.78) (31.6), where 1.78 is Euler's constant, denoted by γ, and 31.6 is a shape factor for a circle with a well at the center, denoted by C_A. Dietz (1965) has shown that all well/reservoir configurations that depend on drainage shape and well position have a characteristic shape factor. Therefore, Equation (2-32) can be generalized for any shape into

$$\bar{p} - p_{wf} = \frac{141.2qB\mu}{kh}\left(\frac{1}{2}\ln\frac{4A}{\gamma C_A r_w^2} + s\right). \tag{2-44}$$

Figure 2-3 (Earlougher, 1977) shows t_{DA} values and shape factors for some commonly encountered (approximate) drainage shapes and well positions.

Example 2-5 Impact of Irregular Well Positioning on Production Rate

Assume that two wells in the reservoir described in Appendix A each drain 640 acres. Furthermore, assume that $\bar{p} = 5651$ psi (same as p_i) and that $s = 0$. The flowing bottomhole pressure in both is 3500 psi. However, well A is placed at the center of a square, whereas well B is at the center of the upper right quadrant of a square drainage shape. Calculate the production rates from the two wells at the onset of pseudosteady state. (This calculation is valid only at very early time. At late time, either drainage shapes will change, if they are artificially induced, or the average reservoir pressure will not decline uniformly within the drainage areas because of different production rates and resulting different rates of depletion.)

	Shape	t_{DApss}	C_A
1.		0.1	31.6
2.		0.1	31.6
3.		0.2	27.6
4.	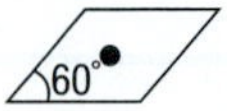 60°	0.2	27.1
5.	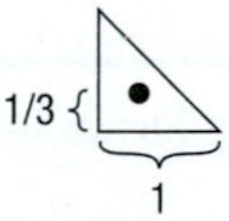 1/3, 1	0.4	21.9
6.	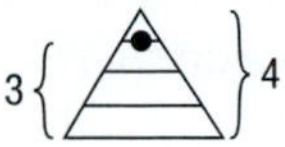 3, 4	0.9	0.098
7.		0.1	30.9
8.	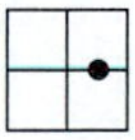	0.7	13
9.	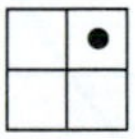	0.6	4.5
10.		0.7	3.3
11.	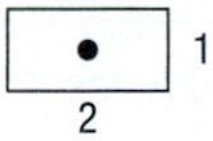 1, 2	0.3	21.8
12.	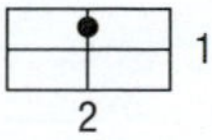 1, 2	0.4	10.8
13.	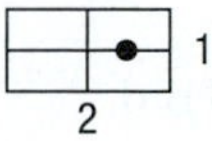 1, 2	1.5	4.5
14.	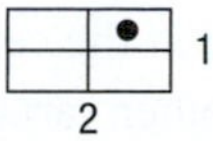 1, 2	1.7	2.08
15.	 1, 2	0.4	3.16
16.	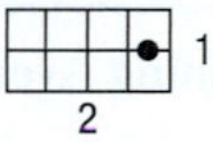 1, 2	2.0	0.58
17.	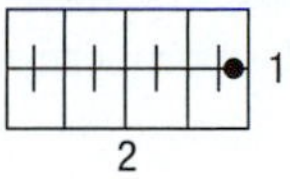 1, 2	3.0	0.011
18.	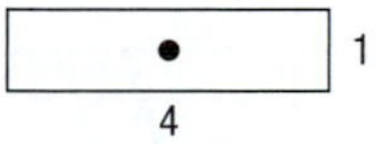 1, 4	0.8	5.38
19.	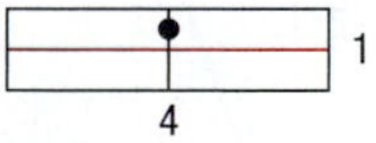 1, 4	0.8	2.69
20.	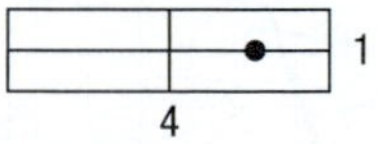 1, 4	4.0	0.23
21.	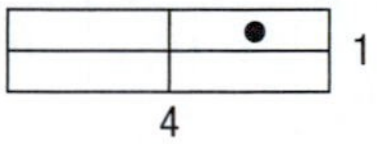 1, 4	1.0	0.12
22.	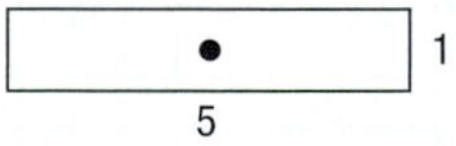 1, 5	1.0	2.36

Figure 2-3 Shape factor for various closed, single-well drainage areas. (From Earlougher, 1977.)

Solution

Well A The shape factor, C_A, from Figure 2-3 is equal to 30.9. Therefore, from Equation (2-44),

$$q = \frac{(8.2)(53)(2151)}{(141.2)(1.2)(1.03)(0.5)\ \ln\left[(4)(640)(43560)/(1.78)(30.9)(0.328)^2\right]}$$

$$= 640\ \text{STB/d}. \tag{2-45}$$

Well B Since it is located at the center of the upper right quadrant, its shape factor (from Figure 2-3) is equal to 4.5. All other variables in Equation (2-44) remain the same. The flow rate calculated is then equal to 574 STB/d, representing a 10% reduction.

Example 2-6 Determining Average Reservoir Pressure within Adjoining Drainage Areas

The following data were obtained on a three-well fault block. A map with well locations is shown in Figure 2-4. (Use properties as for the reservoir described in Appendix A.)

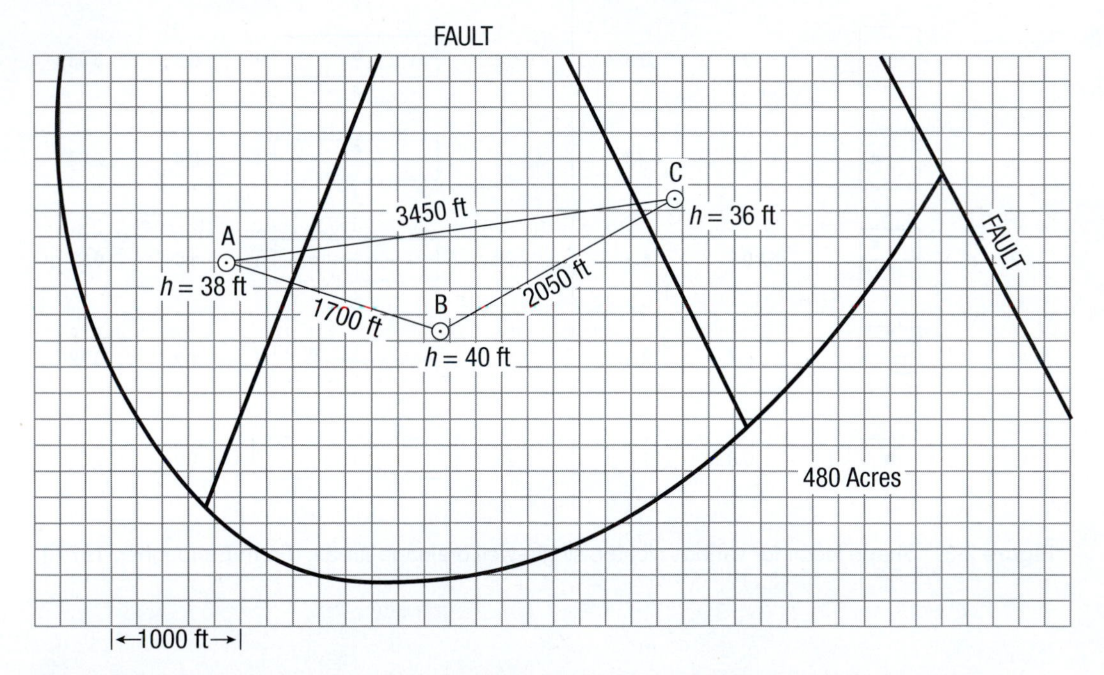

Figure 2-4 Three-well fault block for Example 2-6. (From a problem by H. Dykstra, class notes, 1976.)

Each well produced for 200 days since the previous shut-in. At the end of the 200 days, the following rates and skin effects were obtained from each well:

	A	B	C
q_o (STB/d at time of shut-in)	100	200	80
h (ft)	38	40	36
s (skin effect)	2	0	5

If the bottomhole pressure is 2000 psi for each well, calculate the average reservoir pressure within each drainage area.

Solution

The drainage volumes *formed* by the production rate of each well are related by

$$\frac{V_A}{V_B} = \frac{q_A}{q_B} \tag{2-46}$$

and

$$\frac{V_A}{V_C} = \frac{q_A}{q_C}. \tag{2-47}$$

If the reservoir thickness were the same throughout, then the ratios of the areas would have sufficed. However, since h varies, the volumes can be replaced by the product $h_i A_i$, where i refers to each well. Finally, a third equation is needed:

$$A_A + A_B + A_C = A_{\text{total}} = 480 \text{ acres}. \tag{2-48}$$

Equations (2-46), (2-47), and (2-48) result in $A_A = 129$ acres (5.6×10^6 ft^2), $A_B = 243$ acres (1.06×10^7 ft^2), and $A_C = 108$ acres (4.7×10^6 ft^2). The next step is to sketch these areas on the fault block map.

Each map square represents 40,000 ft^2 (200×200 ft), and therefore the three areas are to be allocated 140, 365, and 118 squares, respectively. The drainage divide must be normal to the tieline between adjoining wells. Thus, counting squares, the approximate drainage areas, shown in Figure 2-4, can be drawn. These describe shapes and therefore approximate shape factors. From Figure 2-3, well A is shape no. 12 ($C_A = 10.8$), well B is shape no. 7 ($C_A = 30.9$), and well C is shape no. 10 ($C_A = 3.3$).

From Equation (2-44) and for well A,

$$\bar{p} = 2000 + \frac{(141.2)(100)(1.2)(1.03)}{(8.2)(38)}\left[0.5\ln\frac{(4)(5.6 \times 10^6)}{(1.78)(10.8)(0.328)^2} + 2\right]$$

$$= 2565 \text{ psi}. \quad \textbf{(2-49)}$$

The average pressures in the drainage areas for wells B and C are calculated to be 2838 and 2643 psi, respectively.

Such uneven depletion is common and is an important variable to know in any reservoir exploitation strategy.

2.6 Inflow Performance Relationship

All well deliverability equations relate the well production rate and the driving force in the reservoir, that is, the pressure difference between the initial, outer boundary or average reservoir pressure and the flowing bottomhole pressure.

If the bottomhole pressure is given, the production rate can be obtained readily. However, the bottomhole pressure *is* a function of the wellhead pressure, which, in turn, depends on production engineering decisions, separator or pipeline pressures, and so on. Therefore, what a well will actually produce must be the combination of what the reservoir can deliver and what the imposed wellbore hydraulics would allow.

It is then useful to present the relationship between the well production rate and the bottomhole pressure as the inflow performance relationship (IPR), introduced in Chapter 1. Usually, the bottomhole pressure, p_{wf}, is graphed on the ordinate and the production rate, q, is graphed on the abscissa.

Equations (2-9), (2-26), and (2-44) can be used for steady-state, transient, and pseudo-steady-state IPR curves. The following examples illustrate these concepts.

Example 2-7 Transient IPR

Using the well and reservoir data in Appendix A, construct transient IPR curves for 1, 6, and 24 months. Assume zero skin.

Solution

Equation (2-24) with substituted variables takes the form

$$q = \frac{2.16(5651 - p_{wf})}{\log t + 4.25}. \quad \textbf{(2-50)}$$

The relationship between q and p_{wf} of course will depend on time, t [which in Equation (2-50) must be entered in hours]. Figure 2-5 is a graph of the *transient* IPR curves for the three different times.

Example 2-8 Steady-State IPR: Influence of the Skin Effect

Assume that the initial reservoir pressure of the well described in Appendix A is also the constant pressure of the outer boundary, p_e (steady state). Draw IPR curves for skin effects equal to 0, 5, 10, and 50, respectively. Use a drainage radius of 2980 ft ($A = 640$ acres).

Solution

Equation (2-9) describes a straight-line relationship between q and p_{wf} for any skin effect. For example, after substitution of variables for a skin equal to 5, Equation (2-9) becomes

$$p_{wf} = 5651 - 5.66q. \tag{2-51}$$

Similarly, the multipliers of the rate for the 0, 10, and 50 skin effects are 3.66, 7.67, and 23.7, respectively.

Figure 2-6 gives the steady-state IPR curves for the four skin effects.

Example 2-9 Pseudosteady-State IPR: Influence of Average Reservoir Pressure

This calculation is the most useful and the one most commonly done for the forecast of well performance. Each IPR curve reflects a "snapshot" of well performance at a given reservoir pressure. This is a time-dependent calculation, done in discrete intervals. Combination with volumetric material balances (treated in Chapter 10) will allow the forecast of rate and cumulative production versus time.

For this exercise, calculate the IPR curves for zero skin effect but for average reservoir pressures in increments of 500 psi from the "initial" 5651 to 3500 psi. Use all other variables from Appendix A. Drainage radius is 2980 ft.

Solution

Equation (2-44) is the generalized pseudosteady-state equation for any drainage shape and well position. For a circular drainage shape, Equation (2-34) is sufficient.

Substituting the variables from Appendix A into Equation (2-34) results (for $\bar{p} = 5651$) in

$$p_{wf} = 5651 - 3.36q. \tag{2-52}$$

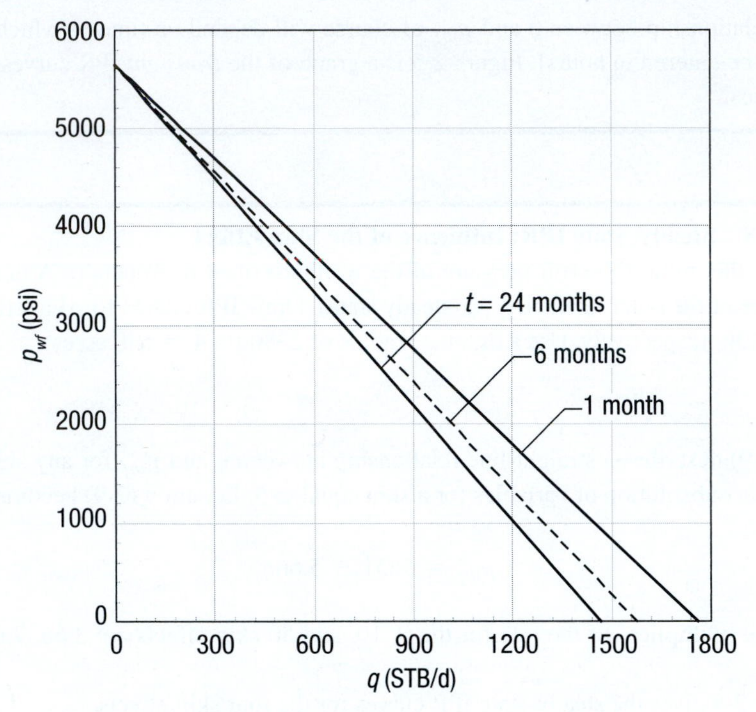

Figure 2-5 Transient IPR curves for Example 2-7.

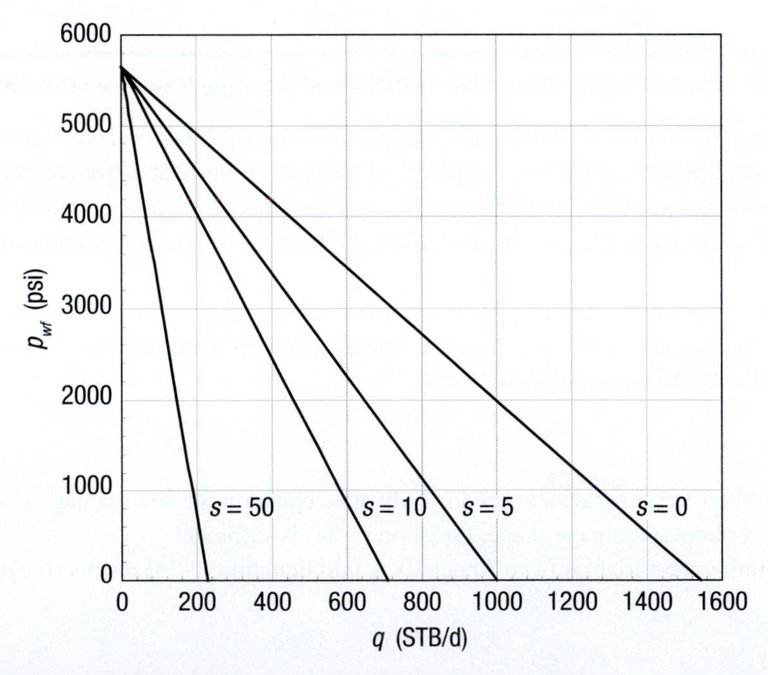

Figure 2-6 Steady-state and impact of skin effect for Example 2-8.

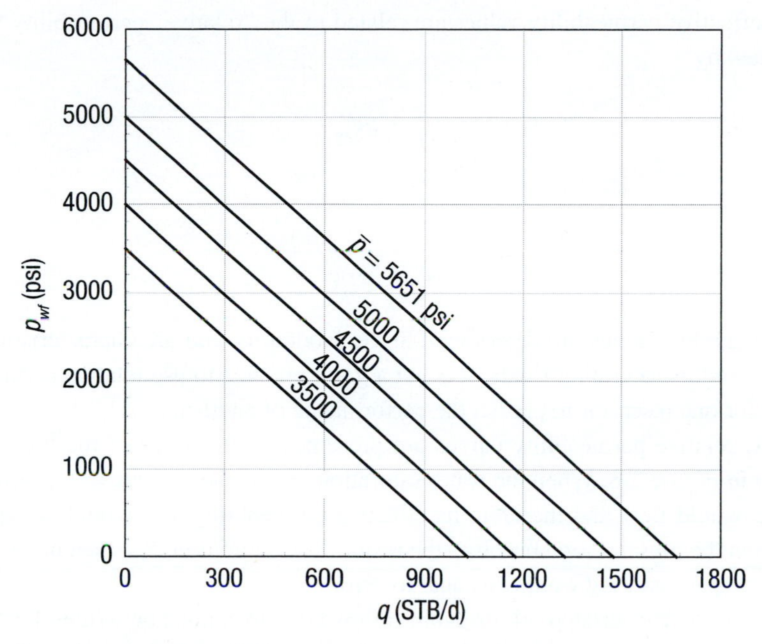

Figure 2-7 Pseudosteady-state IPR curves for a range of average reservoir pressures (Example 2-9).

For all average reservoir pressures, the slope in Equation (2-51) will remain the same. The intercept will simply be the average pressure. Therefore, as shown in Figure 2-7, the pseudosteady-state IPR curves are depicted as parallel straight lines, each reflecting an average reservoir pressure.

For these simplified examples, the slope of the IPR is the reciprocal of the productivity index, J.

2.7 Effects of Water Production, Relative Permeability

All previous sections in this chapter provided volumetric flow rates of undersaturated oil reservoirs as functions of the permeability, k. This permeability was used as a reservoir property. In reality this is only an approximation, since such a use of permeability is correct only if the flowing fluid is *also* the only saturating fluid. In such case the "absolute" and "effective" permeability values are the same.

In petroleum reservoirs, however, water is always present at least as connate water, denoted as S_{wc}. Thus, in all previous equations in this chapter the permeability should be considered as effective, and it would be invariably less (in certain cases significantly less) than the one obtained from core flooding or other laboratory techniques using a single fluid.

If both oil *and* free water are flowing, then effective permeability must be used. The sum of these permeability values is invariably less than the absolute permeability of the formation (to either fluid).

These effective permeability values are related to the "relative" permeability values (also rock properties) by

$$k_o = kk_{ro} \tag{2-53}$$

and

$$k_w = kk_{rw}. \tag{2-54}$$

Relative permeability values are determined in the laboratory and are characteristic of a given reservoir rock and its saturating fluids. It is not a good practice to use relative permeability values obtained for one reservoir to predict the performance of another.

Usually, relative permeability curves are presented as functions of the water saturation, S_w, as shown in Figure 2-8. When the water saturation, S_w, is the connate water saturation, S_{wc}, no free water would flow and therefore its effective permeability, k_w, would be equal to zero. Similarly, when the oil saturation becomes the residual oil saturation, S_{or}, then no oil would flow and its effective permeability would be equal to zero.

Thus, in an undersaturated oil reservoir, inflow equations must be written for both oil and water. For example, for steady-state production,

$$q_o = \frac{kk_{ro}h(p_e - p_{wf})_o}{141.2B_o\mu_o[\ln(r_e/r_w) + s]} \tag{2-55}$$

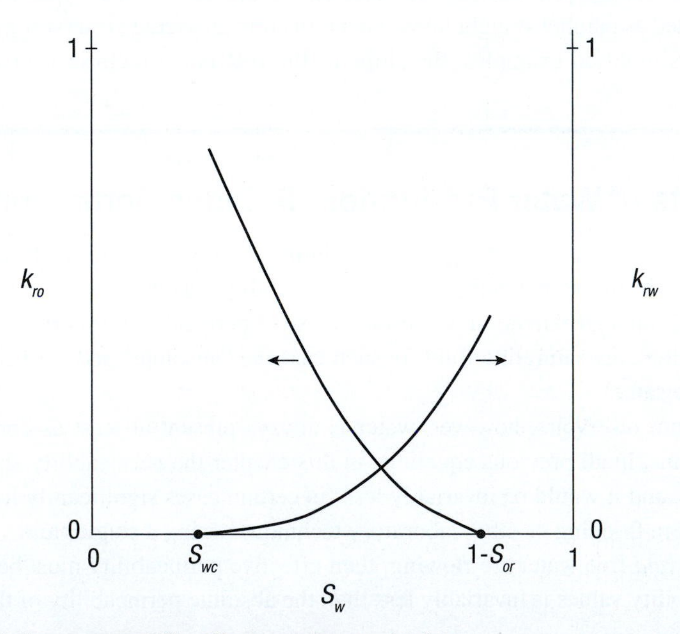

Figure 2-8 Relative permeability effects, water production.

and

$$q_w = \frac{kk_{rw}h(p_e - p_{wf})_w}{141.2B_w\mu_w[\ln(r_e/r_w) + s]} \tag{2-56}$$

with relative permeability values, k_{ro} and k_{rw}, being functions of S_w, as shown in Figure 2-11. Note that the pressure gradients have been labeled with subscripts for oil and water to allow for different pressures within the oil and water phases.

The ratio q_w/q_o is referred to as the water–oil ratio. In an almost depleted reservoir it would not be unusual to obtain water–oil ratios of 10 or larger. Such a well is often referred to as a "stripper" with the production rates of less than 10 STB/d of oil. In mature petroleum areas, stripper wells may constitute the overwhelming majority of producers. Their economic viability is frequently one of the most important questions confronting production engineers.

2.8 Summary of Single-Phase Oil Inflow Performance Relationships

This chapter has presented three inflow performance equations that can be used to analyze the reservoir behavior for single-phase oil production: steady-state, transient, and pseudosteady-state flows. The production engineer selects the most appropriate of these relationships based on the far-field boundary condition for the well of interest. If the pressure, p_e, at the drainage boundary can be approximated as being constant, the steady-state equation should be used. If there is no effect of a boundary felt at the well, transient flow is occurring. Finally, if the boundary at r_e is a no-flow boundary being felt at the well, the well performance is pseudosteady-state.

References

1. Carslaw, H.S., and Jaeger, J.C., *Conduction of Heat in Solids*, 2nd ed., Clarendon Press, Oxford, 1959.
2. Dake, L.P., *Fundamentals of Reservoir Engineering*, Elsevier, Amsterdam, 1978.
3. Darcy, H., *Les Fontaines Publiques de la Ville de Dijon*, Victor Dalmont, Paris, 1856.
4. Dietz, D.N., "Determination of Average Reservoir Pressure from Build-up Surveys," *JPT*, 955–959 (August 1965).
5. Earlougher, R.C., *Advances in Well Test Analysis*, Society of Petroleum Engineers, Dallas, 1977.
6. Van Everdingen, A.F., and Hurst, W., "The Application of the Laplace Transformation to Flow Problems in Reservoirs," *Trans. AIME*, **186**: 305–324 (1949).

Problems

2-1 An oil well produces under steady-state conditions. Assume that there is no skin effect in the well system. The drainage area is 40 acres, and the reservoir pressure is 5000 psi. Generate inflow performance relationship (IPR) curves for three permeability values of 1 md, 10 md, and 100 md, respectively. Obtain other variables from Appendix A.

2-2 Assume a production rate of 150 STB/d and a drainage area of 40 acres. The reservoir pressure is 5000 psi. Use the well and reservoir properties in Appendix A. Calculate the total pressure gradient required for values of skin of 0, 5, 10, and 20. In each case, what fraction of the pressure gradient is across the damage zone? Plot the well IPR for each skin value.

2-3 Equation (2-42) provided the time for beginning of pseudosteady state. Calculate this time for the well described in Appendix A if the drainage area is 80 acres. What would be the average reservoir pressure at that time? The flowing bottomhole pressure is 4000 psi. (Note: Calculate flow rate from transient relationship. Use this flow rate to solve for $\overline{p}$ from the pseudosteady-state relationship.)

2-4 Using the variables for the well in Appendix A, and assuming that the reservoir initial pressure is 5000 psi, develop a cumulative production curve for 1 year. At what time will the well produce 50% of the total? The flowing bottomhole pressure is 4000 psi. Select the appropriate flow equation based on Equation (2-42).

2-5 Continue Problem 4. Assuming a pressure decline rate of 500 psi/year, calculate the discounted present value for the expected cumulative production of the first 3 years. Assume that the oil price is $100. The discount rate (time value of money) is 0.08. Discounted present value $= (\Delta N_p)_n/(1 + i)^n$, where $(\Delta N_p)_n$ is the cumulative production in year n and i is the discount rate. What are the fractional contributions from each year to the 3-year discounted present value?

2-6 The average pressure in a reservoir is 5000 psi. If the well drainage area in another reservoir is half that of the first reservoir, what should be its average pressure to produce the same well flow rate? Use zero skin and 4000 psi bottomhole flowing pressure for both wells. Assume that all other variables are the same. Explain your result.

2-7 Given the well data in Appendix A, and assuming a 40 acre drainage area, calculate:

a. the time required to reach pseudosteady state.

b. the flow rate for transient flow from time $= 0$ to time to pseudosteady state at wellbore flowing pressures ranging from 0 to initial reservoir pressure. Display 5 equal increments for time and 10 for wellbore flowing pressure. Plot the corresponding IPR curves.

c. the flow rate for pseudosteady state, if the flow boundary pressure equals the initial reservoir pressure and the wellbore flowing pressure is 500 psi.

d. the IPR curve for a reservoir at pseudosteady-state condition with average reservoir pressure equal to 5000 psi at the time of transition. Repeat for average reservoir pressures of 4500 and 4000 psi.

2-8 For the well described in Problem 2-7, what should be the impact on the well production rate if the well were placed within the drainage area as depicted in configurations no. 8, no. 9, or no. 11 in Figure 2-3? How would be the presence of a skin effect change this comparison? Use skins equal to 0, 10, and 30.

CHAPTER 3

Production from Two-Phase Reservoirs

3.1 Introduction

The performance relationships presented in Chapter 2 were for single-phase oil wells and, while gas may come out of solution after oil enters the wellbore, the use of those relationships does not consider free gas to be present in the reservoir. Expansion of oil itself as a means of recovery is a highly inefficient mechanism because of the oil's small compressibility. It is likely that even in the best of cases (the issue is addressed in Chapter 10), if the bottomhole pressure is above the bubble-point pressure, as is the case for heavy oil, a very small fraction of the original-oil-in-place would be recovered. Therefore, in most oilfields, oil will be produced along with free gas in the reservoir, either because the reservoir pressure is naturally below the bubble-point pressure (saturated reservoirs) or because the flowing bottomhole pressure is set below that point to provide adequate driving force. In terms of ultimate recovery, expansion of free or solution gas is a much more efficient mechanism than the expansion of oil.

Figure 3-1 is a schematic of a classic phase diagram, plotting pressure versus temperature and identifying the important variables. Depending on the initial and flowing pressures and the reservoir temperature, a diagram such as in Figure 3-1 can indicate whether single-phase deliverability relationships as presented in Chapter 2 apply, or if the two-phase equations presented in this chapter are needed.

In Figure 3-1, the initial reservoir conditions for an undersaturated oil reservoir are marked as p_i, T_R. In this depiction, the reservoir is above the bubble-point pressure. The flowing bottomhole pressure, p_{wf}, is marked within the two-phase region. The flowing bottomhole temperature is taken as equal to the reservoir temperature, reflecting the typically isothermal flow in the reservoir (unless thermal recovery is attempted). The wellhead flowing conditions, with pressure p_{tf} and temperature T_{tf}, are also marked. Thus, the reservoir fluid would follow

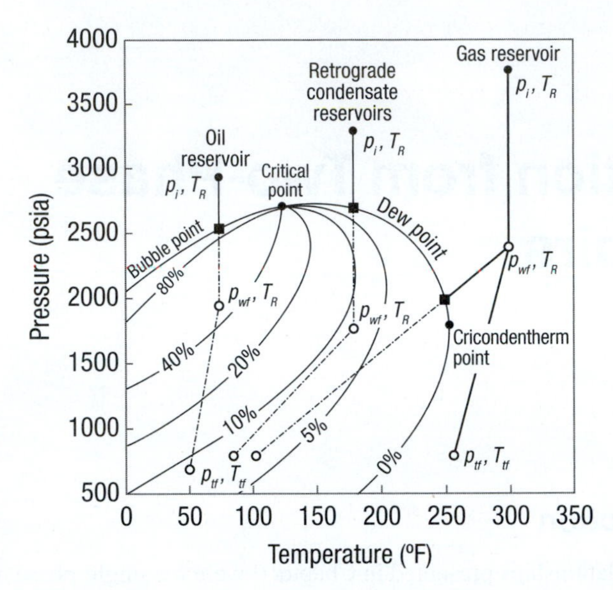

Figure 3-1 Schematic phase diagram of hydrocarbon mixture. Marked are reservoir, bottomhole, and wellhead flowing conditions for an oil reservoir.

a path from the reservoir to the surface, joining these three points. In an initially saturated reservoir, all three points would be within the two-phase envelope. For comparison, the paths for the fluids in a retrograde gas condensate reservoir and in a single-phase gas reservoir are shown. In the gas cases there is a much more pronounced reduction in the temperature along the path, reflecting Joule-Thomson expansion effects associated with gas flow. Solid lines connote single-phase flow.

3.2 Properties of Saturated Oil

3.2.1 General Properties of Saturated Oil

The bubble-point pressure is the important variable in characterizing saturated oil. At pressures above the bubble point, oil behaves like a liquid; below the bubble point, gas comes out of solution, becoming free gas coexisting with oil. The formation volume factor, B_o, measured in res bbl/STB, for oil above the bubble-point pressure includes all of the solution gas. At a pressure below the bubble point, the B_o refers to the liquid phase and its remaining dissolved gas at that pressure.

Figure B-1a in Appendix B shows a plot of B_o versus pressure for an example two-phase well. As oil is produced from a fixed reservoir volume, the reservoir pressure decreases. Above p_b, B_o increases with decreasing pressure, reflecting the expansion of the undersaturated oil. This increase is slight because liquid oil compressibility is small. Below p_b, B_o decreases as pressure decreases, reflecting the loss of gas from the oil and the resulting increase in the oil gravity.

In Figure B-1b the formation volume factor of the gas, B_g, increases nonlinearly with decreasing pressure because the gaseous phase is much more compressible than liquid oil, and this effect is more pronounced at lower pressure. Finally, Figure B-1c shows the solution gas–oil ratio, R_s. Above the bubble-point pressure, R_s is constant as R_{sb}, the solution gas–oil ratio at the bubble-point pressure, because the solution gas remains in the oil. Below the bubble-point pressure, R_s decreases with decreasing reservoir pressure because gas comes out of solution, leaving a decreasing amount of gas still dissolved in the oil. The solution gas–oil ratio is the amount of gas that would be liberated from a unit volume of oil at standard conditions.

When the produced gas–oil ratio observed for a well is greater than the solution gas–oil ratio, this signals that the reservoir pressure near the well is below the bubble-point pressure, and free gas that came out of solution will flow with the oil. The variables B_o, B_g, and R_s are related through the total formation volume factor, B_t, which accounts for both oil and free gas:

$$B_t = B_o + (R_{sb} - R_s)B_g \tag{3-1}$$

where $R_{sb} - R_s$ is produced as free gas. If B_g is given in res ft^3/SCF, then B_g must be divided by 5.615 to convert to res bbl/SCF. These pressure–volume–temperature (PVT) properties are usually obtained in the laboratory and are unique to a given reservoir fluid.

Example 3-1 PVT Properties Below the Bubble-Point Pressure: Impact on Oil Reserves

Calculate the total formation volume factor at 3000 psi for the reservoir fluid described in Appendix B. What would be the reduction in volume of oil (STB) in 4000 acres of the reservoir described in the same Appendix when the average pressure is reduced from the initial pressure to 3000 psi? Assume that the initial pressure is the bubble-point pressure and is equal to 4336 psi.

Solution

From the figures in Appendix B, $R_{sb} = 800$ SCF/STB, and at 3000 psi $B_o = 1.33$ res bbl/STB, $B_g = 5.64 \times 10^{-3}$ res ft^3/SCF, $R_s = 517$ SCF/STB. Then from Equation (3-1),

$$B_t = B_o + (R_{sb} - R_s)B_g = 1.33 + (800 - 517)\frac{5.64 \times 10^{-3}}{5.615} = 1.61 \text{ res bbl/STB} \tag{3-2}$$

From Appendix B, $B_{ob} = 1.46$ res bbl/STB, and from Equation (1-3), the original-oil-in-place at $p_i = p_b$ is (from Figure B-1a)

$$N = \frac{7758Ah\phi(1 - S_w)}{B_{oi}} = \frac{(7758)(4000)(115)(0.21)(0.7)}{1.46} \tag{3-3}$$

$$= 3.6 \times 10^8 \text{ STB} = 360 \text{ MMSTB}$$

For $B_t = 1.61$ res bbl/STB, $N = 330$ MMSTB, a reduction of 30 MMSTB. The remaining original reservoir volume is now filled with gas that came out of solution from the oil.

Figures 3-2 and 3-3, from Standing's (1977) seminal work in oilfield hydrocarbon correlations, relate important variables such as R_s, γ_g, γ_o, T, p_b, and B_o for a large number of hydrocarbons. These correlations should be used only in the absence of experimentally determined *PVT* properties obtained for the specific reservoir. The correlation shown in Figure 3-2 can be computed with the following equation:

$$p_b = 18.2\left[\left(\frac{R_s}{\gamma_g}\right)^{0.83}\left(10^{(0.00091T - 0.0125\gamma_o)}\right) - 1.4\right] \tag{3-4}$$

with T in °F and γ_o in °API.

Example 3-2 Use of Correlations to Obtain PVT Properties of Saturated Oil Reservoirs

Supposing that $R_s = 500$ SCF/STB, $\gamma_g = 0.7$, $\gamma_o = 28$°API, and $T = 160$°F, calculate the bubble-point pressure. What would be the impact if $R_s = 1000$ SCF/STB but all other variables remain the same?

If $R_s = 500$ SCF/STB, $B_{ob} = 1.2$ res bbl/STB, $T = 180$°F, and $\gamma_o = 32$°API, what should be the gas gravity, γ_g?

Solution

Reading from Figure 3-2 in a stairstep manner with the first set of variables, $p_b = 2650$ psi. If $R_s = 1000$ SCF/STB, then $p_b = 4650$ psi. Using Equation (3-4),

$$p_b = 18.2\left[\left(\frac{R_s}{\gamma_g}\right)^{0.83}\left(10^{(0.00091T - 0.0125\gamma_o)}\right) - 1.4\right]$$

$$= 18.2\left[\left(\frac{500}{0.7}\right)^{0.83}\left(10^{(0.00091(160) - 0.0125(28))}\right) - 1.4\right] = 2632 \text{ psi.} \tag{3-5}$$

For $R_s = 1000$ SCF/STB, $p_b = 4698$ psi, Figure 3-3 can be used for the last question. Stating from the right at $B_{ob} = 1.2$ res bbl/STB, $T = 180$°F, and $\gamma_o = 32$°API and then from the left at $R_s = 500$ SCF/STB, the intersection of the two lines results in $\gamma_g = 0.7$.

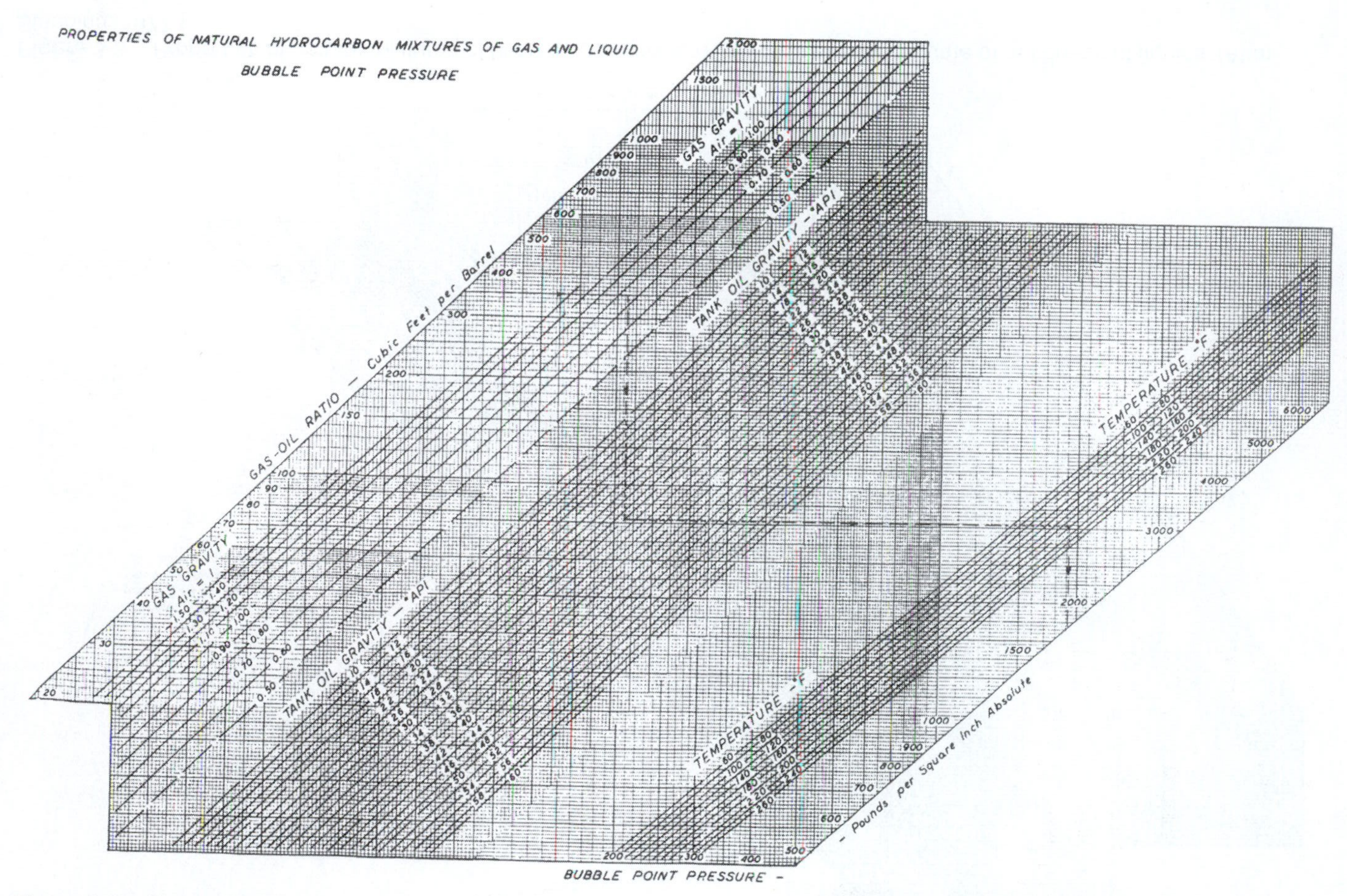

Figure 3-2 Properties of natural mixtures of hydrocarbon gas and liquids, bubble-point pressure. (After Standing, 1977.)

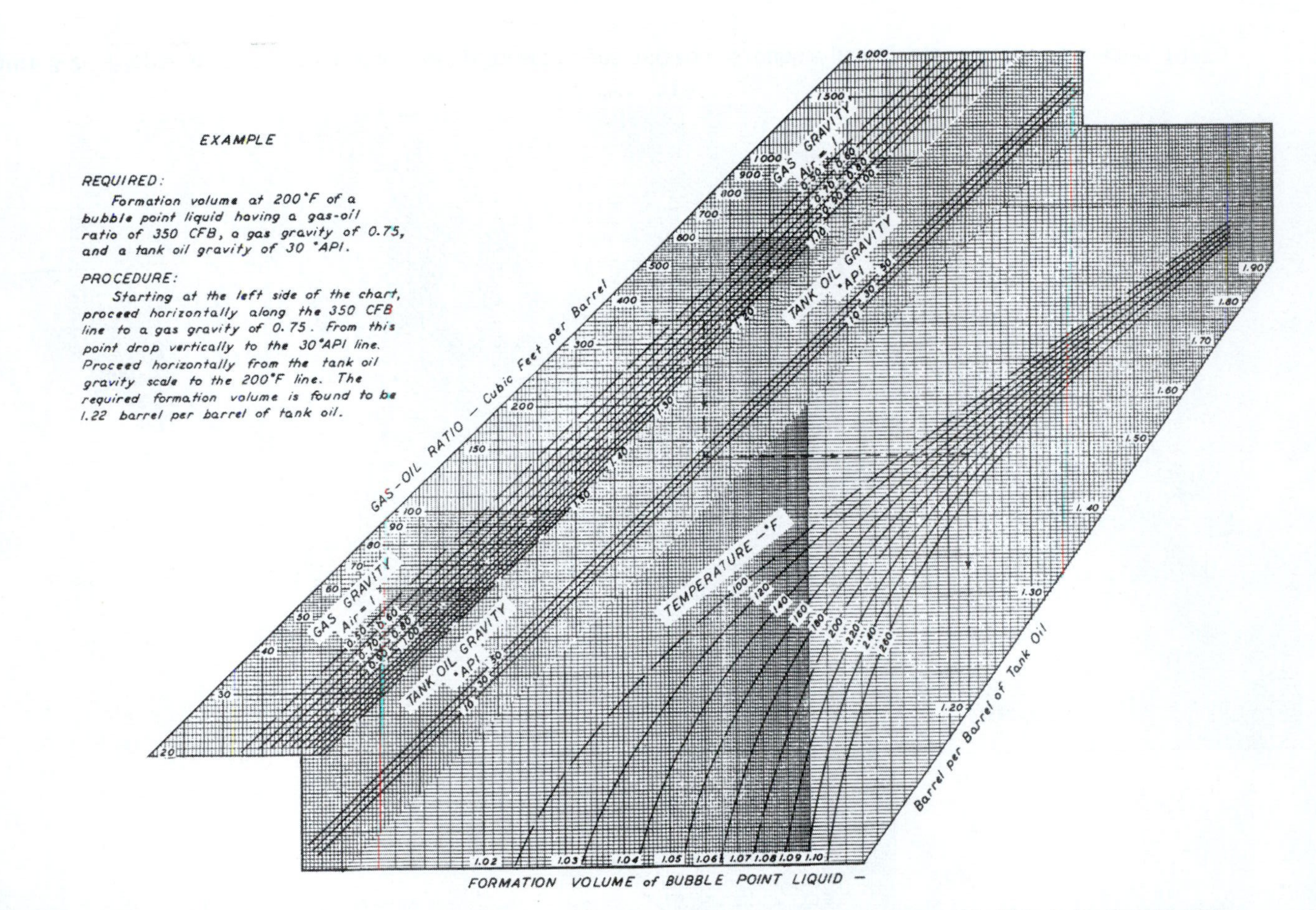

Figure 3-3 Properties of natural mixtures of hydrocarbon gas and liquids, formation volume of bubble-point liquids. (After Standing, 1977.)

3.2.2 Property Correlations for Two-Phase Systems

This subsection presents the most widely used property correlations for two-phase oilfield hydrocarbon systems.

The downhole volumetric flow rate of oil is related to the surface rate through the formation volume factor, B_o:

$$q_l = B_o q_o. \tag{3-6}$$

Here q_l is the actual liquid flow rate at some location in the well or reservoir. The downhole gas rate depends on the solution gas–oil ratio, R_s, according to

$$q_g = B_g(R_p - R_s)q_o \tag{3-7}$$

where B_g is the gas formation volume factor, addressed further in Chapter 4, and R_p is the produced gas–oil ratio in SCF/STB.

The oil-formation volume factor and the solution gas–oil ratio, R_s, vary with temperature and pressure. They can be obtained from laboratory *PVT* data or from correlations. One common correlation is the one of Standing, given in Figures 3-2 and 3-3. Another correlation that is accurate for a wide range of crude oils is that by Vasquez and Beggs (1990), given here.

First, the gas gravity is corrected to the reference separator pressure of 100 psig (114.7 psia):

$$\gamma_{gs} = \gamma_{gsep}\left[1 + 5.912 \times 10^{-5}\,\gamma_o T_{sep} \log\left(\frac{p_{sep}}{114.7}\right)\right] \tag{3-8}$$

where T_{sep} is in °F, p_{sep} is in psia, and γ_o is in °API. The solution gas–oil ratio is then, for $\gamma_o \leq 30$°API,

$$R_s = \frac{\gamma_{gs} p^{1.0937}}{27.64}(10^{11.172A}) \tag{3-9}$$

and for $\gamma_o > 30$°API,

$$R_s = \left(\frac{\gamma_{gs} p^{1.187}}{56.06}\right)(10^{10.393A}) \tag{3-10}$$

where

$$A = \frac{\gamma_o}{T + 460}. \tag{3-11}$$

For pressures below the bubble-point pressure, the oil-formation volume factor for $\gamma_o \leq 30°\text{API}$ is

$$B_o = 1.0 + 4.677 \times 10^{-4} R_s + 0.1751 \times 10^{-4} F - 1.8106 \times 10^{-8} R_s F \tag{3-12}$$

and for $\gamma_o > 30°\text{API}$ is

$$B_o = 1.0 + 4.67 \times 10^{-4} R_s + 0.11 \times 10^{-4} F + 0.1337 \times 10^{-8} R_s F \tag{3-13}$$

where

$$F = (T - 60)\left(\frac{\gamma_o}{\gamma_{gs}}\right) \tag{3-14}$$

and for pressure above the bubble point is

$$B_o = B_{ob} e^{c_o(p_b - p)} \tag{3-15}$$

where

$$c_o = \frac{-1.433 + 5R_s + 17.2T - 1.180\gamma_{gs} + 12.61\gamma_o}{p \times 10^5} \tag{3-16}$$

and B_{ob} is the formation volume factor at the bubble point. At the bubble point, the solution gas–oil ratio is equal to the produced gas–oil ratio, R_p, so the bubble-point pressure can be estimated from Equations (3-9) or (3-10) by setting $R_s = R_p$ and solving for p. Then Equation (3-12) or (3-13) may be used to calculate B_{ob}.

3.2.2.1 Liquid Density

The oil density at pressures below the bubble point is

$$\rho_o = \frac{[8830/(131.5 + \gamma_o)] + 0.01361\gamma_{gd} R_s}{B_o} \tag{3-17}$$

where ρ_o is in lb_m/ft^3 and γ_{gd} is the dissolved gas gravity because of the changing gas composition with temperature. It can be estimated from Figure 3-4 (Katz et al., 1959). Above the bubble point, the oil density is

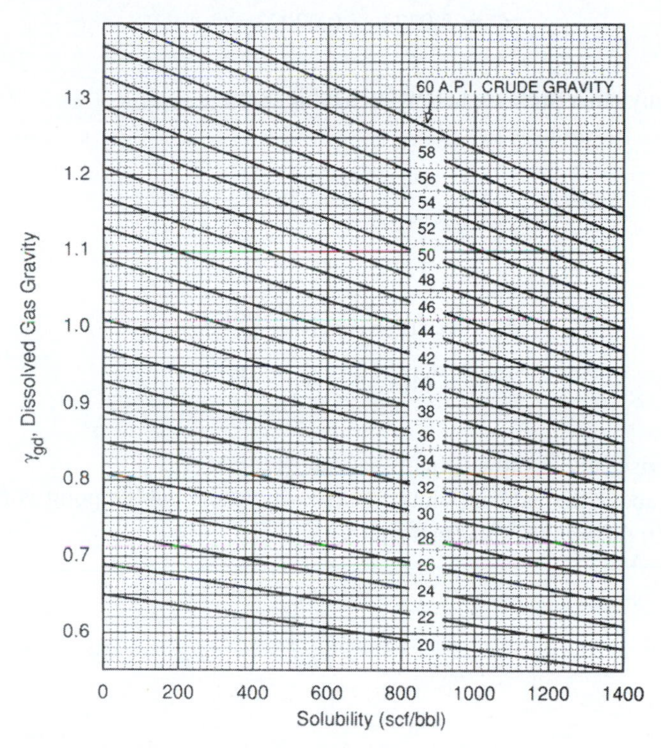

Figure 3-4 Prediction of gas gravity from solubility and crude-oil gravity. (After Katz et al., Handbook of Natural Gas Engineering, Copyright 1959, McGraw-Hill, reproduced with permission of McGraw-Hill.)

$$\rho_o = \rho_{ob}\left(\frac{B_{ob}}{B_o}\right) \quad \textbf{(3-18)}$$

where B_o is calculated from Equation (3-13) and B_{ob} from Equation (3-12) or (3-13), with $R_s = R_p$.

3.2.2.2 Oil Viscosity

Oil viscosity can be estimated with the correlations of Beggs and Robinson (1975) and Vasquez and Beggs (1980). The "dead" oil viscosity is

$$\mu_{od} = 10^A - 1 \quad \textbf{(3-19)}$$

where

$$A = BT^{-1.163} \quad \textbf{(3-20)}$$

$$B = 10^C \quad \textbf{(3-21)}$$

$$C = 3.0324 - 0.02023\gamma_l. \tag{3-22}$$

The oil viscosity at any other pressure below the bubble point is

$$\mu_o = a\mu_{od}^b \tag{3-23}$$

where

$$a = 10.715(R_s + 100)^{-0.515} \tag{3-24}$$

$$b = 5.44(R_s + 150)^{-0.338}. \tag{3-25}$$

If the stock tank oil viscosity is known, this value can be used for μ_{od}.

For pressures above the bubble point, the viscosity at the bubble point is first computed with Equations (3-19) through (3-25). Then

$$\mu_o = \mu_{ob}\left(\frac{p}{p_b}\right)^m \tag{3-26}$$

where

$$m = 2.6p^{1.187}e^{(-11.513-8.98\times10^{-5}p)}. \tag{3-27}$$

Gas viscosity can be estimated with the correlation that will be given in Chapter 4.

3.2.2.3 Accounting for the Presence of Water

When water is produced, the liquid flow properties are generally taken to be averages of the oil and water properties. If there is no slip between the oil and water phases, the liquid density is the volume fraction-weighted average of the oil and water densities. The volume fraction-weighted averages will be used to estimate liquid viscosities and surface tension, though there is no theoretical justification for this approach. The reader should note that in the petroleum literature it has been common practice to use volume fraction-weighted average liquid properties in oil–water–gas flow calculations. Also, the formation volume factor for water is normally assumed to be 1.0 because of low compressibility and gas solubility. Thus, when water and oil are flowing,

$$q_l = q_o(WOR + B_o) \tag{3-28}$$

$$\rho_l = \frac{WOR\rho_w + B_o\rho_o}{WOR + B_o} \tag{3-29}$$

$$\mu_l = \left(\frac{WOR\rho_w}{WOR\rho_w + B_o\rho_o}\right)\mu_w + \left(\frac{B_o\rho_o}{WOR\rho_w + B_o\rho_o}\right)\mu_o \quad \text{(3-30)}$$

$$\sigma_l = \left(\frac{WOR\rho_w}{WOR\rho_w + B_o\rho_o}\right)\sigma_w + \left(\frac{B_o\rho_o}{WOR\rho_w + B_o\rho_o}\right)\sigma_o \quad \text{(3-31)}$$

where WOR is the water–oil ratio, and σ is the surface tension.

Example 3-3 Estimating Downhole Properties

Suppose that 500 bbl/d of the oil described in Appendix B is being produced at WOR = 1.5 and R_p = 500 SCF/STB. The separator conditions for properties given in Appendix B are 100 psig and 100°F. Using the correlations presented in Section 3.2.2, estimate the volumetric flow rates of the gas and liquid and the density and viscosity of the liquid at a point in the tubing where the pressure is 2000 psia and the temperature is 150°F.

Solution

The first step is to calculate R_s and B_o. Since the separator is at the reference condition of 100 psig, $\gamma_{gs} = \gamma_g$. From Equations (3-9) to (3-14),

$$A = \frac{\gamma_o}{T + 460} = \frac{32}{150 + 460} = 0.0525 \quad \text{(3-32)}$$

$$R_s = \left(\frac{\gamma_{gs}p^{1.187}}{56.06}\right)(10^{10.393A}) = \left[\frac{(0.71)(2000)^{1.187}}{56.06}\right](10^{10.393\times 0.0525}) = 369 \text{ SCF/STB} \quad \text{(3-33)}$$

$$F = (T - 60)\left(\frac{\gamma_o}{\gamma_{gs}}\right) = (150 - 60)\left(\frac{32}{0.71}\right) = 4.056 \times 10^3 \quad \text{(3-34)}$$

$$\begin{aligned} B_o &= 1.0 + 4.67 \times 10^{-4}R_s + 0.11 \times 10^{-4}F + 0.1337 \times 10^{-8}R_sF \\ &= 1 + (4.67 \times 10^{-4})(369) + 0.11 \times 10^{-4}(4.056 \times 10^3) \\ &\quad + 0.1337 \times 10^{-8}(369)(4.056 \times 10^3) = 1.22 \text{ res bbl/STB.} \end{aligned} \quad \text{(3-35)}$$

The gas-formation volume factor, B_g, can be calculated from the real gas law. For T = 150°F and p = 2000 psi, it is 6.97×10^{-3} res ft^3/SCF. (This calculation is shown explicitly in Chapter 4.)

The volumetric flow rates are [Equations (3-28) and (3-7)]

$$q_l = q_o(WOR + B_o) = (500)(1.5 + 1.22) = 1360 \text{ bbl/day} = 7640 \text{ ft}^3\text{/d} \quad \text{(3-36)}$$

$$q_g = B_g(R_p - R_s)q_o = (6.9710 \times 10^{-3})(500 - 369)(500) = 457 \text{ ft}^3\text{/day.} \quad \text{(3-37)}$$

To calculate the oil density, the dissolved gas gravity, γ_{gd}, must be estimated. From Figure 3-4 it is found to be equal to 0.85. Then, from Equation (3-17),

$$\rho_o = \frac{[8830/(131.5 + \gamma_o)] + 0.01361\gamma_{gd}R_s}{B_o}$$

$$= \frac{[8830/(131.5 + 32)] + (0.01361)(0.85)(369)}{1.22} = 47.8 \text{ lb}_m/\text{ft}^3 \quad \textbf{(3-38)}$$

and, from Equation (3-29),

$$\rho_l = \frac{WOR\rho_w + B_o\rho_o}{WOR + B_o} = \frac{1.5(62.4) + 1.22(47.8)}{1.5 + 1.22} = 55.9 \text{ lb}_m/\text{ft}^3. \quad \textbf{(3-39)}$$

The oil viscosity can be estimated with Equation (3-19) through (3-25):

$$C = 3.0324 - 0.02023\gamma_o = 3.0324 - (0.02023)(32) = 2.385 \quad \textbf{(3-40)}$$

$$B = 10^C = 10^{2.385} = 242.7 \quad \textbf{(3-41)}$$

$$A = BT^{-1.163} = (242.7)(150)^{-1.163} = 0.715 \quad \textbf{(3-42)}$$

$$\mu_{od} = 10^A - 1 = 10^{0.715} - 1 = 4.19 \text{ cp} \quad \textbf{(3-43)}$$

$$a = 10.715(R_s + 100)^{-0.515} = 10.715(369 + 100)^{-0.515} = 0.451 \quad \textbf{(3-44)}$$

$$b = 5.44(R_s + 150)^{-0.338} = 5.44(369 + 150)^{-0.338} = 0.657 \quad \textbf{(3-45)}$$

$$\mu_o = a\mu_{od}^b = 0.451(4.19)^{0.657} = 1.16 \text{ cp}. \quad \textbf{(3-46)}$$

The liquid viscosity is then found from Equation (3-30), assuming $\mu_w = 1$ cp:

$$\mu_l = \left(\frac{WOR\rho_w}{WOR\rho_w + B_o\rho_o}\right)\mu_w + \left(\frac{B_o\rho_o}{WOR\rho_w + B_o\rho_o}\right)\mu_o$$

$$= \frac{(1.5)(62.4)}{(1.5)(62.4) + (1.22)(4.78)}(1)$$

$$+ \frac{(1.22)(4.78)}{(1.5)(62.4) + (1.22)(47.8)}(1.16) = 1.06 \text{ cp}. \quad \textbf{(3-47)}$$

3.3 Two-Phase Flow in a Reservoir

Although a rigorous treatment of two-phase flow in a reservoir is outside the scope of this textbook, it is necessary to understand the impact of competing phases on the flow of a fluid through the porous medium.

If there are two or three fluids flowing at the same time in a porous medium, the absolute reservoir permeability, k, is necessarily divided into "effective" permeability values, one for each fluid. Therefore, in a multiphase flow, oil flows with an effective permeability, k_o, water with k_w, and gas with k_g.

Even the presence of a nonflowing phase, such as connate water, whose saturation is denoted by S_{wc}, would cause some reduction in the effective permeability to oil when compared to a rock fully saturated with oil. Thus, laboratory-derived core permeability values with air or water should not be used automatically for reservoir calculations. Pressure transient test-derived permeability values are far more reliable (not only because of the reason outline above but also because they account for reservoir heterogeneities; cores reflect only local permeability values). The effective permeability values are related to the relative permeability values by simple expressions:

$$k_{ro} = \frac{k_o}{k}, \qquad k_{rw} = \frac{k_w}{k}, \qquad k_{rg} = \frac{k_g}{k}. \tag{3-48}$$

Relative permeability values are laboratory-derived relationships, are functions of fluid saturations, and, although frequently misapplied, are functions of the specific reservoir rock. Thus, relative permeability data developed for a reservoir cannot be readily transferred to another, even seemingly similar, formation.

Figure 3-5 is a schematic diagram of laboratory-derived oil and gas relative permeability data. Frequently relative permeability values are represented by a fit for laboratory data using Corey equations given by the following:

$$k_{rg} = k_{rg}^{o}\left(\frac{S_g - S_{gr}}{1 - S_{or} - S_{gr}}\right)^{n}, \qquad k_{ro} = k_{ro}^{o}\left(\frac{1 - S_g - S_{or}}{1 - S_{or} - S_{gr}}\right)^{m} \tag{3-49}$$

where k_{rg}^{o} and k_{ro}^{o} are water (wetting phase) and oil (nonwetting phase) endpoint relative permeability values at residual saturations S_{or} and S_{gr}. The effective permeability values are analogous, being simply the relative permeability values multiplied by the absolute permeability, k.

The rate equations developed in Chapter 2 for single-phase flow of oil must then be adjusted to reflect effective permeability values in the presence of gas. Furthermore, formation volume factor and viscosity vary significantly with pressure below the bubble-point pressure. The

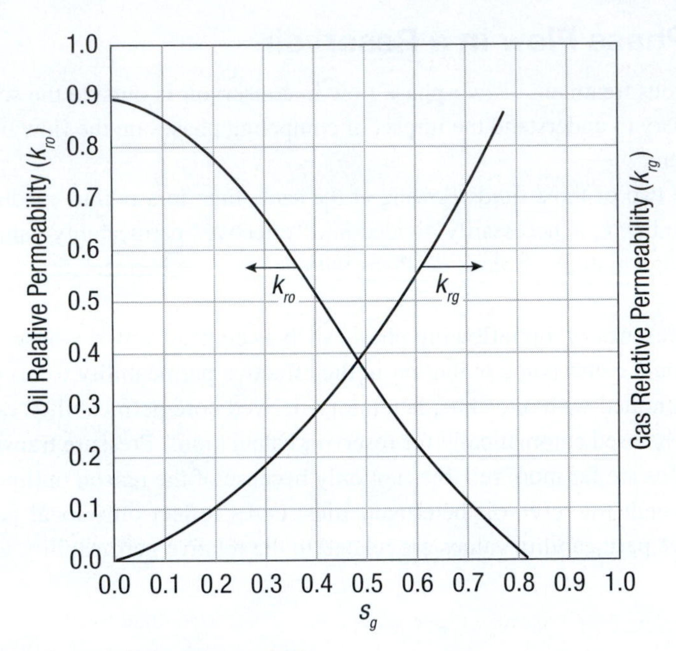

Figure 3-5 Oil and gas relative permeability.

generalized expression for the flow of oil, taking into account property variation and relative permeability effects, can be written for steady state as

$$q_o = \frac{kh}{141.2[\ln(r_e/r_w) + s]} \int_{p_{wf}}^{p_e} \frac{k_{ro}}{\mu_o B_o} dp. \tag{3-50}$$

For pseudosteady state, $\ln(r_e/r_w)$ and p_e can be changed to $\ln(0.472r_e/r_w)$ and $\overline{p}$, respectively. Finally, in the brackets in the denominator of Equation (3-50), the term Dq_o can be added to account for turbulence effects in high-rate wells, where D is the turbulence coefficient.

Example 3-4 The Effect of Relative Permeability on the Flow of Oil in a Two-Phase Reservoir

Using the experimentally determined relative permeability data in Appendix B, calculate the well flow rate if the flowing bottomhole pressure is 3000 psi. Compare the single-phase flow rate with the flow rate accounting for flow below bubble-point pressure. The drainage radius is 1490 ft and the skin effect is zero.

Solution

From a rearrangement of Equation (2-9),

$$q_{o,\text{singlephase}} = \frac{kh(p_i - p_{wf})}{141.2 B_o \mu \ln(r_e/r_w)} = \frac{(13)(115)(4336 - 3000)}{(141.2)(1.5)(0.45)[\ln(1490/0.406)]} = 2553 \text{ STB/d.} \quad \textbf{(3-51)}$$

Strictly speaking, relative permeability is a function of saturation, not pressure. However, a gas saturation value can be assumed for each pressure below bubble-point pressure. For the calculation below bubble point, $S_g(p)$ is assumed to be linear and ranging from 0 for $p = 4336$ psia to 0.3 for $p = 3000$ psia.

$$q_o = \frac{kh}{141.2[\ln(r_e/r_w) + s]} \int_{p_{wf}}^{p_e} \frac{k_{ro}}{\mu_o B_o} dp$$

$$= \frac{(13)(115)}{141.2[\ln(1490/0.406)]} \int_{3000}^{4350} \frac{k_{ro}(p)}{\mu_o(p) B_o(p)} dp = 1.29(1563) = 2016 \text{ STB/d.} \quad \textbf{(3-52)}$$

This is a significant reduction from the single-phase value. The integration is computed numerically using the trapezoidal rule.

3.4 Oil Inflow Performance for a Two-Phase Reservoir

Vogel (1968) introduced an empirical relationship for q_o based on a number of history-matching simulations. The relationship, normalized for the absolute open flow potential, $q_{o,\max}$ is

$$\frac{q_o}{q_{o,\max}} = 1 - 0.2\frac{p_{wf}}{\overline{p}} - 0.8\left(\frac{p_{wf}}{\overline{p}}\right)^2 \quad \textbf{(3-53)}$$

where, for pseudosteady state,

$$q_{o,\max} = \left(\frac{1}{1.8}\right) \frac{k_o h \overline{p}}{141.2 B_o(\overline{p}) \mu_o(\overline{p})[\ln(0.472 r_e/r_w) + s]} \quad \textbf{(3-54)}$$

where k_o is the effective permeability to oil that might be estimated from a pressure buildup test. Therefore,

$$q_o = \frac{k_o h \overline{p}[1 - 0.2(p_{wf}/\overline{p}) - 0.8(p_{wf}/\overline{p})^2]}{254.2 B_o(\overline{p}) \mu_o(\overline{p})[\ln(0.472 r_e/r_w) + s]}. \quad \textbf{(3-55)}$$

The convenience of the Vogel correlation is that it allows the use of the properties of only oil in a two-phase system.

Example 3-5 Calculation of Inflow Performance Using Vogel's Correlation

Develop an IPR curve for the well described in Appendix B. The drainage radius is 1490 ft and the skin effect is zero.

Solution

Equation (3-55) for $\overline{p} = 4336$ psi becomes

$$q_o = \frac{kh\overline{p}[1 - 0.2(p_{wf}/\overline{p}) - 0.8(p_{wf}/\overline{p})^2]}{254.2B_o(\overline{p})\mu_o(\overline{p})[\ln(0.472r_e/r_w) + s]}$$

$$= \frac{13(115)(4336)[1 - 0.2(p_{wf}/\overline{p}) - 0.8(p_{wf}/\overline{p})^2]}{254.2(1.5)(0.45)[\ln(0.472(1490)/0.406) + 0]}$$

$$= 5067\left[1 - 0.2\frac{p_{wf}}{\overline{p}} - 0.8\left(\frac{p_{wf}}{\overline{p}}\right)^2\right]. \tag{3-56}$$

The flow-rate prediction with Equation (3-55) is simple. For example, if $p_{wf} = 3000$ psi, from Equation (3-56), $q_o = 2427$ STB/d.

Figure 3-6 is a plot of the IPR curve for this well. Note that the shape of the curve is no longer a straight line as for the single-phase oil in Chapter 2.

3.5 Generalized Vogel Inflow Performance

If the reservoir pressure is above the bubble point and yet the flowing bottomhole pressure is below, a generalized inflow performance can be written. The following approach enables generation of an IPR that has a straight portion for $p_{wf} \geq p_b$, and follows the Vogel equation for $p_{wf} < p_b$ adapted to the straightforward logic found in Standing (1971). This can be done for transient, steady state, and pseudosteady state. First, q_b, the flow rate where $p_{wf} = p_b$, can be written as

$$q_b = \frac{kh(p_i - p_b)}{141.2B\mu(p_D + s)} \tag{3-57}$$

where p_D is the transient dimensionless pressure drop or is equal to $\ln(r_e/r_w)$ for steady state or $\ln(0.472\, r_e/r_w)$ for pseudosteady state.

The productivity index above the bubble point is simply

$$J = \frac{q_b}{p_i - p_b} \tag{3-58}$$

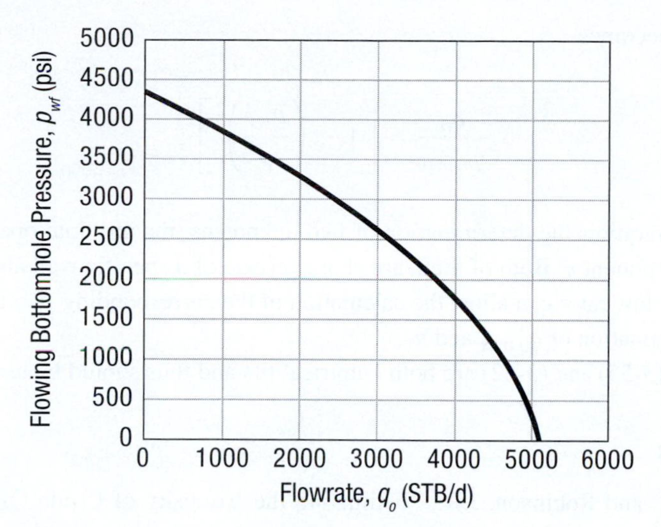

Figure 3-6 Inflow performance curve for a two-phase well. IPR Curve

and is related to q_v (denoted here as "Vogel" flow, qV) by

$$q_v = \frac{p_b J}{1.8}. \tag{3-59}$$

For $p_{wf} \geq p_b$,

$$q_o = J(p_i - p_{wf}). \tag{3-60}$$

For $p_{wf} < p_b$,

$$q_o = q_b + q_v\left[1 - 0.2\frac{p_{wf}}{p_b} - 0.8\left(\frac{p_{wf}}{p_b}\right)^2\right]. \tag{3-61}$$

When $\overline{p}$ is less than or equal to the original bubble-point pressure, use Equation 3-55.

3.6 Fetkovich's Approximation

Vogel's correlation, normalizing q_o by $q_{o,\,\max}$, is frequently not in close accordance with field data. Fetkovich (1973) suggested a normalization with $q_{o,\,\max} = C\overline{p}^{2n}$, and in a flow equation of the form

$$q_o = C(\overline{p}^2 - p_{wf}^2)^n \tag{3-62}$$

the relationship becomes

$$\frac{q_o}{q_{o,\,\max}} = \left[1 - \left(\frac{p_{wf}}{\overline{p}}\right)^2\right]^n. \tag{3-63}$$

Equation (3-63) requires the determination of two unknowns, the absolute open flow potential, $q_{o,\max}$, and the exponent n. Both of them are characteristic of a specific well and therefore a test at two stabilized flow rates can allow the calculation of the corresponding p_{wf} and therefore can lead to the determination of $q_{o,\max}$ and n.

Equations (3-55) and (3-62) are both empirical fits and thus should be used accordingly.

References

1. Beggs, H.D., and Robinson, J.R., "Estimating the Viscosity of Crude Oil Systems," *JPT*, 1140–1141 (September 1975).
2. Fetkovich, M.J., "The Isochronal Testing of Oil Wells," SPE Paper 4529, 1973.
3. Katz, D.L., Cornell, D., Kobayashi, R.L., Poettmann, F.H., Vary, J.A., Elenbaas, J.R., and Weinang, C.F., *Handbook of Natural Gas Engineering*, McGraw-Hill, New York, 1959.
4. Standing, M.B., "Concerning the Calculation of Inflow Performance of Wells Producing from Solution Gas Drive Reservoirs," *JPT*, 1141–1142 (September 1971).
5. Standing, M.B., *Volumetric and Phase Behavior of Oil Field Hydrocarbon Systems*, Society of Petroleum Engineers, Dallas, 1977.
6. Vasquez, M., and Beggs, H.D., "Correlations for Fluid Physical Property Predictions," *JPT*, 968–970 (June 1990).
7. Vogel, J.V., "Inflow Performance Relationships for Solution-Gas Drive Wells," *JPT*, 83–92 (January 1968).

Problems

3-1 A reservoir at pressure 5000 psi and temperature 160°F, with 35°API oil and gas–oil ratio 1 Mscf/stb produces no water. Estimate the bubble-point pressure. What are the values for B_o and oil viscosity at the bubble point pressure? Plot B_o versus pressure. The separator is 100°F and 100 psi, and $\gamma_g = 0.7$.

3-2 For the 30°API oil at 180°F, for gas–oil ratio values ranging from 0 to 2000 scf/stb, plot the bubble point B_o versus R_s, and the bubble point pressure p_b versus R_s. Use $\gamma_g = 0.7$.

3-3 Assume $\gamma_o = 25°\text{API}$, $\gamma_g = 0.7$ and $R_s = 600$ scf/stb. For temperature ranging from 120°F to 210°F, plot bubble point B_o versus T, and the bubble-point pressure p_b versus T. On the same plot axes, graph oil viscosity versus pressure for temperature values 120°F, 150°F, 180°F, and 210°F for pressure ranging from 14.7 psi to 4000 psi.

3-4 Suppose that 500 bbl/d of the oil described in Appendix B is being produced at a WOR (water–oil ratio) of 1.5 and GOR (gas–oil ratio) of 500. The separator conditions for properties given in Appendix B are 100 psig and 100°F. Plot the liquid volumetric flow rate at constant temperature 150°F for pressure ranging from 0 to the reservoir pressure.

3-5 Assume that each 0.02 in the gas saturation in Figure B-2 represents a reduction in the bottomhole pressure of 200 psi. With these relative permeability data, the PVT properties in Figure B-1, and the data in Appendix B, use Equation (3-50) to forecast the well oil flow rate as a function of bottomhole flowing pressure p_{wf}. Assume that the skin factor is equal to zero, and the well drainage area is 40 acres.

3-6 For a well draining 80 acres, with thickness 80 ft and permeability 15 md, assume $B_o = 1.2$, $\mu = 1.8$ cp, $r_w = 0.3$ ft, and the flow is steady-state flow following the Vogel correction. Plot IPRs in one figure for skin values of $s = 0$, 5, 10, and 20. Assume the reservoir and bubble-point pressure are 5000 psi.

3-7 A well drains 40 acres with thickness 100 ft and permeability 20 md. The initial reservoir pressure is 6000 psi, and the bubble-point pressure is 4500 psi. Assume the initial bottom-hole pressure is 5000 psi and decreases at a rate of 500 psi/year for 3 years. What is the cumulative oil production for the 3 years? Assume pseudosteady-state flow, and that $B_o = 1.1$, $\mu = 1.7$ cp, $r_w = 0.328$ ft.

3-8 Assume that a well produces oil following Fetkovich's approximation given by Equation 3-63. Several tests show that the well produces 1875 STB/d at $p_{wf} = 3000$ psi, 1427 STB/d at $p_{wf} = 3500$ psi, and 860 STB/d at $p_{wf} = 4000$ psi. Determine the $q_{o,\max}$, n, and the average pressure of the reservoir. Plot the IPR curve for this well.

3-9 An operator is trying to decide how much drawdown to apply to his new well producing from a reservoir with properties as in Appendix B. He knows from experience that wells in the area that are produced above the bubble point produce at a 30% exponential decline for 5 years. Wells that are produced with a p_{wf} of 3000 psi produce at a much higher rate but for only 3 years at a decline rate of 30%. After that only gas is produced and there is no pipeline to market the gas. Determine the p_{wf} that will maximize the NPV of this well. Assume pseudosteady-state flow, oil price \$100/bbl, and discount rate 10%. The average S_g in the reservoir when $p_{wf} = 3000$ is 0.15.

CHAPTER 4

Production from Natural Gas Reservoirs

4.1 Introduction

Natural gas reservoirs produce hydrocarbons that exist primarily in the gaseous phase at reservoir conditions. To predict the gas production rate from these reservoirs, there is a need to review some of the fundamental properties of hydrocarbon gases. This is particularly important (more so than in the case of oil reservoirs) because certain physical properties of gases and gas mixtures vary significantly with pressure, temperature, and gas composition. Following is a brief review of gas gravity, the real gas law, gas compressibility, the impact of nonhydrocarbon gases, gas viscosity, and gas isothermal compressibility.

4.1.1 Gas Gravity

Gas gravity, as used in natural gas production and reservoir engineering, is the ratio of the molecular weight of a natural gas mixture to that of air, itself a mixture of gases. Gas gravity is perhaps the most important defining property of a natural gas because almost all properties and, in fact, the actual description of the natural gas itself, are intimately related to it. The molecular weight of air is usually taken as equal to 28.97 (approximately 79% nitrogen and 21% oxygen). Therefore the gas gravity, symbolized by γ_g, is

$$\gamma_g = \frac{MW}{28.97} = \frac{\sum y_i MW_i}{28.97} \tag{4-1}$$

where y_i and MW_i are the mole fraction and molecular weight, respectively, of an individual component in the natural gas mixture.

Table 4-1 Molecular Weights and Critical Properties of Pure Components of Natural Gas

Compound	Chemical Composition	Symbol (for calculations)	Molecular Weight	Critical Pressure (psi)	Critical Temperature R
Methane	CH_4	C_1	16.04	673	344
Ethane	C_2H_6	C_2	30.07	709	550
Propane	C_3H_8	C_3	44.09	618	666
iso-Butane	C_4H_{10}	$i\text{-}C_4$	58.12	530	733
n-Butane	C_4H_{10}	$n\text{-}C_4$	58.12	551	766
iso-Pentane	C_5H_{12}	$i\text{-}C_5$	72.15	482	830
n-Pentane	C_5H_{12}	$n\text{-}C_5$	72.15	485	847
n-Hexane	C_6H_{14}	$n\text{-}C_6$	86.17	434	915
n-Heptane	C_7H_{16}	$n\text{-}C_7$	100.2	397	973
n-Octane	C_8H_{18}	$n\text{-}C_8$	114.2	361	1024
Nitrogen	N_2	N_2	28.02	492	227
Carbon Dioxide	CO_2	CO_2	44.01	1072	548
Hydrogen Sulfide	H_2S	H_2S	34.08	1306	673

Table 4-1 gives the molecular weights and critical properties for most hydrocarbon and nonhydrocarbon gases likely to be found in a natural gas reservoir. A *light* gas reservoir is one that contains primarily methane with some ethane and traces of higher molecular-weight gases. Pure methane has a gravity equal to (16.04/28.97) = 0.55. A *rich* or *heavy* gas reservoir may have a gravity equal to 0.75 or, in some rare cases, higher than 0.9. Such gravity would mean a significant concentration of ethane plus propane and butane. In commercial practice, the latter two components are removed and sold as natural gas liquids (NGLs).

Example 4-1 Calculation of the Gravity of a Natural Gas

A natural gas consists of the following (molar) composition: $C_1 = 0.880$, $C_2 = 0.082$, $C_3 = 0.021$, and $CO_2 = 0.017$. Calculate the gas gravity to air.

Solution

With the data in Table 4-1 and the given composition, the contributions to this natural gas molecular weight are

Compound	Composition	y_iMW_i
C_1	0.880	14.115
C_2	0.082	2.466
C_3	0.021	0.926
CO_2	0.017	0.748
		18.255

Therefore, the gas gravity is 18.225/28.97 = 0.63.

4.1.2 Real Gas Law

The behavior of natural gas mixtures can be approximated by the real gas law

$$pV = ZnRT \tag{4-2}$$

where Z is the compressibility factor, also called the gas deviation factor in the petroleum engineering literature. The universal gas constant, R, is equal to 10.73 psi ft^3/lb-mol-°R. Equation (4-2) is a general equation of state for gases. The gas compressibility factor for mixtures of hydrocarbon gases can be obtained from Figure 4-1 (Standing and Katz, 1942). This well-known graph was constructed for hydrocarbon gas mixtures. In the presence of large amounts of nonhydrocarbon gases, the gas compressibility factor must be adjusted, as is demonstrated later in the chapter.

To use Figure 4-1 it is necessary to calculate the pseudoreduced properties (pressure and temperature) of the mixture. These properties are simply

$$p_{pr} = \frac{p}{p_{pc}} \tag{4-3}$$

and

$$T_{pr} = \frac{T}{T_{pc}} \tag{4-4}$$

where p_{pc} and T_{pc} are the pseudocritical pressure and temperature of the mixture, respectively. The temperature must be absolute (R or K), which is simply °F + 460 or °C + 273.

Several researchers have developed analytical equations to reproduce the Standing and Katz Z-factor charts. Comparative studies have shown that the Dranchuk and Abou-Kassem

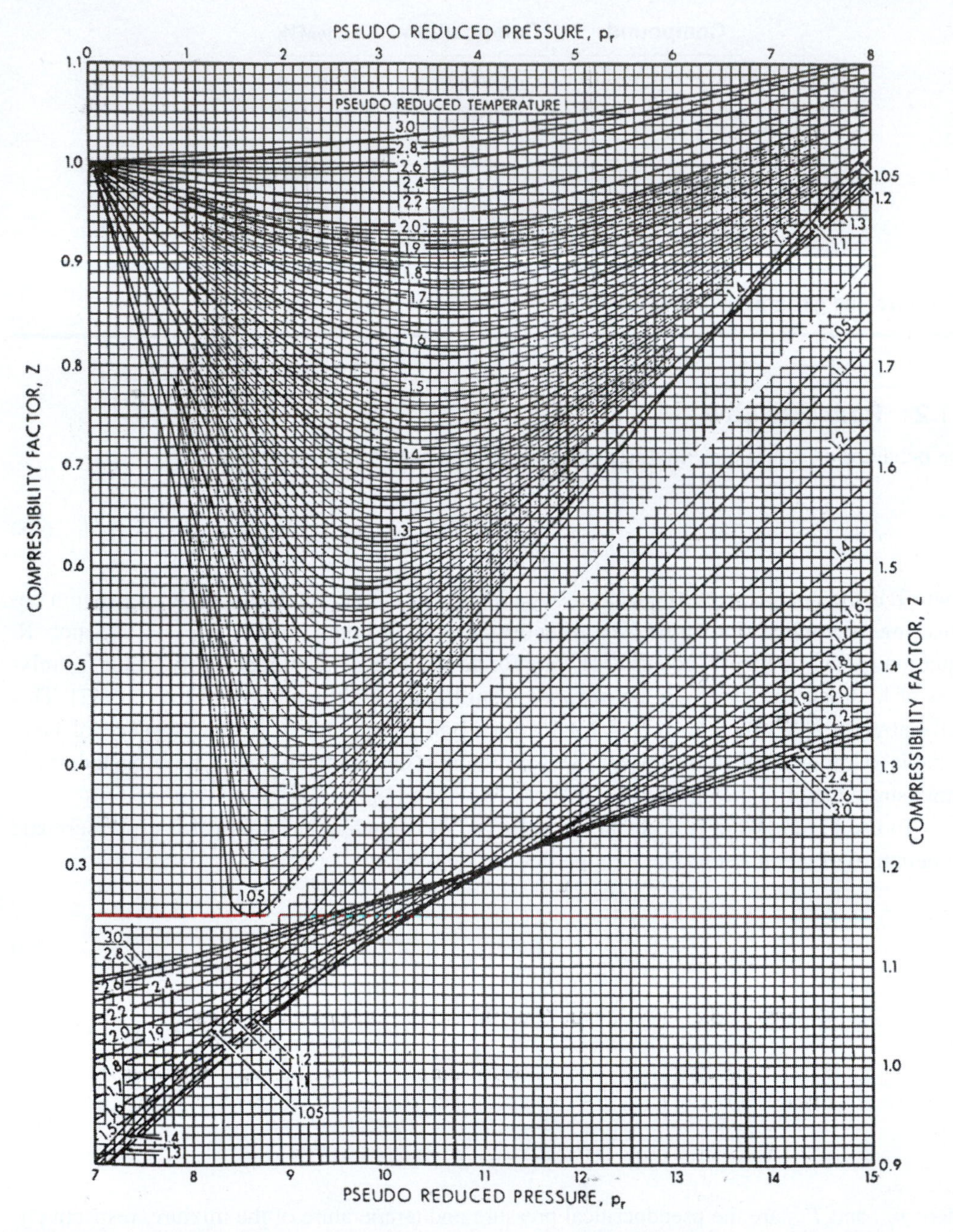

Figure 4-1 The gas deviation factor for natural gases. (From Standing and Katz, 1942.)

(1975) correlation is accurate to better than 1% for a wide range of temperatures and pressures. This correlation, as presented by McCain (1990), is:

$$Z = 1 + (A_1 + A_2/T_{pr} + A_3/T_{pr}^3 + A_4/T_{pr}^4 + A_5/T_{pr}^5)\rho_{pr}$$
$$+ (A_6 + A_7/T_{pr} + A_8/T_{pr}^2)\rho_{pr}^2$$
$$- A_9(A_7/T_{pr} + A_8/T_{pr}^2)\rho_{pr}^5$$
$$+ A_{10}(1 + A_{11}\rho_{pr}^2)(\rho_{pr}^2/T_{pr}^3)\,\text{EXP}(-A_{11}\rho_{pr}^2) \quad \textbf{(4-5)}$$

where

$$\rho_{pr} = 0.27[p_{pr}/(ZT_{pr})]. \quad \textbf{(4-6)}$$

The constants are as follows: $A_1 = 0.3265$, $A_2 = -1.0700$, $A_3 = -0.5339$, $A_4 = 0.01569$, $A_5 = -0.05165$, A6 = 0.5475, A7 = −0.7361, A8 = 0.1844, A9 = 0.1056, $A_{10} = 0.6134$, $A_{11} = 0.7210$.

The range of applicability is

$$0.2 \leq p_{pr} < 30 \text{ for } 1.0 < T_{pr} \leq 3.0$$

and

$$p_{pr} < 1.0 \text{ for } 0.7 < T_{pr} < 1.0.$$

Finally, as can be seen from Figure 4-1, at the standard conditions of $p_{sc} = 14.7$ psi and $T_{sc} = 60°F = 520°R$, the gas compressibility factor, Z_{sc}, can be taken as equal to 1.

Example 4-2 Use of the Real Gas Law to Calculate the Volume of a Gas Mixture at Reservoir Conditions

Assume that a natural gas has the following molar composition (this is the fluid in the reservoir described in Appendix C): $C_1 = 0.875$, $C_2 = 0.083$, $C_3 = 0.021$, i-$C_4 = 0.006$, n-$C_4 = 0.002$, i-$C_5 = 0.003$, n-$C_5 = 0.008$, n-$C_6 = 0.001$, and $C_7+ = 0.001$.

Calculate the volume of 1 lb-mol of this mixture at reservoir conditions of $T = 180°F$ and $p = 4000$ psi.

Solution

The first step is to calculate the pseudocritical properties of the mixture. These properties are simply the summation of the individual contributions of the component gases, weighted by their

Table 4-2 Calculation of Pseudocritical Properties for Example 4-2

Compound	y_i	MW_i	$y_i\,MW_i$	P_{ci}	y_iP_{ci}	T_{ci}	y_iT_{ci}
C_1	0.875	16.04	14.035	673	588.87	344	301
C_2	0.083	30.07	2.496	709	58.85	550	45.65
C_3	0.021	44.1	0.926	618	12.98	666	13.99
i-C_4	0.006	58.12	0.349	530	3.18	733	4.4
n-C_4	0.002	58.12	0.116	551	1.1	766	1.53
i-C_5	0.003	72.15	0.216	482	1.45	830	2.49
n-C_5	0.008	72.15	0.577	485	3.88	847	6.78
n-C_6	0.001	86.18	0.086	434	0.43	915	0.92
C_7+	0.001	114.23 [a]	0.114	361 [a]	0.36	1024 [a]	1.02
	1.000		18.92		$671 = p_{pc}$		$378 = T_{pc}$

[a]Use Properties of n-octane

molar fractions. This is based on the classical thermodynamics law for ideal mixtures and Dalton's law of partial pressures. Table 4-2 gives the results of this calculation.

The pseudoreduced properties are then $p_{pr} = 4000/671 = 5.96$ and $T_{pr} = (180 + 460)/378 = 1.69$. From Figure 4-1, $Z = 0.855$. Then, from Equation (4-2) and rearrangement,

$$V = \frac{(0.855)(1)(10.73)(640)}{4000} = 1.47 \text{ ft}^3. \quad \textbf{(4-7)}$$

Equation (4-5) gives $Z = 0.89$ and $V = 1.53$ ft^3.

4.2 Correlations and Useful Calculations for Natural Gases

Several important works have presented correlations for natural gas properties. Following is a summary of these, with brief descriptions of the use of these correlations.

4.2.1 Pseudocritical Properties from Gas Gravity

In the absence of detailed composition of a natural gas, Figure 4-2 can be used to relate the gas gravity (to air) with the pseudocritical properties of gas mixtures. Using the results of Example 4-2, the calculated molecular weight is 18.92, leading to $\gamma_g = 18.92/28.97 = 0.65$. From Figure 4-2, $p_{pc} = 670$ psi and $T_{pc} = 375°$R, which compare with 671 psi and 378°R, calculated explicitly in Example 4-2.

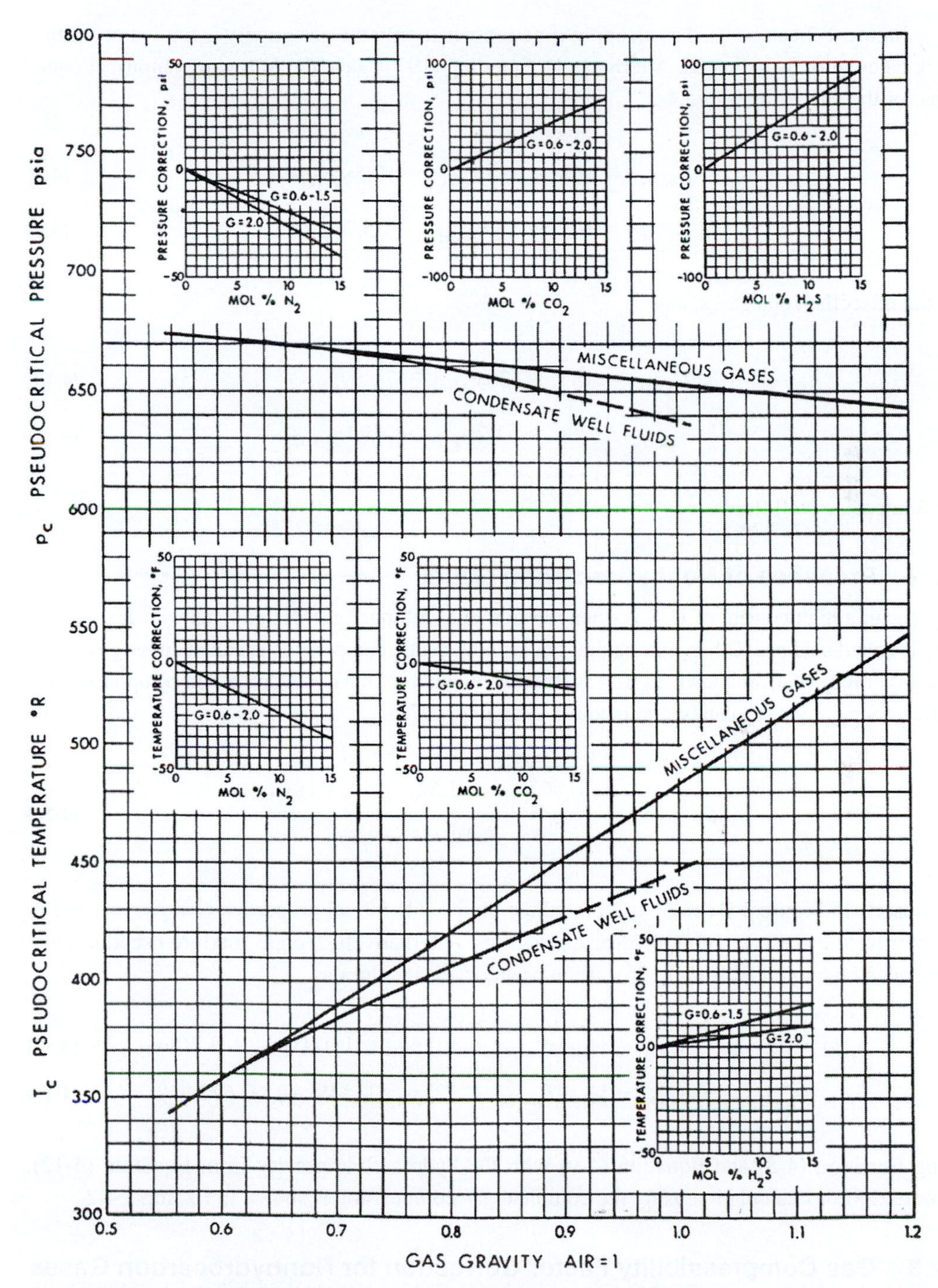

Figure 4-2 Pseudocritical properties of natural gases. (From Brown, Katz, Oberfell, and Alden, 1948; inserts from Carr, Kobayashi, and Burrows, 1954.)

Figure 4-2 can be used as an approximation when only the gas gravity is known or when a quick calculation is indicated. Alternatively, Standing (1977) gave the following empirical equations for the curves in Figure 4-2:

$$T_{pcHC} = 168 + 325\gamma_{gHC} - 12.5\gamma_{gHC}^2 \quad \textbf{(4-8)}$$

$$p_{pcHC} = 677 + 15.0\gamma_{gHC} - 35.7\gamma_{gHC}^2 \quad \textbf{(4-9)}$$

for the miscellaneous gases, and

$$T_{pcHC} = 187 + 330\gamma_{gHC} - 71.5\gamma_{gHC}^2 \quad \textbf{(4-10)}$$

$$p_{pcHC} = 706 + 51.7\gamma_{gHC} - 11.1\gamma_{gHC}^2 \quad \textbf{(4-11)}$$

for condensate well fluids.

4.2.2 Presence of Nonhydrocarbon Gases

A gas with a high content of H_2S is often referred to as a "sour" gas. Natural gas with H_2S and/or CO_2 is called acid gas. Otherwise, natural gas without acidic components is called sweet gas. If the total gas gravity, γ_{gM}, is known, as well as the mole fractions of the nonhydrocarbon components, the hydrocarbon gas gravity can be computed as (Standing, 1977)

$$\gamma_{gHC} = \frac{\gamma_{gM} - 0.967y_{N_2} - 1.52y_{CO_2} - 1.18y_{H_2S}}{1\text{-}y_{N_2} - y_{CO_2} - y_{H_2S}}. \quad \textbf{(4-12)}$$

The inserts in Figure 4-2 can be used to adjust the pseudocritical properties of a gas mixture to account for the presence of nonhydrocarbon gases. Alternatively, the gas mixture pseudocritical values can be computed from hydrocarbon pseudocritical values as

$$p_{pcM} = (1\text{-}y_{N_2} - y_{CO_2} - y_{H_2S})p_{pcHC} + 493y_{N_2} + 1{,}071y_{CO_2} + 1.306y_{H_2S} \quad \textbf{(4-13)}$$

$$T_{pcM} = (1\text{-}y_{N_2} - y_{CO_2} - y_{H_2S})T_{pcHC} + 227y_{N_2} + 548y_{CO_2} + 672y_{H_2S} \quad \textbf{(4-14)}$$

using Equation (4-8) and Equation (4-9) with the hydrocarbon gravity from Equation (4-12). A more rigorous calculation using the complete gas composition is shown in Example 4-2.

4.2.3 Gas Compressibility Factor Correction for Nonhydrocarbon Gases

Wichert and Aziz (1972) have presented a correlation that allows the use of the Standing-Katz graph (Figure 4-1) in the presence of nonhydrocarbon gases. In this case corrected pseudocritical values are

$$T'_{pc} = T_{pcM} - \varepsilon_3 \tag{4-15}$$

and

$$p'_{pc} = \frac{p_{pcM} T'_{pc}}{T_{pcM} + y_{H_2S}(1 - y_{H_2S})\varepsilon_3} \tag{4-16}$$

where the term ε_3 is a function of the H_2S and CO_2 concentrations given by

$$\varepsilon_3 = 120[(y_{CO_2} + y_{H_2S})^{0.9} - (y_{CO_2} + y_{H_2S})^{1.6}] + 15(y_{CO_2}^{0.5} - y_{H_2S}^{4}) \tag{4-17}$$

that can also be obtained graphically from Figure 4-3.

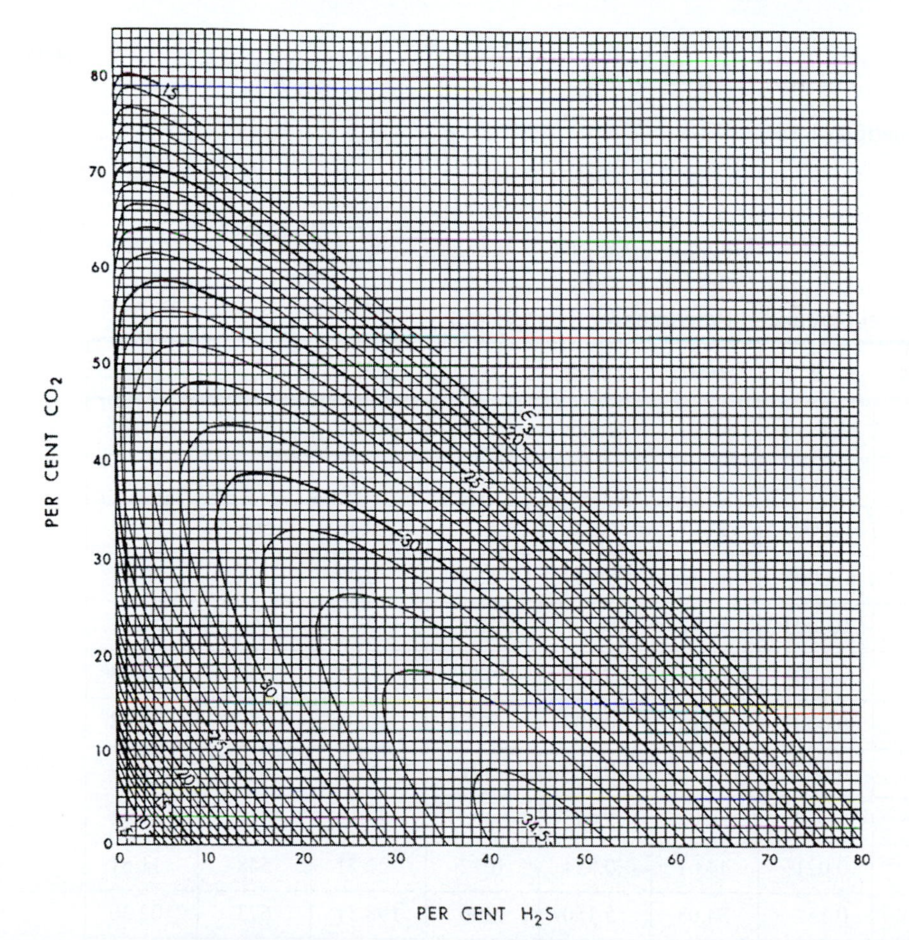

Figure 4-3 Pseudocritical temperature adjustment factor, ε_3. (From Wichert and Aziz, 1972.)

Example 4-3 Calculation of the Gas Compressibility Factor of a Sour Gas

A natural gas has the following composition: C_1 = 0.7410, C_2 = 0.0245, C_3 = 0.0007, i-C_4 = 0.0005, n-C_4 = 0.0003, i-C_5 = 0.0001, n-C_5 = 0.0001, C_6+ = 0.0005, N_2 = 0.0592, CO_2 = 0.021, and H_2S = 0.152. Calculate the gas compressibility factor, Z, at 180°F and 4000 psi.

Solution

The pseudoreduced properties are calculated as shown in Table 4-3. From Figure 4-3 and using the compositions of CO_2 and H_2S, the adjustment factor $\varepsilon_3 = 23.5°R$, or using Equation (4-17),

$$\varepsilon_3 = 120[(0.021 + 0.152)^{0.9} - (0.021 + 0.152)^{1.6}] + 15(0.021^{0.5} - 0.152^4) = 19.7°R. \quad (4\text{-}18)$$

Therefore, from Equation (4-15),

$$T'_{pc} = 397.4 - 23.5 = 373.9°R \quad (4\text{-}19)$$

and from Equation (4-16),

$$p'_{pc} = \frac{(767.4)(373.9)}{397.4 + (0.152)(1 - 0.152)(23.5)} = 716.6 \text{ psi}. \quad (4\text{-}20)$$

Table 4-3 Pseudoreduced Properties for Example 4-3

Compound	y_i	MW_i	$y_i\,MW_i$	p_{ci}	y_iP_{ci}	T_{ci}	y_iT_{ci}
C_1	0.7410	16.04	11.886	673	498.69	344	254.90
C_2	0.0246	30.07	0.740	709	17.44	550	13.53
C_3	0.0007	44.01	0.031	618	0.43	666	0.47
i-C_4	0.0005	58.12	0.029	530	0.26	733	0.37
n-C_4	0.0003	58.12	0.017	551	0.17	766	0.23
i-C_5	0.0001	72.15	0.007	482	0.05	830	0.08
n-C_5	0.0001	72.15	0.007	485	0.05	847	0.08
C_6+	0.0005	100.2	0.050	397	0.20	973	0.49
N_2	0.0592	28.02	1.659	492	29.13	227	13.44
CO_2	0.021	44.01	0.924	1072	22.51	548	11.51
H_2S	0.152	34.08	5.180	1306	198.51	673	102.30
	1.0000		20.53		767.44		397.4

In addition to the calculation shown in this example, it is possible to calculate the pseudocritical properties from Figure 4-2. The molecular weight is 20.53, so $\gamma_g = 20.53/28.97 = 0.709$. Therefore, from Figure 4-2, $T_{pc} = 394°R$ and $p_{pc} = 667$ psi. These must be corrected by the inserts in Figure 4-2 or by using Equations (4-13) and (4-14). Thus,

$$T_{pc} = 394 - 15 - 2 + 20 = 397°R \tag{4-21}$$

and

$$p_{pc} = 667 - 10 + 5 + 92 = 754 \text{ psi.} \tag{4-22}$$

The adjustments are for N_2, CO_2, and H_2S, respectively. The values from Equations (4-21) and (4-22) compare with 397.4 and 767.4, as calculated explicitly. To use the Z graph, these values must be adjusted again using Equations (4-19) and (4-20) before computing the reduced pseudocritical values.

4.2.4 Gas Viscosity

Gas viscosity correlations have been presented by a number of authors. The Carr et al. (1954) correlation has been the most popular. It is presented in Figures 4-4 and 4-5. Figure 4-4 allows the calculation of the viscosity at any temperature and at a pressure of 1 atm. Figure 4-5 provides the estimation of μ/μ_{1atm}, which is the ratio of the viscosity at an elevated pressure to the viscosity at 1 atm.

A widely used correlation for gas viscosity that gives results comparable to those in Figures 4-4 and 4-5 is that of Lee, Gonzales, and Eakin (1966):

$$\mu_g = A(10^{-4})\, EXP(B\rho_g^c) \tag{4-23}$$

where

$$A = \frac{(9.379 + 0.01607M_a)T^{1.5}}{209.2 + 19.26M_a + T} \tag{4-24}$$

$$B = 3.448 + \frac{986.4}{T} + 0.01009M_a \tag{4-25}$$

$$C = 2.447 - 0.2224B \tag{4-26}$$

ρ_g is gas density in gm/cc, M_a is apparent molecular weight, T is temperature in °R, and μ_g is gas viscosity in cp.

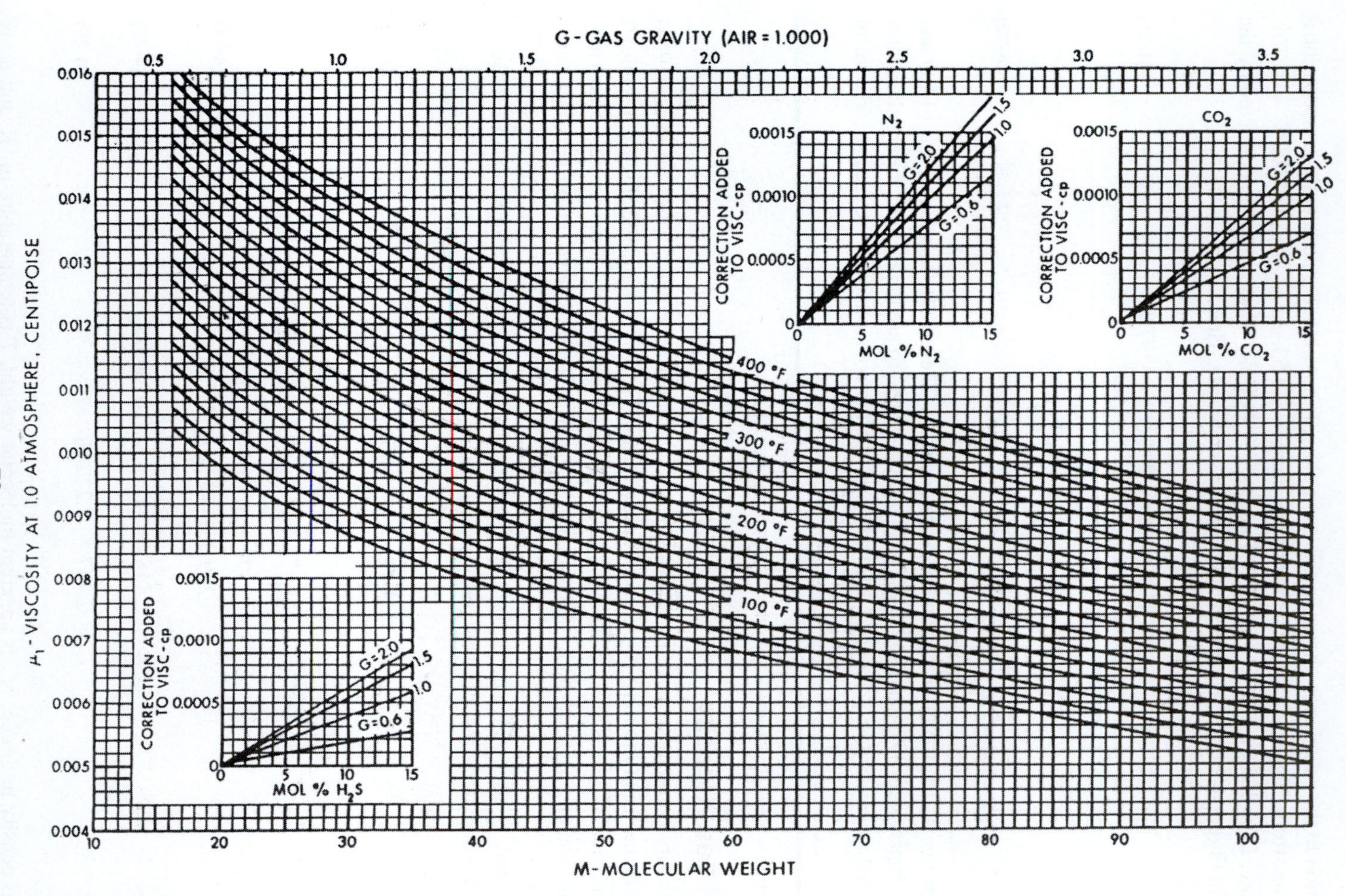

Figure 4-4 Viscosity of natural gases at 1 atm. (From Carr et al., 1954.)

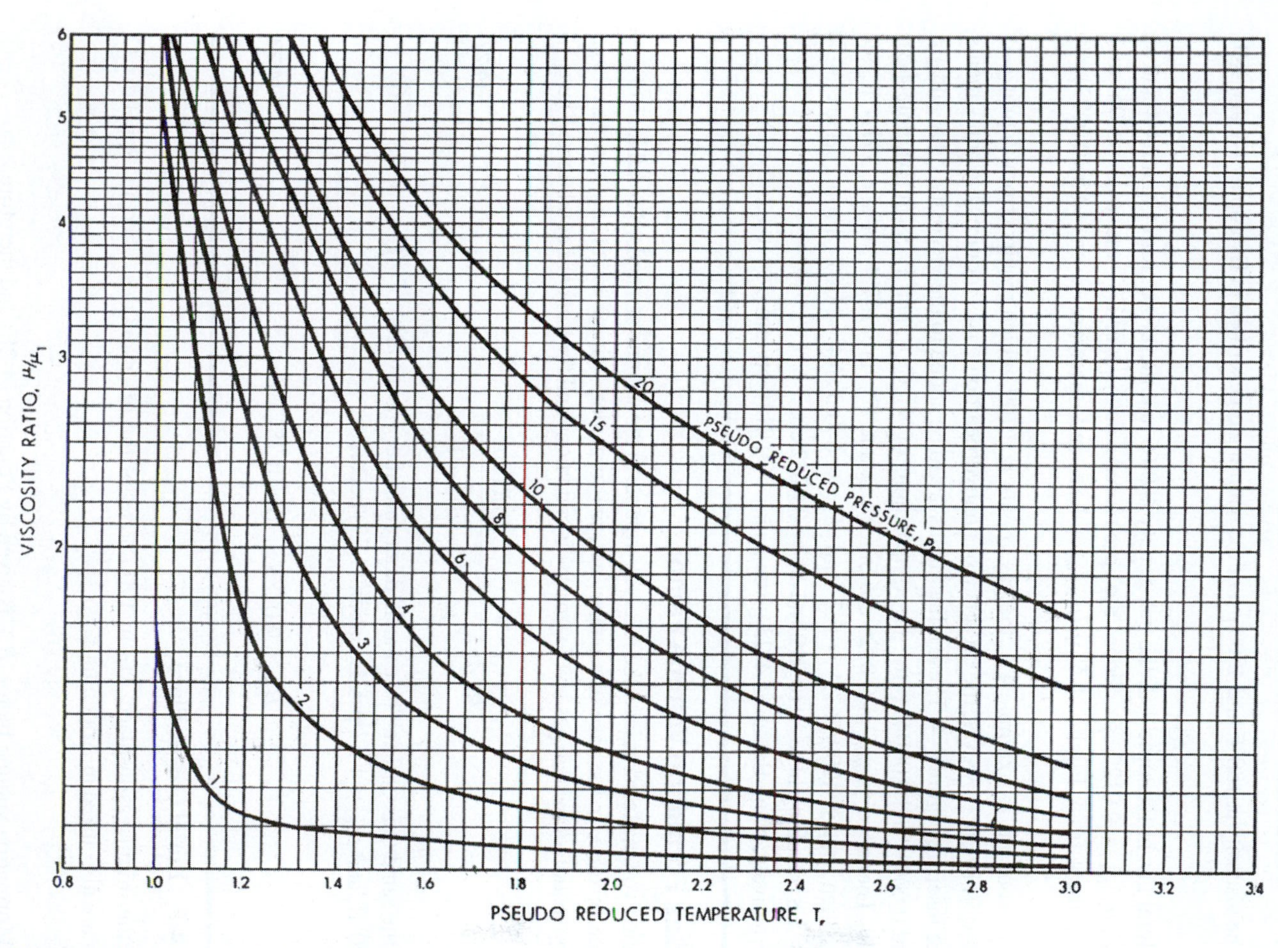

Figure 4-5 Viscosity ratio at elevated pressures and temperatures. (From Carr et al., 1954.)

Example 4-4 Calculation of the Viscosity of a Natural Gas and a Sour Gas

Calculate the viscosity at 180°F and 4000 psi of the natural gas described in Examples 4-2 and 4-3.

Solution

For the natural gas in Example 4-2, the gas gravity is 0.65 and therefore from Figure 4-4 and at $T = 180°\text{F}$, $\mu_{1atm} = 0.0122$ cp. Since the pseudoreduced properties are $p_{pr} = 5.96$ and $T_{pr} = 1.69$, then from Figure 4-5, $\mu/\mu_{1atm} = 1.85$ and therefore $\mu = (1.85)(0.0122) = 0.0226$ cp.

For the sour gas in Example 4-3, the gas gravity is 0.709, which results (from Figure 4-4) in $\mu_{1atm} = 0.0119$ cp. However, the presence of nonhydrocarbon gases requires the adjustments given in the insets in Figure 4-4. These adjustments are to be added to the viscosity value and are 0.0005, and 0.0001, and 0.0004 cp for the compositions of N_2, CO_2, and H_2S (in Example 4-3), respectively. Therefore, $\mu_{1atm} = 0.0129$ cp. Since p_{pc} and T_{pc} are 767.4 psi and 397.4°R, respectively, then $p_{pr} = 4000/767.4 = 5.2$ and $T_{pr} = 640/397.4 = 1.61$. From Figure 4-5, $\mu/\mu_{1atm} = 1.8$, resulting in $\mu = (0.0129)(1.8) = 0.0232$ cp.

Alternatively, using Equation 4-23 gives $\mu_{1atm} = 0.0119$ cp and $\mu = 0.0232$ cp.

4.2.5 Gas Formation Volume Factor

The formation volume factor relates the reservoir volume to the volume at standard conditions of any hydrocarbon mixture. In the case of a natural gas, the formation volume factor, B_g, can be related with the application of the real gas law for reservoir conditions and for standard conditions. Thus,

$$B_g = \frac{V}{V_{sc}} = \frac{ZnRT/p}{Z_{sc}nRT_{sc}/p_{sc}}. \tag{4-27}$$

For the same mass, nR can be cancelled out and, after substitution of $Z_{sc} \approx 1$, $T_{sc} = 60 + 460 = 520°\text{R}$, and $p_{sc} = 14.7$ psi, Equation (4-27) becomes

$$B_g = 0.0283\frac{ZT}{p}(\text{res ft}^3/\text{SCF}). \tag{4-28}$$

Example 4-5 Initial Gas-in-Place

Calculate the initial gas-in-place, G_i, in 1900 acres of the reservoir described in Appendix C. Properties are also listed in Appendix C.

Solution

The initial formation volume factor, B_{gi}, is given by Equation (4-28) and therefore

$$B_{gi} = \frac{(0.0283)(0.945)(640)}{4613} = 3.71 \times 10^{-3} \text{ res ft}^3/\text{SCF}. \tag{4-29}$$

Then

$$G_i = \frac{Ah\phi S_g}{B_{gi}} \tag{4-30}$$

and

$$G_i = \frac{(43{,}560)(1{,}900)(78)(0.14)(0.73)}{3.71 \times 10^{-3}} = 1.78 \times 10^{11} \text{ SCF.} \tag{4-31}$$

4.2.6 Gas Isothermal Compressibility

The gas compressibility, c_g, often referred to as isothermal compressibility, has an exact thermodynamic expression:

$$c_g = -\frac{1}{V}\left(\frac{\partial V}{\partial p}\right)_T. \tag{4-32}$$

For an ideal gas, it can be shown readily that c_g is exactly equal to $1/p$. For a real gas, using Equation (4-2), the derivative $\partial V/\partial p$ can be evaluated:

$$\frac{\partial V}{\partial p} = -\frac{ZnRT}{p^2} + \frac{nRT}{p}\left(\frac{\partial Z}{\partial p}\right)_T. \tag{4-33}$$

Substitution of the volume, V, by its equivalent from Equation (4-2) and the derivative $\partial V/\partial p$ from Equation (4-33) into Equation (4-32) results in

$$c_g = \frac{1}{p} - \frac{1}{Z}\left(\frac{\partial Z}{\partial p}\right)_T \tag{4-34}$$

or, more conveniently, remembering the relationship between pseudoreduced and pseudocritical pressure (Equation 4-3),

$$c_g = \frac{1}{p} - \frac{1}{Z\,p_{pc}}\left(\frac{\partial Z}{\partial p_{pr}}\right)_T. \tag{4-35}$$

Equation (4-35) is useful because it allows the calculation of the compressibility of a real gas at any temperature and pressure. Needed are the gas compressibility factor Z and the *slope* of the Standing-Katz correlation, $\partial Z/\partial p_{pr}$, at the corresponding temperature (i.e., the associated pseudoreduced temperature curve). The slope can be determined by differentiating Equation 4-35 with respect to the pseudoreduced pressure holding the temperature constant.

4.3 Approximation of Gas Well Deliverability

The steady-state relationship developed from Darcy's law for an incompressible fluid (oil) was presented as Equation (2-9) in Chapter 2. A similar relationship can be derived for a natural gas well by converting the flow rate from STB/d to Mscf/d and using the real gas law to describe the PVT behavior of the gas. Beginning with the differential form of Darcy's law for radial flow,

$$q_{act} = \frac{2\pi rkh}{\mu}\frac{dp}{dr} \tag{4-36}$$

where q_{act} denotes the actual volumetric flow rate at some temperature and pressure in the reservoir. From the real gas law, the volumetric flow rate, q_{act}, at any temperature and pressure is related to the volumetric flow rate at standard conditions, q, by

$$q_{act} = q\left(\frac{p_{sc}}{p}\right)\left(\frac{T}{T_{sc}}\right)Z. \tag{4-37}$$

Bringing this expression to Equation 4-36 and separating variables yields

$$\frac{q\mu ZT}{2\pi r}dr = \left(\frac{T_{sc}}{p_{sc}}\right)khpdp. \tag{4-38}$$

For steady-state flow with a pressure p_e at the drainage boundary, r_e, and a bottomhole flowing pressure, p_{wf}, at r_w, integrating Equation 4-38 for average values of viscosity and compressibility factor yields

$$q = \frac{kh\left(\frac{T_{sc}}{p_{sc}}\right)\left(p_e^2 - p_{wf}^2\right)}{4\pi\bar{\mu}\bar{Z}T\left(\ln\frac{r_e}{r_w}\right)}. \tag{4-39}$$

Expressing the flow rate in Mscf/d, reconciling the other usual oilfield units, and adding the skin factor gives the steady-state gas inflow equation

$$q = \frac{kh\left(p_e^2 - p_{wf}^2\right)}{1424\bar{\mu}\bar{Z}T\left(\ln\frac{r_e}{r_w} + s\right)}, \tag{4-40}$$

which also results in

$$p_e^2 - p_{wf}^2 = \frac{1424q\bar{\mu}\bar{Z}T}{kh}\left(\ln\frac{r_e}{r_w} + s\right). \tag{4-41}$$

Equation (4-41) suggests that a gas well production rate is approximately proportional to the pressure squared difference. The properties $\bar{\mu}$ and $\bar{Z}$ are average properties evaluated at a pressure between p_e and p_{wf}. (Henceforth the bars will be dropped for simplicity).

A similar approximation can be developed for pseudosteady state. It has the form

$$\bar{p}^2 - p_{wf}^2 = \frac{1424q\mu ZT}{kh}\left(\ln 0.472\frac{r_e}{r_w} + s\right). \tag{4-42}$$

Equations (4-41) and (4-42) are not only approximations in terms of properties but also because they assume Darcy flow in the reservoir. For reasonably small gas flow rates this approximation is acceptable. A common presentation of Equation (4-42) [or Equation (4-41)] is

$$q = C(\bar{p}^2 - p_{wf}^2). \tag{4-43}$$

For larger flow rates, where non-Darcy flow is evident in the reservoir,

$$q = C(\bar{p}^2 - p_{wf}^2)^n \tag{4-44}$$

where $0.5 < n < 1$.

A log-log plot of q versus $(\bar{p}^2 - p_{wf}^2)$ would yield a straight line with slope equal to n and intercept C.

Example 4-6 Flow Rate versus Bottomhole Pressure for a Gas Well

Graph the gas flow rate versus flowing bottomhole pressure for the well described in Appendix C. Use the steady-state relationship given by Equation (4-40). Assume that $s = 0$ and $r_e = 1490$ ft ($A = 160$ acres).

Solution

Equation (4-40) after substitution of variables becomes

$$2.128 \times 10^7 - p_{wf}^2 = (5.79 \times 10^5) q\bar{\mu}\bar{Z}. \tag{4-45}$$

For p_{wf} increments of 500 psi from 1000 psi to 4000 psi, calculations are shown in Table 4-4. (Note that $\mu_{1atm} = 0.0122$ cp and $T_{pr} = 1.69$ throughout, since the reservoir is considered isothermal.)

As an example calculation, for $p_{wf} = 3000$, Equation (4-45) yields

$$q = \frac{2.128 \times 10^7 - 3000^2}{(5.79 \times 10^5)(0.022)(0.903)} = 1.07 \times 10^3 \text{ MSCF/d}. \tag{4-46}$$

If the initial μ_i and Z_i were used (i.e., not averages) the flow rate q would be 9.22×10^2 MSCF/d, a deviation of 14%. Figure 4-6 is a graph of p_{wf} versus q for this example.

Table 4-4 Viscosity and Gas Deviation Factor for Example 4-6

p_{wf}(psi)	$p_{pr,w}$	μ/μ_{1atm} [a]	μ(cp)	$\bar{\mu}$(cp) [b]	Z [c]	$\bar{Z}$ [d]
1000	1.49	1.10	0.0134	0.0189	0.920	0.933
1500	2.24	1.20	0.0146	0.0195	0.878	0.912
2000	2.98	1.35	0.0165	0.0205	0.860	0.903
2500	3.73	1.45	0.0177	0.0211	0.850	0.898
3000	4.47	1.60	0.0195	0.022	0.860	0.903
3500	5.22	1.70	0.0207	0.0226	0.882	0.914
4000	5.96	1.85	0.0226	0.0235	0.915	0.930

[a] From Figure 4-5, following $T_{pr} = 1.69$

[b] Arithmetic average with $\mu_i = 0.0244$ cp (at $p_e = 4613$ psi).

[c] From Figure 4-1, following $T_{pr} = 1.69$.

[d] Arithmetic average with $Z_i = 0.945$ (at $p_e = 4613$ psi).

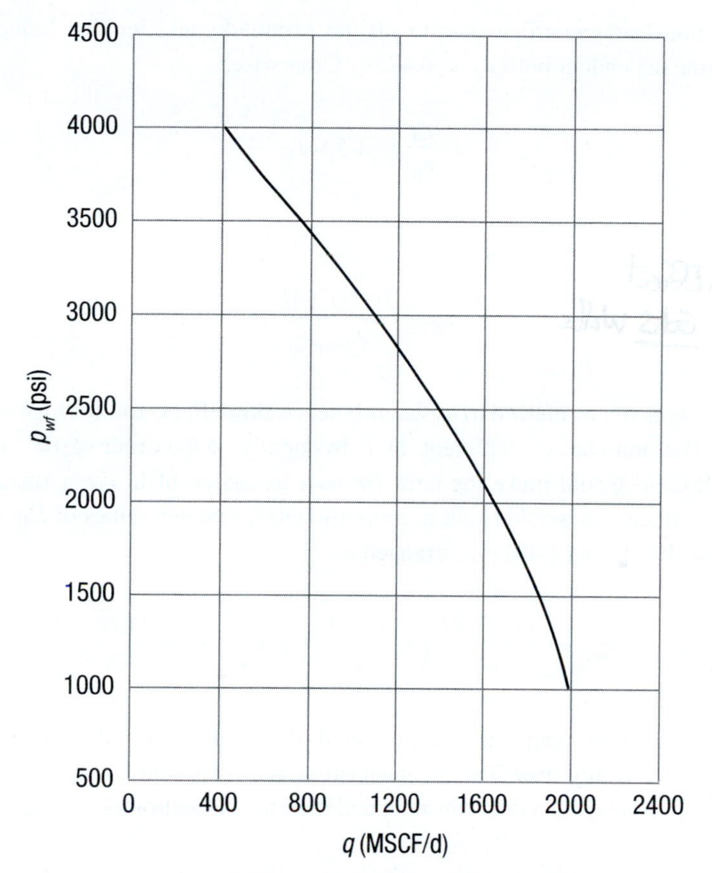

Figure 4-6 Production rate versus flowing bottomhole pressure (IPR) for the gas well in Example 4-6.

Although this is done in the style of IPR curves as shown in the previous two chapters, this is not a common construction for gas reservoirs, as is shown later in the chapter.

The "irregular" shape of the curve in Figure 4-6 reflects the changes in the Z factor, which for this example reaches a minimum between 2500 and 3000 psi.

4.4 Gas Well Deliverability for Non-Darcy Flow

A more "exact" deliverability relationship for stabilized gas flow was developed by Aronofsky and Jenkins (1954) from the solution of the differential equation for gas flow through porous media using the Forchheimer (rather than the Darcy) equation for flow. This solution is

$$q(\text{MSCF/d}) = \frac{kh(\overline{p}^2 - p_{wf}^2)}{1424\overline{\mu}\overline{Z}T[\ln(r_d/r_w) + s + Dq]} \tag{4-47}$$

where D is the non-Darcy coefficient and r_d is the Aronofsky and Jenkins "effective" drainage radius, and is time dependent until $r_d = 0.472r_e$. Otherwise,

$$\frac{r_d}{r_w} = 1.5\sqrt{t_D} \tag{4-48}$$

where

Turbulence Skin Effect only applies to GAS Wells

$$t_D = \frac{0.000264kt}{\phi\mu c_t r_w^2}. \tag{4-49}$$

The term Dq is often referred to as the turbulence skin effect, and for high rate wells it can be substantial. The non-Darcy coefficient, D, is frequently of the order of 10^{-3} and therefore a rate of 10 MMSCF/d would make the term Dq near the value of $\ln r_d/r_w$ (usually between 7 and 9). Smaller values of q would result in proportionately smaller values of Dq.

Frequently, Equation (4-47) is rearranged as

$$\overline{p}^2 - p_{wf}^2 = \frac{1424\overline{\mu}\overline{Z}T}{kh}\left(\ln\frac{0.472r_e}{r_w} + s\right)q + \frac{1424\overline{\mu}\overline{Z}TD}{kh}q^2. \tag{4-50}$$

The first term on the right-hand side of Equation (4-50) is identical to the one developed earlier [Equation (4-42)] for Darcy flow. The second term accounts for non-Darcy effects. All multipliers of q and q^2 can be considered as constant, and therefore Equation (4-50) may take the form

$$\overline{p}^2 - p_{wf}^2 = aq + bq^2. \tag{4-51}$$

In field applications the constant a and b in Equation (4-51) can be calculated from a "four-point test" where $(\overline{p}^2 - p_{wf}^2)/q$ is graphed on Cartesian coordinates against q. The flowing bottomhole pressure, p_{wf}, is calculated for four different stabilized flow rates. The intercept of the straight line is a, and the slope is b. From b and its definition [see Equation (4-50)], the non-Darcy coefficient, D, can be obtained.

Approximations for the non-Darcy coefficient have been given by a number of authors. In the absence of field measurements, an empirical relationship for a cased, perforated well is proposed by Jones (1987) as:

$$D = \frac{6 \times 10^{-5}\,\gamma k_s^{-0.1} h}{\mu r_w h_{\text{perf}}^2} \tag{4-52}$$

where γ is the gas gravity; k_s is the near-wellbore permeability in md; h and h_{perf} the net and perforated thickness, both in ft; and μ is the gas viscosity in cp, evaluated at the flowing bottomhole pressure.

Katz et al. (1959) have presented a more comprehensive expression for natural gas flow with turbulence:

$$p_e^2 - p_{wf}^2 = \frac{1{,}424\mu ZT}{kh}\left[\ln\left(\frac{r_e}{r_w}\right) + s\right]q + \frac{3.16 \times 10^{-12}\, \beta \gamma_g ZT\left(\frac{1}{r_w} - \frac{1}{r_e}\right)}{h^2}q^2 \tag{4-53}$$

where k equals the horizontal permeability, k_H. β is the non-Darcy coefficient in the Forcheimer equation and can be calculated by using the Tek, Coats, and Katz (1962) correlation:

$$\beta = \frac{5.5 \times 10^9}{k^{1.25}\phi^{0.75}}. \tag{4-54}$$

Equation (4-53) is for open-hole completions. Other correlations for non-Darcy flow coefficients for flow in the formation are listed in Table 4-5.

Table 4-5 Correlations for the non-Darcy Coefficient for Flow in the Formation

Reference	Formation Correlation	Unit	
		β	**k**
Geerstma (1974)	$\beta = \frac{0.005}{k^{0.5}\phi^{5.5}}$	1/cm	cm^2
Tek et al. (1962)	$\beta = \frac{5.5 \times 10^9}{k^{1.25}\phi^{0.75}}$	1/ft	md
Jones (1987)	$\beta = \frac{6.15 \times 10^{10}}{k^{1.55}}$	1/ft	md
Coles & Hartman (1998)	$\beta = \frac{1.07 \times 10^{12} \times \phi^{0.449}}{k^{1.88}}$	1/ft	md
Coles & Hartman (1998)	$\beta = \frac{2.49 \times 10^{11}\phi^{0.537}}{k^{1.79}}$	1/ft	md
Li et al. (2001)	$\beta = \frac{11500}{k\phi}$	1/cm	darcy
Thauvin & Mohanty (1998)	$\beta = \frac{3.1 \times 10^4 \tau^3}{k}$	1/cm	darcy

(continued)

Table 4-5 Correlations for the non-Darcy Coefficient for Flow in the Formation *(continued)*

Reference	Formation Correlation	Unit	
		β	k
Tek et al. (1962)	$\beta = \dfrac{5.5 \times 10^9}{k^{1.25}\phi^{0.75}}$	1/ft	md
Liu et al. (1995)	$\beta = \dfrac{8.91 \times 10^8 \tau}{k\phi}$	1/ft	md
Janicek & Katz (1955)	$\beta = \dfrac{1.82 \times 10^8}{k^{1.25}\phi^{0.75}}$	1/cm	md
Pascal et al. (1980)	$\beta = \dfrac{4.8 \times 10^{12}}{k^{1.176}}$	1/m	md
Wang et al. (1999) Wang (2000)	$\beta = \dfrac{(10)^{-3.25}\tau^{1.943}}{k^{1.023}}$ τ is tortuosity	1/cm	cm^2

Example 4-7 Development of a Gas Well Deliverability Curve

Use the data in Appendix C and $r_e = 1490$ ft ($A = 160$ acres), $\overline{\mu} = 0.022$ cp, $\overline{Z} = 0.93$, and $c_t = 1.5 \times 10^{-4}$ psi^{-1}. Develop a deliverability relationship, and graph the Darcy and non-Darcy components and the correct deliverability curve. Show the absolute open-flow (AOF) potential, that is, what the flow rate would be if $p_{wf} = 0$. The skin effect is equal to 3, and the non-Darcy coefficient, *D*, is equal to 4.9×10^{-2} (this is a very large coefficient).

Solution

The first step is to calculate the time required for stabilized flow. From Equations (4-48) and (4-49),

$$\left(\frac{0.472r_e}{1.5r_w}\right)^2 = \frac{0.000264kt}{\phi\mu c_t r_w^2}, \quad \textbf{(4-55)}$$

which leads to

$$t = \left[\frac{(0.472)(1490)}{(1.5)(0.328)}\right]^2 \frac{(0.14)(0.0.22)(1.5 \times 10^{-4})(0.328)^2}{(0.000264)(0.17)}$$

$$= 2260 \text{ hr} = 94 \text{ d} \quad \textbf{(4-56)}$$

Therefore, the stabilized relationship implied by Equation (4-50) will be in effect after this time.

The coefficients a and b in Equation (4-51) can be calculated next:

$$a = \frac{(1424)(0.022)(0.93)(640)}{(0.17)(78)}\left[\ln\frac{(0.472)(1490)}{0.328} + 3\right] = 1.5 \times 10^4 \tag{4-57}$$

and

$$b = \frac{(1424)(0.022)(0.93)(640)(4.9 \times 10^{-2})}{(0.17)(78)} = 68.9. \tag{4-58}$$

The deliverability relationship for this well is then

$$\overline{p}^2 - p_{wf}^2 = 1.5 \times 10^4 q + 68.9q^2 \tag{4-59}$$

Figure 4-7 is a log–log graph of the Darcy flow result and the correct curve incorporating non-Darcy effects.

The AOF potential for the correct curve (at $p_{wf} = 0$ and therefore $\overline{p}^2 = 2.13 \times 10^7$ psi^2) is 460 MSCF/d, whereas this AOF, when taking into account only the Darcy flow contribution, would be 1420 MSCF/d.

Example 4-8 Estimation of the Non-Darcy Coefficient

Calculate the non-Darcy coefficient for the well in Appendix C. Assume that k_s is the same as the reservoir permeability, but that $h_{perf} = 39$ ft (half the net thickness). Use a viscosity $\mu = 0.02$ cp. What would happen if the near-wellbore permeability is reduced by damage to one-tenth of the reservoir permeability?

Solution

From Equation (4-52) and substitution of the variables in Appendix C,

$$D = \frac{(6 \times 10^{-5})(0.65)(0.17^{-0.1})(78)}{(0.02)(0.328)(39)^2} = 3.6 \times 10^{-4}\,(\text{MSCF/d})^{-1}. \tag{4-60}$$

If $k_s = 0.017$ md, then $D = 4.5 \times 10^{-4}\,(\text{MSCF/d})^{-1}$.

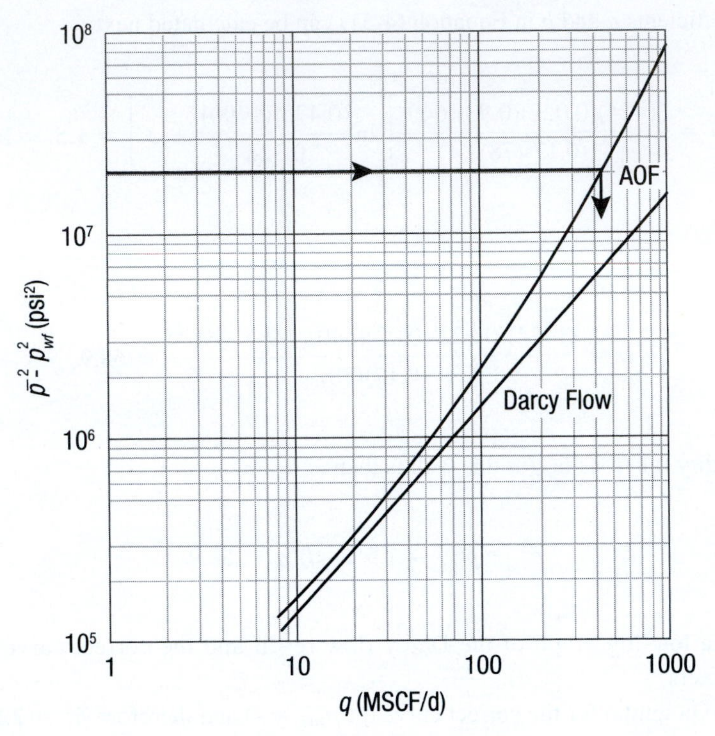

Figure 4-7 Deliverability relationship for the gas well in Example 4-7. Non-Darcy and Darcy effects.

4.5 Transient Flow of a Gas Well

Gas flow in a reservoir under transient conditions can be approximated by the combination of Darcy's law (rate equation) and the continuity equation. In general,

$$\phi \frac{\partial p}{\partial t} = \nabla\left(\rho \frac{k}{\mu} \nabla p\right), \tag{4-61}$$

which in radial coordinates reduces to

$$\phi \frac{\partial p}{\partial t} = \frac{1}{r} \frac{\partial}{\partial r}\left(\rho \frac{k}{\mu} r \frac{\partial p}{\partial r}\right). \tag{4-62}$$

From the real gas law,

$$\rho = \frac{m}{V} = \frac{pMW}{ZRT} \tag{4-63}$$

and therefore

$$\phi\frac{\partial}{\partial t}\left(\frac{p}{Z}\right)=\frac{1}{r}\frac{\partial}{\partial r}\left(\frac{k}{\mu Z}rp\frac{\partial p}{\partial r}\right). \tag{4-64}$$

If the permeability k is considered constant, then Equation (4-64) can be approximated further:

$$\frac{\phi}{k}\frac{\partial}{\partial t}\left(\frac{p}{Z}\right)=\frac{1}{r}\frac{\partial}{\partial r}\left(\frac{p}{\mu Z}r\frac{\partial p}{\partial r}\right). \tag{4-65}$$

Performing the differentiation on the right-hand side of Equation (4-65) and assuming that Z and μ are either constant or that they change uniformly and slowly with pressure, then

$$\frac{1}{\mu Z}\left[\frac{p}{r}\frac{\partial p}{\partial r}+p\frac{\partial^2 p}{\partial r^2}+\left(\frac{\partial p}{\partial r}\right)^2\right]=\text{RHS}. \tag{4-66}$$

Via rearrangement and remembering that

$$\frac{1}{2}\frac{\partial^2 p^2}{\partial r^2}=p\frac{\partial^2 p}{\partial r^2}+\left(\frac{\partial p}{\partial r}\right)^2 \tag{4-67}$$

it becomes

$$\frac{1}{2\mu Z}\left(\frac{1}{r}\frac{\partial p^2}{\partial r}+\frac{\partial^2 p^2}{\partial r^2}\right)=\text{RHS}. \tag{4-68}$$

Therefore, Equation (4-65) can be written as

$$\frac{\phi\mu}{kp}\frac{\partial p^2}{\partial t}=\frac{\partial^2 p^2}{\partial r^2}+\frac{1}{r}\frac{\partial p^2}{\partial r}. \tag{4-69}$$

For an ideal gas, $c_g = 1/p$ and, as a result, Equation (4-55) leads to

$$\frac{\partial^2 p^2}{\partial r^2}+\frac{1}{r}\frac{\partial p^2}{\partial r}=\frac{\phi\mu c}{k}\frac{\partial p^2}{\partial t}. \tag{4-70}$$

This approximation is in the form of the diffusivity equation [see Equation (2-18) in Chapter 2] and its solution, under the assumptions listed in this section, could have the shape of the solutions of the equation for oil, presuming that p^2 instead of p is used. This approximation was used in earlier sections of this chapter and was developed independently. Pressure squared differences can be used as a reasonable approximation. However, the assumptions used to derive Equation (4-70) are limiting and, in fact, they can lead to large errors in high-rate wells unless using the Al-Hussainy and Ramey (1966) real gas pseudopressure function.

The real gas pseudopressure function, $m(p)$, is defined as

$$m(p) = 2\int_{p_o}^{p} \frac{p}{\mu Z} dp \tag{4-71}$$

where p_o is some arbitrary reference pressure (could be zero). The differential pseudopressure $\Delta m(p)$, defined as $m(p) - m(p_{wf})$, is then the driving force in the reservoir.

For low pressure it can be shown that

$$2\int_{p_{wf}}^{p_i} \frac{p}{\mu Z} dp \approx \frac{p_i^2 - p_{wf}^2}{\mu Z} \tag{4-72}$$

whereas for high pressure (both p_i and p_{wf} higher than 3000 psi),

$$2\int_{p_{wf}}^{p_i} \frac{p}{\mu Z} dp \approx 2\frac{\overline{p}}{\overline{\mu Z}}(p_i - p_{wf}). \tag{4-73}$$

The real gas pseudopressure can be used instead of pressure squared difference in any gas well deliverability relationship (properly adjusted for the viscosity and the gas deviation factor). For example, Equation (4-47) would be of the form

$$q(\text{MSCF/d}) = \frac{kh[m(\overline{p}) - m(p_{wf})]}{1424T[\ln(0.472r_e/r_w) + s + Dq]}. \tag{4-74}$$

Of course, more appropriately, the real gas pseudopressure can be used as an integrating factor for an exact solution of the diffusivity equation for a gas. Beginning with Equation (4-65), and using the definition of the real gas pseudopressure [Equation (4-71)] and the chain rule, it can be written readily that

$$\frac{\partial m(p)}{\partial t} = \frac{\partial m(p)}{\partial p}\frac{\partial p}{\partial t} = \frac{2p}{\mu Z}\frac{\partial p}{\partial t}. \tag{4-75}$$

Similarly,

$$\frac{\partial m(p)}{\partial r} = \frac{2p}{\mu Z}\frac{\partial p}{\partial r}. \tag{4-76}$$

Therefore, Equation (4-65) becomes

$$\frac{\partial^2 m(p)}{\partial r^2} + \frac{1}{r}\frac{\partial m(p)}{\partial r} = \frac{\phi \mu c_t}{k}\frac{\partial m(p)}{\partial t}. \tag{4-77}$$

The solution of Equation (4-77) is exactly similar to the solution for the diffusivity equation in terms of pressure. Dimensionless time has (by convention) been defined as

$$t_D = \frac{0.000264kt}{\phi(\mu c_t)_t r_w^2} \tag{4-78}$$

and dimensionless pressure as

$$p_D = \frac{kh[m(p_i) - m(p_{wf})]}{1424qT}. \tag{4-79}$$

All solutions (such as the line source solution and the wellbore storage and skin effect solution) that have been developed for oil wells using the diffusivity equation in terms of pressure are applicable for a gas well using the real gas pseudopressure. For example, the logarithmic approximation to the exponential integral [compare Equations (2-19) and (2-20)] would lead to an analogous expression for a natural gas. Thus,

$$q(\text{Mscf/d}) = \frac{kh[m(p_i) - m(p_{wf})]}{1638T}\left[\log t + \log\frac{k}{\phi(\mu c_t)_i r_w^2} - 3.23\right]^{-1}. \tag{4-80}$$

This expression can be used for transient IPR curves for a gas well.

Example 4-9 Transient IPR for a Gas Well

With the data for the well in Appendix C, calculate transient IPR curves for 10 days, 3 months, and 1 year.

Solution

Table 4-6 gives the viscosity, gas deviation factor, and real gas pseudopressure for the reservoir fluid described in Appendix C. For the initial condition of $p_i = 4613$ psi, the real gas pseudopressure, viscosity, and gas compressibility factor are 1.265×10^9 psi^2/cp, 0.0235 cp, and 0.968, respectively. (These values, derived from numerical fits, are slightly different from the ones in Appendix C.)

Equation (4-80) can be used for the transient IPR calculations if real gas pseudopressures are to be used. If pressure squared differences are to be used, then the denominator $1638T$ must be replaced by $1638\mu ZT$.

The gas compressibility can be calculated from Equation (4-35). At initial conditions,

$$c_g = \frac{1}{4613} - \frac{0.045}{(0.968)(671)} = 1.475 \times 10^{-4}\ \text{psi}^{-1}. \tag{4-81}$$

The slope 0.045 was obtained from Figure 4-1 at $T_{pr} = 1.69$ and $p_{pr} = 6.87$. Thus the total system compressibility is

$$c_t \approx S_g c_g \approx 0.73 \times 1.475 \times 10^{-4} = 1.08 \times 10^{-4}\ \text{psi}^{-1}. \tag{4-82}$$

Hence for time equal to 10 days (240 hr), Equation (4-80) becomes

$$q = \frac{(0.17)(78)[1.265 \times 10^9 - m(p_{wf})]}{(1638)(640)} \left[\log 240 + \log \frac{(0.17)}{(0.14)(0.0235)(1.08 \times 10^{-4})(0.328)^2} - 3.23\right]^{-1} \tag{4-83}$$

and finally,

$$q = 2.18 \times 10^{-6}[1.265 \times 10^9 - m(p_{wf})]. \tag{4-84}$$

Table 4-6 Calculated Viscosity, Gas Deviation Factor, and Real Gas Pseudopressure for Example 4-9

p(psi)	μ(cp)	Z	m(p)(psi^2/cp)
100	0.0113	0.991	8.917×10^5
200	0.0116	0.981	3.548×10^6
300	0.0118	0.972	7.931×10^6
400	0.0120	0.964	1.401×10^7
500	0.0123	0.955	2.174×10^7

p(psi)	μ (cp)	Z	m(p)(psi^2/cp)
600	0.0125	0.947	3.108×10^7
700	0.0127	0.939	4.202×10^7
800	0.0130	0.931	5.450×10^7
900	0.0132	0.924	6.849×10^7
1000	0.0135	0.917	8.396×10^7
1100	0.0137	0.910	1.009×10^8
1200	0.0140	0.904	1.192×10^8
1300	0.0142	0.899	1.389×10^8
1400	0.0145	0.894	1.598×10^8
1500	0.0147	0.889	1.821×10^8
1600	0.0150	0.885	2.056×10^8
1700	0.0153	0.881	2.303×10^8
1800	0.0155	0.878	2.562×10^8
1900	0.0158	0.876	2.831×10^8
2000	0.0161	0.874	3.111×10^8
2100	0.0163	0.872	3.401×10^8
2200	0.0166	0.872	3.700×10^8
2300	0.0169	0.871	4.009×10^8
2400	0.0171	0.871	4.326×10^8
2500	0.0174	0.872	4.651×10^8
2600	0.0177	0.873	4.984×10^8
2700	0.0180	0.875	5.324×10^8
2800	0.0183	0.877	5.670×10^8

(*continued*)

Table 4-6 Calculated Viscosity, Gas Deviation Factor, and Real Gas Pseudopressure for Example 4-9 *(continued)*

p(psi)	μ (cp)	Z	m(p)(psi^2/cp)
2900	0.0185	0.879	6.023×10^8
3000	0.0188	0.882	6.381×10^8
3100	0.0191	0.885	6.745×10^8
3200	0.0194	0.889	7.114×10^8
3300	0.0197	0.893	7.487×10^8
3400	0.0200	0.897	7.864×10^8
3500	0.0203	0.902	8.245×10^8
3600	0.0206	0.907	8.630×10^8
3700	0.0208	0.912	9.018×10^8
3800	0.0211	0.917	9.408×10^8
3900	0.0214	0.923	9.802×10^8
4000	0.0217	0.929	1.020×10^9
4100	0.0220	0.935	1.059×10^9
4200	0.0223	0.941	1.099×10^9
4300	0.0223	0.947	1.139×10^9
4400	0.0226	0.954	1.180×10^9
4500	0.0232	0.961	1.220×10^9
4600	0.0235	0.967	1.260×10^9
4700	0.0238	0.974	1.301×10^9
4800	0.0241	0.982	1.342×10^9
4900	0.0244	0.989	1.382×10^9
5000	0.0247	0.996	1.423×10^9

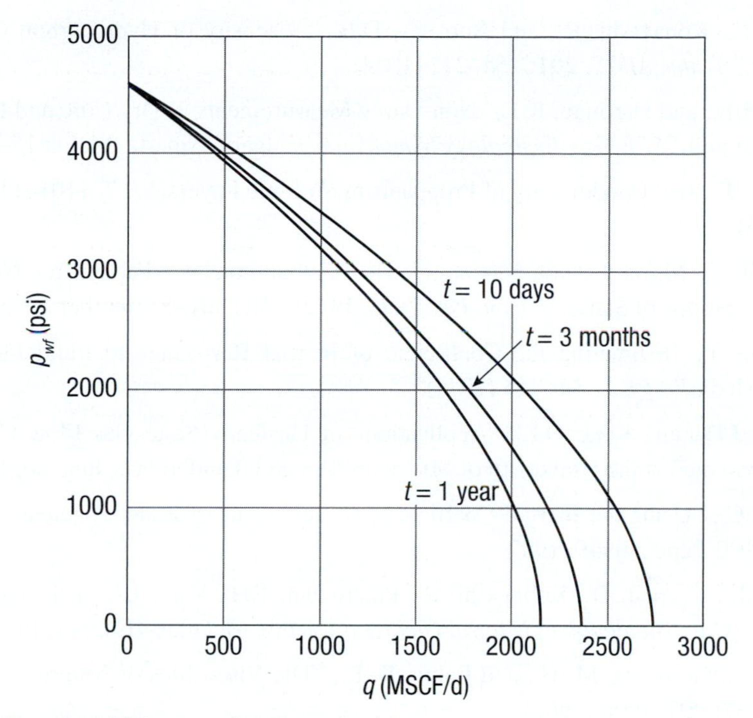

Figure 4-8 Transient IPR curves for Example 4-9.

Similar expressions can be developed readily for the other times.

Figure 4-8 is a graph of the transient IPR curves for 10 days, 3 months, and 1 year.

Note: A large fraction of all natural gas-producing wells are hydraulically fractured, which makes the inflow relationships presented in this chapter, implying radial flow, sometimes invalid. If the reservoir permeability is relatively large (i.e., $k > 10$ md), then the fracture lengths are relatively short, and the well performance can be described with a fractured well skin factor and a radial flow equation is appropriate. However, in low-permeability reservoirs fracture lengths are very large and radial flow-equivalent skin effects are not appropriate. Fracture-performance models are then indicated. In such cases, non-Darcy flow effects are likely to occur in a hydraulic fracture so these effects need to be included. Reservoir inflow to hydraulically fractured gas wells is covered in Chapters 17 and 18.

References

1. Al-Hussainy, R., and Ramey, H.J., Jr., "Application of Real Gas Theory to Well Testing and Deliverability Forecasting," *JPT,* 637–642 (May 1966).
2. Aronofsky, J.S., and Jenkins, R., "A Simplified Analysis of Unsteady Radial Gas Flow," *Trans. AIME,* **201:** 149–154 (1954).
3. Brown, G.G., Katz, D.L., Oberfell, C.G., and Alden, R.C., "Natural Gasoline and the Volatile Hydrocarbons," NGAA Paper, Tulsa, OK, 1948.

4. Carr, N.L., Kobayashi, R., and Burrows, D.B., "Viscosity of Hydrocarbon Gases Under Pressure," *Trans. AIME,* **201:** 264–272 (1954).

5. Coles, M.E., and Hartman, K.J., "Non-Darcy Measurements in Dry Core and Effect of Immobile Liquid," *SPE Gas Technology Symposium,* Calgary, Alberta, March 15–18, 1998.

6. Cooke, C.E., Jr., "Conductivity of Proppants in Multiple Layers," *JPT,* 1101–1107 (September 1973).

7. Dranchuk, P. M., and Abou-Kassem, J. H., "Calculation of z-Factors for Natural Gases Using Equations of State," *J. Can. Pet. Tech.,* **14:** 34–36 (July–September 1975).

8. Geertsma, J., "Estimating the Coefficient of Inertial Resistance in Fluid Flow Through Porous Media," *SPE J.,* **14:** 445 (1974).

9. Janicek, J.D., and Katz, D.L., "Applications of Unsteady State Gas Flow Calculations," Paper presented at the University of Michigan Research Conference, June 20, 1955.

10. Jones, S.C., "Using the Inertial Coefficient, β, to Characterize Heterogeneity in Reservoir Rock," SPE Paper 16949, 1987.

11. Katz, D.L., Cornell, D., Kobayashi, R., Poettmann, F.H., Vary, J.A., Ellenbaas, J.R., and Weinang, C.F., *Handbook of Natural Gas Engineering,* McGraw-Hill, NY, 1959.

12. Lee, A. L., Gonzales, M. H., and Eakin, B. E., "The Viscosities of Natural Gases," *Trans. AIME,* **237:** 997–1000 (1966).

13. Li, D., and Engler, T. W., "Literature Review on Correlations of the Non-Darcy Coefficient," SPE Paper 70015, 2001.

14. Liu, X., Civan, F., and Evans, R.D., "Correlation of the non-Darcy flow coefficient," *JCPT,* **43:** 50 (1995).

15. McCain, William D, Jr., *The Properties of Petroleum Fluids,* 2nd edition, PennWell Publishing Co., Tulsa, OK, 1990.

16. Pascal, H., and Quillian, R. G., "Analysis of Vertical Fracture Length and non-Darcy Flow Coefficient Using Variable Rate Tests," SPE Paper 9348, 1980.

17. Standing, M.B., "Volumetric and Phase Behavior of Oil Field Hydrocarbon Systems," SPE, Dallas (1977).

18. Standing, M.B., and Katz, D.L., "Density of Natural Gases," *Trans. AIME,* **146:** 140–149 (1942).

19. Tek, M.R., Coats, K.H., and Katz, D.L., "The Effect of Turbulence on Flow of Natural Gas Through Porous Reservoir," *J. Pet. Tech.*, **799** (July 1962).

20. Thauvin, F., and Mohanty, K.K., "Network Modeling of Non-Darcy Flow Through Porous Media," *Transport in Porous Media,* **31**(1): 19 (1998).

21. Wichert, E., and Aziz, K., "Calculation of Z's for Sour Gases," *Hydrocarbon Processing,* **51**(5) (1972).

22. Wang, X., Thauvin, F., and Mohanty, K.K., "Non-Darcy Flow Through Anisotropic Porous Media," *Chem. Eng. Sci.,* **54:** 1859 (1999).

23. Wang, X., "Pore-Level Modeling of Gas-Condensate Flow in Porous Media," PhD dissertation, University of Houston (May 2000).

Problems

4-1 For two natural gases, A and B, the only differences between their compositions are C_1 and C_2. Assume the composition in gas A: $C_1 = 0.8, C_2 = 0.15$ and gas B: $C_1 = 0.85, C_2 = 0.1$. The pseudocritical pressure of gas A and gas B are 663 psi and 661.2 psi, respectively. What is the specific gravity of gas B if the gravity of gas A is 0.6815?

4-2 A closed container has 10 ft^3 volume and is filled with a gas of $\gamma_g = 0.7$. The temperature of the container is 200°F and the pressure is 4000 psi. If the pressure in the container increases to 4100 psi at the same temperature, how would the volume change? What is the slope of the Standing-Katz correlation $\partial Z/\partial p_{pr}$ at this temperature?

4-3 For the gas compositions below and conditions in Appendix C, compute the gas density.

Methane = 0.875
Ethane = 0.075
Propane = 0.025
Case 1: Nitrogen = 0.025, other gases = 0
Case 2: Carbon dioxide = 0.025, other gases = 0
Case 3: Hydrogen sulfide = 0.025, other gases = 0

4-4 The pseudocritical temperature and pressure of a sour gas are 397°R and 720 psi, respectively. To use the Standing-Katz graph, T_{pc} and p_{pc} are corrected to 367°R and 657.6 psi. What are the percentages of H_2S and CO_2 in this gas?

4-5 For the well described in Appendix C, plot the IPR curve for bottomhole pressure $500 < p_{wf} < 3500$ at steady-state conditions, for both Darcy and Non-Darcy Flow [$D = 7.6 \times 10^{-4}$ (Mscf/d)$^{-1}$)].

4-6 A well has stabilized gas production following Aronofsky and Jenkins's relation. Assume the reservoir average pressure is 3000 psi, drainage area is 80 acres, and net thickness is 80 ft. The wellbore radius is 0.328 ft and the perforated thickness is 20 ft. The well produces 2 MMscf/d at $p_{\text{wf}} = 2000$ psi and 5.2 MMscf/day at $p_{\text{wf}} = 1000$ psi. What is the permeability? Use $\gamma_g = 0.7$, $\mu_g = 0.01$ for gas, and the skin factor is 5. Generate IPR curves for this well at the specified condition.

4-7 For a gas well at steady-state flow, use the following data to

a. generate the IPR equation in a standard quadratic format ($ax^2 + bx + c = 0$) for a gas well.
b. What is the gas flow rate if $p_{wf} = 300$ psi?
c. generate an IPR plot for this well.

d. Discussion: At what flow rate can the non-Darcy flow effect be neglected and why?

Reservoir: $k = 0.17$ md, $h = 78$ ft, $p_e = 4350$ psi, $T = 180$ °F, skin factor = 5, $r_e = 1000$ ft, and $r_w = 0.328$ ft

Fluid: $\gamma_g = 0.65$, and at average condition, $\mu_g = 0.02$ cp, $Z = 0.95$

Non-Darcy flow coefficient $D = 10^{-3}$

4-8 Assume a well in a gas reservoir is under pseudosteady state and the production rate is following Equation (4-47). The reservoir average pressure decrease rate is 500 psi/year. The non-Darcy coefficient is 10^{-3} (Mscf/d)$^{-1}$.

a. What is the cumulative production for 3 years if the skin factor is zero?

b. If the skin factor is 10, what is the cumulative production for 3 years?

c. At what flow rate is the non-Darcy flow effect on production reduction equal to the effect of skin factor of 10?

Use data in Appendix C. The bottomhole pressure is 3000 psi. $r_e = 1490$ ft. Ignore the transient to pseudosteady state process.

4-9 Calculate a cumulative production curve for the gas well in Appendix C, assuming infinite-acting behavior [Equation (4-80)] for 1 year. Bottomhole pressure is 2000 psi. What fraction of the original gas-in-place will be produced if $A = 4000$ acres?

CHAPTER 5

Production from Horizontal Wells

5.1 Introduction

Starting in the 1980s, horizontal wells began capturing an ever-increasing share of hydrocarbon production. They proved to be successful in a great number of applications which include thin reservoirs ($h < 50$ ft), heavy oil, tight formations ($k < 0.1$ md) with multiple hydraulic fractures, highly heterogeneous reservoirs, or even thick reservoirs with good vertical permeability, k_V. Figure 5-1 presents the drainage pattern of a horizontal well with a length, L. The concept of horizontal wells is extending the contact with the reservoir, thus changing the drainage pattern from radial flow in vertical wells to a combination of radial, linear, and elliptical flows. Reducing water or gas coning is another main benefit of horizontal wells.

Of particular importance in the production from horizontal wells is the horizontal-to-vertical permeability anisotropy. Sufficient vertical permeability is required for a horizontal well to be beneficial. Low vertical permeability or discontinuities in vertical permeability may render horizontal wells unattractive in relatively thick formations. Another often-ignored issue is the permeability anisotropy in the *horizontal* plane. A well that is drilled normal to the larger horizontal permeability direction would be a better producer than one drilled in an arbitrary direction or normal to the smaller horizontal permeability. The larger this permeability anisotropy, the more important the well azimuth becomes.

In a new reservoir development, measurements are always recommended before a horizontal well is drilled. These measurements should be done in a vertical pilot hole. Testing techniques to estimate horizontal and vertical permeability values are described in Chapter 13. Stress measurements in the pilot hole can identify the maximum and minimum horizontal stress directions. While different correlations exist between stress and permeability, almost always stress and permeability directions coincide: the maximum horizontal stress orientation is

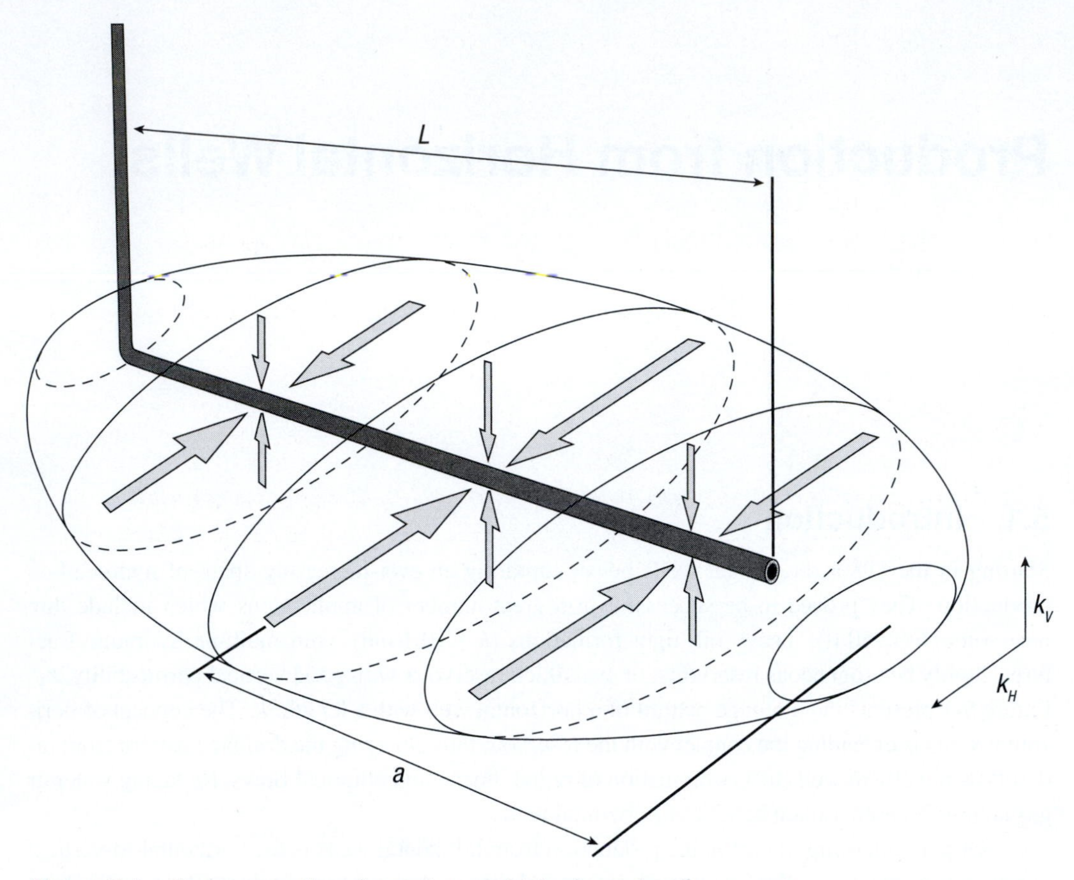

Figure 5-1 Drainage pattern formed around a horizontal well.

along maximum permeability and the minimum horizontal stress is along the minimum permeability. Therefore, a horizontal well that is not intended to be hydraulically fractured should be drilled along the direction of minimum horizontal stress.

The drainage regime of a horizontal well is principally different from a vertical well. Deviation from a simple radial flow pattern in a vertical well makes the inflow models for horizontal wells more complex. Models for horizontal well inflow performance were developed based on assumptions of flow patterns. Most of them are combinations of more than one standard pattern (radial, linear, elliptical, or spherical). With assumed boundary conditions—for example, steady-state or pseudosteady-state—analytical models are available to predict well inflow performance relationships. For infinite-acting radial flow (transient boundary), analytical equations exist (Goode and Kuchuk, 1991; Ozkan, Yildiz, and Kuchuk, 1998) that are useful for testing purposes. But most models contain infinite summation terms and are not convenient for inflow performance calculations. We discuss only the flow equations for steady-state and pseudosteady-state conditions in this chapter.

5.2 Steady-State Well Performance

Two models developed by Joshi (1988) and Furui, Zhu, and Hill (2003) for horizontal well inflow performance under the steady-state flow condition are introduced in this section. In the Joshi model the well is centered in an elliptical-shaped drainage boundary. The Furui et al. model is based on a geometry of a horizontal well centered in a box-shaped drainage boundary. Reservoir pressure is assumed constant at the boundary for both models.

5.2.1 The Joshi Model

The steady-state flow model developed by Joshi (1988) was one of the first analytical models for horizontal well inflow. Joshi derived an equation for the flow rate to a horizontal well of length L by adding a solution for the flow resistance in the horizontal plane with the solution for the flow resistance in the vertical plane, and taking into account vertical-to-horizontal anisotropy.

Considering a horizontal well extending in the x-direction in a reservoir of thickness h, Joshi treated the horizontal well inflow as a combination of the flow in the x-y plane and the flow in the y-z plane separately, as shown in Figure 5-2. Two-dimensional flow in the x-y plane to a sink of length L will have elliptical isobars at steady state (the right-hand picture in Figure 5-2), so presuming a drainage ellipse with a major axis length of $2a$ and a constant pressure at the drainage boundary gives

$$q_h = \frac{2\pi k \Delta p}{\mu B_o \ln\left(\dfrac{a + \sqrt{a^2 - (L/2)^2}}{L/2}\right)}, \tag{5-1}$$

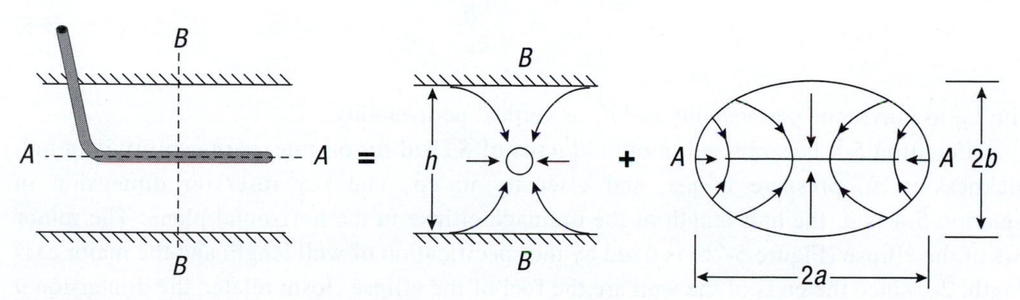

Figure 5-2 Schematic of the Joshi model.

which was multiplied by reservoir thickness to approximate the production from a planar sink (a stack of horizontal wells or a fully penetrating, infinitely conductive fracture). Flow in the vertical plane *y-z* plane (the center picture in Figure 5-2) was approximated to be radial flow from the vertical boundary located at a distance *h*/2 from the well, where, interestingly, the pressure is presumed to be the same as at the elliptical horizontal boundary. This assumption yields

$$q_v = \frac{2\pi k \Delta p}{\mu B_o \ln\left(\frac{h}{2r_w}\right)}, \tag{5-2}$$

which was multiplied by the total well length, *L*, to sum the *y-z* flow contributions of the entire well. The flow resistances $\Delta p/q$ for the *x-y* and *y-z* planes were then added together and equated to $\Delta p/q$ for the well to give the inflow for an isotropic formation as:

$$q = \frac{2\pi k_H h \Delta p}{\mu B_o \left(\ln\left(\frac{a + \sqrt{a^2 - (L/2)^2}}{L/2} \right) + \frac{h}{L} \ln\left(\frac{h}{2r_w} \right) \right)}. \tag{5-3}$$

For an anisotropic reservoir, Equation 5-3 becomes (Economides, Deimbacher, Brand, and Heinemann, 1991):

$$q = \frac{k_H h (p_e - p_{wf})}{141.2\, \mu B_o \left(\ln\left(\frac{a + \sqrt{a^2 - (L/2)^2}}{L/2} \right) + \frac{I_{ani} h}{L} \ln\left(\frac{I_{ani} h}{r_w (I_{ani} + 1)} \right) \right)} \tag{5-4}$$

where the anisotropy ratio, I_{ani}, is defined as

$$I_{ani} = \sqrt{\frac{k_H}{k_V}} \tag{5-5}$$

with k_H as horizontal permeability and k_V as vertical permeability.

Equation 5-4 is expressed in oilfield units of STB/d for oil rate, permeability is in md, thickness in ft, pressure in psi, and viscosity in cp. The key reservoir dimension in Equation 5-4 is *a,* the half length of the drainage ellipse in the horizontal plane. The minor axis of the ellipse (Figure 5-2b) is fixed by the specification of well length and the major axis length, 2*a,* since the ends of the well are the foci of the ellipse. Joshi related the dimension *a* to an equivalent cylindrical drainage radius by equating the areas of the ellipse to that of a cylinder of radius r_e, yielding

$$a = \frac{L}{2}\left\{0.5 + \left[0.25 + \left(\frac{r_{eH}}{L/2}\right)^4\right]^{0.5}\right\}^{0.5} \tag{5-6}$$

Equation 5-4 is derived for a well that is centered in the drainage volume, both vertically and horizontally. Selecting the appropriate value of the parameter, *a,* is an important part of applying this equation. It should be selected based on the best information available about the drainage area extent either in the direction of the well (x-direction) or in the horizontal direction perpendicular to the well (y-direction). Joshi presented a modification to the model (not presented here) to account for eccentricity in the vertical plane.

Example 5-1 Well Performance with the Joshi Model

A 2000-ft long horizontal lateral is producing from a 100-ft thick reservoir, having a horizontal permeability of 10 md and a vertical permeability of 1 md. The lateral is 6 inches in diameter and is draining a region that is 4000-ft long in the direction of the well. The pressure at the drainage boundary is 4000 psia, the oil has a viscosity of 5 cp, and the formation volume factor is 1.1. Generate an equation for the inflow performance relationship and plot the IPR curve for this well. What is the production rate for a bottomhole pressure of 2000 psia?

Solution

For the Joshi model, from Equation 5-4, the inflow performance relationship is

$$p_{wf} = p_e - \frac{141.2qB_o\mu}{k_H h}\left(\ln\left(\frac{a + \sqrt{a^2 - (L/2)^2}}{L/2}\right) + \frac{I_{ani}h}{L}\ln\left(\frac{I_{ani}h}{r_w(I_{ani} + 1)}\right)\right). \tag{5-7}$$

From Equation 5-5, we have

$$I_{ani} = \sqrt{\frac{10}{1}} = 3.162$$

and a is 2000 ft, one-half of the reservoir extent in the direction of the well. Substituting in Equation 5-7,

$$p_{wf} = 4000 - 0.78q(1.32 + 0.16(5.72)) \tag{5-8}$$

or

$$p_{wf} = 4000 - 1.74q. \tag{5-9}$$

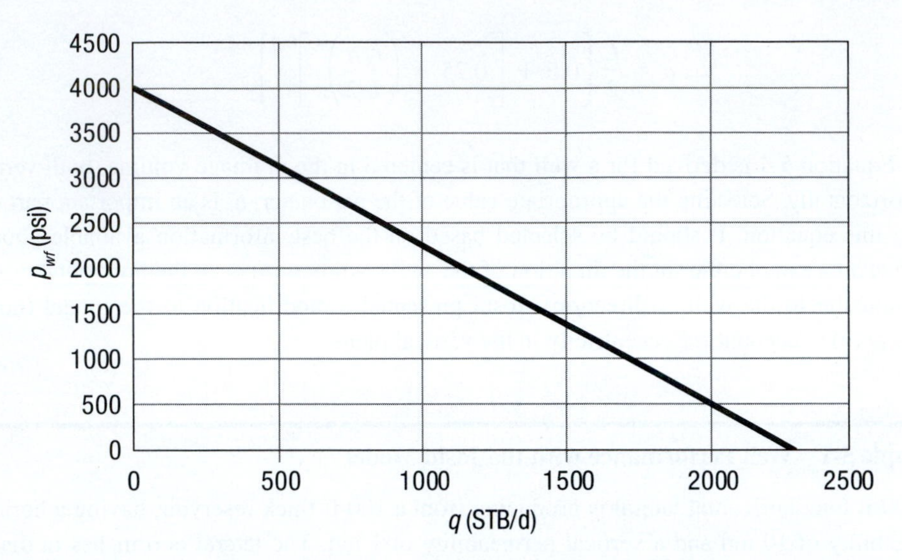

Figure 5-3 Inflow performance of a horizontal well by Joshi's equation.

Equation 5-9 is the inflow performance relationship for this well, and the IPR curve is shown in Figure 5-3. For a p_{wf} of 2000 psi, the flow rate is 1149 STB/d.

5.2.2 The Furui Model

Furui et al. (2003) solved the flow problem in the cross-sectional area perpendicular to the wellbore (Figure 5-4). The model assumes that the flow near the well is radial and becomes linear farther from the well. Thus, the total pressure drop can simply be added as follows:

$$\Delta p = \Delta p_r + \Delta p_l \tag{5-10}$$

where Δp_r and Δp_l are the pressure drops for the radial and linear flow regions, respectively. From Darcy's law in radial coordinates, the pressure drop is

$$\Delta p_r = \frac{q\mu}{2\pi kL}\ln\left(\frac{r_t}{r_w}\frac{2I_{ani}}{(I_{ani}+1)}\right) \tag{5-11}$$

where r_t is the outer extent of the radial flow region (the location of the transition to linear flow), and k is the geometric mean permeability,

$$k = \sqrt{k_H k_V} \tag{5-12}$$

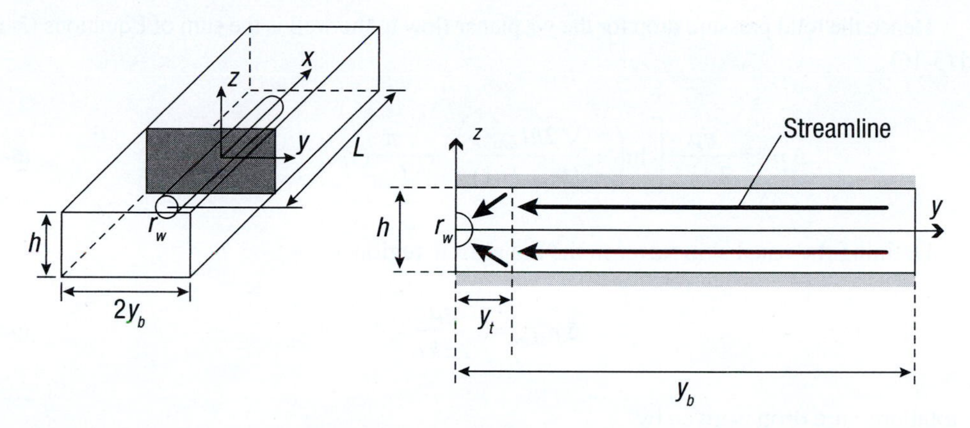

Figure 5-4 Schematic of Furui et al. (2003) model.

and the ratio $2I_{ani}/(I_{ani} + 1)$ represents the Peaceman transform for radial flow in an anisotropic medium. Similarly, the pressure drop in the linear flow region is

$$\Delta p_l = \frac{(q/2)\mu(y_b - y_t)}{kI_{ani}hL} \tag{5-13}$$

where y_t is the location of the beginning of the linear flow region and y_b is the distance to the drainage boundary in the y-direction. Realize that k_H is replaced with kI_{ani} for the anisotropic case in Equation (5-13). Finite element modeling indicates that a suitable empirical geometric relationship for r_t and y_t is

$$r_t = y_t\sqrt{2} = \frac{\sqrt{2}I_{ani}}{2}h. \tag{5-14}$$

Substituting the above equations into Equations (5-11) and (5-13) gives

$$\Delta p_r = \frac{q\mu}{2\pi kL}\ln\left(\frac{\sqrt{2}h}{r_w}\frac{I_{ani}}{(I_{ani} + 1)}\right) \tag{5-15}$$

and

$$\Delta p_l = \frac{q\mu(y_b/h - I_{ani}/2)}{2kI_{ani}L}. \tag{5-16}$$

Hence the total pressure drop for the *y-z* planar flow to the well is the sum of Equations (5-15) and (5-16)

$$\Delta p = \frac{q\mu}{2\pi kL}\left[\ln\left(\frac{\sqrt{2}hI_{ani}}{r_w(I_{ani}+1)}\right) + \frac{\pi}{I_{ani}}(y_b/h - I_{ani}/2)\right]. \tag{5-17}$$

Defining the usual skin factor in the radial flow region as

$$\Delta p_{skin} = \frac{q\mu}{2\pi kL}s, \tag{5-18}$$

the total pressure drop is given by

$$\Delta p = \frac{q\mu}{2\pi kL}\left[\ln\left(\frac{\sqrt{2}hI_{ani}}{r_w(I_{ani}+1)}\right) + \frac{\pi}{I_{ani}}(y_b/h - I_{ani}/2) + s\right]. \tag{5-19}$$

Solving for q, and incorporating conversions for oilfield units, the equation becomes

$$q = \frac{kL(p_e - p_{wf})}{141.2\,\mu B_o\left(\ln\left[\frac{hI_{ani}}{r_w(I_{ani}+1)}\right] + \frac{\pi y_L}{hI_{ani}} - 1.224 + s\right)}. \tag{5-20}$$

The constant in the above equation, 1.224, comes from $\ln(\sqrt{2}) - \pi/2$. To account for the drainage area beyond the horizontal well length, we introduce a partial penetration skin, s_R, for a non-fully penetrating wellbore. The calculation of the partial penetration skin factor is discussed in the next section. For a partially-penetrating well, Equation 5-20 becomes

$$q = \frac{kb(p_e - p_{wf})}{141.2\,\mu B_o\left(\ln\left[\frac{hI_{ani}}{r_w(I_{ani}+1)}\right] + \frac{\pi y_b}{hI_{ani}} - 1.224 + s + s_R\right)} \tag{5-21}$$

where k is the geometry mean permeability (same as Equation 5-12).

Example 5-2 Well Performance with the Furui et al. Model

Consider again the reservoir described in Example 5-1. Using the Furui et al. model, calculate the production rate for a bottomhole flowing pressure of 2000 psi if the well is 2000 ft in length as before. Assume the partial penetration skin factor is 14. Using $k_H = 10$ md as a reference, study the effect of vertical permeability on production rate for I_{ani} equal to 1 and 10.

Solution

For the conditions described in Example 5-1, with a 2000-ft long well in a 4000-ft long reservoir, the length $2b$ of the minor axis of the drainage ellipse assumed by Joshi is $\sqrt{3}(1000)$, or 1732 ft. This will be used as the distance between the reservoir boundaries, $2y_b$. The geometric mean permeability to be used in the Furui equation is $k = \sqrt{k_y k_z} = \sqrt{10} = 3.162$ md.

For case 1, applying Equation 5-21, we have

$$q = \frac{(3.162)(4000)(2000)}{141.2(5)(1.1)\left(\ln\left[\frac{(100)(3.162)}{(0.25)(3.162+1)}\right] + \frac{\pi(866)}{(100)(3.162)} - 1.224 + 14\right)}$$

$$= 1276 \text{ STB}/d \tag{5-22}$$

Under the same condition, if I_{ani} is 1 ($k_V = 10$ md) then $q = 2275$ STB/d, and if I_{ani} is 10 ($k_V = 0.1$ md) then $q = 482$ STB/d. For small k_V, q is very sensitive to I_{ani}.

5.3 Pseudosteady-State Flow

Pseudosteady-state models of inflow performance, as mentioned in Chapter 2, presumes that the reservoir is bounded by no-flow boundaries and that pressure declines in a uniform fashion in the reservoir.

5.3.1 The Babu and Odeh Model

In the Babu and Odeh (1988, 1989) model, the physical system is a box-shaped drainage area with a horizontal well with radius r_w and length $L = x_2 - x_1$, placed parallel to the x-direction, as shown in Figure 5-5. The reservoir has a length in the x-direction of b, a width in the horizontal direction perpendicular to the well (y-direction) of a, and a thickness of h. The well can be in any arbitrary location in this reservoir, except that the well must lie in the x-direction and cannot be too close to any boundary. The well location is defined by specifying the location of the heel of the well as being at x_1, y_o, and z_o, relative to the origin located at one corner of the reservoir.

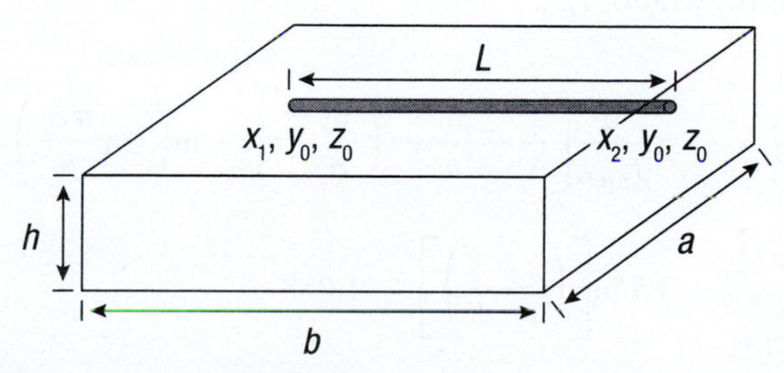

Figure 5-5 Schematic of the Babu and Odeh model.

The Babu and Odeh model is based on radial flow in the *y-z* plane, with the deviation of the drainage area from a circular shape in this plane accounted for with a geometry factor, and flow from beyond the wellbore in the *x* direction accounted for with a partial penetration skin factor. Note that the Babu and Odeh geometry factor is related inversely to the commonly used Dietz shape factor (Dietz, 1965). Thus, the Babu and Odeh inflow equation is

$$q = \frac{\sqrt{k_y k_z} b(\bar{p} - p_{wf})}{141.2 B_o \mu \left[\ln\left(\frac{A^{0.5}}{r_w}\right) + \ln C_H - 0.75 + s_R + s \right]}. \tag{5-23}$$

In Equation 5-23, A is the cross-section area (***ah,*** Figure 5-5), C_H is the shape factor, s_R is the partial penetration skin, and s includes any other skin factors, such as completion or damage skin effects. The shape factor, C_H, accounts for the deviation of the shape of the drainage area from cylindrical and the departure of the wellbore location from the center of the system (Figure 5-5). The partial penetration skin, s_R, accounts for the flow from the reservoir located beyond the ends of the well in the *x*-direction, and is equal to zero for a fully penetrating horizontal well.

The heart of the Babu and Odeh model are procedures for calculating the shape factor and the partial penetration skin factor. These parameters were obtained by simplifying the solution of the diffusivity equation for the parallelepiped reservoir geometry and comparing it with the assumed inflow equation (Equation 5-23). Babu and Odeh solved the 3-D diffusivity equation with a wellbore boundary condition of constant flow rate (uniform flux) at the well and no flow across the reservoir boundaries using the Green's function approach. In this manner, the following correlations for the shape factor and partial penetration skin factor were obtained.

$$\ln C_H = 6.28 \frac{a}{h} \sqrt{\frac{k_z}{k_y}} \left[\frac{1}{3} - \frac{y_0}{a} + \left(\frac{y_0}{a}\right)^2 \right] - \ln\left(\sin \frac{\pi z_0}{h}\right) - 0.5 \ln\left[\left(\frac{a}{h}\right)\sqrt{\frac{k_z}{k_y}}\right] - 1.088 \tag{5-24}$$

or in terms of the anisotropy ratio, I_{ani},

$$\ln C_H = 6.28 \frac{a}{I_{ani} h} \left[\frac{1}{3} - \frac{y_0}{a} + \left(\frac{y_0}{a}\right)^2 \right] - \ln\left(\sin \frac{\pi z_0}{h}\right) - 0.5 \ln\left[\left(\frac{a}{I_{ani} h}\right)\right] - 1.088 \tag{5-25}$$

s_R is evaluated for two different cases, depending on the horizontal dimensions of the reservoir. The first case is for a reservoir that is relatively wide [i.e., the reservoir extends farther in the horizontal direction perpendicular to the well than in the direction of the wellbore trajectory ($a > b$)]. The second case is for a long reservoir ($b > a$). The particular criteria for Case 1 are

$$\frac{a}{\sqrt{k_x}} \geq 0.75\frac{b}{\sqrt{k_y}} > 0.75\frac{h}{\sqrt{k_z}},$$

then

$$s_R = P_{xyz} + P'_{xy}, \tag{5-26}$$

where

$$P_{xyz} = \left(\frac{b}{L} - 1\right)\left[\ln\frac{h}{r_w} + 0.25\ln\frac{k_y}{k_z} - \ln\left(\sin\left(\frac{\pi z_0}{h}\right)\right) - 1.84\right] \tag{5-27}$$

and

$$P'_{xy} = \frac{2b^2}{Lh}\sqrt{\frac{k_z}{k_x}}\left\{F\left(\frac{L}{2b}\right) + 0.5\left[F\left(\frac{4x_{mid} + L}{2b}\right) - F\left(\frac{4x_{mid} - L}{2b}\right)\right]\right\} \tag{5-28}$$

where x_{mid} is the x-coordinate of the midpoint of the well,

$$x_{mid} = \frac{x_1 + x_2}{2} \tag{5-29}$$

and

$$F\left(\frac{L}{2b}\right) = -\left(\frac{L}{2b}\right)\left[0.145 + \ln\left(\frac{L}{2b}\right) - 0.137\left(\frac{L}{2b}\right)^2\right]. \tag{5-30}$$

$F((4x_{mid} + L)/2b)$ and $F((4x_{mid} - L)/2b)$ in Equation 5-28 are evaluated as follows, taking the argument, $(4x_{mid} + L)/2b$ or $(4x_{mid} - L)/2b$, as X. If the values of X are less than or equal to 1, $F(X)$ is calculated by Equation 5-30 with the argument of $L/2b$ replaced by X. Otherwise, if X is greater than 1, then $F(X)$ is calculated by

$$F(X) = (2 - X)[0.145 + \ln(2 - X) - 0.137(2 - X)^2] \tag{5-31}$$

with X either $(4x_{mid} + L)/2b$ or $(4x_{mid} - L)/2b$.

The criteria for Case 2 are

$$\frac{b}{\sqrt{k_y}} \geq 1.33\frac{a}{\sqrt{k_x}} > \frac{h}{\sqrt{k_z}}.$$

For this case,

$$s_R = P_{xyz} + P_y + P_{xy}, \tag{5-32}$$

where

$$P_y = \frac{6.28b^2}{ah}\frac{\sqrt{k_y k_z}}{k_x}\left[\left(\frac{1}{3} - \frac{x_{mid}}{b} + \frac{x_{mid}^2}{b^2}\right) + \frac{L}{24b}\left(\frac{L}{b} - 3\right)\right] \tag{5-33}$$

and

$$P_{xy} = \left(\frac{b}{L} - 1\right)\left(\frac{6.28a}{h}\sqrt{\frac{k_z}{k_y}}\right)\left(\frac{1}{3} - \frac{y_0}{a} + \left(\frac{y_0}{a}\right)^2\right) \tag{5-34}$$

where P_{xyz} in Equation 5-32 is the same as defined in Equation 5-27.

Example 5-3 Well Performance with the Babu and Odeh Model

Consider again the 4000-ft long reservoir described in Examples 5-1 and 5-2. For a 2000-ft long horizontal well centered in the reservoir as in Example 5-1, and with width, a, equal to 1414 ft, what is the production rate predicted by the Babu and Odeh model if average reservoir pressure is 4000 psi and the bottomhole flowing pressure is 2000 psi? Assume all other parameters are the same as in Examples 5-1 and 5-2.

Solution

For the well centered in the Babu and Odeh box-shaped reservoir of the dimensions given, the length of the reservoir, b, is 4000 ft; the width of the reservoir, a, is 1414 ft; the height of the reservoir, h, is 100 ft; the ends of the well are at $x_1 = 1000$ ft and $x_2 = 3000$ ft, $x_{mid} = 2000$ ft,

$z_0 = 50$ ft, and $y_0 = 707$ ft. Other necessary data from the previous examples are horizontal permeability, which is 10 md ($k_x = k_y$) and vertical permeability (k_z) of 1 md, the lateral is 6 in. in diameter, the oil viscosity is 5 cp, and the formation volume factor is 1.1. I_{ani} is 3.16.

First, we calculate the shape factor, ln C_H, using Equation 5-25:

$$\ln C_H = 6.28\frac{1414}{(3.16)(100)}\left[\frac{1}{3} - \frac{707}{1414} + \left(\frac{707}{1414}\right)^2\right] - \ln\left(\sin\frac{\pi(50)}{100}\right)$$
$$- 0.5\ln\left[\left(\frac{1414}{(3.16)(100)}\right)\right] - 1.088 = 0.5. \quad \text{(5-35)}$$

Checking for which case to use for calculating the partial penetration skin factor, a is 1414 ft and b is 4000 ft, thus

$$\frac{4000}{\sqrt{10}} \geq 1.33\frac{1414}{\sqrt{10}} > \frac{100}{\sqrt{1}},$$

and therefore Case 2 applies (long reservoir).

Using Equations 5-27, 5-33, and 5-34, then

$$P_{xyz} = \left(\frac{4000}{2000} - 1\right)\left[\ln\frac{100}{0.25} + 0.25\ln 10 - 1.05\right] = 4.73 \quad \text{(5-36)}$$

$$P_y = \frac{6.28(4000)^2}{(1414)(100)}\frac{\sqrt{(10)(1)}}{(10)}\left[\left(\frac{1}{3} - \frac{2000}{4000} + \left(\frac{2000}{4000}\right)^2\right)\right.$$
$$\left. + \frac{2000}{24(4000)}\left(\frac{2000}{4000} - 3\right)\right] = 7.02 \quad \text{(5-37)}$$

$$P_{xy} = \left(\frac{4000}{2000} - 1\right)\left(\frac{6.28(1414)}{100}\sqrt{\frac{1}{10}}\right)\left(\frac{1}{3} - \frac{707}{1414} + \left(\frac{707}{1414}\right)^2\right) = 2.34 \quad \text{(5-38)}$$

so, from Equation 5-32,

$$s_R = 4.73 + 7.02 + 2.3 = 14.1. \quad \text{(5-39)}$$

The flow rate for the given conditions is then calculated with Equation 5-23:

$$q = \frac{\sqrt{(10)(1)}(4000)(4000 - 2000)}{141.2(1.1)(5)\left[\ln\left(\frac{\{(1414)(100)\}^{0.5}}{0.25}\right) + 0.5 - 0.75 + 14.1\right]} = 1527 \text{ STB/d} \quad \text{(5-40)}$$

There are more parameters in horizontal well performance calculations than for vertical wells, such as the wellbore length and permeability anisotropy. One common question about horizontal well design is how long the wellbore length should be (notice that we are comparing producing length, which is the formation thickness in the vertical well case). The length of a horizontal well should be designed based on the geometry of the drainage area (length, width, formation thickness), the type of reservoir fluid, and the well structure (new well or reentry sidetrack).

In general, for an effectively unlimited drainage area, the longer the wellbore, the higher the flow rate. However, this conclusion could be misleading if the pressure drop along the horizontal well is large compared to the pressure drop from the formation to the well. For large-permeability formations, selecting horizontal well length should consider the flow diameter in the horizontal well. If the well length is too long, the production rate will be choked back by the pressure drop in the tubing. Detailed discussion about pressure drop in pipe flow is presented in Chapters 7 and 8, and methods to estimate the conditions for which the pressure drop in the wellbore are important relative to reservoir pressure drop is presented in Section 8.5.1.

Horizontal wells sometimes may not be the optimal plan to develop a field. For example, if the formation payzone is relatively thick, then a vertical well or a hydraulically fractured vertical well may be more beneficial, especially when vertical connectivity is low (small vertical permeability). Productivity indices for horizontal, vertical, and hydraulically fractured wells should be compared to select the well architecture.

Example 5-4 Comparison of Productivity Index for Horizontal and Vertical Wells

For the same reservoir that was presented in Example 5-3, calculate the ratio of productivity index for a 2000-ft long horizontal well to a vertical well. Horizontal permeability, k_x, is 10 md, vertical permeability, k_z, is 1 md, wellbore radius is 0.25 ft, viscosity is 5 cp, and the formation volume factor is 1.1.

Solution

For pseudosteady state, the vertical well productivity index is

$$J_V = \frac{q}{(\bar{p} - p_{wf})} = \frac{k_x h}{141.2 B\mu \ln\left(\dfrac{0.472 r_e}{r_w}\right)}. \tag{5-41}$$

And using the Babu and Odeh model, the horizontal well productivity index is

$$J_H = \frac{q}{(\bar{p} - p_{wf})} = \frac{\sqrt{k_x k_z} b}{141.2 B\mu \left(\ln\left(\dfrac{A^{0.5}}{r_w}\right) + \ln C_H - 0.75 + s_R\right)}. \tag{5-42}$$

For the horizontal well, using the result from Example 5.3, the productivity index is the flow rate, 1527 STB/d, dividing by the pressure drawdown, 2000 psi, which gives 0.76 STB/d/psi. For the vertical well, the equivalent drainage radius, r_e, can be calculated from the drainage area, 4000-ft by 1414-ft. Thus, the productivity index for the vertical well is

$$J_V = \frac{(10)(100)}{141.2(1.1)(5)\ln\left(\dfrac{0.472\sqrt{\dfrac{(4000)(1414)}{\pi}}}{0.25}\right)} = 0.164 \text{ STB/d/psi} \tag{5-43}$$

The productivity index ratio is 4.6, showing that drilling a horizontal well in this case is beneficial from a production point of view.

5.3.2 The Economides et al. Model

Economides et al. (1991; Economides, Brand, and Frick, 1996) developed a completely general model for one or more horizontal wells or laterals by integrating unit length point sources in no-flow boundary "boxes" to create any arbitrary well trajectory or trajectories. Figure 5-6 shows a line source with arbitrary trajectory inside a parallelepiped drainage volume. This modeling approach has enabled many useful models for both transient and pseudosteady-state flow behavior of vertical, deviated, and horizontal wells with and without fractures.

The productivity index, J, is related to the dimensionless pressure (in oilfield units):

$$J = \frac{q}{\bar{p} - p_{wf}} = \frac{\bar{k}x_e}{887.22B\mu\left(p_D + \dfrac{x_e}{2\pi L}\Sigma s\right)} \tag{5-44}$$

$\bar{k}$ is the geometric average reservoir permeability given by the cubed root of the three directional permeability values, $\bar{k} = \sqrt[3]{k_x k_y k_z}$. Σs is the sum of all damage and pseudo-skin factors. Dimensioned calculations are based on the reservoir length, x_e, and L is the horizontal well length.

The generalized solution to the dimensionless pressure, p_D, starts with early-time transient behavior and ends with pseudosteady state if all drainage boundaries are felt. At that moment, the three-dimensional (3-D) p_D is decomposed into one two-dimensional (2-D) and one one-dimensional (1-D) part,

$$P_D = \frac{x_e C_H}{4\pi h} + \frac{x_e}{2\pi L}s_x \tag{5-45}$$

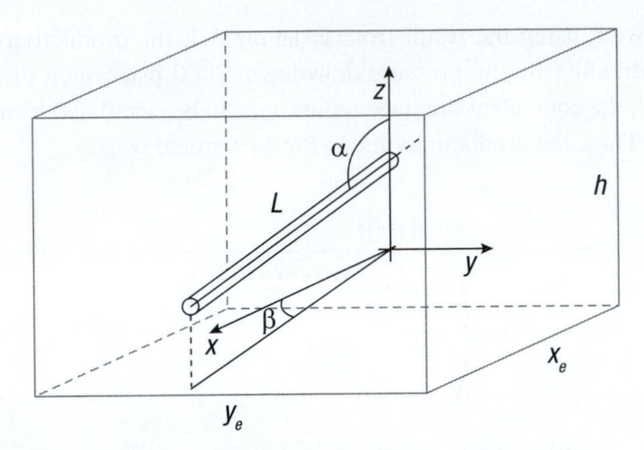

Figure 5-6 Basic parallelepiped model with appropriate coordinates.

where C_H is a "shape" factor, characteristic of well and reservoir configurations in the horizontal plane, and s_x is the skin accounting for vertical flow effects.

Table 5-1 contains a library of approximate shape factors.

The expression for this skin effect (after Kuchuk, Goode, Brice, Sherrard, and Thambynayagam, 1990) is

$$s_x = \ln\left(\frac{h}{2\pi r_w}\right) + \frac{h}{6L} + s_e \tag{5-46}$$

and s_e, describing eccentricity effects in the vertical direction, is

$$s_e = \frac{h}{L}\left[\frac{2z_w}{h} - \frac{1}{2}\left(\frac{2z_w}{h}\right)^2 - \frac{1}{2}\right] - \ln\left[\sin\left(\frac{\pi z_w}{h}\right)\right], \tag{5-47}$$

where z_w is the vertical distance from the wellbore to the bottom of the payzone. s_e is negligible if the well is placed near the vertical middle of the reservoir.

The way the Economides et al. (1996) model works is that lengths, radii, and azimuthal departures are adjusted with permeability anisotropy and then the adjusted variables are used for the calculation. Thus,

Length:

$$L' = La^{-1/3}\beta \tag{5-48}$$

Table 5-1 Shape Factors for Various Single, Multilateral, and Multibranch Well Configurations

Configuration		L/x_e	C_H
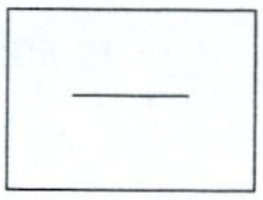 Single horizontal well	$x_e = 4y_e$	0.25	3.77
		0.5	2.09
		0.75	1.00
		1	0.26
	$x_e = 2y_e$	0.25	3.19
		0.5	1.80
		0.75	1.02
		1	0.52
	$x_e = y_e$	0.25	3.55
		0.4	2.64
		0.5	2.21
		0.75	1.49
		1	1.04
	$2x_e = y_e$	0.25	4.59
		0.5	3.26
		0.75	2.53
		1	2.09
	$4x_e = y_e$	0.25	6.69
		0.5	5.35
		0.75	4.63
		1	4.18
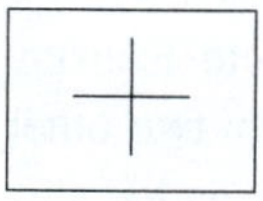 Cross (two laterals)	$x_e = y_e$	0.25	2.77
		0.5	1.47
		0.75	0.81
		1	0.46
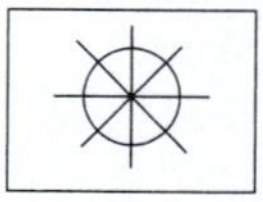 Radial (multiple laterals)	$x_e = y_e$	0.25	2.66
		0.5	1.36
		0.75	0.69
		1	0.32

Configuration			C_H
Inclined well at angle ϕ	$x_e = y_e$, $L/x_e = 0.75$	ϕ = 0	1.49
		30	1.48
		45	1.48
		75	1.49
		90	1.49
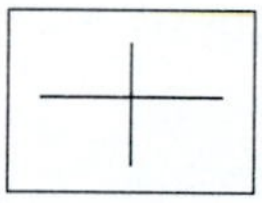 One branch pair	$x_e = y_e$, $L_x/X_e = 0.4$	$L_y = 2L_x$	1.10
		$L_y = L_x$	1.88
		$L_y = 0.5L_x$	2.52
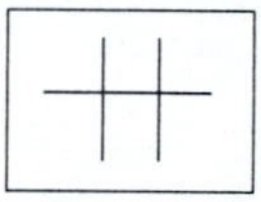 Two branch pairs	$x_e = y_e$, $L_x/x_e = 0.4$	$L_y = 2L_x$	0.79
		$L_y = L_x$	1.51
		$L_y = 0.5L_x$	2.04
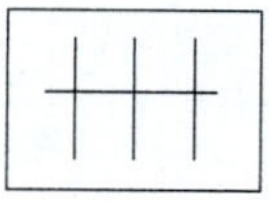 Three branch pairs	$x_e = y_e$, $L_x/x_e = 0.4$	$L_y = 2L_x$	0.66
		$L_y = L_x$	1.33
		$L_y = 0.5L_x$	1.89
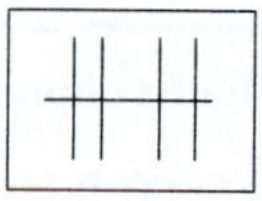 Four branch pairs	$x_e = y_e$, $L_x/x_e = 0.4$	$L_y = 2L_x$	0.59
		$L_y = L_x$	1.22
		$L_y = 0.5L_x$	1.79

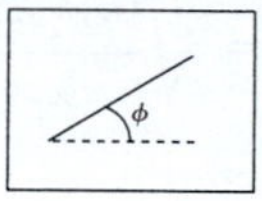

(From Economides et al., 1994; Retnanto and Economides, 1996.)

<u>Wellbore radius:</u>

$$r'_w = r_w \frac{a^{2/3}}{2} \left(\frac{1}{a\beta} + 1 \right) \tag{5-49}$$

with

$$a = \sqrt{\frac{(k_x k_y)^{1/2}}{k_z}} \quad \text{(5-50)}$$

and

$$\beta = \left(\sqrt{\frac{k_y}{k_x}} \cos^2 \phi + \sqrt{\frac{k_x}{k_y}} \sin^2 \phi \right)^{1/2} \quad \text{(5-51)}$$

where the angle is the departure between the well trajectory and the x-direction of the reservoir drainage.

Similarly, for the reservoir dimensions:

$$x' = x\frac{\sqrt{k_y k_z}}{k} \; y' = y\frac{\sqrt{k_x k_z}}{k} \; z' = z\frac{\sqrt{k_x k_y}}{\bar{k}} \quad \text{(5-52)}$$

Example 5-5 Well Performance with the Economides et al. Model

Consider a 4000-ft square reservoir with a 2000-ft long horizontal well centered in the reservoir and parallel to the reservoir boundary. Assume all other parameters are the same as in Examples 5-1 and 5-2 (i.e., the reservoir thickness is 100 ft, the well radius is 0.25 ft, the oil has a viscosity of 5 cp, and the formation volume factor is 1.1). First, calculate the PI under complete isotropy ($k_H = k_V = 10$ md). Then repeat the calculation with $k_z = 1$ md and finally perform two other calculations with 5:1 horizontal permeability anisotropy (keeping $k_H = 10$ md but k_x and k_y different), the first with the well perpendicular to the minimum permeability (i.e., $k_x > k_y$) and the second with the well perpendicular to the maximum permeability. Assume no damage skins.

Solution

1. **Perfect isotropy** Start with Equation 5-46. The skin effect s_x, with $s_e = 0$, using unadjusted, h, r_w and L with Equation 5-46 is 4.15. From Table 5-1, because $x_e = y_e$ and $L/x_e = 0.5$, then $C_H = 2.21$. From Equation 5-45, $p_D = 8.36$ and, finally, from Equation 5-44, $J = 0.98$ STB/d/psi.
2. **Vertical-to-horizontal anisotropy** If $k_z = 1$ md, $k_x = k_y = 10$ md, using Equation (5-53) $\bar{k} = 4.63$ md. With Equation (5-52), $x' = 2725$ ft, $y' = 2725$ ft, and $h' = 215$ ft. From Equations 5-48 and 5-49, $L' = 1363$ ft and $r'_w = 0.354$ ft. From Equation 5-46 and using the adjusted variables, $s_x = 4.54$. The C_H will not change because both the L' and x' will be adjusted similarly and because the permeability in the

horizontal plane is isotropic, y' is also adjusted similarly. From Equation 5-45, $p_D = 3.67$ and, finally, from Equation 5-44, $J = 0.7$ STB/d/psi, an almost 30% reduction in the PI from the isotropic case.

3. **Horizontal anisotropy, unfavorable orientation** Let $k_H = 10$ md but let $k_x = 5k_y$. This means that their product, square root, should be equal to 10 and therefore $k_x = 22.36$ md and $k_y = 4.47$ md. The average permeability is still 4.63 md. The adjustment to h' is still the same. However, all other variables will change. This means that (from Equation 5-52) $x' = 1822$ ft, $y' = 4076$ ft. This leads to $2.24x' = y'$. From Equations 5-48 to 5-49, $L' = 911$ ft and $r'_w = 0.397$ ft. As should be expected since x and L are in the same direction, L/x does not change and is equal to 0.5. Then, by interpolation of Table 5-1, C_H is equal to 3.5. From Equation 5-46 and using the adjusted variables, $s_x = 4.42$. Finally, from Equation 5-45, $p_D = 3.77$ and from Equation 5-44, $J = 0.46$ STB/d/psi, more than 50% reduction in the PI from the isotropic case.
4. **Horizontal anisotropy, favorable orientation** The results from the calculation above are repeated but now $k_y = 22.36$ md and $k_x = 4.47$ md (i.e., the well is rotated to be perpendicular to the favorable permeability, although the average reservoir permeability remains constant). The adjustment to h' is still the same. However, from Equation 5-52, $y' = 1822$ ft, $x' = 4076$ ft. This leads to $x' = 2.24y'$. From Equation 5-49 and 5-49, $L' = 2038$ ft (much more elongated than the previous case) and $r'_w = 0.326$ ft. From Table 5-1 and by interpolation, $C_H = 1.83$.From Equation 5-46 and using the adjusted variables, $s_x = 4.64$. Finally, from Equation 5-45, $p_D = 4.24$, and from Equation 5-44, $J = 0.91$ STB/d/psi, twice the PI of the previous case, showing the importance of proper well orientation.

Goode and Kuckuk (1991) presented an inflow model obtained by solving the 2-D problem of flow to a fracture that is the full height of the reservoir, then accounting for the flow convergence in the z-direction with a z-direction partial penetration skin factor. Uniform flux along the well is assumed. The original Goode and Kuchuk model contains infinite summations, so it is somewhat more unwieldy than the Babu and Odeh model.

Table 5-2 Summary of Results for Example 5-5

	k_{avg}	h'	r'_w	L'	s_x	x'_e	y'_e	L'/x'_e	C_H	J	p_D
1	10	100	0.25	2000	4.15	4000	4000	0.5	2.21	0.98	8.36
2	4.63	215	0.354	1363	4.54	2725	2725	0.5	2.21	0.70	3.67
3	4.63	215	0.397	911	4.42	1822	4076	0.5	3.5	0.46	3.77
4	4.63	215	0.326	2038	4.64	4076	1822	0.5	1.83	0.91	4.24

5.4 Inflow Performance Relationship for Horizontal Gas Wells

IPR equations for gas wells can be directly developed from the equations for oil wells. Analogous to the derivation for vertical well equations presented in Chapter 4, the horizontal well equations are developed by replacing the formation volume factor of oil with the one for gas in Equations 5-4, 5-20, and 5-23, and then applying the real gas law to the definition of the gas formation volume factor, as shown in Equation 4-28 For steady state, using the Furui et al. model (Equation 5-20), for example, the inflow equation for a horizontal gas well can be expressed as

$$q_g = \frac{kL(p_e^2 - p_{wf}^2)}{1424\overline{Z}\,\overline{\mu}_g T\left(\ln\left[\frac{hI_{ani}}{r_w(I_{ani}+1)}\right] + \frac{\pi y_b}{hI_{ani}} - 1.224 + s\right)}. \tag{5-53}$$

This equation averages Z and μ_g as a constant over the pressure range from p_{wf} to p_e. To more accurately account for the effects of pressure on these physical properties, as in Chapter 4, the real gas pseudopressure function presented by Al-Hussainy and Ramey (1966) can be used

$$m(p) = 2\int_{p_o}^{p} \frac{p}{\mu_g z}\,dp \tag{5-54}$$

where p_o is the reference pressure and can be any convenient base pressure. The equation for horizontal gas wells in terms of the real gas pseudopressure is

$$q_g = \frac{kL(m(p_e) - m(p_{wf}))}{1424T\left(\ln\left[\frac{hI_{ani}}{r_w(I_{ani}+1)}\right] + \frac{\pi y_b}{hI_{ani}} - 1.224 + s\right)}. \tag{5-55}$$

Non-Darcy flow can be added to Equation 5-55 as

$$q_g = \frac{kL(m(p_e) - m(p_{wf}))}{1424T\left(\ln\left[\frac{hI_{ani}}{r_w(I_{ani}+1)}\right] + \frac{\pi y_b}{hI_{ani}} - 1.224 + s + Dq_g\right)}. \tag{5-56}$$

The non-Darcy coefficient, *D,* is the same as the one presented and discussed in detail in Chapter 4.

Similarly, for pseudosteady-state conditions, the IPR equation for gas wells can be obtained from Babu and Odeh's model (Equation 5-32). The equation for a horizontal gas well is

$$q_g = \frac{b\sqrt{k_y k_z}(\overline{p}^2 - p_{wf}^2)}{1424\overline{Z}\overline{\mu}_g T\left[\ln\left(\frac{A^{0.5}}{r_w}\right) + \ln C_H - 0.75 + s_R + s + Dq_g\right]}. \tag{5-57}$$

The gas properties are estimated at the average pressure between the flowing bottomhole pressure and the reservoir pressure. To use the real gas pseudopressure, the equation becomes

$$q_g = \frac{b\sqrt{k_y k_z}(m(\overline{p}) - m(p_{wf}))}{1424T\left[\ln\left(\frac{A^{0.5}}{r_w}\right) + \ln C_H - 0.75 + s_R + s + Dq_g\right]}. \tag{5-58}$$

One of the advantages of using horizontal wells is to reduce the drawdown for production. This actually reduces the non-Darcy flow effect in horizontal wells compared with vertical wells. The non-Darcy flow effect can sometimes also be masked by the partial penetration skin, s_R. When the ratio of wellbore length to the reservoir size, L/b, is low, s_R can be significant, diminishing the effect of non-Darcy flow.

5.5 Two-Phase Correlations for Horizontal Well Inflow

Similar to vertical wells, analytical inflow relationships for two-phase flow to horizontal wells are unavailable because of the complexity of relative permeability and the variable phase distribution in the reservoir. Correlations, led by Vogel's equation presented in Chapter 3, have been used for two-phase IPR calculations for vertical wells. There have been publications for correlations of two-phase horizontal wells (e.g., Bendakhlia and Aziz, 1989; Cheng, 1990; Retnanto and Economides, 1996; Kabir, 1992). Here we use the original Vogel equation to build IPR relationships for two-phase horizontal wells. From Equation 3-51 for two-phase flow wells,

$$\frac{q_o}{q_{o,\max}} = 1 - 0.2\left(\frac{p_{wf}}{\overline{p}}\right) - 0.8\left(\frac{p_{wf}}{\overline{p}}\right)^2. \tag{5-59}$$

To modify the above equation for horizontal wells, we use a model of single-phase oil flow in horizontal wells to calculate maximum open-flow potential, $q_{o,max}$ (the flow rate at p_{wf} is zero). Since the original correlation was developed for a pseudosteady-state condition, we should use pseudosteady-state equations in the modification. For example, Babu and Odeh's equation can be employed here,

$$q_{\max,o} = \frac{\sqrt{k_y k_z} b(\overline{p})}{254.2 B_o \mu \left[\ln\left(\frac{A^{0.5}}{r_w}\right) + \ln C_H - 0.75 + s_R + s \right]}. \quad \text{(5-60)}$$

Example 5-6 Well Performance in Two-phase Flow Horizontal Wells

Using the data in Appendix B, calculate the well flow rate if the average reservoir pressure is 4000 psi for a horizontal well draining from the reservoir, assuming no oil relative permeability reduction. The horizontal well length is 2500 ft. Assume that the vertical permeability is 10% of horizontal permeability.

Solution

Using Equation 5-60, we first calculate the maximum open flow potential of the horizontal well. Assume that the length of the drainage area, *b,* for horizontal well case is $2r_e(1490 \times 2 = 2980)$, then the width of the drainage, *a,* is 2339 ft ($\pi r_e^2/b$). Using the Babu and Odeh model, in this case, ln*Ch* is 1.35 and s_R is 1.92. With absolute permeability,

$$q_{o,\max} = \frac{\sqrt{(13)(1.3)}(2960)(4000)}{(254.2)(1.42)(0.5)[\ln(\sqrt{(2339)(115)}/0.406 + 1.35 + 1.92 - 0.75)]}$$

$$= 27{,}900 \text{ STB/d} \quad \text{(5-61)}$$

Then

$$q_0 = q_{o,\max}\left(1 - 0.2\left(\frac{p_{wf}}{\overline{p}}\right) - 0.8\left(\frac{p_{wf}}{\overline{p}}\right)^2\right)$$

$$= 27{,}900\left(1 - 0.2\left(\frac{3000}{4350}\right) - 0.8\left(\frac{3000}{4350}\right)^2\right) = 13{,}400\ STB/d. \quad \text{(5-62)}$$

5.6 Multilateral Well Technology

Built from the horizontal well concept, multilateral well technology has become an important part of oil and gas production. A multilateral well is a well that has two or more producing/injecting laterals contacting different parts of the reservoir. Multilateral wells have a wide range of applications, including coal production, heavy oil, unconventional low-permeability gas formation (shale gas formation), and locations that have limited access, such as offshore fields. Figure 5-7 shows a complex multilateral well used in heavy oil recovery in Venezuela.

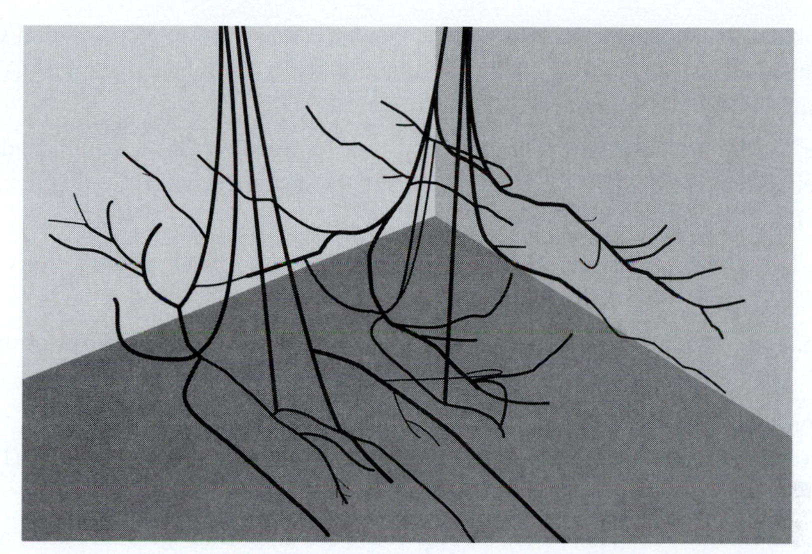

Figure 5-7 An example of multilateral well for heavy oil production. (After Robles, 2001.)

Multilateral wells can be more attractive than horizontal wells in some cases. One example would be a wellbore pressure-drop limitation in long horizontal wells. To use a multilateral well, we can still ensure the reservoir contact, but reduce the wellbore frictional pressure drop.

The well performance of a multilateral well is not simply adding the individual horizontal lateral's production rate. Since the laterals are linked to the mother bore through the junctions, the flow conditions from each lateral have to be balanced at the junctions, which means that junction pressure has to be equal from each lateral. This condition regulates the flow rates from each lateral, and that makes flow in the wellbore important to multilateral well performance. The details of calculating well performance is presented in *Multilateral Wells* (Hill, Economides, and Zhu, 2007).

References

1. Al-Hussainy, R., and Ramey, H.J., Jr., "Application of Real Gas Theory to Well Testing and Deliverability Forecasting," *JPT,* 637–642, May 1966.
2. Babu, D.K., and Odeh, A.S., Appendices A and B of SPE 18298, "Productivity of a Horizontal Well," SPE 18334 presented at the 1988 SPE Annual Technical Conference and Exhibition, Houston, TX, October 2–5, 1988.
3. Babu, D.K., and Odeh, A.S., "Productivity of a Horizontal Well," *SPE Reservoir Engineering,* p. 417, November 1989.

4. Bendakhlia, H., and Aziz, K., "Inflow Performance Relationships for Solution-Gas Drive Horizontal Wells," Paper SPE 19823, presented at the 64th Annual Technical Conference, San Antonio, TE, October 8–11, 1989.

5. Cheng, A.M., "Inflow Performance Relationships for Solution-Gas-Drive Slanted/Horizontal Wells," Paper SPE 20720, presented at the 65th Annual Technical Conference, New Orleans, LA, September 23–26, 1990.

6. Dietz, D.N., "Determination of Average Reservoir Pressure from Build-up Surveys," *JPT,* 955–959 (August 1965).

7. Economides, M.J., Brand, C.W., and Frick, T.P., "Well Configurations in Anisotropic Reservoirs," *SPEFE,* 257–262 (December 1996).

8. Economides, M.J., Deimbacher, F.X., Brand, C.W., and Heinemann, Z.E., "Comprehensive Simulation of Horizontal Well Performance," *SPEFE,* 418–426 (December 1991).

9. Furui, K., Zhu, D., and Hill, A. D., "A Rigorous Formation Damage Skin Factor and Reservoir Inflow Model for a Horizontal Well," *SPE Production and Facilities,* 151–157, August 2003.

10. Goode, P. A. and Kuchuk, F. J., "Inflow Performance of Horizontal Wells," *SPE Reservoir Engineering,* 319–323 (August 1991).

11. Hill, A. D., Economides, M. J., and Zhu, D., *Multilateral Wells,* Society of Petroleum Engineers, Richardson, Texas, 2007.

12. Joshi, S.D., "Augmentation of Well Productivity with Slant and Horizontal Wells," *JPT,* 729–739 (June 1988).

13. Kabir, C.S., "Inflow Performance of Slanted and Horizontal Wells in Solution-Gas-Drive Reservoir," Paper SPE 24056, presented at the 1992 SPE WESTERN REGIONAL Meeting, Bakersfield, CA, March 30–April 1.

14. Kuchuk, F.J., Goode, P.A., Brice, B.W., Sherrard, D.W., and Thambynayagam, R.K.M., "Pressure Transient Analysis and Inflow Performance for Horizontal Well," JPT, 974–1031 (August 1990).

15. Ozkan, E., Yildiz, T., and Kuckuk, F. J., "Transient Pressure Behavior of Dual Lateral Wells," *SPEJ,* 181–190 (June 1998).

16. Robles, Jorge, "Application of Advanced Heavy-Oil-Production Technologies in the Orinoco Heavy-Oil-Belt, Venezuela," SPE 69848 presented at the 2001 SPE International Thermal Operations and Heavy Oil Symposium, Margarita Island, VA, March 12–14, 2001.

17. Retnanto, A. and Economides, M.J., "Performance of Multiple Horizontal Well Laterals in Low-To Medium-Permeability Reservoirs," *SPERE,* 73–77 (May 1996).

Problems

5-1 Consider the reservoir described in Example 5-1, having a horizontal permeability of 100 md. Generate equations for inflow performance relationships with Joshi's model and plot the IPR curves for vertical permeabilities of 1, 10, and 100 md, respectively. Study the effect of the vertical permeability on the results considering the anisotropy index. Repeat the problem with a net reservoir thickness of 25 ft (one-fourth of the previous case). Comment on the relative difference in the results.

5-2 For each thickness, graph the dimensionless productivity index, J_D, versus the anisotropy index, I_{ani}, for the cases considered in Problem 5-1. Use J_D defined as

$$J_D = \frac{141.2qB\mu}{kh(p_e - p_{wf})}.$$

5-3 Suppose that the correlation for the location of the beginning of the linear flow region, y_t, and the outer extent of the radial flow region, r_t, is

$$r_t = \sqrt{2}y_t = \frac{1}{2}h.$$

Re-derive Equation 5-19.

5-4 A 4000-ft long horizontal lateral is producing from a reservoir having a horizontal permeability of 10 md and a vertical permeability of 1 md. The lateral is 6 inches in diameter and is draining an area that is 4000-ft long and 2000-ft wide (in the direction perpendicular to the well). Other conditions remain the same as in Example 5-1. Assume the skin factor is 10. Generate equations for inflow performance relationships with Furui's model and plot the IPR curves for reservoirs with thicknesses of 50, 250, and 500 ft, respectively. Examine the benefit of a horizontal well over a vertical well by graphing J_H/J_V for the three thicknesses.

5-5 Consider a reservoir with a square drainage area of 4000-ft × 4000-ft. The thickness is 200 ft. The average reservoir pressure is 4000 psi. The horizontal permeability is 50 md and the vertical permeability is 8 md. Assume no damage skin. A 3500-ft horizontal well is placed at the center of the reservoir. Generate the IPR plot for pseudosteady-state inflow and calculate the production rate for a bottomhole pressure of 2000 psi with the Babu and Odeh model. The oil viscosity is 5 cp and the formation volume factor is 1.1. The wellbore radius is 6 in.

5-6 For Problem 5-5, examine the effect of the following parameters on production performance and provide conclusions. Assume p_{wf} = 2000 psi.

a. For a vertical well drilled at the center of the reservoir, calculate J_H/J_V.

b. To avoid water or gas contact, the well is sometimes placed close to the boundary. Plot q versus z_0 and comment.

c. When a well is moved away from the center location, the production rate decreases. Plot q versus y_0 and comment.

d. When the vertical permeability is too small, horizontal wells become unfavorable. Plot q versus k_V and draw observations.

5-7 Repeat Problems 5-5 and 5-6 using the Economides et al. model.

5-8 Use the data in Appendix C. A 2000-ft long horizontal lateral is producing from a 78-ft thick reservoir, having a horizontal permeability of 2 md and a vertical permeability of 0.2 md. The reservoir boundary pressure is 4600 psi. The lateral is 0.328-ft in radius and is draining a region that is 4000-ft long in the direction of the well and 2000-ft long in the direction perpendicular to the well. Neglect the non-Darcy flow effect. Calculate the gas flow rate with a 3000 psi bottomhole pressure using the Furui et al. model.

5-9 Develop an inflow performance relationship equation for horizontal gas wells analogous to Equation (5-61) for the Economides et al. model.

5-10 Calculate the gas well flow rate for a 4000-ft long horizontal lateral and compare the results with a vertical well for the 3000 psi bottomhole pressure under pseudosteady-state conditions. Both wells are centered in each direction. The drainage area is a square with 4000-ft length sides. The average reservoir pressure is 4600 psi. Use the data in Appendix C and Problem 5-8. Neglect the non-Darcy flow effect and assume no damage skin. Also, calculate the productivity index ratio of the horizontal well to the vertical one.

5-11 Use the data in Problem 5-5. Generate an IPR curve for a two-phase flow horizontal well if the average drainage volume pressure is 3000 psi.

CHAPTER 6

The Near-Wellbore Condition and Damage Characterization; Skin Effects

6.1 Introduction

Radial flow to a vertical well results in increasingly higher flow velocity near the well and accounts for the semilogarithmic nature of the pressure–distance relationship presented in Chapter 2 [Equation (2-4)]. Van Everdingen and Hurst (1949) introduced the concept of a pressure difference, Δp_s, occurring over an infinitesimal distance at the wellbore radius proportional to the skin factor, s, as in Equation (2-5), to account for a nonideal flow profile.

Mathematically, the Van Everdingen and Hurst skin effect has no physical dimension and is analogous to the film coefficient in heat transfer. When added to $\ln(r_e/r_w)$ for steady state, $\ln(0.472\, r_e/r_w)$ for pseudosteady state, or p_D for transient solutions, the sum becomes proportional to the total (reservoir plus near-wellbore) pressure drop.

The well skin factor can be positive or negative. Because well productivity decreases with positive skin and increases with negative skin, it is useful to determine the reasons for skin and to devise ways to remove positive skin or, better yet, to induce negative skin. Negative skin mainly results from well stimulation, such as matrix acidizing discussed in Chapters 14–16 or hydraulic fracturing discussed in Chapters 17 and 18, and it can also be observed in a highly inclined wellbore. This chapter is mainly about positive skin effects.

Positive skin effects can be created by causes such as partial completion (i.e., a perforated thickness that is less than the reservoir thickness), by inadequate number of flowing perforations or slots, by phase changes (relative permeability reduction to the main fluid), by turbulence, and, of course, by near-wellbore alteration of the natural reservoir permeability. The well skin effect is a composite variable. In general, any phenomenon that causes a distortion of the flow lines from the perfectly normal to the well direction or a restriction to flow (which could be viewed as a distortion at the pore-throat scale) would result in a positive value of the skin effect.

In this chapter, the skin effect, its components, and the estimation of the contributions of each component are presented in detail. Skin effects characterizing the damage around a horizontal well are also introduced. Finally, the nature of formation damage and the types of formation damage are outlined as a precursor to the appropriate choice of matrix stimulation treatments.

It is important to realize that there may be high contrasts in skin along the length of the productive interval. This is particularly likely in wells in which production from two or more dissimilar and vertically separated intervals is commingled. Different formation properties (permeability, stress, mechanical stability, fluids) and differential pressure create an environment for uneven damage due to drilling fluid invasion, poor completion cleanup, and other causes. Similar reasoning also implies the likelihood of uneven damage along a horizontal borehole. Chapter 15 describes stimulation techniques for commingled and horizontal wells.

6.2 Hawkins' Formula

Figure 6-1 is a depiction of near-wellbore permeability alteration, or damage, with r_s and k_s being the penetration of damage and permeability in the damaged region, respectively. Outside this zone the reservoir remains undisturbed, with permeability k. A well-known equation relating the skin effect and the above variables has been presented by Hawkins (1956) and is frequently referred to as Hawkins' formula. Figure 6-2 provides an easy means for the development of this relationship.

If the near-wellbore permeability is the reservoir permeability (i.e., no damage), then a steady-state pressure drop between the outer boundary pressure (p_s) and the well would result in a $p_{wf,\,\mathrm{ideal}}$ given by

$$p_s - p_{wf,\,\mathrm{ideal}} = \frac{q\mu}{2\pi kh} \ln \frac{r_s}{r_w}. \tag{6-1}$$

If, though, the near-wellbore permeability is altered to k_s, then the real bottomhole pressure is related by

$$p_s - p_{wf,\,\mathrm{real}} = \frac{q\mu}{2\pi k_s h} \ln \frac{r_s}{r_w}. \tag{6-2}$$

The difference between $p_{wf,\,\mathrm{ideal}}$ and $p_{wf,\,\mathrm{real}}$ is exactly the pressure drop due to the skin effect, Δp_s, which was given in Chapter 2 by Equation (2-5). Therefore, from Equations (6-1), (6-2), and (2-5),

$$\frac{q\mu}{2\pi kh} s = \frac{q\mu}{2\pi k_s h} \ln \frac{r_s}{r_w} - \frac{q\mu}{2\pi kh} \ln \frac{r_s}{r_w} \tag{6-3}$$

and simply

$$s = \left(\frac{k}{k_s} - 1 \right) \ln \frac{r_s}{r_w}. \tag{6-4}$$

Equation (6-4) is Hawkins' formula and is useful in assessing the relative effects of permeability impairment and the penetration of damage.

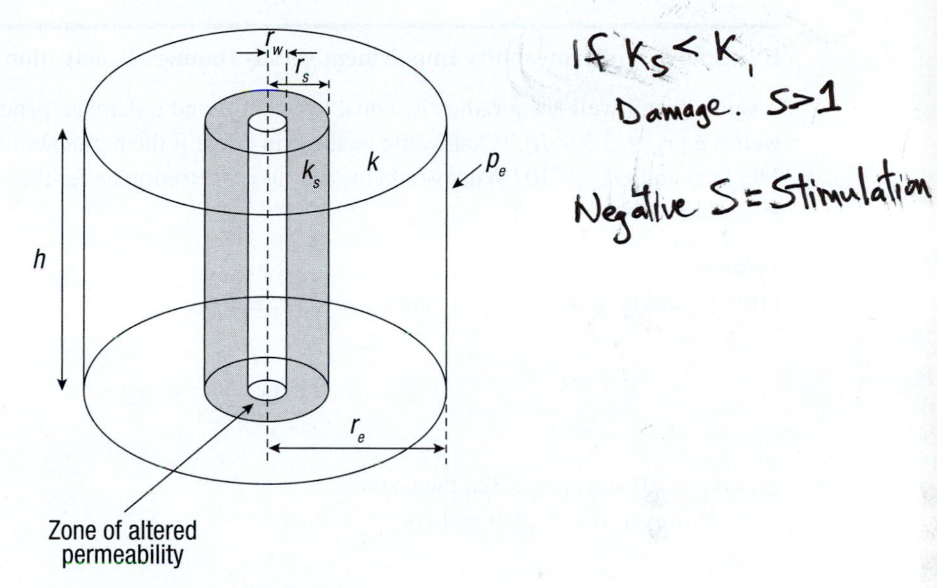

Figure 6-1 Near-wellbore zone with altered permeability.

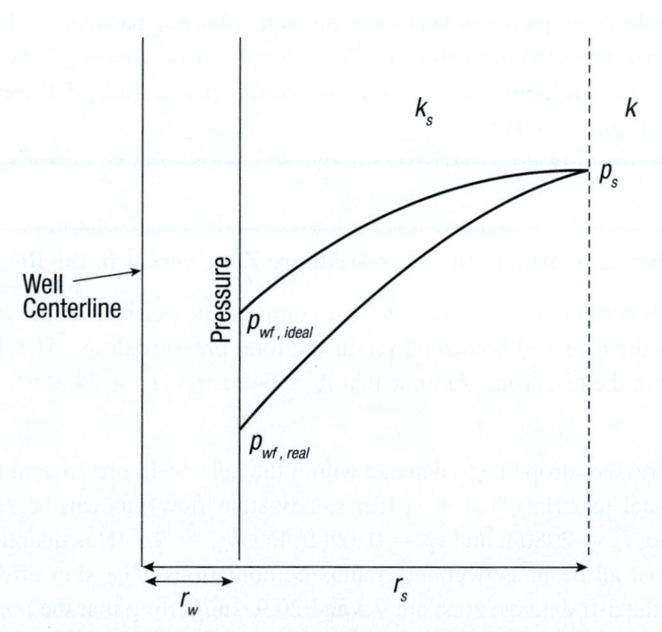

Figure 6-2 Near-wellbore zone. Ideal and real flowing bottomhole pressures.

Example 6-1 Permeability Impairment versus Damage Penetration

Assume that a well has a radius r_w equal to 0.328 ft and a damage penetration 3 ft beyond the well (i.e., $r_s = 3.328$ ft). What would be the skin effect if the permeability impairment results in $k/k_s = 5$ and $k/k_s = 10$? What would the damage penetration be for the second value of the skin effect but with $k/k_s = 5$?

Solution

From Equation (6-4), $k/k_s = 5$, and the given r_s and r_w,

$$s = (5 - 1)\ln\frac{3.328}{0.328} = 9.3 \quad \textbf{(6-5)}$$

for $k/k_s = 10$ and $r_s = 3.328$ then, similarly, $s = 20.9$.

However, if $s = 20.9$ and $k/k_s = 5$, then

$$r_s = r_w e^{20.9/4} = 61 \text{ ft.} \quad \textbf{(6-6)}$$

This exercise suggests that permeability impairment has a much larger effect on the value of the skin effect than the damage penetration. Except for a phase change-dependent skin effect, a damage penetration such as the one calculated in Equation (6-6) is impossible. Thus, skin effects derived from well tests (frequently ranging between 5 and 20) are likely to be caused by substantial permeability impairment very near the well. This is a particularly important point in the design of matrix stimulation treatments. For example, it is often likely that a skin of 20.9, calculated in the example above, can also result from a penetration of damage of just 0.5 ft (i.e., $r_s = 0.828$) if $k/k_s = 0.042$.

Example 6-2 Pressure Drop in the Near-Wellbore Zone versus in the Reservoir

With the skin effects calculated in Example 6-1, compare the portions of the pressure drop due to damage within the near-wellbore zone versus the total pressure drop. (The difference will be the pressure drop in the reservoir.) Assume that $A = 640$ acres ($r_e = 2980$ ft).

Solution

The ratio of the pressure drop due to damage within the near-wellbore zone to the total pressure drop is proportional to $s/[\ln(r_e/r_w) + s]$ (for steady-state flow), as can be seen readily from Equation (2-8). For $r_e = 2980$ ft and $r_w = 0.328$ ft, $\ln r_e/r_w = 9.1$. This quantity will be largely constant for almost all drainage/wellbore radius combinations. The skin effects calculated in Example 6-1 for the 3-ft damage zone are 9.3 and 20.9, suggesting that the portions of the total pressure drop due to damage would be 0.51 and 0.70, respectively. Possible elimination of these skin effects, at a constant $p_e - p_{wf}$, would result in production rate increases of 102% and 230%, respectively.

Although it is unlikely that the permeability in the damaged zone is constant as assumed in Hawkins' formula, the relationships developed are valid as long as an appropriate average permeability is used for the damaged zone permeability. More generally, for any distribution of permeability, $k_s(r)$ between r_w and r_s, the damage skin factor is

$$s = k \int_{r_w}^{r_w} \frac{dr}{rk_s(r)} - \ln\frac{r_s}{r_w}. \tag{6-7}$$

Alternatively, the following average damaged zone permeability can be used in Hawkins' formula:

$$\bar{k}_s = \frac{\ln\dfrac{r_s}{r_w}}{\displaystyle\int_{r_w}^{r_s} \frac{dr}{rk_s(r)}}. \tag{6-8}$$

Equations 6-7 or 6-8 can be used to calculate the damage skin effect when the permeability profile in the damaged zone has been predicted with a model of the damaging process.

Example 6-3 Skin Factor for Concentric Radial Damage Zones

Damage in some cases may be the result of two concentric damage penetrations. For example, in addition to the filtrate loss and resulting damage, drilling mud particles penetrate into the formation, causing an additional, shorter but far more damaged zone. What is the skin factor resulting from the combined particle and filtrate damage? Assume that the permeability in the particle-invaded zone is k_p and the permeability in the filtrate invaded zone is k_f.

Solution

Equation 6-7 can be used to solve this problem. For the composite damage described, the damage permeability distribution, $k_s(r)$ is

$$k_s(r) = \begin{matrix} k_p & for\ r_w < r < r_p \\ k_f & for\ r_p < r < r_f. \end{matrix} \tag{6-9}$$

Substituting this function in Equation 6-7 yields

$$s = k\left\{\int_{r_w}^{r_p} \frac{dr}{rk_p} + \int_{r_p}^{r_f} \frac{dr}{rk_f}\right\} - \ln\frac{r_f}{r_w}. \tag{6-10}$$

Integrating

$$s = \left(\frac{k}{k_p}\right)\ln\frac{r_p}{r_w} + \left(\frac{k}{k_f}\right)\ln\frac{r_f}{r_p} - \ln\frac{r_f}{r_w}. \tag{6-11}$$

6.3 Skin Components for Vertical and Inclined Wells

The total skin effect, s, for a vertical or inclined well consists of a number of components. Generally these can be added together, and therefore

$$s = (s_{comp})_d + s_c + s_\theta + \sum s_{\text{pseudo}} \tag{6-12}$$

where $(s_{comp})_d$ is the combined effect of the completion and permeability damage surrounding the completion, s_c is the skin due to partial completion, and s_θ is the deviated well skin effect. For an openhole completion, $(s_{comp})_d$ is simply the damaged skin as given by Hawkins' formula. For any other completion type the interaction of the completion and damage skin must be considered. All "pseudoskins" are grouped together within the summation sign. These pseudoskins include all phase- and rate-dependent effects. Following is a discussion of these skin effects. Subsequent sections discuss the other skin components.

The rate-dependent effect was discussed in Chapter 4 in conjunction with the turbulence in high-rate gas producers. (It can also affect very high-rate oil wells with large gas–oil ratios.) This skin effect is equal to Dq, where D is the non-Darcy coefficient (see Section 4.4). The skin effect extracted from a well test in a high-rate gas well is likely to be larger and, in certain instances, much larger than the sum of all other skin effects. Thus, from a well test, an apparent skin, s', can be obtained that is equal to

$$s' = s + Dq. \tag{6-13}$$

Tests performed at several different rates can be used to isolate the skin effect, s. A plot such as the one shown in Figure 6-3 of s' versus q suggests that s is the intercept and D is the slope. This is a straightforward manner for field determination of D that can be sure to forecast the impact of the rate-dependent skin on future well production.

Phase-dependent skin effects are associated with phase changes because of the near-wellbore pressure gradient. In the case of oil wells, if the flowing bottomhole pressure is below the bubble-point pressure, then a gas saturation will form, causing a reduction in the effective permeability to oil even if the gas phase is not mobile. A version of Hawkins' formula with k/k_s substituted by the ratio of the effective (or relative) permeabilities can be used.

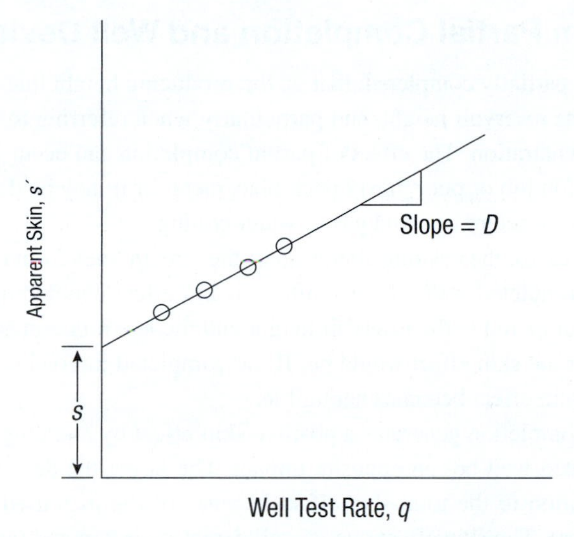

Figure 6-3 Field determination of skin effect and non-Darcy coefficient from multiple-rate well tests.

A similar phenomenon can be observed in the case of gas retrograde condensate reservoirs, where liquid is formed around the well, causing a reduction in gas permeability. This is a particularly adverse occurrence. While the gas that is formed in an oil reservoir will reenter solution at an elevated pressure (e.g., as the pressure builds up when the well is shut in), in the case of a gas condensate reservoir much of the formed condensate will not reenter the gas. Several authors (e.g., Fussell, 1973; Cvetkovic, Economides, Omrcen, and Longaric, 1990) have studied the process of liquid condensate deposition with time and have shown that permeability impairment to gas in gas condensate reservoirs is not eliminated following a shut-in. Thus, after reopening the well, the gas flow rate is still affected by the near-wellbore permeability reduction. A method to combat this skin effect is by the injection of neat natural gas, which may re-dissolve the condensate and displace it into the reservoir. This "huff-and-puff" operation can be repeated periodically.

The skin components in Equation 6-12 cause alterations of the flow or pressure-drop behavior from that which occurs for radial flow to vertical or inclined wells having uniform permeability along any streamline. Other skin components will be introduced for flow configurations associated with fractures (choke, fracture face) or fracture-to-well contact, as in the case of a largely vertical fracture intersecting an inclined or horizontal well transversely. These skin components are addressed in Chapter 17. However, the reader must be alerted here that once a hydraulic fracture is generated, most pretreatment skin effects ($(s_{comp})_d$, s_c, s_θ) are bypassed and have no impact on the post-treatment well performance. Phase- and rate-dependent skin effects are either eliminated or contribute in the calculation of the fracture skin effects. In general, it is not correct to add pretreatment skin effects to any post-fracture skin effects.

6.4 Skin from Partial Completion and Well Deviation

Frequently, wells are partially completed; that is, the producing height that is open to the formation is smaller than the reservoir height, and particularly when referring to horizontal wells, this is known as partial penetration. The effect of partial completion can occur inadvertently as a result of a bad perforation job or poor gravel pack placement, or it may be done deliberately in an often misguided effort to retard or avoid gas or water coning.

In any of these cases, the ensuing alteration of the stream lines from those that occur in radial flow to a fully completed well would result in a skin effect denoted by s_c. The smaller the perforated interval, compared to the reservoir height and the less it is centered in the total formation height, the larger the skin effect would be. If the completed interval is 75% of the reservoir height or more, this skin effect becomes negligible.

While partial completion generates a positive skin effect by reducing the well exposure to the reservoir, a deviated well has an opposite impact. The larger the deviation angle, the larger the negative contribution to the total skin effect because of the increased amount of reservoir contact by the wellbore. The skin effect due to well deviation is denoted by s_θ.

The effect of partial completion on the productivity of vertical wells has been considered in numerous studies, beginning with the work of Muskat (1946), who derived an analytical solution for the flow to a well penetrating only partially into the upper part of a reservoir (a common completion practice in the time of Muskat's work).

For a completion like that shown in Figure 6-4, Muskat presented the following inflow equation:

$$q = \frac{2\pi\, kh_w\Delta p\left(1 + 7\sqrt{\frac{r_w}{2h_w}}\cos\frac{\pi h_w}{2h}\right)}{\mu \ln\frac{r_e}{r_w}}. \tag{6-14}$$

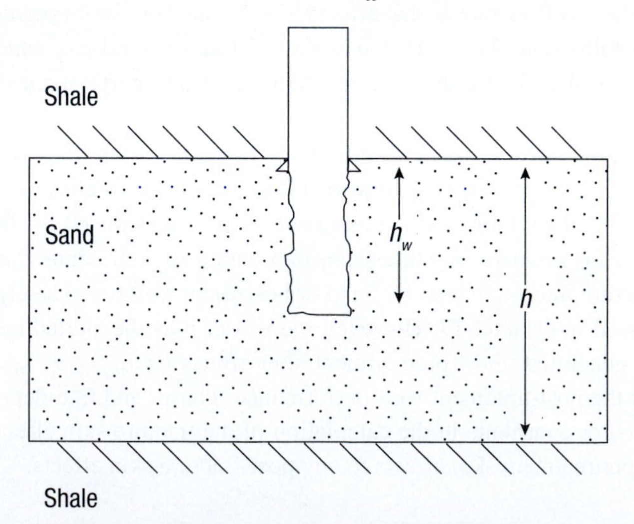

Figure 6-4 Well completed openhole in the top of the reservoir.

In this equation, h_w is the thickness of the completed interval. By comparing this equation with a steady-state inflow equation, including a skin factor (e.g., Equation 2-7), the partial completion skin factor from Muskat's relationship is

$$s_c = \left\{ \frac{\frac{h}{h_w}}{1 + 7\sqrt{\frac{r_w}{2h_w}}\cos\frac{\pi h_w}{2h}} - 1 \right\} \ln\frac{r_e}{r_w}. \tag{6-15}$$

Because of symmetry, Muskat's partial completion inflow equation can also be used to derive a partial completion skin factor for a well completed in the center of a reservoir zone. Thus, for a well with a completed thickness of h_w in the middle of a reservoir of thickness h, the partial completion skin factor is

$$s_c = \left\{ \frac{\frac{h}{h_w}}{1 + 7\sqrt{\frac{r_w}{h_w}}\cos\frac{\pi h_w}{2h}} - 1 \right\} \ln\frac{r_e}{r_w}. \tag{6-16}$$

Several studies of partial completion effects have been presented since Muskat's work, including those of Brons and Marting (1961), Cinco-Ley, Ramey, and Miller (1975), Strelstova-Adams (1979), Odeh (1980), and Papatzacos (1987). Because it is relatively simple, includes the effect of permeability anisotropy, allows for any arbitrary location of the completed interval in the reservoir zone, and reproduces other partial completion models well, we present here the Papatzacos model. To describe the completion geometry shown in Figure 6-5, Papatzacos uses the following dimensionless variables:

$$h_{wD} = \frac{h_w}{h} \tag{6-17}$$

$$r_D = \frac{r_w}{h}\sqrt{\frac{k_V}{k_H}} \tag{6-18}$$

and

$$h_{1D} = \frac{h_1}{h}. \tag{6-19}$$

In terms of these dimensionless variables, the partial completion skin factor is

$$s_c = \left(\frac{1}{h_{wD}} - 1\right)\ln\frac{\pi}{2r_D} + \frac{1}{h_{wD}}\ln\left(\frac{h_{wD}}{2 + h_{wD}}\sqrt{\frac{A - 1}{B - 1}}\right) \tag{6-20}$$

where

$$A = \frac{1}{h_{1D} + \dfrac{h_{wD}}{4}} \tag{6-21}$$

and

$$B = \frac{1}{h_{1D} + \dfrac{3h_{wD}}{4}}. \tag{6-22}$$

The effect of wellbore deviation through the producing reservoir has also been studied frequently. A deviated well has higher productivity than a vertical well through the same reservoir because of the longer length of wellbore in contact with the formation in the deviated well case. Thus, a skin factor accounting for this effect will always be negative. Besson (1990) presented analytical equations for the deviated well skin effect for both isotropic and anisotropic reservoirs for a wellbore deviated at an angle θ from the vertical. For the isotropic case,

$$s_\theta = \ln\left(\frac{4r_w \cos\theta}{h}\right) + \cos\theta \ln\left(\frac{h}{4r_w\sqrt{\cos\theta}}\right). \tag{6-23}$$

For the anisotropic case,

$$s_\theta = \ln\left\{\frac{1}{I_{ani}\gamma}\left(\frac{4r_w \cos\theta}{h}\right)\right\} + \frac{\cos\theta}{\gamma}\ln\left\{\frac{2I_{ani}\sqrt{\gamma}}{1+\dfrac{1}{\gamma}}\left(\frac{h}{4r_w\sqrt{\cos\theta}}\right)\right\} \tag{6-24}$$

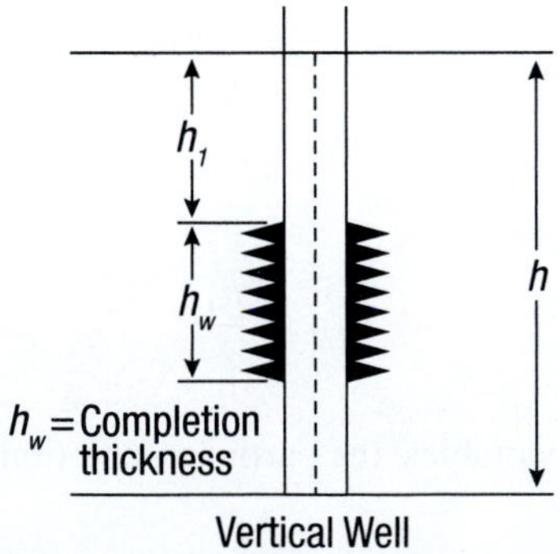

Figure 6-5 Partially completed well configuration.

where

$$\gamma = \sqrt{\frac{1}{I_{ani}^2} + \cos^2\theta\left(1 - \frac{1}{I_{ani}^2}\right)}. \tag{6-25}$$

This correlation for the isotropic case reproduces the work of Cinco-Ley et al. (1975).

For a partially completed, deviated well, we use the Papatzacos correlation to calculate the partial completion skin factor, but using the true vertical thickness of the completed interval for h_w, not the length of the completed interval measured along the wellbore. The Besson equation applies for the deviated well skin for such a well.

Partial completion effects (often called partial penetration) are a very important aspect of horizontal well inflow behavior, and have been discussed in Chapter 5.

Example 6-4 Partial Completion and Well Deviation Skin Effect

Two factors that are likely to have important influences on partial completion skin effects are the reservoir anisotropy and the location of the completed interval in the reservoir, because the convergence of flow to the completed interval in the vertical direction is the cause of the positive partial completion skin. Consider a vertical well with a radius of 0.25 ft completed along 20 ft of the wellbore in a 100-ft thick reservoir. Using the Papatzacos partial completion skin model, show how the partial completion skin factor depends on the anisotropy ratio and on the location of the completed interval.

Repeat this exercise using the Besson correlation for the well deviation skin factor for wells completed in 20 vertical ft of the reservoir with deviations of 30 and 60 degrees from the vertical.

Solution

First, we calculate the dimensionless size and position of the completion:

$$h_{wD} = \frac{h_w}{h} = \frac{20\,ft}{100\,ft} = 0.2. \tag{6-26}$$

For the completion being at the top of the reservoir,

$$h_{1D} = \frac{h_1}{h} = \frac{0}{100} = 0 \tag{6-27}$$

while for a completion in the middle of the reservoir,

$$h_{1D} = \frac{h_1}{h} = \frac{40\,ft}{100\,ft} = 0.4. \tag{6-28}$$

The dimensionless radius, r_D, is

$$r_D = \frac{r_w}{h}\sqrt{\frac{k_V}{k_H}} = \frac{0.25\ ft}{(100\ ft)I_{ani}} = \frac{0.0025}{I_{ani}}. \tag{6-29}$$

Now, for the case of the completion in the middle of the reservoir, for example ($h_{1D} = 0.4$),

$$A = \frac{1}{h_{1D} + \dfrac{h_{wD}}{4}} = \frac{1}{0.4 + (0.25 * 0.2)} = 2.222 \tag{6-30}$$

$$B = \frac{1}{h_{1D} + 0.75h_{wD}} = \frac{1}{0.4 + (0.75 * 0.2)} = 1.8181 \tag{6-31}$$

and

$$s_c = \left(\frac{1}{0.2} - 1\right)\ln\frac{\pi}{2r_D} + \frac{1}{0.2}\ln\left(\frac{0.2}{2 + 0.2}\sqrt{\frac{2.222 - 1}{1.8181 - 1}}\right) = 4\ln\frac{\pi}{2r_D} - 11 \tag{6-32}$$

The results of this calculation for I_{ani}, ranging from 1 to 20, and for the completion being at the top of the reservoir and in the middle are shown in Figure 6-6. The partial completion skin factor

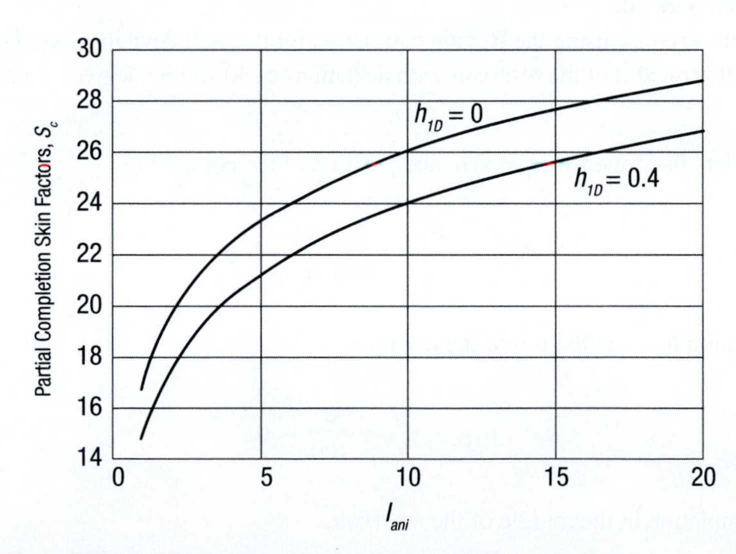

Figure 6-6 The effect of anisotropy on the partial completion skin factor.

depends strongly on the fraction of the reservoir thickness completed and strongly on reservoir anisotropy. The vertical location of the completion has a smaller effect.

For the deviated well skin calculation, using Equations 6-25 and 6-24, for the 30° deviation case,

$$\gamma = \sqrt{\frac{1}{I_{ani}^2} + \cos^2(30)\left(1 - \frac{1}{I_{ani}^2}\right)} = \sqrt{\frac{0.25}{I_{ani}^2} + 0.75} \tag{6-33}$$

$$s_\theta = \ln\left\{\frac{1}{I_{ani}\gamma}\left(\frac{4(0.25\ ft)\cos(30)}{100\ ft}\right)\right\} + \frac{\cos(30)}{\gamma}\ln\left\{\frac{2I_{ani}\sqrt{\gamma}}{1 + \dfrac{1}{\gamma}}\left(\frac{100\ ft}{4(0.25\ ft)\sqrt{\cos(30)}}\right)\right\}$$

$$= \ln\left\{\frac{0.00866}{I_{ani}\gamma}\right\} + \frac{0.866}{\gamma}\ln\left\{\frac{215I_{ani}\sqrt{\gamma}}{1 + \dfrac{1}{\gamma}}\right\} \tag{6-34}$$

while for the 60° case,

$$\gamma = \sqrt{\frac{1}{I_{ani}^2} + \cos^2(60)\left(1 - \frac{1}{I_{ani}^2}\right)} = \sqrt{\frac{0.75}{I_{ani}^2} + 0.25} \tag{6-35}$$

$$s_\theta = \ln\left\{\frac{1}{I_{ani}\gamma}\left(\frac{4(0.25\ ft)\cos(60)}{100\ ft}\right)\right\} + \frac{\cos(60)}{\gamma}\ln\left\{\frac{2I_{ani}\sqrt{\gamma}}{1 + \dfrac{1}{\gamma}}\left(\frac{100\ ft}{4(0.25\ ft)\sqrt{\cos(60)}}\right)\right\}$$

$$= \ln\left\{\frac{0.005}{I_{ani}\gamma}\right\} + \frac{0.5}{\gamma}\ln\left\{\frac{283I_{ani}\sqrt{\gamma}}{1 + \dfrac{1}{\gamma}}\right\}. \tag{6-36}$$

Using these equations, the variation of deviated well skin with anisotropy ratio is generated (Figure 6-7).

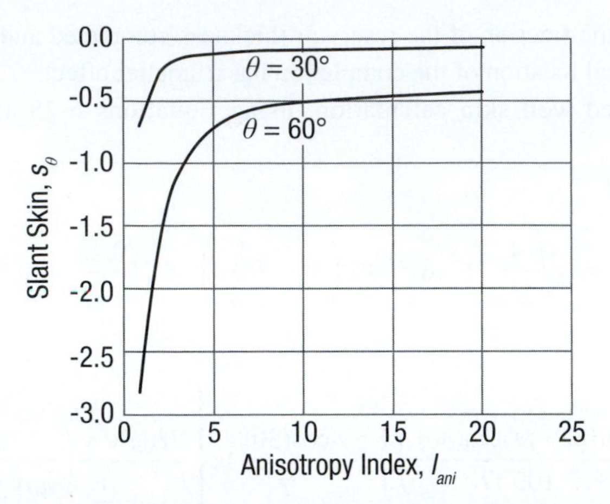

Figure 6-7 Dependence of deviated well skin factor on reservoir anisotropy.

6.5 Horizontal Well Damage Skin Effect

A horizontal well creates a drainage pattern that is quite different from that for a vertical well. The flow geometry in a reservoir containing a horizontal lateral is radial near the well and can be predominantly linear far from the well, while for a vertical well (with no hydraulic fracture), only radial flow is dominant. As a result, while the fully penetrated vertical well is totally insensitive to permeability anisotropy, both vertical and lateral permeability anisotropy is important to the horizontal well productivity, and both impact damage mechanisms. The conventional Hawkins' formula cannot be applied for the estimation of formation damage skin for horizontal wells because radial flow to the horizontal well is subject to horizontal-to-vertical permeability anisotropy. Furthermore, formation damage along the lateral is most likely nonuniform because of the reservoir heterogeneity and varying exposure time to drilling and completion fluid.

Furui, Zhu, and Hill (2003) presented a general damage skin factor model for horizontal wells. This model assumes that the damage cross section perpendicular to the well (Figure 6-8) mimics the isobars given by Peaceman's (1983) solution for flow through an anisotropic permeability field to a cylindrical wellbore. Because formation damage is often directly related to flux or velocity, it was assumed that the permeability impairment is distributed similarly to the pressure field. With this assumption about the damage distribution in the *y-z* plane normal to the well axis, Hawkins' formula can be transformed for the anisotropic space, and the local skin, $s_d(x)$, is expressed as

$$s_d(x) = \left[\frac{k}{k_s(x)} - 1\right] \ln\left[\frac{1}{I_{ani} + 1}\left(\frac{r_{sH}(x)}{r_w} + \sqrt{\left(\frac{r_{sH}(x)}{r_w}\right)^2 + I_{ani}^2 - 1}\right)\right]. \tag{6-37}$$

The local skin factor in Equation (6-37) describes skin effects for 2-D flow in a plane perpendicular to the horizontal wellbore axis.

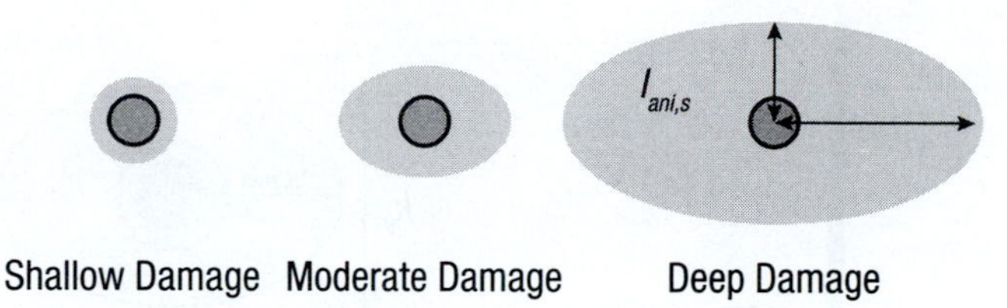

Figure 6-8 Cross sections of damage around a horizontal well in an anisotropic reservoir.

In Equation (6-37), r_{sH} is the half-length of the horizontal axis of the damage ellipse, k_s is the permeability in the damaged zone, and k is the undamaged formation permeability. Because ultimately flow to the horizontal well becomes pseudoradial as an ellipse about the entire well length, the overall damage skin factor for a horizontal lateral is obtained by integrating a 2-D inflow equation over the length of the lateral (Furui et al., 2003), yielding

$$s_h = \frac{L}{\int_0^L \left\{ \left[\frac{I_{ani} h}{r_w (I_{ani} + 1)} \right] + s_d(x) \right\}^{-1} dx} - \ln\left[\frac{I_{ani} h}{r_w (I_{ani} + 1)} \right] \tag{6-38}$$

where L is the length of the horizontal lateral, r_w is the wellbore radius, h is the thickness of the payzone, I_{ani} is the anisotropic ratio, and $s_d(x)$ is the local skin factor distribution given by Equation (6-37).

The local skin factor can be caused by any arbitrary distribution of damage along the horizontal well. To obtain an overall damage skin factor for a horizontal well, the distribution of damage along the well, $s_d(x)$, must be known, which in turn requires knowledge of the depth of damage distribution, $r_{sH}(x)$. Two limiting cases have been presented in the literature. Frick and Economides (1991) postulated that damage caused by drilling fluid filtrate invasion will be deepest at the heel of a horizontal well and shallowest at the toe because of the variation in the length of time the formation has been exposed to drilling fluid. This would lead to a conical damaged region tapering from the heel to the toe, as shown in Figure 6-9. For this situation, if the damage depth in the horizontal direction is $r_{sH,\,max}$ at the heel, tapering to zero damage at the toe, $r_{sH}(x)$ is

$$r_{sH}(x) = r_{sH,\,max}\left\{1 - \frac{x}{L}\right\}. \tag{6-39}$$

Incorporating this expression into Equation (6-37), the overall skin factor for this conical distribution of damage is obtained by placing the resulting expression for $s_d(x)$ into Equation (6-38) and numerically integrating.

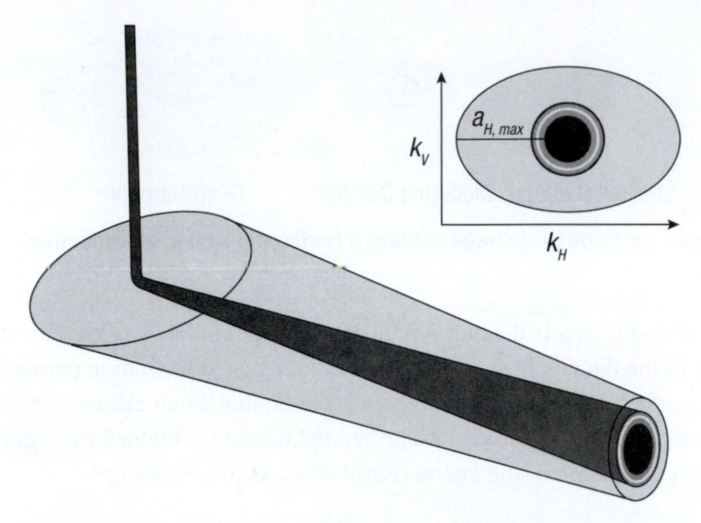

Figure 6-9 Distribution of filtrate-caused damage along a horizontal well.

Another approach is to assume that the damage is distributed uniformly along the entire length of the well. This assumption is justified when drilling mud filtrate or completion fluid invasion occurs rapidly and retards further invasion. Plugging by particles can lead to damage of this type. For this distribution of damage, $r_{sH}(x)$ and $s_d(x)$ are constants, and combining Equations (6-37) and (6-38) yields

$$s_h = \left[\frac{k_H}{k_{sH}} - 1\right] \ln\left[\frac{1}{I_{ani} + 1}\left(\frac{r_{sH}}{r_w} + \sqrt{\left(\frac{r_{sH}}{r_w}\right)^2 + I_{ani}^2 - 1}\right)\right]. \tag{6-40}$$

In general, the effect of near well formation damage for a horizontal well completion is relatively small compared with vertical wells. However, if the reservoir thickness is large and/or vertical permeability is small, radial or elliptical flow becomes dominant and the impact of formation damage on a horizontal lateral can be significant. The importance of formation damage in a horizontal lateral can be determined by comparing the magnitude of the damage and completion skin factor with other terms in a horizontal inflow equation (Hill and Zhu, 2008). For more information about completion and damage skin effects in horizontal wells and the laterals of multilateral wells, the reader is referred to Hill, Zhu, and Economides (2008).

Example 6-5 Effect of Damage Distribution on Horizontal Well Skin Factor

For a 2000-ft long horizontal well with $I_{ani} = 3$, $h = 53$ ft, $r_w = 0.328$ ft, calculate the equivalent skin factor, s_{eq}, for damage depths up to 5 ft and for k_H/k_{sH} ratios of 5, 10, and 20 for

a. conical-shaped damage tapering from $r_{sH,\,max}$ at the heel to zero at the toe
b. uniform damage along the entire well

Solution

a. The equivalent skin for the conical distribution of damage is obtained by numerically integrating Equation (6-38) from $x = 0$ at the heel of the well to $x = L$ at the toe, using the expression in Equation (6-39) for $r_{sH,max}(x)$. The results are shown in Figure 6-10.

b. For the uniform damage case, Equation (6-40) applies and the results are shown in Figure 6-11.

For the same level of permeability impairment (i.e., the same k_s/k), the overall skin factor is higher for the uniform damage case than the tapered damage when the damage penetration at the heel is the same in both cases.

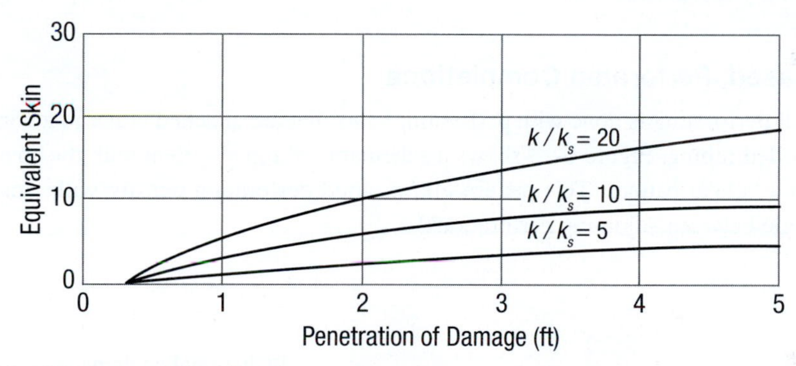

Figure 6-10 Equivalent skin effect for a conical damage distribution along a horizontal well.

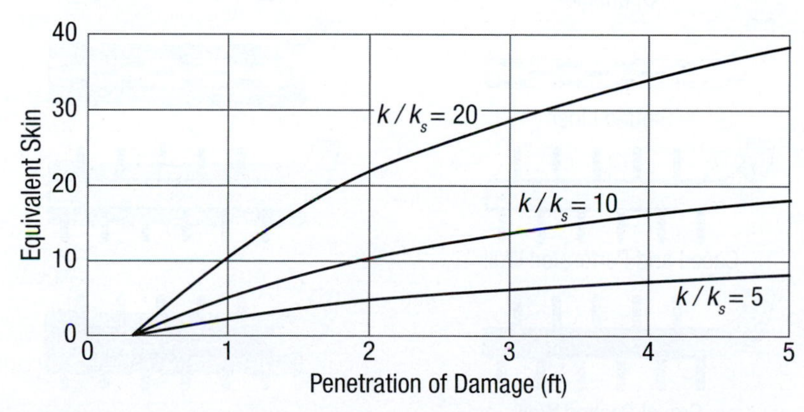

Figure 6-11 Equivalent skin effect for a uniform damage distribution along a horizontal well.

6.6 Well Completion Skin Factors

Oil and gas wells are completed in a variety of ways, with the most common completions being open-hole completions; cased, perforated completions; slotted or perforated liner completions; and gravel-pack completions. Figure 6-12 illustrates the completion types for the horizontal well trajectory, but they apply as well for vertical wells. Any well completion using hardware in the wellbore that changes the flow path from strictly radial flow in the near well vicinity, or through the completion itself, can lead to mechanical skin effects that can have a significant impact on well performance. Thus, all completions except for undamaged openhole completions may restrict production. These effects can be described with a completion skin factor. Another critical factor concerning well completion performance is whether or not there is formation damage in the region immediately surrounding the well. For some completion types, combined effects of damage and flow convergence caused by the completion can lead to very high skin factors, and correspondingly low productivity.

6.6.1 Cased, Perforated Completions

Modern well perforating is done with perforating guns that are attached either to a wireline or to tubing or coiled tubing. Figure 6-13 shows a schematic of a gun system with the shape charges arranged in a helical pattern. This pattern allows good perforation density with small phasing (i.e., the angle between adjoining perforations).

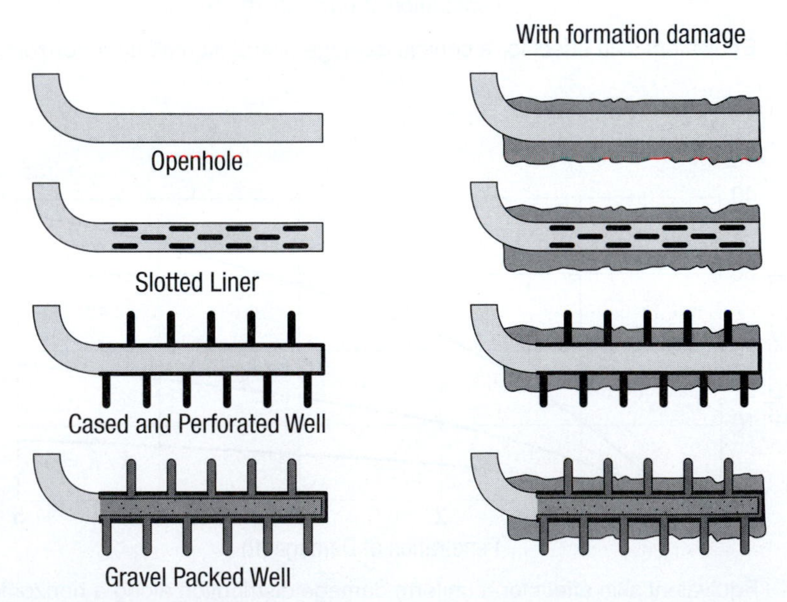

Figure 6-12 Common well completions.

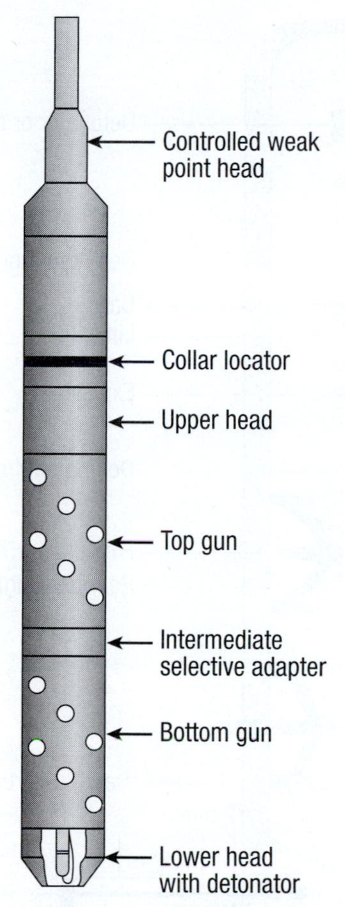

Figure 6-13 Perforating gun string. (Courtesy of Schlumberger.)

The perforating string contains a cable head, a correlation device, a positioning device, and the perforation guns. The cable head connects the string to the wireline and at the same time provides a weak point at which to disconnect the cable if problems arise. The correlation device is used to identify the exact position with a previously run correlation log, and frequently it locates casing collars. The positioning device orients the shots toward the casing for more optimum perforation geometry. The perforating guns are loaded with shape charges, which consist of the case, the explosive, and the liner, as shown in Figure 6-14. Electric current initiates an explosive wave; the sequences of the detonation process are shown in Figure 6-14. Perforations with a diameter between 0.25 and 0.4 in. and a length between 6 and 12 in. are typically created. Significantly longer perforations can be created with special charges in some formations.

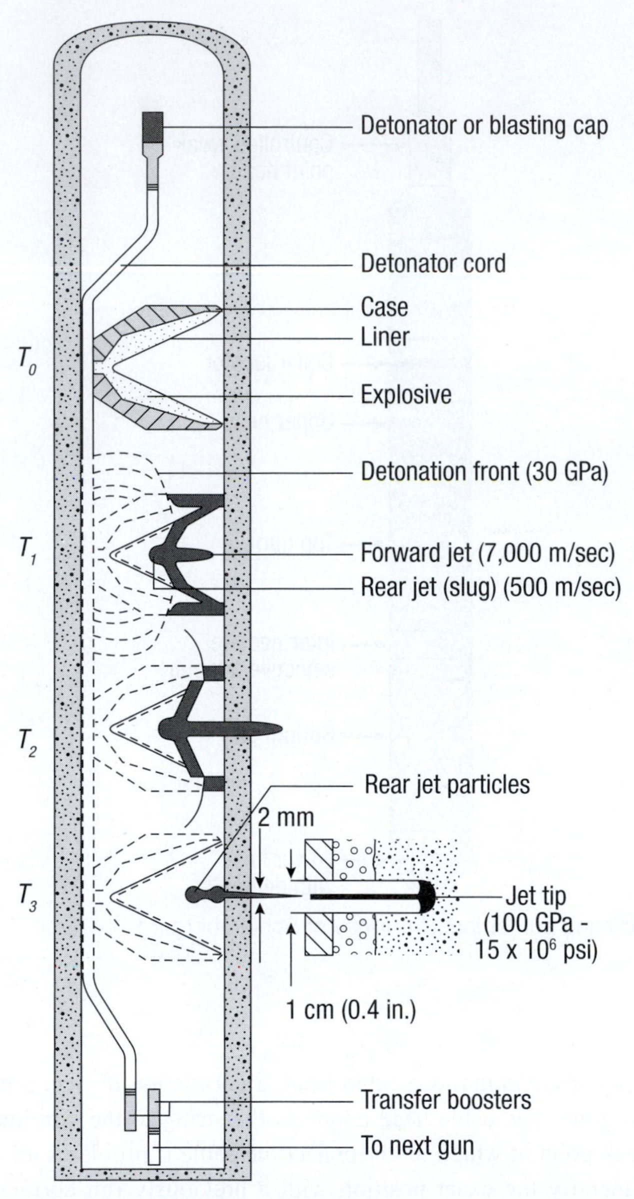

Figure 6-14 Detonation process of a shaped charge. (Courtesy of Schlumberger.)

Perforating is often done underbalanced; that is, the pressure in the well is less than the reservoir pressure at the moment the perforations are created. This facilitates immediate flow-back following the detonation, carrying the debris out of the perforations and resulting in a cleaner perforation cavity. The dimensions, number, and phasing of perforations have a controlling role in well performance.

6.6.1.1 Calculation of the Perforation Skin Effect

Based on detailed finite element simulations, Karakas and Tariq (1988) presented an empirical model of the perforation skin effect, which they divide into components: the planar flow effect, s_H; the vertical convergence effect, s_V; and the wellbore blockage effect, s_{wb}. The total perforation skin effect is then

$$s_p = s_H + s_V + s_{wb}. \tag{6-41}$$

Figure 6-15 gives all relevant variables for the calculation of the perforation skin. These include the well radius, r_w, the perforation radius, r_{perf}, the perforation length, l_{perf}, the perforation phase angle, θ, and, very importantly, the distance between the perforations, h_{perf}, which is exactly inversely proportional to the perforation density (e.g., two shots per foot (SPF), result in $h_{perf} = 0.5$ ft). Below, the method of estimating the individual components of the perforation skin is outlined.

6.6.1.2 Calculation of s_H

$$s_H = \ln \frac{r_w}{r'_w(\theta)} \tag{6-42}$$

where $r'_w(\theta)$ is the effective wellbore radius and is a function of the phase angle θ:

$$r'_w(\theta) = \begin{cases} \dfrac{l_{perf}}{4} & \text{for } \theta = 0 \\ a_\theta(r_w + l_{perf}) & \text{for } \theta \neq 0. \end{cases} \tag{6-43}$$

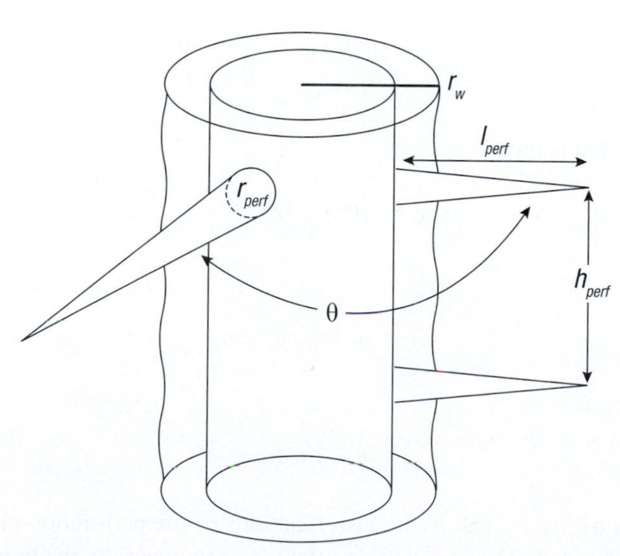

Figure 6-15 Well variables for perforation skin calculation. (From Karakas and Tariq, 1988.)

Table 6-1 Constants for Perforation Skin Effect Calculation

Perforation Phasing	a_θ	a_1	a_2	b_1	b_2	c_1	c_2
0° (360°)	0.250	−2.091	0.0453	5.1313	1.8672	1.6E–1	2.675
180°	0.500	−2.025	0.0943	3.0373	1.8115	2.6E–2	4.532
120°	0.648	−2.018	0.0634	1.6136	1.7770	6.6E–3	5.320
90°	0.726	−1.905	0.1038	1.5674	1.6935	1.9E–3	6.155
60°	0.813	−1.898	0.1023	1.3654	1.6490	3.0E–4	7.509
45°	0.860	−1.788	0.2398	1.1915	1.6392	4.6E–5	8.791

From Karakas and Tariq (1988).

The constant a_θ depends on the perforation phase angle and can be obtained from Table 6-1. This skin effect is negative (except for $\theta = 0$), but its total contribution is usually small.

6.6.1.3 Calculation of s_V

To obtain s_V, two dimensionless variables must be calculated:

$$h_D = \frac{h_{\text{perf}}}{l_{\text{perf}}}\sqrt{\frac{k_H}{k_V}} \tag{6-44}$$

where k_H and k_V are the horizontal and vertical permeability values, respectively, and

$$r_D = \frac{r_{\text{perf}}}{2h_{\text{perf}}}\left(1 + \sqrt{\frac{k_V}{k_H}}\right). \tag{6-45}$$

The vertical pseudoskin is then

$$s_V = 10^a h_D^{b-1} r_D^b \tag{6-46}$$

with

$$a = a_1 \log r_D + a_2 \tag{6-47}$$

and

$$b = b_1 r_D + b_2. \tag{6-48}$$

The constants a_1, a_2, b_1, and b_2 are also functions of the perforation phase angle and can be obtained from Table 6-1. The vertical skin effect, s_V, is potentially the largest contribution to s_p; for small perforation densities (i.e., large h_{perf}), s_V can be very large.

6.6.1.4 Calculation of s_{wb}

For the calculation of s_{wb}, a dimensionless quantity is calculated first:

$$r_{wD} = \frac{r_w}{l_{perf} + r_w}. \quad (6\text{-}49)$$

Then

$$s_{wb} = c_1 e^{c_2 r_{wD}}. \quad (6\text{-}50)$$

The constants c_1 and c_2 can also be obtained from Table 6-1.

Example 6-6 Perforation Skin Effect

Assume that a well with $r_w = 0.328$ ft is perforated with 2 SPF, $r_{perf} = 0.25$ in. (0.0208 ft), $l_{perf} = 8$ in. (0.667 ft), and $\theta = 180°$. Calculate the perforation skin effect if $k_H/k_V = 10$.

Repeat the calculation for $\theta = 0°$ and $\theta = 60°$.

If $\theta = 180°$, show the effect of the horizontal-to-vertical permeability anisotropy with $k_H/k_V = 1$.

Solution

From Equation (6-43) and Table 6-1 ($\theta = 180°$),

$$r'_w(\theta) = (0.5)(0.328 + 0.667) = 0.5. \quad (6\text{-}51)$$

Then, from Equation (6-42),

$$s_H = \ln\frac{0.328}{0.5} = -0.4. \quad (6\text{-}52)$$

From Equation (6-44) and remembering that $h_{perf} = 1/\text{SPF}$,

$$h_D = \frac{0.5}{0.667}\sqrt{10} = 2.37 \quad (6\text{-}53)$$

and

$$r_D = \frac{0.0208}{(2)(0.5)}\left(1 + \sqrt{0.1}\right) = 0.027 \quad (6\text{-}54)$$

From Equations (6-47) and (6-48) and the constants in Table 6-1,

$$a = -2.025\log(0.027) + 0.0943 = 3.271 \quad (6\text{-}55)$$

and

$$b = (3.0373)(0.027) + 1.8115 = 1.894. \quad (6\text{-}56)$$

From Equation (6-46),

$$s_V = 10^{3.271} 2.37^{0.894} 0.027^{1.894} = 4.3. \quad \textbf{(6-57)}$$

Finally, from Equation (6-49),

$$r_{wD} = \frac{0.328}{0.667 + 0.328} = 0.33 \quad \textbf{(6-58)}$$

and with the constants in Table 6-1 and Equation (6-50),

$$s_{wb} = (2.6 \times 10^{-2}) e^{(4.532)(0.33)} = 0.1 \quad \textbf{(6-59)}$$

The total perforation skin effect is then

$$s_p = -0.4 + 4.3 + 0.1 = 4. \quad \textbf{(6-60)}$$

If $\theta = 0°$, then $s_H = 0.7$, $s_V = 3.6$, $s_{wb} = 0.4$, and therefore $s_p = 4.7$.

If $\theta = 60°$, then $s_H = -0.9$, $s_V = 4.9$, $s_{wb} = 0.004$, and $s_p = 4$.

For $\theta = 180°$ and $k_H/k_V = 1$, s_H and s_{wb} do not change; s_V, though, is only 1.2, leading to $s_p = 0.9$, reflecting the beneficial effects of good vertical permeability even with relatively unfavorable perforation density (only 2 SPF).

Example 6-7 Perforation Density

Using typical perforation characteristics such as $r_{perf} = 0.25$ in. (0.0208 ft), $l_{perf} = 8$ in. (0.667 ft), $\theta = 120°$, in a well with $r_w = 0.328$ ft, develop a table of s_V versus perforation density for permeability anisotropies $k_H/k_V = 10$, 5, and 1.

Solution

Table 6-2 presents the skin effect s_V for perforation densities from 0.5 SPF to 4 SPF using Equations (6-44) to (6-48). For the higher perforation densities (3–4 SPF), this skin contribution

Table 6-2 Vertical Contribution to Perforation Skin Effect

	s_V		
SPF	$k_H/k_V = 10$	$k_H/k_V = 5$	$k_H/k_V = 1$
0.5	21.3	15.9	7.7
1	10.3	7.6	3.6
2	4.8	3.5	1.6
3	3.0	2.1	0.9
4	2.1	1.5	0.6

becomes small. For low shot densities, this skin effect in normally anisotropic formations can be substantial. For the well in this problem, $s_H = -0.7$, and $s_{wb} = 0.04$.

6.6.1.5 Near-Well Damage and Perforations

When there is formation damage around a cased, perforated completion, the combined effect of the perforation skin factor and the damage skin factor can be much greater than the sum of these separate effects. This is because the converging flow to the perforations creates a high-pressure drop if the permeability is reduced in this region. Karakas and Tariq (1988) have also shown that damage and perforations can be characterized by a composite skin effect.

$$(s_p)_d = \left(\frac{k}{k_s} - 1\right)\left(\ln\frac{r_s}{r_w} + s_p\right) + s_p = (s_d)_o + \frac{k}{k_s}s_p \tag{6-61}$$

if the perforations terminate within the damage zone ($l_{\text{perf}} < r_s$). In Equation (6-61), $(s_d)_o$ is the open-hole equivalent skin effect given by Hawkins' formula [Equation (6-4)]. If the perforations terminate outside the damage zone, then

$$(s_p)_d = s'_p \tag{6-62}$$

where s'_p is evaluated at a modified perforation length, l'_{perf}, and a modified radius, r'_w. These are

$$l'_{\text{perf}} = l_{\text{perf}} - \left(1 - \frac{k_s}{k}\right)(r_s - r_w) \tag{6-63}$$

and

$$r'_w = r_s - \frac{k_s}{k}(r_s - r_w). \tag{6-64}$$

These variables are used in Equations (6-41) to (6-50) for the calculation of the skin effects contributing to the composite skin effect in Equation (6-62).

6.6.1.6 Perforation Skin Factor for Horizontal Wells

The impact of horizontal to vertical permeability anisotropy is different for perforations in a horizontal well than in a vertical well. In both cases the pressure drop associated with flow to the perforation cavity depends on the orientation of the perforation relative to the orientation of the permeability anisotropy. Because of the vertical permeability often being significantly lower than the horizontal permeability, the skin factor for a perforated horizontal well completion can be different from that of a vertical well. Perhaps more significantly, a horizontal perforated completion's productivity depends on the orientation of the perforations relative to the permeability field. In a formation with low vertical permeability (high I_{ani}), perforations oriented up and down will be more productive than perforations oriented horizontally. Following a similar approach to that of Karakas and Tariq, Furui et al. (2008) developed skin factor models for perforated horizontal wells that include this perforation orientation effect. The reader is referred to their work for more details.

6.6.2 Slotted or Perforated Liner Completions

Slotted or perforated liners are commonly used to complete horizontal and multilateral wells. In this type of completion, typically a small fraction of the pipe area is slotted or pre-perforated to communicate with the formation, so if these slots or perforations are plugged, or if there is formation damage in the few inches near the openings where the flow is converging, very high skin factors can result. Slotted or perforated liners are placed in horizontal wellbores, but not cemented in place. If the borehole is very stable so that the formation does not collapse around the liner, the well should behave like an openhole completion, and the simple Hawkins' formula can describe the skin factor if there is formation damage present. If the formation deforms to contact the liner, flow convergence to the slots or perforations can lead to much higher completion skin factors.

Figure 6-16 shows common geometries of slots used in slotted liners. Because only a small portion of the liner is actually open to flow, the flow converges radially to the slots, as illustrated in Figure 6-17. Furui, Zhu, and Hill (2005) developed skin factor models for both slotted and perforated liners that include non-Darcy effects. These models are complex and beyond the scope of this book.

Figure 6-18 from Furui et al. illustrates the skin factor predictions of these models. This figure compares the skin factors for various well completions: an open-hole completion; a cased and perforated well with an excellent perforation geometry ($l_p = 12$ in. and $s_p = -1.20$), a cased and perforated well with a good perforation geometry ($l_p = 12$ in. $s_p = 0.00$), and a slotted liner ($s_{SL} = 1.54$). For cased and perforated well completions with efficient perforating (i.e., $s_p < 0$), skin factors are lower than for an open-hole completion. The perforations extending beyond the damage zone create flowpaths through the damage zone so that the effect of formation damage becomes less important. For such completions, the combined perforation/damage skin factor is zero or even negative if sufficient perforation density is used. Even if all the perforations terminate inside the damaged zone, the increase of skin due to damage can be less than that of open-hole completions. On the other hand, slotted (perforated) liner completions are unsuitable for formations where severe permeability reduction is observed. The skin factor may significantly increase even for shallow penetration of damage since the reduced permeability magnifies the slotted liner geometry skin factor. To lower the damage effects requires either an excellent slot design to achieve $s_{SL} \approx 0$ or appropriate damage removal operations to recover the damaged permeability (e.g., acidizing).

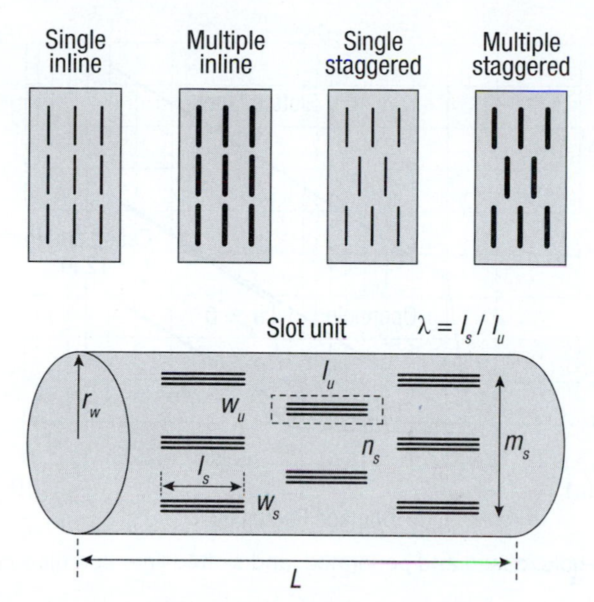

Figure 6-16 Slot configurations used in slotted liners. (From Furui et al., 2005.)

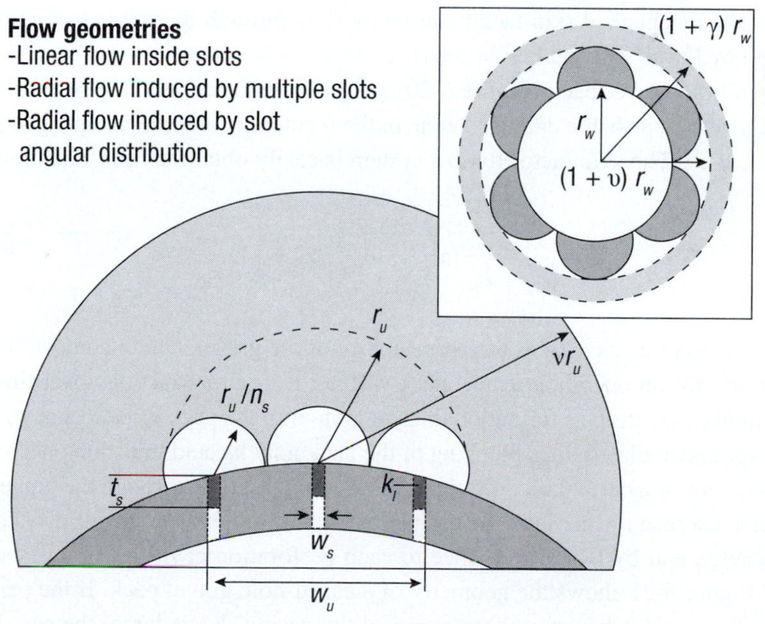

Figure 6-17 Flow regions around a slotted liner. (From Furui, 2004.)

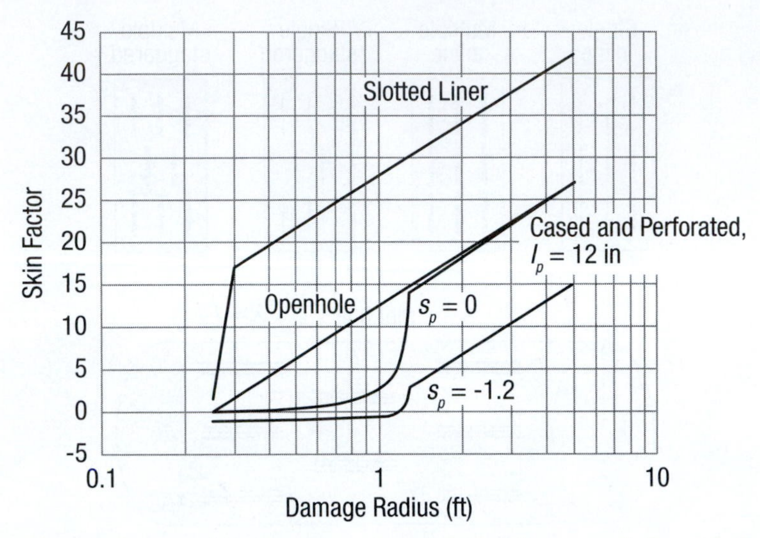

Figure 6-18 Open-hole, cased and perforated, and slotted liner completion skin factors. (From Furui et al., 2005.)

6.6.3 Gravel Pack Completions

Gravel pack completions are commonly used in poorly consolidated formations to prevent sand production. Two types are common, open-hole and cased-hole gravel packs, as shown in Figure 6-19. The mechanical skin factor caused by flow through these types of completions has been modeled by Furui, Zhu, and Hill (2004).

An open-hole gravel pack (Figure 6-20) is simply modeled as three permeability regions in series: the gravel, a possible damaged zone in the formation, and the undamaged permeability away from the well. The skin factor for this system is easily obtained from Equation 6-7 as

$$s_{g,o} = \frac{k}{k_g} \ln \frac{r_w}{r_g} + \left(\frac{k}{k_s} - 1 \right) \ln \frac{r_s}{r_w}. \tag{6-65}$$

This equation shows that as long as the permeability of the gravel is high compared with the formation permeability, an open-hole gravel pack will not restrict production. Over time, the gravel pack permeability may drop as formation fines migrate into the pack or collect at its outer radius. The remedy for accumulated fines plugging of the pack may be acid injection on a regular basis.

The behavior of a cased-hole gravel pack is affected by flow through the gravel-packed annulus between a screen or liner and the casing, by flow through each gravel-packed perforation tunnel and cavity, and by the convergence to each perforation cavity as in a cased, perforated completion. Figure 6-21 shows the geometry of a cased-hole gravel pack. If the permeability of the gravel in the annulus between the screen and the casing, k_g, is large, the cased-hole gravel pack skin factor can be separated into components related to the flow through the perforation

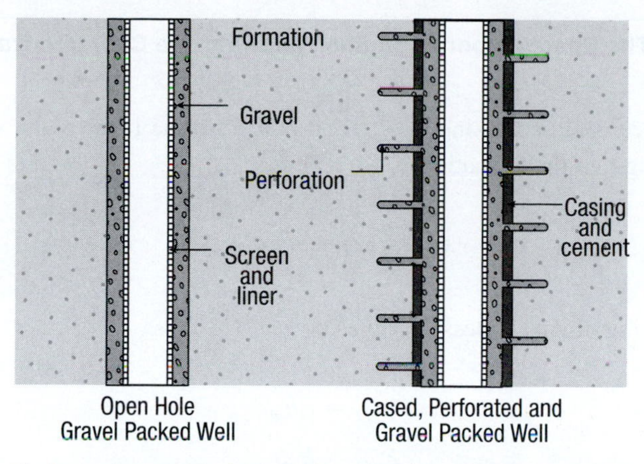

Figure 6-19 Open-hole and cased-hole gravel pack completions. (From Furui et al., 2004.)

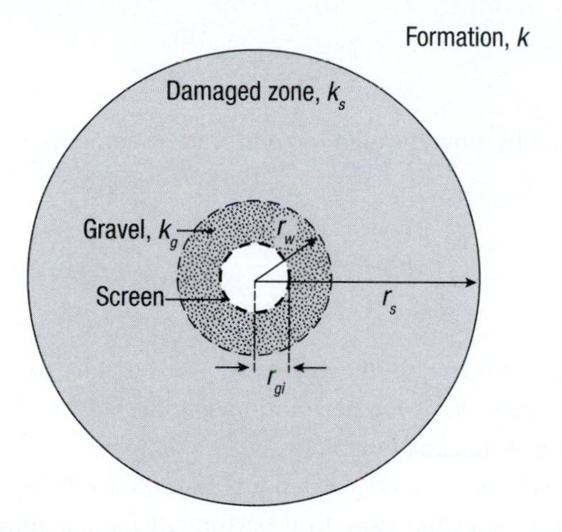

Figure 6-20 Cross section of an open-hole gravel pack completion. (From Furui et al., 2004.)

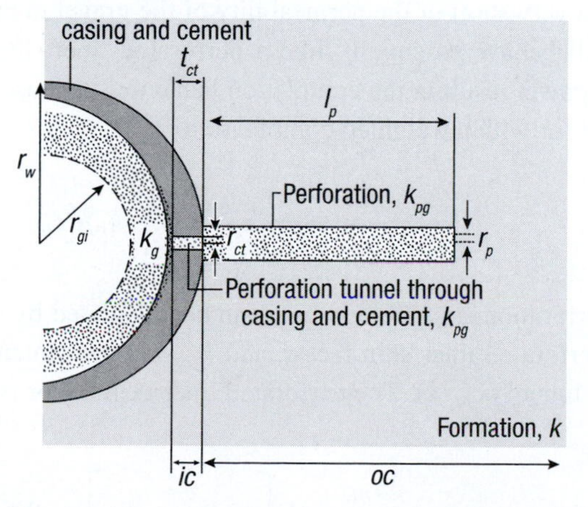

Figure 6-21 Cross section of a cased-hole gravel pack completion. (From Furui et al., 2004.)

tunnel through the cement and casing, $s_{CG,ic}$, and related to the flow to and within the perforation cavity extending into the formation, $s_{CG,oc}$. Thus,

$$s_{CG} = s_{CG,ic} + s_{CG,oc}. \quad \textbf{(6-66)}$$

Defining the following dimensionless variables,

$$r_{tD} = r_t/r_w \quad \textbf{(6-67)}$$

$$h_{pD} = h_p/r_w \quad \textbf{(6-68)}$$

$$l_{tD} = l_t/r_w \quad \textbf{(6-69)}$$

$$k_{gD} = k_g/k \quad \textbf{(6-70)}$$

the skin component caused by flow through the perforation tunnel through the casing is

$$s_{CG,ic} = \left(\frac{2h_{pD}}{r_{tD}^2}\right)\frac{l_{tD}}{k_{gD}}. \quad \textbf{(6-71)}$$

In Equations 6-67 through 6-71, r_t is the radius of the tunnel, l_t is the length of the tunnel through the casing and cement, h_p is the perforation spacing (inverse of shot density), and k_g is the permeability of the gravel-packed tunnel.

The flow to the gravel-packed cavity extending into the formation is analogous to the flow to a perforated completion or to the flow to a perforated liner completion, depending on the lengths of the perforations and the permeability of the gravel packed in the perforation tunnel. If the perforations are very short or the permeability of the gravel in the perforation tunnels is low, the completion will behave essentially like a perforated liner; longer perforations containing high permeability gravel result in the completion behavior of a cased, perforated well. These effects are accounted for with a weighted combination,

$$s_{CG,oc} = (1 - k_{gD}^{-0.5})s_p + k_{gD}^{-0.5}s_{pl}. \quad \textbf{(6-72)}$$

Where s_p is the perforation skin factor which can be calculated by the Karakas and Tariq correlation, s_{pl} is the perforated liner skin factor, and k_{pgD} is the dimensionless permeability of the gravel packing the tunnel (k_{pg}/k). The perforated liner skin factor is

$$s_{pl} = \frac{3h_{pD}}{2r_{pD}} + \ln\left[\frac{v^2}{h_{pD}^2(1 + v)}\right] - 0.61 \quad \textbf{(6-73)}$$

and

$$v = \begin{cases} 1.5 & \theta = 360°, 0° \\ \sin\left(\dfrac{\pi}{360/\theta}\right) & \theta \neq 360°, 0° \end{cases}. \tag{6-74}$$

In gravel-packed completions, it is very important that the perforation cavities are filled with gravel. If a perforation tunnel is not packed, over time fines may migrate into the cavity and plug it. Successful packing of gravel pack tunnels can require considerable energy, especially for completions in overpressured formations. Plugged perforations can result in non-uniform flow through the pack. Too much flow through a few perforations can cause completion failure.

6.7 Formation Damage Mechanisms

In the next two sections, the underlying causes of formation damage and the source of damage during well operations are described. Formation damage can be caused by plugging of the pore spaces by solid particles, by mechanical crushing or disaggregation of the porous media, or by fluid effects such as the creation of emulsions or changes in relative permeability. Plugging of pores by solid particles is the most pervasive of these mechanisms and can result from many sources, including the injection of solids into formation, the dispersion of clays present in the rock, precipitation, and the growth of bacteria.

6.7.1 Particle Plugging of Pore Spaces

A porous medium is a complex assembly of irregularly shaped mineral grains with void spaces (pores), which are likewise irregularly shaped and distributed, providing the pathway for fluid transport. Scanning electron microscope (SEM) photographs, such as that shown in Figure 6-22 (Krueger, 1986), have illustrated the tortuous nature of the pore spaces and the common presence of small particles, called fines, in naturally occurring porous media. This complicated structure can be idealized as a collection of relatively large chambers, the pore bodies, connected by narrower openings, the pore throats. The permeability of the medium is controlled largely by the number and conductivity of the pore throats.

When fines are moving through a porous medium, they will often be deposited, and if this deposition occurs in the pore throats, a severe reduction in the permeability may result. Figure 6-23 (Schechter, 1992) illustrates possible modes of particle entrapment. Large particles transported to the surface of the porous medium will bridge over the surface pores and form a filter cake external to the porous medium. An example of surface bridging and filter cake formation is the mud cake formed on the wellbore wall during drilling operations. Such a filter cake greatly reduces the ability to transport fluids to or from the porous medium, but is also relatively easily removed or bypassed.

Figure 6-22 Scanning electron micrograph of pore space in a sandstone. (From Krueger, 1986.)

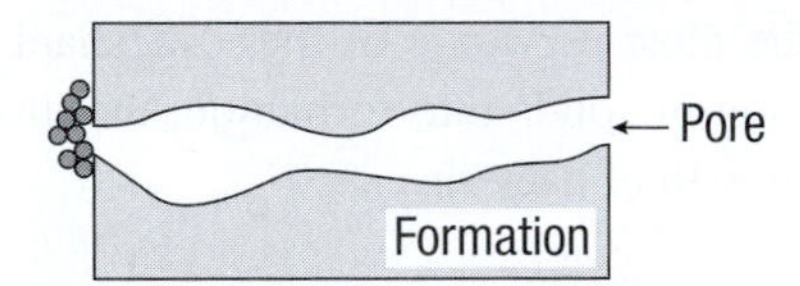

Cake Formation by Large Particles

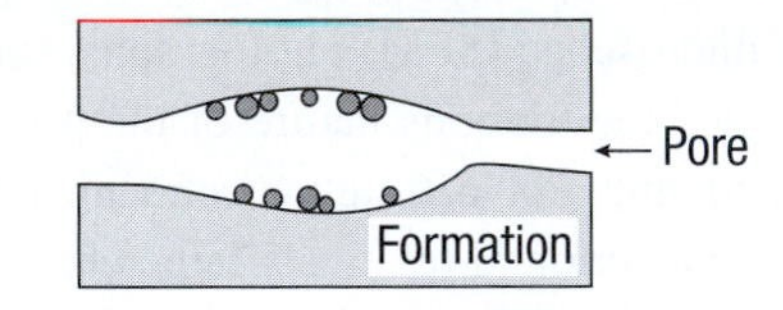

Surface Deposits of Adhering Particles

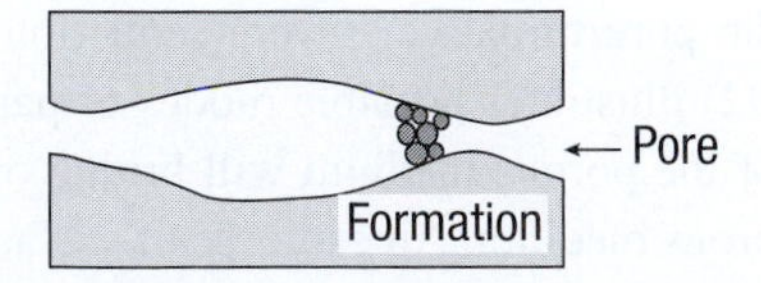

Plugging Type Deposits

Figure 6-23 Modes of particle entrapment. (From Schechter, 1992.)

Small fines passing through the porous medium may adhere to the surfaces of the pore bodies, resulting in little permeability impairment, or may bridge in the pore throats, effectively plugging the pore. Bridging can occur when the particles are on the order of one-third to one-seventh the size of the pore throat, or larger; thus, the relative sizes of the fines and the pore throats is a primary factor in determining whether formation damage due to fines movement is likely.

Example 6-8 Change in Permeability Due to Pore Throat Plugging

A capillary model of pore structure (Schechter and Gidley, 1969) describes the porous medium as a collection of capillary tubes, with their sizes given by a pore size density function, $\eta(A)$, defined as the number of pores per unit volume with a cross-sectional area between A and $A + dA$. With this model, the permeability is

$$k = F \int_0^{\infty} A^2 \eta(A)\, dA \qquad \textbf{(6-75)}$$

where F is the average pore length multiplied by a tortuosity factor. For Berea sandstone, the pore size density function is approximately

$$\eta = 0 \quad \text{for } A < 10^{-10}\ \text{cm}^2 \qquad \textbf{(6-76)}$$

$$\eta = A^{-2} \quad \text{for } 10^{-10}\ \text{cm}^2 < A < 10^{-4}\ \text{cm}^2 \qquad \textbf{(6-77)}$$

$$\eta = 0 \quad \text{for } A > 10^{-4}\ \text{cm}^2 \qquad \textbf{(6-78)}$$

where η is the number of pores per cubic centimeter. It can be assumed that this describes the distribution of pore throats.

Particles with radii no more than 5μm (5×10^{-4} cm) are injected into the porous medium described above until the permeability no longer changes. If particles bridge over and completely plug all pore throats that are less than seven times larger than the particles, what would be the ratio of the permeability after injection of the particles to that before?

Solution

All pores small enough to be plugged will be eliminated from the pore size distribution contributing to permeability. The smallest pore that would be plugged is seven times the size of the particles or has a radius of $(7)(5 \times 10^{-4}\ \text{cm})$ or 3.5×10^{-3} cm. Thus, the area of the smallest effective pore in the damaged porous medium is

$$A_d = \pi(r_d)^2 = \pi(3.5 \times 10^{-3}\ \text{cm})^2 = 3.85 \times 10^{-5}\ \text{cm}^2. \qquad \textbf{(6-79)}$$

The ratio of the damaged to the original permeability is then

$$\frac{k_d}{k} = \frac{F\int_{A_d}^{A_{max}} A^2 \eta dA}{F\int_{A_{min}}^{A_{max}} A^2 \eta dA} = \frac{\int_{A_d}^{A_{max}} A^2(A^{-2})dA}{\int_{A_{min}}^{A_{max}} A^2(A^{-2})dA} = \frac{A_{max} - A_d}{A_{max} - A_{min}}, \tag{6-80}$$

so

$$\frac{k_d}{k} = \frac{10^{-4} - (3.85 \times 10^{-5})}{10^{-4} - 10^{-10}} = 0.61. \tag{6-81}$$

Most of the pores would be plugged by the particles; however, since the largest pores are not affected, the permeability impairment is not severe. Using Hawkins' formula and assuming that $r_s = 3$ ft and $r_w = 0.5$ ft, this permeability impairment would result in a skin effect approximately equal to 1.

6.7.2 Mechanisms for Fines Migration

The fines responsible for particle plugging may come from external sources, such as conventional drilling mud, or may originate in the porous medium itself. Fines in the porous medium may be mobilized by a change in the chemical composition of the water or simply mechanically entrained due to the shear forces applied by the moving fluid. Formation damage is often caused by the dispersion of fine clay particles when the salinity of the interstitial water is reduced or ionic composition is changed. Thus, any fluids that may come in contact with the producing formation (drilling fluid filtrate, completion fluids, stimulation fluids, etc.) should have an ionic composition that will be nondamaging.

Numerous studies have shown that a sudden decrease in salinity of the brine flowing through a sandstone will cause formation damage by dispersing clay particles. This phenomenon, called water sensitivity, depends on the cations present in the brine, the pH, and the rate of salinity change. In general, monovalent cations are much more damaging than divalent or trivalent cations; the water sensitivity is greatest for NaCl brines and decreases in the order $Na^+ > K^+ > NH_4^+$. The higher the pH, the more sensitive the porous medium will be to salinity changes. To prevent clay dispersion because of a salinity change, any aqueous fluid that may come in contact with the formation should have a minimum concentration of a monovalent ion or a sufficient fraction of divalent ions. Commonly used criteria to prevent damage are for brines to contain at least 2 wt% of KCl or that at least one-tenth of the cations are divalent cations.

6.7.3 Chemical Precipitation

Precipitation of solids from the brine or the crude oil in the formation can cause severe formation damage when these solids plug the pore spaces. The formed precipitates can be either inorganic compounds precipitating from brine or organic species precipitating from the oil; in either case, the precipitation can be due to changes in temperature or pressure near the wellbore or from alterations in the composition of the phase by injected fluids.

Inorganic precipitates causing formation damage are usually divalent cations, such as calcium or barium, combined with carbonate or sulfate ions. The ionic species in solution in the connate water in a reservoir are initially in chemical equilibrium with the formation minerals. A change in the composition of the brine may lead to precipitation.

For example, the equilibrium reaction between calcium and bicarbonate ions can be expressed as

$$Ca^{2+} + 2HCO_3^- \leftrightarrow CaCO_3(s) + H_2O + CO_2(g). \tag{6-82}$$

If the brine is initially saturated with respect to calcium bicarbonate, an increase in the concentration of species on the left side of the equation or a decrease in the concentration of any species on the right side of the equation will drive the reaction to the right and calcium carbonate will precipitate. Addition of calcium ions will cause calcium carbonate to precipitate; likewise, removal of CO_2 will lead to precipitation. Thus, in a reservoir with high bicarbonate concentrations, injection of fluids high in calcium, such as $CaCl_2$ completion fluids, may cause severe formation damage. Likewise, as the pressure decreases near a production well, CO_2 will be liberated from the brine, and again precipitation may occur. Precipitation of $CaCO_3$ from the bicarbonate-rich connate water is a common source of formation damage at the Prudhoe Bay field (Tyler, Metzger, and Twyford, 1984).

The most common organic species that cause formation damage are waxes (paraffins) and asphaltenes. Waxes are long-chain hydrocarbons that precipitate from certain crude oils when the temperature is reduced, or the oil composition changes because of the liberation of gas as the pressure is reduced. Asphaltenes are high-molecular-weight aromatic and napthenic compounds that are thought to be colloidally dispersed in crude oils (Schechter, 1992). This colloidal state is stabilized by the presence of resins in the crude oil; when these resins are removed, the asphaltenes can flocculate, creating particles large enough to cause formation damage. Chemical changes to the crude oil that reduce the resin concentration can thus lead to asphaltene deposition in the formation.

6.7.4 Fluid Damage: Emulsions, Relative Permeability, and Wettability Changes

Formation damage can be caused by changes in the fluids themselves rather than a change in the permeability of the rock. The damage caused by fluids is due either to a change in the apparent viscosity of the oil phase or to a change in relative permeability. These types of damage can be thought of as temporary because the fluids are mobile and theoretically can all be removed from the near-well vicinity. However, such removal is sometimes difficult.

The formation of water-in-oil emulsions in the reservoir rock around the wellbore can cause damage because the apparent viscosity of the emulsion may be more than an order or magnitude higher than that of the oil. In addition, emulsions are often non-Newtonian and may require sufficient force to overcome a yield stress to be mobilized. Emulsions are most commonly caused by mechanical mixing of oil and water, which breaks one of the phases into small droplets dispersed in the other phase. In the formation, it is more likely that emulsions are formed chemically, through the introduction of surfactants or fines that tend to stabilize small droplets.

Apparent formation damage can also be due simply to an increase in the water saturation around the wellbore, resulting in a reduction of the permeability to oil. This effect, called water block, can occur any time aqueous fluids are introduced into the formation.

Finally, certain chemicals can alter the wettability of the formation, changing the relative permeability characteristics of the formation entirely. If a water-wet formation is changed to oil-wet around the wellbore, the oil-relative permeability may be greatly reduced in the near-wellbore region. This was addressed also in Section 6.3.

6.7.5 Mechanical Damage

The formation near the wellbore can also be damaged by physical crushing or compaction of the rock. Pulverization and compaction of the rock in the vicinity of a perforation is an unavoidable consequence of perforating, leading to a damaged region around a perforation like that shown schematically in Figure 6-24 (Krueger, 1986). Based on laboratory testing of perforating of sandstone cores, Krueger reported the damaged zone around a perforation to be $^1/_4$- to $^1/_2$-in. thick with a permeability of 7–20% of the undamaged permeability. Because of the convergent flow to a perforation, this small layer of damage around a perforation can significantly impair the productivity of the perforation. For example, Lea, Hill, and Sepehrnoori (1991) showed that a crushed zone with a permeability of 10% of the original permeability results in a perforation skin factor of about 15.

Mechanical damage around the wellbore can also result from the collapse of weak formation material around the wellbore. This type of damage is possible in friable formations or those weakened by acidizing in the near-wellbore region.

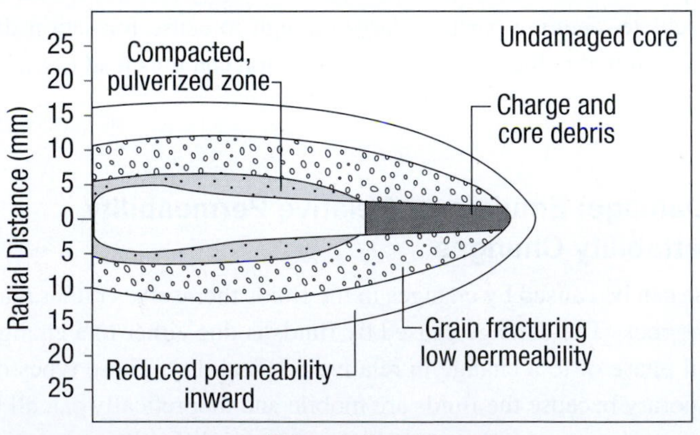

Figure 6-24 Damaged regions around a perforation. (From Krueger, 1986.)

6.7.6 Biological Damage

Some wells, particularly water injection wells, are susceptible to damage caused by bacteria in the near-wellbore environment. Bacteria injected into the formation, particularly anaerobic bacteria, may grow rapidly in the formation, plugging the pore spaces with the bacteria themselves or with precipitates that result from biological activity of the organisms. The permeability reduction caused by bacteria can be so significant that the injection of bacteria to intentionally reduce the permeability of thief zones has been studied as an enhanced oil recovery method (Zajic, Cooper, Jack, and Kosaric 1983). Biological damage is best prevented by treating injection waters with bactericides.

6.8 Sources of Formation Damage During Well Operations

6.8.1 Drilling Damage

The most common source of formation damage in wells is the drilling process. Drilling damage results from the invasion of the formation by drilling fluid particles and drilling fluid filtrate. The damage caused by drilling fluid particles is likely to be the more severe.

The deposition of drilling mud particles in the formation around the wellbore can severely reduce the permeability in this critical region; fortunately, however, the depth of particle invasion is usually small, ranging from less than an inch to a maximum of about 1 ft. To minimize this damage, the mud particles should be larger than the pores; Abrams (1977) suggests that having 5 vol% of the mud particles with a diameter greater than one-third of the mean pore size will prevent significant particle invasion. Since the invasion depth is small, it is often possible to overcome drilling mud particle damage by perforating through the damaged region or by acidizing.

Drilling mud filtrate will invade the formation to a greater depth than drilling mud particles, with depths of invasion of 1–6 ft being common (Hassen, 1980). As filtrate enters the formation, a filter cake of drilling mud solids is built up on the formation face, decreasing the rate of filtrate invasion. However, the filter cake will also be eroded by the shear force of the drilling fluid. The dynamic filtration rate accounts for this balance between filter cake formation and erosion and is given by

$$u_f = \frac{C}{\sqrt{t}} + 3600b\dot{\gamma} \tag{6-83}$$

where u_f is the filtrate flux in cm/hr, C is the dynamic fluid loss coefficient for the filter cake in $\text{cm}^3/\text{cm}^2\text{-hr}^{1/2}$, t is the exposure time (hr), b is a constant accounting for the mechanical stability of the filter cake, and $\dot{\gamma}$ is the shear rate at the wall (sec^{-1}). Hassen (1980) reports values of b of $2 \cdot 10^{-8}$ to $5 \cdot 10^{-7}$ cm^3/cm^2. The fluid loss coefficient can be obtained from a laboratory dynamic fluid loss test.

The drilling of horizontal wells with reported horizontal lengths up to 8000 ft poses new problems of significant penetrations of damage because of the long exposures to drilling fluids while the horizontal well is drilled. The shape of damage along a horizontal well is likely to reflect the longer exposure near the vertical section. This issue was addressed earlier in this chapter.

Example 6-9 The Depth of Filtrate Invasion

Calculate the depth of drilling mud filtrate invasion after 10 hr and after 100 hr of drilling mud exposure to the formation for a mud with a dynamic fluid loss coefficient of 5 in.^{3}/in.^{2}-hr$^{1/2}$. The wellbore radius is 6 in. and the formation porosity is 0.2. Assume that b is $5 \cdot 10^{-7}$ cm^3/cm^2 and the shear rate at the wall is 20 sec^{-1}.

Solution

Equation (6-83) gives the filtrate flux at the wellbore; the volumetric flow rate per unit thickness of formation is just the flux multiplied by the borehole circumference. The volume of filtrate is then the integral of the volumetric flow rate with time:

$$q_f = 2\pi r_w u_f = 2\pi r_w \left(\frac{C}{\sqrt{t}} + 3600 b\dot{\gamma} \right) \tag{6-84}$$

for q_f in cm^2/hr, r_w in cm, and C in cm^3/cm^2-hr$^{1/2}$.

$$V = \int_0^t q_f dt \tag{6-85}$$

$$V = \int_0^t 2\pi r_w \left(\frac{C}{\sqrt{t}} + 36000 b\dot{\gamma} \right) dt = 2\pi r_w (2C\sqrt{t} + 3600 b\dot{\gamma} t). \tag{6-86}$$

The volume of filtrate injected per unit thickness is related to the depth of penetration of the filtrate by

$$V = \pi\phi(r_p^2 - r_w^2). \tag{6-87}$$

Equating V from (6-86) and (6-87), we obtain

$$r_p = \sqrt{r_w^2 + \frac{2r_w}{\phi}(2Ct^{1/2} + 3600 b\dot{\gamma} t)} \tag{6-88}$$

and for the data given with appropriate unit conversions,

$$r_p = \sqrt{6^2 + \frac{(2)(6)}{0.2}\left[(2)(5)t^{1/2} + (3600)\left(\frac{5 \times 10^{-7}}{2.54} \right)(20)t \right]}. \tag{6-89}$$

From Equation (6-89), we find that for $t = 10$ hr, $r_p = 44$ in., and for $t = 100$ hr, $r_p = 78$ in. Thus, the filtrate has invaded 38 in. into the formation after 10 hr and 72 in. into the formation after 100 hr. The actual filtrate invasion is likely to be somewhat higher than that calculated here because of higher fluid loss rates that occur before a stable filter cake is formed.

Drilling fluid filtrate can damage the formation by fines movement, by precipitation, or by water blocking, as discussed in Section 6.7. Fines migration and precipitation damage can be minimized by tailoring the ionic composition of the drilling fluid to be compatible with that of the formation. If water blocking is a serious problem, water-based muds may have to be avoided.

6.8.2 Completion Damage

Damage to the formation during well completion operations can be caused by invasion of completion fluids into the formation, by cementing, by perforating, or by well stimulation techniques. Since a primary purpose of the completion fluid is to maintain a higher pressure in the wellbore than in the formation (overbalance), completion fluids will be forced into the formation. Thus, if the completion fluids contain solids or are chemically incompatible with the formation, damage can result similar to the damage caused by drilling mud. It is particularly important that completion fluids be filtered well to prevent injection of solids into the formation. It is recommended that completion fluids contain no more than 2 ppm of solids of a size less than 2 μm (Millhone, 1982).

Cement filtrate is another potentially damaging fluid when it enters the formation. Since cement filtrate will usually contain a high concentration of calcium ions, precipitation damage may occur. However, the small volume of the cement filtrate limits such damage to very near the well.

Perforating will inevitably result in some crushing of the formation in the immediate vicinity of the perforation. This damage is minimized by perforating underbalanced, that is, with the wellbore pressure lower than the formation pressure. Guidelines for the amount of underbalance needed in gas and oil zones are given in Figures 6-25 and 6-26 (King, Anderson, and Bingham, 1985). The minimum overbalance needed for a given formation permeability can be read from the correlation lines drawn on these figures.

An alternative to underbalanced perforating to obtain clean perforations is to perforate with extreme overbalance (Handren, Jupp, and Dees, 1993). In this technique the pressure in the wellbore is well above the fracturing pressure at the moment the perforations are created, with wellbore pressure gradients typically greater than 1.0 psi/ft. Additionally, the wellbore and/or tubing are partially filled with gas so that the high pressure is maintained for a short duration after the perforations are created. The well configuration for extreme overbalanced perforating is shown in Figure 6-27. This process is thought to create a network of short fractures emanating from the perforation (Figure 6-28), providing a place for the perforating debris to move away from the perforation tunnels.

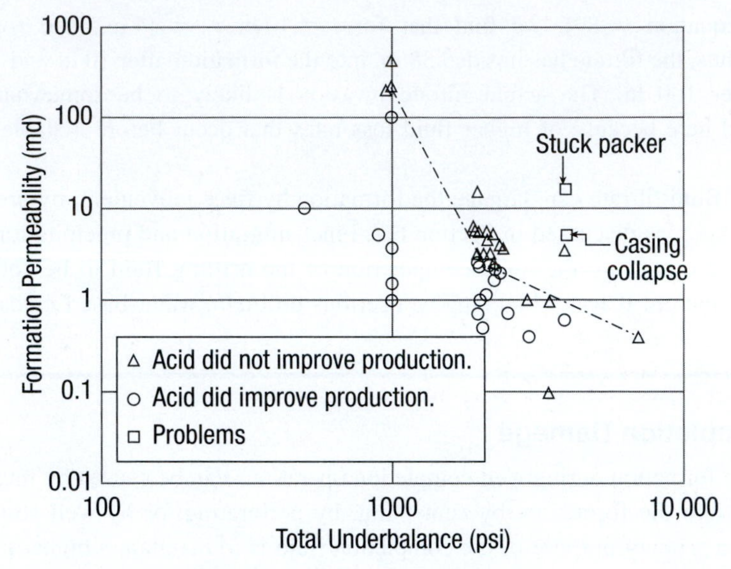

Figure 6-25 Underbalance needed to minimize perforation damage in gas wells. (From King et al., 1985.)

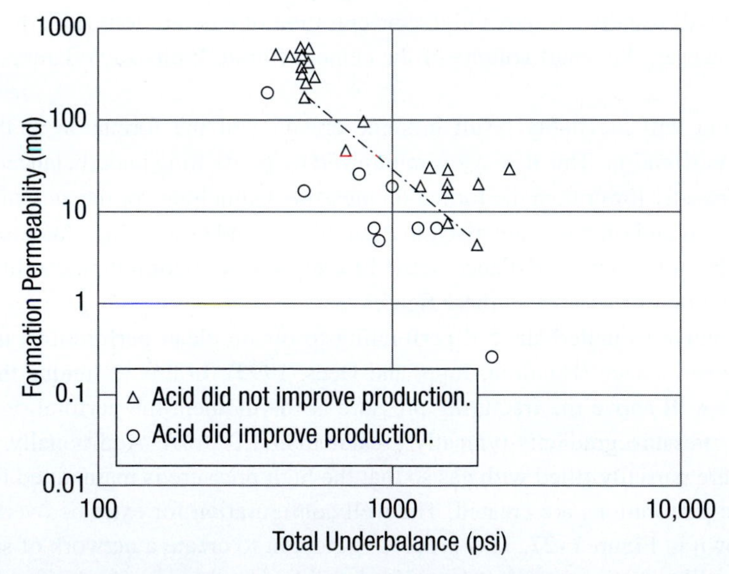

Figure 6-26 Underbalance needed to minimize perforation damage in oil wells. (From King et al., 1985.)

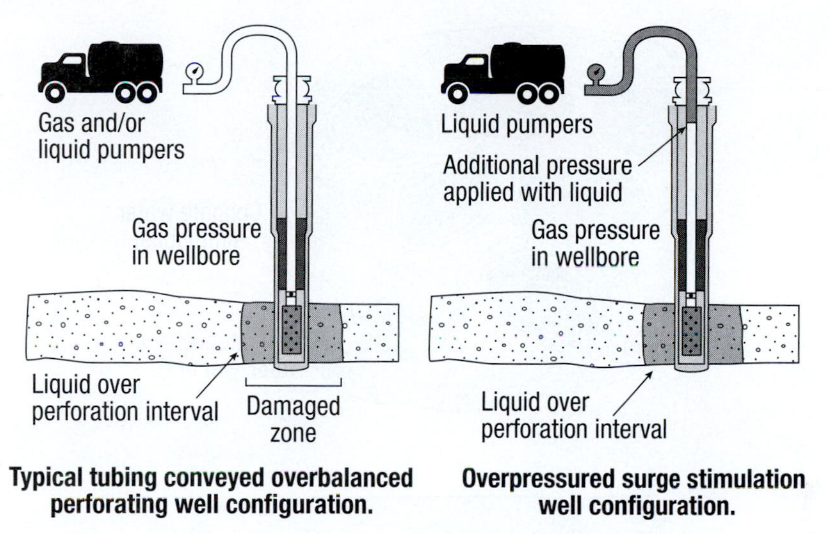

Figure 6-27 Well configuration to perforate with extreme overbalance. (From Handren et al., 1993.)

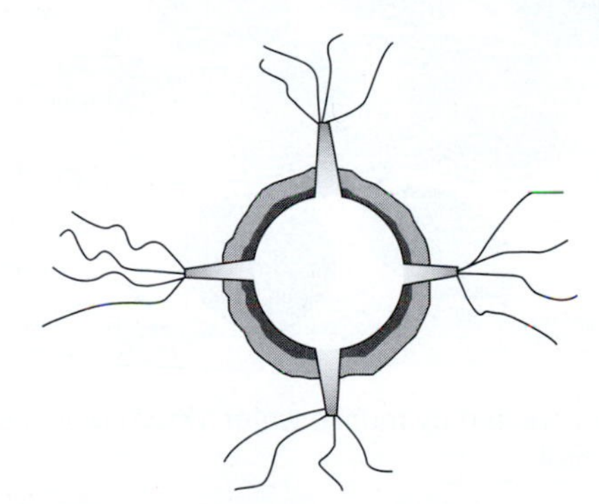

Figure 6-28 Short fractures from perforations created by extreme overbalanced perforating. (After Handren et al., 1993.)

Well stimulation fluids, though intended to increase the productivity of a well, can cause formation damage themselves by solids invasion of the formation or precipitation. The potential for damage from stimulation fluids is discussed in the chapters on well stimulation.

6.8.3 Production Damage

Formation damage during production can be caused by fines migration in the formation or by precipitation. The high velocity in the porous medium near the well is sometimes sufficient to mobilize fines that can then plug pore throats. Numerous studies have shown that a critical velocity exists, above which formation damage by particle migration occurs (Schechter, 1992). Unfortunately, this critical velocity depends on the particular rock and fluids in a complex manner; the only means of determining the critical velocity is through laboratory coreflood tests.

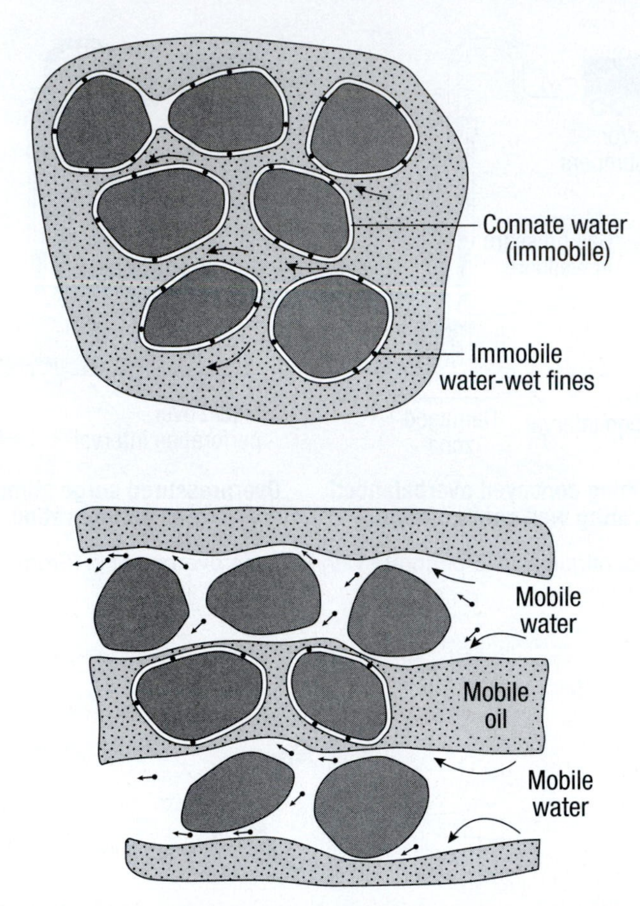

Figure 6-29 Fines migration caused by mobile water. (From Muecke, 1979.)

Fines may be mobilized in the vicinity of a production well when water production begins. Figure 6-29 (Muecke, 1979) illustrates this mechanism. Fines are most likely to move when the phase they wet is mobile, and since most formation fines are water-wet, the presence of a mobile water phase can cause fines migration and subsequent formation damage.

Precipitation of solids, either inorganic material from the brine or organic solids from the crude oil, may occur near a production well because of the reduced pressure in the near-wellbore region. These sources of formation damage can often be overcome with occasional stimulation treatments (e.g., acids to remove carbonate precipitates or solvents to remove waxes) or can be prevented with chemical squeeze treatments using sequestering agents.

6.8.4 Injection Damage

Injection wells are susceptible to formation damage caused by the injection of solid particles, by precipitation due to incompatibility of injected and formation water, or by the growth of bacteria. Solids injection is always a danger if the injected water is not well filtered; filtration to remove all particles larger than 2 μm is recommended.

Damage by precipitated solids can occur whenever mixing of the injected water with the formation water leads to supersaturation of one or more chemical species. The most common problem of this type is the injection of water with relatively high concentrations of sulfate or carbonate ions into formations with divalent cations, such as calcium, magnesium, or barium present. Because cation exchange with clays in the formation can release divalent cations into solution when a water with a different ionic composition is injected, precipitation may occur in the formation even when the injected water is apparently compatible with the formation water. In other words, the fact that no precipitation occurs when a sample of formation water is mixed with a sample of injection water is not sufficient to guarantee that no precipitation will occur in the formation. Dynamic processes such as cation exchange must be considered. Modeling of these processes is beyond the scope of this book.

Injection water may contain bacteria, which can plug the formation just like any other solid particles. Injected bacteria may also grow in the near-wellbore vicinity, causing severe formation damage. Injection water should be tested for the presence of bacteria, and bactericides should be added to the water if the potential for formation damage is indicated.

References

1. Abrams, A., "Mud Design to Minimize Rock Impairment Due to Particle Invasion," *JPT,* 586–592 (May 1977).
2. Besson, J., "Performance of Slanted and Horizontal Wells on an Anisotropic Medium," SPE Paper 20965 presented at Europec 90, The Hague, The Netherlands, October 22–24, 1990.
3. Brons, F., and Marting, V.E., "The Effect of Restricted Flow Entry on Well Productivity," *JPT,* 172–174 (February 1961).
4. Cinco-Ley, H., Ramey, H.J., Jr., and Miller, F.G., "Pseudoskin Factors for Partially Penetrating Directionally Drilled Wells," SPE Paper 5589, 1975.
5. Cvetkovic, B., Economides, M.J., Omrcen, B., and Longaric, B., "Production from Heavy Gas Condensate Reservoirs," SPE Paper 20968, 1990.
6. Frick, T.P., and Economides, M.J., "Horizontal Well Damage Characterization and Removal," SPE Paper 21795, 1991.
7. Furui, K., "A Comprehensive Skin Factor Model for Well Completions based on Finite Element Simulations," Ph.D. dissertation, University of Texas at Austin, May 2004.
8. Furui, K., Zhu, D., and Hill, A.D., "A Rigorous Formation Damage Skin Factor and Reservoir Inflow Model for a Horizontal Well," *SPE Production and Facilities,* 151–57 (August 2003).
9. Furui, K., Zhu, D., and Hill, A.D., "A New Skin Factor Model for Gravel-Packed Completions," SPE Paper 90433 presented at the SPE Annual Technical Conference and Exhibition, Houston, Texas, September 26–29, 2004.
10. Furui, K., Zhu, D., and Hill, A.D., "A Comprehensive Skin Factor Model of Horizontal Well Completion Performance," *SPE Production and Facilities,* 207–220 (May 2005).

11. Furui, K., Zhu, D., and Hill, A.D., "A New Skin Factor Model for Perforated Horizontal Wells," *SPE Drilling and Completion,* Vol. 23, No. 3, pp. 205–215 (September 2008).

12. Fussell, D.D., "Single-Well Performance Predictions for Gas Condensate Reservoirs," *JPT,* 860–870 (July 1973).

13. Handren, P.J., Jupp, T.B., and Dees, J.M., "Overbalance Perforating and Stimulation of Wells," SPE Paper 26515, 1993.

14. Hassen, B.R., "New Technique Estimates Drilling Fluid Filtrate Invasion," SPE Paper 8791, 1980.

15. Hawkins, M.F., Jr., "A Note on the Skin Effect," *Trans. AIME,* **207:** 356–357 (1956).

16. Hill A.D., and Zhu, D., "The Relative Importance of Wellbore Pressure Drop and Formation Damage in Horizontal Well," *SPE Production and Operations,* **23**(2): 232–240 (May 2008).

17. Hill, A.D., Zhu, D., and Economides, M.J., *Multilateral Wells,* Society of Petroleum Engineers, Richardson, TX, 2008.

18. Karakas, M., and Tariq, S., "Semi-Analytical Production Models for Perforated Completions," SPE Paper 18247, 1988.

19. King, G.E., Anderson, A., and Bingham, M., "A Field Study of Underbalance Pressures Necessary to Obtain Clean Perforations Using Tubing-Conveyed Perforating," SPE Paper 14321, 1985.

20. Krueger, R.F., "An Overview of Formation Damage and Well Productivity in Oilfield Operations," *JPT,* 131–152 (February 1986).

21. Lea, C.M., Hill, A.D., and Sepehrnoori, K., "The Effect of Fluid Diversion on the Acid Stimulation of a Perforation," SPE Paper 22853, 1991.

22. Millhone, R.S., "Completion Fluids—Maximizing Productivity," SPE Paper 10030, 1982.

23. Muecke, T.W., "Formation Fines and Factors Controlling their Movement in Porous Media," *JPT,* 144–150 (February 1979).

24. Muskat, M., *The Flow of Homogeneous Fluids Through Porous Media,* J. W. Edwards, Inc., Ann Arbor, MI, 1946.

25. Odeh, A.S., "An Equation for Calculating Skin Factor Due to Restricted Entry," *JPT,* 964–965 (June 1980).

26. Peaceman, D.W., "Interpretation of Well-Block Pressure in Numerical Reservoir Simulation with Nonsquare Grid Blocks and Anisotropic Permeability," *SPEJ,* 531–543 (June 1983).

27. Papatzacos, Paul, "Approximate Partial Penetration Pseudoskin for Infinite-Conductivity Wells," *SPERE,* 227–234 (May 1987).

28. Schechter, R.S., *Oil Well Stimulation,* Prentice Hall, Englewood Cliffs, NJ, 1992.

29. Schechter, R.S., and Gidley, J.L., "The Change in Pore Size Distributions from Surface Reactions in Porous Media," *AICHE J.,* 339–350 (May 1969).

30. Strelstova-Adams, T.D., "Pressure Drawdown in a Well with Limited Flow Entry," *JPT,* 1469–1476 (November 1979).

31. Tyler, T.N., Metzger, R.R., and Twyford, L.R., "Analysis and Treatment of Formation Damage at Prudhoe Bay, AK," SPE Paper 12471, 1984.

32. Van Everdingen, A.F., and Hurst, N., "The Application of the Laplace Transformation to Flow Problems in Reservoirs," *Trans. AIME,* **186:** 305–324 (1949).

33. Zajic, J.E., Cooper, D.G., Jack, T.R., and Kosaric, N., *Microbial Enhanced Oil Recovery,* PennWell Publishing Co., Tulsa, OK, 1983.

Problems

6-1 Consider a well that is completed in two zones. Zone 1 is 20 ft thick, has a permeability of 500 md, and has a damaged region with a permeability of 50 md extending one foot into the reservoir. Zone 2 is 40 ft thick, has a permeability of 200 md, and its damaged region having a permeability of 40 md extending 6 inches into the reservoir. The well radius is 0.25 ft, the well drains a 40-acre region, and there are no other completion skin effects. What is the overall skin factor for this well? If we injected acid into this well to restore the permeability in the damaged region, what would be the initial distribution of the injected acid into the two layers?

6-2 A general equation for the skin factor (without turbulence) was given in SPE 84401 as:

$$s^0 = \left(\int_{\xi_{D0}}^{\xi_{D1}} k_D^{-1} A_D^{-1} d\xi_D - \int_{\xi'_{D0}}^{\xi'_{D1}} A_D^{-1} d\xi'_D \right) \tag{1}$$

where

$$A_D = A/(2\pi r_w L)$$
$$\xi_D = \xi/r_w$$
$$k_D = k(\xi)/k.$$

Using this general formula, derive the standard Hawkins' formula for damage skin factor. Do not present the usual derivation of Hawkins' formula, such as that in this text; derive it directly from Equation 1 above.

6-3 A slanted well is being drilled and the degree of slant is not yet known, but it is expected to be between 45 and 65 degrees from the vertical. Determine the expected slant skin factors for 45, 50, 55, 60, and 65 degrees using the Besson equation. The reservoir is to be fully completed through the pay zone. The reservoir is 120 ft thick and has a horizontal permeability of 10 md and a vertical permeability of 1 md. Use $r_w = 0.328$ ft. Repeat the exercise assuming that the vertical and horizontal permeabilities are equal (isotropic). For both cases calculate the lowest value of the slant skin effect.

6-4 If $r_w = 0.328$ ft, $h = 165$ ft, $h_w = 75$ ft, and $z_w = 82.5$ ft (midpoint of the reservoir), calculate the partial completion skin effect for a vertical well. At which angle of well deviation would the contribution from s_θ nullify the effects of s_c (i.e., $s_{c+\theta} \approx 0$)?

6-5 A vertical well with a wellbore radius of 0.4 ft has been completed through the top half of the pay zone. The pay zone is 80 ft thick. The perforation density is 2 shots/ft and the perforations are 6 in. long, resulting in a perforation skin factor, not considering any damage or other effects, of 6 ($s_p = 6$). There also exists a damage zone around the well extending one foot into the formation from the wellbore, in which the permeability is 10% of the undamaged permeability.

a. What is the total skin factor for this well?

b. If the well is reperforated, adding the same type of perforations (2 spf, 6 in.-long perforations) along the remaining 40 ft of reservoir zone (fully completed), what is the new total skin factor?

c. If the well is originally perforated in the same zone (top half of the pay zone) to 2 shots/ft and 1.5 ft long, resulting in a perforation skin of 1, what is the total skin?

6-6 A fully penetrating 3000-ft long horizontal well is drilled in a reservoir with Appendix A properties. The distance to the reservoir boundary in the horizontal direction perpendicular to the well is 1000 feet, the pressure at the drainage boundary is 4000 psi, and the well flowing pressure is 2000 psi. The well has uniform damage along its entire length, with the damage extending 12 inches beyond the well in the horizontal direction. The damage permeability is 10% of the undamaged permeability. Calculate the damage skin factor for the well and productivity index ratio before and after an acid stimulation treatment that completely removed the damage.

6-7 A drilling mud has a dynamic fluid loss coefficient of 10 $\text{in.}^3/\text{in.}^2\text{-hr}^{1/2}$. During the 20 hr that the formation is exposed to this mud during drilling, the shear rate at the wall is 50 sec^{-1} and b is 5×10^{-7} cm^3/cm^2. What is the depth of drilling filtrate invasion assuming that $\phi = 0.19$? What is the skin factor if the filtrate reduces the permeability to 50% of the original value? The drainage radius is 660 ft, and $r_w = 0.328$ ft.

6-8 In addition to the filtrate loss and resulting damage, assume that, in the well of Problem 6-7, drilling mud particles penetrate 5 in. into the formation, reducing the permeability to 10% of the original reservoir permeability. What is the skin factor resulting from the combined particle and filtrate damage?

CHAPTER 7

Wellbore Flow Performance

7.1 Introduction

Wellbore flow can be divided into several broad categories, depending on the flow geometry, the fluid properties, and the flow rate. First, the flow in a wellbore is either single phase or multiphase; in most production wells, the flow is multiphase, with at least two phases (e.g., gas and liquid) present. Some production wells and most injection wells experience single-phase flow. The flow geometry of interest in the wellbores is usually flow through a circular pipe, though flow in an annular space, such as between tubing and casing, sometimes occurs. Furthermore, the flow of interest may be in any direction relative to the gravitational field. The properties of the fluids, both their PVT behavior and their rheological characteristics, must be considered in describing wellbore flow performance. Finally, depending on the flow rate and the fluid properties, flow in a wellbore may be either laminar or turbulent, and this will strongly influence the flow behavior.

In considering wellbore flow performance, the objective is usually to predict the pressure as a function of position between the bottomhole location and the surface. In addition, the velocity profile and the distribution of the phases in multiphase flow are sometimes of interest, particularly when interpreting production logs.

In this chapter the fluid flow relationships assume Newtonian behavior, which is appropriate for most hydrocarbon fluids. However, when injecting gelled fracturing fluids, the flow behavior is non-Newtonian, which is covered in Chapter 18.

7.2 Single-Phase Flow of an Incompressible, Newtonian Fluid

7.2.1 Laminar or Turbulent Flow

Single-phase flow can be characterized as being either laminar or turbulent, depending on the value of a dimensionless group, the Reynolds number, N_{Re}. The Reynolds number is the ratio of the inertial forces to the viscous forces in a flowing fluid, and for flow in a circular pipe is given by

$$N_{Re} = \frac{Du\rho}{\mu}. \tag{7-1}$$

When the flow is laminar, the fluid moves in distinct laminae, with no fluid motion transverse to the bulk flow direction. Turbulent flow, on the other hand, is characterized by eddy currents that cause fluctuating velocity components in all directions. Whether the flow is laminar or turbulent will strongly influence the velocity profile in the pipe, the frictional pressure drop behavior, and the dispersion of solutes contained in the fluid, among other factors; all of these attributes are considerations at times in production operations.

The transition from laminar to turbulent flow in circular pipes generally occurs at a Reynolds number of 2100, though this value can vary somewhat depending on the pipe roughness, entrance conditions, and other factors (Govier and Aziz, 1977). To calculate the Reynolds number, all variables must be expressed in consistent units so that the result is dimensionless.

Example 7-1 Determining the Reynolds Number for Flow in an Injection Well

For the injection of 1.03-specific gravity water ($\rho = 64.3$ lb$_m$/ft^3) in an injection well with 7-in., 32-lb/ft casing, construct a graph of Reynolds number versus volumetric flow rate (in bbl/d). The viscosity of the water at bottomhole conditions is 0.6cp. At what volumetric flow rate will the transition from laminar to turbulent flow occur?

Solution

Equation (7-1) relates the Reynolds number to the average velocity, pipe size, and fluid properties. The average velocity is simply the volumetric flow rate divided by the cross-sectional area of flow,

$$u = \frac{q}{A} \tag{7-2}$$

and for flow in a circular pipe, the cross-sectional area is

$$A = \frac{\pi}{4}D^2, \tag{7-3}$$

so

$$u = \frac{4q}{\pi D^2}. \tag{7-4}$$

Substituting for u in Equation (7-1),

$$N_{\text{Re}} = \frac{4q\rho}{\pi D\mu}. \tag{7-5}$$

The units must now be converted to a consistent set. For this problem, English engineering units will be most convenient. The appropriate conversion factors are given in Table 1-1.

$$N_{\text{Re}} = \frac{4(q \text{ bbl/d})(5.615 \text{ ft}^3\text{/bbl})(\text{day}/86{,}400 \text{ s})(64.3 \text{ lb}_m/\text{ft}^3)}{\pi(6.094 \text{ in.})(\text{ft}/12 \text{ in.})(0.6 \text{ cp})(6.72 \times 10^{-4} \text{ lb}_m/\text{ft-s-cp})} = 26.0q \tag{7-6}$$

where q is in bbl/d, ρ in lb_m/ft^3, D in in., and μ in cp. Of course, the constant 26.0 is specific to the pipe size and fluid properties of this example. For these oilfield units, the Reynolds number can be expressed in general as

$$N_{\text{Re}} = \frac{1.48q\rho}{D\mu}. \tag{7-7}$$

We see that the Reynolds number varies linearly with the volumetric flow rate for a given pipe size and fluid (Figure 7-1). The transition from laminar to turbulent flow occurs at $N_{\text{Re}} = 2100$, so for this example $2100 = 26.0q$ and therefore $q = 81$ bbl/d. Below about 81 bbl/d, the flow will be laminar, at rates higher than 81 bbl/d, flow is turbulent.

Example 7-1 illustrates that laminar flow occurs at low flow rates, generally less than about 100 bbl/d, when the fluid is a low-viscosity liquid. As the viscosity increases, the likelihood of laminar flow also increases. Figure 7-2 shows the variation of Reynolds number with flow rate, with pipe size and viscosity as parameters. Laminar flow occurs at relatively high flow rates in wells in which a viscous fluid is being produced or injected.

7.2.2 Velocity Profiles

The velocity profile, or variation of velocity with radial position, is sometimes an important consideration when analyzing flow in wellbores, particularly with regard to certain production logging

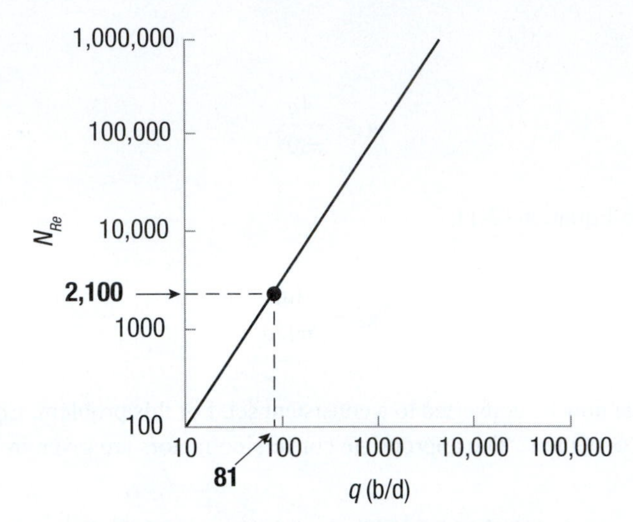

Figure 7-1 Variation of Reynolds number with volumetric flow rate.

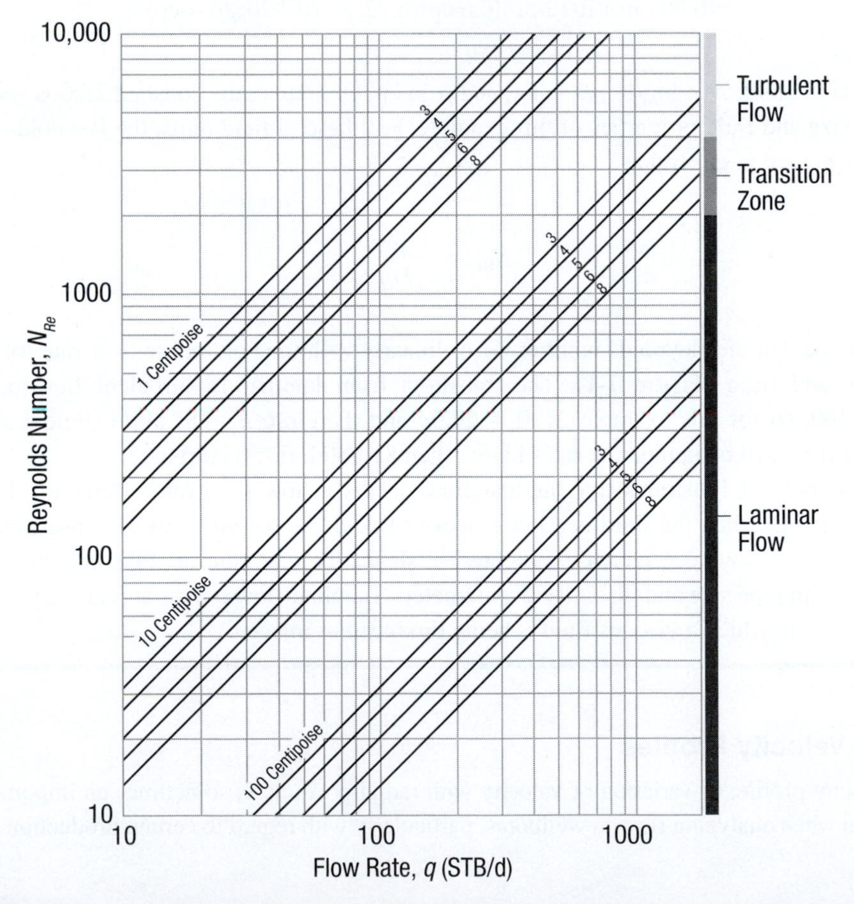

Figure 7-2 Variation of Reynolds number with volumetric flow rate, viscosity, and pipe size.

measurements. In laminar flow, the velocity profile in a circular pipe can be derived analytically and is given by

$$u(r) = \frac{(\Phi_0 - \Phi_L)R^2}{4\mu L}\left[1 - \left(\frac{r}{R}\right)^2\right] \tag{7-8}$$

where $\Phi_0 = p_0 + \rho g z_0$; $\Phi_L = p_L + pgz_L$; p_0 and p_L are the pressures at longitudinal positions a distance L apart; z_0 and z_L are the heights above some datum at these axial positions; R is the inside pipe radius; r is the radial distance from the center of the pipe; and $u(r)$ is velocity as a function of radial position. This equation shows that the velocity profile is parabolic in laminar flow, with the maximum velocity occurring at the center of the pipe as shown in Figure 7-3. It can be readily shown from Equation (7-8) that the average velocity, u, and maximum or centerline velocity, u_{max}, are

$$\bar{u} = \frac{1}{\pi R^2}\int_0^R 2\pi u(r)dr = \frac{(\Phi_0 - \Phi_L)R^2}{8\mu L} \tag{7-9}$$

$$u_{\text{max}} = \frac{(\Phi_0 - \Phi_L)R^2}{4\mu L} \tag{7-10}$$

so that the ratio of mean velocity to maximum velocity in laminar flow in a circular pipe is

$$\frac{u}{u_{\text{max}}} = 0.5. \tag{7-11}$$

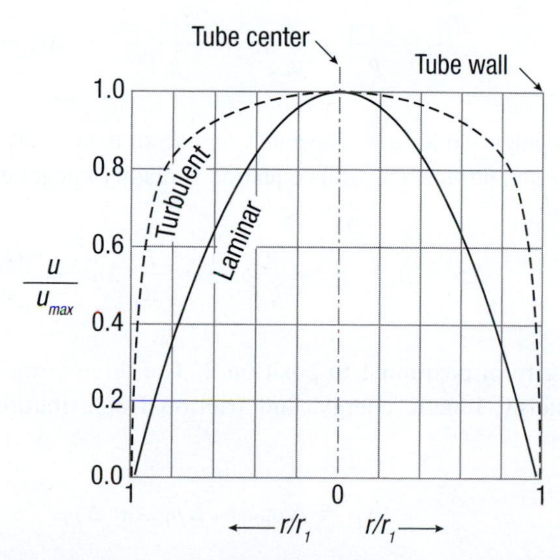

Figure 7-3 Velocity profiles in laminar and turbulent flow.

Equation (7-9) is a form of the Hagen-Poiseuille equation for laminar flow in a pipe.

Turbulent flow does not lend itself to a simple analytical treatment as applied to laminar flow because of the fluctuating nature of turbulent flow. From experiments, empirical expressions have been developed to describe the velocity profiles in turbulent flow. One such expression that is fairly accurate for $10^5 > N_{Re} > 3000$ is the power law model

$$\frac{u(r)}{u_{max}} = \left[1 - \left(\frac{r}{R}\right)\right]^{1/7}. \tag{7-12}$$

From this expression, it can be shown that

$$\frac{u}{u_{max}} \approx 0.8. \tag{7-13}$$

Thus, in turbulent flow, the velocity profile is much flatter than that found in laminar flow and the average velocity is closer to the maximum velocity (Figure 7-3). The ratio u/u_{max} in turbulent flow varies with Reynolds number and the roughness of the pipe, but is generally in the range 0.75–0.86.

7.2.3 Pressure-Drop Calculations

The pressure drop over a distance, L, of single-phase flow in a pipe can be obtained by solving the mechanical energy balance equation, which in differential form is

$$\frac{dp}{\rho} + \frac{u\,du}{g_c} + \frac{g}{g_c}dz + \frac{2f_f u^2 dL}{g_c D} + dW_s = 0. \tag{7-14}$$

If the fluid is incompressible (ρ = constant), and there is no shaft work device in the pipeline (a pump, compressor, turbine, etc.), this equation is readily integrated to yield

$$\Delta p = p_1 - p_2 = \frac{g}{g_c}\rho\Delta z + \frac{\rho}{2g_c}\Delta u^2 + \frac{2f_f \rho u^2 L}{g_c D} \tag{7-15}$$

for fluid moving from position 1 to position 2. The three terms on the right-hand side are the potential energy, kinetic energy, and frictional contributions to the overall pressure drop, or

$$\Delta p = \Delta p_{PE} + \Delta p_{KE} + \Delta p_F. \tag{7-16}$$

7.2.3.1 Δp_{PE}, the Pressure Drop Due to Potential Energy Change

Δp_{PE} accounts for the pressure change due to the weight of the column of fluid (the hydrostatic head); it will be zero for flow in a horizontal pipe. From Equation (7-15), the potential energy pressure drop is given by

$$\Delta p = \frac{g}{g_c}\rho\Delta z. \tag{7-17}$$

In this equation, Δz is the difference in elevation between positions 1 and 2, with z increasing upward. θ is defined as the angle between horizontal and the direction of flow. Thus, θ is $+90°$ for upward, vertical flow, $0°$ for horizontal flow, and $-90°$ for downward flow in a vertical well (Figure 7-4). For flow in a straight pipe of length L with flow direction θ,

$$\Delta z = z_2 - z_1 = L \sin\theta. \tag{7-18}$$

Example 7-2 Calculation of the Potential Energy Pressure Drop

Suppose that 1000 bbl/d of brine ($\gamma_w = 1.05$) is being injected through 2 7/8-in., 8.6-lb$_m$/ft tubing in a well that is deviated 50° from vertical. Calculate the pressure drop over 1000 ft of tubing due to the potential energy change.

Solution

Combining Equations (7-17) and (7-18),

$$\Delta p_{PE} = \frac{g}{g_c}\rho L \sin\theta. \tag{7-19}$$

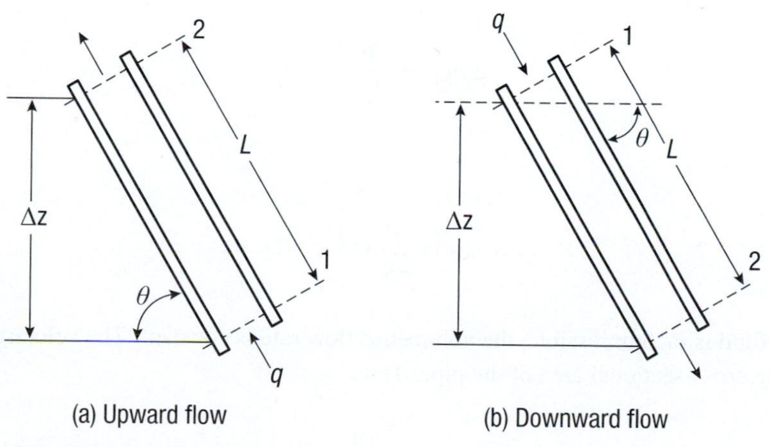

Figure 7-4 Flow geometry for pipe flow.

For downward flow in a well deviated 50° from vertical, the flow direction is −40° from horizontal, so θ is −40°. Converting to oilfield units, $\rho = (1.05)(62.4)\ \text{lb}_m/\text{ft}^3 = 65.5\ \text{lb}_m/\text{ft}^3$ and $\Delta p_{PE} = -292$ psi from Equation (7-19).

Shortcut Solution

For fresh water with $\gamma_w = 1(\rho = 62.4\ \text{lb}_m/\text{ft}^3)$, the potential energy pressure drop per foot of vertical distance is

$$\frac{dp}{dz} = \frac{g}{g_c}\rho = \left(1\frac{\text{lb}_f}{\text{lb}_m}\right)\left(62.4\frac{\text{lb}_m}{\text{ft}^3}\right)\left(\frac{1\ \text{ft}^2}{144\ \text{in.}^2}\right) = 0.433\ \text{psi/ft}. \tag{7-20}$$

For a fluid of any other specific gravity,

$$\frac{dp}{dz} = 0.433\gamma_w \tag{7-21}$$

where γ_w is the specific gravity. Thus,

$$\Delta p_{PE} = 0.433\gamma_w \Delta z. \tag{7-22}$$

For the example given, $\gamma_w = 1.05$ and $\Delta z = L \sin\theta$, so $\Delta p_{PE} = 0.433\ \gamma_w L \sin\theta = -292$ psi.

7.2.3.2 Δp_{KE}, the Pressure Drop Due to Kinetic Energy Change

Δp_{KE} is the pressure drop resulting from a change in the velocity of the fluid between positions 1 and 2. It will be zero for an incompressible fluid unless the cross-sectional area of the pipe is different at the two positions of interest. From Equation (7-15),

$$\Delta p_{KE} = \frac{\rho}{2g_c}(\Delta u^2) \tag{7-23}$$

or

$$\Delta p_{KE} = \frac{\rho}{2g_c}(u_2^2 - u_1^2). \tag{7-24}$$

If the fluid is incompressible, the volumetric flow rate is constant. The velocity then varies only with the cross-sectional area of the pipe. Thus,

$$u = \frac{q}{A} \tag{7-25}$$

and since $A = \pi D^2/4$, then

$$u = \frac{4q}{\pi D^2}. \tag{7-26}$$

Combining Equations (7-24) and (7-26), the kinetic energy pressure drop due to a pipe diameter change for an incompressible fluid is

$$\Delta p_{KE} = \frac{8\rho q^2}{\pi^2 g_c}\left(\frac{1}{D_2^4} - \frac{1}{D_1^4}\right). \tag{7-27}$$

Example 7-3 Calculation of the Kinetic Energy Pressure Drop

Suppose that 2000 bbl/d of oil with a density of 58 lb_m/ft^3 is flowing through a horizontal pipeline having a diameter reduction from 4 in. to 2 in., as illustrated in Figure 7-5. Calculate the kinetic energy pressure drop caused by the diameter change.

Solution

Equation (7-27) can be used if the fluid is incompressible. First, the volumetric flow rate must be converted to ft^3/sec:

$$q = (2000 \text{ bbl/d})(5.615 \text{ ft}^3\text{/bbl})\left(\frac{\text{day}}{86400 \text{ sec}}\right) = 0.130 \text{ ft}^3\text{/sec} \tag{7-28}$$

and from Equation (7-27),

$$\Delta p_{KE} = \frac{8(58 \text{ lb}_m\text{/ft}^3)(0.130 \text{ ft}^3\text{/sec})^2}{\pi^2(32.17 \text{ ft-lb}_m\text{/lb}_f - \text{sec}^2)}\left[\frac{1}{\left(\frac{2}{12}\text{ ft}\right)^4} - \frac{1}{\left(\frac{4}{12}\text{ ft}\right)^4}\right]\left(\frac{1 \text{ ft}^2}{144 \text{ in.}^2}\right)$$

$$= 0.208 \text{ psi}. \tag{7-29}$$

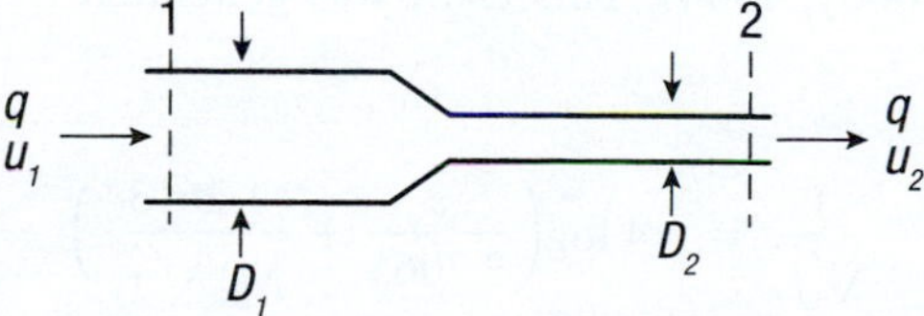

Figure 7-5 Flow with a reduction in pipe size (Example 7-3).

For oilfield units of bbl/d for flow rate, lb_m/ft^3 for density, and in. for diameter, the constants in Equation (7-27) and unit conversions can be combined to yield

$$\Delta p_{KE} = 1.53 \times 10^{-8} q^2 \rho \left(\frac{1}{D_2^4} - \frac{1}{D_1^4} \right) \tag{7-30}$$

where q is in bbl/d, ρ in lb_m/ft^3, and D in in.

7.2.3.3 Δp_F, The Frictional Pressure Drop

The frictional pressure drop is obtained from the Fanning equation,

$$\Delta p_F = \frac{2 f_f \rho u^2 L}{g_c D} \tag{7-31}$$

where f_f is the Fanning friction factor. In laminar flow, the friction factor is a simple function of the Reynolds number,

$$f_f = \frac{16}{N_{Re}} \tag{7-32}$$

whereas in turbulent flow, the friction factor may depend on both the Reynolds number and the relative pipe roughness, ε. The relative roughness is a measure of the size of surface features on the pipe wall protruding into the flow stream compared with the pipe diameter, or

$$\varepsilon = \frac{k}{D} \tag{7-33}$$

where k is the length of the protrusions on the pipe wall. The relative roughness of some common types of pipes are given in Figure 7-6 (Moody, 1944). However, it should be remembered that the pipe roughness may change with service, so that the relative roughness is essentially an empirical parameter that can be obtained through pressure-drop measurements.

The Fanning friction factor is most commonly obtained from the Moody friction factor chart (Figure 7-7; Moody, 1944). This chart was generated from the Colebrook-White equation,

$$\frac{1}{\sqrt{f_f}} = -4 \log \left(\frac{\varepsilon}{3.7065} + \frac{1.2613}{N_{Re}\sqrt{f_f}} \right). \tag{7-34}$$

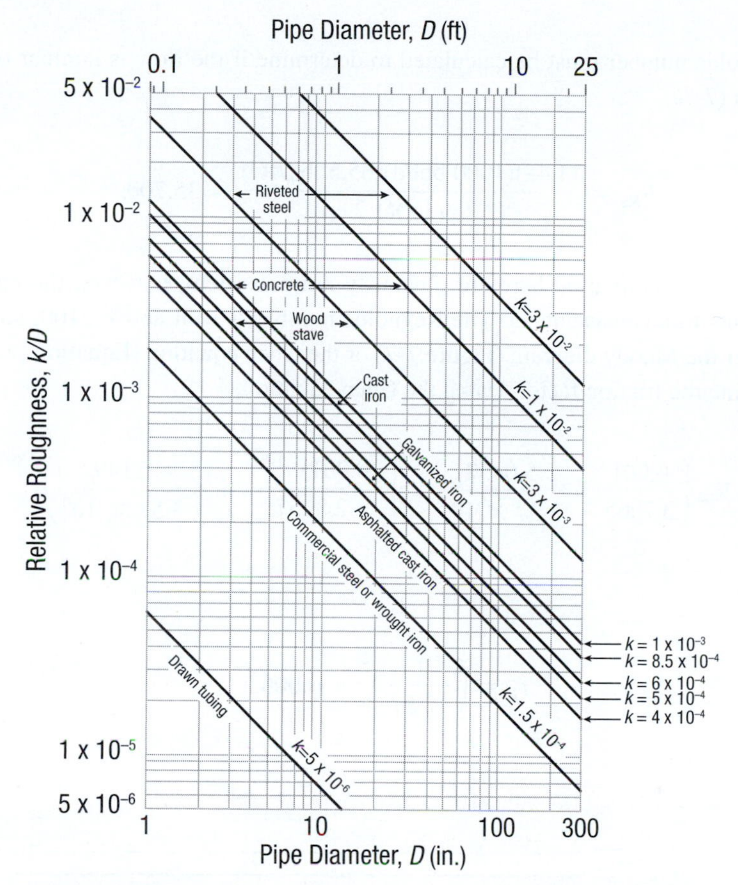

Figure 7-6 Relative roughness of common piping material. (From Moody, 1944.)

The Colebrook-White equation is implicit in f_f, requiring an iterative procedure, such as the Newton-Raphson method, for solution. An explicit equation for the friction factor with similar accuracy to the Colebrook-White equation (Gregory and Fogarasi, 1985) is the Chen equation (Chen, 1979):

$$\frac{1}{\sqrt{f_f}} = -4\log\left\{\frac{\varepsilon}{3.7065} - \frac{5.0452}{N_{\mathrm{Re}}}\log\left[\frac{\varepsilon^{1.1098}}{2.8257} + \left(\frac{7.149}{N_{\mathrm{Re}}}\right)^{0.8981}\right]\right\}. \tag{7-35}$$

Example 7-4 Calculating the Frictional Pressure Drop

Calculate the frictional pressure drop for the 1000 bbl/d of brine injection described in Example 7-2. The brine has a viscosity of 1.2 cp, and the pipe relative roughness is 0.001.

Solution

First, the Reynolds number must be calculated to determine if the flow is laminar or turbulent. Using Equation (7-7),

$$N_{\text{Re}} = \frac{(1.48)(1000 \text{ bbl/d})(65.5 \text{ lb}_m/\text{ft}^3)}{(2.259 \text{ in.})(1.2 \text{ cp})} = 35{,}700. \tag{7-36}$$

Note that the oilfield units used here are obviously not consistent; however, the constant 1.48 converts the units to a consistent set. The Reynolds number is well above 2100, so the flow is turbulent. Either the Moody diagram (Figure 7-7) or the Chen equation [Equation (7-35)] can be used to determine the friction factor. Using the Chen equation,

$$\frac{1}{\sqrt{f_f}} = -4\log\left\{\frac{0.001}{3.7065} - \frac{5.0452}{3.57 \times 10^4}\log\left[\frac{(0.001)^{1.1098}}{2.8257} + \left(\frac{7.149}{3.57 \times 10^4}\right)^{0.8981}\right]\right\} \tag{7-37}$$

and

$$f_f = \left(\frac{1}{12.57}\right)^2 = 0.0063. \tag{7-38}$$

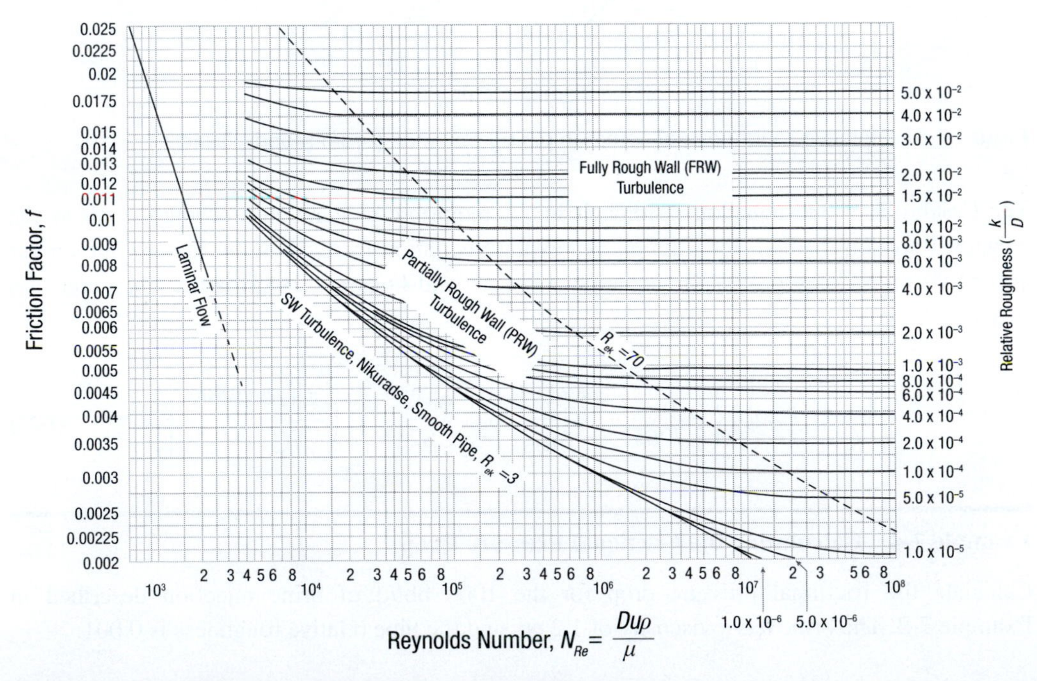

Figure 7-7 Moody friction factor diagram. (From Moody, 1944.)

Now using Equation (7-31), and noting that 2 7/8-in., 8.6-lb$_m$/ft tubing has an I.D. of 2.259 in.,

$$u = \frac{q}{A} = \frac{4q}{\pi D^2} = \frac{4(1000 \text{ bbl/day})(5.615 \text{ ft}^3/\text{bbl})(1 \text{ d}/86{,}400 \text{ s})}{\pi[(2.259/12) \text{ ft}]^2} = 2.33 \text{ ft/s} \quad \textbf{(7-39)}$$

then,

$$\Delta p_F = \frac{2(0.0063)(65.5 \text{ lb}_m/\text{ft}^3)(2.33 \text{ ft/s})^2(1000 \text{ ft})}{(32.17 \text{ ft-lb}_m/\text{lb}_f - \text{s}^2)[(2.259/12) \text{ ft}]}$$
$$= (740 \text{ lb}_f/\text{ft}^2)(\text{ft}^2/144 \text{ in.}^2) = 5.14 \text{ psi}. \quad \textbf{(7-40)}$$

Notice that the frictional pressure drop is considerably less than the potential energy or hydrostatic pressure drop, which we calculated to be –292 psi in Example 7-2.

7.2.4 Annular Flow

There are many instances in production operations in which flow in the annular space between a tubing string and a casing string occurs, including injection of gas down the annulus in a gas lift completion, injection of workover fluids down the annulus, and production of gas up the annulus. The frictional pressure drop for flow in an annulus can be calculated with Equation 7-31 by using the equivalent diameter in this equation and in the Reynolds' number calculation. The equivalent diameter for annular flow is the inside diameter of the large pipe minus the outside diameter of the small pipe.

7.3 Single-Phase Flow of a Compressible, Newtonian Fluid

To calculate the pressure drop in a gas well, the compressibility of the fluid must be considered. When the fluid is compressible, the fluid density and fluid velocity vary along the pipe, and these variations must be included when integrating the mechanical energy balance equation.

To derive an equation for the pressure drop in a gas well, we begin with the mechanical energy balance, Equation (7-14). With no shaft work device and neglecting for the time being any kinetic energy changes, the equation simplifies to

$$\frac{dp}{\rho} + \frac{g}{g_c}dz + \frac{2f_f u^2 dL}{g_c D} = 0. \quad \textbf{(7-41)}$$

Since dz is $\sin\theta dL$, the last two terms can be combined as

$$\frac{dp}{\rho} + \left(\frac{g}{g_c}\sin\theta + \frac{2f_f u^2}{g_c D}\right)dL = 0. \quad \textbf{(7-42)}$$

From the real gas law (Chapter 4), the density is expressed as

$$\rho = \frac{MWp}{ZRT} \tag{7-43}$$

or, in terms of gas gravity,

$$\rho = \frac{28.97\gamma_g p}{ZRT}. \tag{7-44}$$

The velocity can be written in terms of the volumetric flow rate at standard conditions, q,

$$u = \frac{4}{\pi D^2} qZ\left(\frac{T}{T_{sc}}\right)\left(\frac{p_{sc}}{p}\right). \tag{7-45}$$

Then, substituting for ρ and u from Equations (7-44) and (7-45), Equation (7-42) becomes

$$\frac{ZRT}{28.97\gamma_g p}dp + \left\{\frac{g}{g_c}\sin\theta + \frac{32f_f}{\pi^2 g_c D^5}\left[\left(\frac{T}{T_{sc}}\right)\left(\frac{p_{sc}}{p}\right)qZ\right]^2\right\}dL = 0. \tag{7-46}$$

This equation still contains three variables that are functions of position: Z, the compressibility factor, temperature, and pressure. To solve Equation (7-46) rigorously, the temperature profiles can be provided and the compressibility factor replaced by a function of temperature and pressure using an equation of state. This approach will likely require numerical integration.

Alternatively, single, average values of temperature and compressibility factor over the segment of pipe of interest can be assumed. If the temperature varies linearly between upstream position 1 and downstream position 2, the average temperature can be estimated as the mean temperature $(T_1 + T_2)/2$ or the log-mean temperature (Bradley, 1987), given by

$$T_{1m} = \frac{T_2 - T_1}{\ln(T_2/T_1)}. \tag{7-47}$$

An estimate of the average compressibility factor, $\overline{Z}$, can be obtained as a function of average temperature, $\overline{T}$, and the known pressure, p_1, as described in Chapter 4. Once the pressure, p_2, has been calculated, $\overline{Z}$ can be checked using $\overline{T}$ and the mean pressure, $(p_1 + p_2)/2$. If the new estimate differs significantly, the pressure calculation can be repeated using a new estimate of $\overline{Z}$.

Using average values of Z and T, Equation (7-46) can be integrated for nonhorizontal flow to yield

$$p_2^2 = e^s p_1^2 + \frac{32 f_f}{\pi^2 D^5 g_c \sin\theta}\left(\frac{\overline{Z}\overline{T} q p_{sc}}{T_{sc}}\right)^2 (e^s - 1) \tag{7-48}$$

where s is defined as

$$s = \frac{-(2)(28.97)\gamma_g (g/g_c)\sin\theta L}{\overline{Z}R\overline{T}}. \tag{7-49}$$

For the special case of horizontal flow, $\sin\theta$ and s are zero; integration of Equation (7-46) gives

$$p_1^2 - p_2^2 = \frac{(64)(28.97)\gamma_g f_f \overline{Z}\overline{T}}{\pi^2 g_c D^5 R}\left(\frac{p_{sc} q}{T_{sc}}\right)^2 L. \tag{7-50}$$

To complete the calculation, the friction factor must be obtained from the Reynolds number and the pipe roughness. Since the product, ρq, is a constant for flow of a compressible fluid, N_{Re} can be calculated based on standard conditions as

$$N_{Re} = \frac{4(28.97)\gamma_g q p_{sc}}{\pi D \overline{\mu} R T_{sc}}. \tag{7-51}$$

The viscosity should be evaluated at the average temperature and pressure.

The constants and conversion factors for oilfield units for Equations (7-48) through (7-51) can be combined to give

For vertical flow or inclined flow:

$$p_2^2 = e^s p_1^2 + 2.685 \times 10^{-3} \frac{f_f(\overline{Z}\overline{T}q)^2}{\sin\theta\, D^5}(e^s - 1) \tag{7-52}$$

where

$$s = \frac{-0.0375\gamma_g \sin\theta L}{\overline{Z}\overline{T}}. \tag{7-53}$$

For horizontal flow:

$$p_1^2 - p_2^2 = 1.007 \times 10^{-4} \frac{\gamma_g f_f \overline{Z}\overline{T} q^2 L}{D^5} \tag{7-54}$$

$$N_{Re} = 20.09 \frac{\gamma_g q}{D\mu}. \tag{7-55}$$

In Equations (7-52) through (7-55), p is in psia, q is in MSCF/d, D is in in., L is in ft, μ is in cp, T is in °R, and all other variables are dimensionless.

Frequently, in production operations, the unknown pressure may be the upstream pressure, p_1. For example, in a gas production well, in calculating the bottomhole pressure from the surface pressure, the upstream pressure is the unknown. Rearranging Equation (7-52) to solve for p_1, we have

$$p_1^2 = e^{-s} p_2^2 - 2.685 \times 10^{-3} \frac{f_f (\overline{Z}\,\overline{T} q)^2}{\sin \theta D^5} (1 - e^{-s}). \tag{7-56}$$

Equations (7-51) through (7-56) are the working equations for computing the pressure drop in gas wells. Remember that these equations are based on the use of an average temperature, compressibility factor, and viscosity over the pipe segment of interest. The longer the flow distance, the larger will be the error due to this approximation. It is advantageous to divide the well into multiple segments and calculate the pressure drop for each segment if the length (well measured depth) is large. We have also neglected changes in kinetic energy to develop these equations, even though we know that velocity will be changing throughout the pipe. The kinetic energy pressure drop can be checked after using these equations to estimate the pressure drop and corrections made, if necessary.

Example 7-5 Calculation of the Bottomhole Flowing Pressure in a Gas Well

Suppose that 2 MMSCF/d of natural gas is being produced through 10,000 ft of 2 7/8-in. tubing in a vertical well. At the surface, the temperature is 150°F and the pressure is 800 psia; the bottomhole temperature is 200°F. The gas has the composition given in Example 4-3, and the relative roughness of the tubing is 0.0006 (a common value used for new tubing).

Calculate the bottomhole flowing pressure directly from the surface pressure. Repeat the calculation, but this time dividing the well into two equal segments. Show that the kinetic energy pressure drop is negligible.

Solution

Equations (7-53), (7-55), and (7-56) are needed to solve this problem. From Example 4-3, T_{pc} is 374°R, p_{pc} is 717 psia, and γ_g is 0.709. Using the mean temperature, 175°F, the pseudoreduced temperature is $T_{pr} = (175 + 460)/374 = 1.70$; and using the known pressure at the surface to approximate the average pressure, $p_{pr} = 800/717 = 1.12$. From Figure 4-1, $\overline{Z} = 0.935$. Following Example 4-4, the gas viscosity is estimated: from Figure 4-4, $\mu_{1\text{ atm}} = 0.012$ cp; from Figure 4-5, $\mu/\mu_{1\text{ atm}} = 1.07$, and therefore $\mu = (0.012 \text{ cp})(1.07) = 0.013$ cp.

The Reynolds number is, from Equation (7-55),

$$N_{\text{Re}} = \frac{(20.09)(0.709)(2000 \text{ MSCF/d})}{(2.259 \text{ in.})(0.013 \text{ cp})} = 9.70 \times 10^5 \tag{7-57}$$

and $\varepsilon = 0.0006$, so, from the Moody diagram (Figure 7-7), $f_f = 0.0044$. Since the flow direction is vertical upward, $\theta = +90°$.

Now, using Equation (7-53),

$$s = \frac{-(0.0375)(0.709)[\sin(90)](10{,}000)}{(0.935)(635)} = -0.448. \tag{7-58}$$

The bottomhole pressure is calculated from Equation (7-56):

$$p_1^2 = e^{0.448}(800)^2 - 2.685 \times 10^{-3}\left\{\frac{(0.0044)[(0.935)(635)(2000)]^2}{\sin(90)(2.259)^5}\right\}(1 - e^{0.448}) \tag{7-59}$$

and $p_1 = p_{wf} = 1078$ psia.

The well is next divided into two equal segments and the calculation of bottomhole pressure repeated. The first segment is from the surface to a depth of 5000 ft. For this segment, $\overline{T}$ is 162.5°F, T_{pr} is 1.66, and p_{pr} is 1.12 as before. From Figure 4-1, $\overline{Z} = 0.93$. The viscosity is essentially the same as before, 0.0131cp. Thus, the Reynolds number and friction factor will be the same as in the previous calculation. From Equations (7-53) and (7-56),

$$s = \frac{(0.0375)(0.709)[\sin(90)](5000)}{(0.93)(622.5)} = -0.2296 \tag{7-60}$$

$$p_1^2 = e^{0.2296}(800)^2 - 2.685 \times 10^{-3}\left\{\frac{(0.0044)[(0.93)(622.5)(2000)]^2}{2.259^5}\right\}(1 - e^{0.2296}) \tag{7-61}$$

and $p_{5000} = 935$ psia.

For the second segment, from a depth of 5000 ft to the bottomhole depth of 10,000 ft, we use $\overline{T} = 187.5°F$ and $p = 935$ psia. Thus, $T_{pr} = 1.73$, $p_{pr} = 1.30$, and from Figure 4-1, $\overline{Z} = 0.935$. Viscosity is again 0.0131 cp. So, for this segment,

$$s = -\frac{(0.0375)(0.709)[\sin(90)](5000)}{(0.935)(647.5)} = -0.2196 \tag{7-62}$$

$$p_1^2 = e^{0.2196}(935)^2 - 2.685 \times 10^{-3}\left\{\frac{(0.0044)[(0.935)(647.5)(2000)]^2}{2.259^5}\right\}(1 - e^{0.2196}) \tag{7-63}$$

and $p_1 = p_{wf} = 1078$ psia.

Since neither temperature nor pressure varied greatly throughout the well, little error resulted from using average T and Z for the entire well. It is not likely that kinetic energy changes are significant, also because of the small changes in temperature and pressure, but this can be checked.

The kinetic energy pressure drop in this well can be estimated by

$$\Delta p_{\mathrm{KE}} \approx \frac{\bar{\rho}}{2g_c}\Delta u^2 \approx \frac{\bar{\rho}}{2g_c}(u_2^2 - u_1^2). \tag{7-64}$$

This calculation is approximate, since an average density is being used. The velocities at points 1 and 2 are

$$u_1 = \frac{Z_1(p_{\mathrm{sc}}/p_1)(T_1/T_{\mathrm{sc}})q}{A} \tag{7-65}$$

and

$$u_2 = \frac{Z_2(p_{\mathrm{sc}}/p_2)(T_2/T_{\mathrm{sc}})q}{A} \tag{7-66}$$

and the average density is

$$\bar{\rho} = \frac{28.97\gamma_g\bar{p}}{\bar{Z}R\bar{T}}. \tag{7-67}$$

At position 2 (surface), T is 150°F, $T_{pr} = 1.63$, $p = 800$ psia, $p_{pr} = 1.12$, and $Z = 0.925$; while at 1 (bottomhole), $T = 200$°F, $T_{pr} = 1.76$, $p = 1078$ psia, $p_{pr} = 1.50$, and $Z = 0.93$. For 2 7/8-in., 8.6-lb$_m$/ft tubing, the I.D. is 2.259 in., so the cross-sectional area is 0.0278 ft^2. To calculate average densities, the average pressure, 939 psia, average temperature, 175°F, and average compressibility factor, 0.93, are used. We then calculate $\Delta p_{\mathrm{KE}} = 0.06$ psia. The kinetic energy pressure drop is negligible compared with the potential energy and frictional contributions to the overall pressure drop.

7.4 Multiphase Flow in Wells

Multiphase flow—the simultaneous flow of two or more phases of fluid—will occur in almost all oil production wells, in many gas production wells, and in some types of injection wells. In an oil well, whenever the pressure drops below the bubble point, gas will evolve, and from that point to the surface, gas–liquid flow will occur. Thus, even in a well producing from an undersaturated reservoir, unless the surface pressure is above the bubble point, two-phase flow will occur in the wellbore and/or tubing. Many oil wells also produce significant amounts of water, resulting in oil–water flow or oil–water–gas three-phase flow.

Two-phase flow behavior depends strongly on the distribution of the phases in the pipe, which in turn depends on the direction of flow relative to the gravitational field. In this chapter, upward vertical and inclined flow are described; horizontal and near-horizontal flow are treated in Chapter 8.

7.4.1 Holdup Behavior

In two-phase flow, the amount of the pipe occupied by a phase is often different from its proportion of the total volumetric flow rate. As an example of a typical two-phase flow situation, consider the upward flow of two phases, α and β, where α is less dense than β, as shown in Figure 7-8. Typically, in upward two-phase flow, the lighter phase (α) will be moving faster than the denser phase (β). Because of this fact, called the *holdup phenomenon,* the in-situ volume fraction of the denser phase will be greater than the input volume fraction of the denser phase—that is, the denser phase is "held up" in the pipe relative to the lighter phase.

This relationship is quantified by defining a parameter called *holdup, y,* as

$$y_\beta = \frac{V_\beta}{V} \tag{7-68}$$

where V_β = volume of denser phase in pipe segment and V = volume of pipe segment. The holdup, y_β, can also be defined in terms of a local holdup, $y_{\beta l}$, as

$$y_\beta = \frac{1}{A}\int_0^A y_{\beta l}\, d\, A. \tag{7-69}$$

The local holdup, $y_{\beta l}$, is a time-averaged quantity—that is, $y_{\beta l}$ is the fraction of the time a given location in the pipe is occupied by phase β.

The holdup of the lighter phase, y_α, is defined identically to y_β as

$$y_\alpha = \frac{V_\alpha}{V} \tag{7-70}$$

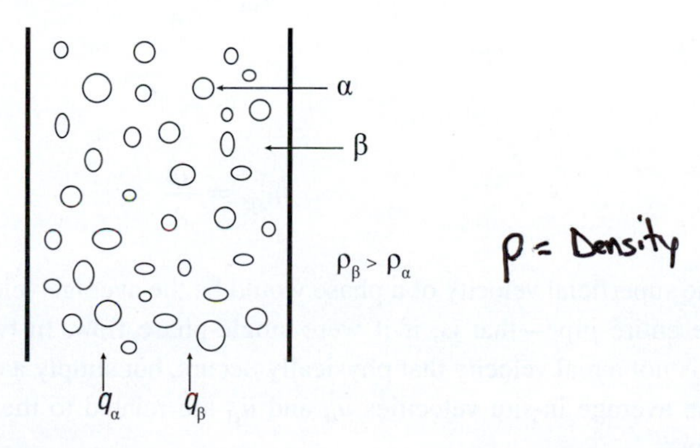

Figure 7-8 Schematic of two-phase flow.

or, because the pipe is completely occupied by the two phases,

$$y_\alpha = 1 - y_\beta. \tag{7-71}$$

In gas–liquid flow, the holdup of the gas phase, y_α, is sometimes called the *void fraction.*

Another parameter used in describing two-phase flow is the input fraction of each phase, λ, defined as

$$\lambda_\beta = \frac{q_\beta}{q_\alpha + q_\beta} \tag{7-72}$$

and

$$\lambda_\alpha = 1 - \lambda_\beta \tag{7-73}$$

where q_α and q_β are the volumetric flow rates of the two phases. The input volume fractions, λ_α and λ_β, are also referred to as the "no-slip holdups."

Another measure of the holdup phenomenon that is commonly used in production log interpretation is the "slip velocity," u_s. Slip velocity is defined as the difference between the average velocities of the two phases. Thus,

$$u_s = \bar{u}_\alpha - \bar{u}_\beta \tag{7-74}$$

where $\bar{u}_\alpha$ and $\bar{u}_\beta$ are the average in-situ velocities of the two phases. Slip velocity is not an independent property from holdup, but is simply another way to represent the holdup phenomenon. In order to show the relationship between holdup and slip velocity, we introduce the definition of superficial velocity, $u_{s\alpha}$ or $u_{s\beta}$, defined as

$$u_{s\alpha} = \frac{q_\alpha}{A} \tag{7-75}$$

and

$$u_{s\beta} = \frac{q_\beta}{A}. \tag{7-76}$$

The superficial velocity of a phase would be the average velocity of the phase if that phase filled the entire pipe—that is, if it were single-phase flow. In two-phase flow, the superficial velocity is not a real velocity that physically occurs, but simply a convenient parameter.

The average in-situ velocities $\bar{u}_\alpha$ and $\bar{u}_\beta$ are related to the superficial velocities and the holdup by

$$\bar{u}_\alpha = \frac{u_{s\alpha}}{y_\alpha} \tag{7-77}$$

and

$$\bar{u}_\beta = \frac{u_{s\beta}}{y_\beta}. \tag{7-78}$$

Substituting these expressions into the equation defining slip velocity (7-74) yields

$$u_s = \frac{1}{A}\left(\frac{q_\alpha}{1 - y_\beta} - \frac{q_\beta}{y_\beta}\right). \tag{7-79}$$

Correlations for holdup are generally used in two-phase pressure gradient calculations; the slip velocity is usually used to represent holdup behavior in production log interpretation.

Example 7-6 Relationship between Holdup and Slip Velocity

If the slip velocity for a gas–liquid flow is 60 ft/min and the superficial velocity of each phase is also 60 ft/min, what is the holdup of each phase?

Solution

From Equation (7-79), since superficial velocity of a phase is q/A,

$$u_s = \frac{u_{sg}}{1 - y_l} - \frac{u_{sl}}{y_l}. \tag{7-80}$$

Solving for y_l, a quadratic equation is obtained:

$$u_s y_l^2 - (u_s - u_{sg} - u_{sl})y_l - u_{sl} = 0. \tag{7-81}$$

For $u_s = u_{sg} = u_{sl} = 60$ ft/min, the solution is $y_l = 0.62$. The holdup of the gas phase is then $y_g = 1 - y_l = 0.38$. The holdup of the liquid is greater than the input fraction (0.5), as is typical in upward gas–liquid flow.

7.4.2 Two-Phase Flow Regimes

The manner in which the two phases are distributed in the pipe significantly affects other aspects of two-phase flow, such as slippage between phases and the pressure gradient. The "flow regime" or flow pattern is a qualitative description of the phase distribution. In gas–liquid, vertical, upward flow, four flow regimes are now generally agreed upon in the two-phase flow literature: bubble, slug, churn, and annular flow. These occur as a progression with increasing gas rate for a given liquid rate. Figure 7-9 (Govier and Aziz, 1977) shows these flow patterns and the approximate regions in which they occur as functions of superficial velocities for air–water flow in a small-diameter pipe. A brief description of these flow regimes is as follows.

1. Bubble flow: Dispersed bubbles of gas in a continuous liquid phase.
2. Slug flow: At higher gas rates, the bubbles coalesce into larger bubbles, called Taylor bubbles, that eventually fill the entire pipe cross section. Between the large gas bubbles are slugs of liquid that contain smaller bubbles of gas entrained in the liquid.
3. Churn flow: With a further increase in gas rate, the larger gas bubbles become unstable and collapse, resulting in churn flow, a highly turbulent flow pattern with both phases dispersed. Churn flow is characterized by oscillatory, up-and-down motion of the liquid.
4. Annular flow: At very high gas rates, gas becomes the continuous phase, with liquid flowing in an annulus coating the surface of the pipe and with liquid droplets entrained in the gas phase.

The flow regime in gas–liquid vertical flow can be predicted with a flow regime map, a plot relating flow regime to flow rates of each phase, fluid properties, and pipe size. One such map that is used for flow regime discrimination in some pressure-drop correlations is that of Duns and Ros (1963), shown in Figure 7-10. The Duns and Ros map correlates flow regime with two dimensionless numbers, the liquid and gas velocity numbers, N_{vl} and N_{vg}, defined as

$$N_{vl} = u_{sl} \sqrt[4]{\frac{\rho_l}{g\sigma}} \tag{7-82}$$

and

$$N_{vg} = u_{sg} \sqrt[4]{\frac{\rho_l}{g\sigma}} \tag{7-83}$$

where ρ_l is liquid density, g is the acceleration of gravity, and σ is the interfacial tension of the liquid–gas system. This flow pattern map does account for some fluid properties; note, however, that for a given gas–liquid system, the only variables in the dimensionless groups are the superficial velocities of the phrases.

Duns and Ros defined three distinct regions on their map, but also included a transition region where the flow changes from a liquid-continuous to a gas-continuous system. Region I contains bubble and low-velocity slug flow, Region II is high-velocity slug and churn flow, and Region III contains the annular flow pattern.

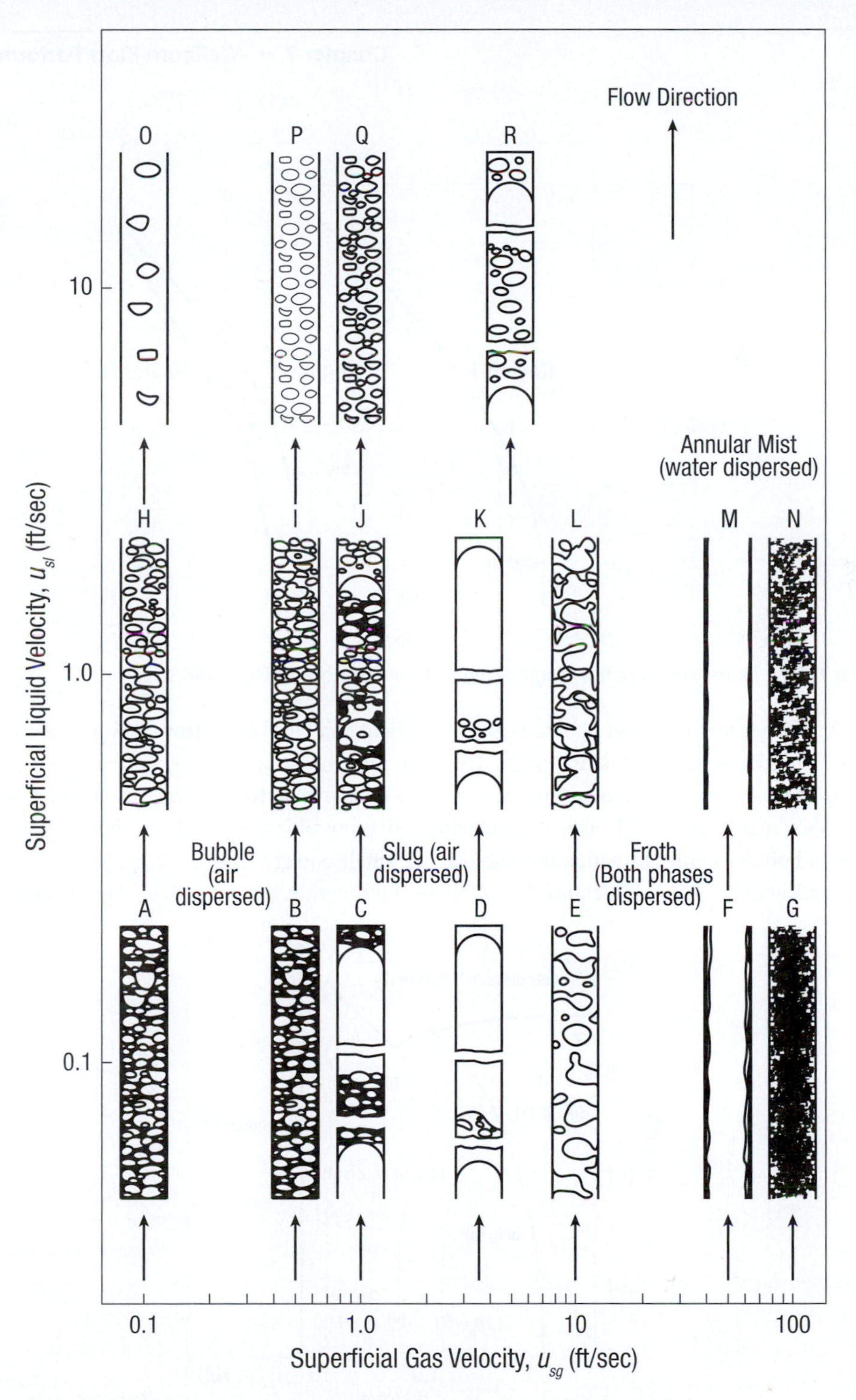

Figure 7-9 Flow regimes in gas–liquid flow. (From Govier and Aziz, 2008.)

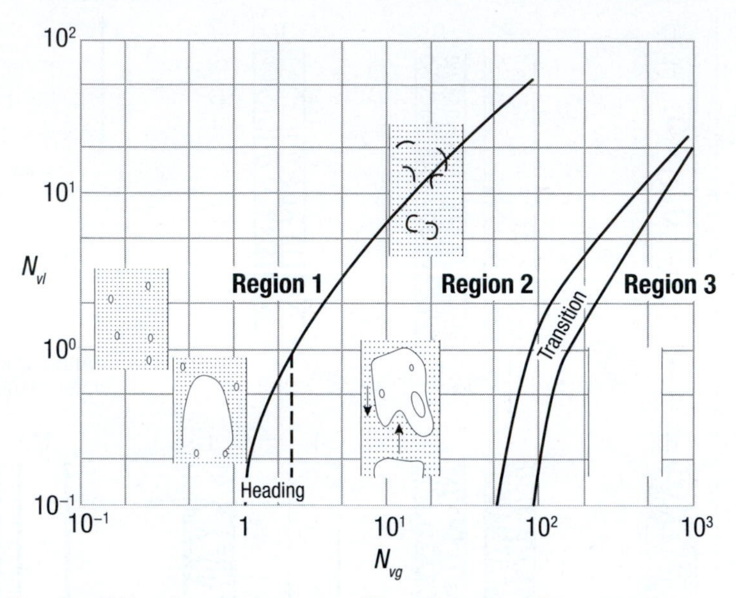

Figure 7-10 Duns and Ros flow regime map. (From Duns and Ros, 1963.)

A flow regime map that is based on a theoretical analysis of the flow regime transitions is that of Taitel, Barnea, and Dukler (1980). This map must be generated for particular gas and liquid properties and for a particular pipe size; a Taitel-Dukler map for air–water flow in a 2-in. I.D. pipe is shown in Figure 7-11. This map identifies five possible flow regimes: bubble, dispersed bubble (a bubble regime in which the bubbles are small enough that no slippage occurs), slug, churn, and annular. The slug/churn transition is significantly different than that of other flow

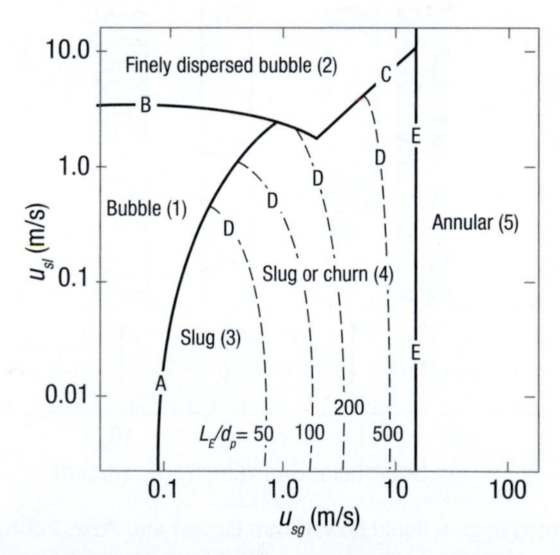

Figure 7-11 Taitel-Dukler flow regime map. (From Taitel et al., 1980.)

regime maps in that churn flow is thought to be an entry phenomenon leading to slug flow in the Taitel-Dukler theory. The D lines show how many pipe diameters from the pipe entrance churn flow is expected to occur before slug flow develops. For example, if the flow conditions fell on the D line labeled $L_E/D = 100$, for a distance of 100 pipe diameters from the pipe entrance, churn flow is predicted to occur; beyond this distance slug flow is the predicted flow regime.

Example 7-7 Predicting Two-Phase Flow Regime

200 bbl/d of water and 10,000 ft^3/day of air are flowing in a 2-in. vertical pipe. The water density is 62.4 lb$_m$/ft^3 and the surface tension is 74 dynes/cm. Predict the flow regime that will occur using the Duns-Ros and the Taitel-Dukler flow regime maps.

Solution

First, the superficial velocities are calculated as

$$u_{sl} = \frac{q_l}{A} = \left[\frac{(200 \text{ bbl/d})(5.615 \text{ ft}^3/\text{bbl})(1 \text{ d}/86{,}400 \text{ s})}{0.02182 \text{ ft}^2}\right] = 0.6 \text{ ft/s} = 0.18 \text{ m/s} \quad \textbf{(7-84)}$$

$$u_{sg} = \frac{q_g}{A} = \left[\frac{(10{,}000 \text{ ft}^3/\text{d})(1 \text{ d}/86{,}400 \text{ s})}{0.02182 \text{ ft}^2}\right] = 5.3 \text{ ft/s} = 1.62 \text{ m/s}. \quad \textbf{(7-85)}$$

For the Duns and Ros map, the liquid and gas velocity numbers must be calculated. For units of ft/s for superficial velocity, lb$_m$/ft^3 for density, and dynes/cm for surface tension, these are

$$N_{vl} = 1.938 u_{sl} \sqrt[4]{\frac{\rho_l}{\sigma}} \quad \textbf{(7-86)}$$

$$N_{vg} = 1.938 u_{sg} \sqrt[4]{\frac{\rho_l}{\sigma}} \quad \textbf{(7-87)}$$

Using the physical properties and flow rates given, we find $N_{vl} = 1.11$ and $N_{vg} = 9.8$. Referring to Figure 7-10, the flow conditions fall in region 2; the predicted flow regime is high-velocity slug or churn flow. Using the Taitel-Dukler map (Figure 7-11), the flow regime is also predicted to be slug or churn, with L_E/D of about 150. Thus, the Taitel-Dukler map predicts that churn flow will occur for the first 150 pipe diameters from the entrance; beyond this position, slug flow is predicted.

7.4.3 Two-Phase Pressure Gradient Models

In this section we consider correlations used to calculate the pressure drop in gas–liquid two-phase flow in wells. As in single-phase flow, the starting point is the mechanical energy balance given by Equation (7-16). Since the flow properties may change significantly along the pipe (mainly the gas density and velocity) in gas–liquid flow, we must calculate the pressure gradient for a particular location in the pipe; the overall pressure drop is then obtained with a pressure traverse calculation procedure (Section 7.4.4). A differential form of the mechanical energy balance equation is

$$\frac{dp}{dz} = \left(\frac{dp}{dz}\right)_{\text{PE}} + \left(\frac{dp}{dz}\right)_{\text{KE}} + \left(\frac{dp}{dz}\right)_{\text{F}}. \quad \textbf{(7-88)}$$

In most two-phase flow correlations, the potential energy pressure gradient is based on the in-situ average density, $\bar{\rho}$,

$$\left(\frac{dp}{dz}\right)_{\text{PE}} = \frac{g}{g_c}\bar{\rho}\sin\theta \quad \textbf{(7-89)}$$

where

$$\bar{\rho} = (1 - y_l)\rho_g + y_l\rho_l. \quad \textbf{(7-90)}$$

Various definitions of the two-phase average velocity, viscosity, and friction factor are used in the different correlations to calculate the kinetic energy and frictional pressure gradients.

There are many different correlations that have been developed to calculate gas–liquid pressure gradients, ranging from simple empirical models to complex deterministic models. The reader is referred to Brill and Mukherjee (1999) for a detailed treatment of several of these correlations. Although the deterministic models have aided our understanding of multiphase flow behavior, they have not shown a great improvement in predictive accuracy compared with some of the simpler models. Table 7-1 from Ansari et al. (1994a and 1994b) compares the relative errors of eight different two-phase flow correlations using several combinations of databases. In this table, the smaller the relative performance factor, the more accurate is the correlation. One of the consistently best correlations was found to be the empirical Hagedorn and Brown correlation.

We consider two of the most commonly used two-phase flow correlations for oil wells: the modified Hagedorn and Brown method (Brown, 1977) and the Beggs and Brill (1973) method with the Payne et al. (1979) correction. The first of these was developed for vertical, upward flow and is recommended only for near-vertical wellbores; the Beggs and Brill correlation can be applied for any wellbore inclination and flow direction. Finally, we will review the Gray (1974) correlation, which is commonly used for gas wells that are also producing liquid.

7.4.3.1 The Modified Hagedorn and Brown Method

The modified Hagedorn and Brown method (mH-B) is an empirical two-phase flow correlation based on the original work of Hagedorn and Brown (1965). The heart of the Hagedorn-Brown method is a correlation for liquid holdup; the modifications of the original method include using the

Table 7.1 Relative Performance Factors[a]

	EDB	VW	DW	VNH	ANH	AB	AS	VS	SNH	VSNH	AAN
n	1712	1086	626	755	1381	29	1052	654	745	387	70
MODEL	0.700	1.121	1.378	0.081	0.000	0.143	1.295	1.461	0.112	0.142	0.000
HAGBR	0.585	0.600	0.919	0.876	0.774	2.029	0.386	0.485	0.457	0.939	0.546
AZIZ	1.312	1.108	2.085	0.803	1.062	0.262	1.798	1.764	1.314	1.486	0.214
DUNRS	1.719	1.678	1.678	1.711	1.792	1.128	2.056	2.028	1.852	2.296	1.213
HASKA	1.940	2.005	2.201	1.836	1.780	0.009	2.575	2.590	2.044	1.998	1.043
BEGBR	2.982	2.908	3.445	3.321	3.414	2.828	2.883	2.595	3.261	3.282	1.972
ORKIS	4.284	5.273	2.322	5.838	4.688	1.226	3.128	3.318	3.551	4.403	6.000
MUKBR	4.883	4.647	6.000	3.909	4.601	4.463	5.343	5.140	4.977	4.683	1.516

EBD = entire databank; VW = vertical well cases; DW = deviated well cases; VNH = vertical well cases without Hagedorn and Brown data; ANH = all well cases without Hagedorn and Brown data; AB = all well cases with 75% bubble flow; AS = all well cases with 100% slug flow; VS = vertical well cases with 100% slug flow; SNH = all well cases with 100% slug flow without Hagedorn and Brown data; VSNH = vertical well cases with 100% slug flow without Hagedorn and Brown data; AAN = all well cases with 100% annular flow; HAGBR = Hagedorn and Brown correlation; AZIZ = Aziz *et al.* correlation; DUNRS = Duns and ROS correlation; Haska = Hasan and Kabir mechanistic model; BEGBR = Beggs and Brill correlation; ORKIS = Orkiszewski correlation; MUKBR = Mukherjee and Brill correlation.

[a]From Ansari *et al.*, 1994.

no-slip holdup when the original empirical correlation predicts a liquid holdup value less than the no-slip holdup and the use of the Griffith correlation (Griffith and Wallis, 1961) for the bubble flow regime.

These correlations are selected based on the flow regime as follows. Bubble flow exists if $\lambda_g < L_B$, where

$$L_B = 1.071 - 0.2218\left(\frac{u_m^2}{D}\right) \tag{7-91}$$

and $L_B \geq 0.13$. Thus, if the calculated value of L_B is less than 0.13, L_B is set to 0.13. If the flow regime is found to be bubble flow, the Griffith correlation is used; otherwise, the original Hagedorn-Brown correlation is used.

7.4.3.2 Flow Regimes Other than Bubble Flow: The Original Hagedorn-Brown Correlation

The form of the mechanical energy balance equation used in the Hagedorn-Brown correlation is

$$\frac{dp}{dz} = \frac{g}{g_c}\overline{\rho} + \frac{2f\overline{\rho}u_m^2}{g_c D} + \overline{\rho}\frac{\Delta(u_m^2/2g_c)}{\Delta z}, \quad \text{(7-92)}$$

which can be expressed in oilfield units as

$$144\frac{dp}{dz} = \overline{\rho} + \frac{f\dot{m}^2}{(7.413 \times 10^{10} D^5)\overline{\rho}} + \overline{\rho}\frac{\Delta(u_m^2/2g_c)}{\Delta z} \quad \text{(7-93)}$$

where f is the friction factor, $\dot{m}$ is the total mass flow rate (lb_m/d), $\overline{\rho}$ is the in-situ average density [Equation (7-90)] (lb_m/ft^3), D is the diameter (ft), u_m is the mixture velocity (ft/sec), and the pressure gradient is in psi/ft. The mixture velocity used in H-B is the sum of the superficial velocities,

$$u_m = u_{sl} + u_{sg}. \quad \text{(7-94)}$$

To calculate the pressure gradient with Equation (7-93), the liquid holdup is obtained from a correlation and the friction factor is based on a mixture Reynolds number. The liquid holdup, and hence, the average density, is obtained from a series of charts using the following dimensionless numbers. Liquid velocity number, N_{vl}:

$$N_{vl} = u_{sl}\sqrt[4]{\frac{\rho_l}{g\sigma}} \quad \text{(7-95)}$$

Gas velocity number, N_{vg}:

$$N_{vg} = u_{sg}\sqrt[4]{\frac{\rho_l}{g\sigma}} \quad \text{(7-96)}$$

Pipe diameter number, N_D,

$$N_D = D\sqrt{\frac{\rho_l g}{\sigma}} \quad \text{(7-97)}$$

Liquid viscosity number, N_L:

$$N_L = \mu_l\sqrt[4]{\frac{g}{\rho_l\sigma^3}} \quad \text{(7-98)}$$

In field units, these are

$$N_{vl} = 1.938u_{sl}\sqrt[4]{\frac{\rho_l}{\sigma}} \quad \text{(7-99)}$$

$$N_{vg} = 1.938 u_{sg} \sqrt[4]{\frac{\rho_l}{\sigma}} \tag{7-100}$$

$$N_D = 120.872 D \sqrt{\frac{\rho_l}{\sigma}} \tag{7-101}$$

$$N_L = 0.15726 \mu_l \sqrt[4]{\frac{1}{\rho_l \sigma^3}} \tag{7-102}$$

where superficial velocities are in ft/sec, density is in lb_m/ft^3, surface tension in dynes/cm, viscosity in cp, and diameter in ft. The holdup is obtained from Figures 7-12 through 7-14 or calculated from equations that fit the correlation curves presented on the charts (Brown, 1977). First, CN_L is read from Figure 7-12 or calculated by

$$CN_L = \frac{0.0019 + 0.0322 N_L - 0.6642 N_L^2 + 4.9951 N_L^3}{1 - 10.0147 N_L + 33.8696 N_L^2 + 277.2817 N_L^3}. \tag{7-103}$$

Then the group

$$H = \frac{N_{vl} p^{0.1} (CN_L)}{N_{vg}^{0.575} p_a^{0.1} N_D} \tag{7-104}$$

is calculated; from Figure 7-13, we get y_l/ψ, or it is calculated by

$$\frac{y_l}{\psi} = \left(\frac{0.0047 + 1123.32 H + 729489.64 H^2}{1 + 1097.1566 H + 722153.97 H^2} \right)^{\frac{1}{2}}. \tag{7-105}$$

Here p is the absolute pressure at the location where pressure gradient is wanted, and p_a is atmospheric pressure. Finally, compute

$$B = \frac{N_{vg} N_L^{0.380}}{N_D^{2.14}} \tag{7-106}$$

and read ψ from Figure 7-14 or calculate it with

$$\psi = \frac{1.0886 - 69.9473 B + 2334.3497 B^2 - 12896.683 B^3}{1 - 53.4401 B + 1517.9369 B^2 - 8419.8115 B^3}. \tag{7-107}$$

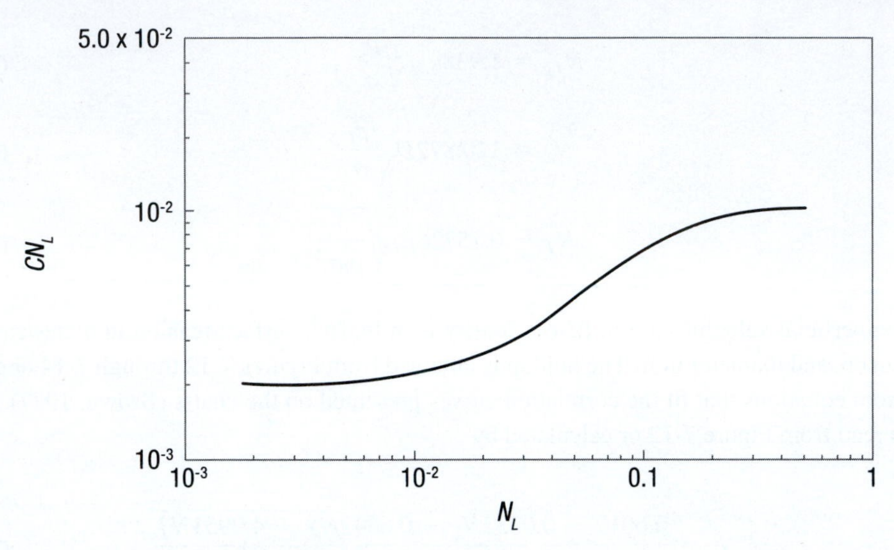

Figure 7-12 Hagedorn and Brown correlation for CN_L. (From Hagedorn and Brown, 1965.)

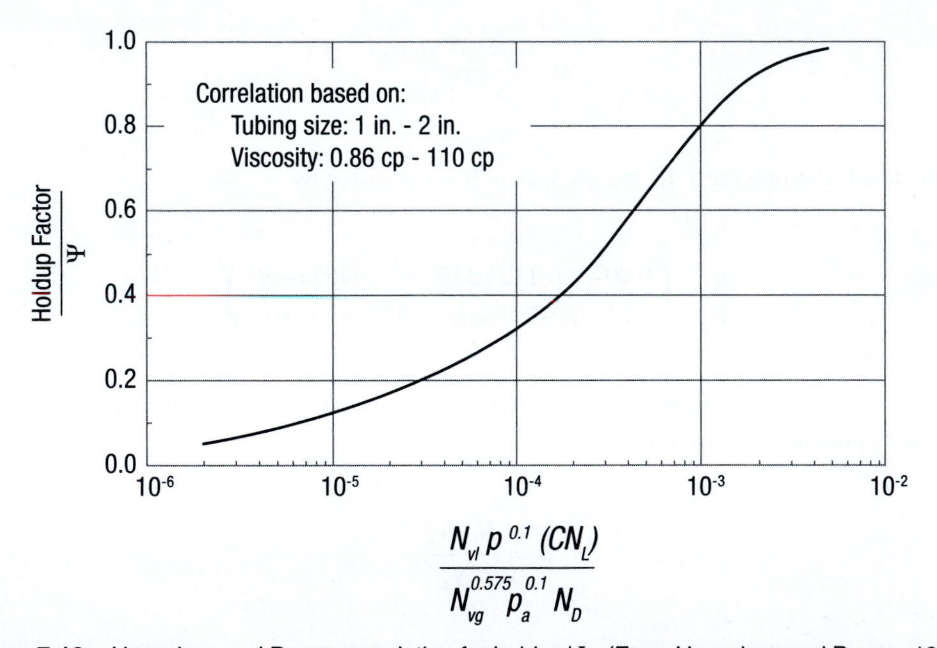

Figure 7-13 Hagedorn and Brown correlation for holdup/Ψ. (From Hagedorn and Brown, 1965.)

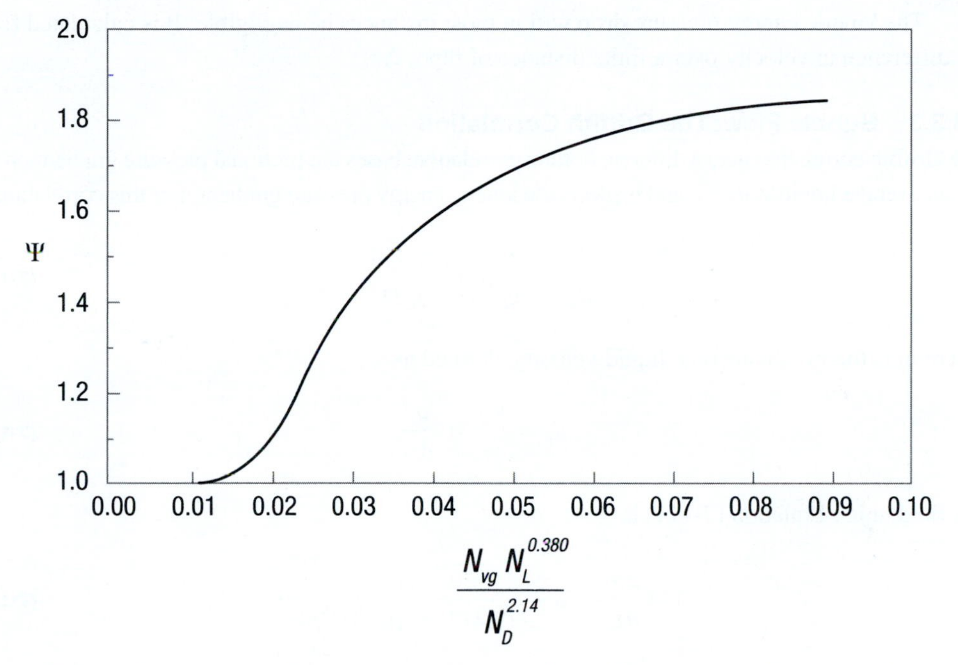

Figure 7-14 Hagedorn and Brown correlation for ψ. (From Hagedorn and Brown, 1965.)

The liquid holdup is then

$$y_l = \left(\frac{y_l}{\psi}\right)\psi. \tag{7-108}$$

The mixture density is then calculated from Equation (7-90).

The frictional pressure gradient is based on a Fanning friction factor using a mixture Reynolds number, defined as

$$N_{\text{Re}} = \frac{Du_m\bar{\rho}}{\mu_l^{y_l}\mu_g^{(l-y_l)}} \tag{7-109}$$

or, in terms of mass flow rate and using field units,

$$N_{\text{Re}} = \frac{2.2 \times 10^{-2}\dot{m}}{D\mu_l^{y_l}\mu_g^{(l-y_l)}} \tag{7-110}$$

where mass flow rate, $\dot{m}$, is in lb_m/day, D is in ft, and viscosities are in cp. The friction factor is then obtained from the Moody diagram (Figure 7-7) or calculated with the Chen equation [Equation (7-35)] for the calculated Reynolds number and the pipe relative roughness.

The kinetic energy pressure drop will in most instances be negligible; it is calculated from the difference in velocity over a finite distance of pipe, Δz.

7.4.3.3 Bubble Flow: The Griffith Correlation

The Griffith correlation uses a different holdup correlation, bases the frictional pressure gradient on the in-situ average liquid velocity, and neglects the kinetic energy pressure gradient. For this correlation,

$$\frac{dp}{dz} = \frac{g}{g_c}\bar{\rho} + \frac{2f\rho_l\bar{u}_l^2}{g_c D} \tag{7-111}$$

where u_l is the in-situ average liquid velocity, defined as

$$\bar{u}_l = \frac{u_{sl}}{y_l} = \frac{q_l}{Ay_l}. \tag{7-112}$$

For field units, Equation (7-111) is

$$144\frac{dp}{dz} = \bar{\rho} + \frac{f\dot{m}_l^2}{(7.413 \times 10^{10})D^5\rho_l y_l^2} \tag{7-113}$$

where m_l is the mass flow of the liquid only. The liquid holdup is

$$y_l = 1 - \frac{1}{2}\left[1 + \frac{u_m}{u_s} - \sqrt{\left(1 + \frac{u_m}{u_s}\right)^2 - 4\frac{u_{sg}}{u_s}}\right] \tag{7-114}$$

where $u_s = 0.8$ ft/sec. The Reynolds number used to obtain the friction factor is based on the in-situ average liquid velocity,

$$N_{\text{Re}} = \frac{D\bar{u}_l\rho_l}{\mu_l} \tag{7-115}$$

or

$$N_{\text{Re}} = \frac{2.2 \times 10^{-2}\dot{m}_l}{D\mu_l}. \tag{7-116}$$

Example 7-8 Pressure Gradient Calculation Using the Modified Hagedorn and Brown Method

Suppose that 2000 bbl/d of oil ($\rho = 0.8$ g/cm^3, $\mu = 2$ cp) and 1 MMSCF/d of gas of the same composition as in Example 7-5 are flowing in 2 7/8-in. tubing. The surface tubing pressure is 800 psia and the temperature is 175°F. The oil–gas surface tension is 30 dynes/cm, and the pipe relative roughness is 0.0006. Calculate the pressure gradient at the top of the tubing, neglecting any kinetic energy contribution to the pressure gradient,

Solution

From Example 7-5, we have $\mu_g = 0.0131$ cp and $Z = 0.935$. Converting volumetric flow rates to superficial velocities with $A = (\pi/4)(2.259/12)^2 = 0.0278\ \text{ft}^2$,

$$u_{sl} = \frac{(2000\ \text{bbl/d})(5.615\ \text{ft}^3/\text{bbl})(\text{d}/86{,}400\ \text{s})}{0.0278\ \text{ft}^2} = 4.67\ \text{ft/s}. \tag{7-117}$$

The gas superficial velocity can be calculated from the volumetric flow rate at standard conditions with Equation (7-45),

$$u_{sg} = \frac{4}{\pi(2.259/12)^2}(10^6\ \text{ft}^3/\text{d})(0.935)\left(\frac{460 + 175}{460 + 60}\right)\left(\frac{14.7}{800}\right)\frac{\text{d}}{86{,}400\ \text{s}} = 8.72\ \text{ft/s}. \tag{7-118}$$

The mixture velocity is

$$u_m = u_{sl} + u_{sg} = (4.67 + 8.72) = 13.39\ \text{ft/s} \tag{7-119}$$

and the input fraction of gas is

$$\lambda_g = \frac{u_{sg}}{u_m} = \frac{8.72}{13.39} = 0.65. \tag{7-120}$$

First, we check whether the flow regime is bubble flow. Using Equation (7-91),

$$L_B = 1.071 - 0.2218\left[\frac{(13.39)^2}{2.259/12}\right] = -210 \tag{7-121}$$

but L_B must be ≥ 0.13, so $L_B = 0.13$. Since λ_g (0.65) is greater than L_B, the flow regime is not bubble flow and we proceed with the Hagedorn-Brown correlation.

We next compute the dimensionless numbers, N_{vl}, N_{vg}, N_D, and N_L. Using Equations (7-99) through (7-102), we find $N_{vl} = 10.28$, $N_{vg} = 19.20$, $N_D = 29.35$, $N_L = 9.26 \times 10^{-3}$. Now, we determine liquid holdup, y_l, from Figures 7-12 through 7-14 or Equations 7-103 through 7-107. From Figure 7-12 or Equation 7-103, $CN_L = 0.0022$. Then

$$H = \frac{N_{vl}}{N_{vg}^{0.575}}\left(\frac{p}{p_a}\right)^{0.1}\frac{CN_L}{N_D} = \frac{10.28}{(19.2)^{0.575}}\left(\frac{800}{14.7}\right)^{0.1}\frac{(0.0022)}{29.35} = 2.1 \times 10^{-4} \tag{7-122}$$

and, from Figure 7-13 or Equation 7-105, $y_l/\psi = 0.46$. Finally, we calculate

$$B = \frac{N_{vg}N_L^{0.38}}{N_D^{2.14}} = \frac{(19.2)(9.26 \times 10^{-3})^{0.38}}{(29.35)^{2.14}} = 2.34 \times 10^{-3} \tag{7-123}$$

and from Figure 7-14 or Equation 7-107, $\psi = 1.0$. Note than ψ will generally be 1.0 for low-viscosity liquids. The liquid holdup is thus 0.46. This is compared with the input liquid fraction, λ_l, which in this case is 0.35. If y_l is less than λ_l, y_l is set to λ_l.

Next, we calculate the two-phase Reynolds number using Equation (7-110). The mass flow rate is

$$\dot{m} = \dot{m}_l + \dot{m}_g = A(u_{sl}\rho_l + u_{sg}\rho_g). \tag{7-124}$$

The gas density is calculated from Equation (7-44),

$$\rho_g = \frac{(28.97)(0.709)(800\text{ psi})}{(0.935)(10.73\text{ psi-ft}^3/\text{lb-mol-°R})(635\text{°R})} = 2.6\text{ lb}_m/\text{ft}^3, \tag{7-125}$$

so

$$\begin{aligned}\dot{m} &= (0.0278\text{ ft}^2)[(4.67\text{ ft/s})(49.9\text{ lb}_m/\text{ft}^3) + (8.72\text{ ft/s})(2.6\text{ lb}_m/\text{ft}^3)](86{,}400\text{ s/d}) \\ &= 614{,}000\text{ lb}_m/\text{d}\end{aligned} \tag{7-126}$$

and

$$N_{\text{Re}} = \frac{(2.2 \times 10^{-2})(6.14 \times 10^5)}{(2.259/12)(2)^{0.46}(0.0131)^{0.54}} = 5.42 \times 10^5. \tag{7-127}$$

From Figure 7-7 or Equation (7-35), $f = 0.0046$. The in-situ average density is

$$\bar{\rho} = y_l\rho_l + (1 - y_l)\rho_g = (0.46)(49.9) + (0.54)(2.6) = 24.4\text{ lb}_m/\text{ft}^3. \tag{7-128}$$

Finally, from Equation (7-93),

$$\begin{aligned}\frac{dp}{dz} &= \frac{1}{144}\left[\bar{\rho} + \frac{f\dot{m}^2}{(7.413 \times 10^{10})D^5\bar{\rho}}\right] \\ &= \frac{1}{144}\left[24.4 + \frac{(0.0046)(614{,}000)^2}{(7.413 \times 10^{10})(2.259/12)^5(24.4)}\right] \\ &= \frac{1}{144}(24.4 + 4.1) = 0.190\text{ psi/ft}.\end{aligned} \tag{7-129}$$

7.4.3.4 The Beggs and Brill Method

The Beggs and Bill (1973) correlation differs significantly from that of Hagedorn and Brown in that the Beggs and Brill correlation is applicable to any pipe inclination and flow direction. This method is based on the flow regime that would occur if the pipe were horizontal; corrections are then made to account for the change in holdup behavior with inclination. It should be kept in mind that the flow regime determined as part of this correlation is the flow regime that would occur if the pipe were perfectly horizontal and is probably not the actual flow regime that occurs at any other angle. The Beggs and Brill method is the recommended technique for any wellbore that is not near vertical.

The Beggs and Brill method uses the general mechanical energy balance [Equation (7-88)] and the in-situ average density [Equation (7-90)] to calculate the pressure gradient and is based on the following parameters:

$$N_{FR} = \frac{u_m^2}{gD} \tag{7-130}$$

$$\lambda_l = \frac{u_{sl}}{u_m} \tag{7-131}$$

$$L_1 = 316\lambda_l^{0.302} \tag{7-132}$$

$$L_2 = 0.0009252\lambda_l^{-2.4684} \tag{7-133}$$

$$L_3 = 0.10\lambda_l^{-1.4516} \tag{7-134}$$

$$L_4 = 0.5\lambda_l^{-6.738} \tag{7-135}$$

The horizontal flow regimes used as correlating parameters in the Beggs-Brill method are segregated, transition, intermittent, and distributed (see Chapter 8 for a discussion of horizontal flow regimes). The flow regime transitions are given by the following.

Segregated flow exists if

$$\lambda_l < 0.01 \text{ and } N_{FR} < L_1 \text{ or } \lambda_l \geq 0.01 \text{ and } N_{FR} < L_2 \tag{7-136}$$

Transition flow occurs when

$$\lambda_l \geq 0.01 \text{ and } L_2 < N_{FR} \leq L_3 \tag{7-137}$$

Intermittent flow exists when

$$0.01 \leq \lambda_l < 0.4 \text{ and } L_3 < N_{FR} \leq L_1 \text{ or } \lambda_l > 0.4 \text{ and } L_3 < N_{FR} \leq L_4 \tag{7-138}$$

Distributed flow occurs if

$$\lambda_l < 0.4 \text{ and } N_{FR} \geq L_1 \text{ or } \lambda_l \geq 0.4 \text{ and } N_{FR} > L_4 \quad \textbf{(7-139)}$$

The same equations are used to calculate the liquid holdup, and hence the average density, for the segregated, intermittent, and distributed flow regimes. These are

$$y_l = y_{lo}\psi \quad \textbf{(7-140)}$$

$$y_{lo} = \frac{a\lambda_l^b}{N_{FR}^c} \quad \textbf{(7-141)}$$

with the constraint that $y_{lo} \geq \lambda_l$ and

$$\psi = 1 + C[\sin(1.8\,\theta) - 0.333\,\sin^3(1.8\,\theta)] \quad \textbf{(7-142)}$$

where

$$C = (1 - \lambda_l)\ln(d\lambda_l^e N_{vl}^f N_{FR}^g) \quad \textbf{(7-143)}$$

where *a, b, c, d, e, f,* and *g* depend on the flow regime and are given in Table 7-2. C must be ≥ 0.

If the flow regime is transition flow, the liquid holdup is calculated using both the segregated and intermittent equations and interpolated using the following:

$$y_l = Ay_l\,(\text{segregated}) + By_l\,(\text{intermittent}) \quad \textbf{(7-144)}$$

Table 7-2 Beggs and Brill Holdup Constants

Flow Regime	***a***	***b***	***c***	
Segregated	0.98	0.4846	0.0868	
Intermittent	0.845	0.5351	0.0173	
Distributed	1.065	0.5824	0.0609	
	d	***e***	***f***	***g***
Segregated uphill	0.011	−3.768	3.539	−1.614
Intermittent uphill	2.96	0.305	−0.4473	0.0978
Distributed uphill	No correction, $C = 0, \psi = 1$			
All regimes downhill	4.70	−0.3692	0.1244	−0.5056

where

$$A = \frac{L_3 - N_{FR}}{L_3 - L_2} \tag{7-145}$$

and

$$B = 1 - A. \tag{7-146}$$

The frictional pressure gradient is calculated from

$$\left(\frac{dp}{dz}\right)_F = \frac{2f_{tp}\rho_m u_m^2}{g_c D} \tag{7-147}$$

where

$$\rho_m = \rho_l \lambda_l + \rho_g \lambda_g \tag{7-148}$$

and

$$f_{tp} = f_n \frac{f_{tp}}{f_n}. \tag{7-149}$$

The no-slip friction factor, f_n, is based on the actual pipe relative roughness and the Reynolds number,

$$N_{Re_m} = \frac{\rho_m u_m D}{\mu_m} = 1488 \frac{\rho_m u_m D}{\mu_m} \tag{7-150}$$

for ρ_m in lb_m/ft^3, u_m in ft/s, D in in., and μ_m in cp, and where

$$\mu_m = \mu_l \lambda_l + \mu_g \lambda_g. \tag{7-151}$$

The two-phase friction factor, f_{tp}, is then

$$f_{tp} = f_n e^s \tag{7-152}$$

where

$$S = \frac{[\ln(x)]}{\{-0.0523 + 3.182\ln(x) - 0.8725[\ln(x)]^2 + 0.01853[\ln(x)]^4\}} \tag{7-153}$$

and

$$x = \frac{\lambda_l}{y_l^2}. \tag{7-154}$$

Since S is unbounded in the interval $1 < x < 1.2$, for this interval,

$$S = \ln(2.2x - 1.2). \tag{7-155}$$

The kinetic energy contribution to the pressure gradient is accounted for with a parameter E_k as follows:

$$\frac{dp}{dz} = \frac{(dp/dz)_{PE} + (dp/dz)_F}{1 - E_k} \tag{7-156}$$

where

$$E_k = \frac{u_m u_{sg} \rho_m}{g_c p} \tag{7-157}$$

In comparisons with extensive measurements with natural gas and water flowing in inclined schedule 40 2-inch I.D. pipe, Payne et al. [1979] found that the Beggs and Brill correlation underpredicted friction factors and overpredicted liquid holdup. To correct these errors, Payne et al. suggest that the friction factor be calculated incorporating pipe roughness (the original correlation assumed smooth pipe), and found the following holdup corrections improved the correlation. Denoting the liquid holdup calculated by the original correlation as y_{lo}, the corrected liquid holdup is

$$y_l = 0.924 y_{lo} \text{ for } \theta > 0 \tag{7-158}$$

or

$$y_l = 0.685 y_{lo} \text{ for } \theta < 0. \tag{7-159}$$

Example 7-9 Pressure Gradient Calculation Using the Beggs and Brill Method

Repeat Example 7-8, using the Beggs and Brill method.

Solution

First, we determine the flow regime that would exist if the flow were horizontal. Using Equations (7-130) through (7-135) and the values of u_m (13.39 ft/s) and λ_l (0.35) calculated in Example 7-8, we find $N_{FR} = 29.6$, $L_1 = 230$, $L_2 = 0.0124$, $L_3 = 0.462$, and $L_4 = 606$. Checking the flow regime limits [Equations (7-136) through (7-139)], we see that

$$0.01 \leq \lambda_l < 0.4 \text{ and } L_3 < N_{FR} \leq L_1. \quad \textbf{(7-160)}$$

Then, using Equations (7-142) and (7-143),

$$C = (1 - 0.35)\ln[2.96(0.35)^{0.305}(10.28)^{-0.4473}(29.6)^{0.0978}] = 0.0351 \quad \textbf{(7-161)}$$

$$\psi = 1 + 0.0351\{\sin[1.8(90)] - 0.333\sin^3[1.8(90)]\} = 1.01 \quad \textbf{(7-162)}$$

so that, from Equation (7-140),

$$y_l = (0.454)(1.01) = 0.459. \quad \textbf{(7-163)}$$

Applying the Payne correction for upward flow,

$$y_l = 0.924(0.459) = 0.424. \quad \textbf{(7-164)}$$

The in-situ average density is

$$\overline{\rho} = y_l\rho_l + y_g\rho_g = (0.424)(49.9) + (1 - 0.424)(2.6) = 22.66 \text{ lb}_m/\text{ft}^3 \quad \textbf{(7-165)}$$

and the potential energy pressure gradient is

$$\left(\frac{dp}{dz}\right)_{PE} = \frac{g}{g_c}\overline{\rho}\sin\theta = \frac{(22.66)\sin(90)}{144} = 0.157 \text{ psi/ft}. \quad \textbf{(7-166)}$$

To calculate the frictional pressure gradient, we first compute the input fraction weighted density and viscosity from Equations (7-148) and (7-151):

$$\rho_m = (0.35)(49.9) + (0.65)(2.6) = 19.1 \text{ lb}_m/\text{ft}^3 \quad \textbf{(7-167)}$$

$$\mu_m = (0.35)(2) + (0.65)(0.0131) = 0.709 \text{ cp} \quad \textbf{(7-168)}$$

The Reynolds number from Equation (7-150) is

$$N_{Re_m} = (1488)\frac{(19.1)(13.39)(2.259/12)}{0.709} = 101{,}000. \quad \textbf{(7-169)}$$

For the relative roughness of 0.0006, from the Moody diagram or the Chen equation, the no-slip friction factor, f_n, is 0.005. Then, using Equations (7-152) through (7-154),

$$x = \frac{0.35}{(0.424)^2} = 1.95 \tag{7-170}$$

$$S = \frac{\ln(1.95)}{\{-0.0523 + 3.182\ln(1.95) - 0.8725[\ln(1.95)]^2 + 0.01853[\ln(1.95)]^4\}}$$
$$= 0.395 \tag{7-171}$$

$$f_{tp} = 0.005e^{0.395} = 0.0068 \tag{7-172}$$

From Equation (7-147), the frictional pressure gradient is

$$\left(\frac{dp}{dz}\right)_F = \frac{(2)(0.0068)(19.1)(13.39)^2}{(32.17)(2.259/12)} = 7.7 \text{ lb}_f/\text{ft}^3 = 0.054 \text{ psi/ft} \tag{7-173}$$

and the overall pressure gradient is

$$\left(\frac{dp}{dz}\right) = \left(\frac{dp}{dz}\right)_{PE} + \left(\frac{dp}{dz}\right)_F = 0.157 + 0.054 = 0.211 \text{ psi/ft.} \tag{7-174}$$

7.4.3.5 The Gray Correlation

The Gray correlation was developed specifically for wet gas wells and is commonly used for gas wells producing free water and/or condensate with the gas. This correlation empirically calculates liquid holdup to compute the potential energy gradient and empirically calculates an effective pipe roughness to determine the frictional pressure gradient.

First, three dimensionless numbers are calculated:

$$N_1 = \frac{\rho_m^2 u_m^4}{g\sigma(\rho_l - \rho_g)} \tag{7-175}$$

$$N_2 = \frac{gD^2(\rho_l - \rho_g)}{\sigma} \tag{7-176}$$

$$N_3 = 0.0814\left[1 - 0.0554\ln\left(1 + \frac{730R_v}{R_v + 1}\right)\right] \tag{7-177}$$

where

$$R_v = \frac{u_{sl}}{u_{sg}}. \quad (7\text{-}178)$$

The liquid holdup correlation is

$$y_l = 1 - (1 - \lambda_l)(1 - \exp(f_l)) \quad (7\text{-}179)$$

where

$$f_l = -2.314\left[N_1\left(1 + \frac{205}{N_2}\right)\right]^{N_3}. \quad (7\text{-}180)$$

The potential energy pressure gradient is then calculated using the in-situ average density.

To calculate the frictional pressure gradient, the Gray correlation uses an effective pipe roughness to account for liquid along the pipe walls. The effective roughness correlation is

$$k_e = k_o \text{ for } R_v > 0.007 \quad (7\text{-}181)$$

or

$$k_e = k + R_v\left\{\frac{k_o - k}{0.007}\right\} \text{ for } R_v < 0.007 \quad (7\text{-}182)$$

where

$$k_o = \frac{12.92738\sigma}{\rho_m u_m^2} \quad (7\text{-}183)$$

and k is the absolute roughness of the pipe. The constant in Equation 7-183 is for all variables in consistent units. For oilfield units of dynes/cm for σ, lb_m/ft^3 for ρ, and ft/sec for u_m,

$$k_o = \frac{0.285\sigma}{\rho_m u_m^2}. \quad (7\text{-}184)$$

The effective roughness is used to calculate the relative roughness by dividing by the pipe diameter. The friction factor is obtained using this relative roughness and a Reynolds number of 10^7.

Example 7-10 Pressure Gradient Calculation Using the Gray Method

An Appendix C gas well is producing 2 MMscf/day of gas with 50 bbl of water produced per MMscf of gas. The surface tubing pressure is 200 psia and the temperature is 100°F. The gas–water surface tension is 60 dynes/cm, and the pipe relative roughness is 0.0006. Calculate the pressure gradient at the top of the tubing, neglecting any kinetic energy contribution to the pressure gradient, At this location, the water density is 65 lb_m/ft^3 and the viscosity is 0.6 cp.

Solution

First, we need to determine the Z factor for this gas. $p_{pr} = 0.298$ $(p/p_{pc} = 200/671)$ and $T_{pr} = 1.49$ $(T/T_{pc} = 560/375)$. From Figure 4-1, $Z = 0.97$. The flowing area $A = 0.0278$ ft^2 if 2 7/8-in. tubing is used.

The superficial velocities are calculated as

$$u_{sl} = \frac{q_l}{A} = \left[\frac{(100 \text{ bbl/d})(5.615 \text{ ft}^3\text{/bbl})(1 \text{ d/86,400 s})}{0.0278 \text{ ft}^2}\right] = 0.2335 \text{ ft/s.} \tag{7-185}$$

The gas superficial velocity is calculated from the volumetric flow rate at standard conditions with Equation (7-45),

$$u_{sg} = \frac{4}{\pi(2.259/12)^2}(2 \times 10^6 \text{ ft}^3\text{/d})(0.97)\left(\frac{460 + 100}{460 + 60}\right)\left(\frac{14.7}{200}\right)\frac{\text{d}}{86{,}400 \text{ s}} = 63.9 \text{ ft/s.} \tag{7-186}$$

The mixture velocity is

$$u_m = u_{sl} + u_{sg} = (0.2335 + 63.9) = 64.09 \text{ ft/s.} \tag{7-187}$$

And the input fraction of liquid is

$$\lambda_l = \frac{u_{sl}}{u_m} = \frac{0.2335}{63.9} = 0.0036. \tag{7-188}$$

The gas density is calculated from Equation (7-44),

$$\rho_g = \frac{(28.97)(0.65)(200 \text{ psi})}{(0.97)(10.73 \text{ psi-ft}^3\text{/lb-mol-°R})(560\text{°R})} = 0.65 \text{ lb}_m\text{/ft}^3. \tag{7-189}$$

The input fraction weighted density is calculated from Equation (7-148),

$$\rho_m = (0.0036)(65) + (1 - 0.0036)(0.65) = 0.88 \text{ lb}_m\text{/ft}^3. \tag{7-190}$$

From Equations (7-175) and (7-176), we calculate N_1 and N_2,

$$N_1 = \frac{\rho_m^2 u_m^4}{g\sigma(\rho_l - \rho_g)}$$

$$= \left[\frac{(0.88\ \text{lb}_\text{m}/\text{ft}^3)^2(64.09\ \text{ft/s})^4}{(32.17\ \text{ft-lb}_\text{m}/\text{lb}_\text{f}\text{-s}^2)(60\ \text{dynes/cm})\left(6.85 \times 10^{-5}\dfrac{\text{lb}_\text{f}/\text{ft}}{\text{dynes/cm}}\right)(65\ \text{lb}_\text{m}/\text{ft}^3 - 0.65\ \text{lb}_\text{m}/\text{ft}^3)}\right]$$

$$= 1.537 \times 10^6 \tag{7-191}$$

$$N_2 = \frac{gD^2(\rho_l - \rho_g)}{\sigma} = \left[\frac{(32.17\text{ft-lb}_\text{m}/\text{lb}_\text{f}\text{-s}^2)(2.259/12\ \text{ft})^2(65\ \text{lb}_\text{m}/\text{ft}^3 - 0.65\ \text{lb}_\text{m}/\text{ft}^3)}{(60\ \text{dynes/cm})\left(6.85 \times 10^{-5}\dfrac{\text{lb}_\text{f}/\text{ft}}{\text{dynes/cm}}\right)}\right]$$

$$= 1.785 \times 10^4 \tag{7-192}$$

We calculate R_v and N_3 with Equations (7-178) and (7-177),

$$R_v = \frac{u_{sl}}{u_{sg}} = \frac{0.2335}{63.9} = 0.0037 \tag{7-193}$$

$$N_3 = 0.0814\left[1 - 0.0554\ \ln\left(1 + \frac{730 \times 0.0037}{0.0037 + 1}\right)\right] = 0.0755 \tag{7-194}$$

With Equations (7-180) and (7-179), f_1 and y_1 are

$$f_l = -2.314\left[N_1\left(1 + \frac{205}{N_2}\right)\right]^{N_3} = -6.7942 \tag{7-195}$$

$$y_l = 1 - (1 - \lambda_l)(1 - \exp(f_l)) = 0.0048 \tag{7-196}$$

The in-situ average density is

$$\overline{\rho} = y_l\rho_l + (1 - y_l)\rho_g = (0.0048)(65) + (1 - 0.0048)(0.65) = 0.9524\ \text{lb}_m/\text{ft}^3. \tag{7-197}$$

The potential energy pressure gradient is

$$\left(\frac{dp}{dz}\right)_{\text{PE}} = \frac{g}{g_c}\overline{\rho}\sin\theta = \frac{(0.9524)\sin(90°)}{144} = 0.0066\ \text{psi/ft}. \tag{7-198}$$

The absolute roughness of the pipe is $k = \varepsilon D = (0.0006)(2.259)/12 \text{ ft} = 0.000113 \text{ ft}$.

Using Equation (7-184),

$$k_o = \frac{0.285(60 \text{ dynes/cm})}{(0.88 \text{ lb}_m/\text{ft}^3)(64.09 \text{ ft/s})^2} = 0.0047 \text{ ft.} \tag{7-199}$$

For $R_v = 0.0037 < 0.007$,

$$k_e = k + R_v\left\{\frac{k_o - k}{0.007}\right\} = 0.0025 \text{ ft.} \tag{7-200}$$

The effective relative roughness is $k_e/D = 0.0134$. With Equation (7-37), the Fanning friction factor is 0.0105.

From Equation (7-147), the frictional pressure gradient is

$$\left(\frac{dp}{dz}\right)_F = \frac{(2)(0.0105)(0.88)(64.09)^2}{(32.17)(2.259/12)(144)} = 0.087 \text{ psi/ft} \tag{7-201}$$

and the overall pressure gradient is

$$\left(\frac{dp}{dz}\right) = \left(\frac{dp}{dz}\right)_{PE} + \left(\frac{dp}{dz}\right)_F = 0.0066 + 0.087 = 0.0936 \text{ psi/ft.} \tag{7-202}$$

7.4.4 Pressure Traverse Calculations

We have examined several methods for calculating the pressure gradient, *dp/dz,* which can be applied at any location in a well. However, our objective is often to calculate the overall pressure drop, Δp, over a considerable distance, and over this distance the pressure gradient in gas–liquid flow may vary significantly as the downhole flow properties change with temperature and pressure. For example, in a well such as that pictured in Figure 7-15, in the lower part of the tubing the pressure is above the bubble point and the flow is single-phase oil. At some point, the pressure drops below the bubble point and gas comes out of solution, causing gas–liquid bubble flow; as the pressure continues to drop, other flow regimes may occur farther up the tubing.

Thus, we must divide the total distance into increments small enough that the flow properties, and hence the pressure gradient, are almost constant in each increment. Summing the pressure drop in each increment, we obtain the overall pressure drop. This stepwise calculation procedure is generally referred to as a pressure traverse calculation.

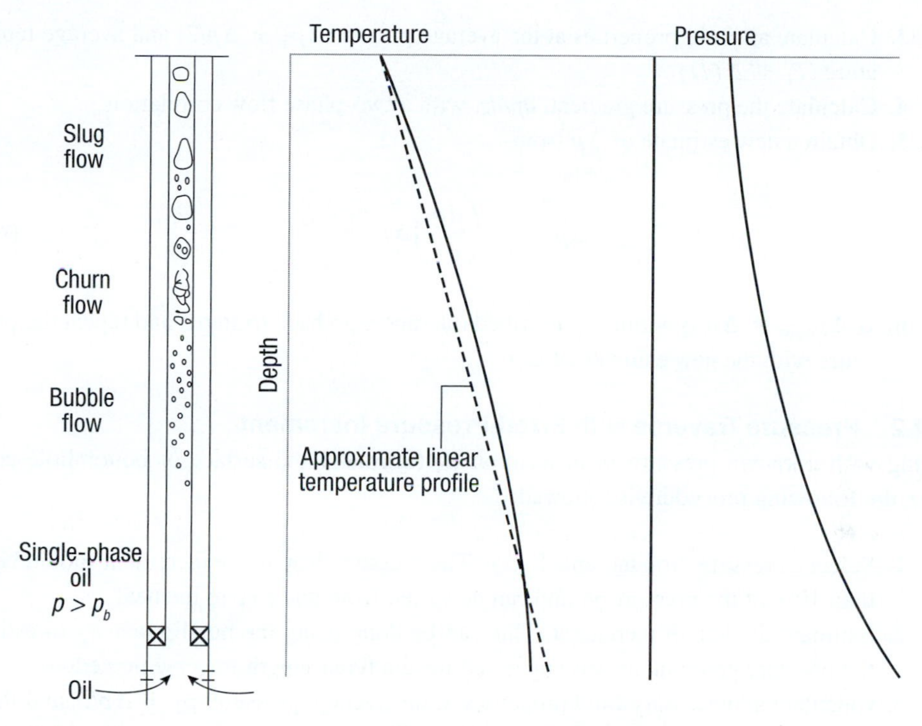

Figure 7-15 Pressure, temperature, and flow regime distribution in a well.

Since both the temperature and pressure will be varying, a pressure traverse calculation is usually iterative. The temperature profile is usually approximated as being linear between the surface temperature and the bottomhole temperature, as shown in Figure 7-15. Pressure traverse calculations can be performed either by fixing the length increment and calculating the pressure drop, or by fixing the pressure drop and finding the depth interval over which this pressure drop would occur (Brill and Beggs, 1978). Fixing the length interval is often more convenient when programming a pressure traverse calculation for computer solution; fixing the pressure drop increment is more convenient for hand calculations.

7.4.4.1 Pressure Traverse with Fixed Length Interval

Starting with a known pressure p_1 at position L_1 (normally the surface or bottomhole conditions), the following procedure is followed:

1. Select a length increment, ΔL. A typical value for flow in tubing is 200 ft.
2. Estimate the pressure drop, Δp. A starting point is to calculate the no-slip average density and from this, the potential energy pressure gradient. The estimated Δp is then the potential energy pressure gradient times the depth increment. This will generally underestimate the pressure drop.

3. Calculate all fluid properties at the average pressure ($p_1 + \Delta p/2$) and average temperature ($T_1 + \Delta T/2$).
4. Calculate the pressure gradient, dp/dz, with a two-phase flow correlation.
5. Obtain a new estimate of Δp from

$$\Delta p_{\text{new}} = \left(\frac{dp}{dz}\right)\Delta L. \tag{7-203}$$

6. If $\Delta p_{\text{new}} \neq \Delta p_{\text{old}}$ within a prescribed tolerance, go back to step 3 and repeat the procedure with the new estimate of Δp.

7.4.4.2 Pressure Traverse with Fixed-Pressure Increment

Starting with a known pressure p_1 at position L_1 (normally the surface or bottomhole conditions), the following procedure is followed:

1. Select a pressure drop increment, Δp. The pressure drop in the increment should be less than 10% of the pressure p_1 and can be varied from one step to the next.
2. Estimate the length increment. This can be done using the no-slip density to estimate the pressure gradient, as was suggested for the fixed-length-traverse procedure.
3. Calculate all necessary fluid properties at the average pressure, $p_1 + \Delta p/2$, and the estimated average temperature, $T_1 + \Delta T/2$.
4. Calculate the estimated pressure gradient, dp/dz, using a two-phase flow correlation.
5. Estimate the length increment by

$$\Delta L_{\text{new}} = \frac{\Delta p}{(dp/dz)}. \tag{7-204}$$

6. If $\Delta L_{\text{new}} \neq \Delta L_{\text{old}}$ within a prescribed tolerance, go back to step 3 and repeat the procedure. In this procedure, since temperature is changing more slowly in a well and the average pressure of the increment is fixed, convergence should be rapid. If the well can be assumed to be isothermal, no iteration is required.

Since the pressure traverse calculations are iterative, and the fluid properties and pressure gradient calculations are tedious, it is most convenient to write computer programs for pressure traverse calculations.

Example 7-11 Pressure Traverse Calculation for a Vertical Well

Using the modified Hagedorn and Brown method, generate plots of pressure versus depth, from the surface to 10,000 ft, for gas–oil ratios ranging from 0 to 4000 for the following flow conditions in a vertical well: q_o = 400 bbl/d, WOR = 1, p_{sep} = 100 psig, average T = 140°F, γ_g = 0.65, γ_o = 35°API, γ_w = 1.074, tubing size = 2.5 in. I.D.

Solution

A series of pressure traverse calculations must be performed, one for each GOR. This is best done with a simple computer program. Using the modified Hagedorn and Brown method and calculating fluid properties with the correlations presented in Section 3.2, the results shown in Figure 7-16 were obtained. Plots such as this one are often called gradient curves; those presented by Brown (1977) were generated with the modified Hagedorn and Brown method, as illustrated here.

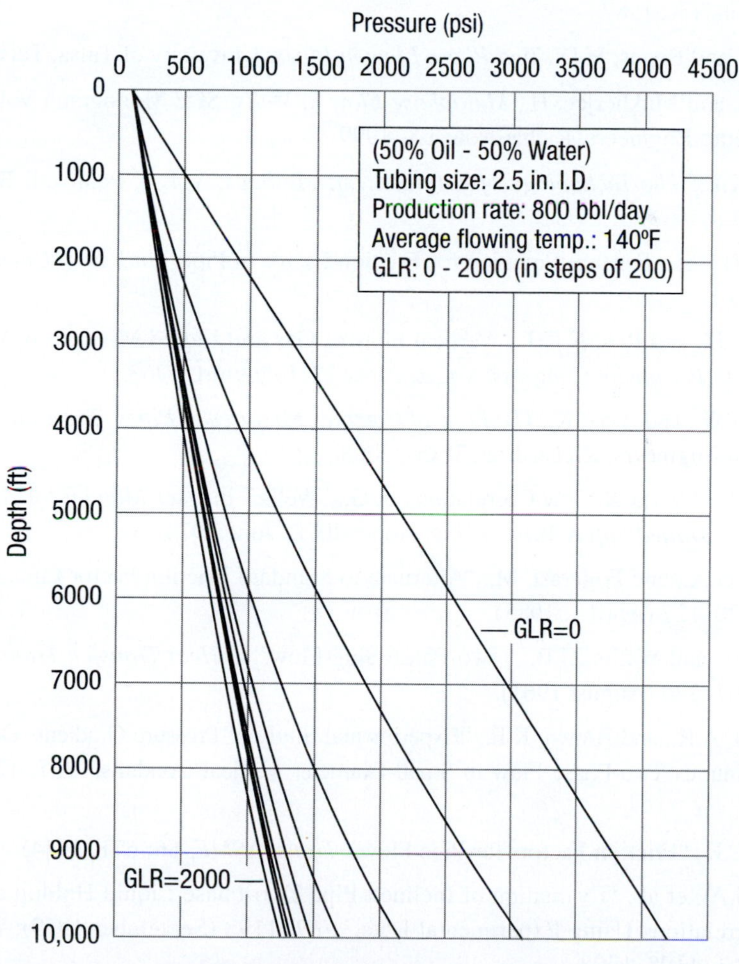

Figure 7-16 Pressure gradient curves generated with the modified Hagedorn and Brown correlation.

References

1. Ansari, A.M., et al., "A Comprehensive Mechanistic Model for Two-Phase Flow in Wellbores," SPEPF Paper, May 1994, p. 143; *Trans. AIME,* p. 285.
2. Ansari, A.M., et al., "Supplement to Paper SPE 20630, A Comprehensive Mechanistic Model for Two-Phase Flow in Wellbores," SPE 28671, May 1994.
3. Beggs, H.D., and Brill, J.P., "A Study of Two-Phase Flow in Inclined Pipes," *JPT,* 607–617 (May 1973).
4. Bradley, H.B., ed., *Petroleum Engineering Handbook,* Society of Petroleum Engineers, Richardson, TX, 1987.
5. Brill, J.P., and Beggs, H.D., *Two-Phase Flow in Pipes,* University of Tulsa, Tulsa, OK, 1978.
6. Brill, J.P., and Mukherjee, H., *Multiphase Flow in Wells,* SPE Monograph Vol. 17, Society of Petroleum Engineers, Richardson, TX, 1999.
7. Brown, K.E., *The Technology of Artificial Lift Methods,* Vol. 1, Pennwell Books, Tulsa, OK, 1977.
8. Chen, N.H., "An Explicit Equation for Friction Factor in Pipe," *Ind. Eng. Chem. Fund.,* **18:** 296 (1979).
9. Duns, H., Jr., and Ros, N.C.J., "Vertical Flow of Gas and Liquid Mixtures in Wells," *Proc., Sixth World Petroleum Congress,* Vol. 2, Paper 22, Frankfurt, 1963.
10. Govier, G.W., and Aziz, K., *The Flow of Complex Mixtures in Pipes,* 2nd edition, Society of Petroleum Engineers, Richardson, Texas, 2008.
11. Gray, H. E., "Vertical Flow Correlations in Gas Wells," in *User Manual for API 14B, Subsurface Controlled Safety Valve Sizing,* Appendix B, June 1974.
12. Gregory, G.A., and Fogarasi, M., "Alternate to Standard Friction Factor Equation," *Oil and Gas J.,* 120–127 (April 1, 1985).
13. Griffith, P., and Wallis, G.B., "Two-Phase Slug Flow," *J. Heat Transfer, Trans. ASME,* Ser. C., **83:** 307–320 (August 1961).
14. Hagedorn, A.R., and Brown, K.E., "Experimental Study of Pressure Gradients Occurring During Continuous Two-Phase Flow in Small-Diameter Vertical Conduits," *JPT,* 475–484 (April 1965).
15. Moody, L.F., "Friction Factors for Pipe Flow," *Trans. ASME,* **66:** 671 (1944).
16. Payne, G.A., et al., "Evaluation of Inclined-Pipe Two-Phase Liquid Holdup and Pressure-Loss Correlations Using Experimental Data," *JPT,* 1198 (September 1979); *Trans. AIME,* **267:** Part 1, 1198–1208.
17. Taitel, Y., Barnea, D., and Dukler, A.E., "Modelling Flow Pattern Transitions for Steady Upward Gas–Liquid Flow in Vertical Tubes," *AIChE J.,* **26**(6): 345–354 (May 1980).

Problems

7-1 Calculate the bottomhole pressure for the injection of 1500 bbl/d of brine with a density of 1.08 g/cm^3 and a viscosity of 1.3 cp in an 8000-ft vertical well having a surface pressure of 200 psi and (a) 2-in., 3.4-lb$_m$/ft tubing; (b) 2 7/8-in., 8.6-lb$_m$/ft tubing; (c) 3 1/2-inch, 12.70-lb$_m$/ft tubing.

7-2 The fluid in Appendix A flows in a well of 10,000-ft deep with a production tubing of new, 2 7/8 in. and 8.6 lb$_m$/ft. The well is 25° deviated, and the production rate is 2000 bbl/day. Calculate:

a. the potential energy pressure change of oil being pumped to the surface.
b. the frictional pressure change if the pipe relative roughness is 0.0006.
c. can the kinetic energy change be neglected and why?
d. generate a plot of frictional pressure drop as a function of flow rate for this well, with the flow rate varying from 500 bbl/day to 10,000 bbl/day.

7-3 In order to supply a pipeline, a particular gas well must maintain a wellhead pressure of 600 psia. The vertical well is 15,000 ft deep, and the production is through new, 2 7/8 in., 8.6 lb$_m$/ft tubing. Generate a plot of gas flow rate as a function of p_{wf}. The gas gravity is 0.65. Assume that the surface temperature is 80°F and the temperature gradient is 0.02° F/ft.

7-4 Construct a curve of bottomhole flowing pressure versus flow rate for the gas well in Appendix C for 500 MSCF/d $< q <$ 10,000 MSCF/d with the following data: $\mu = 0.022$ cp, $Z = 0.93$, $H = 10{,}000$ ft, $p_{tf} = 500$ psi, $D_i = 2$ 7/8 in., $\varepsilon = 0.0006$. Generate the curve with a tubing I.D. equal to 2 in., then 3 in.

7-5 Calculate the slip velocity for the flow of 5-ft^3/sec of oil and 10-ft^3/sec of gas in a 6-in. I.D. pipe when the liquid holdup is 0.8.

7-6 For a flow of 1000 bbl/d of oil with a density of 0.8 g/cm^3 and a surface tension of 30 dynes/cm in a 3-in. I.D. vertical pipe, at what gas rate will the transition from Region I to Region II (Duns and Ros flow regime map) occur?

7-7 Using the modified Hagedorn and Brown method, calculate the pressure gradient for the flow of 1200 STB/d of a vertical oil in Appendix B with GOR of 1000 and a WOR of 1.0 in 2 7/8-in. vertical tubing at (a) the surface, where $T = 100$°F and $p_{tf} = 100$ psig; (b) the bottomhole, where $T = 180$°F and $p_{wf} = 3000$ psia.

7-8 Repeat problem 7-7, but use the Beggs and Brill correlation to calculate pressure gradient at the bottomhole for the same well with 35° deviation, where $T = 180$°F and $p = 3000$ psia.

7-9 Write a computer program to implement a two-phase pressure traverse calculation with fixed-length increment incorporating both correlations presented in this chapter as user options. Assume that the surface pressure is known and the bottomhole pressure is the desired result.

7-10 Given the well in Appendix B with surface temperature ranging from 30°F to 100°F, gas gravity = 0.71, and oil gravity = 32, suppose the separator and other surface equipment for the well is located 2000 ft from the wellhead and up an incline that rises 200 above the wellhead. The flow rate from this well should be no more than 500 BBL/D but the GOR may be between 500 and 1000 SCF/STB with a wellhead pressure of 500 psi, and pressure drop should be no more than 50 psi. What pipe diameter should run from the wellhead to the separator?

CHAPTER 8

Flow in Horizontal Wellbores, Wellheads, and Gathering Systems

8.1 Introduction

This chapter is concerned with the transport of fluids in horizontal or near-horizontal pipes, including flow in horizontal wellbores and the flow from the wellhead to the facility where processing of the fluids begins. For oil production, this facility is typically a two- or three-phase separator; for gas production, the facility may be a gas plant, a compressor station, or simply a transport pipeline; and for injection wells, the surface transport of interest is from a water treating/pumping facility to the wells. We are not concerned here with pipeline transport over large distances; thus we do not consider the effect of hilly terrain or changes in fluid temperature in our calculations.

As in vertical wellbore flow, we are interested primarily in the pressure as a function of position as fluids move through flow lines. In addition to flow through pipes, flows through fittings and chokes are important considerations for surface transport.

8.2 Flow in Horizontal Pipes

8.2.1 Single-Phase Flow: Liquid

Single-phase flow in horizontal pipes is described by the same equations as those for single-phase flow in wellbores presented in Chapter 7, but with the simplification that the potential energy pressure drop is zero. If the fluid is incompressible and the pipe diameter is constant, the kinetic energy pressure drop is also zero, and the mechanical energy balance [Equation (7-15)] simplifies to

$$\Delta p = p_1 - p_2 = \frac{2 f_f \rho u^2 L}{g_c D}. \tag{8-1}$$

The friction factor is obtained as in Chapter 7 by the Chen equation [Equation (7-35)] or the Moody diagram (Figure 7-7).

Example 8-1 Pressure Drop in a Water Injection Supply Line

The 1000 bbl/d of injection water described in Examples 7-2 and 7-4 is supplied to the wellhead through a 3000-ft long, 1 1/2-in. I.D. flow line from a central pumping station. The relative roughness of the galvanized iron pipe is 0.004. If the pressure at the wellhead is 100 psia, what is the pressure at the pumping station, neglecting any pressure drops through valve or other fittings? The water has a specific gravity of 1.05 and a viscosity of 1.2 cp.

Solution

Equation (8-1) applies, with the Reynolds number and the friction factor calculated as in Chapter 7. Using Equation (7-7), the Reynolds number is calculated to be 53,900; from the Chen equation [Equation (7-35)], the friction factor is 0.0076. Dividing the volumetric flow rate by the pipe cross-sectional area, we find the mean velocity to be 5.3 ft/sec. Then, from Equation (8-1),

$$\Delta p = p_1 - p_2 = \frac{(2)(0.0076)(65.5)(5.3)^2(3000)}{(32.17)[(1.5)/(12)]}$$
$$= 20{,}800 \text{ lb}_f/\text{ft}^2 = 145 \text{ psi} \tag{8-2}$$

so

$$p_1 = p_2 + 145 = 100 + 145 = 245 \text{ psi.} \tag{8-3}$$

This is a significant pressure loss over 3000 ft. It can be reduced substantially by using a larger pipe for this water supply, since the frictional pressure drop depends approximately on the pipe diameter to the fifth power.

8.2.2 Single-Phase Flow: Gas

The pressure drop for the horizontal flow of a compressible fluid (gas), neglecting the kinetic energy pressure drop, was given by Equations (7-50) and (7-54). These equations use average values for Z, T, and μ for the entire length of pipe being considered. In a high-rate, low-pressure line, the change in kinetic energy may be significant. In this case, for a horizontal line, the mechanical energy balance is

$$\frac{dp}{\rho} + \frac{u\,du}{g_c} + \frac{2f_f u^2 dL}{g_c D} = 0. \tag{8-4}$$

For a real gas, ρ and u are given by Equations (7-44) and (7-45), respectively. The differential form of the kinetic energy term is

$$u\,du = -\left(\frac{4qZT}{\pi D^2}\frac{p_{sc}}{T_{sc}}\right)^2 \frac{dp}{p^3}. \tag{8-5}$$

Substituting for ρ and $u\,du$ in Equation (8-4), assuming average values of Z and T over the length of the pipeline, and integrating, we obtain

$$p_1^2 - p_2^2 = \frac{32}{\pi^2}\frac{28.97\gamma_g \overline{Z}\,\overline{T}}{Rg_c D^4}\left(\frac{p_{sc}q}{T_{sc}}\right)^2\left(\frac{2f_f L}{D} + \ln\frac{p_1}{p_2}\right), \tag{8-6}$$

which for field units is

$$p_1^2 - p_2^2 = (4.195 \times 10^{-6})\frac{\gamma_g \overline{Z}\,\overline{T} q^2}{D^4}\left(\frac{24 f_f L}{D} + \ln\frac{p_1}{p_2}\right) \tag{8-7}$$

where p_1 and p_2 are in psi, T is in °R, q in Mscf/d, D in in., and L in ft. The friction factor is obtained from the Reynolds number and pipe roughness, with the Reynolds number for field units given by Equation (7-55).

Equation (8-7) is identical to Equation (7-50) except for the additional $\ln(p_1/p_2)$ term, which accounts for the kinetic energy pressure drop. Equation (8-7) is an implicit equation in p and must be solved iteratively. The equation can be solved first by neglecting the kinetic energy term; then, if $\ln(p_1/p_2)$ is small compared with $24 f_f\, L/D$, the kinetic energy pressure drop is negligible.

Example 8-2 Flow Capacity of a Low-Pressure Gas Line

Gas production from a low-pressure gas well (wellhead pressure = 100 psia) is to be transported through 1000 ft of a 3-in. I.D. line ($\varepsilon = 0.001$) to a compressor station, where the inlet pressure must be at least 20 psia. The gas has a specific gravity of 0.7, a temperature of 100°F, and an average viscosity of 0.012 cp. What is the maximum flow rate possible through this gas line?

Solution

We can apply Equation (8-7), solving for q. We need the Reynolds number to find the friction factor. However, we can begin by assuming that the flow rate, and hence the Reynolds number, is high enough that the flow is fully rough wall turbulent so that the friction $f_f = 0.0049$ for high Reynolds number and a relative roughness of 0.001. Then

$$q = \sqrt{\frac{(p_1^2 - p_2^2)D^4}{(4.195 \times 10^{-6})\gamma_g \overline{Z}\,\overline{T}[(24 f_f L/D) + \ln(p_1/p_2)]}}. \tag{8-8}$$

Assuming $Z = 1$ at these low pressures,

$$q = \sqrt{\frac{(100^2 - 20^2)(3)^4}{(4.195 \times 10^{-6})(0.7)(1)(560)\{[(24)(0.0049)(1000)/3] + \ln(100/20)\}}} \tag{8-9}$$

then

$$q = \sqrt{\frac{4.73 \times 10^8}{39.2 + 1.61}} = 3{,}404 \text{ Mscf/d}. \tag{8-10}$$

Checking the Reynolds number using Equation (7-55),

$$N_{\text{Re}} = \frac{(20.09)(0.7)(3{,}404)}{(3)(0.012)} = 1.4 \times 10^6, \tag{8-11}$$

so the friction factor based on fully rough wall turbulence is correct.

We find that we can transport over 3 MMscf/d through this line. Notice that even at this high flow rate and with a velocity five times higher at the pipe outlet than at the entrance, the kinetic energy contribution to the overall pressure drop is still small relative to the frictional pressure drop.

8.2.3 Two-Phase Flow

Two-phase flow in horizontal pipes differs markedly from that in vertical pipes. Except for the Beggs and Brill (1973) correlation, which can be applied for any flow direction, completely different correlations are used for horizontal flow than for vertical flow. In this section, we first consider flow regimes in horizontal gas-liquid flow, then a few commonly used pressure drop correlations.

8.2.3.1 Flow Regimes

The flow regime does not affect the pressure drop as significantly in horizontal flow as it does in vertical flow because there is no potential energy contribution to the pressure drop in horizontal flow. However, the flow regime is considered in some pressure drop correlations and can affect production operations in other ways. Most important, the occurrence of slug flow necessitates designing separators or sometimes special pieces of equipment (slug catchers) to handle the large volume of liquid contained in a slug. Particularly in offshore operations, where gas and liquid from subsea completions are transported significant distances to a platform, the possibility of slug flow, and its consequences, must be considered.

Figure 8-1 (Brill and Beggs, 1978) depicts the commonly described flow regimes in horizontal gas–liquid flow. These can be classified as three types of regimes: *segregated* flows, in which the two phases are for the most part separate; *intermittent* flows, in which gas and liquid are alternating; and *distributive* flows, in which one phase is dispersed in the other phase.

Segregated flow is further classified as being stratified smooth, stratified wavy (ripple flow), or annular. Stratified smooth flow consists of liquid flowing along the bottom of the pipe and gas flowing along the top of the pipe, with a smooth interface between the phases. This flow regime occurs at relatively low rates of both phases. At higher gas rates, the interface becomes wavy, and stratified wavy flow results. Annular flow occurs at high gas rates and relatively high liquid rates and consists of an annulus of liquid coating the wall of the pipe and a central core of gas flow, with liquid droplets entrained in the gas.

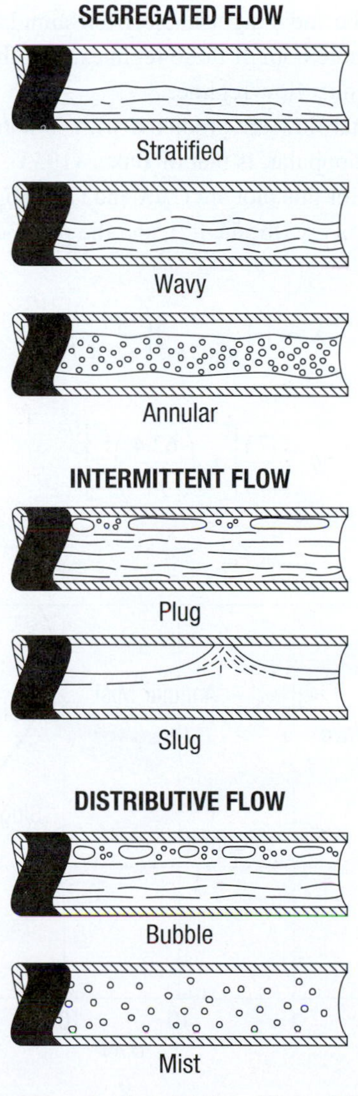

Figure 8-1 Flow regimes in two-phase horizontal flow. (From Brill and Beggs, 1978.)

The intermittent flow regimes are slug flow and plug (also called elongated bubble) flow. Slug flow consists of large liquid slugs alternating with high-velocity bubbles of gas that fill almost the entire pipe. In plug flow, large gas bubbles flow along the top of the pipe, which is otherwise filled with liquid.

Distributive flow regimes described in the literature include bubble, dispersed bubble, mist, and froth flow. The bubble flow regimes differ from those described for vertical flow in that the gas bubbles in a horizontal flow will be concentrated on the upper side of the pipe. Mist flow occurs at high gas rates and low liquid rates and consists of gas with liquid droplets entrained. Mist flow will often be indistinguishable from annular flow, and many flow regime maps use "annular mist" to denote both of these regimes. "Froth flow" is used by some authors to describe the mist or annular mist flow regime.

Flow regimes in horizontal flow are predicted with flow regime maps. One of the first of these, and still one of the most popular, is that of Baker (1953), later modified by Scott (1963), shown in Figure 8-2. The axes for this plot are G_g/λ and $G_l\lambda\phi/G_g$, where G_l and G_g are the mass fluxes of liquid and gas, respectively ($\text{lb}_m/\text{hr-ft}^2$), and the parameters λ and ϕ are

$$\lambda = \left[\left(\frac{\rho_g}{0.075}\right)\left(\frac{\rho_l}{62.4}\right)\right]^{1/2} \tag{8-12}$$

$$\phi = \frac{73}{\sigma_l}\left[\mu_l\left(\frac{62.4}{\rho_l}\right)^2\right]^{1/3} \tag{8-13}$$

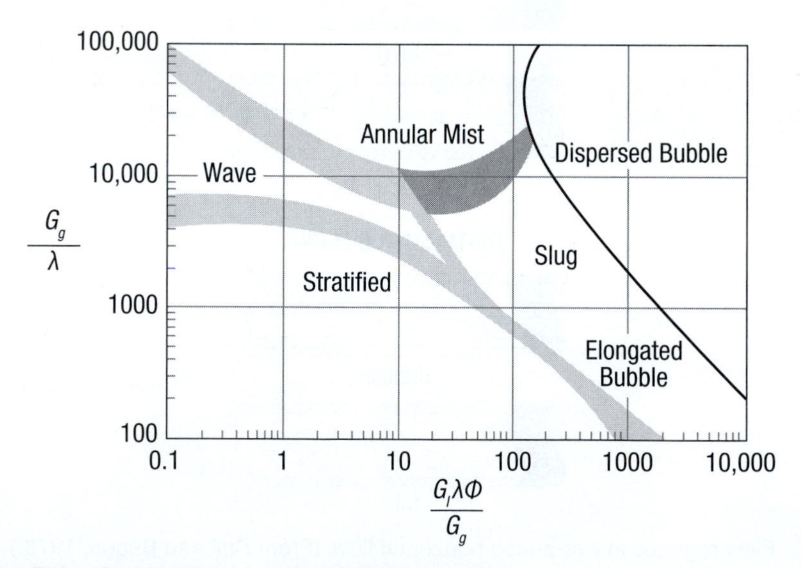

Figure 8-2 Baker flow regime map. (From Baker, 1953.)

where densities are in lb_m/ft^3, μ is in cp, and σ_l is in dynes/cm. The shaded regions on this diagram indicate that the transitions from one flow regime to another are not abrupt, but occur over these ranges of flow conditions.

Another commonly used flow regime map is that of Mandhane, Gregory, and Aziz (1974) (Figure 8-3). Like many vertical flow regime maps, this map uses the gas and liquid superficial velocities as the coordinates.

The Beggs and Brill correlation is based on a horizontal flow regime map that divides the domain into the three flow regime categories: segregated, intermittent and distributed. This map, shown in Figure 8-4, plots the mixture Froude number, defined as

$$N_{\mathrm{Fr}} = \frac{u_m^2}{gD} \tag{8-14}$$

versus the input liquid fraction, λ_l.

Finally, Taitel and Dukler (1976) developed a theoretical model of the flow regime transitions in horizontal gas-liquid flow; their model can be used to generate flow regime maps for particular fluids and pipe size. Figure 8-5 shows a comparison of their flow regime predictions with those of Mandhane et al. for air–water flow in a 2.5-cm diameter pipe.

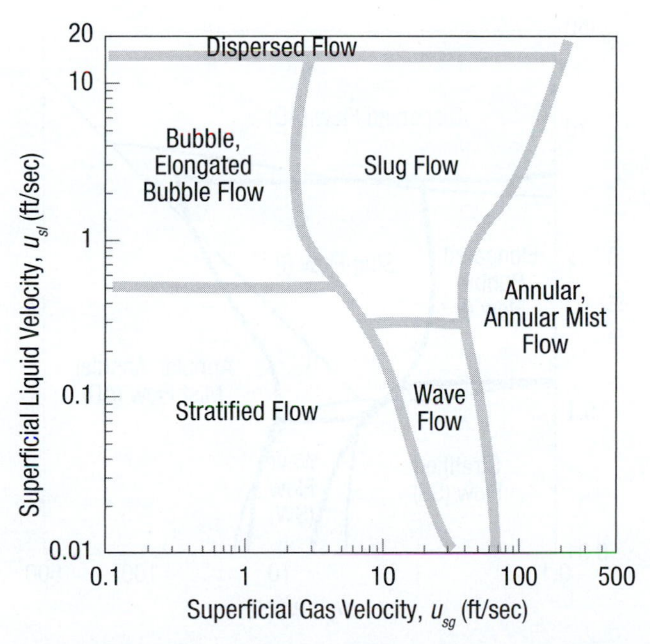

Figure 8-3 Mandhane flow regime map. (From Mandhane et al., 1974.)

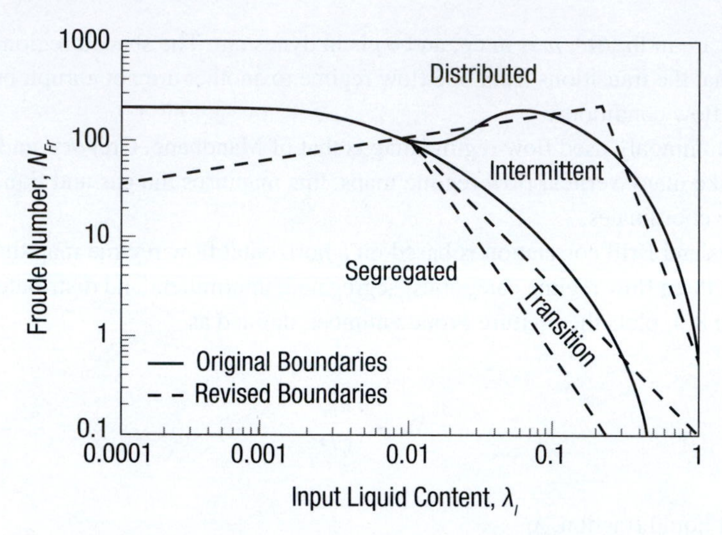

Figure 8-4 Beggs and Brill flow regime map. (From Beggs and Brill, 1973.)

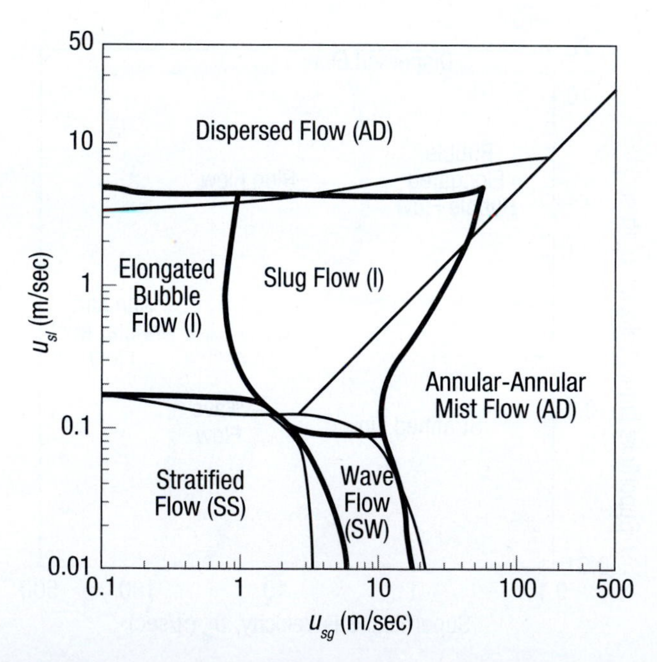

Figure 8-5 Taitel-Dukler flow regime map. (From Taitel and Dukler, 1976.)

Example 8-3 Predicting Horizontal Gas–Liquid Flow Regime

Using the Baker, Mandhane et al., and Beggs and Brill flow regime maps, determine the flow regime for the flow of 2000 bbl/d of oil and 1 MMscf/d of gas at 800 psia and 175°F in a 2 1/2-in. I.D. pipe. The fluids are the same as in Example 7-8.

Solution

From Example 7-8, we find the following properties:

$$\text{Liquid: } \rho = 49.92 \text{ lb}_m/\text{ft}^3;\ \mu_l = 2 \text{ cp};\ \sigma = 30 \text{ dynes/cm};\ q_l = 0.130 \text{ ft}^3/\text{sec}$$

$$\text{Gas: } \rho = 2.6 \text{ lb}_m/\text{ft}^3;\ \mu_g = 0.0131 \text{ cp};\ Z = 0.935;\ q_g = 0.242 \text{ ft}^3/\text{sec}$$

The 2 1/2-in. with pipe has a cross-sectional area of 0.0341 ft^2; dividing the volumetric flow rates by the cross-sectional area, we find $u_{sl} = 3.81$ ft/sec, $u_{sg} = 7.11$ ft/sec, and $u_m = 10.9$ ft/sec.

For the Baker map, we computer the mass fluxes, G_l and G_g, and the parameters, λ and ϕ. The mass flux is just the superficial velocity times density, so

$$G_l = u_{sl}\rho_l = (3.81)(49.92)(3600) = 6.84 \times 10^5 \text{ lb}_m/\text{hr} - \text{ft}^2 \tag{8-15}$$

$$G_g = u_{sg}\rho_g = (7.11)(2.6)(3600) = 6.65 \times 10^4 \text{ lb}_m/\text{hr} - \text{ft}^2 \tag{8-16}$$

Then, from Equations (8-12) and (8-13),

$$\lambda = \left[\left(\frac{2.6}{0.075}\right)\left(\frac{49.92}{62.4}\right)\right]^{1/2} = 5.27 \tag{8-17}$$

$$\phi = \frac{73}{30}\left[2\left(\frac{62.4}{49.92}\right)^2\right]^{1/3} = 3.56 \tag{8-18}$$

The coordinates for the Baker map are

$$\frac{G_g}{\lambda} = \frac{6.65 \times 10^4}{5.27} = 1.26 \times 10^4 \tag{8-19}$$

$$\frac{G_l \lambda \phi}{G_g} = \frac{(6.84 \times 10^5)(5.27)(3.56)}{6.65 \times 10^4} = 193 \tag{8-20}$$

Reading from Figure 8-2, the flow regime is predicted to be dispersed bubble, though the conditions are very near the boundaries with slug flow and annular mist flow.

The Mandhane map (Figure 8-3) is simply a plot of superficial liquid velocity versus superficial gas velocity. For the values of $u_{sl} = 3.81$ ft/sec and $u_{sg} = 7.11$ ft/sec, the flow regime is predicted to be slug flow.

The Beggs and Brill map plots the mixture Froude number, defined by Equation (7-130), against the input fraction of liquid. These parameters are

$$N_{\mathrm{Fr}} = \frac{(10.9)^2}{(32.17)[(2.5/12)]} = 17.8 \tag{8-21}$$

$$\lambda_l = \frac{3.81}{10.9} = 0.35 \tag{8-22}$$

From Figure 8-4, the flow regime is predicted to be intermittent.

The prediction of dispersed bubble flow with the Baker map disagrees with that of Mandhane et al. and Beggs and Brill, which predict slug flow. However, the conditions were very near the flow regime transition on the Baker map, and slug flow is the likely flow regime.

8.2.3.2 Pressure Gradient Correlations

Over the years, several correlations have been developed to calculate the pressure gradient in horizontal gas–liquid flow. The most commonly used in the oil and gas industry today are those of Beggs and Brill (1973), Eaton et al. (1967), and Dukler (1969). These correlations all include a kinetic energy contribution to the pressure gradient. However, this can be considered negligible unless the gas rate is high and the pressure is low. Yuan and Zhou (2009) compared these correlations with a mechanistic model presented by Xiao et al. (1990) for horizontal flow and for upward flow at angles up to 9° from vertical, using an extensive database. They concluded that the Xiao et al. mechanistic model gave the best overall predictions of actual pressure drops. However, both the Dukler and Beggs and Brill correlations also performed well, and Yuan and Zhou did not apply the Payne et al. (1979) corrections to the Beggs and Brill correlation. Xiao et al. also compared their mechanistic model to the other correlations using a pipeline database. The mechanistic model compared better overall, but the improvement was not dramatic. Because of its complexity and only a small advantage in accuracy over the simpler correlations, the Xiao et al. model is not included here. The interested reader is referred to their paper.

8.2.3.3 Beggs and Brill Correlation

The Beggs and Brill correlation presented in Chapter 7 can be applied to horizontal flow, as well as flow in any other direction. For horizontal flow, the correlation is somewhat simplified, since the angle θ is 0, making the factor ψ equal to 1. This correlation is presented in Section 7.4.3, Equations (7-130) through (7-159).

Example 8-4 Pressure Gradient Calculation Using the Beggs and Brill Correlation

Calculate the pressure gradient for the flow of 2000 bbl/d oil and 1 MMscf/d of gas described in Example 8-3.

Solution

The first step in using this correlation is to determine the flow regime using Equations (7-136) through (7-139); however, we have already found the flow regime to be intermittent by using the flow regime map in Example 8-3. Note that the dashed lines in Figure 8-4 correspond to Equations (7-136) through (7-139).

Next, we calculate the holdup for horizontal flow with Equation (7-141). From Example 8-3, $\lambda_l = 0.35$ and $N_{\text{Fr}} = 17.8$, so

$$y_{lo} = \frac{(0.845)(0.35)^{0.5351}}{(17.8)^{0.0173}} = 0.458. \tag{8-23}$$

Even though there is no potential energy contribution to the pressure gradient in horizontal flow, the holdup is used as a parameter in calculating the frictional pressure gradient.

To calculate the frictional gradient, we first obtain the no-slip friction factor based on the mixture Reynolds number. Using Equations (7-148), (7-150), and (7-151),

$$\rho_m = (49.92)(0.35) + (2.6)(0.65) = 19.2 \text{ lb}_m/\text{ft}^3 \tag{8-24}$$

$$\mu_m = (2)(0.35) + (0.0131)(0.65) = 0.71 \text{ cp} \tag{8-25}$$

$$N_{\text{Re}_m} = \frac{(19.2)(10.92)[(2.5/12)]}{((0.71)(6.72 \times 10^{-4})} = 91{,}600 \tag{8-26}$$

and from the Moody diagram (Figure 7-7) or the Chen equation [Equation (7-35)], $f_n = 0.0046$. Now, we calculate the parameters x, S, and f_{tp} using Equations (7-154), (7-153), and (7-152), respectively:

$$x = \frac{0.35}{(0.458)^2} = 1.67 \tag{8-27}$$

$$S = \frac{\ln(1.67)}{\{-0.0523 + 3.182\ln(1.67) - 0.8725[\ln(1.67)]^2 + 0.01853[\ln(1.67)]^4\}} = 0.379 \tag{8-28}$$

$$f_{tp} = (0.0046)e^{0.379} = 0.0067 \tag{8-29}$$

Finally, from Equation (7-147),

$$\frac{dp}{dx} = \frac{(2)(0.0067)(19.2)(10.92)^2}{(32.17)[(2.5/12)]} = 4.6\ \text{lb}_m/\text{ft}^3 = 0.03\ \text{psi/ft}. \tag{8-30}$$

8.2.3.4 Eaton Correlation

The Eaton correlation (Eaton, Andrews, Knowles, Silberberg, and Brown, 1967) was developed empirically from a series of tests in 2- and 4-in.-diameter, 1700-ft-long lines. It consists primarily of correlations for liquid holdup and friction factor. The Eaton correlation is derived by applying mechanical energy balances to both the gas and liquid phases, then adding these together. Integrating this equation over a finite distance of pipe, Δx, yields

$$\frac{\Delta p}{\Delta x} = \frac{f\bar{\rho}_m u_m^2}{2g_c D} + \frac{\lambda_l \bar{\rho}_l\, \Delta(u_l)^2 + \lambda_g \bar{\rho}_g\, \Delta(u_g)^2}{2g_c\, \Delta x}. \tag{8-31}$$

In this equation, the overbars indicate properties evaluated at the average pressure over the distance Δx and the velocities, u_l and u_g, are the in-situ average velocities. If the kinetic energy term is neglected, the equation can be applied at a point, or

$$\left(\frac{dp}{dx}\right)_F = \frac{f\rho_m u_m^2}{2g_c D}. \tag{8-32}$$

The friction factor, f, is obtained from the correlation shown in Figure 8-6 as a function of the mass flow rate of the liquid, m_l, and the total mass flow rate, m_m. For the constant given in this figure to compute the abscissa, mass flow rates are in lb_m/sec, diameter is in ft, and viscosity is in lb_m/ft-sec.

To compute the kinetic energy pressure drop, the liquid holdup is needed so that the in-situ average velocities of the phases can be calculated. Liquid holdup is given by the correlation of Figure 8-7, using the same dimensionless groups as the Hagedorn and Brown correlation (see Section 7.4.3). The base pressure, p_b, is 14.65 psi.

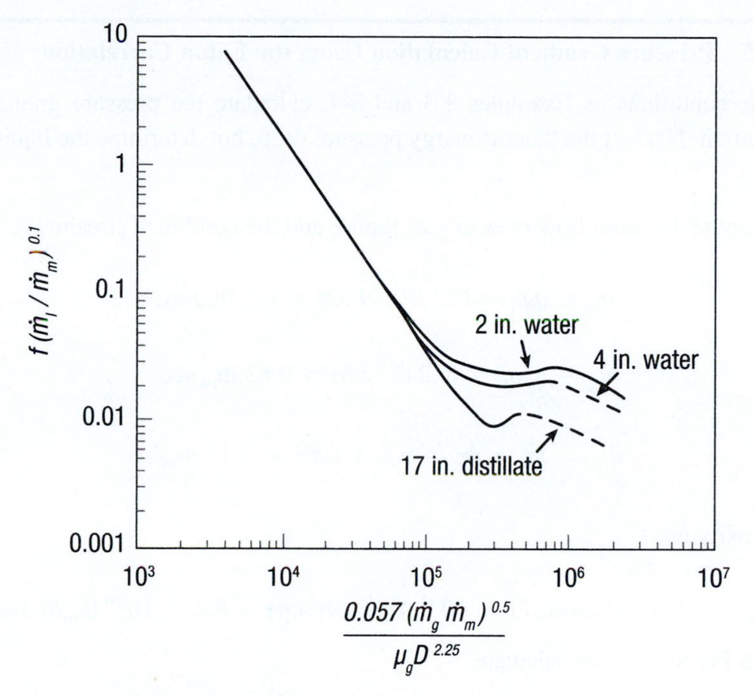

Figure 8-6 Eaton friction factor correlation. (From Eaton et al., 1967.)

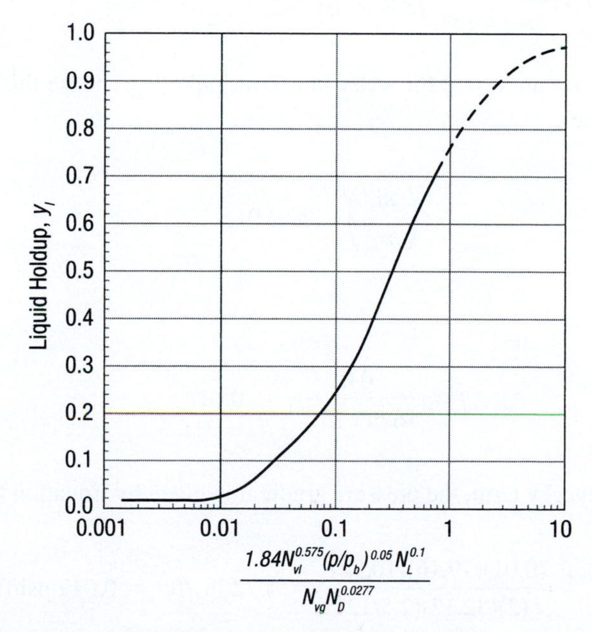

Figure 8-7 Eaton Holdup correlation. (From Eaton et al., 1967.)

Example 8-5 Pressure Gradient Calculation Using the Eaton Correlation

For the same conditions as Examples 8-3 and 8-4, calculate the pressure gradient using the Eaton correlation. Neglect the kinetic energy pressure drop, but determine the liquid holdup.

Solution

First, we compute the mass flow rates of gas, liquid, and the combined stream:

$$\dot{m}_l = q_l \rho_l = (0.130)(49.92) = 6.5 \text{ lb}_m/\text{sec} \tag{8-33}$$

$$\dot{m}_g = q_g \rho_g = (0.242)(2.6) = 0.63 \text{ lb}_m/\text{sec} \tag{8-34}$$

$$\dot{m}_m = \dot{m}_l + \dot{m}_g = 6.5 + 0.63 = 7.13 \text{ lb}_m/\text{sec} \tag{8-35}$$

The gas viscosity is

$$\mu_g = (0.0131 \text{ cp})(6.72 \times 10^{-4} \text{ lb}_m/\text{ft-sec-cp}) = 8.8 \times 10^{-6} \text{ lb}_m/\text{ft-sec.} \tag{8-36}$$

To find f with Figure 8-6, we calculate

$$\frac{(0.057)(\dot{m}_g \dot{m}_m)^{0.5}}{\mu_g D^{2.25}} = \frac{(0.057)[(0.63)(7.13)]^{0.5}}{(8.8 \times 10^{-6})(2.5/12))^{2.25}} = 4.7 \times 10^5 \tag{8-37}$$

and reading from the correlation line for water in a 2-in. pipe (we choose this line because the pipe is closest to this size),

$$f\left(\frac{\dot{m}_l}{\dot{m}_m}\right)^{0.1} = 0.01, \tag{8-38}$$

so

$$f = \frac{0.01}{(6.5/7.13)^{0.1}} = 0.01. \tag{8-39}$$

Neglecting the kinetic energy term, the pressure gradient is given by Equation (8-32),

$$\left(\frac{dp}{dx}\right)_F = \frac{(0.01)(19.16)(10.92)^2}{(2)(32.17)(2.5/12)} = 1.72 \text{ lb}_f/\text{ft}^3 = 0.012 \text{ psi/ft.} \tag{8-40}$$

The liquid holdup is obtained from the correlation given by Figure 8-7. The dimensionless numbers needed are given by Equations (7-99) through (7-82).

$$N_{vl} = (1.938)(3.81)\sqrt[4]{\frac{49.92}{30}} = 8.39 \tag{8-41}$$

$$N_{vg} = (1.98)(7.11)\sqrt[4]{\frac{49.92}{30}} = 15.65 \tag{8-42}$$

$$N_D = (120.872)\left(\frac{2.5}{12}\right)\sqrt{\frac{49.92}{30}} = 32.48 \tag{8-43}$$

$$N_L = (0.15726)(2)\sqrt[4]{\frac{1}{(49.92)(30)^3}} = 0.00923 \tag{8-44}$$

Calculating the abscissa value, we have

$$\frac{(1.84)N_{vl}^{0.575}(p/p_b)^{0.05}N_L^{0.1}}{N_{vg}N_D^{0.0277}} = \frac{(1.84)(8.39)^{0.575}(800/14.65)^{0.05}(0.00923)^{0.1}}{(15.65)(32.48)^{0.0277}} = 0.277 \tag{8-45}$$

and from Figure 8-7, $y_l = 0.45$.

The liquid holdup predictions from the Beggs and Brill and Eaton correlations agree very closely; the Eaton correlation predicts a lower pressure gradient.

8.2.3.5 Dukler Correlation

The Dukler (1969) correlation, like that of Eaton, is based on empirical correlations of friction factor and liquid holdup. The pressure gradient again consists of frictional and kinetic energy contributions:

$$\frac{dp}{dx} = \left(\frac{dp}{dx}\right)_{\mathrm{F}} + \left(\frac{dp}{dx}\right)_{\mathrm{KE}}. \tag{8-46}$$

The frictional pressure drop is

$$\left(\frac{dp}{dx}\right)_{\mathrm{F}} = \frac{f\rho_k u_m^2}{2g_c D} \tag{8-47}$$

where

$$\rho_k = \frac{\rho_l \lambda_l^2}{y_l} + \frac{\rho_g \lambda_g^2}{y_g}. \tag{8-48}$$

Notice that the liquid holdup enters the frictional pressure drop through ρ_k. The friction factor is obtained from the no-slip friction factor, f_n, defined as

$$f_n = 0.0056 + 0.5(N_{\mathrm{Re}_k})^{-0.32} \tag{8-49}$$

where the Reynolds number is

$$N_{\mathrm{Re}_k} = \frac{\rho_k u_m D}{\mu_m}. \tag{8-50}$$

The two-phase friction factor is given by the correlation

$$\frac{f}{f_n} = 1 - \frac{\ln \lambda_l}{1.281 + 0.478 \ln \lambda_l + 0.444(\ln \lambda_l)^2 + 0.094(\ln \lambda_l)^3 + 0.00843(\ln \lambda_l)^4}. \tag{8-51}$$

The liquid holdup is given as a function of the input liquid fraction, λ_l, in Figure 8-8, with N_{Re_k} as a parameter. Since the holdup is needed to calculate N_{Re_k} (ρ_k depends on y_l), determining the liquid holdup is an iterative procedure. We can begin by assuming that $y_l = \lambda_l$; estimates of ρ_k and N_{Re_k} are then calculated. With these estimates, y_l is obtained from Figure 8-8. We then compute new estimates of ρ_k and N_{Re_k}, repeating this procedure until convergence is achieved.

The pressure gradient due to kinetic energy changes is given by

$$\left(\frac{dp}{dx}\right)_{\mathrm{KE}} = \frac{1}{g_c \Delta x} \Delta\left(\frac{\rho_g u_{sg}^2}{y_g} + \frac{\rho_l u_{sl}^2}{y_l}\right). \tag{8-52}$$

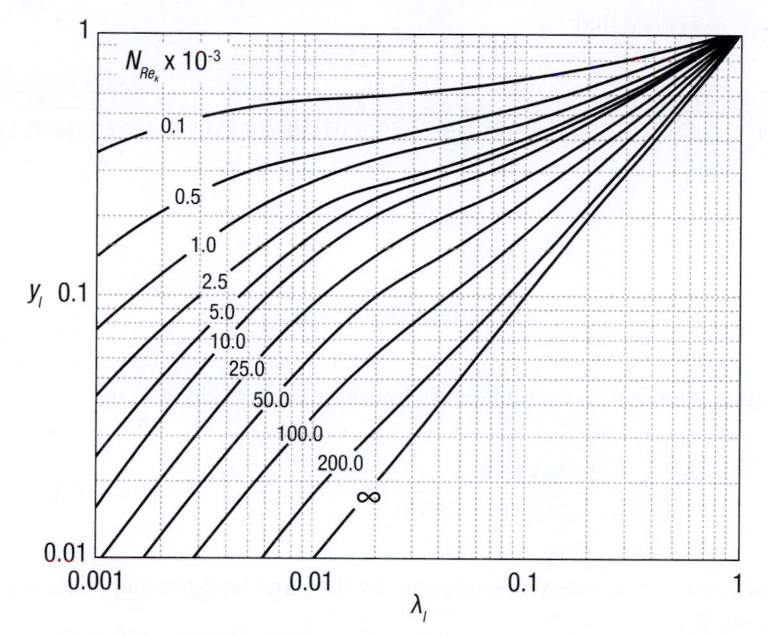

Figure 8-8 Dukler holdup correlation. (From Brill and Beggs, 1978.)

Example 8-6 Pressure Gradient Calculation Using the Dukler Correlation

Repeat Examples 8-4 and 8-5, using the Dukler correlation.

Solution

An iterative procedure is required to find the liquid holdup. We begin by assuming that $y_l = \lambda_l$. In this case, $\rho_k = \rho_m$, found previously to be 19.16 lb_m/ft^3, and $N_{\text{Re}_k} = N_{\text{Re}_m}$, which was 91,600. From Figure 8-8, we estimate y_l to be 0.44. Using this new value of liquid holdup,

$$\rho_k = \frac{(49.92)(0.35)^2}{0.44} + \frac{(2.6)(0.65)^2}{(1 - 0.44)} = 15.86 \text{ lb}_m/\text{ft}^3 \tag{8-53}$$

and

$$N_{\text{Re}_k} = (91{,}600)\left(\frac{15.86}{19.16}\right) = 75{,}800. \tag{8-54}$$

Again reading y_l from Figure 8-8, $y_l = 0.46$. This is the converged value.

The no-slip friction factor from Equation (8-49) is

$$f_n = 0.0056 + (0.5)(75{,}800)^{-0.32} = 0.019. \tag{8-55}$$

From Equation (8-51), we find

$$\frac{f}{f_n} = 1 - \frac{ln(0.35)}{1.281 + 0.478[ln(0.35)] + 0.444[ln(0.35)]^2 + 0.094[ln(0.35)]^3 + 0.00843[ln(0.35)]^4} \tag{8-56}$$

$$= 1.90$$

so

$$f = (1.90)(0.019) = 0.036. \tag{8-57}$$

Finally, from Equation (8-47), we find the frictional pressure gradient to be

$$\left(\frac{dp}{dx}\right)_{\mathrm{F}} = \frac{(0.036)(15.86)(10.92)^2}{(2)(32.17)(2.5/12)} = 5.08\ \mathrm{lb}_f/\mathrm{ft}^3 = 0.035\ \mathrm{psi/ft} \tag{8-58}$$

We see that all three correlations predict essentially the same liquid holdup, but the pressure gradient predictions differ.

8.2.3.6 Pressure Traverse Calculations

The correlations we have just examined provide a means of calculating the pressure gradient at a point along a pipeline; to determine the overall pressure drop over a finite length of pipe, the variation of the pressure gradient as the fluid properties change in response to the changing pressure must be considered. The simplest procedure is to evaluate fluid properties at the mean pressure over the distance of interest and then calculate a mean pressure gradient. For example, integrating the Dukler correlation over a distance L of pipe, we have

$$\Delta p = \frac{\overline{f}\overline{\rho}_k\overline{u}_m^2 L}{2g_c D} + \frac{1}{g_c}\Delta\left(\frac{\rho_g u_{sg}^2}{y_g} + \frac{\rho_l u_{sl}^2}{y_l}\right). \tag{8-59}$$

The overbars indicate that f, ρ_k, and u_m are evaluated at the mean pressure, $(p_1 + p_2)/2$. In the kinetic energy term, the Δ means the difference between conditions at point 1 and point 2. If the overall pressure drop, $p_1 - p_2$, is known, the pipe length, L, can be calculated directly with this equation. When L is fixed, Δp must be estimated to calculate the mean pressure; using this mean pressure, Δp is calculated with Equation (8-59) and, if necessary, the procedure is repeated until convergence is reached.

Since the overall pressure drop over the distance L is being calculated based on the mean properties over this distance, the pressure should not change too much over this distance. In general, if the Δp over the distance L is greater than 8% of p_1, the distance L should be divided into smaller increments and the pressure drop over each increment calculated. The pressure drop over the distance L is then the sum of the pressure drops over the smaller increments.

Example 8-7 Calculating the Overall Pressure Drop

Just downstream from the wellhead choke, 2000 bbl/d of oil and 1 MMscf/d of gas enters a 2.5-in. pipeline at 800 psia and 175°F (the same fluids as in the previous four examples). If these fluids are transported 3000 ft through this flow line to a separator, what is the discharge pressure at the separator? Use the Dukler correlation and neglect kinetic energy pressure losses.

Solution

We found in Example 8-6 that the pressure gradient at the entrance conditions is 0.035 psi/ft. If this value holds over the entire line, the overall pressure drop would be (0.035)(3000) = 105 psi. Since this pressure drop is slightly greater than 10% of the entrance condition, we will divide the pipe into two length increments and calculate a mean pressure gradient over each increment.

It is most convenient to fix the Δp for the first increment and solve for the length of the increment. The remainder of the pipeline will then comprise the second section. We will choose a Δp of 60 psi for the first increment, L_1. Thus, for this section, $\overline{p} = 800 - 60/2 = 770$ psi. We previously calculated the properties of the fluids at 800 psi; the only property that may differ significantly at 700 psi is the gas density.

Following Example 4-3 for this gas, at 770 psi and 175°F, $p_{pr} = 1.07$ and $T_{pr} = 1.70$. From Figure 4-1, $Z = 0.935$, and using Equation (7-67),

$$\overline{\rho}_g = \frac{(28.97)(0.709)(770)}{(0.935)(10.73)(635)} = 2.5 \text{ lb}_m/\text{ft}^3. \tag{8-60}$$

The gas volumetric flow rate is then

$$\overline{q}_g = \frac{\dot{m}_g}{\overline{\rho}_g} = \frac{0.63}{2.5} = 0.252 \text{ ft}^3/\text{sec}. \tag{8-61}$$

Since the liquid density is essentially the same at 770 psi as at the entrance condition, q_l is 0.13 ft^3/sec, as before, and the total volumetric flow rate is $0.252 + 0.13 = 0.382$ ft^3/sec.

The input liquid fraction is $0.13/0.382 = 0.34$, and the mixture velocity ($\overline{u}_m$) is found to be 11.2 ft/sec by dividing the total volumetric flow rate by the pipe cross-sectional area.

We can now use the Dukler correlation to calculate the pressure gradient at the mean pressure. We begin by estimating the liquid holdup at mean conditions to be the same as that found at the entrance, or $y_l = 0.46$. Then

$$\rho_k = \frac{(49.92)(0.34)^2}{0.46} + \frac{(2.5)(0.66)^2}{0.56} = 14.56 \text{ lb}_m/\text{ft}^3 \tag{8-62}$$

and

$$\mu_m = (2)(0.34) + (0.131)(0.66) = 0.689 \text{ cp} \tag{8-63}$$

so,

$$N_{Re_k} = \frac{(14.56)(11.2)(2.5/12)}{(0.689)(6.72 \times 10^{-4})} = 73{,}400. \tag{8-64}$$

Checking Figure 8-8, we find that $y_l = 0.46$, so no iteration is required. Using Equations (8-49) and (8-51), $f_n = 0.019$ and $f/f_n = 1.92$, so $f = 0.036$. From Equation (8-47), we find that *dp/dx* is 0.034 psi/ft. The length of the first increment is

$$L_1 = \frac{\Delta p}{dp/dx} = \frac{60}{0.034} = 1760 \text{ ft.} \tag{8-65}$$

The remaining section of the pipeline is $3000 - 1760 = 1240$ ft long. If the pressure gradient over this section is also 0.034 psi/ft, the pressure drop will be 42 psi; thus we can estimate the mean pressure over the second increment to be $740 - 42/2 \approx 720$ psi.

With this mean pressure, we repeat the procedure to calculate the mean pressure gradient using the Dukler correlation and find that $dp/dx = 0.036$ psi/ft, giving an overall pressure drop for the second segment of 45 psi. Adding the two pressure drops, the pressure drop over the 3000-ft pipe is 85 psi. This happens to be exactly what we estimated using the pressure gradient at the pipe entrance conditions, illustrating that the pressure gradient is not varying significantly at these relatively high pressures.

8.2.4 Pressure Drop through Pipe Fittings

When fluids pass through pipe fittings (tees, elbows, etc.) or valves, secondary flows and additional turbulence create pressure drops that must be included to determine the overall pressure drop in a piping network. The effects of valves and fittings are included by adding the "equivalent length" of the valves and fittings to the actual length of straight pipe when calculating the pressure drop. The equivalent lengths of many standard valves and fittings have been determined experimentally (Crane, 1957) and are given in Table 8-1. The equivalent lengths are given in pipe diameters; this value is multiplied by the pipe diameter to find the actual length of pipe to be added to account for the pressure drop through the valve or fitting.

8.3 Flow through Chokes

The flow rate from almost all flowing wells is controlled with a wellhead choke, a device that places a restriction in the flow line (Figure 8-9). A variety of factors may make it desirable to restrict the production rate from a flowing well, including the prevention of coning or sand production, satisfying production rate limits set by regulatory authorities, and meeting limitations of rate or pressure imposed by surface equipment.

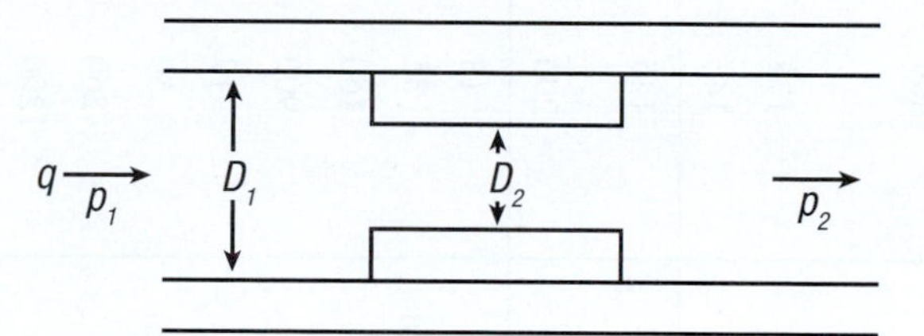

Figure 8-9 Choke schematic.

When gas or gas–liquid mixtures flow through a choke, the fluid may be accelerated sufficiently to reach sonic velocity in the throat of the choke. When this condition occurs, the flow is called "critical," and changes in the pressure downstream of the choke do not affect the flow rate because pressure disturbances cannot travel upstream faster than the sonic velocity. (*Note:* Critical flow is not related to the critical point of the fluid.) Thus, to predict the flow rate–pressure drop relationship for compressible fluids flowing through a choke, we must determine whether or not the flow is critical. Figure 8-10 shows the dependence of flow rate through a choke on the ratio of the downstream to upstream pressure for a compressible fluid, with the rate being independent of the pressure ratio when the flow is critical.

In this section, we examine the flow of liquid, gas, and gas–liquid mixtures through chokes.

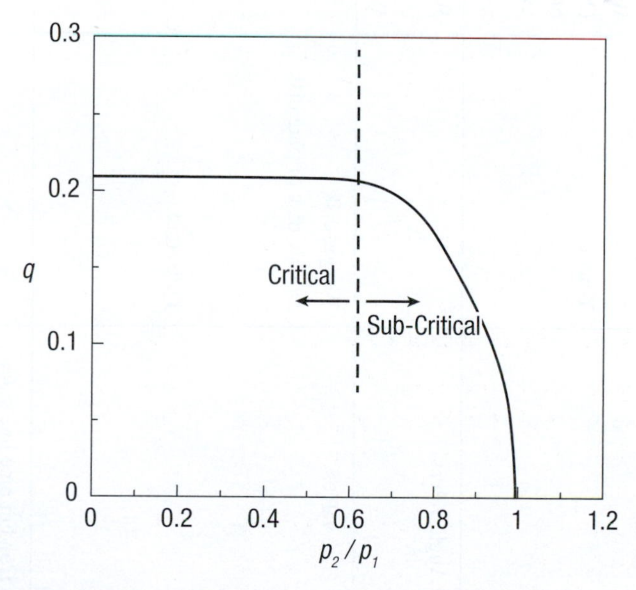

Figure 8-10 Dependence of flow rate through a choke on the ratio of the upstream to the downstream pressure.

Table 8-1 Equivalent Lengths of Valves and Fittings

		Description of Fitting		Equivalent Length in Pipe Diameters
Globe valves	Stem perpendicular to run	With no obstruction in flat-, bevel-, or plug-type seat	Fully open	340
		With wing- or pin-guided disk	Fully open	450
	Y-pattern	(No obstruction in flat-, bevel-, or plug-type seat)		
		–With stem 60° from run of pipe line	Fully open	175
		–With stem 45° from run of pipe line	Fully open	145
Angle valves		With no obstruction in flat-, bevel-, or plug-type seat	Fully open	145
		With wing- or pin-guided disk	Fully open	200
Gate valves	Wedge, disk		Fully open	13
	Double disk or plug disk		Three-quarters open	35
			One-half open	160
			One-quarter open	900
	Pulp stock		Fully open	17
			Three-quarters open	50
			One-half open	260
			One-quarter open	1200
Conduit pipe line gate, ball, and plug valves			Fully open	3

Check valves	Conventional swing		Fully open	135
	Clearaway swing		Fully open	50
	Globe life or stop; stem perpendicular to run or Y-pattern		Fully open	Same as globe
	Angle lift or stop		Fully open	
	Same as angle			
	In-line ball		Fully open	150
Foot valves with strainer		With poppet lift-type disk	Fully open	420
		With leather-hinged disk	Fully open	75
Butterfly valves (8 in. and larger)			Fully open	40
Cocks	Straight-through	Rectangular plug port area equal to 100% of pipe area	Fully open	18
	Three-way	Rectangular plug port area equal to 80% of pipe area (fully open)	Flow straight through	44
			Flow through branch	140
Fittings	90° standard elbow			30
	45° standard elbow			16
	90° long radius elbow			20
	90° street elbow			50
	45° street elbow			26
	Square corner elbow			57
	Standard tee	With flow-through run		20
		With flow-through branch		60
	Close-pattern return bend			50

From Crane (1957).

8.3.1 Single-Phase Liquid Flow

The flow through a wellhead choke rarely consists of single-phase liquid, since the flowing tubing pressure is almost always below the bubble point. However, when this does occur, the flow rate is related to the pressure drop across the choke by

$$q = CA\sqrt{\frac{2g_c\,\Delta p}{\rho}} \tag{8-66}$$

where C is the flow coefficient of the choke and A is the cross-sectional area of the choke. The flow coefficient for flow through nozzles is given in Figure 8-11 (Crane, 1957) as a function of the Reynolds number in the choke and the ratio of the choke diameter to the pipe diameter. Equation (8-66) is derived by assuming that the pressure drop through the choke is equal to the kinetic energy pressure drop divided by the square of a drag coefficient. This equation applies for subcritical flow, which will usually be the case for single-phase liquid flow.

For oilfield units, Equation (8-66) becomes

$$q = 22{,}800C(D_2)^2\sqrt{\frac{\Delta p}{p}} \tag{8-67}$$

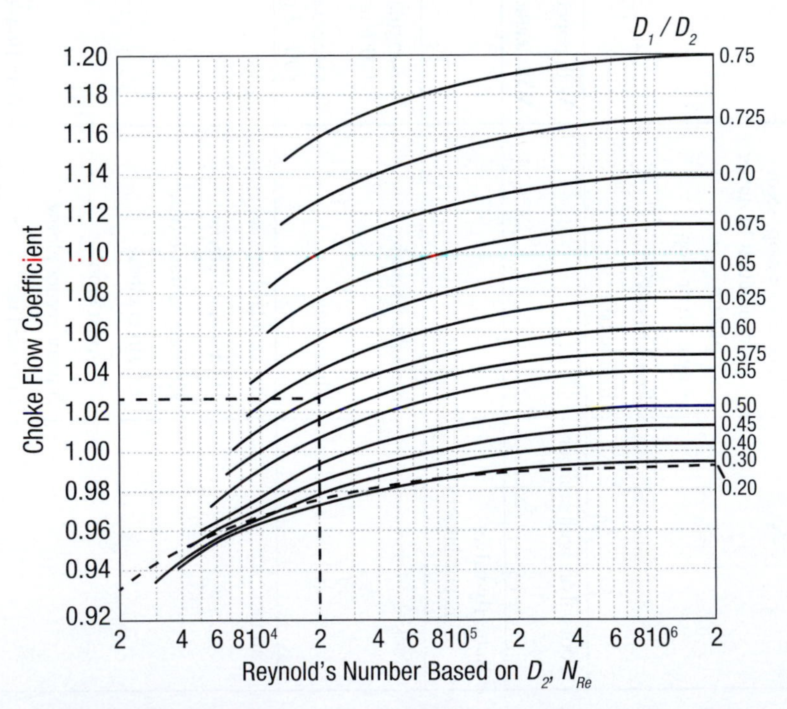

Figure 8-11 Flow coefficient for liquid flow through a choke. (From Crane, 1957.)

where q is in bbl/d, D_2 is the choke diameter in in., Δp is in psi, and ρ is in lb_m/ft^3. The choke diameter is often referred to as the "bean size" because the device in the choke that restricts the flow is called the bean. Bean sizes are usually given in 64ths of an inch.

Example 8-8 Liquid Flow through a Choke

What will be the flow rate of a 0.8-specific gravity, 2-cp oil through a 20/64-in. choke if the pressure drop across the choke is 20 psi and the line size is 1 in.?

Solution

Figure 8-11 gives the flow coefficient as a function of the diameter ratio and the Reynolds number through the choke. Since we do not know the Reynolds number until we know the flow rate, we assume that the Reynolds number is high enough that the flow coefficient is independent of the Reynolds number. For $D_2/D_1 = 0.31$, C is approximately 1.00. Then, from Equation (8-67),

$$q = (22{,}800)(1.00)\left(\frac{20}{64}\right)^2\sqrt{\frac{20}{49.92}} = 1410 \text{ bbl/d}. \tag{8-68}$$

Checking the Reynolds number through the choke [Equation (7-7)], we find $N_{Re} = 1.67 \times 10^5$. From Figure 8-11, $C = 0.99$ for this N_{Re}; using this value, the flow rate is 1400 bbl/d.

8.3.2 Single-Phase Gas Flow

When a compressible fluid passes through a restriction, the expansion of the fluid is an important factor. For isentropic flow of an ideal gas through a choke, the rate is related to the pressure ratio, p_2/p_1, by (Szilas, 1975)

$$q_g = \frac{\pi}{4}D_2^2 p_1 \frac{T_{sc}}{p_{sc}}\alpha\sqrt{\left(\frac{2g_cR}{28.97\gamma_g T_1}\right)\left(\frac{\gamma}{\gamma-1}\right)\left[\left(\frac{p_2}{p_1}\right)^{2/\gamma} - \left(\frac{p_2}{p_1}\right)^{(\gamma+1)/\gamma}\right]}, \tag{8-69}$$

which can be expressed in oilfield units as

$$q_g = 3.505 D_{64}^2\left(\frac{p_1}{p_{sc}}\right)\alpha\sqrt{\left(\frac{1}{\gamma_g T_1}\right)\left(\frac{\gamma}{\gamma-1}\right)\left[\left(\frac{p_2}{p_1}\right)^{2/\gamma} - \left(\frac{p_2}{p_1}\right)^{(\gamma+1)/\gamma}\right]} \tag{8-70}$$

where q_g is in Mscf/d, D_{64} is the choke diameter (bean size) in 64ths of an inch (e.g., for a choke diameter of 1/4 in., $D_2 = 16/64$ in. and $D_{64} = 16$), T_1 is the temperature upstream of the choke in °R, γ is the heat capacity ratio, C_p/C_v, α is the flow coefficient of the choke, γ_g is the gas gravity, p_{sc} is standard pressure, and p_1 and p_2 are the pressures upstream and downstream of the choke, respectively.

Equations (8-69) and (8-70) apply when the pressure ratio is equal to or greater than the critical pressure ratio, given by

$$\left(\frac{p_2}{p_1}\right)_c = \left(\frac{2}{\gamma + 1}\right)^{\gamma/(\gamma-1)}. \tag{8-71}$$

When the pressure ratio is less than the critical pressure ratio, p_2/p_1 should be set to $(p_2/p_1)_c$ and Equation (8-70) used, since the flow rate is insensitive to the downstream pressure whenever the flow is critical. For air and other diatomic gases, γ is approximately 1.4, and the critical pressure ratio is 0.53; in petroleum engineering operations, it is commonly assumed that flow through a choke is critical whenever the downstream pressure is less than about half of the upstream pressure.

Example 8-9 The Effect of Choke Size on Gas Flow Rate

Construct a chart of gas flow rate versus pressure ratio for choke diameters (bean sizes) of 8/64, 12/64, 16/24, 20/64, and 24/64 of an inch. Assume that the choke flow coefficient is 0.85, the gas gravity is 0.7, γ is 1.25, and the wellhead temperature and flowing pressure are 100°F and 600 psia.

Solution

From Equation (8-71), we find that the critical pressure ratio is 0.56 for this gas. Using Equation (8-70),

$$q_g = 3.505 D_{64}^2 \left(\frac{600}{14.7}\right)(0.85)\sqrt{\left(\frac{1}{(0.7)(560)}\right)\left(\frac{1.25}{1.25 - 1}\right)\left[\left(\frac{p_2}{p_1}\right)^{2/1.25} - \left(\frac{p_2}{p_1}\right)^{(1.25+1)/1.25}\right]} \tag{8-72}$$

or

$$q_g = 13.73 D_{64}^2 \sqrt{\left(\frac{p_2}{p_1}\right)^{1.6} - \left(\frac{p_2}{p_1}\right)^{1.8}}. \tag{8-73}$$

The maximum gas flow rate will occur when the flow is critical, that is, when $(p_2/p_1) = 0.56$. For any value of the pressure ratio below the critical value, the flow rate will be the critical flow rate. Using values of p_2/p_1 from 0.56 to 1 for each choke size, Figure 8-12 is constructed.

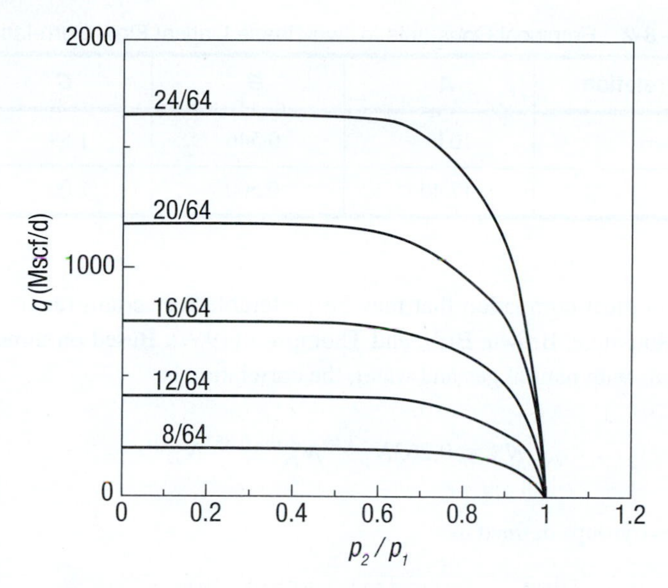

Figure 8-12 Gas flow performance for different choke sizes.

8.3.3 Gas–Liquid Flow

Two-phase flow through a choke has not been described well theoretically. To determine the flow rate of two phases through a choke, empirical correlations for critical flow are generally used. Some of these correlations are claimed to be valid up to pressure ratios of 0.7 (Gilbert, 1954). One means of estimating the conditions for critical two-phase flow through a choke is to compare the velocity in the choke with the two-phase sonic velocity given by Wallis (1969) for homogeneous mixtures as

$$v_c = \left\{[\lambda_g \rho_g + \lambda_l \rho_l]\left[\frac{\lambda_g}{\rho_g v_{gc}^2} + \frac{\lambda_l}{\rho_l v_{lc}^2}\right]\right\}^{-1/2} \tag{8-74}$$

where v_c is the sonic velocity of the two-phase mixture and v_{gc} and v_{lc} are the sonic velocities of the gas and liquid, respectively.

The empirical correlations of Gilbert (1954) and Ros (1960) have the same form, namely,

$$p_1 = \frac{A q_l (\text{GLR})^B}{D_{64}^C}, \tag{8-75}$$

differing only in the empirical constants *A, B,* and *C,* given in Table 8-2. The upstream pressure, p_1, is in psig in the Gilbert correlation and psia in Ros's correlation. In these correlations, q_l is the liquid rate in bbl/d, GLR is the producing gas-liquid ratio in SCF/bbl, and D_{64} is the choke diameter in 64ths of an inch.

Table 8-2 Empirical Constants in Two-Phase Critical Flow Correlations

Correlation	*A*	*B*	*C*
Gilbert	10.0	0.546	1.89
Ros	17.40	0.500	2.00

Another empirical correlation that may be preferable for certain ranges of conditions is that of Omana, Houssiere, Brown, Brill, and Thompson (1969). Based on dimensional analysis and a series of tests with natural gas and water, the correlation is

$$N_{ql} = 0.263 N_{\rho}^{-3.49} N_{pl}^{3.19} \lambda_l^{0.657} N_D^{1.80} \quad \textbf{(8-76)}$$

with dimensionless groups defined as

$$N_{\rho} = \frac{\rho_g}{\rho_l} \quad \textbf{(8-77)}$$

$$N_{pl} = 1.74 \times 10^{-2} p_1 \left(\frac{1}{\rho_l \sigma} \right)^{0.5} \quad \textbf{(8-78)}$$

$$N_D = 0.1574 D_{64} \sqrt{\frac{\rho_l}{\sigma}} \quad \textbf{(8-79)}$$

$$N_{ql} = 1.84 q_l \left(\frac{\rho_l}{\sigma} \right)^{1.25} \quad \textbf{(8-80)}$$

For the constants given, the oilfield units are q, bbl/d; ρ, lb_m/ft^3; σ, dynes/cm; D_{64}, 64ths of an inch; and p_1, psia. The Omana correlation is restricted to critical flow, requiring that $q_g/q_l > 1.0$ and $p_2/p_1 < 0.546$. It is best suited to low-viscosity liquids (near the viscosity of water) and choke diameters (bean sizes) of 14/64 in. or less. The fluid properties are evaluated at the upstream conditions.

Example 8-10 Finding the Choke Size for Gas–Liquid Flow

For the flow of 2000 bbl/d of oil and 1 MMscf/d of gas at a flowing tubing pressure of 800 psia as described in Example 8-3, find the choke diameter (bean size) using the Gilbert, Ros, and Omana correlations.

Solution

For the Gilbert and Ros correlations, solving Equation (8-75) for the choke diameter, we have

$$D_{64} = \left(\frac{Aq_l(\mathrm{GLR})^B}{p_1} \right)^{1/C}. \tag{8-81}$$

For the given flow rate of 2000 bbl/d, a GLR of 500 SCF/bbl, and an absolute pressure of 800 psi upstream of the choke, from the Gilbert correlation,

$$D_{64} = \left(\frac{(10)(2000)(500)^{0.546}}{800 - 14.7} \right)^{1/1.89} = 33 \text{ 64ths of an inch} \tag{8-82}$$

and, from the Ros correlation,

$$D_{64} = \left(\frac{(17.4)(2000)(500)^{0.5}}{800} \right)^{0.5} = 31 \text{ 64ths of an inch.} \tag{8-83}$$

With the Omana correlation, we solve Equation (8-76) for N_D:

$$N_D = \left(\frac{1}{0.263} N_{ql} N_\rho^{3.49} N_{p_1}^{-3.19} \lambda_l^{-0.657} \right)^{1/1.8}. \tag{8-84}$$

From Example 8-3, $\lambda_l = 0.35$, $\rho_l = 49.92 \text{ lb}_m/\text{ft}^3$, $\rho_g = 2.6 \text{ lb}_m/\text{ft}^3$, and $\sigma = 30$ dynes/cm. From Equations (8-77), (8-78), and (8-80),

$$N_\rho = \frac{2.6}{49.92} = 0.0521 \tag{8-85}$$

$$N_{pl} = (1.74 \times 10^{-2})(800) \left[\frac{1}{(49.92)(30)} \right]^{0.5} = 0.36 \tag{8-86}$$

$$N_{ql} = (1.84)(2000) \left(\frac{49.92}{30} \right)^{1.25} = 6.95 \times 10^3 \tag{8-87}$$

Then

$$N_D = \left[\left(\frac{1}{0.263} \right) (6.95 \times 10^3)(0.0521)^{3.49}(0.36)^{-3.19}(0.35)^{-0.657} \right]^{1/1.8} = 8.35. \tag{8-88}$$

Solving Equation (8-79) for the choke diameter, we have

$$D_{64} = \frac{N_D}{(0.1574)\sqrt{\frac{\rho_l}{\sigma_l}}} \quad \text{(8-89)}$$

so

$$D_{64} = \frac{8.35}{(0.1574)\sqrt{49.92/30}} = 41 \text{ 64ths of an inch.} \quad \text{(8-90)}$$

The Gilbert and Ros correlations predict a choke size of about 1/2 in. (32/64 in.), whereas the Omana correlation predicts a larger choke size of 41/64 in. Since the Omana correlation was based on liquid flow rates of 800 bbl/d or less, the Gilbert and Ros correlations are probably the more accurate in this case.

When a well is being produced with critical flow through a choke, the relationship between the wellhead pressure and the flow rate is controlled by the choke, since downstream pressure disturbances (such as a change in separator pressure) do not affect the flow performance through the choke. Thus, the attainable flow rate from a well for a given choke can be determined by matching the choke performance with the well performance, as determined by a combination of the well IPR and the vertical lift performance. The choke performance curve is a plot of the flowing tubing pressure versus the liquid flow rate, and can be obtained from the two-phase choke correlations, assuming that the flow is critical.

Example 8-11 Choke Performance Curves

Construct performance curves for 16/64-, 24/64-, and 32/64-in. chokes for a well with a GLR of 500, using the Gilbert correlation.

Solution

The Gilbert correlation predicts that the flowing tubing pressure is a linear function of the liquid flow rate, with an intercept at the origin. Using Equation (8-75), we find

$$p_{tf} = 1.57q_l \quad \text{for 16/64-in. choke} \quad \text{(8-91)}$$

$$p_{tf} = 0.73q_l \quad \text{for 24/64-in. choke} \quad \text{(8-92)}$$

$$p_{tf} = 0.43q_l \quad \text{for 32/64-in. choke} \quad \text{(8-93)}$$

These relationships are plotted in Figure 8-13, along with a well performance curve. The intersections of the choke performance curves with the well performance curve are the flow rates that would occur with these choke sizes. Note that the choke correlation is valid only when the flow

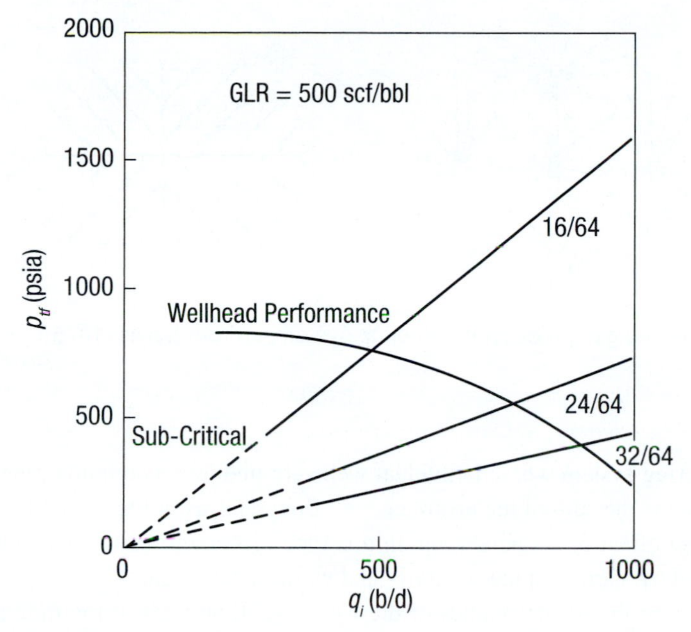

Figure 8-13 Choke performance curves (Example 8-11).

through the choke is critical; for each choke, there will be a flow rate below which flow through the choke is subcritical. This region is indicated by the dashed portions of the choke performance curves—the predictions are not valid for these conditions.

8.4 Surface Gathering Systems

In most oil and gas production installations, the flow from several wells will be gathered at a central processing station or combined into a common pipeline. Two common types of gathering systems were illustrated by Szilas (1975) (Figure 8-14). When the individual well flow rates are controlled by critical flow through a choke, there is little interaction among the wells. However, when flow is subcritical, the downstream pressure can influence the performance of the wells, and the flow through the entire piping network may have to be treated as a system.

When individual flow lines all join at a common point (Figure 8-14, left) the pressure at the common point is equal for all flow lines. The common point is typically a separator in an oil production system. The flowing tubing pressure of an individual well i is related to the separator pressure by

$$p_{\text{tf}\,i} = p_{\text{sep}} + \Delta p_{Li} + \Delta p_{Ci} + \Delta p_{fi} \tag{8-94}$$

where Δp_{Li} is the pressure drop through the flow line, Δp_{Ci} is the pressure drop through the choke (if present), and Δp_{fi} is the pressure drop through fittings.

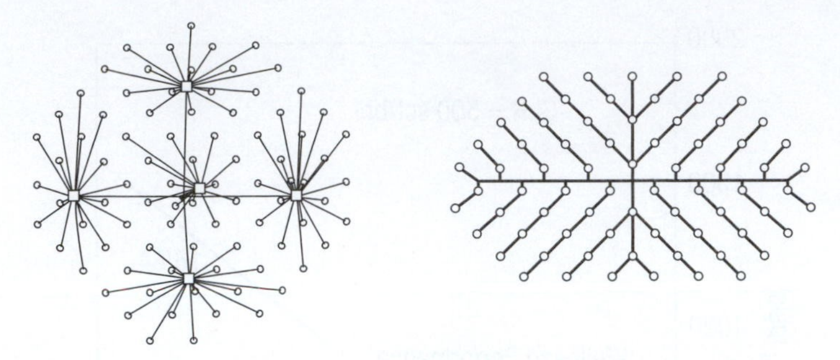

Figure 8-14 Oil and gas production gathering systems. (From Szilas, 1975.)

In a gathering system where individual wells are tied into a common pipeline, so that the pipeline flow rate is the sum of the upstream well flow rates as in Figure 8-14 (right), each well has a more direct effect on its neighbors. In this type of system, individual wellhead pressures can be calculated by starting at the separator and working upstream.

Depending on the lift mechanics of the wells, the flow rates of the individual wells may depend on the flowing tubing pressures. In this case, the IPRs and the vertical lift performance characteristics of the wells and surface gathering system must all be considered together to predict the performance of the well network.

Example 8-12 Analysis of a Surface Gathering System

The liquid production from three rod-pumped wells is gathered in a common 2-in. line, as shown in Figure 8-15. One-inch flow lines connect each well to the gathering line, and each well line contains a ball valve and a conventional check valve. Well 1 is tied into the gathering line with a standard 90° elbow, whereas wells 2 and 3 are connected with standard tees. The oil density is 0.85 g/cm^3 (53.04 lb$_m$/ft^3), and its viscosity is 5 cp. The separator pressure is 100 psig. Assuming the relative roughness of all lines to be 0.001, calculate the flowing tubing pressures of the three wells.

Solution

Since the flow rates are all known (and are assumed independent of the wellhead pressures for these rod-pumped wells), the pressure drop for each pipe segment can be calculated independently using Equation (8-1). The friction factors are obtained from the Chen equation [Equation (7-35)] or the Moody diagram (Figure 7-7). The pressure drops through the fittings and valves in the well flow lines are considered by adding their equivalent lengths from Table 8-1 to the well flow line lengths. For example, for well flow line 2, the ball valve adds three pipe diameters, the check valve 135 pipe diameters, and the tee (with flow through a branch) 60 pipe diameters.

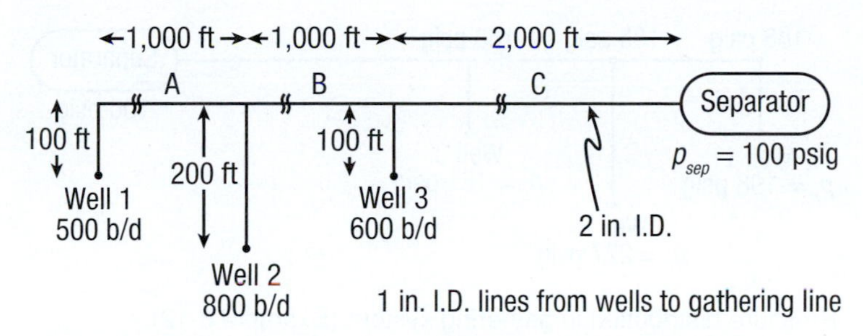

Figure 8-15 Surface gathering system (Example 8-12).

The equivalent length of flow line 2 is then (3 + 135 + 60)(1/12 ft) + 200 ft = 216.5 ft. A summary of the calculated results is given in Table 8-3.

The pressure at each point in the pipe network is obtained by starting with the known separator pressure and adding the appropriate pressure drops. The resulting system pressures are shown in Figure 8-16. The differences in the flowing tubing pressures in these wells would result in different liquid levels in the annuli, if the IPRs and elevations are the same in all three wells.

Table 8-3 Pressure Drop Calculation Results

Gathering Line					
Segment	***q* (bbl/d)**	**N_{Re}**	***f***	***u* (ft/sec)**	**Δp (psi)**
A	500	3,930	0.0103	1.49	3
B	1,300	10,200	0.0081	3.88	17
C	1,900	14,900	0.0077	5.66	68

Well Flow Lines						
Well No.	**N_{Re}**	***f***	***u* (ft/sec)**	**$(L/D)_{fittings}$**	***L* (ft)**	**Δp (psi)**
1	7,850	0.0086	5.96	168	114	10
2	12,600	0.0077	9.54	198	216.5	42
3	9,420	0.0082	7.16	198	116.5	13

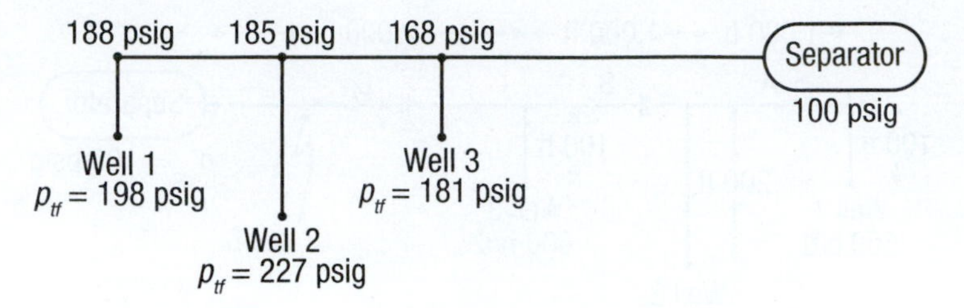

Figure 8-16 Pressure distribution in gathering system (Example 8-12).

8.5 Flow in Horizontal Wellbores

Analytical models of horizontal well inflow, such as those presented in Chapter 5, assume that the wellbore pressure is uniform along the horizontal well (i.e., there is no significant pressure drop along the well). However, in some high productivity reservoirs and/or very long horizontal sections, this assumption is not valid and the pressure drop in the horizontal well itself must be considered. In this section, we illustrate the conditions for which horizontal well pressure drop is significant, review correlations for wellbore pressure drop including simultaneous inflow, and discuss multiphase flow effects that occur in nominally horizontal wells that are not perfectly horizontal.

8.5.1 Importance of Wellbore Pressure Drop

The importance of the pressure drop in the wellbore on the overall performance of a horizontal well has been considered in several previous studies (Dikken, 1990; Folefac, Archer, and Issa, 1991; Novy, 1995; Ozkan, Sarica, and Haciislamogly, 1995; Penmatcha et al., 1999; Ankalm and Wiggins, 2005), all of which show that the wellbore pressure drop is important only if it is of significance relative to the reservoir drawdown. Whether or not the pressure drop in a horizontal well is important depends on the magnitude of the pressure drop in the lateral relative to the pressure drop in the reservoir (the drawdown). Hill and Zhu (2008) presented a method to easily determine whether the pressure drop in a horizontal well is important compared with the reservoir pressure drop. For single-phase liquid flow in the reservoir and in the wellbore, they showed that the ratio of wellbore frictional pressure drop to the reservoir drawdown is

$$\frac{\Delta p_f}{\Delta p_r} = 4 f_f N_{\mathrm{Re}} N_{\mathrm{H}} \tag{8-95}$$

where N_{Re} is the Reynolds number for pipe flow, using the total production rate from the well, f_f is the friction factor evaluated at this Reynolds number, and N_{H} is the horizontal well dimensionless number defined as

$$N_{\mathrm{H}} = \frac{k L_w^2}{D^4 F_g}. \tag{8-96}$$

In this dimensionless number, F_g represents the geometric terms in the reservoir inflow equation. For example, for the Furui equation (Equation 5-20),

$$F_g = \ln\left[\frac{hI_{ani}}{r_w(I_{ani} + 1)}\right] + \frac{\pi y_b}{hI_{ani}} - 1.224 + s. \quad \text{(8-97)}$$

When the $\Delta p_f/\Delta p_r$ ratio is small, the effect of pressure drop in the lateral can be neglected.

Example 8-13 Relative Pressure Drops

A 4000-ft-long 4-in.-diameter horizontal well is completed in an oil reservoir with a viscosity of 1 cp, a density of 60 lb_m/ft^3, and a formation volume factor of 1.1. The net thickness of the formation is 50 ft, horizontal-to-vertical permeability anisotropy ratio is 10, and the distance to the drainage boundary perpendicular to the well (y_b) is 2000 ft. The relative roughness of the liner is 0.001. Assume the skin factor is 0. The pressure at the outer drainage boundary is 4000 psi. For a range of horizontal permeability values and reservoir pressure drawdowns, find the conditions for which the wellbore pressure drop is greater than 10% of the drawdown.

Solution

We first use Equation 5-20 to calculate the well flow rate for each combination of wellbore pressure (and hence drawdown) and horizontal permeability. We then apply Equation 8-95 to find the ratio of the wellbore pressure drop to the reservoir drawdown. Results for several combinations of permeability values and drawdowns are given in Table 8-4.

Table 8-4 Comparison of Wellbore and Reservoir Pressure Drops

k_x	Δp_r	q	Δp_{ratio}
50	500	4633	0.01
50	1000	9266	0.02
100	500	9266	0.04
100	1000	18533	0.08
1000	50	9266	0.41
1000	100	18533	0.78
500	50	4633	0.11
500	100	9266	0.21
500	200	18533	0.39

Table 8-4 shows that for a 4000-ft horizontal well, the wellbore pressure drop is relatively small for permeability values of 100 md or less. For example, for a horizontal permeability of 100 md and a drawdown of 1000 psi, the wellbore pressure drop is about 8% of the drawdown. Any combination of lower permeability or lower drawdown results in a smaller wellbore pressure drop relative to the reservoir drawdown. However, the wellbore pressure drop becomes important in high permeability reservoirs. With 500 md horizontal permeability and only 50 psi drawdown, the wellbore pressure drop is 11% of the drawdown. In a 1000 md reservoir with 100 psi drawdown, the wellbore pressure drop is 78% of the reservoir drawdown. This amount of wellbore pressure drop is an unnecessary limiting factor on the productivity of such a well. A larger-diameter wellbore or shorter lateral would more efficiently produce such a reservoir.

The calculation in this example uses one-half of the total well flow rate to calculate the wellbore pressure drop in a single calculation. This has been shown to be a reasonable approximation for the purpose of comparing wellbore Δp to reservoir drawdown (Hill and Zhu, 2008).

8.5.2 Wellbore Pressure Drop for Single-Phase Flow

A rigorous method to calculate the actual wellbore pressure drop for a horizontal well with single-phase flow is to divide the well into a number of segments and use the average flow rate in the segment to compute the frictional pressure drop in that segment. The simplest approach is to calculate the pressure drop for the entire horizontal wellbore in one step by using one-half of the total rate of the well in the calculation. With these approaches, the standard equations for pressure drop in pipe flow (Equation 8-1 for liquid, Equations 8-6 or 8-7 for gas) are used.

There is some effect of the inflow occurring along the horizontal well on the pressure drop in the wellbore, as shown by Yuan, Sarica, and Brill (1996, 1998), Ouyang et al. (1998), and others. These studies led to empirical methods for calculating wellbore pressure drop. These effects are usually small (Hill et al., 2008).

8.5.3 Wellbore Pressure Drop for Two-Phase Flow

The pressure drop behavior in a horizontal well producing two or more phases can be calculated using the two-phase flow correlations presented previously for horizontal two-phase pipe flow. The wellbore should be divided into a series of increments so that the assumption of steady flow conditions for each segment (notably, constant flow rates of each phase) is reasonable.

Example 8-14 Pressure Drop in a High-Rate Horizontal Well

A horizontal well with a 5-in. I.D. is producing 15,000 bbl/d at a GOR of 1000 along a 3000-ft horizontal well from the saturated reservoir. A large inflow of 5000 STB/d occurs at about 1000 feet from the heel, another large inflow of 6000 STB/d occurs at 2000 feet, and the remaining 4000 STB/d are produced near the toe at 3000 feet. The pressure at the beginning of the horizontal section is 3000 psia. Estimate the pressure profile along the horizontal well.

Solution

The Beggs and Brill method is appropriate to calculate the pressure drops between each fluid entry location, since this method is applicable to horizontal flow. This may underestimate the overall pressure drop because there will be additional turbulence at the fluid entry locations that is not considered in our calculations.

Beginning with the known pressure at location 1, the total liquid rate at this point is 15,000 STB/d and the temperature is 180°F (the temperature is assumed constant throughout the well and equal to the reservoir temperature). First, we calculate the downhole flow conditions. Following the correlations in Sections 3.2 and 4.2, and Examples 3-3, 4-2, and 4-4, we find:

$$R_s = 562 \text{ ft}^3/\text{bbl} \qquad B_o = 1.29 \qquad B_g = 5.071 \times 10^{-3}$$

$$\rho_o = 46.8 \text{ lb}_m/\text{ft}^3 \qquad \rho_g = 10.7 \text{ lb}_m/\text{ft}^3$$

$$\mu_o = 0.69 \text{ cp} \qquad \mu_g = 0.02 \, cp$$

$$q_l = 19{,}350 \text{ bbl/d} = 1.26 \text{ ft}^3/\text{sec} \qquad q_g = 3.33 \times 10^4 \text{ ft}^3/\text{d} = 0.39 \text{ ft}^3/\text{sec}$$

From the downhole volumetric flow rates and the well cross-sectional area, we calculate

$$\lambda_l = 0.766 \qquad \lambda_g = 0.234$$

$$u_m = 12.05 \text{ ft/sec}$$

From Equations (7-125) through (7-130), we find $N_{FR} = 10.8$, $L_1 = 292$, $L_2 = 1.79 \times 10^{-3}$, $L_3 = 0.147$, and $L_4 = 3.01$. Since $\lambda_l \geq 0.4$ and $N_{FR} \geq L_4$, the flow regime is distributed.

Since the well is horizontal, the potential energy pressure drop is zero. However, we still must calculate the holdup, y_l, for the frictional pressure gradient. Using the constants for distributed flow, from Equation (7-136),

$$y_{lo} = \frac{1.065(0.766)^{0.5824}}{(10.8)^{0.0609}} = 0.788. \quad \textbf{(8-98)}$$

Since the well is horizontal, $\psi = 1$ and $y_l = y_{lo}$. Next, we calculate the mixture density and viscosity from Equations (7-143) and (7-146):

$$\rho_m = (0.766)(46.8) + (0.234)(10.7) = 38.4 \text{ lb}_m/\text{ft}^3 \quad \textbf{(8-99)}$$

$$\mu_m = (0.766)(0.69) + (0.234)(0.02) = 0.533 \text{ cp} \quad \textbf{(8-100)}$$

The mixture Reynolds number is [from Equation (7-145)]

$$N_{\mathrm{Re}_m} = \frac{(38.4)(12.05)(5/12)(1488)}{0.533} = 538{,}000 \tag{8-101}$$

and from the Moody diagram, the no-slip friction factor, $f_{n,} = 0.0032$. Using Equations (7-147) through (7-149), the two-phase friction factor, f_{tp}, is computed to be 0.0046. Finally,

$$\frac{dp}{dz} = \left(\frac{dp}{dz}\right)_{\mathrm{F}} = \frac{2f\rho_m u_m^2}{g_c D} = \frac{(2)(0.0046)(38.4)(12.05)^2}{(32.17)/(5/12)}$$

$$= 382 \text{ lb}_f/\text{ft}^3 = 0.0265 \text{ psi/ft} \tag{8-102}$$

and the pressure drop over the first 1000 ft of wellbore is

$$\Delta p = \left(\frac{dp}{dz}\right)(L) = (0.0265)(1000) = 26.5 \text{ psi} \tag{8-103}$$

For the next 1000 ft of wellbore, the oil flow rate is 10,000 STB/d. Since the pressure has changed so little, we can use the same downhole fluid properties as before. Thus, the gas and liquid flow rates will be in the same proportion, and the only significant difference is that u_m is now 8.03 ft/sec. Repeating the calculation procedure, the flow regime is also distributed flow between points 2 and 3 and the pressure drop is 11 psi.

In the last 1000 ft of wellbore, the oil flow is 4000 STB/d, so $u_m = 3.21$ ft/sec. The flow regime in this section is predicted to be intermittent and the overall pressure drop is 1.7 psi. The total pressure drop over the entire horizontal section is predicted to be only 39 psi. For these calculations to be valid, the well must be perfectly horizontal; slight changes in inclination would lead to some potential energy contributions to the pressure drop and may significantly change the pressure profile in a "horizontal" well.

Many nominally horizontal wells are not truly horizontal and slight variations in inclination can have profound effects on multiphase flow. Some "horizontal" wells are angled upward from the heel to the toe (a "toe-up" well); some are angled slightly downward from the heel to the toe (a "toe-down" well); and others may have both upward and downward inclined sections (an undulating well). These trajectories are sometimes intentional for purposes such as augmenting artificial lift of liquids from gas wells. Other times, the final trajectory of the well being not perfectly horizontal is simply a result of the directional drilling process or because of a dip in the reservoir. For wells with constant inclination near horizontal, any of the horizontal pipe two-phase flow correlations presented here are applicable.

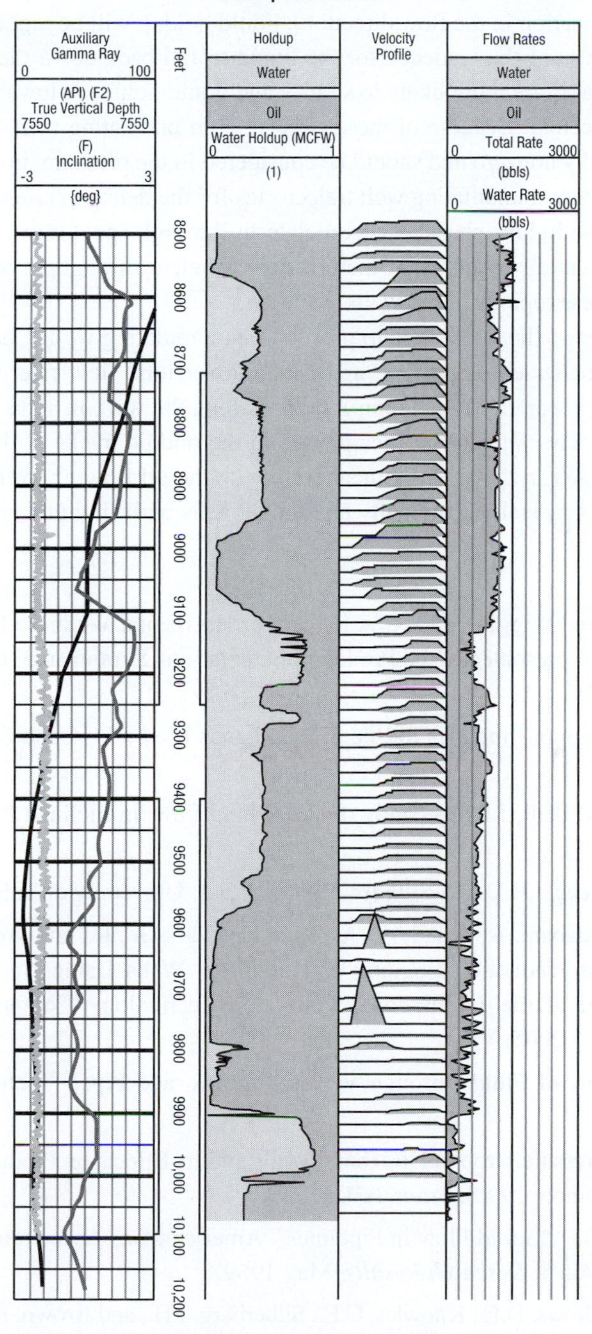

Figure 8-17 Horizontal well flow profile with small choke size.

For undulating wells, the upward and the downward sections must be separated. For upward flow (an upward inclination in the flow direction), liquid holdup will be higher than in a perfectly horizontal pipe because of the tendency for the liquid to fall back down the well. In downward flow sections, the phases are more likely to stratify and liquid holdup is low compared with a perfectly horizontal trajectory. Because of these effects, in an undulating well, the potential energy pressure drop is usually nonzero and should be considered in the pressure drop calculations.

It is also common in undulating well trajectories for the denser phase to accumulate in low spots (sumps) and the lighter phase to accumulate in the high spots (crests). These accumulations can remain essentially stagnant if there is not sufficient throughput to carry the accumulated fluids out of these regions of the well.

Figure 8-17 shows the water holdup profile of an undulating well (Chandran et al., 2005). Producing with a smaller choke setting, and hence lower total flow rate, the log shows large water holdups in the lower elevation (sump) regions along the well, in spite of the low produced water cut of 7–10%. The velocity profiles plotted in the middle tracks of these logs shows that the velocity in the water in the sumps is near zero, with the oil flowing over the standing water. This type of profile is typical of nominally horizontal wells with slight undulations in trajectory.

References

1. Ankalm, E.G., and Wiggins, M.L., "A Review of Horizontal Wellbore Pressure Equations," SPE Paper 94314 presented at the Production Operations Symposium, Oklahoma City, OK, April 17–19, 2005.
2. Baker, O., "Design of Pipelines for the Simultaneous Flow of Oil and Gas," *Oil and Gas J.,* **53:** 185 (1953).
3. Beggs, H.D., and Brill, J.P., "A Study of Two-Phase Flow in Inclined Pipes," *JPT,* 607–617 (May 1973).
4. Brill, J.P., and Beggs, H.D., *Two-Phase Flow in Pipes,* University of Tulsa, Tulsa, OK, 1978.
5. Chandran, T., Talabani, S., Jehad, A., Al-Anzi, E., Clark, Jr., R., and Wells, J.C., "Solutions to Challenges in Production Logging of Horizontal Wells Using New Tool," SPE Paper 10248 presented at the International Petroleum Technology Conference, Doha, Qatar, November 21–23, 2005.
6. Crane Co., "Flow of Fluids through Valves, Fittings, and Pipe," Technical Paper No. 48, Chicago, 1957.
7. Dikken, B.J., "Pressure Drop in Horizontal Wells and Its Effects on Production Performance," *JPT,* 1426 (November 1990); *Trans. AIME,* 289.
8. Dukler, A.E., "Gas–Liquid Flow in Pipelines," American Gas Association, American Petroleum Institute, Vol. 1, *Research Results,* May 1969.
9. Eaton, B.A., Andrews, D.E., Knowles, C.E., Silberberg, I.H., and Brown, K.E., "The Prediction of Flow Patterns, Liquid Holdup, and Pressure Losses Occurring during Continuous Two-Phase Flow in Horizontal Pipelines," *Trans. AIME,* **240:** 815–828 (1967).

10. Folefac, A.N., Archer, J.S., and Issa, R.I.: "Effect of Pressure Drop Along Horizontal Wellbores on Well Performance," SPE Paper 23094 presented at the Offshore Europe Conference, Aberdeen, September 3–6, 1991.

11. Gilbert, W.E., "Flowing and Gas-Lift Well Performance," *API Drilling and Production Practice,* 143 (1954).

12. Hill, A. D., Zhu, D., and Economides, M. J., *Multilateral Wells,* Society of Petroleum Engineers, Richardson, TX, 2008.

13. Mandhane, J.M., Gregory, G.A., and Aziz, K., "A Flow Pattern Map for Gas–Liquid Flow in Horizontal Pipes," *Int. J. Multiphase Flow,* **1:** 537–553 (1974).

14. Novy, R.A., "Pressure Drop in Horizontal Wells: When Can They Be Ignored?" *SPERE,* 29 (February 1995); *Trans. AIME,* 299.

15. Omana, R., Houssiere, C., Jr., Brown, K.E., Brill, J.P., and Thompson, R.E., "Multiphase Flow through Chokes," SPE Paper 2682, 1969.

16. Ouyang, L.-B., and Aziz, K., "A Simplified Approach to Couple Wellbore Flow and Reservoir Inflow for Arbitrary Well Configurations," SPE Paper 48936 presented at the SPE Annual Technical Conference and Exhibition, New Orleans, LA., September 27–30, 1998.

17. Ozkan, E., Sarica, C., and Haciislamoglu, M.: "Effect of Conductivity on Horizontal-Well Pressure-Behavior," *SPE Advanced Technology Series,* p. 85 (March 1995).

18. Penmatcha, V.R., Arbabi, S., and Aziz, K., "Effects of Pressure Drop in Horizontal Wells and Optimum Well Length," *SPE Journal,* **4**(3): 215–223 (September 1999).

19. Ros, N.C.J., "An Analysis of Critical Simultaneous Gas/Liquid Flow through a Restriction and Its Application to Flowmetering," *Appl. Sci. Res.,* **9**, Sec. A, 374 (1960).

20. Scott, D.S., "Properties of Concurrent Gas–Liquid Flow," *Advances in Chemical Engineering, Volume 4,* Dres, T.B., Hoopes, J.W., Jr., and Vermeulen, T., eds., Academic Press, New York, pp. 200–278, 1963.

21. Szilas, A.P., *Production and Transport of Oil and Gas,* Elsevier, Amsterdam, 1975.

22. Taitel, Y., and Dukler, A.E., "A Model for Predicting Flow Regime Transitions in Horizontal and Near Horizontal Gas-Liquid Flow," *AICHE J.,* **22**(1): 47–55 (January 1976).

23. Wallis, G.B., *One Dimensional Two-Phase Flow,* McGraw-Hill, New York, 1969.

24. Xiao, J.J., Shoham, O., and Brill, J.P., "A Comprehensive Mechanistic Model of Two-Phase Flow in Pipes," SPE Paper 20631 presented at the SPE ATCE, New Orleans, LA, September 23–26, 1990.

25. Yuan, H., and Zhou, D., "Evaluation of Two Phase Flow Correlations and Mechanistic Models for Pipelines at Horizontal and Inclined Upward Flow," SPE Paper 120281 presented at the SPE Production and Operations Symposium, Oklahoma City, OK, April 4–8, 2009.

26. Yuan, H., Sarica, C., and Brill, J.P., "Effect of Perforation Density on Single Phase Liquid Flow Behavior in Horizontal Wells," SPE Paper 37109 presented at the SPE International Conference on Horizontal Well Technology, Calgary, Canada, 1996.

27. Yuan, H., Sarica, C., and Brill, J.P., "Effect of Completion Geometry and Phasing on Single Phase Liquid Flow Behavior in Horizontal Wells," SPE Paper 48937 presented at the 1998 SPE Annual Technical Conference and Exhibition, New Orleans, LA, September 27–30, 1998.

Problems

8-1 Suppose that 3000 bbl/d of injection water is supplied to a well by a central pumping station located 2000 ft away, where the pressure is 400 psig. The water has a specific gravity of 1.02 and a viscosity of 1 cp. Determine the smallest-diameter flow line (within the nearest 1/2 in.) that can be used to maintain a wellhead pressure of at least 300 psig if the pipe relative roughness is 0.001.

8-2 Create a plot of pressure drop as a function of flow rate in a pipeline for the fluid defined in Appendix A. The pipe is 4000 ft long and the I.D. is 2 in. with relative roughness 0.006. Vary flow rate from 500 bbl/day to 10,000 bbl/day.

8-3 Suppose that 20 MMscf/d of natural gas with specific gravity of 0.7 is connected to a pipeline with 4000 ft of 2-in. flow line. The pipeline pressure is 200 psig and the gas temperature is 150°F. Calculate the wellhead pressure assuming: (a) smooth pipe; (b) $\varepsilon = 0.001$; and (c) a 20/64-in. choke ($\alpha = 0.9$) is added to the flowline of the smooth pipe.

8-4 Using the Baker, Mandhane et al., and Beggs and Brill flow regime maps, find the flow regime for the flow of 500 bbl/d oil and 800 scf/bbl of associated gas in a 2-in. flow line. The oil and gas are those described in Appendix B, $\sigma_l = 20$ dynes/cm, the temperature is 120°F, and the pressure is 800 psia.

8-5 Using the Beggs and Brill, Eaton, and Dukler correlations, generate a plot of the pressure gradient versus oil flow rate for the fluid in Appendix B. The oil and gas flow in a 3-in. I.D. line with a relative roughness of 0.001. $T = 150$°F, $p = 200$ psia, and $\sigma_l = 20$ dynes/cm. Neglect the kinetic energy pressure drop. Assume that no water is present, and the GOR is 600 scf/stb.

8-6 For the flow described in Problem 8-5, assume that the pressure given is at the wellhead. What is the maximum possible length of this flow line?

8-7 Construct choke performance curves for flowing tubing pressures, up to 800 psi for the well in Appendix A for choke sizes of 8/64, 12/64, and 16/64 in.

8-8 Construct choke performance curves for flowing tubing pressures up to 800 psi for the well in Appendix B for choke sizes of 8/64, 12/64, and 16/64 in.

8-9 Construct choke performance curves for flowing tubing pressures up to 800 psi for the gas well in Appendix C for choke sizes of 8/64, 12/64, and 16/64.

8-10 The liquid production from several rod-pumped wells producing from the reservoir of Appendix A are connected to a separator with the piping network shown in Figure 8-18. The relative roughness of all pipes is 0.001. For a separator pressure of 150 psig and

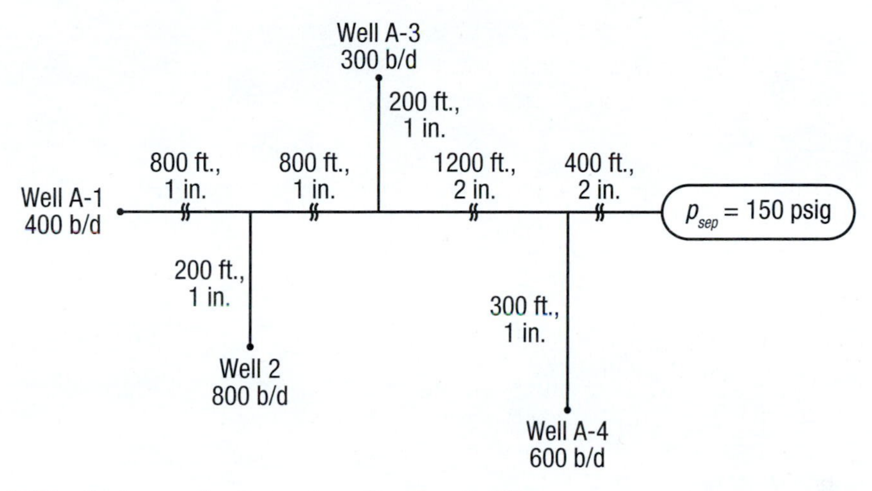

Figure 8-18 Surface gathering system (Problem 8-10).

assuming that the temperature is approximately 80°F throughout the system, find the wellhead pressure of the wells.

8-11 Redesign the piping network of Problem 8-10 so that no wellhead pressure is greater than 225 psig by changing the pipe diameters of as few pipe segments as possible.

8-12 An undulating wellbore can be simplified as a series of up- and down-inclined pipes. Assume the fluid in Appendix B with GOR of 500 scf/stb is uniformly distributed along the well shown in Figure 8-19, that each section has a 1000 bbl/day oil entering the section at the start, and that each section is 1000 ft long. If the pipe has an I.D. of 3 in., and $\varepsilon = 0.0006$, calculate and plot the pressure at each low point of the well. The start point has a temperature $T = 150°F$ and pressure $p = 1000$ psia.

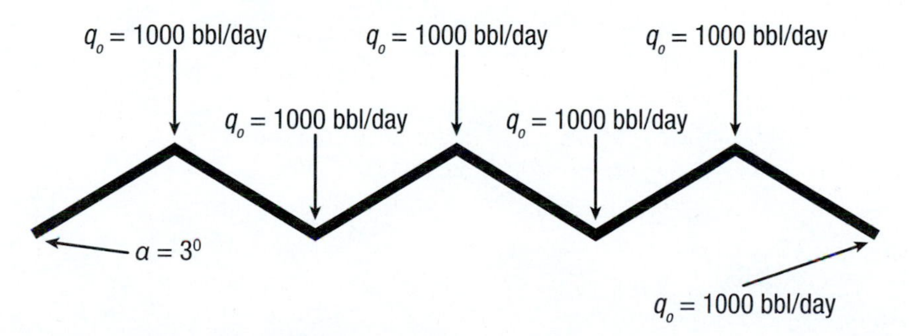

Figure 8-19 Well trajectory for Problem 8-12.

CHAPTER 9

Well Deliverability

9.1 Introduction

Chapters 2, 3, 4, and 5 described well inflow performance of oil, two-phase, and gas reservoirs, for vertical, inclined, and complex well architectures. The inflow performance relationship (IPR) was presented in a traditional and standardized manner, with the bottomhole flowing pressure on the ordinate of a graph and the corresponding production rate on the abscissa. This type of depiction gives a comprehensive picture of what the reservoir can deliver into the well at a specific time. Except for the case of steady-state flow, the well IPR would change with time. This would be both under infinite-acting conditions and under pseudosteady-state conditions where the *reservoir* pressure declines with time.

The reservoir inflow performance must be combined with the well's vertical flow performance (VFP). Wellbore flow was addressed in Chapters 7 and 8. For a required wellhead flowing pressure, p_{tf}, there exists a corresponding bottom hole flowing pressure, p_{wf}, which is a function of the hydrostatic pressure difference and the frictional pressure losses in the tubing string. Both of these variables are related implicitly to the pressure values themselves. Density differences and phase changes affect both the hydrostatic and the frictional pressure drops. For two-phase flow an increase in the imposed wellhead pressure ordinarily would result in an even larger increase in the corresponding bottomhole pressure because gas will be redissolved, increasing the density of the fluid in the wellbore. Two-phase flow in the well is common for almost all oil reservoirs because even if the bottomhole flowing pressure is above the bubble point, the wellhead pressure is likely to be significantly below the bubble point. Also, many gas wells have a free liquid phase, either condensate or water or both, flowing in the tubing. Thus, it is common to combine a single-phase oil or gas IPR with two-phase VFP.

Increasing the flowing gas–liquid ratio (GLR) would result in a reduction in the bottomhole pressure. This is the purpose of gas lift. However, there exists a limiting GLR where the decrease in the hydrostatic pressure with be offset by the increase in the frictional pressure drop. Gas lift performance is outlined in Chapter 11, while pump-assisted lift is discussed in Chapter 12.

In this chapter natural vertical flow performance is combined with the IPR to estimate the well deliverability. The material consists largely of examples, each intended to demonstrate an important issue.

9.2 Combination of Inflow Performance Relationship (IPR) and Vertical Flow Performance (VFP)

The traditional manner of solving the problem is depicted graphically in Figure 9-1. The IPR curve is shown in the p_{wf} versus q plot. Then, for a given wellhead pressure, p_{tf}, the bottomhole flowing pressure, p_{wf}, is calculated for each flow rate through an application of the mechanical energy balance. (See Chapter 7 and, especially, Section 7.2.3.) This is the VFP curve, plotted as shown in Figure 9-1. The intersection of the two curves provides the expected production rate and the bottomhole flowing pressure. Usually, the VFP curve is largely linear, with a relatively small slope. For low-GLR fluids the hydrostatic pressure would comprise the overwhelming portion of the pressure gradient in the well. Therefore, the frictional pressure drop would be relatively small and, since it is the pressure component affected most by the flow rate, the associated VFP curve is likely to be flat. For higher GLR values or for a gas well, the VFP curve is not expected to be linear. When two-phase gas–liquid flow is occurring in the tubing, there is usually a minimum p_{wf} on the VFP curve. This minimum occurs because at low liquid flow rates, the liquid holdup, and hence the hydrostatic pressure drop, decreases with increasing flow rate. The minimum occurs at the flow rate at which the increase in frictional pressure drop for rates above the minimum is greater than the decrease in hydrostatic pressure drop. At this value of flow rate the change in the frictional pressure drop is equal in magnitude to the change in hydrostatic pressure drop.

Example 9-1 Calculation of VFP for Single-Phase Fluid and Combination with Single-Phase IPR

Use the well in Appendix A with 160-acre well spacing. Assume that the depth is 8000 ft, the oil gravity, γ_o, is equal to 0.88 (API° $= 28$, $p_o = 55$ lb/ft^3), the tubing size is 2 3/8-in. (I.D. $\approx$ 2 in.) with a roughness $\epsilon = 0.0006$. What is the expected production rate and the corresponding bottomhole pressure if the wellhead pressure is 0 psi? The reservoir operates under steady-state conditions (use the IPR curve developed in Example 2-8 for $s = 0$).

Solution

The IPR curve for this problem is [from Equation (2-9)]

$$p_{wf} = 5651 - 5.58q. \tag{9-1}$$

Ignoring the kinetic energy pressure drop, as indicated by Example 7-3, the potential and frictional pressure drops must be calculated for each flow rate.

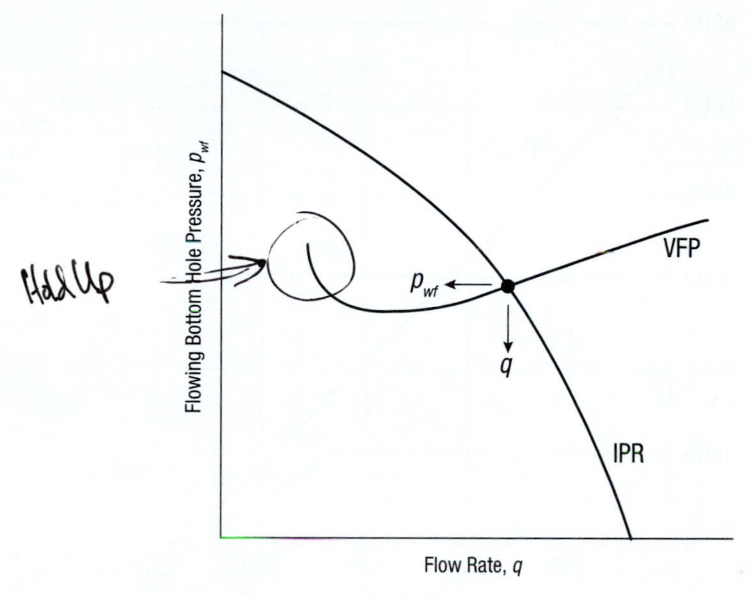

Figure 9-1 Well deliverability: Combination of inflow performance relationship (IPR) and vertical flow performance (VFP).

Since the fluid in this example is considered single phase, and largely incompressible, the potential energy pressure drop (hydrostatic pressure) would be the same, regardless of the well-head pressure and rate. Therefore, from Equation (7-22),

$$\Delta p_{\mathrm{PE}} = (0.433)(0.88)(8000) = 3048 \text{ psi}. \quad \textbf{(9-2)}$$

The frictional pressure drop must then be calculated for each flow rate. For example, if $q_o = 200$ STB/d, the Reynolds number is [from Equation (7-7) and using the properties in Appendix A]

$$N_{\mathrm{Re}} = \frac{(1.48)(200)(55)}{(2)(1.72)} = 4732, \quad \textbf{(9-3)}$$

which implies turbulent flow.

From Equation (7-35) and $\epsilon = 0.0006$, the Fanning friction factor, f_f, is 0.0096. The velocity, u, is

$$u = \frac{(4)(200)(5.615)(1/86{,}400)}{\pi(2/12)^2} = 0.6 \text{ ft/sec} \quad \textbf{(9-4)}$$

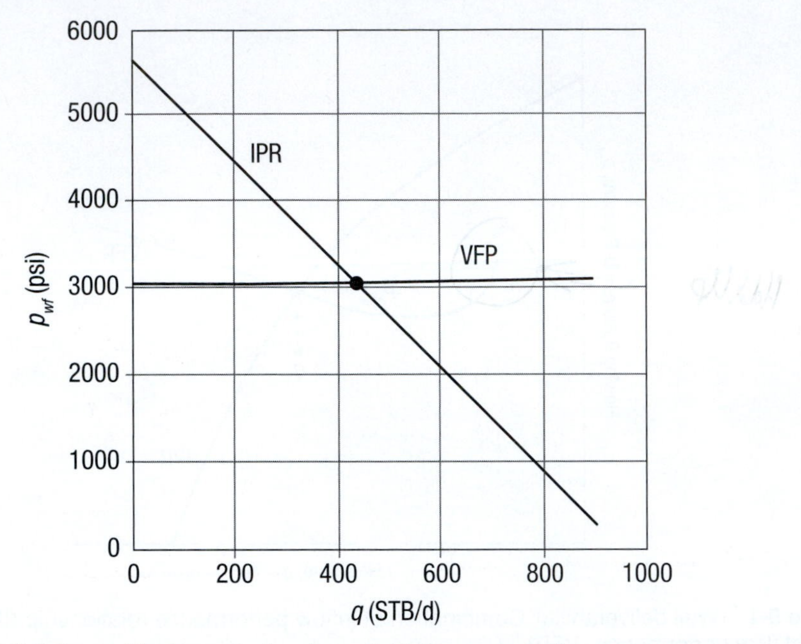

Figure 9-2 Well deliverability for a single-phase fluid (Example 9-1).

and therefore, from Equation (7-31),

$$\Delta p_F = \frac{(2)(0.0096)(55)(0.6)^2(8000)}{(32.17)(2/12)} = 568 \text{ lb}_f/\text{ft}^2 = 3.9 \text{ psi}, \tag{9-5}$$

which is insignificant when compared to the hydrostatic pressure drop.

The total pressure drop is the sum of the results of Equations (9-2) and (9-5), that is, 3052 psi. p_{wf} is equal to this pressure drop, since $p_{tf} = 0$. Figure 9-2 shows the combination of the IPR and VFP (mainly a flat, straight line) for this problem. The intersection is at $p_{wf} = 3055$ psi with $q = 430$ STB/d.

Example 9-2 Impact of Gas–Liquid Ratio (GLR) on Vertical Flow Performance (VFP)

Assume that the reservoir in Appendix A ($R_s = 250$ SCF/STB) is completed with a well with 2 3/8-in. (I.D. $\approx$ 2 in.) tubing. If the wellhead flowing pressure, p_{tf}, is 100 psi, develop gradient curves for $q = 500$ STB/d and GLR values from 100 to 800 SCF/STB. All GLRs larger than 250 must be supplied through gas lift. What are the bottomhole flowing pressures and the flowing pressure gradients (between 5000 and 8000 ft) for GLRs equal to 300 and 800 SCF/STB, respectively? The average flowing temperature is 130°F.

Solution

Using the modified Hagedorn and Brown method (see Section 7.4.3 and Example 7-11), Figure 9-3 can be constructed readily. The bottomhole flowing pressures with GLR values of 300 and 800 SCF/STB are 1976 and 1270 psi, respectively. The corresponding flowing gradients (between 5000 and 8000 ft) are 0.30 and 0.16 psi/ft.

Example 9-3 Increasing the Wellhead Flowing Pressure

For the well in Appendix A (GLR = 250 SCF/STB and tubing-size I.D. ≈ 2 in.) develop VFP curves for p_{tf} = 0, 100, and 500 psi, respectively, for the same depth of 8000 ft. What are the production rates with the steady-state IPR used in Example 9-1?

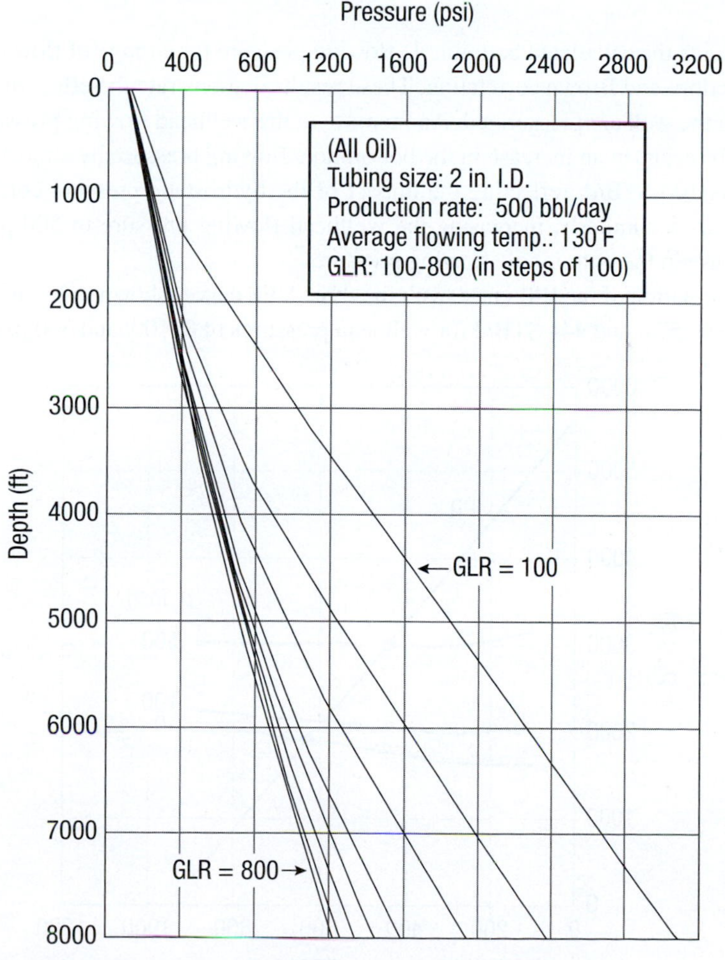

Figure 9-3 Gradient curves for a range of GLR values (Example 9-2).

Table 9-1 Flowing Bottomhole Pressures for Various Production Rates for Example 9-3

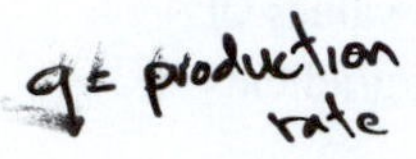

	p_{wf} (psi)		
q(STB/d)	***pw$_f$* (psi)**	***pw$_f$* (psi)**	***pw$_f$* (psi)**
100	1631.0	2077.0	3182.0
300	1668.0	2011.0	3018.0
500	1827.0	2099.0	3025.0
700	1976.0	2175.0	3040.0
900	2118.0	2251.0	3064.0

Solution

Table 9-1 contains the calculated bottomhole flowing pressure for a range of flow rates using the modified Hagedorn and Brown correlation. These results demonstrate the effect of phase change as the fluids in the well are pressurized. An increase in the wellhead flowing pressure by 100 psi (from 0 to 100) results in an increase in the bottomhole flowing pressure by almost 500 psi when the flow rate is 100 STB/d, reflecting the impact of the hydrostatic pressure component in the total pressure drop. Similarly, increasing the wellhead flowing pressure to 500 psi results in a 1400 psi increase in the bottomhole flowing pressure.

The combination of the IPR curve with the three VFP curves shown in Figure 9-4 results in flow rates of 608, 577, and 444 STB/d for wellhead pressures of 0, 100, and 500 psi, respectively.

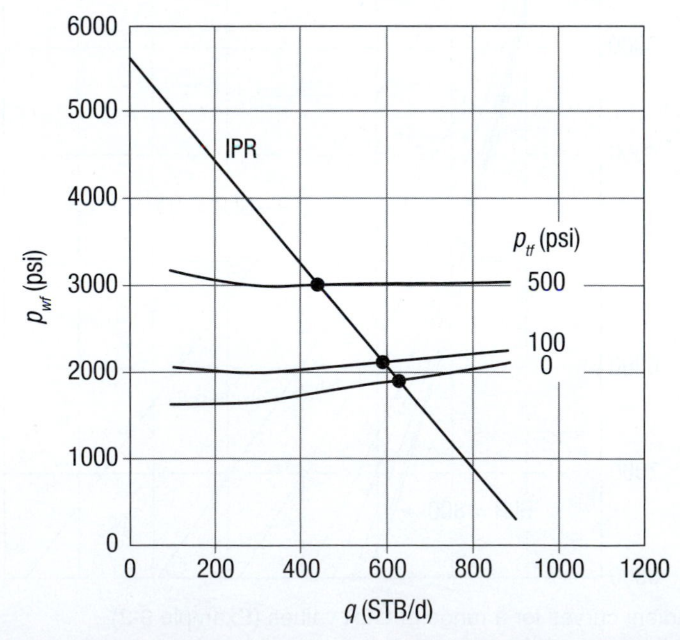

Figure 9-4 Effect of wellhead flowing pressure on well deliverability (Example 9-3).

This problem demonstrates that the design of wellhead and surface equipment is inherently related to the well deliverability.

Example 9-4 Changes in the IPR

Well deliverability is affected both by the well flow performance and the reservoir inflow. The previous two examples examined the influence of GLR and the wellhead pressure on the well performance and, hence, well deliverability. The reservoir inflow is affected by the near-well situation (damage or stimulation) and the evolution of the average reservoir pressure.

In this example, use the IPR curves of Example 2-8 (different skin effects) and Example 2-9 (declining reservoir pressure) with the VFP curve for $p_{tf} = 100$ psi, developed in Example 9-3, to show the effects of skin and declining reservoir pressure on well deliverability.

Solution

Figure 9-5 is the graphical solution of the skin effect impact on the IPR and the resulting changes in well deliverability. While the production rate would be 577 STB/d with zero skin, it would be reduced to 273 and less than 73 STB/d with skins equal to 10 and 50, respectively. However, because VFP values computed to the left of the pressure minimum are considered to be unstable, the flow rate of 73 STB/d is uncertain.

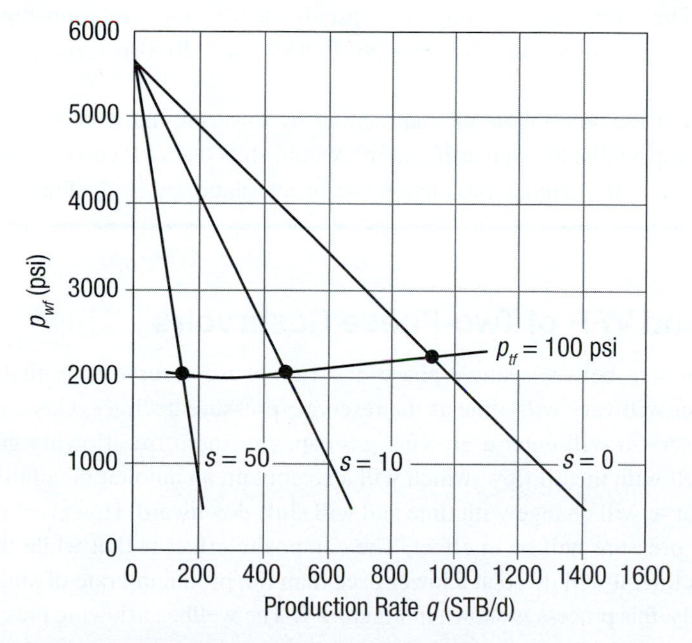

Figure 9-5 Skin effect and well deliverability (Example 9-4).

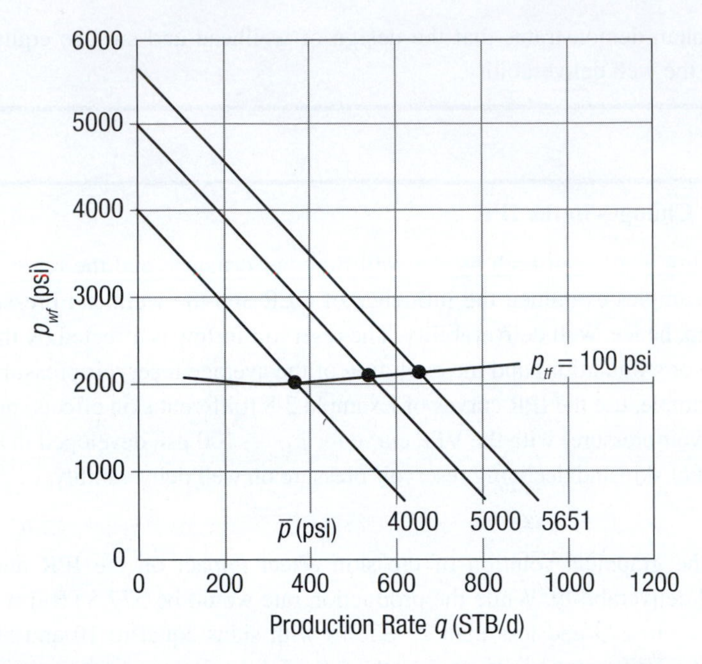

Figure 9-6 Well deliverability for declining reservoir pressure (Example 9-4).

The declining reservoir pressure and its impact on well deliverability are shown in Figure 9-6. [The IPR curve using the pseudosteady-state relationship is given by Equation (2-50).] The flow rates for $\overline{p}$ = 5651, 5000, and 4000 psi are 614, 496, and 319 STB/d, respectively.

Reduction in the bottomhole pressure, either by lowering the wellhead pressure (usually not feasible) or, especially, through artificial lift, would shift the VFP curve downward, resulting in the maintenance of the original production rate or retardation of its decline.

9.3 IPR and VFP of Two-Phase Reservoirs

The main difference between single-phase and two-phase reservoirs is that the producing GLR in the latter will vary with time as the reservoir pressure declines. Gas coming out of solution in the reservoir will enter a growing gas cap, causing a free-flowing gas. This will be mixed in the well with the oil flow, which will also contain an amount of solution gas. As a result, the VFP curve will change with time and will shift downward. However, a new IPR at the lower reservoir pressure will be in effect. The composite effect is that while the well producing rate will decline, it will do so at a lower pace than the producing rate of single-phase reservoirs. Graphically, this process is shown in Figure 9-7. The wellhead flowing pressure is constant.

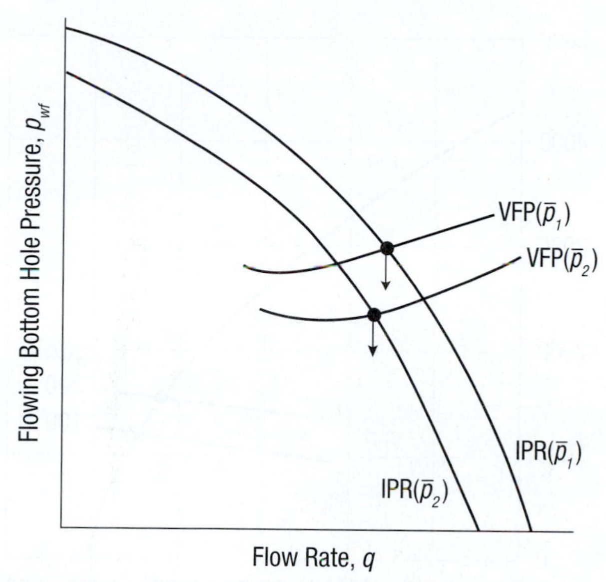

Figure 9-7 IPR and VFP curves for two average reservoir pressures *(p2 < p1)*. The flowing well head pressure is constant.

A comprehensive example of a forecast of well performance, taking into account the reservoir decline behavior, is presented in Chapter 10.

The behavior depicted in Figure 9-7 will be observed under natural flow conditions. This process will be more pronounced at early time in reservoirs with high initial pressure and that contain fluids with high solution gas–oil ratios. At a later time, as the producing GLR increases with time, the gap between the corresponding VFP curves is reduced. Ultimately, they are likely to overlap.

Example 9-5 Well Deliverability from a Two-Phase Reservoir

Using the well in Appendix B, a drainage radius equal to 1490 ft, and a skin effect equal to 0, develop an IPR curve for the (initial) average reservoir pressure of 4336 psi. For a 4 in. tubing I.D., show the influence of the wellhead flowing pressure by using $p_{tf} = 100$ and 300 psi, respectively.

Solution

The IPR expression for a two-phase reservoir was given in Section 3.4 [Equation (3-55)]. With the variables in Appendix B and the ones given in this example, it becomes

$$q_o = 5317\left[1 - 0.2\frac{p_{wf}}{\overline{p}} - 0.8\left(\frac{p_{wf}}{\overline{p}}\right)^2\right]. \tag{9-6}$$

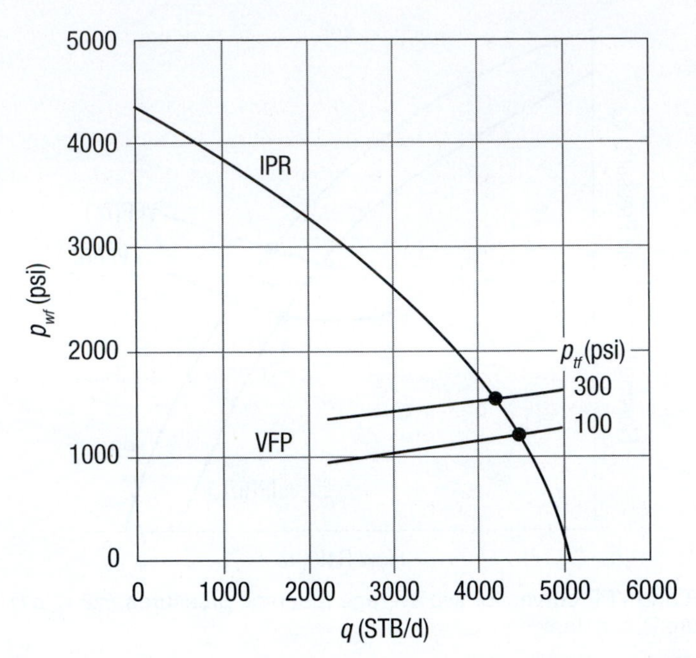

Figure 9-8 IPR and VFP curves for two wellhead flowing pressures in a two-phase reservoir (Example 9-5).

The IPR curve with average values of the formation volume factor, B_o, is plotted in Figure 9-8. Also on Figure 9-8 are the two VFP curves for the two wellhead pressures (100 and 300 psi). The corresponding production rates are 4180 and 4493 STB/d, respectively. The bottomhole flowing pressures are approximately 1549 and 1177 psi, a difference of 372 psi, compared to the 200 psi wellhead pressure difference, reflecting the more liquid-like phase distribution at the higher wellhead pressure.

9.4 IPR and VFP in Gas Reservoirs

Equations (4-56) and (4-62) in Chapter 4 describe the inflow performance of a gas well under pseudosteady-state and transient conditions, respectively. Both of these expressions use the real-gas pseudo-pressure function $m(p)$, and they are the most appropriate forms of the gas IPR.

Approximate expressions can be obtained by substituting for the real-gas pseudo-pressures with the pressure-squared difference divided by the product of the average values of the viscosity and the gas deviation factor. For example, the pseudosteady-state IPR is given by Equation (4-32).

The same expression contains the non-Darcy coefficient, D, which accounts for turbulence effects evident in high-rate gas wells.

The vertical flow performance of a gas well consists of the hydrostatic and friction pressure drop components, just as for any other well. However, the friction pressure contribution is likely to be large, in contrast to largely liquid reservoirs. Hence, the tubing diameter and its selection become critical for gas wells. This, of course, is true for all high-rate reservoirs.

Section 7.3 and Example 7-5 presented the methodology for the calculation of the flowing bottomhole pressure for a gas well. A graph of bottomhole flowing pressure versus the gas rate is the VFP curve for the well.

Example 9-6 Well Deliverability for a Gas Reservoir

With the data for the gas well in Appendix C, $r_e = 1490$ ft, $\overline{\mu} = 0.0249$ cp, $\overline{Z} = 0.96$, $c_t = 8.610^{-5}$ psi^{-1}, $s = 0$, and $D = 1.5 \times 10^{-4}$(MSCF/d)$^{-1}$, construct an IPR curve using Equations (4-50) and (4-51) (i.e., assume that the well is stabilized).

What is the well deliverability if the wellhead flowing pressure is 500 psi? The depth of the well is 10,000 ft, the inside tubing diameter is 2.259 in., and the relative roughness is 0.0006. Assume that the wellhead temperature is 140°F.

Solution

The coefficients a and b in Equation (4-51) are

$$a = \frac{(1424)(0.0249)(0.96)(640)}{(0.17)(78)}\left[\ln\frac{(0.472)(1490)}{0.328}\right] = 1.3 \times 10^4 \quad \text{(9-7)}$$

$$b = \frac{(1424)(0.0249)(0.96)(640)(1.5 \times 10^{-4})}{(0.17)(78)} = 0.25, \quad \text{(9-8)}$$

leading to an IPR given by

$$\overline{p}^2 - p_{wf}^2 = (1.3 \times 10^4)q + 0.25q^2. \quad \text{(9-9)}$$

The IPR curve is plotted in Figure 9-9 ($\overline{p} = 4613$ psi). The absolute open-flow potential (at $p_{wf} = 0$) is 1588 MSCF/d. For such a small D the non-Darcy effects are minimal. Ignoring the second term in Equation (9-9) would result in an absolute open-flow potential equal to 1640 MSCF/d.

From the intersection of the VFP curve, developed as shown in Example 7-5, the expected gas production rate would be 1550 MSCF/d.

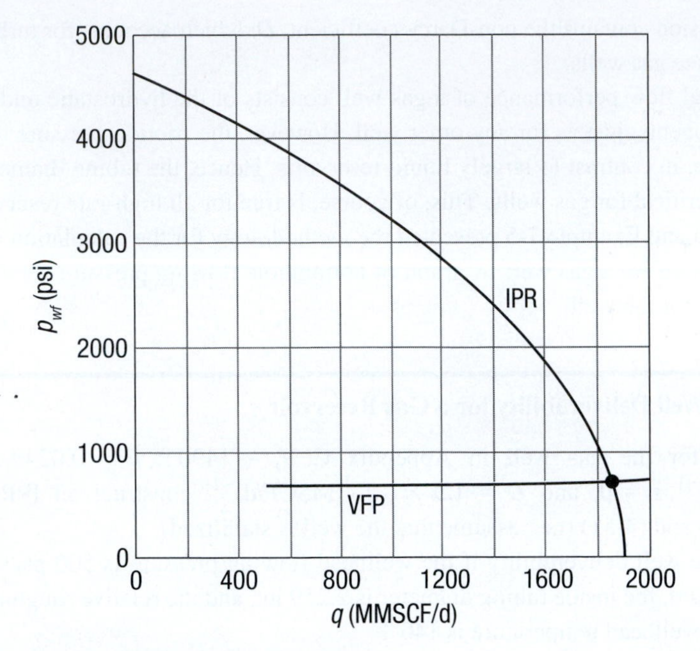

Figure 9-9 Gas well deliverability for Example 9-6.

Example 9-7 High-Rate Gas Well: Effect of Tubing Size

A gas well in a reservoir with $\overline{p} = 4613$ psi has an IPR given by

$$\overline{p}^2 - p_{wf}^2 = 25.8q + 0.115q^2. \quad \textbf{(9-10)}$$

Assuming that all physical properties are as for the gas well in Appendix C, shown with combinations of IPR and VFP the impact of the tubing size.

Solution

The absolute open-flow potential for this well is 13.6×10^3 MSCF/d. The IPR curve for this problem is shown in Figure 9-10, along with the VFP curves for four common tubing sizes (2 3/8, 2 7/8, 3 1/2, and 4 in. with corresponding I.D.s of 2, 2.44, 3, and 3.48 in.). As can be extracted readily from Figure 9-10, production rates from the well would be 10.7, 12.3, 13, and 13.2 MMSCF/d for the four tubing sizes, respectively. This exercise shows the relative benefits from

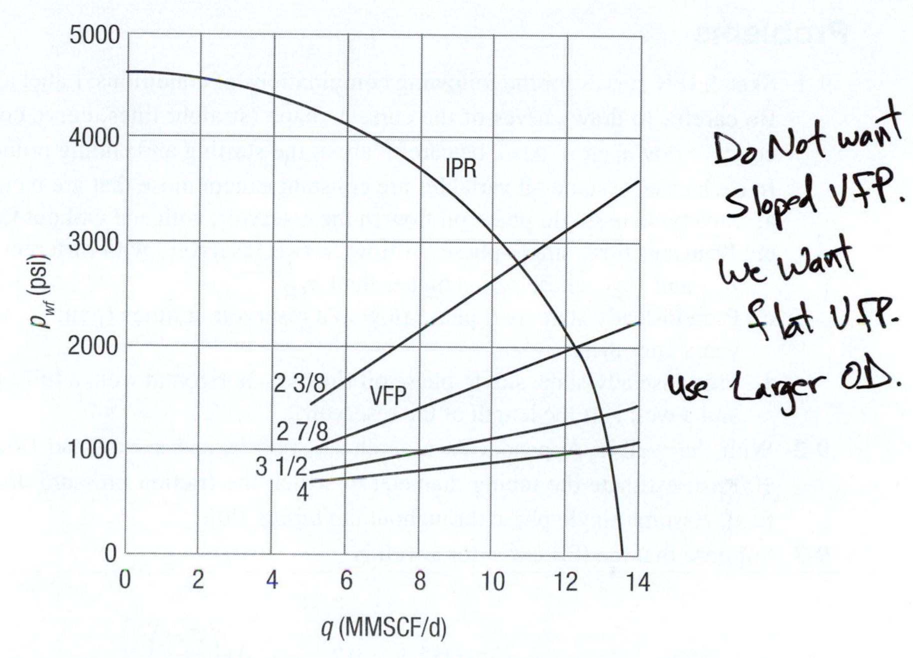

Figure 9-10 Deliverability from a high-rate gas well with a range of tubing diameters (Example 9-7).

the appropriate tubing size selection in a gas well. This is particularly important for gas wells that are likely candidates for hydraulic fracturing (such as the well in Appendix C with permeability $k = 0.17$ md). The anticipated benefits from the improved IPR, following a successful fracturing treatment, may be reduced if the tubing size is not selected on the basis of the expected post-treatment rather than the pretreatment inflow.

This example illustrates the general interrelation among reservoir inflow performance and vertical flow performance on well deliverability. When the VFP curve intersects a very steep portion of the IPR curve, like the 3 1/2 in. and 4 in. VFP curves, the production rate is insensitive to the tubing size. In this case, to significantly increase the production rate, the well must be stimulated and the production rate is *reservoir-limited.* If, on the other hand, the VFP curves intersect a portion of the IPR that has a shallower slope, or the VFP curves themselves have a significant positive slope, as did the 2 3/8 in. and 2 7/8 in. VFP curves, changing the tubing conditions (or adding artificial lift) can significantly increase the production rate, while well stimulation will have less benefit. Such wells are referred to as *tubing-limited.* In general, a tubing-limited condition should be avoided if possible.

Problems

9-1 Sketch IPR curves for the following combinations of conditions. Label all curves clearly. Be careful to draw curves of the correct shape (straight lines, curve concave up, curve concave down, etc.). Also, be careful about the starting and ending points of each curve. In each case, assume all variables are constant except those that are mentioned.

- **a.** Steady-state single-phase oil flow in the reservoir; with and without formation damage
- **b.** Transient flow, single-phase oil flow in two reservoirs with different initial pressures p_{re1} and p_{re2}, where p_{re1} is higher than p_{re2}
- **c.** Pseudosteady-state, two-phase flow in a reservoir at times t_1 and t_2, with t_2 being two years later than t_1
- **d.** Pseudosteady-state, single-phase oil flow to a horizontal well, a fully-penetrating well, and a well half the length of the reservoir.

9-2 With the well in Appendix A, a depth of 8000 ft, and a wellhead flowing pressure of 100 psi, estimate the tubing diameter at which the friction pressure drop is 20% of the total. Assume single phase throughout the tubing flow.

9-3 Suppose that the IPR curve for a well is

$$q_o = 10{,}000\left[1 - 0.2\frac{p_{wf}}{\overline{p}} - 0.8\left(\frac{p_{wf}}{\overline{p}}\right)^2\right] \quad \textbf{(9-11)}$$

with $\overline{p} = 500$ psi (and using an average B_o). If GOR = 300 SCF/STB, $\overline{T} = 150°F$, $p_{tf} = 100$ psi, and tubing I.D. = 3 in., calculate the well deliverability for depths of 4000, 6000, and 8000 ft. For all other properties use the well in Appendix B.

9-4 Is it possible for the tubing diameter in Problem 9-3 to be 2 in.? If so, what would be the well deliverability if the depth is 6000 ft?

9-5 Suppose that a horizontal well with $L = 2000$ ft and $I_{ani} = 3$ were to be drilled in the reservoir described in Problem 9-3. What would be the well deliverability? (Assume that the well has an average trajectory of 45° and that there is negligible pressure drop in the horizontal portion of the well.)

9-6 The gas well in Appendix C is hydraulically fractured, and the treatment results in an equivalent skin effect equal to –6. Develop a gas well IPR with the data in Appendix C and Example 9-6. The flowing wellhead pressure is 500 psi. How does the performance compare with the results of Example 9-6 for the well in the pretreatment state and with a tubing I.D. equal to 1 in.? What would be the performance for a 2-in. tubing I.D.?

9-7 Assume that a horizontal well with $L = 2500$ ft is drilled in the reservoir described in Appendix C and Example 9-6. If the horizontal well $r_w = 0.328$ ft, what is the well deliverability? The vertical well has a 3-in. tubing I.D. Calculate the pressure drop in the horizontal section.

CHAPTER 10

Forecast of Well Production

10.1 Introduction

In the previous chapter the well inflow and vertical flow performances (IPR and VFP) were combined to provide the well deliverability. The intersection of the plots of flow rate versus bottomhole flowing pressure of those two components of the reservoir and well system is the expected well deliverability. This intersection point describes a specific instant in the life of the well. It depends greatly on the type of flow regime controlling the well performance.

For infinite-acting behavior, transient IPR curves can be drawn from a single pressure value on the p_{wf} axis. This is the initial reservoir pressure. The intersections with the VFP curve at different times are the expected flow rates (and corresponding p_{wf} values).

By definition, for steady-state flow, a single IPR curve would intersect the VFP curve. For pseudosteady-state flow the situation is more complicated because the average reservoir pressure will change with time. Given the values of the average reservoir pressure, $\overline{p}$, a family of IPR curves can be drawn, each intersecting the p_{wf} axis at the individual reservoir pressures. What is needed is the element of time. Material balance calculations, relating underground withdrawal and reservoir pressure depletion, can provide the crucial pressure-time relationship. Therefore, the associated well deliverability would be the forecast of well production.

In all cases, integration of the flow rate versus time would lead to the cumulative production, which is an essential element in any economic decision making about the viability of any petroleum production engineering project.

10.2 Transient Production Rate Forecast

In an infinite-acting reservoir, the well IPR can be calculated from the solution to the diffusivity equation. The classic approximations for the well flow rate were given in Chapter 2 for an oil well [Equation (2-23)] and in Chapter 4 for a gas reservoir [Equation (4-62)]. For a two-phase reservoir, Equation (2-23) can be adjusted further, as shown in Section 3.5. These expressions would lead to transient IPR curves, one for each production time. Example 2-7 and associated Figure 2-5 demonstrate the method of calculation and presentation for the transient IPR curves.

Unless the wellhead pressure changes, there will be only one VFP curve for the well, and its intersections with the IPR curves will be the well production rate. The cumulative production under any flow regime is simply

$$N_p = \int_0^t q(t)\,dt. \tag{10-1}$$

Example 10-1 Forecast of Well Production with Transient IPR

Use the transient IPR curves for the well in Appendix A as developed in Example 2-7. Then, with the VFP curve of Example 9-1, calculate and plot the expected well flow rate and cumulative production versus time.

Solution

Equation (2-25) is the transient IPR curve relationship for this well. If t = 1 month (t must be converted into hours), Equation (2-25) becomes

$$q = 0.304(5651 - p_{wf}). \tag{10-2}$$

Figure 10-1 shows on an expanded scale (all IPR curves would merge at $p_{wf} = p_i = 5651$ psi) the intersections between the IPR curves at 1, 6, and 12 months and the VFP curve. The

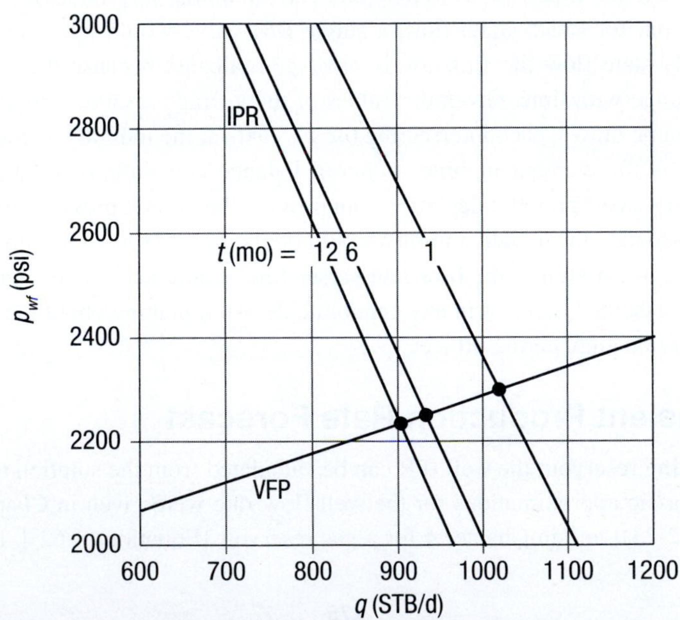

Figure 10-1 Transient IPR and VFP curves for Example 10-1.

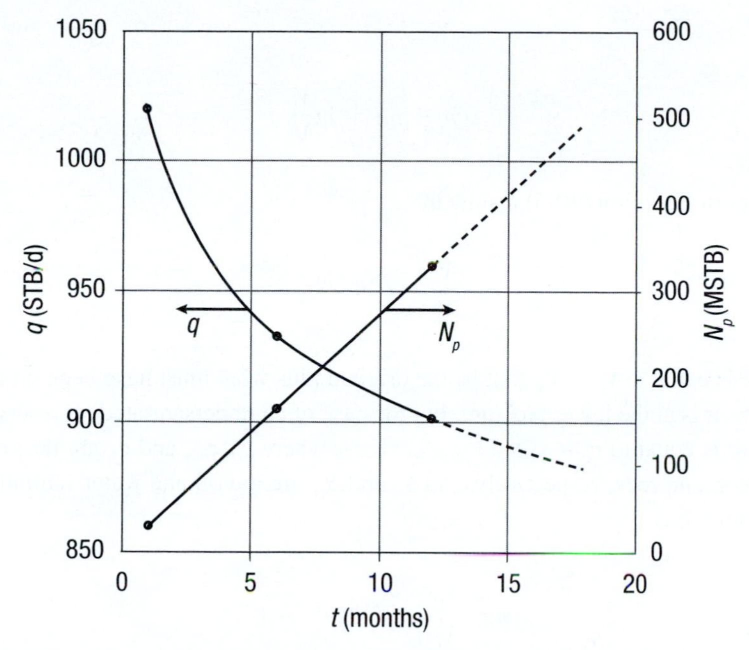

Figure 10-2 Flow rate and cumulative production forecast for Example 10-1.

corresponding production rates (and flowing pressures) would be 1020 (2284), 933 (2233), and 901 (2216) STB/d (psi), respectively.

Figure 10-2 is a graph of the well production rate versus time. The associated cumulative production is also shown on the same figure.

10.3 Material Balance for an Undersaturated Reservoir and Production Forecast Under Pseudosteady-State Conditions

The fluid recovery from an undersaturated oil reservoir depends entirely on the fluid expansion as a result of underground withdrawal and associated pressure reduction. This can be evaluated by starting with the expression for isothermal compressibility.

The isothermal compressibility is defined as

$$c = -\frac{1}{V}\frac{\vartheta V}{\vartheta p} \tag{10-3}$$

where V is the volume of the fluid. By separation of variables, assuming that c is small and largely constant for an oil reservoir,

$$\int_{p_i}^{\bar{p}} c dp = \int_{V_i}^{V}\left(-\frac{dV}{V}\right) \tag{10-4}$$

and therefore

$$c(\overline{p} - p_i) = \ln\frac{V_i}{V}. \tag{10-5}$$

Rearrangement of Equation (10-5) results in

$$\frac{V}{V_i} = e^{c(p_i - \overline{p})}. \tag{10-6}$$

The volume V is equal to $V_i + V_p$, that is, the original plus what must have been produced by the time the reservoir is at the lower pressure. For the case of an undersaturated oil reservoir, the total compressibility is equal to $c_t = c_o S_o + c_w S_w + c_f$, where c_o, c_w, and c_f are the compressibilities of oil, water, and rock, respectively, and S_o and S_w are the oil and water saturations. Finally, the recovery ratio, r, is

$$r = \frac{V_p}{V_i} = e^{c_t(p_i - \overline{p})} - 1. \tag{10-7}$$

If the original oil-in-place, N, is known, the cumulative recovery, N_p, is simply rN.

The general approach to the problem of forecasting the well performance under pseudosteady-state conditions is as follows.

- Assume an average reservoir pressure, $\overline{p}$. From Equation (10-7), the recovery ratio (and the corresponding cumulative recovery) can be calculated. Within each average reservoir pressure interval, an incremental cumulative production ΔN_p can be calculated.
- With the average reservoir pressure (a mean value within the pressure interval can be considered), the pseudosteady-state IPR relationship can be used. Equation (2-33) is for an undersaturated oil reservoir.
- The time is simply $t = \Delta N_p / q$, where q is the calculated average flow rate within the interval.

Thus, the well flow rate and the cumulative recovery are related to time.

Example 10-2 Well Production from an Oil Reservoir Under Pseudosteady-State Conditions

Assume that the well in Appendix A operates under pseudosteady-state conditions draining a 40-acre spacing. Calculate the well production rate and cumulative production versus time. If artificial lift is required when $\overline{p} = 4000$, calculate the time of its onset.

What is the maximum possible cumulative production from this drainage area at $\overline{p} = p_b = 1697$ psi? Use the VFP curve of Example 9-1.

Solution

The pseudosteady-state IPR for an undersaturated oil reservoir is given by Equation (2-33). With the variables in Appendix A, the 40-acre drainage area ($r_e = 745$ ft), $s = 0$, and allowing for the average reservoir pressure to vary, Equation (2-33) becomes

$$q = 0.357(\overline{p} - p_{wf}). \tag{10-8}$$

Figure 10-3 is the combination of IPR curves at different reservoir pressures $\overline{p}$ and the single VFP for the well. At $\overline{p} = 5651$, 5000, 4500, and 4000 psi, the corresponding well production rates would be approximately 1135, 935, 788, and 635 STB/d, respectively.

The next step is to calculate the cumulative production after each pressure interval. At first, the initial oil in place must be estimated. Using the variables in Appendix A, the initial oil in place within 40 acres is

$$N = \frac{(7758)(40)(53)(0.19)(1 - 0.34)}{1.17} = 1.76 \times 10^6 \text{ STB}. \tag{10-9}$$

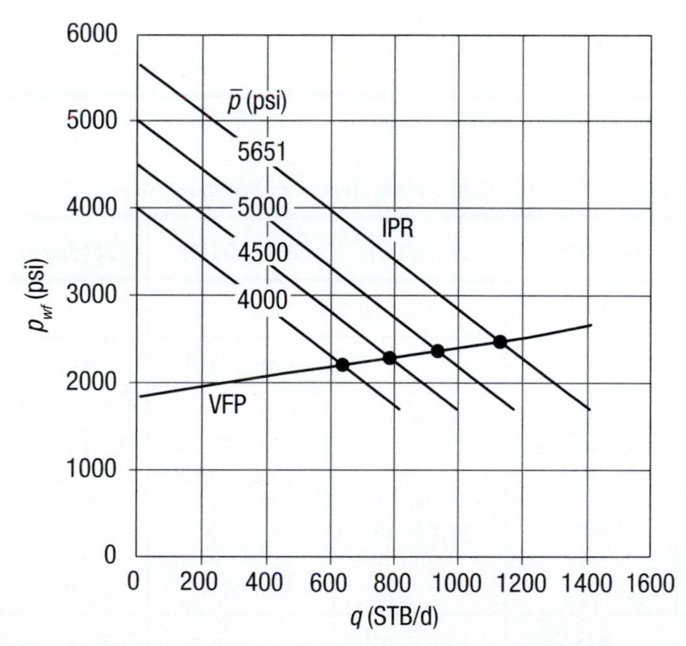

Figure 10-3 IPR and VFP curves for an oil well under pseudosteady-state conditions (Example 10-2).

Then, following the approach described in Section 10.3, a sample calculation is presented below. If $\overline{p} = 5000$ psi, from Equation (10-7) the recovery ratio can be obtained,

$$r = e^{1.29 \times 10^{-5}(5651-5000)} - 1 = 8.43 \times 10^{-3} \quad \text{(10-10)}$$

and therefore

$$N_p = (1.76 \times 10^6)(8.43 \times 10^{-3}) = 14{,}840 \text{ STB}. \quad \text{(10-11)}$$

If the average production rate in this interval (from Figure 9-3) is 1035 STB/d, the time required to produce the cumulative production of Equation (10-11) would be (14840/1035) $\approx$ 14 days.

Table 10-1 gives the production rate, incremental, and total cumulative recovery for the well in this example.

After 44 days the reservoir pressure would decline by 1650 psi, the rate would decline from 1135 to 635 STB/d, while the total recovery would be only 2.15% of the original oil-in-place. This calculation shows the inefficient recovery from highly undersaturated oil reservoirs. Also, after 44 days is when artificial lift must start, since this time coincides with $\overline{p} = 4000$ psi.

Finally, if $\overline{p} = p_b$, the maximum recovery ratio [Equation (10-7)] would be

$$r = e^{1.29 \times 10^{-5}(5651-1697)} - 1 = 0.052 \quad \text{(10-12)}$$

Table 10-1 Production Rate and Cumulative Recovery Forecast for a Well in Example 10-2

$\bar{p}$ (psi)	q (STB/d)	N_p (STB)	ΔN_p (STB)	Δt (days)	t (days)
5651	1135				
			1.48×10^4	14	
5000	935	1.48×10^4			14
			1.15×10^4	13	
4500	788	2.63×10^4			28
			1.16×10^4	16	
4000	635	3.79×10^4			44

10.4 The General Material Balance for Oil Reservoirs

Havlena and Odeh (1963, 1964) introduced the application of material balance to oil reservoirs that have initial oil-in-place N and a ratio m relating the initial hydrocarbon volume in the gas cap to the initial hydrocarbon volume in the oil zone.

10.4.1 The Generalized Expression

The generalized expression presented by Dake (1978) (and ignoring water influx and production) is

$$N_p[B_o + (R_p - R_s)B_g] = NB_{oi}\left[\frac{(B_o - B_{oi}) + (R_{si} - R_s)B_g}{B_{oi}} + m\left(\frac{B_g}{B_{gi}} - 1\right) + (1 + m)\left(\frac{c_w S_{wc} + c_f}{1 - S_{wc}}\right)\Delta p\right]. \quad \text{(10-13)}$$

In Equation (10-13), N_p is the cumulative oil production, R_p is the producing gas–oil ratio, and S_{wc} is the interstitial water saturation. All other variables are the usual thermodynamic and physical properties of a two-phase system. Groups of variables, as defined below, correspond to important components of production.

- Underground withdrawal:

$$F = N_p[B_o + (R_p - R_s)B_g] \quad \text{(10-14)}$$

- Expansion of oil and originally dissolved gas:

$$E_o = (B_o - B_{oi}) + (R_{si} - R_s)B_g \quad \text{(10-15)}$$

- Expansion of gas cap:

$$E_g = B_{oi}\left(\frac{B_g}{B_{gi}} - 1\right) \quad \text{(10-16)}$$

- Expansion of connate water and reduction of pore volume:

$$E_{f,w} = (1 + m)B_{oi}\left(\frac{c_w S_{wc} + c_f}{1 - S_{wc}}\right)\Delta p. \quad \text{(10-17)}$$

Finally, from Equations (10-13) through (10-17),

$$F = N(E_o + mE_g + E_{f,w}). \tag{10-18}$$

10.4.2 Calculation of Important Reservoir Variables

As shown by Havlena and Odeh (1963), Equation (10-18) can be used in the form of straight lines for the calculation of important reservoir variables by observing the well performance and using the properties of the fluids produced.

10.4.2.1 Saturated Reservoirs

For saturated reservoirs,

$$E_{f,w} \approx 0 \tag{10-19}$$

since the compressibility terms can be ignored, and hence

$$F = NE_o + NmE_g. \tag{10-20}$$

With no original gas cap ($m = 0$),

$$F = NE_o \tag{10-21}$$

a plot of F versus E_o would result in a straight line with slope N, the original oil-in-place.

With an original gas cap (N and m are unknown),

$$F = NE_o + NmE_g \tag{10-22}$$

or

$$\frac{F}{E_o} = N + Nm\left(\frac{E_g}{E_o}\right). \tag{10-23}$$

A graph of F/E_o versus E_g/E_o would result in a straight line with N the intercept and Nm the slope.

10.4.2.2 Undersaturated Reservoirs

In undersaturated reservoirs, since $m = 0$ and $R_p = R_s$, Equation (10-13) becomes

$$N_pB_o = NB_{oi}\left[\frac{B_o - B_{oi}}{B_{oi}} + \left(\frac{c_wS_{wc} + c_f}{1 - S_{wc}}\right)\Delta p\right]. \tag{10-24}$$

Equation (10-6) can be written in terms of the formation volume factors and simplified as a Taylor series dropping all terms with powers of 2 and larger:

$$B_o = B_{oi}e^{c_o\Delta p} \simeq B_{oi}(1 + c_o\Delta p) \tag{10-25}$$

therefore

$$\frac{B_o - B_{oi}}{B_{oi}} = c_o\Delta p. \tag{10-26}$$

Multiplying and dividing by S_o results in

$$\frac{B_o - B_{oi}}{B_{oi}} = \frac{c_o S_o \Delta p}{S_o}. \tag{10-27}$$

Since $S_o = 1 - S_{wc}$, then also

$$\frac{B_o - B_{oi}}{B_{oi}} = \frac{c_o S_o \Delta p}{1 - S_{wc}} \tag{10-28}$$

and therefore Equation (10-24) becomes

$$N_p B_o = N B_{oi}\left[\frac{c_o S_o \Delta p}{1 - S_{wc}} + \frac{c_w S_{wc} + c_f}{1 - S_{wc}}\Delta p\right]. \tag{10-29}$$

From the definition of the total compressibility, $c_t = c_o S_o + c_w S_{wc} + c_f$, this is

$$N_p B_o = \frac{N B_{oi} c_i \Delta p}{1 - S_{wc}}. \tag{10-30}$$

A graph of $N_p B_o$ versus $B_{oi}c_t\Delta p/(1 - S_{wc})$ would result in a straight line with a slope equal to N.

Equations (10-21), (10-23), and (10-30) describe different straight lines, the slope and/or intercepts of which provide the original oil-in-place and, if present, the size of the gas cap. The groups of variables to be plotted include production history, reservoir pressure, and associated fluid properties. Hence, well performance history can be used for the calculation of these important reservoir variables.

Example 10-3 Material Balance for Saturated Reservoirs[1]

An oilfield produced as in the schedule in Table 10-2. Fluid properties are also given. The reservoir pressure in the oil and gas zones varied somewhat; therefore, fluid properties are given for

[1] After an example from class notes by M.B. Standing (1978).

Table 10-2 Production and Fluid Data for Example 10-3

Date	N_p (STB)	G_p (MSCF)
5/1/09		
1/1/11	492,500	751,300
1/1/12	1,015,700	2,409,600
1/1/13	1,322,500	3,901,600

Date	$\bar{p}_o$ (psia)	$\bar{p}_g$ (psia)	B_t at $\bar{p}_o$ (res bbl/STB)	B_g at $\bar{p}_g$ (res bbl/SCF)
5/1/09	4415	4245	1.6291	0.00431
1/1/11	3875	4025	1.6839	0.00445
1/1/12	3315	3505	1.7835	0.00490
1/1/13	2845	2985	1.9110	0.00556

p_b = 4290 psia
R_{si} = 975 SCF/STB
B_{ob} = 1.6330 res bbl/STB
B_{oi} = 1.6921 res bbl/STB

the simultaneous pressures in the two zones. Calculate the best value of the initial oil in place (STB) and the initial *total* gas in place (MMSCF).

The oil zone thickness is 21 ft, the porosity is 0.17, and the water saturation is 0.31. Calculate the areal extent of the initial gas cap and the reservoir pore volume.

Reservoir simulation has shown that a good time for a waterflood start is when 16% of the original oil-in-place is produced. When will this happen if a constant flow rate is maintained?

Since B_t data are given in this example, the already developed equations for F, E_o, and E_g are developed below in terms of B_t. Because

$$B_t = B_o + (R_{si} - R_s)B_g \quad \textbf{(10-31)}$$

then

$$B_o = B_t - (R_{si} - R_s)B_g \quad \textbf{(10-32)}$$

and therefore

$$F = N_p[B_t + (R_p - R_{si})B_g] \tag{10-33}$$

$$E_o = B_t - B_{oi} \tag{10-34}$$

$$E_g = B_{oi}\left(\frac{B_g}{B_{gi}} - 1\right) \tag{10-35}$$

Solution

Table 10-3 contains the calculated $R_p(= G_p/N_p)$ and the variables F, E_o, and E_g as given by Equations (10-33) through (10-35). Also, in the manner suggested by Equation (10-23), the variables F/E_o and E_g/E_o are listed.

Figure 10-4 is a plot of the results, providing an intercept of 9×10^6 STB (= N, the initial oil-in-place) and a slope equal to $3.1 \times 10^7 (= Nm)$. This leads to $m = 3.44$. [If there were no initial gas cap, the slope would be equal to zero. In general, the use of Equation (10-23) instead of Equation (10-21) is recommended, since the latter is included in the former.]

The oil pore volume is then

$$V_{po} = NB_{oi} = (9 \times 10^6)(1.6291) = 1.47 \times 10^7 \text{ res bbl} \tag{10-36}$$

and the pore volume of the gas cap is

$$V_{pg} = mV_{po} = (3.44)(1.47 \times 10^7) = 5.04 \times 10^7 \text{ res bbl.} \tag{10-37}$$

The drainage area of the oil zone in acres is

$$A = \frac{V_{po}}{7758\phi h(1 - S_w)} \tag{10-38}$$

Table 10-3 Calculated Variables for the Material Balance Straight Line of Example 10-3

Date	R_p (SCF/STB)	F	E_o	E_g	F/E_o	E_g/E_o
1/1/11	1525	2.04×10^6	5.48×10^{-2}	5.29×10^{-2}	3.72×10^7	0.96
1/1/12	2372	8.77×10^6	1.54×10^{-1}	2.22×10^{-1}	5.69×10^7	1.44
1/1/13	2950	17.05×10^6	2.82×10^{-1}	4.72×10^{-1}	6.0×10^7	1.67

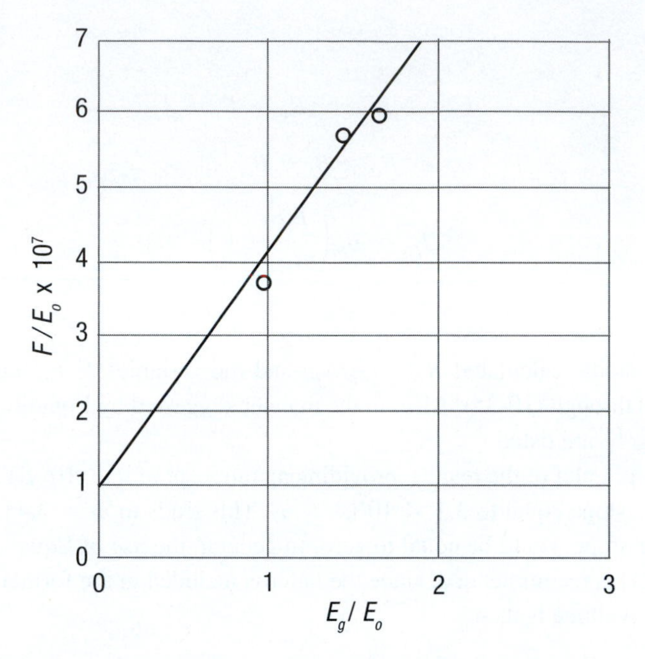

Figure 10-4 Material balance graph for a two-phase reservoir (Example 10-3).

and therefore

$$A = \frac{(1.47 \times 10^7)}{(7758)(0.17)(21)(1 - 0.31)} = 770 \text{ acres.} \tag{10-39}$$

If 16% of initial oil-in-place is to be produced, then

$$N_p = (0.16)(9 \times 10^6) = 1.44 \times 10^6 \text{ STB.} \tag{10-40}$$

From Table 10-2 the flow rate between 1/1/12 and 1/1/13 is approximately 840 STB/d. From 1/1/13 until the required date for water flooding, there will be a need for an additional 1.12×10^5 STB of production. With the same production rate it would require 140 days of additional production. Thus water flooding should begin at about May 13.

10.5 Production Forecast from a Two-Phase Reservoir: Solution Gas Drive

The general material balance of Equation (10-13) can be simplified further if the reservoir is assumed to have no initial gas cap but rapidly passes below the bubble-point pressure after production commences. These assumptions encompass a large number of reservoirs. This mode of recovery is known as solution gas drive, and the calculation method outlined below is referred to as Tarner's (1944) method. It was described further by Craft and Hawkins (1991).

With the assumption that $m=0$ and with gas evolving out of solution immediately (i.e., the water and rock compressibility terms can be neglected), Equation (10-13) becomes

$$N_p[B_o + (R_p - R_s)B_g] = N[(B_o - B_{oi}) + (R_{si} - R_s)B_g]. \tag{10-41}$$

Since G_p (gas produced) $= N_pR_p$, then

$$N_p(B_o - R_sB_g) + G_pB_g = N[(B_o - B_{oi}) + (R_{si} - R_s)B_g] \tag{10-42}$$

and therefore

$$N = \frac{N_p(B_o - R_sB_g) + G_pB_g}{(B_o - B_{oi}) + (R_{si} - R_s)B_g}. \tag{10-43}$$

Equation (10-43) has two distinct components: the oil cumulative production, N_p, and the gas cumulative production, G_p. Multiplying these components are groups of thermodynamic variables. With the following definitions,

$$\Phi_n = \frac{B_o - R_sB_g}{(B_o - B_{oi}) + (R_{si} - R_s)B_g} \tag{10-44}$$

and

$$\Phi_g = \frac{B_g}{(B_o - B_{oi}) + (R_{si} - R_s)B_g}. \tag{10-45}$$

Equation (10-43) becomes

$$N = N_p\Phi_n + G_p\Phi_g. \tag{10-46}$$

For the instantaneous, producing gas–oil ratio, R_p, given by

$$R_p = \frac{\Delta G_p}{\Delta N_p}. \tag{10-47}$$

If $N = 1$ STB, Equation (10-46) in terms of increments of production becomes

$$1 = (N_p + \Delta N_p)\Phi_n + (G_p + \Delta G_p)\Phi_g$$
$$= (N_p + \Delta N_p)\Phi_n + (G_p + R_p \Delta N_p)\Phi_g. \quad \textbf{(10-48)}$$

The instantaneous value of R_p is the average produced gas–oil ratio, $R_{p,av}$, within a production interval with incremental cumulative production ΔN_p. Therefore, in a stepwise fashion between intervals i and $i + 1$,

$$1 = (N_{pi} + \Delta N_{pi\to i+1})\Phi_{n,av} + (G_{pi} + R_{p,av}\Delta N_{pi\to i+1})\Phi_{g,\text{av}} \quad \textbf{(10-49)}$$

and solving for $\Delta N_{pi\to i+1}$,

$$\Delta N_{pi\to i+1} = \frac{1 - N_{pi}\Phi_{n,av} - G_{pi}\Phi_{g,av}}{\Phi_{n,av} + R_{p,av}\Phi_{g,av}}. \quad \textbf{(10-50)}$$

This is an important finding. It can predict incremental production for a decline in the reservoir pressure in the same manner as shown in Section 10.3 for an undersaturated reservoir. The following steps are needed:

1. Define Δp for calculation.
2. Calculate Φ_n and Φ_g using properties obtained for the average pressure in the pressure interval.
3. Assume a value of $R_{p,av} = R_{p,guess}$ in the interval.
4. Calculate $\Delta N_{pi\to i+1}$ from Equation (10-50).
5. Calculate $G_{pi\to i+1}(\Delta G_{pi\to i+1} = \Delta N_{pi\to i+1} R_{p,\, guess})$.
6. Calculate the oil saturation from

$$S_o = \left(1 - \frac{N_p}{N}\right)\frac{B_o}{B_{oi}}(1 - S_w). \quad \textbf{(10-51)}$$

7. Obtain the relative permeability k_g/k_o from a plot versus S_o. Relative permeability curves are usually available.
8. Calculate $R_{p,calc}$ from

$$R_{p,calc} = R_s + \frac{k_g \mu_o B_o}{k_o \mu_g B_g}. \quad \textbf{(10-52)}$$

9. Compare $R_{p,guess}$ to $R_{p,calc}$. If they are equal, both are equal to $R_{p,av}$ for this interval. If not, repeat steps 3–10 with a new value for $R_{p,guess}$.

With the average pressure within the interval, the Vogel correlation [Equation (3-53)] can be used for the calculation of the well IPR. Combination with VFP will lead to the well production rate. Finally, with the material balance outlined above, rate and cumulative production versus time can be established.

Example 10-4 Forecast of Well Performance in a Two-Phase Reservoir

The well described in Appendix B is 8000 ft deep, completed with a 3-in. I.D. tubing and with 100 psi flowing wellhead pressure. If the well drainage area is 40 acres, develop a forecast of oil rate and oil and gas cumulative productions versus time until the average reservoir pressure declines to 3336 psi (i.e., $\Delta \overline{p} = 1000$).

Solution

Since $p_i = p_b = 4336$, the solution gas drive material balance and Tarner's method outlined in Section 10.5 are appropriate. With the properties in Figure B-1 in Appendix B, the variables Φ_n and Φ_g can be calculated for a range of reservoir pressures.

From Example 10-2, the initial oil in place in the well drainage volume is 1.76×10^6 STB.

The following is a sample calculation for a 200 psi pressure interval between 4350 and 4150. The average B_o, B_g, and R_s are 1.45 res bbl/STB, 7.6×10^{-4} res bbl/SCF, and 778 SCF/STB, respectively. Noting that B_{oi} and R_{si} (at p_i) are 1.46 res bbl/STB and 800 SCF/STB, Equation (10-44) yields (for properties at the average pressure during the interval of 4236 psi)

$$\Phi_n = \frac{1.451 - (778.2)(7.607 \times 10^{-4})}{(1.451 - 1.462) + (800 - 778.2)(7.607 \times 10^{-4})} = 154. \quad \textbf{(10-53)}$$

From Equation (10-45),

$$\Phi_g = \frac{7.607 \times 10^{-4}}{(1.451 - 1.462) + (800 - 778.2)(7.607 \times 10^{-4})} = 0.136. \quad \textbf{(10-54)}$$

Similarly, these variables are calculated for four additional 200-psi intervals and are shown in Table 10-4.

For the first interval, assume that the producing $R_{p,guess} = 820$ SCF/STB. Then, from Equation (10-50),

$$\Delta N_{pi\to i+1} = \frac{1}{154 + (820)(0.136)} = 3.77 \times 10^{-3}\text{STB}. \quad \textbf{(10-55)}$$

which, for the first interval, is the same as $\Delta N_{pi\to i+1}$. The incremental gas cumulative production is $(820)(3.77 \times 10^{-3}) = 3.09$ SCF which, for the first interval, also coincides with G_p.

Table 10-4 Physical Properties for Tarner's Calculations (Example 10-4) for Well in Appendix B

$\bar{p}$ (psi)	B_o (res bbl/STB)	B_g (res bbl/SCF)	R_s (SCF/STB)	Φ_n	Φ_g
4336	1.462	7.48×10^{-4}	800		
	1.451	7.607×10^{-4}	779.6	154	0.136
4136					
	1.431	7.874×10^{-4}	734.7	42.5	0.0393
3936					
	1.410	8.170×10^{-4}	691.7	22.4	0.0232
3736					
	1.390	8.520×10^{-4}	649.1	14.9	0.0152
3536					
	1.370	8.935×10^{-4}	607.0	10.3	0.0112
3336					

From Equation (10-51),

$$S_o = (1 - 3.77 \times 10^{-3})\left(\frac{1.451}{1.462}\right)(1 - 0.3) = 0.692 \quad \textbf{(10-56)}$$

and from the relative permeability curves in Appendix B for $S_g = 0.008$, $k_{rg} = 0.000018$ and $k_{ro} = 0.46$. From Equation (10-52), using $\mu_o = 0.45$ cp and $\mu_g = 0.0246$ cp,

$$R_{p,calc} = 778.2 + \frac{0.000018}{0.46}\frac{0.45}{0.0246}\frac{1.451}{(7.607 \times 10^{-4})} = 779.6 \text{ SCF/STB}, \quad \textbf{(10-57)}$$

which is not equal to the assumed value. However, repeating the calculations in Equations (10-55), (10-56), and (10-57) with $R_{p,guess} = 779.6$ SCF/STB leads to the same value for $R_{p,calc}$. Therefore, for the first 200 psi change in average reservoir pressure, the resulting values are

$\Delta N_p = 3.68 \times 10^{-3}$ STB, $\Delta G_p = 3.18$ SCF, and $R_p = 779.6$ SCF/STB. Interestingly, the produced gas–oil ratio drops for this pressure interval. This phenomenon frequently occurs. Effectively, there is a critical gas saturation below which gas cannot flow.

The calculation is repeated for all intervals and appears in summary form in Table 10-5. There is an interesting comparison between the results in Table 10-5 and those in Table 10-1 for the undersaturated case. In the latter, after about 1000 psi reservoir pressure drop (from 5651 to 4500 psi), the recovery ratio would be less than 1.5% (i.e., $2.8 \times 10^4/1.87 \times 10^6$). For a similar pressure drop, the recovery ratio for the two-phase well in this example is about seven times higher, demonstrating the far more efficient recovery in a solution gas–drive reservoir.

For $A = 40$ acres, the initial oil-in-place (using the variables in Appendix B) is

$$N = \frac{(7758)(40)(115)(0.21)(1 - 0.3)}{1.46} = 3.59 \times 10^6 \text{ STB}. \quad \textbf{(10-58)}$$

Table 10-5 Cumulative Production with Depleting Pressure for Well in Example 10-4 (N = 1 STB)

$\bar{p}$ (psi)	ΔN_p (STB)	N_p (STB)	R_p (SCF/STB)	ΔG_p (SCF)	G_p (SCF)
4336					
	3.85×10^{-3}		780	3.0	
4136		3.85×10^{-3}			3.0
	1.37×10^{-2}		775	10.6	
3936		1.75×10^{-2}			13.6
	2.38×10^{-2}		839	20.0	
3736		4.14×10^{-2}			33.6
	3.36×10^{-2}		982	32.9	
3536		7.49×10^{-2}			66.5
	4.17×10^{-2}		1226	51.1	
3336		11.7×10^{-2}			118

The IPR expression for this two-phase reservoir can be obtained from the Vogel correlation [Equation (3-53)]. With $k_o = 8.2$ md, $h = 115$ ft, $r_e = 1490$ ft, $r_w = 0.328$ ft, $\mu_o = 1.7$ cp, $s = 0$, and allowing B_o to vary with the pressure, Equation (3-53) becomes

$$q_o = q_{o,\max}\left[1 - 0.2\frac{p_{wf}}{\overline{p}} - 0.8\left(\frac{p_{wf}}{\overline{p}}\right)^2\right], \tag{10-59}$$

with $q_{o,max} = 2632$ STB/d, computed using values for B_o and μ_o at $\overline{p}$, and the effective permeability, k_o, used for the Tarner calculation. The IPR curves for six average reservoir pressures (from 4350 to 3350 in increments of 200 psi) are shown in expanded form in Figure 10-5. At $q_o = 0$, each curve would intersect the p_{wf} axis at the corresponding $\overline{p}$.

Also plotted are five VFP curves for this well, each for the average producing gas–liquid ratio within the pressure interval.

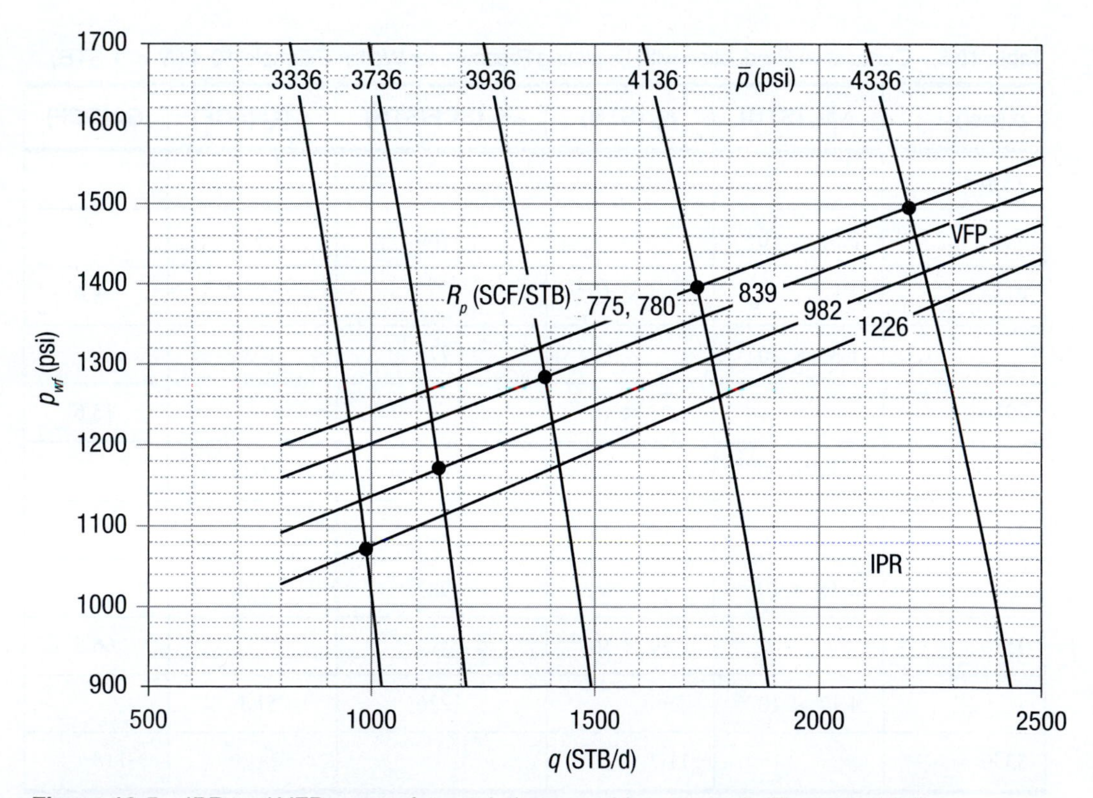

Figure 10-5 IPR and VFP curves for a solution gas–drive reservoir (Example 10-4).

All necessary variables are now available for the calculation of the time duration of each interval. For example, the recovery ratio (= ΔN_p with $N = 1$ STB) in the interval between 4350 and 4150 psi is 3.77×10^{-3}, and since $N = 3.7 \times 10^6$ STB,

$$\Delta N_p = (3.84 \times 10^{-3})(3.59 \times 10^6) = 1.38 \times 10^4 \text{ STB}. \quad \textbf{(10-60)}$$

Also,

$$\Delta G_p = (1.38 \times 10^4)(779.6) = 10.8 \times 10^6 \text{ SCF}. \quad \textbf{(10-61)}$$

The average production rate within the interval is 2200 STB/d (from Figure 10-5), and therefore

$$t = \frac{1.38 \times 10^4}{2200} = 6 \text{ days}. \quad \textbf{(10-62)}$$

The results for all intervals are shown in Table 10-6. Figure 10-6 shows the behavior of the oil and gas flow rates, and the produced gas–oil ratio versus the pressure depletion, $\Delta p = p_i - \bar{p}$.

Table 10-6 Production Rate and Oil and Gas Cumulative Recovery Forecast for Well in Example 10-4

$\bar{p}$ (psi)	q_o(STB/d)	ΔN_p (10^3 STB)	N_p (10^3 STB)	ΔG_p (10^6 SCF)	G_p (10^6 SCF)	Δt (d)	t(d)
4336							
	2200	13.8		10.8		6	
4136			13.8		10.8		6
	1730	49.1		38.1		28	
3936			62.9		48.9		35
	1380	85.5		71.7		62	
3736			148		121		97
	1150	120		118		104	
3536			269		239		201
	980	149		183		152	
3336			418		422		353

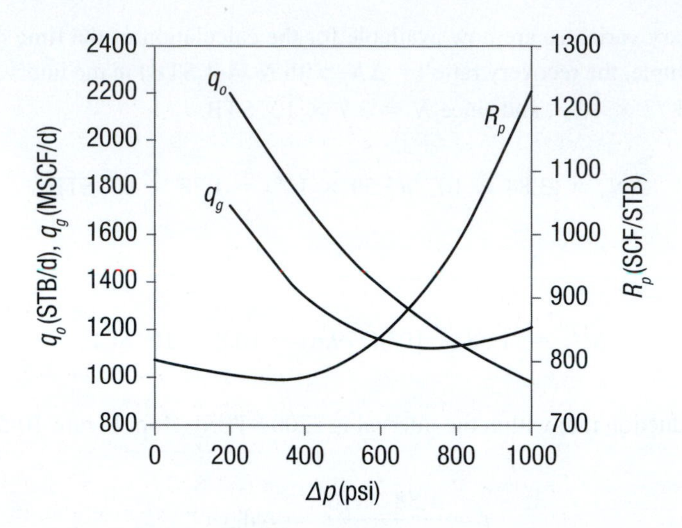

Figure 10-6 Produced oil and gas, and gas–oil ratio versus pressure depletion (Example 10-4).

(*Note:* This example was set up to illustrate both the recovery and time contrasts with the undersaturated reservoir in Example 10-2. Thus a drainage area of 40 acres was used for both cases. Ordinarily, a well such as the one in Example 10-5 would be assigned a much larger drainage area [e.g., 640 acres]. In such a case, while the production rate would not be affected markedly—being inversely proportional to the respective ln r_e—the time values in Table 10-6 would be more than 16 times larger.)

10.6 Gas Material Balance and Forecast of Gas Well Performance

If G_i and G are the initial and current gas-in-place within a drainage area, the cumulative production from a gas reservoir, considering the expansion of the fluid, is

$$G_p = G_i - G = G_i - G_i \frac{B_{gi}}{B_g} \quad \textbf{(10-63)}$$

when B_{gi} and B_g are the corresponding formation volume factors.

Equation (4-27) in Chapter 4 provides B_g in terms of pressure, temperature, and the gas deviation factor. Substitution in Equation (10-63), assuming isothermal operation throughout, and rearrangement results in

$$G_p = G_i\left(1 - \frac{\bar{p}/Z}{p_i/Z_i}\right). \quad \textbf{(10-64)}$$

This expression suggests that if G_p, the cumulative production, is plotted against $\overline{p}/Z$, it should form a straight line. Usually, the variable $\overline{p}/Z$ is plotted on the ordinate and the cumulative production on the abscissa. At $G_p = 0$, then, $\overline{p}/Z = p_i/Z_i$, and at $\overline{p}/Z = 0$, $G_p = G_i$. For any value of the reservoir pressure (and associated Z), there exists a corresponding G_p. Coupled with the gas well IPR expressions for pseudosteady-state flow [Equation (4-42)] or accounting for non-Darcy flow [Equation (4-50)], or using the more exact form [Equation (4-74)], a forecast of well performance versus time can be developed readily.

Example 10-5 Forecast of Gas Well Performance

Assume that a well in the gas reservoir in Appendix C drains 40 acres. The calculation of the initial gas-in-place with $p_i = 4613$ psi was presented in Example 4-5 (for 1900 acres). Develop a forecast of this well's performance versus time until the average reservoir pressure declines to 3600 psi. Use five pressure intervals of 200 psi each. The bottomhole flowing pressure is 1500 psi. Estimate the time to pseudosteady state using the approximation of Equation (2-42). The total compressibility, c_t, is 9.08×10^{-5} psi^{-1}. For the solution to this problem, assume pseudosteady-state conditions throughout and ignore the non-Darcy effects on the flow rate.

Solution

From Equation (2-42) and $r_e = 745$ ft (for $A = 40$ acres),

$$t_{\mathrm{pss}} = 1200\frac{(0.14)(0.0249)(9.08 \times 10^{-5})(745)^2}{0.17} = 1240 \text{ hr.} \tag{10-65}$$

After about 52 days the well will "feel" the no-flow boundaries.

Similar to Example 4-5, the initial gas-in-place within 40 acres (and using the calculated B_{gi}) is

$$G_i = \frac{(43{,}560)(40)(78)(0.14)(0.73)}{3.76 \times 10^{-3}} = 3.69 \times 10^9 \text{ SCF.} \tag{10-66}$$

With $Z_i = 0.96$,

$$\frac{p_i}{Z_i} = \frac{4613}{0.96} = 4805 \text{ psi.} \tag{10-67}$$

From Equation (10-64), the gas cumulative production is

$$G_p = 3.69 \times 10^9\left(1 - \frac{\overline{p}/Z}{4805}\right) \tag{10-68}$$

at any reservoir pressure $\overline{p}$.

Table 10-7 Production Rate and Cumulative Production Forecast for Gas Well in Example 10-5

$\bar{p}$(psi)	Z	G_p(BCF)	ΔG_p(BCF)	q(MMSCF/d)	Δt(d)	t(d)
4613	0.96	0				
			0.0956	2.16	44	
4400	0.94	0.0956				44
			0.127	2.0	63	
4200	0.93	0.222				108
			0.129	1.83	71	
4000	0.92	0.352				178
			0.132	1.66	80	
3800	0.91	0.484				258
			0.135	1.5	90	
3600	0.90	0.619				348

Table 10-7 lists the average reservoir pressures, the calculated gas deviation factors (see Table 4-5), and the corresponding cumulative production.

The production rate is obtained from Equation (4-71), cast in real-gas pseudopressures, and calculated with the average $\overline{p}$ within each pressure interval. Assuming that $s = 0$, Equation (4-42), rearranged, becomes

$$q = \frac{m(\overline{p}) - 1.821 \times 10^8}{4.8 \times 10^5} \tag{10-69}$$

where $m(p_{wf}) = 1.821 \times 10^8$ psi^2/ cp.

The values for $m(\overline{p})$ are computed using Equation (4-71).

References

1. Craft, B.C., and Hawkins, M. (revised by Terry, R.E.), *Applied Petroleum Reservoir Engineering,* 2nd ed., Prentice Hall, Englewood Cliffs, NJ, 1991.
2. Dake, L.P., *Fundamentals of Reservoir Engineering,* Elsevier, Amsterdam, 1978.
3. Havlena, D., and Odeh, A.S., "The Material Balance as an Equation of a Straight Line," *JPT,* 896–900 (August 1963).

4. Havlena, D., and Odeh, A.S., "The Material Balance as an Equation of a Straight Line. Part II—Field Cases," *JPT,* 815–822 (July 1964).

5. Tarner, J., "How Different Size Gas Caps and Pressure Maintenance Programs Affect Amount of Recoverable Oil," *Oil Weekly,* **144:** 32–34 (June 12, 1944).

Problems

10-1 Estimate the time during which the oil well in Appendix A would be infinite acting. Perform the calculations for drainage areas of 40, 80, 160, and 640 acres.

10-2 Develop a forecast of well performance using the VFP curve in Example 9-1 and transient IPR curves for the 640-acre drainage area for two times, when boundaries are first felt and for half that time. Use the data in Appendix A.

10-3 Calculate the average reservoir pressure of the well in Appendix A after 2 years. Use the VFP curve in Example 9-1. Plot production rate and cumulative production versus time for 40-acre and 640-acre spacings.

10-4 Suppose that the drainage area of the well in Appendix B is 640 acres. Assuming that the bottomhole flowing pressure is held constant at 1500 psi, calculate the average reservoir pressure after 3 years. Graph the oil production rate and the oil and gas cumulative productions versus time.

10-5 A well in a gas reservoir has been producing 3.1×10^3 MSCF/d for the last 470 days. The rate was held constant. The reservoir temperature is 624°R, and the gas gravity is 0.7. Assuming that $h = 31$ ft, $S_g = 0.75$, and $\phi = 0.21$, calculate the areal extent of the reservoir.

The initial reservoir pressure, p_i, is 3650 psi. Successive pressure buildup analysis reveals the following average reservoir pressures at the corresponding times from the start of production: 3424 psi (209 days), 3310 psi (336 days), and 3185 psi (470 days).

10-6 Assume that the production rate of the well in Problem 10-5 is reduced after 470 days from 3.1×10^3 MSCF/d to 2.1×10^3 MSCF/d. What would be the average reservoir pressure 600 days after the start of production?

10-7 Repeat the calculation of Example 10-5 but use the non-Darcy coefficient calculated in Example 4-8 [$D = 3.6 \times 10^{-4}$ (MSCF/d)$^{-1}$]. After how many days would the well cumulative production compare with the cumulative production after 345 days obtained in Example 10-5 and listed in Table 10-7?

CHAPTER 11

Gas Lift

11.1 Introduction

In Chapter 9 the relationship between tubing and inflow performance led to well deliverability. If the required bottomhole flowing pressure to deliver a certain inflow is smaller than the tubing pressure difference caused by a fluid with a given flowing gradient, then artificial lift may be employed.

Gas lift is one method of artificial lift. The other is mechanical lift, discussed in Chapter 12. For gas lift, gas is injected continuously or intermittently at selected location(s) along the production string, resulting in a reduction in density of the mixture in the tubing, and thus reducing the hydrostatic component of the pressure difference from the bottom to the top of the well. The purpose is to bring the fluids to the top at a desirable wellhead pressure, while keeping the bottomhole pressure at a value that is small enough to provide good driving force in the reservoir. This pressure drawdown must not violate restrictions for sand control and water or gas coning.

Gas lift can only be applied when hydrostatic pressure drop is the problem of wellbore flow, not frictional pressure drop. Two other considerations must enter the design. First, large amounts of gas injected into the well will affect the separation facilities at the top. Second, there exists a limit gas–liquid ratio (GLR) above which the pressure difference in the well will begin to increase because the reduction in the hydrostatic pressure will be offset by the increase in the friction pressure.

11.2 Well Construction for Gas Lift

Figure 11-1 is a schematic of a typical gas-lift system showing the well, a blowup of a gas-lift valve, and the associated surface process equipment.

Typical completions of a gas-lift system are shown in Figure 11-2 (Brown, Day, Byrd, and Mach, 1980). In Figure 11-2A, the liquid level in the annual isolates injected gas from the

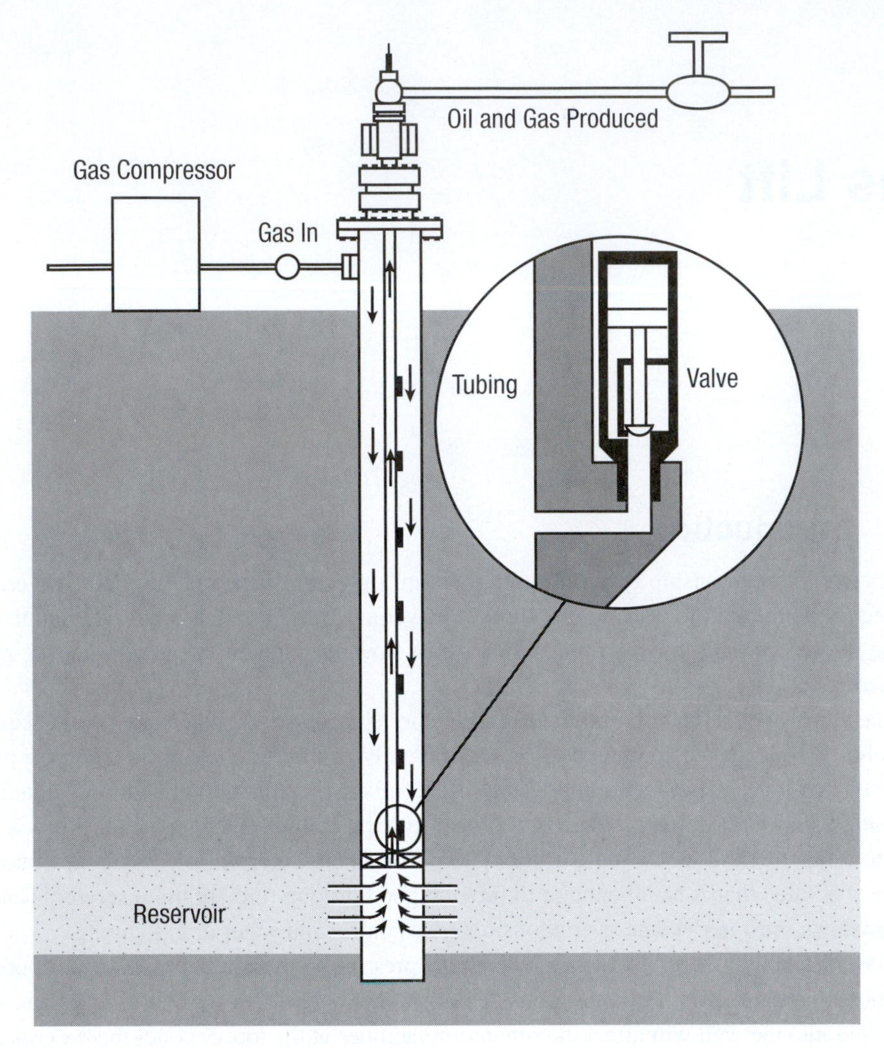

Figure 11-1 Gas-lift concept.

formation, in Figure 11-2B, packers are set to provide the isolation, and in Figure 11-2C, packers are used in the annulus, and a ball valve is used in the production tubing to prevent injected gas from flowing into the formation.

The positioning of the gas-lift valves and their number is a matter of wellbore hydraulics optimization. For continuous gas lift, an "operating valve" is used to inject the appropriate amount of gas at the desirable tubing pressure at the injection point. Other valves may be placed below the injection point and may be put into service during the life of the well as the reservoir pressure declines or if the water–oil ratio increases.

There are two types of gas lift: intermittent lift and continuous lift. For intermittent gas lift, both a single injection point and multiple injection points can be employed. First, a liquid slug

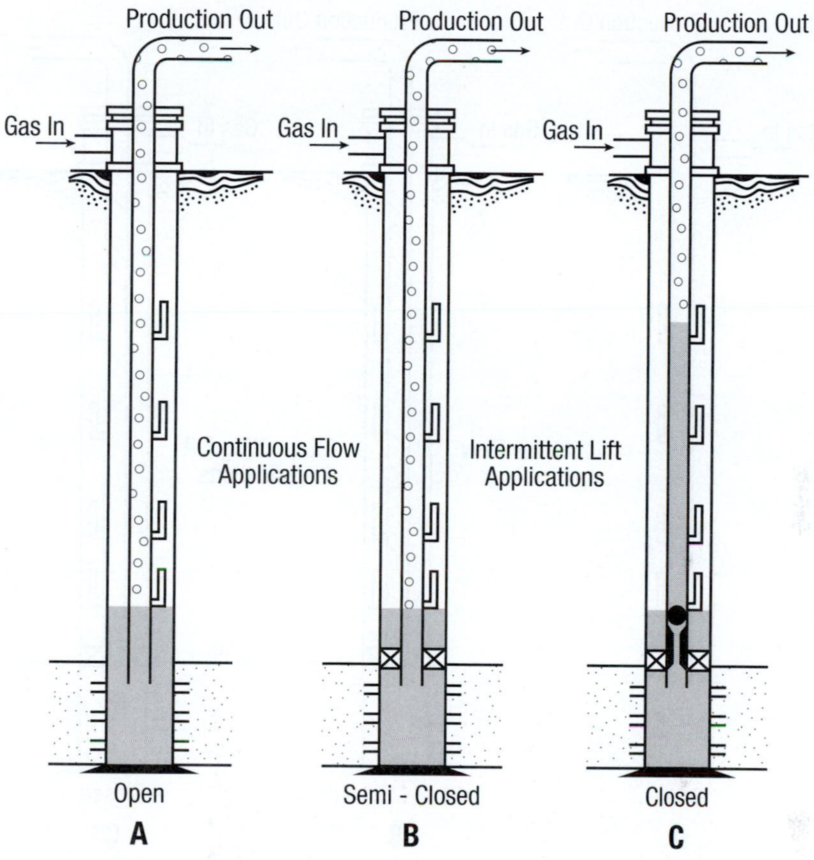

Figure 11-2 Gas-lift completions. (From Brown et al., 1980.)

must be built in the tubing above the bottom valve. Then, the valve opens, displacing the liquid slug upward. The valve remains open until the slug reaches the top, and then the valve closes. It reopens when a new liquid slug builds in the bottom of the well. Type C completion in Figure 11-2C is required for intermittent lift.

For multiple injection points, the bottom valve opens as described for the single-injection-point operation, but as the liquid slug moves upward, valves below the slug open. The valves close after the slug reaches the top. The actuation of the valves for intermittent gas lift can be done with a timing device or can depend on the pressure.

Valves can open and close based on the value of the casing or the tubing pressure. Valves are preset to a certain pressure to operate. Preset pressure is provided by a pressure dome with a diaphragm, springs, or a combination of pressure dome and springs. When casing or tubing pressure is higher than the preset pressure of the valves, the valve opens; otherwise, it remains closed. Figure 11-3 illustrates the principle of how these valves work. A discussion of the operating details of these valves can be found in Brown et al. (1980).

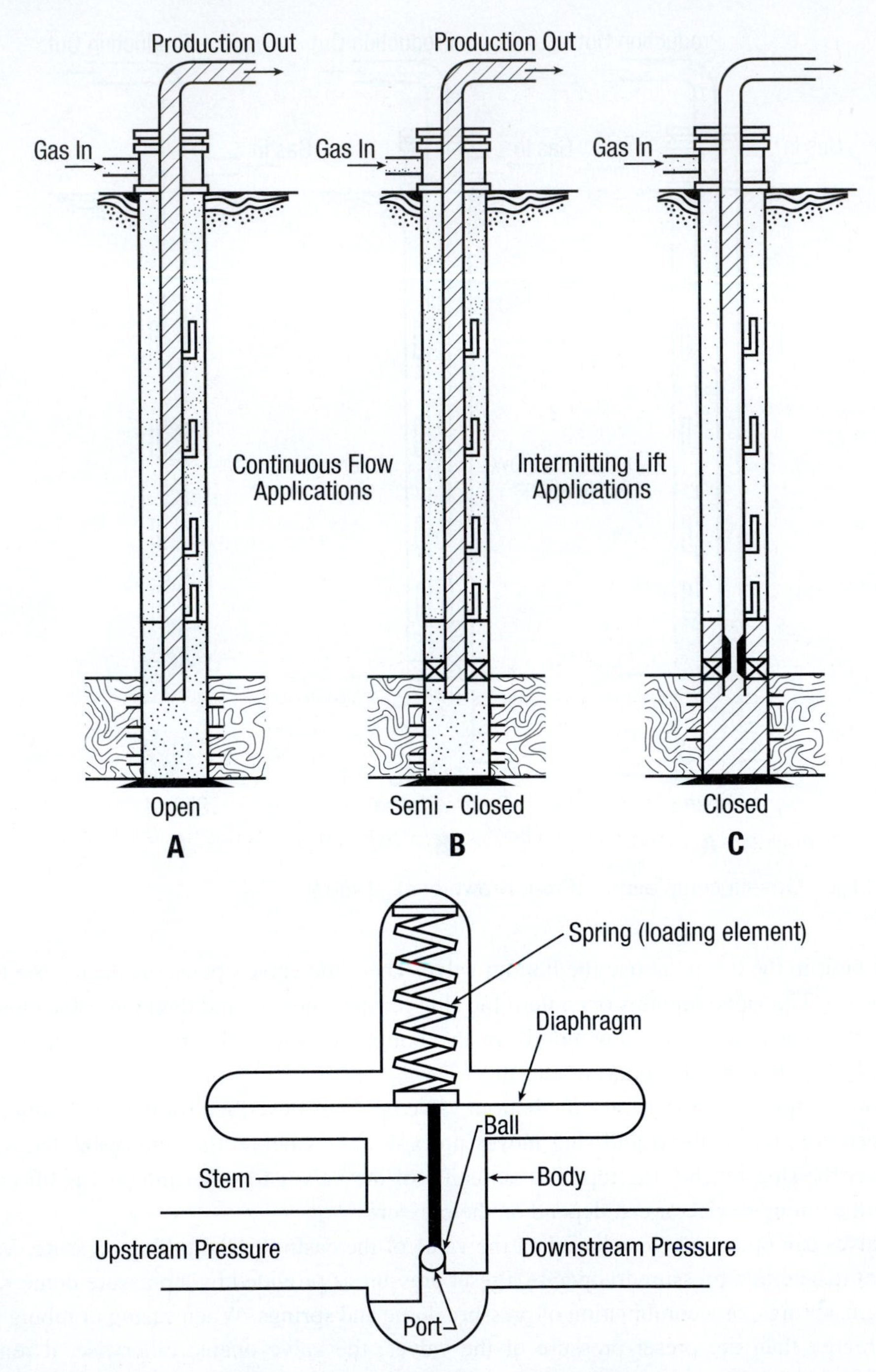

Figure 11-3 Pressure regulator valve. (From Brown et al., 1980.)

11.3 Continuous Gas-Lift Design

11.3.1 Natural versus Artificial Flowing Gradient

In Chapter 7 a number of correlations were introduced for pressure traverse calculations in two- and three-phase oil wells. For each production rate, these calculations result in gradient curves for a range of GLRs. Pressure differences are graphed against depth. For a given flowing well-head pressure, the required bottomhole pressure can be calculated. Thus, simply,

$$p_{tf} + \Delta p_{\text{trav}} = p_{wf} \tag{11-1}$$

where Δp_{trav} is the pressure traverse and is a function of the flow rate, the GLR, the depth, and the properties and composition of the fluid. The methodology is outlined in Chapter 7, especially Section 7.4.4. Example 7-11 (and associated Figure 7-16) shows a typical calculation.

Equation (11-1) can be rewritten in terms of the average pressure gradient in the well. Thus,

$$p_{tf} + \frac{dp}{dz}H = p_{wf} \tag{11-2}$$

where *dp/dz* in Equation (11-2) is considered constant. As can be seen from Figure 7-16, this gradient is not constant throughout the depth of the well. For small values (<100) or relatively large values (>1500) of GLR, the gradient can usually be considered as largely constant for common well depths ($\approx$10,000 ft). For intermediate values, *dp/dz* is not constant but, instead, is a function of depth.

Suppose that the natural GLR_1 would result in a pressure traverse requiring an unacceptably large p_{wf}. Then, for a more desirable p_{wf}, injecting gas at a rate q_g would lead to a higher GLR_2. If the liquid rate is q_l, then

$$q_l(\text{GLR}_2 - \text{GLR}_1) = q_g. \tag{11-3}$$

Example 11-1 Calculation of Required GLR and Gas-Lift Rate

Suppose the well described in example 7-11 (with q_o = 400 STB/d and q_w = 400 STB/d) is in a reservoir where H = 8000 ft and the GLR = 300 SCF/STB. If the indicated bottomhole pressure is 1500 psi, what should be the working GLR and how much gas should be injected at the bottom of the well? Would it be possible to inject (instead) at 4000 ft and still produce the same liquid rates? What would be the GLR above this alternative injection point?

Solution

At 8000 ft and GLR = 300 SCF/STB (see Figure 7-16), the required flowing bottomhole pressure would be 1900 psi. Since the indicated bottomhole pressure is 1500 psi, then (again from Figure 7-16) the required GLR would be 400 SCF/STB. Therefore, from Equation (11-3),

$$q_g = 800(400 - 300) = 8 \times 10^4 \text{ SCF/d}. \tag{11-4}$$

Now for the alternative option. From Equation (11-2) it can be written that

$$100 + \left(\frac{dp}{dz}\right)_a 4000 + 4000(0.3) = 1500 \tag{11-5}$$

where 0.3 is the pressure gradient for the GLR = 300 SCF/STB curve (from Figure 7-16) between depths of 4000 and 8000 ft. $(dp/dz)_a$ should be the pressure gradient above the injection point. Solution of Equation (11-5) results in $(dp/dz)_a = 0.05$ psi/ft, which is not possible, as shown in Figure 7-16. No fluid in this well can provide such a low-flowing gradient. In addition to the artificial GLR, two other variables must be reconciled—the possible gas injection pressure and the required compressor power demand—which is a function of both the gas injection rate and pressure. These are outlined below.

11.3.2 Pressure of Injected Gas

The gas added to the flow stream in the tubing in a gas-lifted well is injected down the casing-tubing annulus. The pressure change from the surface to the gas injection point can be calculated with Equation (7-52) by using the equivalent hydraulic diameter for the annulus and the appropriate cross-sectional area of flow. Also, the Reynolds number uses the equivalent diameter, which is the casing I.D. minus the tubing O.D. instead of the pipe diameter. Thus, to use Equation (7-52) for annular flow, D^5 should be replaced with $(D_c^2 - D_t^2)^2(D_c - D_t)$, and in the Reynold's number (Equation 7-55), D should be replaced with $D_c + D_t$, where D_c is the casing I.D. and D_t is the tubing I.D.

In many cases, the frictional pressure drop is small, so from the mechanical energy balance [Equation (7-15)], ignoring changes in kinetic energy and the friction pressure drop in the casing (i.e., relatively small gas flow rates) and changing into oilfield units,

$$\int_{\text{surf}}^{\text{inj}} \frac{dp}{\rho} + \frac{1}{144}\int_0^H dH = 0. \tag{11-6}$$

From the real-gas law,

$$\rho = \frac{28.97\gamma p}{ZRT} \tag{11-7}$$

where γ is the gas gravity to air and 28.97 is the molecular weight of air. The gas constant R is equal to 10.73 psi ft^3/lb-mole-°R. The temperature is in °R.

Substitution of Equation (11-7) into Equation (11-6) and integration (using average values of $\overline{Z}$ and $\overline{T}$) results in

$$p_{\text{inj,ann}} = p_{\text{surf}} e^{0.01875\gamma H_{\text{inj}}/\overline{Z}\overline{T}}. \tag{11-8}$$

Example 11-2 Injection-Point Pressure

If gas of $\gamma = 0.7$ (to air) is injected at 8000 ft and if $p_{surf} = 900$ psi, $T_{surf} = 80°F$ and $T_{inj} = 160°F$, calculate the pressure in the annulus at the injection point, $p_{inj,ann}$.

Solution

Trial and error is required for this calculation. Assume that $p_{inj} = 1100$ psi. From Chapter 4 (Figure 4-2), $p_{pc} = 668$ psi and $T_{pc} = 390°R$ and therefore

$$p_{pr} = \frac{(900 + 1100)/2}{668} = 1.5 \quad \text{(11-9)}$$

and

$$T_{pr} = \frac{(80 + 160)/2 + 460}{390} = 1.49 \quad \text{(11-10)}$$

From Figure 4-1, $\overline{Z} = 0.86$ and therefore

$$p_{inj,ann} = 900e^{(0.01875)(0.7)(8000)/(0.86)(580)} = 1100 \text{ psi}, \quad \text{(11-11)}$$

which agrees well with the assumed value.

11.3.3 Point of Gas Injection

If Equation (11-8) is expanded as a Taylor series and if typical conditions and fluid properties for a natural gas and reservoir are considered, such as $\gamma = 0.7$, $\overline{Z} = 0.9$, and $\overline{T} = 600R$, then

$$p_{inj,ann} = p_{surf}\left(1 + \frac{H_{inj}}{40{,}000}\right) \quad \text{(11-12)}$$

where the pressures are in psi and H_{inj} is in feet. This relationship was offered first by Gilbert (1954). In current practice this type of approximation is usually not necessary, since the actual downhole injection pressure can be calculated readily with a simple computer program that considers the actual annulus geometry and accounts for both friction and hydrostatic pressure components. Equation (11-12), though, can be useful for first design approximations.

The p_{inj} in the production tubing is an additional 100 to 150 psi less than the annulus pressure because of the pressure drop across the gas-lift valve. The value of this pressure drop is supplied by the valve manufacturer.

The point of gas injection, H_{inj}, creates two zones in the well: one below, with an average flowing pressure gradient $(dp/dz)_b$, and one above, with an average flowing pressure gradient $(dp/dz)_a$. Thus,

$$p_{wf} = p_{tf} + H_{\text{inj}}\left(\frac{dp}{dz}\right)_a + (H - H_{\text{inj}})\left(\frac{dp}{dz}\right)_b. \quad \textbf{(11-13)}$$

It should be obvious that the downhole gas injection pressure in the annulus, $p_{\text{inj,ann}}$, must not be greater than the pressure within the tubing for gas to enter the tubing at the injection point.

Figure 11-4 shows the concept of continuous gas lift in terms of the pressure values, pressure gradients, well depth, and depth of injection. With the available flowing bottomhole pressure and the natural flowing gradient $(dp/dz)_b$, the reservoir fluids would ascend only to the point indicated by the projection of the pressure profile in the well. This would leave a partially filled wellbore. Addition of gas at the injection point would lead to a significant alteration of the pressure gradient $(dp/dz)_a$, which would lift the fluids to the surface. The higher the required wellhead pressure, the lower the flowing gradient should be.

Also marked on Figure 11-4 is the balance point, that is, the position in the tubing where the downhole pressure of the injected gas is equal to the pressure in the tubing. This pressure is related to the surface pressure as shown by Equation (11-8) or its approximation, Equation (11-12). The actual point of injection is placed a few hundred feet higher to account for the pressure drop across the gas-lift valve, Δp_{valve}.

Example 11-3 Relating Gas Injection Pressure, Point of Injection, and Well Flow Rate

Suppose that the well at a depth of 8000 ft and a GLR = 300 SCF/STB (as in Example 11-1) drains a reservoir with an IPR given by

$$q_l = 0.39(\overline{p} - p_{wf}). \quad \textbf{(11-14)}$$

What should be the surface gas injection pressure if the gas-lift valve is at the bottom of the well and $p_{\text{inj}} - \Delta p_{\text{valve}} = p_{wf} = 1000$ psi? [The average reservoir pressure $\overline{p}$ is 3050 psi for $q_l = 800$ STB/d, as can be calculated readily from Equation (11-14)].

What should be the point of gas injection for a production rate of 500 STB/d? Figure 11-5 is the tubing performance curve for $q_l = 500$ STB/d with 50% water and 50% oil. Note from Figure 11-5 that $(dp/dz)_b = 0.33$ psi/gt for GLR = 300 SCF/STB between $H = 5000$ and 8000 ft. Use $\Delta p_{\text{valve}} = p_{inj,ann} - p_{inj} = 100$ psi.

Solution

From Equation (11-12) and rearrangement and noting that $p_{\text{inj,ann}} = 1100$ psi,

$$p_{surf} = 1100/\left(1 + \frac{8000}{40{,}000}\right) = 915 \text{ psi}. \quad \textbf{(11-15)}$$

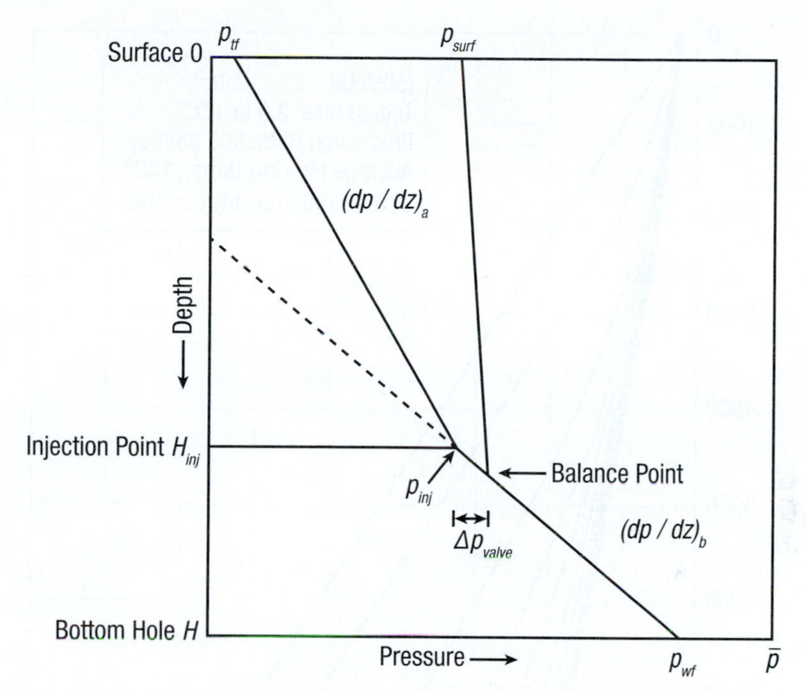

Figure 11-4 Continuous gas lift with pressures and pressure gradients versus depth.

For $q_l = 500$ STB/d, then from the IPR relationship [Equation (11-14)],

$$p_{wf} = 3050 - \frac{500}{0.39} = 1770 \text{ psi}. \tag{11-16}$$

The injection point must be where the pressure between the injected gas and the pressure in the production string must be balanced. Thus, the injection pressure in the wellbore, p_{inj}, is related to the casing gradient by Equation (11-12) and to the wellbore gradient below the injection point:

$$p_{inj} = p_{surf}\left(1 + \frac{H_{\text{inj}}}{40{,}000}\right) - \Delta p_{\text{valve}} = p_{wf} - \left(\frac{dp}{dz}\right)_b (8000 - H_{\text{inj}})$$

$$= 915\left(1 + \frac{H_{\text{inj}}}{40{,}000}\right) - 100 = 1770 - 0.33(8000 - H_{\text{inj}}). \tag{11-17}$$

Solving for the injection depth gives $H_{\text{inj}} = 5490$ ft.

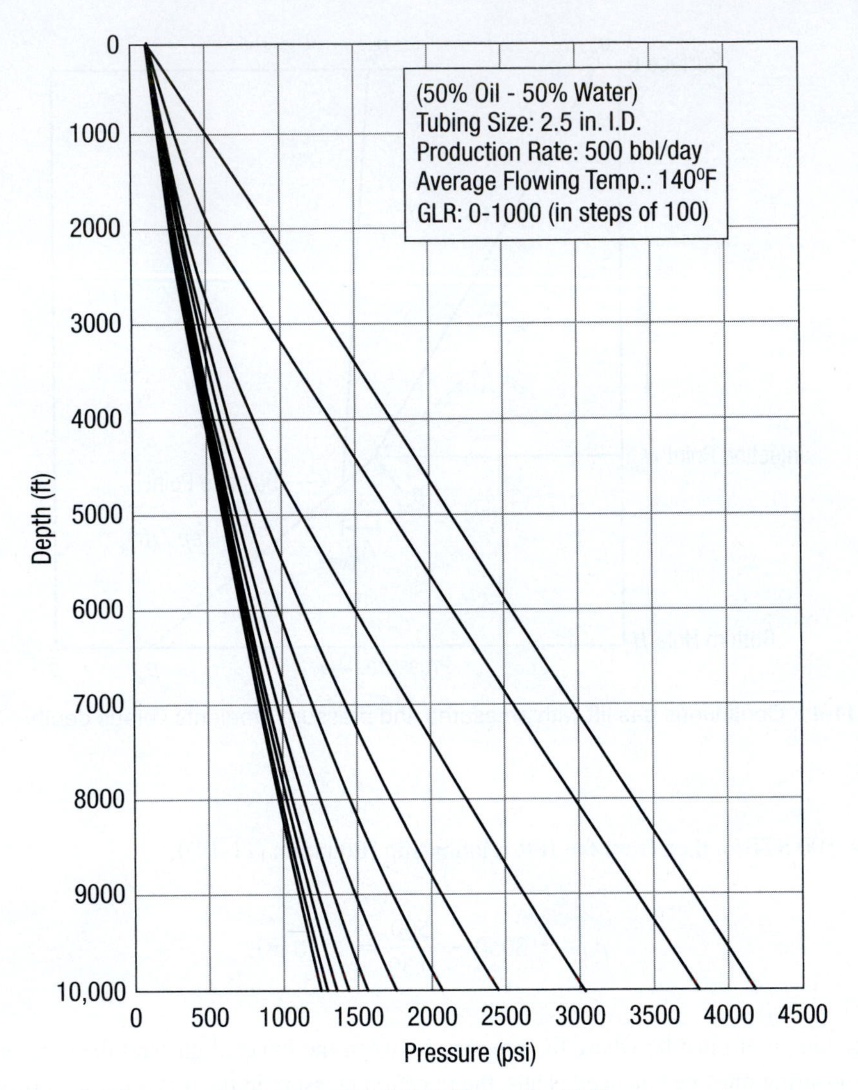

Figure 11-5 Gradient curves for q_l = 500 STB/d (Example 11-3).

Finally, from Equation (11-12), the $p_{\text{inj,ann}}$ at H_{inj} is

$$p_{\text{inj,ann}} = 915\left(1 + \frac{5490}{40{,}000}\right) \approx 1040 \text{ psi.} \tag{11-18}$$

Inside the tubing the pressure is 940 psi and H_{inj} = 5490 is at GLR = 340 SCF/STB (see Figure 11-6).

Therefore, the gas injection from Equation (11-3) would be $q_g = 2 \times 10^4$ SCF/d.

11.3.4 Power Requirements for Gas Compressors

The horsepower requirements can be estimated from the following equation:

$$\text{HHP} = 2.23 \times 10^{-4}\, q_g \left[\left(\frac{p_{\text{surf}}}{p_{\text{in}}} \right)^{0.2} - 1 \right] \qquad \textbf{(11-19)}$$

where p_{in} is the inlet compressor pressure. Thus, for a given gas injection rate in a well and the surface pressure, the calculated horsepower would be an important variable in gas-lift optimization design.

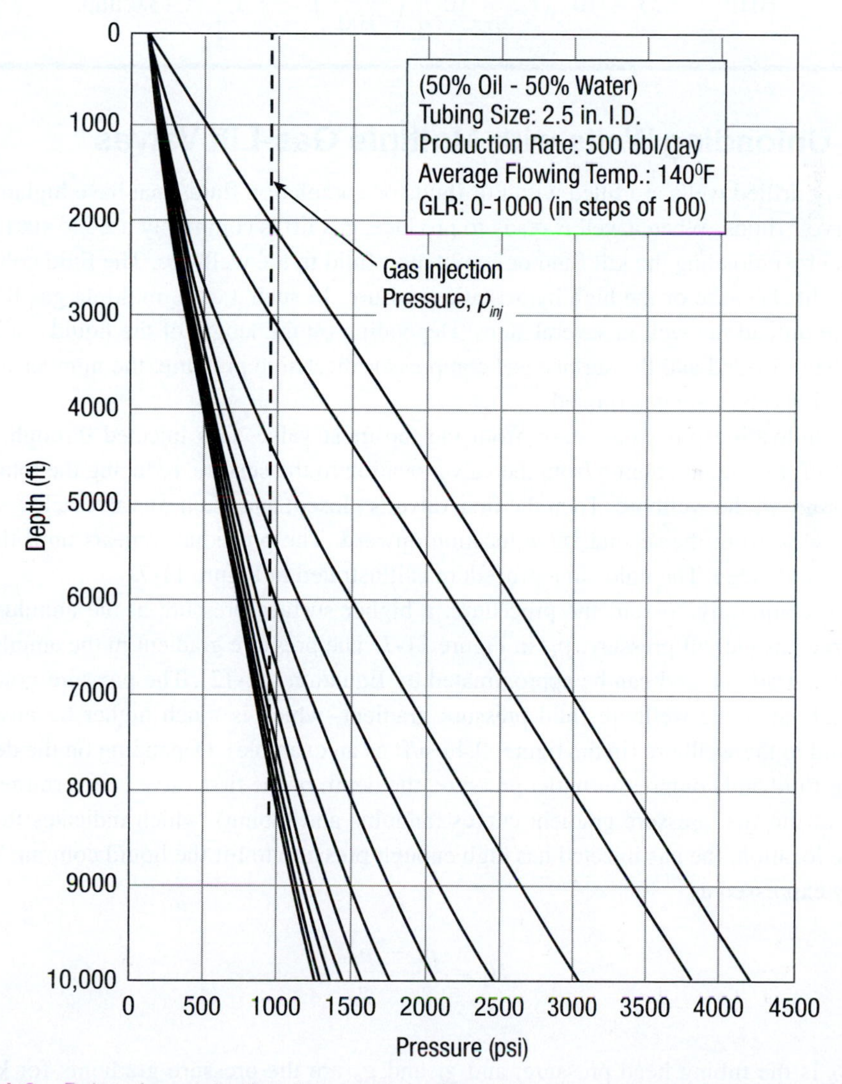

Figure 11-6 Point of gas injection. Intersection of gas injection pressure with gradient curves (Example 11-3).

Example 11-4 Calculation of Gas-Lift Compressor Horsepower

Suppose that the gas injection rate is 1.2×10^5 SCF/d, p_{surf} is 1330 psi, and the compressor inlet pressure, p_{in}, is 100 psi. Calculate the horsepower requirements.

Solution

From Equation (11-19),

$$HHP = 2.23 \times 10^{-4}(1.2 \times 10^5)\left[\left(\frac{1330}{100}\right)^{0.2} - 1\right] = 18.4 \text{ hhp.} \tag{11-20}$$

11.4 Unloading Wells with Multiple Gas-Lift Valves

Most newly drilled wells are filled with kill fluids or completion fluids that have higher density than reservoir fluids. When a well is ready to produce, gas lift is commonly used to start the well production by unloading the kill fluid or completion fluid in the wellbore. The fluid column can be hard to lift because of the high hydrostatic pressure. In such a case, multiple gas-lift valves are used to unload the well in several steps. Depending on the length of the liquid column that needs to be unloaded and the surface gas compressor discharge pressure, the number and location of the lift valves are determined.

The unloading procedure starts from the top-most valve. Gas injected through the first valve will lift the liquid column from the valve location to the surface, reducing the total hydrostatic pressure in the wellbore. Then the first valve is closed, and the next valve below opens to lift the column from the second valve location upward. The procedure repeats until the entire wellbore is unloaded. The unloading procedure is illustrated in Figure 11-7.

Very commonly, to start the procedure, a higher surface pressure at the annulus side is used, marked as kickoff pressure, p_k, in Figure 11-7. The pressure gradient in the annulus is the injected gas gradient, and can be approximated by Equation (11-12). The pressure gradient inside the tubing is the wellbore fluid pressure gradient, which is much higher because of the denser fluid in the wellbore (in the figure, 0.4 psi/ft as an example). Depending on the density of the tubing fluid and compressor outlet pressure, the depth of the first valve is determined by the intercept of the two pressure gradient curves (annulus and tubing), which indicates that at the first valve location, the gas injected has high enough pressure to lift the liquid column. This can be simply expressed as

$$H_1 = \frac{p_k - p_{th}}{g_k - g_g} \tag{11-21}$$

where p_{th} is the tubing head pressure, and g_k and g_g are the pressure gradients for kill fluid and injected gas. Once the first column is lifted, the surface injection pressure can be reduced

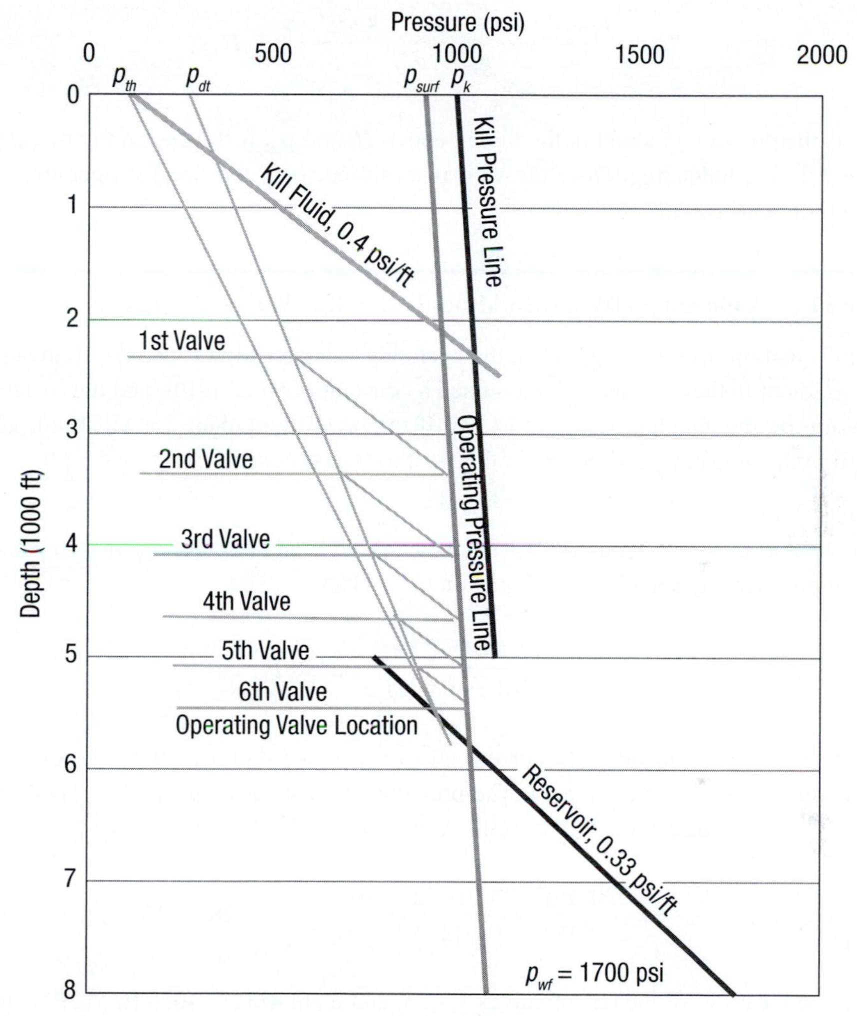

Figure 11-7 Designing the depths of unloading valves.

to a smaller value, and this pressure is used as the injection pressure at the surface on the annulus side, p_{surf}, in Figure 11-7. At this condition, the pressure gradient inside the tubing is divided into two sections by the first valve location, H_1. Above H_1, there is a new pressure gradient of mostly gas with some liquid (two-phase), and below H_1 is the original kill fluid gradient. The intercept of the tubing fluid gradient curve with the annulus gas pressure gradient curve defines the second valve's location. Notice that after the first column is lifted, a higher tubing head pressure is used in the calculation for the locations of the rest of the valves to provide a safety factor. Practically, 100–200 psi is added in unloading design. Figure 11-7 shows a 200-psi safety factor. For each additional valve below the first one, the location can be calculated by

$$H_i = \frac{p_{surf} - p_{dt} + (g_g - g_{dt})H_{i-1}}{(g_k - g_g)} + H_{i-1} \tag{11-22}$$

where g_t is the pressure gradient in the tubing above H_i and p_{dt} is the design tubing pressure at the surface during unloading. Once the well is unloaded, only the deepest operating valve is open for continuous gas-lift production.

Example 11-5 Unloading a Well with Multiple Gas-Lift Valves

Use the information from Example 11-3, the operating valve location is at 5490 ft, the operating pressure gradient in the annulus, g_g, is assumed a constant of 0.022 psi/ft, and the surface injection pressure on the annulus, p_{surf}, is 915 psi. If the pressure gradient for kill fluid, g_k, is 0.4 psi/ft, design the unloading procedure for this well to start production.

Solution

Using the kickoff pressure, p_k, as 1000 psi and the tubing head pressure, p_{th}, of 120 psi, the first valve location is readily calculated by Equation (11-21) as

$$H_1 = \frac{1000 - 120}{(0.4 - 0.022)} = 2328\,ft. \tag{11-23}$$

After the first column is unloaded, the annulus pressure is reduced to p_{surf} (915 psi), and the tubing design pressure is increased to 320 psi. The pressure gradient for unloading, g_{dt}, is 0.113 psi/ft with a straight line assumption, and the section valve location is calculated by Equation (11-22),

$$H_2 = \frac{915 - 320 + (2328)(0.022 - 0.113)}{0.4 - 0.022} + 2328 = 3342\,ft. \tag{11-24}$$

Similarly, the locations of the rest of valves 3, 4, 5, and 6 are 4111ft, 4696 ft, 5139 ft, and 5476 ft. Valve 6 can be used as the operating valve during production. Figure 11-7 presents this unloading valve design procedure graphically.

11.5 Optimization of Gas-Lift Design

11.5.1 Impact of Increase of Gas Injection Rate, Sustaining of Oil Rate with Reservoir Pressure Decline

Figure 11-6 can be used to investigate the effect of increasing the gas injection rate, and therefore increasing the GLR in the wellbore. First, increasing the gas injection rate allows the injection at a lower point in the tubing without increasing the injection pressure.

From Equation (11-13), as H_{inj} increases [with the associated smaller $(dp/dz)_a$] and since $(dp/dz)_b$ remains largely constant within a smaller interval $H - H_{inj}$, then p_{wf} will necessarily decrease. Therefore larger oil production rates are likely to be achieved with higher GLR and a lower injection point.

Similarly, to *sustain* a production rate while the reservoir pressure depletes, the flowing bottomhole pressure must be lowered. This can be accomplished by lowering the injection point and by increasing the gas injection rate. The initial operating valve may not be the deepest valve of the set, and the lower valve can be used in later life of the well production. This has a limit, addressed in the next section.

Example 11-6 Effect of Injection Rate and Injection Point on the Flowing Bottomhole Pressure

Assuming that Figure 11-6 represents the tubing flow behavior, calculate the flowing bottomhole pressure at 8000 ft for the injection points described by the intersection of the gas injection curve and the pressure traverse curves for GLR values equal to 500, 600, and 700 SCF/STB, respectively. The flowing gradient below the injection point is 0.33 psi/ft (GLR = 300 SCF/STB).

Solution

From Figure 11-6, the wellhead pressure is 100 psi. Also, the following table can be constructed from the intersection of the gas and pressure traverse curves:

GLR	Injection Point	Pressure at Injection Point
500	6700 ft	968 psi
600	7050 ft	976 psi
700	7500 ft	987 psi

The pressures at the injection points shown above are obtained from Equation (11-12) using $p_{surf} = 915$ psi and subtracting 100 psi across the gas-lift valve. Therefore the bottomhole pressure can be calculated readily by

$$p_{wf} = p_{inj} + (H - H_{inj})\left(\frac{dp}{dz}\right)_b \tag{11-25}$$

where p_{inj} is the pressure at the injection point.

Substituting the values for p_{inj} and the corresponding values of H_{inj} results in p_{wf} equal to 1358, 1261, and 1137 psi, for GLRs equal to 500, 600, and 700 SCF/STB, respectively. Assuming that Figure 11-6 describes approximately the pressure traverse of flow rates larger than 500 STB/d, these lower bottomhole pressures would lead to larger oil production rates. If Equation (11-14) in Example 11-3 describes the IPR for this well, then the lower bottomhole pressures (using $\overline{p} = 3050$ psi as in Example 11-3) would result in q_l equal to 660, 680, and 746 STB/d, respectively.

Example 11-7 Sustaining a Production Rate While the Reservoir Pressure Depletes

Repeat Example 11-3 with the IPR described by Equation (11-14). Where should the injection point be after the reservoir pressure drops by 500 psi? What should be the gas injection rate to sustain a liquid production rate equal to 500 STB/d? Use Figure 11-6 and the data of Example 11-3 for the analysis.

Solution

From Equation (11-14) and rearrangement if $\overline{p} = 2550$ psi (500 psi less than the original 3050 psi),

$$p_{wf} = 2550 - \frac{500}{0.39} = 1268 \text{ psi.} \quad \textbf{(11-26)}$$

The injection point must then be obtained from the solution of

$$915\left(1 + \frac{H_{\text{inj}}}{40{,}000}\right) - 100 = 1268 - 0.33(8000 - H_{\text{inj}}), \quad \textbf{(11-27)}$$

leading to $H_{\text{inj}} = 7120$ ft and (calculated) $p_{\text{in}} = 980$ psi, the GLR must be equal to 650 SCF/STB.

Finally, the gas injection rate is

$$q_g = (650 - 300)500 = 1.75 \times 10^5 \text{ SCF/d.} \quad \textbf{(11-28)}$$

This injection rate is about 9 times the gas injection rate calculated in Example 11-3, which would be required at the reservoir pressure of 3050 psi. The injection point is also deeper: from 5490 to 7120 ft.

11.5.2 Maximum Production Rate with Gas Lift

As stated in the introduction to this chapter, there exists a limit value of GLR, referred to here as "limit" GLR (Golan and Whiston, 1991, have referred to it as "favorable"), where the flowing pressure gradient is the minimum. For GLR values that are larger than the minimum, the flowing gradient begins to increase. As discussed already, the well flowing pressure gradient is a combination of the hydrostatic pressure head and the friction pressure drop. While increasing the GLR from small values would result in a fluid density reduction with modest friction pressures, continued GLR increase would lead to a disproportionate increase in the friction pressure. Thus, the limit GLR is that value at which the increase in frictional pressure will offset the decrease in the hydrostatic pressure.

Figure 11-8 represents pressure traverse calculations for a specific liquid production rate ($q_l = 800$ STB/d) in a given tubing size. The values of the GLR have been selected to be large enough to observe the behavior reversal at the limit GLR value. For this example, the limit GLR is approximately equal to 4000 SCF/STB.

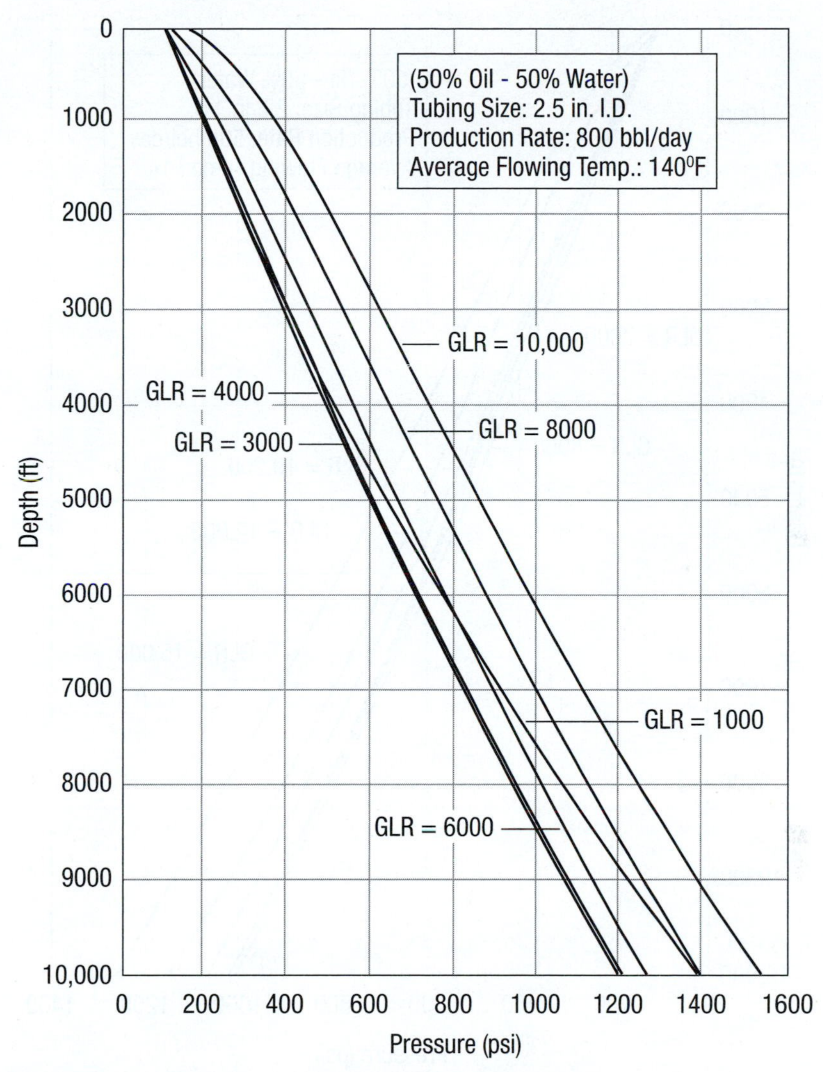

Figure 11-8 Large GLR values and limit GLR for q_l = 800 STB/d.

The limit GLR increases as the production rate decreases. Figure 11-9 is for a smaller liquid production rate (q_l = 500 STB/d) in the same size well. In this example the limit GLR is approximately 7000 SCF/STB. This limit GLR for which the minimum p_{wf} will be obtained is rarely the natural GLR of the reservoir fluids. Thus, it can be reached only through artificial gas lift. While the above applies to an already completed well, appropriate production engineering must take into account these concepts before the well is completed. A larger well size would lead to a larger limit GLR, with, of course, a larger gas injection rate. These design issues, balancing benefits and costs, is dealt with in Section 11.8.

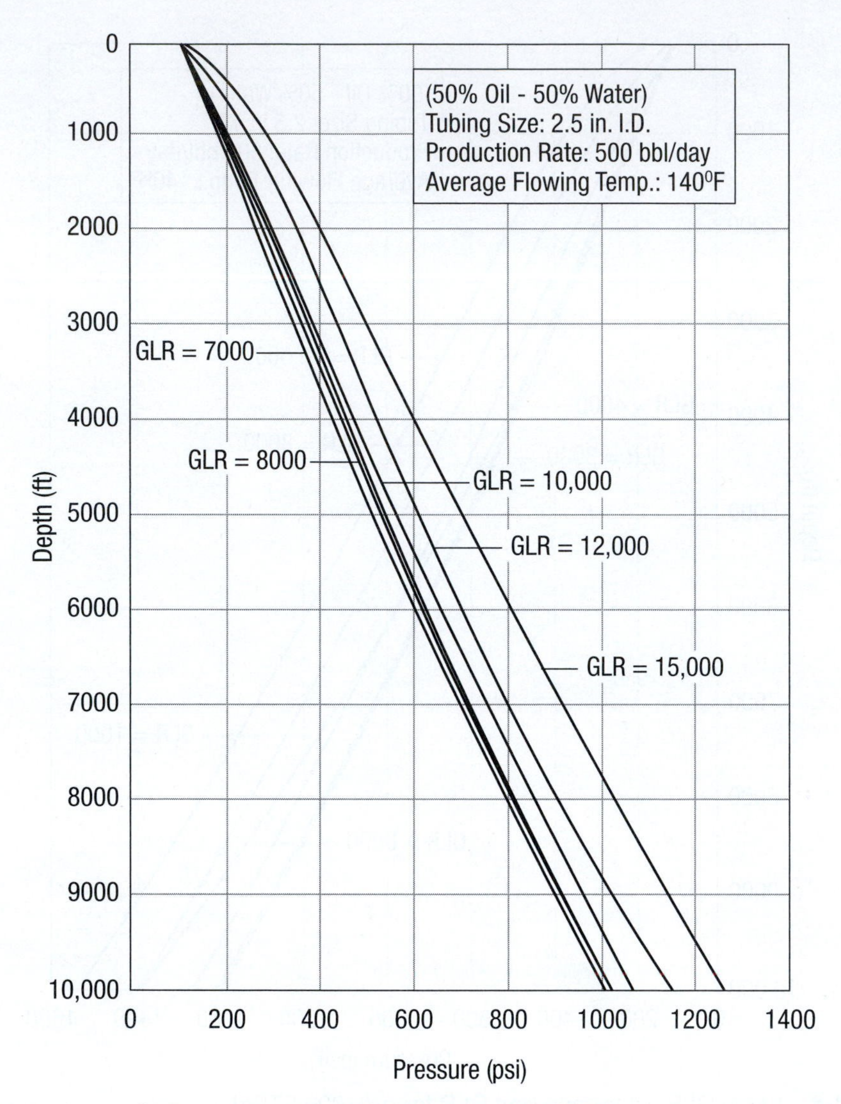

Figure 11-9 Large GLR values and limit GLR for q_l = 500 STB/d.

11.6 Gas-Lift Performance Curve

Poettmann and Carpenter (1952) and Bertuzzi, Welchon, and Poettman (1953), in developing design procedures for gas-lift systems, introduced the "gas-lift performance curve," which has been used extensively in the petroleum industry ever since.

Optimized gas-lift performance would have an important difference from natural well performance. Based on the information presented in the previous section, for each liquid production rate there exists a limit GLR where the minimum p_{wf} will be observed. Thus, a plot similar to the one shown in Figure 11-10 can be drawn. The IPR curve, as always, characterizes the ability of the reservoir to deliver certain combinations of production rates and flowing bottomhole pressures. Intersecting this IPR curve is the optimum gas-lift performance curve of minimum flowing bottomhole pressures (minimum pressure intake curve), and for each production rate they correspond to the limit GLR. Other values of GLR *above* or *below* the limit GLR will result in smaller production rates. This is shown conceptually in Figure 11-11. Each gas-lift performance curve in the *vicinity* of the maximum production rate will result in a larger pressure intake and therefore lower production. In summary, the intersection of the IPR curve

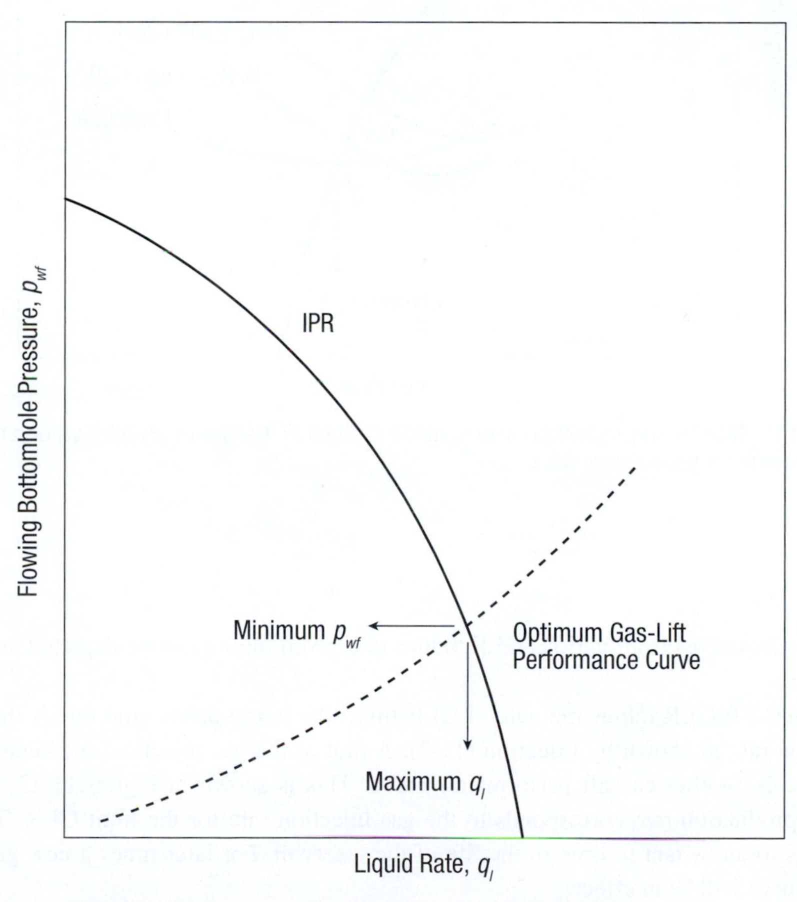

Figure 11-10 IPR and optimum gas-lift performance (at limit GLR).

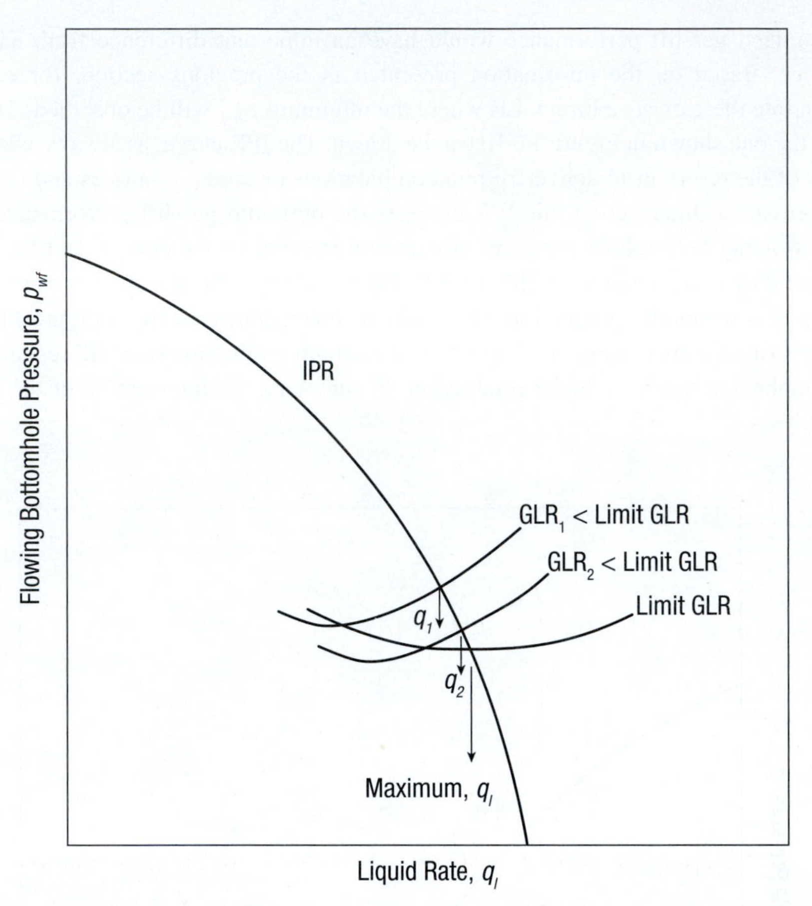

Figure 11-11 Maximum production rate is at the limit GLR. Larger or smaller values of GLR will result in smaller production rates.

with the intake curves from various GLR values in a given tubing can be depicted as shown in Figure 11-12.

The gas-lift GLR minus the natural GLR times the liquid production rate is the required gas injection rate as shown by Equation (11-3). A plot of the gas injection rate versus the production rate is another gas-lift performance curve. This is shown in Figure 11-13, where the maximum production rate corresponds to the gas injection rate for the limit GLR. This graph corresponds to an instant in time in the life of the reservoir. For later times a new gas-lift performance curve will be in effect.

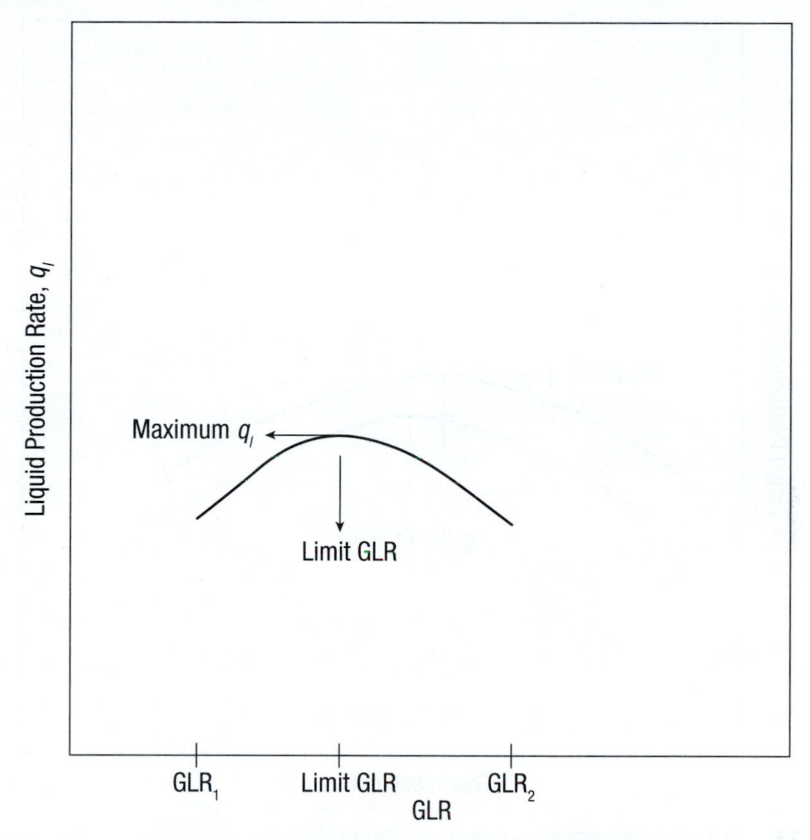

Figure 11-12 Liquid production rate versus GLR. Limit GLR corresponds to maximum q_l.

Example 11-8 Development of Gas-Lift Performance Curve

With the IPR curve described by Equation (11-14) (Example 11-3) and a well at 8000 ft depth whose gradient curves for 800 and 500 STB/d were given in Figures 11-8 and 11-9, develop a gas-lift performance curve assuming that the gas injection valves are at the bottom of the well. The average reservoir pressure $\overline{p}$ is 3050 psi, and the natural GLR is 300 SCF/STB.

Solution

Figure 11-14 is a graph of the intersection of the IPR and the optimum gas-lift performance curve, that is, the locus of the points corresponding to the limit GLR for each production rate. Maximum production rate is 825 STB/d, which corresponds to a limit GLR of 3800 SCF/STB (gradient curves are not shown for this rate).

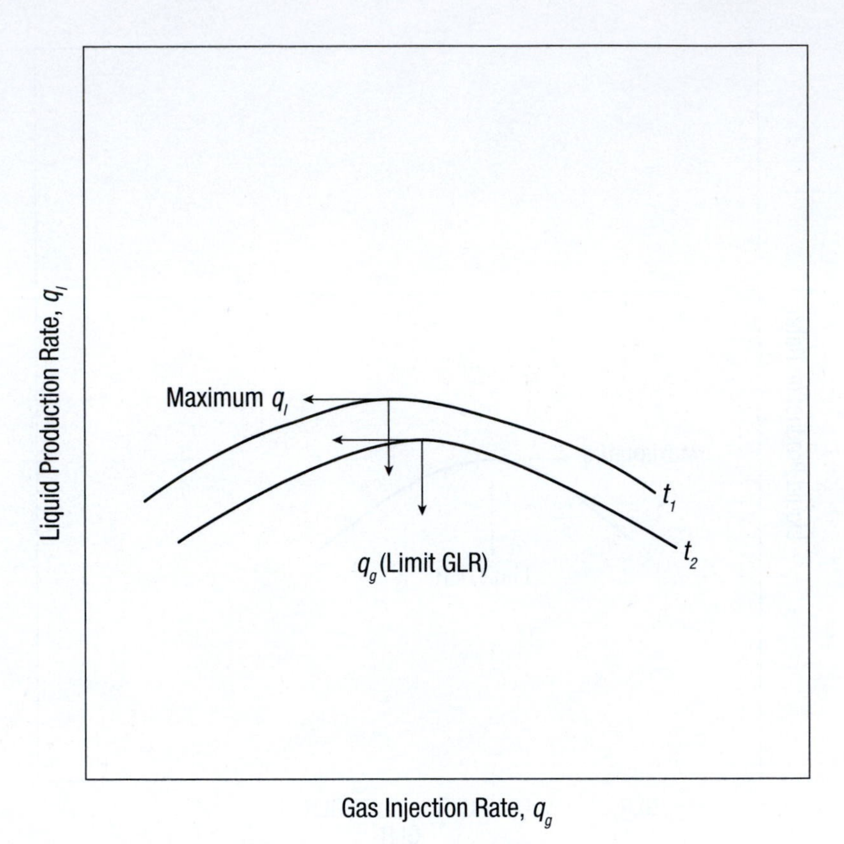

Figure 11-13 Gas-lift performance curve. Time dependency.

From Equation (11-3), the gas injection rate for the maximum production rate of 825 STB/d and limit GLR = 3800 SCF/STB is

$$q_g = 825(3800 - 300) = 2.89 \times 10^6 \text{ SCF/d}. \quad \textbf{(11-29)}$$

Figure 11-15 is the gas-lift performance curve for this well. For production rates other than the maximum, the gas injection rate could be larger or smaller. This performance curve will change with time, as is shown in Example 11-13.

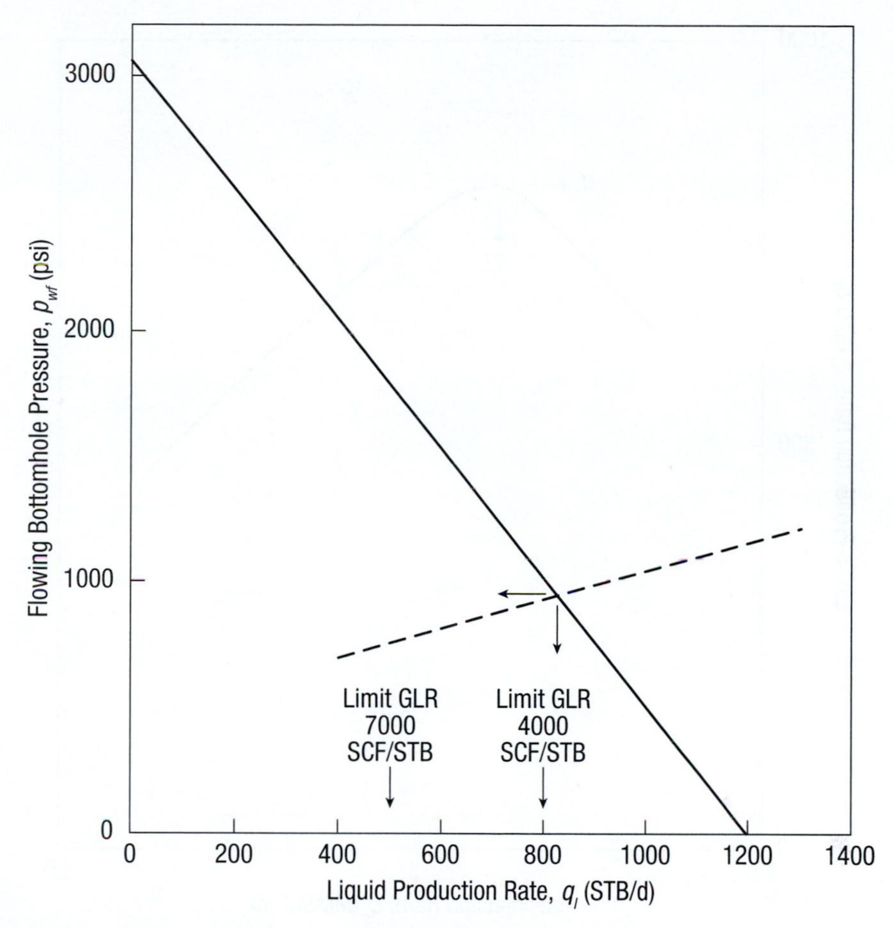

Figure 11-14 IPR and optimum gas lift (Example 11-8).

Example 11-9 Near-Wellbore Damage and Gas Lift

Suppose that wellbore damage ($s \approx +9$) would reduce the IPR curve used in Examples 11-3 and 11-8 to

$$q_l = 0.22(\overline{p} - p_{wf}). \tag{11-30}$$

What would be the maximum production rate and the associated gas-lift requirements?

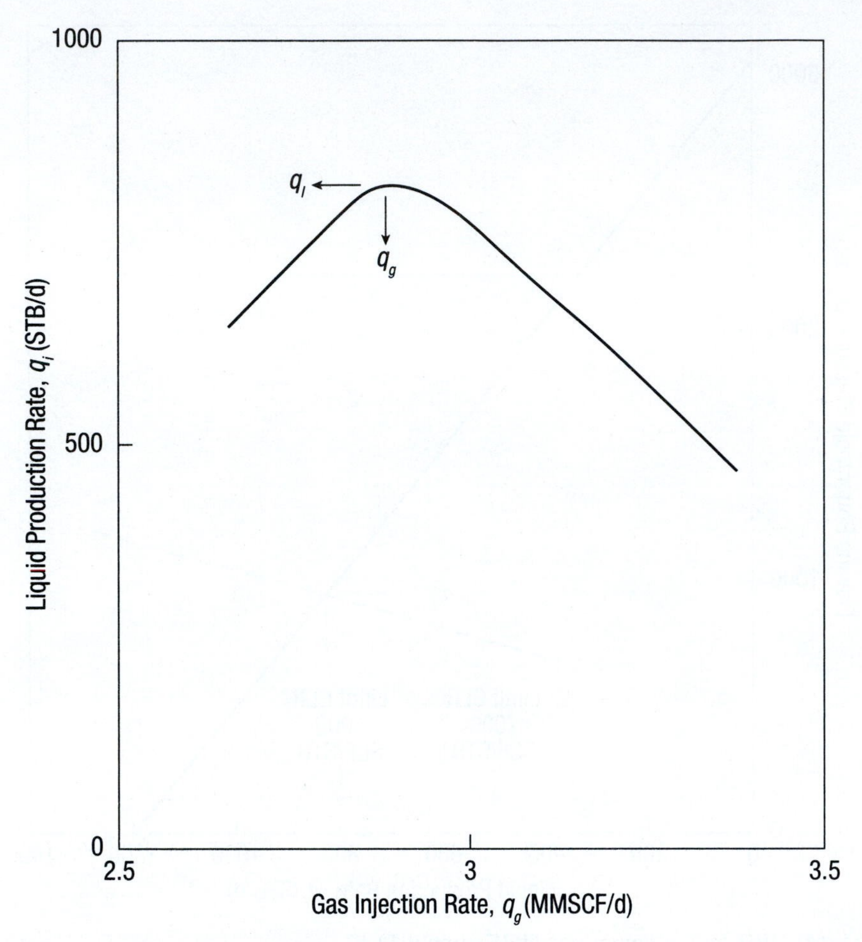

Figure 11-15 Gas-lift performance curve (Example 11-8).

Solution

Figure 11-16 shows the new IPR curve and its intersection with the optimum gas-lift curve. The point of intersection is at $q_l = 500$ STB/d. From Figure 11-15 (done for Example 11-8) at 500 STB/d, the gas injection rate, q_g, is 3.35×10^6 SCF/d. If $q_l = 500$ STB/d, then, from Equation (11-30),

$$p_{wf} = 3050 - \frac{500}{0.22} = 780 \text{ psi.} \quad \textbf{(11-31)}$$

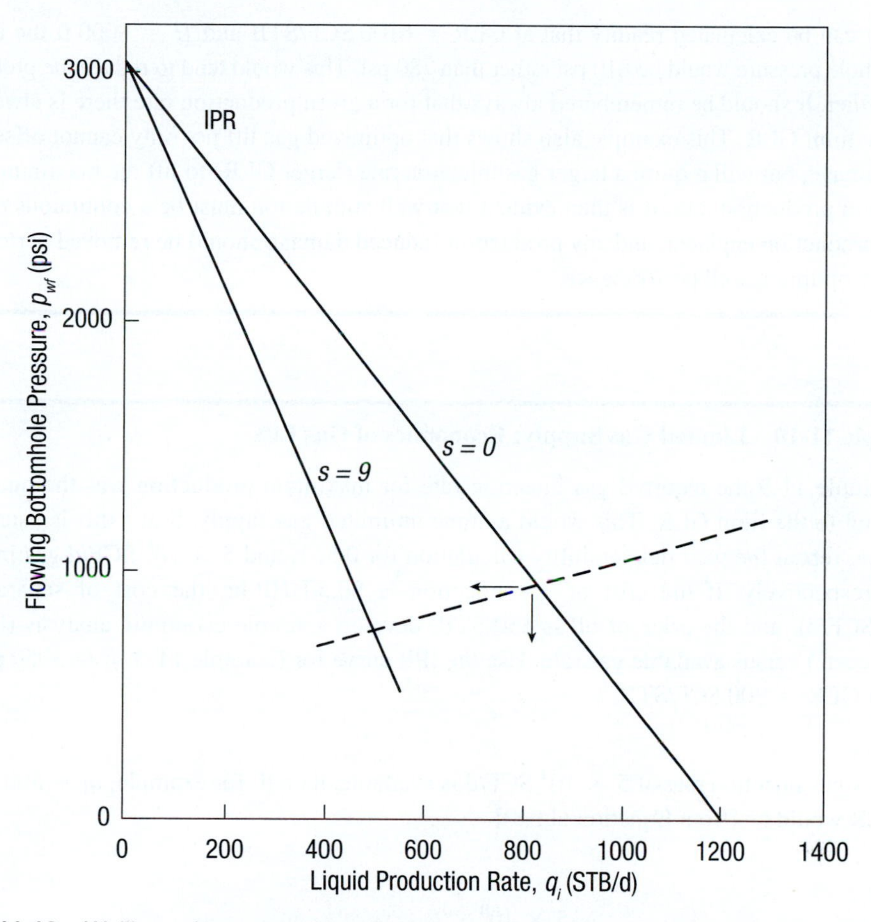

Figure 11-16 Wellbore damage and its effect on gas-lift performance (Example 11-9).

If, though, the gas injection rate for the undamaged well were maintained ($q_g = 2.89 \times 10^6$ SCF/d for $q_l = 825$ STB/d; see Example 11-8), then the GLR would be

$$\text{GLR} = \frac{q_g}{q_l} + (\text{GLR})_{\text{natural}} \tag{11-32}$$

and therefore

$$\text{GLR} = \frac{2.89 \times 10^6}{500} + 300 = 6100 \text{ SCF/STB}. \tag{11-33}$$

It can be calculated readily that at GLR = 6100 SCF/STB and H = 8000 ft the flowing bottomhole pressure would be 810 psi rather than 780 psi. This would tend to reduce the production rate further. It should be remembered always that for a given production rate there is always one specific limit GLR. This example also shows that optimized gas lift not only cannot offset wellbore damage, but will require a larger gas injection rate (larger GLR) to lift the maximum possible liquid production rate. It is then evident that well stimulation must be a continuous concern of the production engineer, and any production-induced damage should be removed periodically to allow optimum well performance.

Example 11-10 Limited Gas Supply: Economics of Gas Lift

In Example 11-9 the required gas injection rate for maximum production was the one corresponding to the limit GLR. This would assume unlimited gas supply. Since this is usually not the case, repeat the well deliverability calculation for 0.5, 1, and 5×10^5 SCF/d gas injection rates, respectively. If the cost of gas injection is \$0.5/HHP-hr, the cost of separation is \$1/(MSCF/d), and the price of oil is \$50/STB, develop a simple economic analysis (benefits minus costs) versus available gas rate. Use the IPR curve for Example 11-3, $\overline{p}$ = 3050 psi, and natural GLR = 300 SCF/STB.

Solution

If only a gas injection rate of 5×10^4 SCF/d is available, then if, for example, q_l = 500 STB/d, the GLR would be [from Equation (11-8)]

$$\text{GLR} = \frac{5 \times 10^4}{500} + 300 = 400 \text{ SCF/STB}. \tag{11-34}$$

From Figure 11-5 at GLR = 400 SCF/STB, the flowing bottomhole pressure (with p_{tf} = 100 psi) is 1500 psi. Similar calculations can be done with the gradient curves for other flow rates. The resulting gas-lift performance curve is shown in Figure 11-17. In addition to $q_g = 5 \times 10^4$ SCF/d, the two other curves, corresponding to 1 and 5×10^5 SCF/d, are also drawn. Intersections with the IPR curve would lead to the expected production rates and flowing bottomhole pressures. With the help of Equation (11-12), the surface pressure can be calculated (assuming 100 psi pressure drop across the valve). Finally, from Equation (11-19), the horsepower requirements are calculated (using p_{in} = 100 psi). The results are shown in Table 11-1.

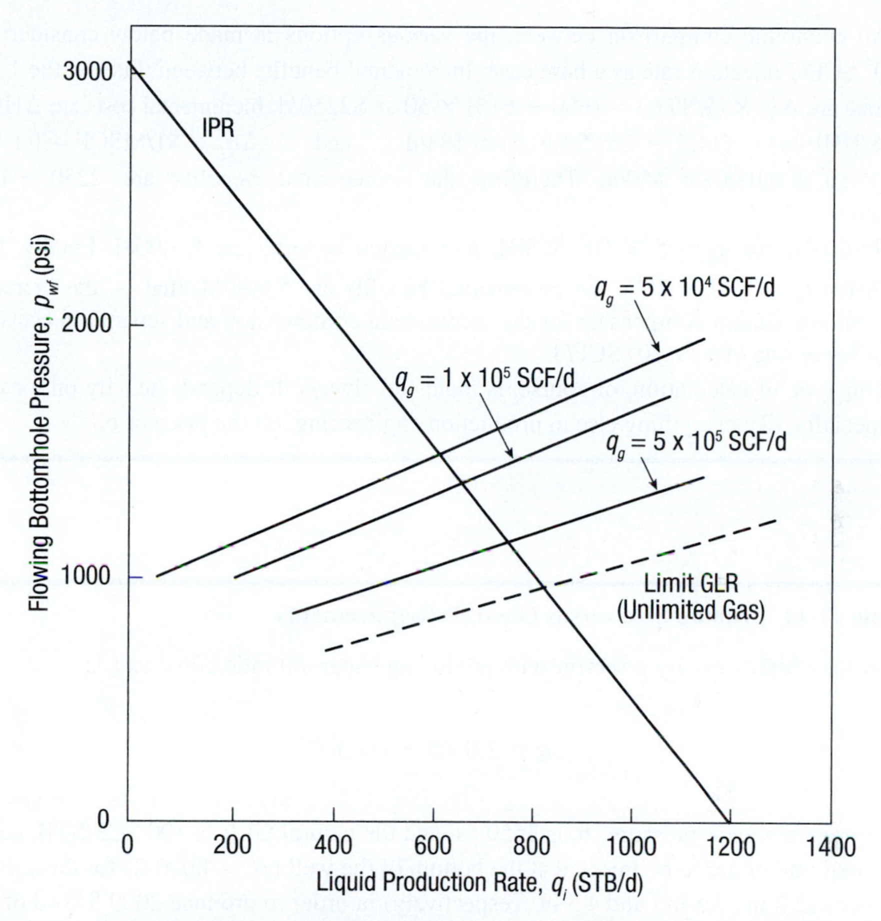

Figure 11-17 Production with gas lift with unlimited and limited gas injection rate (Example 11-10).

Table 11-1 Well Performance with Unlimited and Limited Gas Injection Rates

q_g (SCF/d)	q_l (STB/d)	p_{wf} (psi)	p_{surf} (psi)	HHP
5×10^4	615	1475	1310	7.5
1×10^5	660	1360	1220	14.5
5×10^5	750	1130	1025	66
2.89×10^6	825	935	865	348

An economic comparison between the various options is made below considering the 5×10^4 SCF/d injection rate as a base case. Incremental benefits between this and the 1×10^5 SCF/d rate are $\Delta q_l \times$ (\$/STB) = (660 − 615) × 50 = \$2250/d. Incremental costs are ΔHHP × 24 × (\$/HHP-hr) = (14.5 − 7.5)(24)(0.5) = \$84/d and $\Delta q_g \times$ \$1/MSCF = $[(1 \times 10^5) - (5 \times 10^4)](1/1000)$ = \$50/d. Therefore the incremental benefits are 2250 − 134 = \$2016/d.

Similarly, for $q_g = 5 \times 10^5$ SCF/d, incremental benefits are \$5600/d. Finally, for the limit GLR ($q_g = 2.89 \times 10^6$), the incremental benefits are \$3480/d; that is, the incremental production rate cannot compensate for the incremental compression and separation costs, compared to lower rate of 5×10^5 SCF/d.

This type of calculation, of course, is indicated always. It depends heavily on local costs and, especially, like everything else in production engineering, on the price of oil.

Example 11-11 Tubing Size versus Gas-Lift Requirements

The IPR for a high-capacity reservoir with producing water–oil ratio equal to 1 is

$$q = 1.03(\overline{p} - p_{wf}). \tag{11-35}$$

If the average reservoir pressure, $\overline{p}$, is 3550 psi and the natural GLR is 300 SCF/STB, calculate the required rate of gas to be injected at the bottom of the well (H = 8000 ft) for three different tubing sizes (2.5 in., 3.5 in., and 4.5 in., respectively) in order to produce 2000 STB/d of liquid. The wellhead pressure is 100 psi.

Solution

If q = 2000 STB/d, then, from Equation (11-31), p_{wf} must be equal to 1610 psi. Figures 11-18, 11-19, and 11-20 are the gradient curves for q_l = 2000 STB/d (50% water, 50% oil) for the three tubing sizes.

From Figure 11-18, at H = 8000 ft and p_{wf} = 1610 psi, the GLR must be 800 SCF/STB. Since the natural GLR is 300 SCF/STB, the required gas injection rate, q_g, must be 2000 (800 − 300) = 1×10^6 SCF/d.

From Figure 11-19 (for the 3.5-in. well), the required GLR for the same H and p_{wf} would be only 350 SCF/STB and therefore the gas injection rate would be only 2000 (350 − 300) = 1×10^5 SCF/d, an order of magnitude smaller. Finally, as can be seen readily from Figure 11-20,

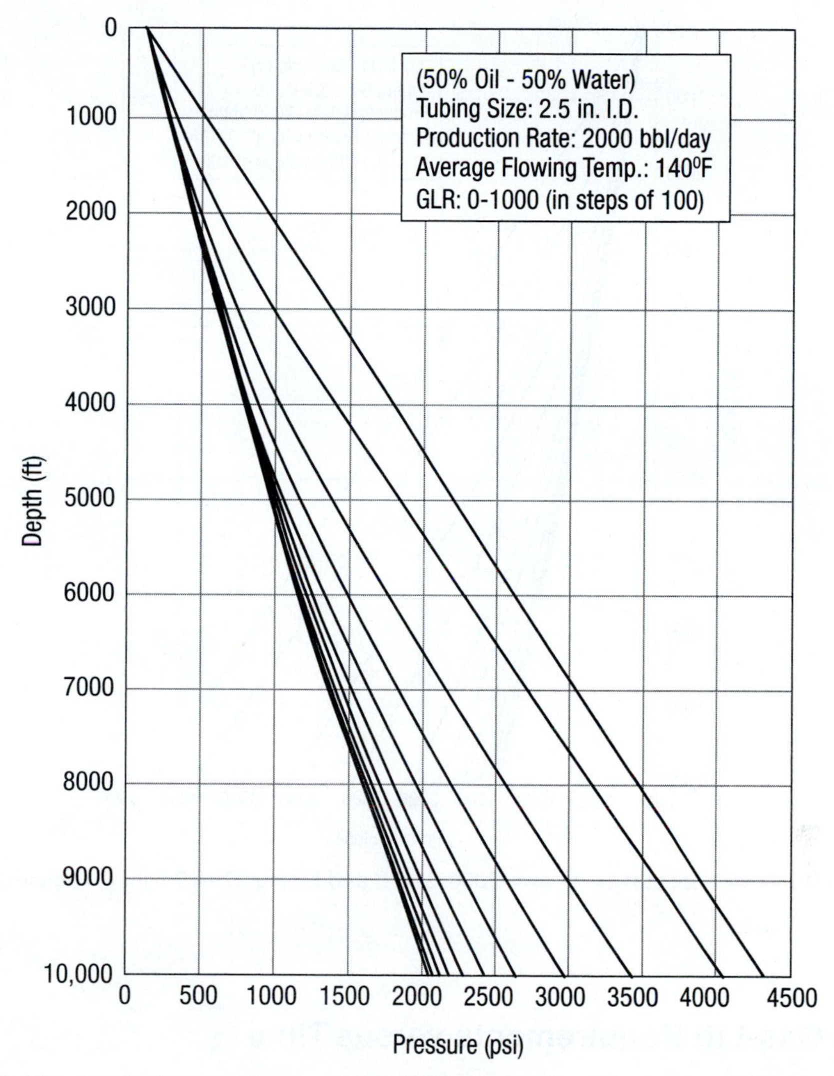

Figure 11-18 Gradient curves for $q_l = 2000$ STB/d and tubing ID $= 2.5$ in. (Example 11-11).

the natural GLR would be sufficient to lift the liquid production rate if the tubing diameter is 4.5 in.

This calculation shows the importance of the tubing size selection (and the associated economics of well completion) versus gas-lift requirements (i.e., volume of gas, compression and separation costs).

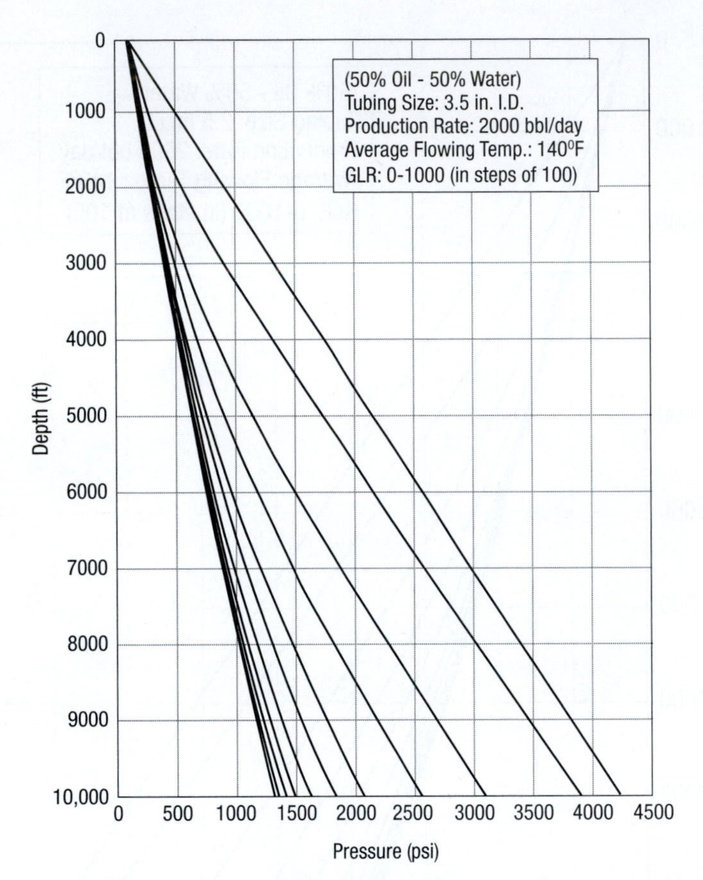

Figure 11-19 Gradient curves for q_l = 2000 STB/d and tubing ID = 3.5 in. (Example 11-11).

11.7 Gas-Lift Requirements versus Time

As the reservoir pressure declines, the driving force in the reservoir is reduced and therefore, in order to sustain a given production rate, the flowing bottomhole pressure must be reduced accordingly. This was discussed in Section 11.5 and was shown in Example 11-6. In that example, it was concluded that sustaining a production rate would require an increase in the gas injection rate as the reservoir pressure declines. In the same example the production rate was not the maximum possible for the given reservoir pressure (500 STB/d versus 825 STB/d, as shown in Example 11-7, where maximized gas-lift performance was calculated).

Maximizing the liquid production rate with gas lift requires an increase in the gas injection rate, since the limit GLR is larger for smaller production rates in a given tubing size. Again, design of gas lift versus time must be the subject of economic evaluation.

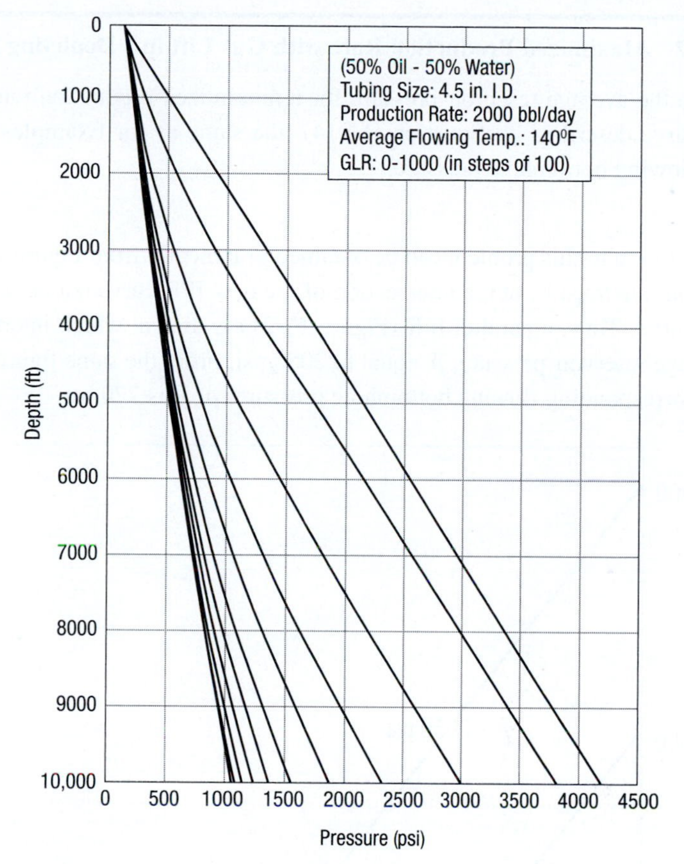

Figure 11-20 Gradient curves for $q_l = 2000$ STB/d and tubing ID $= 4.5$ in. (Example 11-11).

An appropriate method for forecasting gas-lift requirements versus time is as follows.

1. Establish an optimum gas-lift performance curve for the given tubing and reservoir fluid.
2. For each average reservoir pressure, develop a new IPR curve.
3. The intersection of the IPR curves with the optimum gas-lift performance curve is the maximum production rate.
4. This maximum production rate is at the limit GLR, which is also associated with the minimum p_{wf}.
5. Material balance calculations as shown in Chapter 10 can provide the reservoir cumulative production within each reservoir pressure decline interval, $\Delta\overline{p}$.
6. The production rate within this interval and the calculated cumulative production easily allow the calculation of the time required for this pressure decline.
7. Well performance and gas injection rate requirements versus time can then be forecasted.

Example 11-12 Maximized Production Rate with Gas Lift in a Depleting Reservoir

What would be the average reservoir pressure for a maximized production rate of 500 STB/d? Use the IPR curve described by Equation (11-14) (the same as for Examples 11-3 and 11-8). Calculate the flowing bottomhole pressure.

Solution

The simplest solution to this problem can be obtained graphically from Figure 11-14. The maximum production rate must be at the intersection of the new IPR curve (slope 0.39) and the optimum gas-lift curve. Thus, a parallel IPR (Figure 11-21) is drawn, which intersects the vertical axis at an average reservoir pressure, $\bar{p}$, equal to 2050 psi. From the same figure or from the IPR equation, the corresponding flowing bottomhole pressure, p_{wf}, is 770 psi.

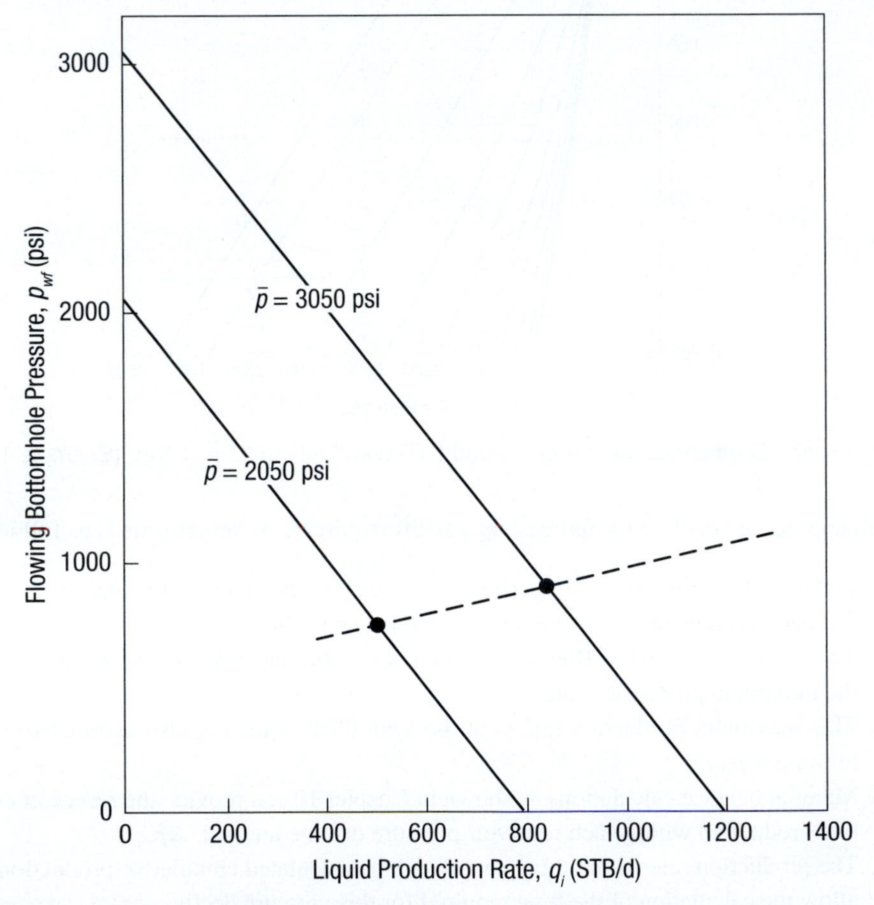

Figure 11-21 IPR curves with declining reservoir pressure for optimized gas lift (Example 11-12).

Example 11-13 Onset of Gas Lift in the Life of a Well, Need for Gas Lift

Use the well described in Appendix A. Suppose that the depth is 8000 ft and the drainage area is 160-acre spacing. If there is to be an attempt to sustain the production rate at 300 STB/d, could gas lift play a role? If so, *when* should it begin? Assume that the well is completed with 2 7/8-in. tubing ($\approx$ 2.259-in. I.D.) and that the wellhead pressure requirement is 300 psi.

Solution

From the data in Appendix A, the IPR curves for this well, using Equation (2-34) and allowing for reservoir pressure to decline from 5651 to 3000 psi, appear in Figure 11-22.

Figure 11-22 shows the series of IPR curves for this problem for average reservoir pressures ranging from the initial value (5651 psi) to 3000 psi. The corresponding required flowing bottomhole pressures for q = 300 STB/d can be readily extracted from this figure.

Figure 11-23 is an expanded scale of the flowing gradients for this well for different GLR values. The natural GLR for this well is 250 SCF/STB.

At the natural GLR (250 SCF/STB) the required flowing bottomhole pressure for 300 STB/d is 2640 psi. This pressure is lower than the required flowing botomhole pressure (3000 psi) for average reservoir pressure of 4500 psi and a rate of 300 STB/d (Figure 11-22)

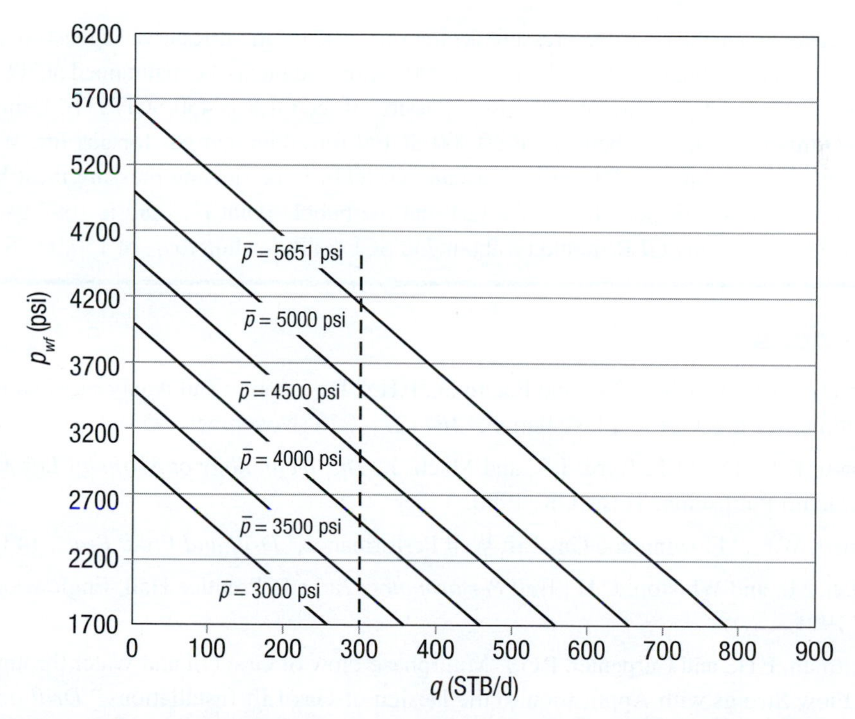

Figure 11-22 IPR curves at different reservoir pressures for Example 11-13.

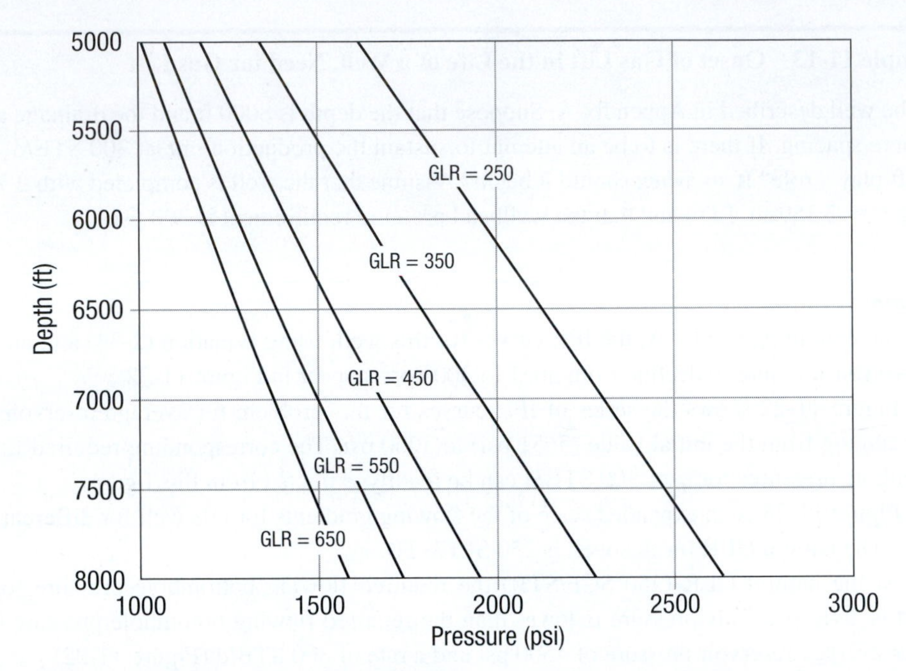

Figure 11-23 Expanded-scale gradient curves for well in Example 11-13.

However, when the reservoir pressure declines to 3500 psi the IPR curve suggests that the required flowing bottomhole pressure should be 2000 psi for the rate to be maintained at 300 STB/d. From Figure 11-23 the pressure of 2000 psi is possible if the GLR is 450 SCF/STB. Using Equation (11-3) this translates to about q_g of 60,000 SCF/d for additional gas for gas lift. When the reservoir pressure declines to 3000 psi to sustain 300 STB/d, the flowing pressure must be about 1500 psi (in this case affected also by the fact that the bubble-point pressure is 1697 psi). From Figure 11-23 the working GLR should be about 750 SCF/STB, leading to q_g of 150,000 SCF/d.

References

1. Bertuzzi, A.F., Welchon, J.K, and Poettman, F.H., "Description and Analyses of an Efficient Continuous Flow Gas-Lift Installation," *JPT,* 271–278 (November 1953).
2. Brown, K.E., Day, J.J., Byrd, J.P., and Mach, J., *The Technology of Artificial Lift Methods,* Petroleum Publishing, Tulsa, OK, 1980.
3. Gilbert, W.E., "Flowing and Gas-Lift Well Performance," *Drill and Prod. Prac.,* 143 (1954).
4. Golan, M., and Whiston, C.H., *Well Performance,* 2nd ed., Prentice Hall, Englewood Cliffs, NJ, 1991.
5. Poettman, F.H., and Carpenter, P.G., "Multiphase Flow of Gas, Oil and Water through Vertical Flow Strings with Application to the Design of Gas-Lift Installations," *Drill and Prod. Prac.,* 257–317 (1952).

Problems

11-1 Consider an 8000-ft deep well that is to produce 500 b/d with gradient curves given in Figure 11-5. The natural GLR is 200 SCF/STB, the average reservoir pressure is 3750 psi, and the IPR for the well is

$$q_o = 0.25(\overline{p} - p_{wf}).$$

The pressure drop across the gas-lift valve is 100 psi. If the gas is to be injected at 8,000 feet,

a. what is the GLR in the tubing above the injection point?
b. what is p_{surf}?
c. what is q_g?

Suppose the maximum possible surface pressure was limited to 100 psi less than the value calculated in part b. In order to produce the same liquid rate with this lower gas pressure, assuming the pressure gradient in the tubing below the gas injection point is 0.33 psi/ft,

a. what is H_{inj}?
b. what is GLR in the tubing above the injection point?
c. what is p_{inj}?
d. what is q_g?

11-2 Use Figure 11-5. Gas surface injection pressure is 1000 psi. Calculate the bottomhole pressures at 10,000 ft if the points of injection are at the curves for GLR equal to 400, 600, and 800 SCF/STB, respectively, and if the flowing gradient below the injection points is 0.37 psi/ft. At what average reservoir pressures should these points be activated to sustain a production rate of 500 STB/d? Use the IPR from Problem 11-1.

11-3 Consider a well to be gas lifted with the following characteristics:

Depth = 8000 ft	p_r = 2900 psi
2-3/8 in tubing (ID = 1.995 in)	p_{th} = 100 psi
T_{bh} = 210°F	T_s = 150°F

Surface operating pressure available = 900 psi
Kick-off pressure = 950 psi
Kill fluid gradient = 0.5 psi/ft
Pressure gradient above kill fluid = 0.2 psi/ft
Δp_{val} = 100 psi
Well loaded to top; first valve only unloaded against zero surface pressure
Point of gas injection = 3820 ft

Select the locations of the unloading valves for this well.

11-4 Suppose that the following limit GLR values and bottomhole pressures (at H = 10,000 ft) correspond to the three production rates listed below.

What would be the maximum production rate if the IPR multiplier were 0.6 and if $\overline{p}$ = 3500 psi? What would be the maximum production rate after the reservoir pressure declines to 3000 and 2500 psi, respectively?

Table 11-2 Production Data

q_l (STB/d)	GLR (SCF/STB)	p_{wf} (psi)
500	7000	1020
1000	2500	1200
2000	1000	2000

11-5 Calculate the gas injection rates, injection pressure, and horsepower requirements for Problem 11-4. Assume that the natural GLR is 300 SCF/STB.

11-6 Consider a vertical gas-lift well with the fluid and reservoir properties given in Appendix A. In addition, assume the following:

$\overline{p}$ = 3000 psi; max gas supply = 2 MMscf/d; depth = 8000 ft; tubing I.D. = 2.44 in.; pipe relative roughness = 0.0006; wellbore radius = 0.328 ft; drainage radius = 2000 ft; skin factor = 10; tubing head flowing pressure = 125 psia; surface temperature = 140°F

The compressor inlet pressure is 100 psia and the maximum gas supply is 1 MMscf/d. The pressure drop across the gas-lift valve is 100 psia.

a. For these conditions, and a gas-lift valve at 8000 feet, find the maximum liquid rate that can be lifted. Plot q_l versus q_{inj} for liquid flow rates ranging from 250 bbl/d up to the maximum liquid rate that can be lifted.

b. Change the skin factor to 0 and again find the maximum liquid rate that can be lifted.

c. Can the maximum liquid rate that can be lifted be increased by increasing the maximum gas supply? What happens to the net lift as the gas injection rate is increased? Why?

CHAPTER 12

Pump-Assisted Lift

12.1 Introduction

Downhole pumps are a common means of boosting the productivity of a well by lowering the bottomhole flowing pressure. Rather than lowering the pressure gradient in the tubing to reduce the bottomhole pressure, as is the case for gas lift, downhole pumps increase the pressure at the bottom of the tubing a sufficient amount to lift the liquid stream to the surface. In fact, the pressure gradient in the tubing is actually higher in a pumped well than it would be without the pump because most of the gas produced with the liquids is vented through the casing-tubing annulus. Furthermore, much of the free gas in the bottom of the well would be redissolved in oil at the higher pump outlet pressure, increasing the fluid density and gradient. A typical well configuration and pressure profiles in the well are shown in Figure 12-1.

Two types of pumps are used: positive-displacement pumps, which include sucker rod pumps and progressing cavity pumps (PCPs), and dynamic displacement pumps, the most common of which is the electrical submersible pump (ESP). Another artificial lift method similar to a positive-displacement pump is plunger lift, commonly used to unload liquids from gas wells.

For any well with a downhole pump, the work supplied by the pump is related to the increase in pressure across the pump by the mechanical energy balance equation, which for incompressible fluid is

$$W_s = \frac{p_2 - p_1}{\rho} + \frac{u_2^2 - u_1^2}{2g_c} + F. \tag{12-1}$$

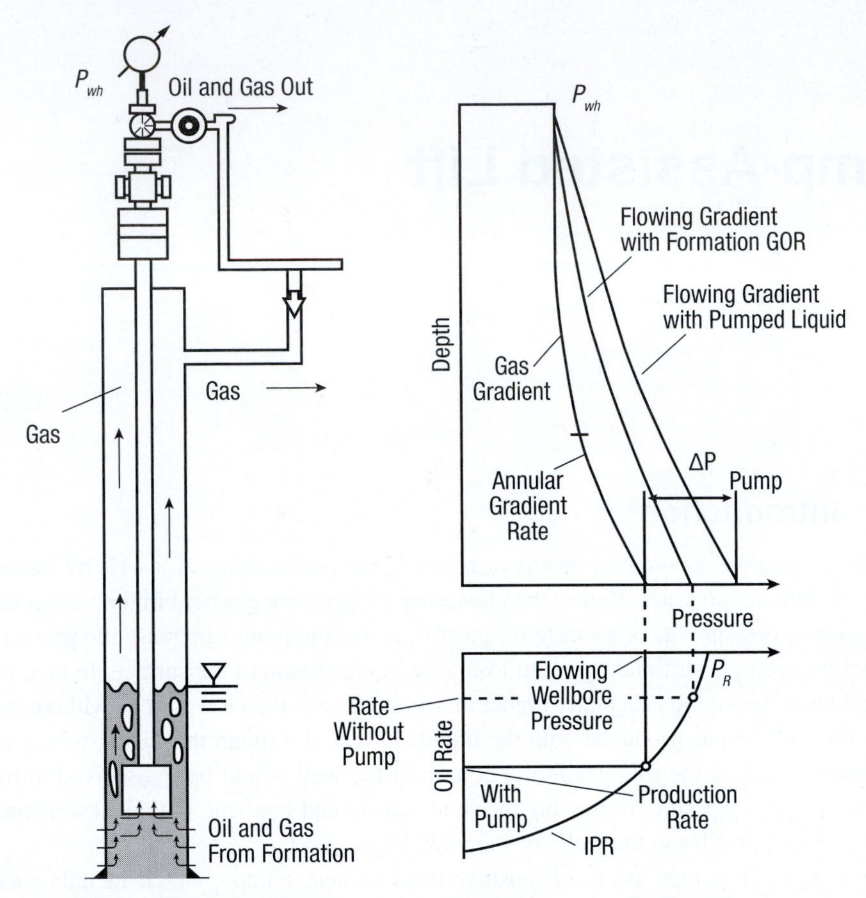

Figure 12-1 Well configuration and pressure profiles for an oil well. (After Golan and Whitson, 1991.)

For liquids, the kinetic energy term is usually small compared to the other terms, so the equation simplifies to

$$W_s = \frac{p_2 - p_1}{\rho} + F \tag{12-2}$$

where W_s is the work supplied by the pump, p_2 is the pressure in the tubing just above the pump, p_1 is the pressure just below the pump, and F is the frictional loss in the pump.

To determine the size and power requirements for a downhole pump, the pressures on either side of the pump are related to the bottomhole flowing pressure by the pressure gradient in the gas-liquid stream below the pump and to the surface pressure by the single-phase liquid gradient in the tubing. Thus, a design procedure is as follows: From the IPR relationship, the p_{wf} needed for a desired production rate is determined from a two-phase flow calculation, the pressure just below the pump, p_1, is calculated from p_{wf} (when the pump is near the production interval, $p_1 \approx p_{wf}$); from the surface tubing pressure, p_2 is determined based on single-phase liquid flow at the desired rate. Once the pressure increase required from the pump is known, the work required from the pump is found, usually based on empirical knowledge of the frictional losses in the pump (the pump efficiency). The flow relationships presented in Chapter 7 can be used to calculate the needed pressure profiles in the wellbore below the pump and in the tubing.

Example 12-1 The Pressure Increase Needed from a Downhole Pump

For the well described in Appendix A, it is desired to obtain a production rate of 500 STB/d at a time when the average reservoir pressure has declined to 3500 psi. Calculate the pressure increase needed from a downhole pump, assuming the pump is set at 9800 ft, just above the production interval, and that the surface tubing pressure is 100 psig. The well has 2 7/8-in., 8.6-lb$_m$/ft tubing (I.D. = 2.259 in.) and a relative roughness of 0.001.

Solution

IPR curves for this well for different reservoir pressures are presented in Figure 2-7. From this figure, when the average reservoir pressure is 3500 psi, for a production rate of 500 STB/d, p_{wf} is 1820 psi. Since the pump is set near the production interval, $p_1 = 1820$ psi. p_2 is calculated from Equation (7-15) for single-phase liquid flow. From Appendix A, the API gravity of the oil is 28°, and calculating the fluid properties using correlations in Chapter 3, at this pressure and the reservoir temperature of 220°F, the oil density is 47 lb$_m$/ft^3 and the viscosity is 1.04 cp. Neglecting the small change in oil density with pressure change, from Equation (7-22), we calculate the potential energy pressure drop to be 3196 psi. Since we can assume the oil to be incompressible, the kinetic energy pressure drop will be 0. To determine the frictional pressure drop, the Reynolds number is found with Equation (7-7) to be 14,804 and the friction factor from the Moody diagram (Figure 7-7) or the Chen equation [Equation (7-35)] is 0.0075. The average velocity is 1.165 ft/s. Using Equation (7-31),

$$\Delta p_{\mathrm{F}} = \frac{(2)(0.0075)(47)(1.165)^2(9{,}800)}{(32.17)(2.259/12)} = 1548 \text{ lb}_f/\text{ft}^2 = 10.7 \text{ psi} \tag{12-3}$$

so

$$\Delta p = \Delta p_{\mathrm{PE}} + \Delta p_{\mathrm{F}} = 3196 + 10.7 = 3207 \text{ psi} \tag{12-4}$$

and

$$p_2 = p_{\mathrm{surf}} + \Delta p = 100 + 14.7 + 3207 = 3322 \text{ psi}, \tag{12-5}$$

so $p_2 - p_1$, the pressure increase needed from the pump, is $3322 - 1820 = 1502$ psi.

12.2 Positive-Displacement Pumps

12.2.1 Sucker Rod Pumping

12.2.1.1 Sucker Rod Pump Equipment

The surface and downhole equipment for a rod-pumped well are shown in Figure 12-2. The rotary motion of the crank is translated to a reciprocating motion of the polished rod by the Pitman and the walking beam; the sucker rods transmit the reciprocating motion from the polished rod to the downhole pump. The pump (Figure 12-3) consists of a barrel with a ball-and-seat check valve at its bottom (the standing valve) and a plunger containing another ball-and-seat check valve (the traveling valve). When the plunger moves up, the standing valve opens, the traveling valve closes, and the barrel fills with fluid. On a down stroke, the traveling valve opens, the standing valve closes, and the fluid in the barrel is displaced into the tubing. For a detailed review of rod pump equipment, see Brown (1980a).

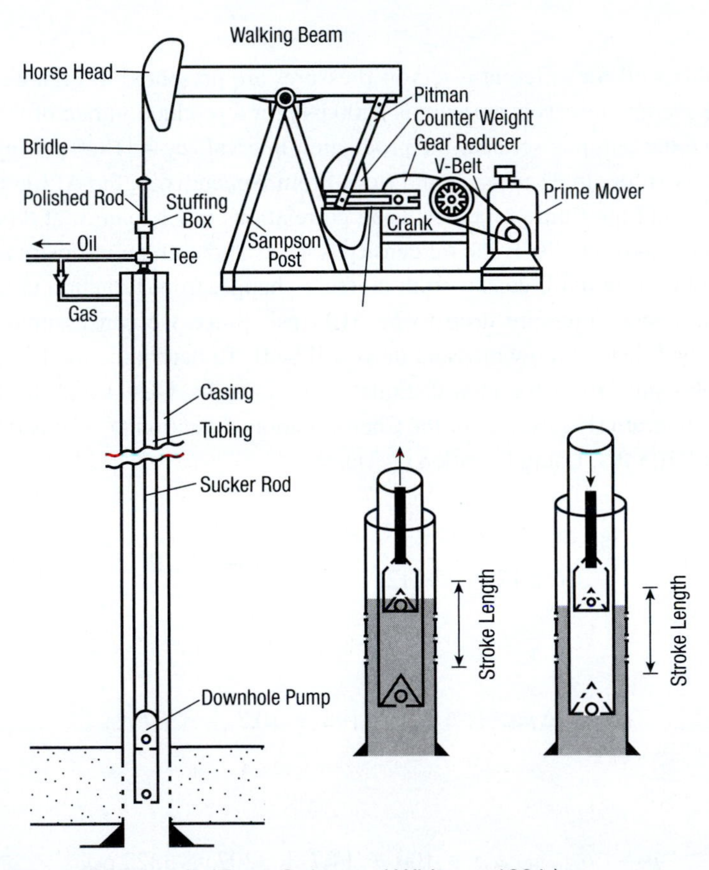

Figure 12-2 Rod-pumped well. (From Golan and Whitson, 1991.)

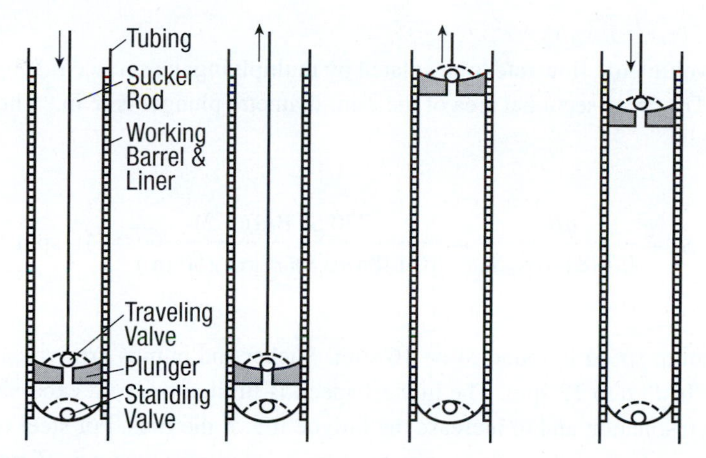

Figure 12-3 Sucker rod pump. (From Brown, 1980a.)

12.2.1.2 Volumetric displacement with Sucker Rod Pumps

Positive-displacement pump performance is evaluated based on the volume of fluid displaced, not the pressure increase generated by the pump, since the compression of the wellbore fluid in the pump will create enough pressure to displace the fluid in the tubing. The volumetric flow rate displaced by a rod pump is

$$q = 0.1484 N E_v A_p S_p \quad (12\text{-}6)$$

where q is the downhole volumetric flow rate (bbl/d), N is the pump speed (strokes per minute, spm), E_v is the volumetric efficiency, A_p is the plunger cross-sectional area (in.^{2}), and S_p is the effective plunger stroke length (in.). The surface production rate is the downhole rate divided by the formation volume factor.

The volumetric efficiency is less than 1 because of leakage of the fluid around the plunger. The volumetric efficiency is usually 0.7–0.8 for a properly working rod pump.

Example 12-2 Calculation of Required Pump Speed

Determine the pump speed (spm) needed to produce 250 STB/d at the surface with a rod pump having a 2-in.-diameter plunger, a 50-in. effective plunger stroke length, and a volumetric efficiency of 0.8. The oil formation volume factor is 1.2.

Solution

The downhole volumetric flow rate is calculated by multiplying the surface rate by the formation volume factor. The cross-sectional area of the 2-in.-diameter plunger is π in.^{2} Then, using Equation (12-6),

$$N = \frac{qB_o}{0.1484E_v A_p S_p} = \frac{(250 \text{ STB/d})(1.2)}{(0.1484)(0.8)(\pi \text{ in.}^2)(50 \text{ in.})} = 16 \text{ spm}. \quad \textbf{(12-7)}$$

The required pump speed is found to be 16 spm. Sucker rod pumps are typically operated at speeds ranging from 6 to 12 spm. The highest speed is limited to avoid excessive vibration of the rods due to resonation and to increase the fatigue life of the rods. For steel rods, the lowest pump speed that can cause the rods to vibrate at their natural frequency is (Craft, Holden, and Graves, 1962)

$$N = \frac{237{,}000}{L} \quad \textbf{(12-8)}$$

where L is the length of the rod string (in ft). Pump speed should be kept below this limiting value, and in fact is usually constrained to well below this speed to decrease rod fatigue.

12.2.1.3 Effective Plunger Stroke Length

The effective plunger stroke length will differ from and generally be less than the polished rod stroke length because of the stretching of the tubing and the rod string and because of overtravel caused by the acceleration of the rod string. Thus,

$$S_p = S + e_p - (e_t + e_r) \quad \textbf{(12-9)}$$

where S is the polished rod stroke length, e_p is the plunger overtravel, e_t is the length the tubing is stretched, and e_r is the length the rod string is stretched. Note that if the tubing is anchored, the tubing stretch will be zero. Assuming harmonic motion of the polished rod and elastic behavior of the rod string and tubing, the effective plunger stroke length is

$$S_p = S + \frac{(5.79 \times 10^{-4})SL^2N^2}{E} - \frac{5.20\gamma_l H A_p L}{E}\left(\frac{1}{A_t} + \frac{1}{A_r}\right). \quad \textbf{(12-10)}$$

In this equation, E is Young's modulus (about 30×10^6 psi for steel), γ_l is the specific gravity of the liquid, H is the liquid level depth in the annulus ($H = L$ if the liquid level is at the pump), A_t is the cross-sectional area of the tubing, A_r is the cross-sectional area of the rods, and all other variables are defined as before. Specifications of common sizes of rods and tubing and some commonly used sucker rod combinations are given in Table 12-1.

Example 12-3 Effective Plunger Stroke Length

Calculate the effective plunger stroke length for a well with a rod pump set at 3600 ft. The well has 3/4-in. sucker rods and 2 7/8-in. tubing, and the specific gravity of the produced liquid is 0.90. The pump speed is 12 spm, the plunger is 2 in. in diameter, and the polished rod stroke length is 64 in. The well is pumped off, so the liquid level is at the pump depth.

Solution

From Table 12-1, the cross-sectional areas of the rods and the tubing are 0.442 and 1.812 in.^{2}, respectively. Since the liquid level is at the pump, $H = L = 3600$ ft. Applying Equation (12-10),

$$S_p = 64 + \frac{(5.79 \times 10^{-4})(64)(3600)^2(12)^2}{30 \times 10^6} - \frac{(5.2)(0.9)(3600)(3.14)(3600)}{30 \times 10^6}\left(\frac{1}{1.812} + \frac{1}{0.442}\right)$$

$$= 64 + 2.31 - 17.9 = 48.44 \tag{12-11}$$

12.2.1.4 Prime Mover Power Requirements

The next step in designing a sucker rod pump installation is the determination of the power requirements for the prime mover. The prime mover must supply sufficient power to provide the useful work needed to lift the fluid; to overcome frictional losses in the pump, the polished rod, and the rod string; and to allow for inefficiencies in the prime mover and the surface mechanical system. Thus the required prime mover power is

$$P_{\text{pm}} = F_s(P_h + P_f) \tag{12-12}$$

Table 12-1 Steel Sucker Rod and Tubing Properties

Steel Sucker Rod Dimensions

Rod Diameter (in.)	Air Weight (lb_m/ft)	Area (in.^{2})
5/8	1.135	0.307
3/4	1.63	0.442
7/8	2.224	0.601
1	2.904	0.785
1 1/8	3.676	0.994

Steel Tubing Dimensions

Nominal Diameter (in.)	Weight (lb_m/ft)	O.D. (in.)	I.D. (in.)	Area (in.^{2})
2	2.90	1.900	1.610	0.800
2 3/8	4.70	2.375	1.995	1.304
2 7/8	6.50	2.875	2.441	1.812
3 1/2	9.30	3.500	2.992	2.590
4	11.00	4.000	3.476	3.077
4 1/2	12.75	4.500	3.958	3.601

Steel Sucker Rod Strengths

API Rod Grade	AISI Grade	T_m Minimum Tensile Strength (psi)
C	C-1536M	60,000
K	A-4621M	60,000
D	A-4320M	90,000
E	A-4330M1	140,000

Steel Sucker Rod Combinations

API Code	Design Plunger Diameter (in.)	Combined Rod Air Weight (lb_m/ft)	1 in.	% of each rod size 7/8 in.	3/4 in.
76	1.25	1.814		30.6	69.4
76	1.5	1.833		33.8	66.2

Table 12-1 Steel Sucker Rod and Tubing Properties *(continued)*

76	1.75	1.855		37.5	62.5
76	2	1.88		41.7	58.3
86	1.25	2.087	24.3	24.5	51.2
86	1.5	2.133	26.8	27	46.2
86	1.75	2.185	29.4	30	40.6
86	2	2.247	32.9	33.2	33.9
86	2.25	2.315	36.9	36	27.1

where P_{pm} is the prime mover horsepower, P_h is the hydraulic horsepower needed to lift the fluid, P_f is the power dissipated as friction in the pump, and F_s is a safety factor accounting for primer mover inefficiency. The safety factor can be estimated as 1.25 to 1.5 (Craft et al., 1962); for several empirical approaches to determining this factor, the reader is referred to Brown (1980a).

The hydraulic power is usually expressed in terms of net lift, L_N,

$$P_h = (7.36 \times 10^{-6})q\gamma_l L_N \tag{12-13}$$

where the flow rate is in bbl/d and the net lift is in ft of produced fluid. The net lift is the height to which the work provided by the pump alone can lift the produced fluid. If the tubing and casing pressure is zero at the surface and the liquid level in the annulus is at the pump, the net lift is simply the depth at which the pump is set. More generally, the fluid in the annulus above the pump exerts the force of its weight in helping to lift the fluid in the tubing and the tubing pressure is an additional force that the pump must work against. In this case, the net lift is

$$L_N = H + \frac{p_{tf}}{0.433\gamma_l} \tag{12-14}$$

where H is the depth to the liquid level in the annulus and p_{tf} is the surface tubing pressure in psig. To obtain Equation (12-14), we have assumed that the pressure at the liquid surface in the annulus is atmospheric pressure (i.e., the casing surface pressure is atmospheric and the hydrostatic head of the column of gas in the annulus is negligible) and that the average density of the liquid in the annulus is the same as that in the tubing (we are neglecting the gas bubbling through the liquid in the annulus).

The horsepower needed to overcome frictional losses is empirically estimated as

$$P_f = 6.31 \times 10^{-7} W_r S N \tag{12-15}$$

where W_r is the weight of the rod string (lb_m), S is the polished rod stroke length (in.), and N is the pump speed (spm).

Example 12-4 Prime Mover Power Requirements

For the well described in Example 12-3, calculate the power requirement for the prime mover if the surface tubing pressure is 100 psig and the liquid level in the annulus is 200 ft above the pump.

Solution

Since the liquid level is 200 ft above the pump, the liquid level, H, is $3600 - 200 = 3400$ ft. Then, from Equation (12-14), the net lift is

$$L_N = 3400 + \frac{100}{(0.433)(0.9)} = 3657 \text{ ft.} \tag{12-16}$$

The flow rate, obtained from Equation (12-6) using a volumetric efficiency of 0.8 and the results from Example 12-3, is

$$q = (0.1484)(20)(0.8)(3.14)(52.5) = 391 \text{ bbl/d.} \tag{12-17}$$

The hydraulic horsepower is then calculated with Equation (12-13):

$$P_h = (7.36 \times 10^{-6})(391)(0.9)(3657) = 9.5 \text{ hp.} \tag{12-18}$$

Using the data in Table 12-1, the weight of 3600 ft of 3/4-in. sucker rods is 5868 lb_m. The frictional horsepower from Equation (12-15) is

$$P_f = (6.31 \times 10^{-7})(5868)(64)(20) = 4.7 \text{ hp.} \quad (12\text{-}19)$$

Finally, using a safety factor of 1.25 in Equation (12-12), the prime mover horsepower requirement is

$$P_{pm} = (1.25)(9.5 + 4.7) = 18 \text{ hp.} \quad (12\text{-}20)$$

12.2.1.5 Selecting Beam Pumping Units

The standard classification of a rod pump unit consists of three specifications. The first one is for gear box rating; the second is for peak beam loading; and the last one for maximum stroke length. Table 12-2 lists some commonly used pumping unit specifications.

12.2.1.6 Sucker Rod and Beam Loads

The maximum load on the top sucker rod and on the beam of the pumping unit occurs during the upstroke when the rod weight, buoyed in the liquid, the liquid weight, and the upstroke dynamic loads are being lifted. Thus,

$$W_{max} = W_{rb} + F_l + W_D \quad (12\text{-}21)$$

Table 12-2 Selected Conventional Pumping Unit Sizes

Size Code	Gear Box Rating (1000 in.-lb_f)	Peak Beam Load (100 lb_f)	Maximum Stroke Length (in.)
160-173-86	160	173	86
228-246-86	228	246	86
320-256-100	320	256	100
456-305-144	456	305	144
640-365-144	640	365	144
912-365-168	912	365	168
1280-427-192	1280	427	192
1840-305-240	1840	305	240

where W_{max} is the maximum rod and beam load, W_{rb} is the buoyed rod weight, F_l is the liquid load, and W_D is the upstroke dynamic load. The liquid load, F_l, is the weight of the column of liquid in the tubing minus the weight of the liquid column in the annulus, so

$$F_l = 0.433\gamma L_N A_p. \tag{12-22}$$

The buoyed rod weight is

$$W_{rb} = W_r(1 - 0.127\gamma) \tag{12-23}$$

where W_r is the rod weight.

The minimum load carried by the rods and beam occurs during the downstroke when the rods are allowed to fall back into the tubing, resisted by the dynamic loads during the downstroke.

$$W_{\min} = W_{rb} - W_D. \tag{12-24}$$

The dynamic loads are best computed using a simulation of the motion of the rods, but an estimate can be made if simple harmonic motion of the rod string is assumed:

$$W_D = \alpha W_{rb} \tag{12-25}$$

where

$$\alpha = \frac{SN^2}{70{,}542}. \tag{12-26}$$

The parameter α is the acceleration constant.

The rods pass through a load cycle on each stroke, with the maximum load occurring during the upstroke and the minimum occurring during the downstroke. Cyclical loads promote failure of the rods due to fatigue. In order to give the rods a reasonable life, the API created a modified Goodman diagram that is used to determine the maximum allowable load for a given minimum load (American Petroleum Institute, 1988). The equation for the maximum acceptable load is:

$$W_{\max} = (0.25T_m A + 0.5625W_{\min})SF \tag{12-27}$$

where T_m is the minimum rod tensile strength (psi), A is the rod cross-sectional area (in^2), W_{min} is the minimum load (lb$_f$), and SF is a service factor. The service factor is given in Table 12-3.

Table 12-3 Rod Load Service Factors

	API Rod Grade			
Service	**C**	**K**	**D**	**High Strength**
Noncorrosive	1.00	1.00	1.00	1.00
Salt water	0.65	0.90	0.90	0.70
Hydrogen sulfide	0.45*	0.70	0.65	0.5*

* Not recommended without adequate chemical inhibition.

The net gear box torque carried by the pumping unit is the difference between the torque created by the well load during the stroke and the offsetting counterbalance torque. These loads vary throughout the stroke and are best estimated with a simulation of the rod loads and pumping unit. The maximum torque can be estimated by

$$T_{max} = \left(W_{max} - 0.95\left(\frac{W_{max} + W_{min}}{2}\right)\right)\frac{S}{2}. \tag{12-28}$$

In Equation (12-28) the value of 0.95 is a service factor that essentially expresses the likelihood of keeping the unit perfectly counterbalanced.

12.2.1.7 Analysis of Rod Pump Performance from Dynamometer Cards

The performance characteristics of sucker rod pumps are commonly monitored by measuring the load on the polished rod with a dynamometer. A recording of the polished rod load over one complete pump cycle is referred to as a "dynamometer card." The dynamometer card plots polished rod load as a function of rod position.

An ideal dynamometer card for elastic rods is shown in Figure 12-4. At point *a,* an upstroke begins and the polished rod load gradually increases as the rods stretch until at point *b* the polished rod supports the weight of the rods in the fluid and the weight of the fluid. The load remains constant until the downstroke begins at point *c.* At this time, the standing valve closes and the standing valve supports the weight of the fluid; the polished rod load decreases as the rods recoil until, at point *d,* the polished rod supports only the weight of the rods in the fluid. The load then remains constant until another cycle begins at point *a.*

Numerous factors will cause an actual dynamometer card to differ from this idealization—a dynamometer card from a properly working rod pump is shown in Figure 12-5. The acceleration and deceleration of the rod string accounts for most of the difference between the ideal load history and the actual history of the polished rod in a properly operating rod-pumped well. The

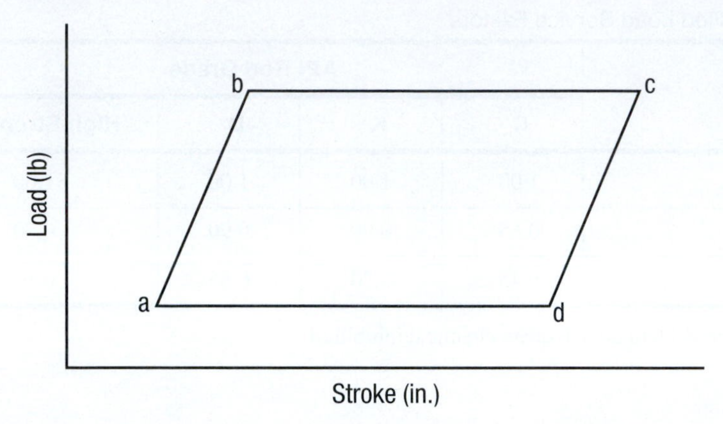

Figure 12-4 Ideal dynamometer card for elastic rods.

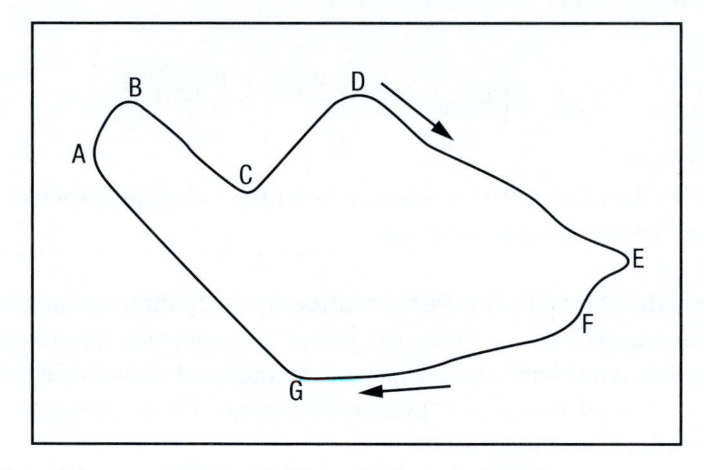

Figure 12-5 Actual dynamometer card. (From Craft et al., 1962.)

character of the dynamometer card can sometimes be used to diagnose abnormal characteristics of the pump or well behavior.

Figure 12-6 illustrates characteristic dynamometer card shapes for a number of adverse conditions that can be encountered in a rod-pumped well (Craft et al., 1962). A well being pumped at synchronous speed will exhibit the distinctive card shown in Figure 12-6a. Notice the decreasing load on the upstroke and the loop in the curve at the end of the upstroke. A restriction

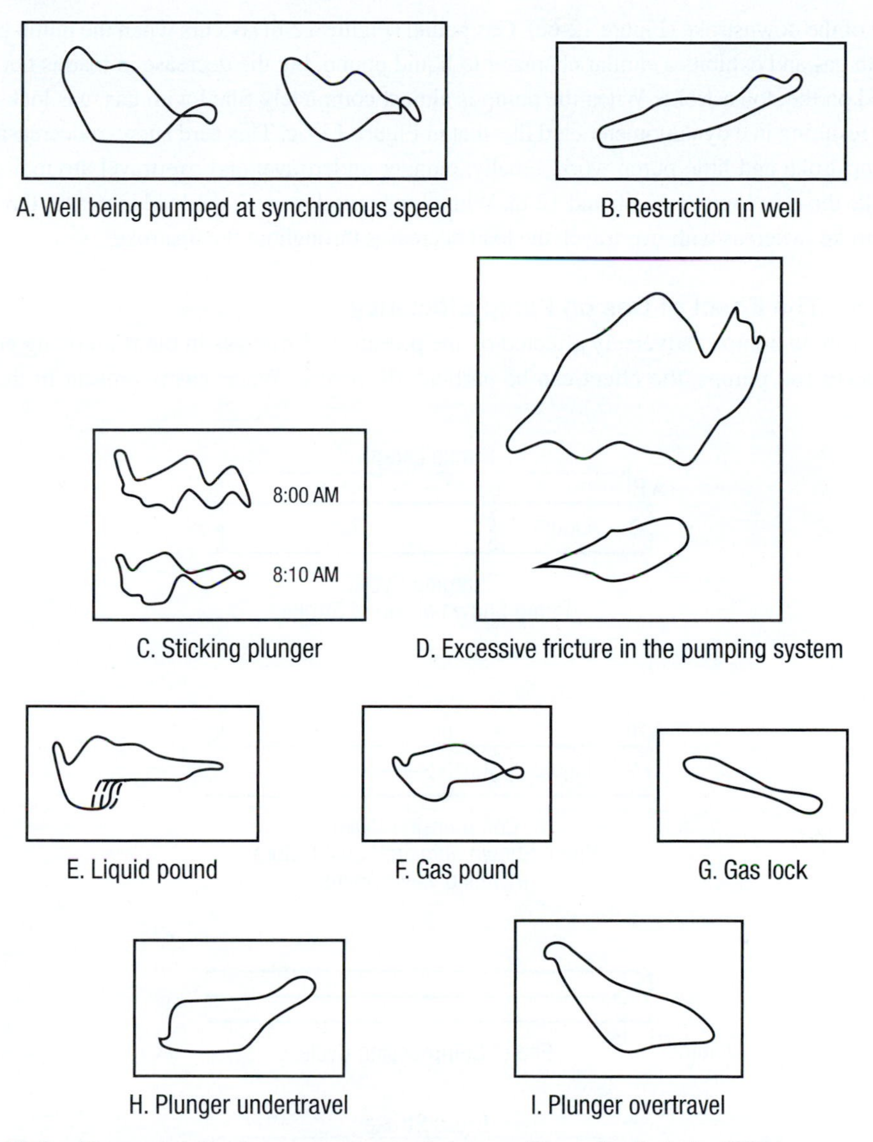

Figure 12-6 Characteristic dynamometer card shapes. (From Craft et al., 1962.)

in the well leads to an increasing load on the upstroke and little pump work, as indicated by the area of the region enclosed by the load curve (Figure 12-6b).

Excessive friction in the pumping system results in erratic dynamometer responses, like those shown in Figure 12-6c (sticking plunger) or Figure 12-6d. Liquid pound occurs when the pump barrel does not fill completely on the upstroke and is characterized by a sudden decrease in load near the end of the downstroke (Figure 12-6e). Gas pound (Figure 12-6f) occurs when the pump partially fills with gas and exhibits a similar character to liquid pound, but the decrease in load is not as pronounced on the downstroke. When the pump is almost completely filled with gas, gas lock has occurred, resulting in a dynamometer card like that in Figure 12-6g. This card shows a decreasing load on the upstroke and little pump work. Finally, plunger undertravel and overtravel are indicated by cards like those in Figures 12-6h and 12-6i. With plunger undertravel, the load increases throughout the upstroke, whereas with overtravel, the load decreases throughout the upstroke.

12.2.1.8 The Effect of Gas on Pump Efficiency

Any downhole pump is adversely affected by the presence of free gas in the fluid being pumped; with sucker rod pumps, the effect can be particularly severe. When gas is present in the pump

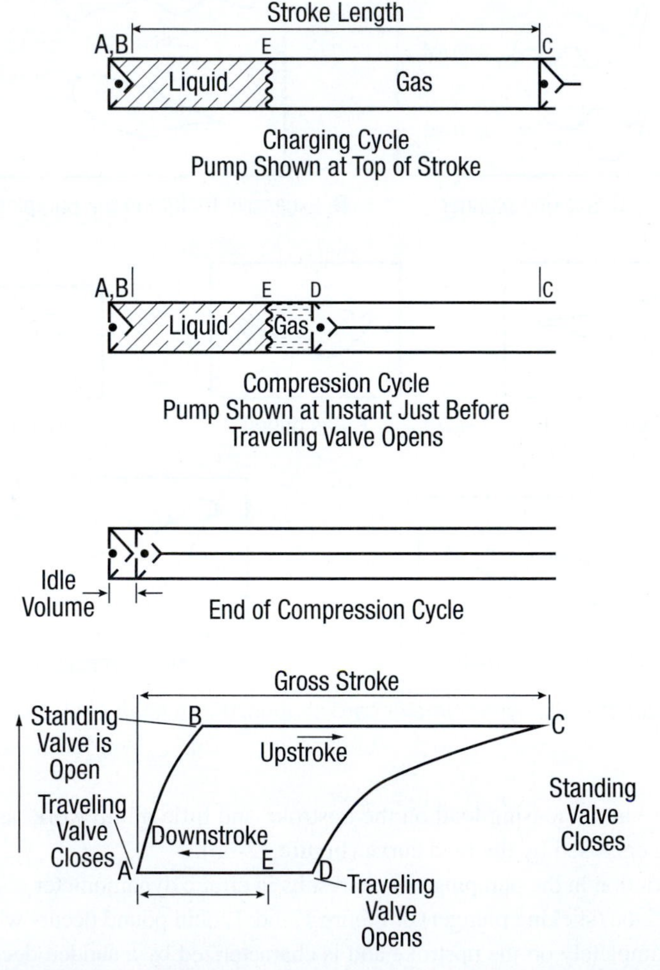

Figure 12-7 Effect of gas on rod pump performance. (From Golan and Whitson, 1991.)

barrel, much of the pump energy is expended in compressing the gas instead of lifting the liquid. Figure 12-7 illustrates this effect. When gas is present in the pump, on a downstroke, the gas must be compressed until the pressure in the barrel is equal to the pressure in the tubing above the pump before the traveling valve will open and allow fluid to pass into the tubing. On an upstroke, the gas must expand until the pressure is below p_1, the pressure in the casing just below the pump, before the standing valve will open and let wellbore fluids enter the barrel. In extreme cases, essentially nothing will occur in the pump except the expansion and compression of gas; in this instance, the pump is said to be "gas-locked."

Because of these deleterious effects, some means must be employed to exclude most, if not all, of the free gas from entering a sucker rod pump. This is accomplished by setting the pump below the perforations so that the gas will rise out of the liquid stream moving to the pump or by employing various mechanical means to separate the gas from the liquid

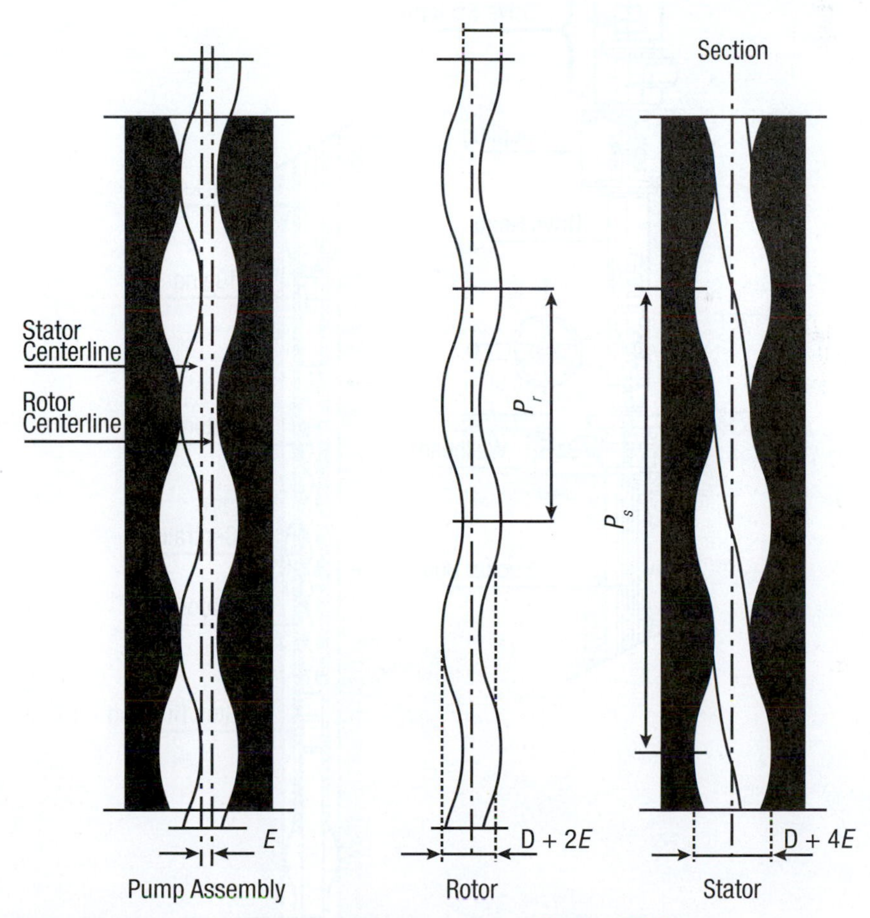

Figure 12-8 Progressing cavity pump geometry. (From Cholet, 2008.)

stream. Downhole devices installed on rod pumps to separate gas from the liquid are called "gas anchors."

12.2.2 Progressing Cavity Pumps

A positive displacement pump that is being increasingly used instead of a traditional rod pump is the progressing cavity pump (PCP). This pump consists of a helically spiraling rotor that rotates inside of a doubly helical stator made of an elastomeric material (Figure 12-8; Cholet, 2008). The PCP geometry creates sealed cavities that progress from the pump inlet to the pump outlet as the rotor turns, thus displacing the fluid in the cavities.

A progressing cavity pump is commonly driven by a rotating rod string, as is illustrated in Figure 12-9. Alternatively, a PCP can be driven by an electrical submersible motor, which

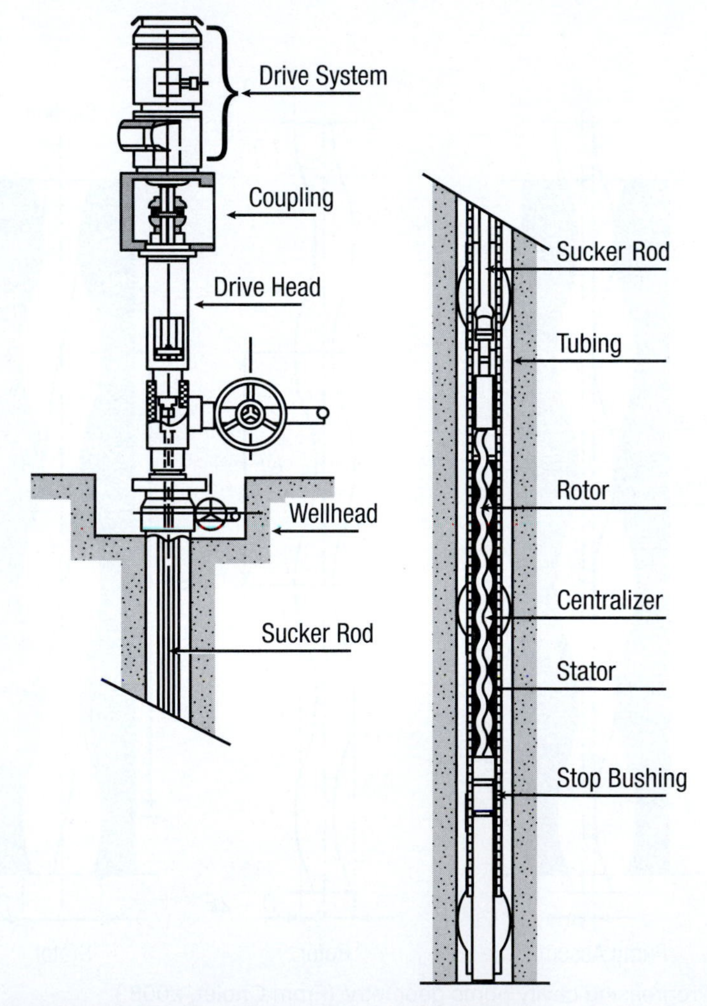

Figure 12-9 Rod-driven PCP completion. (From Cholet, 2008.)

makes a PCP practical in deviated and/or deep wells in which rod strings are impractical (Figure 12-10).

Progressing cavity pumps are better suited to handling fluids such as sand-laden or high-viscosity fluids than are rod pumps. For a thorough review of progressing cavity pumps, the reader is referred to Cholet (2008).

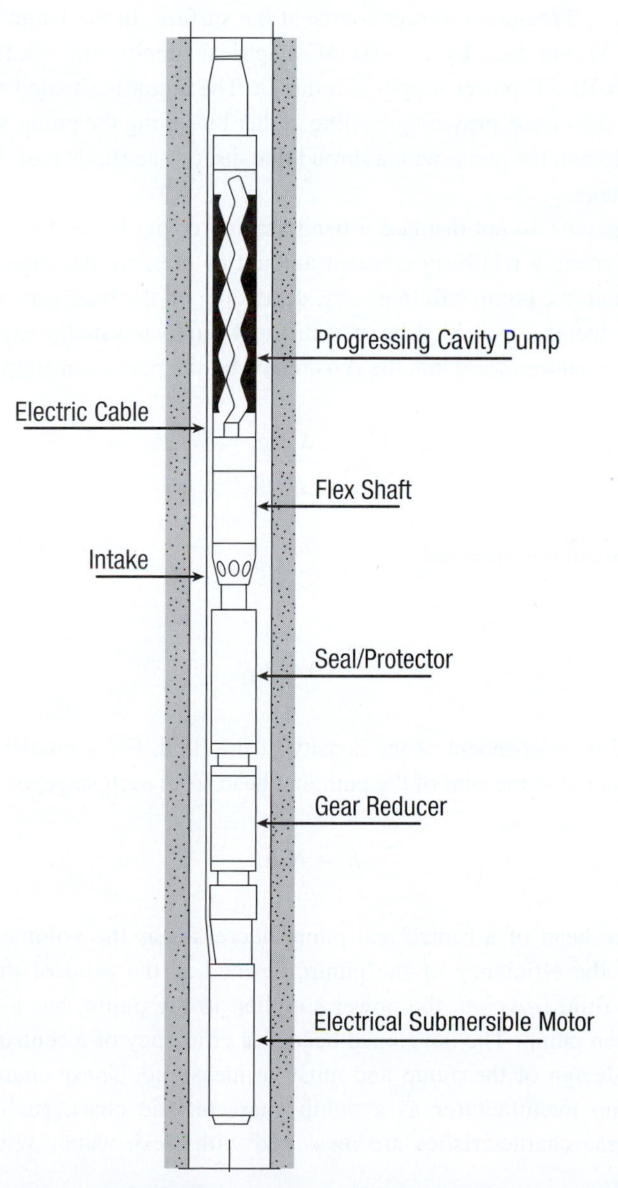

Figure 12-10 Electrical submersible motor-driven PCP completion. (From Cholet, 2008.)

12.3 Dynamic Displacement Pumps

12.3.1 Electrical Submersible Pumps

An electrical submersible pump (ESP) is a multistage centrifugal pump that offers a great deal of flexibility. ESPs are capable of producing very high volumes of liquid, can be used efficiently in deeper wells than sucker rod pumps, and are able to handle some free gas in the pumped fluid. A typical ESP completion is shown in Figure 12-11. The pump is driven by an electric motor connected by cables to a three-phase power source at the surface. In the United States, ESPs typically operate at 3500 rpm drive by a 60-Hz AC electrical supply, while in Europe, operation at 2815 rpm with a 50-Hz AC power supply is common. The motor is situated so that the produced fluids flow around the motor, providing cooling, either by setting the pump above the producing interval, or by equipping the pump with a shroud that directs the fluids past the motor before entering the pump intake.

Centrifugal pumps do not displace a fixed amount of liquid, as do positive-displacement pumps, but rather create a relatively constant amount of pressure increase to the flow stream. The flow rate through the pump will thus vary, depending on the back pressure held on the system. The pressure increase provided by a centrifugal pump is usually expressed as pumping head, the height of produced fluid that the Δp created by the pump can support:

$$h = \frac{\Delta p}{\rho} \frac{g_c}{g}, \tag{12-29}$$

which in field units can be expressed as

$$h = \frac{\Delta p}{0.433\gamma_l}. \tag{12-30}$$

The pumping head is independent of the density of the fluid. For a multistage pump, the total head developed is equal to the sum of the pumping head from each stage, or

$$h = N_s h_s. \tag{12-31}$$

The pumping head of a centrifugal pump decreases as the volumetric throughput increases; however, the efficiency of the pump, defined as the ratio of the hydraulic power transferred to the fluid ($q\Delta p$) to the power supplied to the pump, has a maximum at some flow rate for a given pump. The developed head and efficiency of a centrifugal pump depend on the particular design of the pump and must be measured. These characteristics are provided by the pump manufacturer as a pump characteristic chart, such as that shown in Figure 12-12. These characteristics are measured with fresh water. With another fluid of

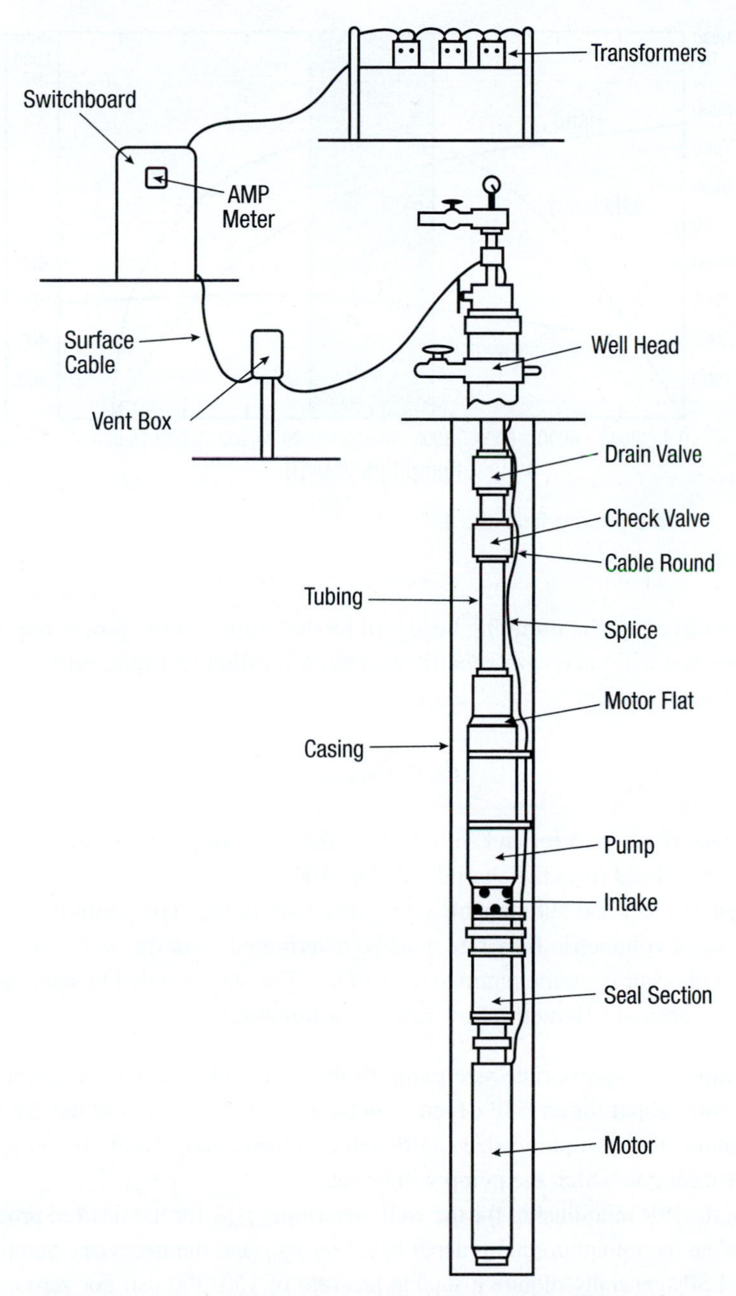

Figure 12-11 Electrical submersible pump completion. (From Centrilift, 1987.)

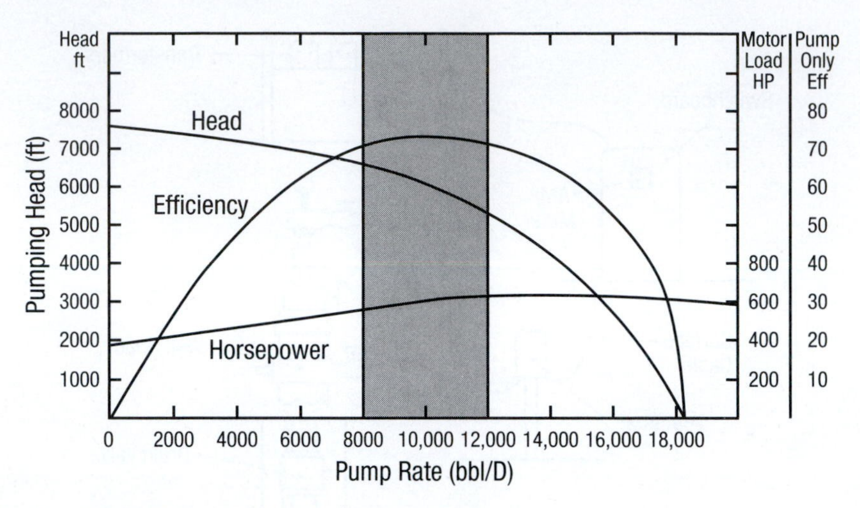

Figure 12-12 Pump characteristic chart.

about the same viscosity, the pumping head will be the same, but the power requirements will differ, since the Δp will vary with specific gravity, according to Equation (12-30). Thus, for a fluid of a different density,

$$P_h = P_{h,\,\text{water}}\gamma_l. \tag{12-32}$$

The pump characteristic chart for an ESP is usually for a 100-stage pump, so the head developed per stage is the total head from the chart divided by 100.

To design an electrical submersible pump installation, the Δp (pumping head) needed to produce the desired volumetric flow rate must be determined from the well's IPR and the pressure drop that will occur from the pump to the surface. The steps needed to select an appropriate ESP to produce a desired volumetric flow rate are as follows:

1. Determine the appropriate size pump from the manufacturer's specifications. An efficient throughput for an ESP depends on the size of the pump, not the Δp developed by the pump. Thus the pump size can be selected based only on the flow rate and the size of the casing in which the pump will be set.
2. From the IPR relationship for the well, determine p_{wf} for the desired production rate.
3. Calculate the minimum pump depth based on p_{wf} and the necessary pump suction pressure. ESPs generally require a suction pressure of 150–300 psi. For zero casing pressure and neglecting the hydrostatic pressure of the gas column in the annulus, the pump depth is

$$H_{\text{pump}} = H\left(\frac{p_{wf} - p_{\text{suction}}}{0.433\gamma_l}\right) \tag{12-33}$$

where H is the depth of the producing interval, H_{pump} is the pump depth, and $p_{suction}$ is the required suction pressure of the pump. The pump can be set at any depth below this minimum depth, and will often be located nearer the production interval. The pump suction pressure can be calculated from Equation (12-33) for any pump depth.

4. Determine the required pump discharge pressure from a pressure traverse calculation for the flow in the tubing, using the equations presented in Chapter 7.
5. The Δp needed from the pump is then

$$\Delta p_{pump} = p_{discharge} - p_{suction}. \tag{12-34}$$

6. From the pump characteristic curve, the head per stage is read. The number of stages needed is then calculated with Equation (12-31).
7. The total power requirement for the pump is obtained by multiplying the power per stage from the pump characteristic chart by the number of stages.

Example 12-5 Electrical Submersible Pump Design

A 10,000-ft-deep well in the reservoir described in Appendix B is to be produced at a rate of 3000 STB/d with an ESP when the average reservoir pressure is 4336 psi. The well will be equipped with 3 1/2 in. tubing ($\epsilon = 0.001$, I.D. $= 2.992$ in.), the surface tubing pressure is 100 psig, and the well casing is 7 in. Assume that the minimum suction pressure required by the pump is 120 psi. Determine the required specifications for an electrical submersible pump for this application.

Solution

The first step is to choose a pump with a capacity range suitable for the desired flow rate. The flow rate through the pump is the downhole volumetric oil rate, assuming that all free gas is excluded from the pump. To determine the formation volume factor for this saturated oil, we must know the pressure.

From Vogel's equation [Equation (3-53)] or Figure 3-6, p_{wf} for a rate of 3000 STB/d is 2600 psi. Using the correlations in Chapter 3, at the p_{wf} of 2600 psi, the density of the oil is 43 lb_m/ft^3, the viscosity is 0.63 cp, and the formation volume factor is 1.29. We will use these values for all calculations, although they will of course change some with pressure and temperature changes in the tubing string and in the annulus. For the B_o of 1.29, the flow rate through the pump is (3000 STB/d)(1.29) $=$ 3870 bbl/d. We now choose a pump that will fit in the 7-in. casing and that has a suitable capacity. Figure 12-13 is the pump characteristic chart for one such pump.

Next, we check the minimum depth for setting the pump. Using Equation (12-33),

$$H_{pump} = 10{,}000 - \frac{2600 - 200}{\left(\frac{43}{144}\right)} = 1963 \text{ ft.} \tag{12-35}$$

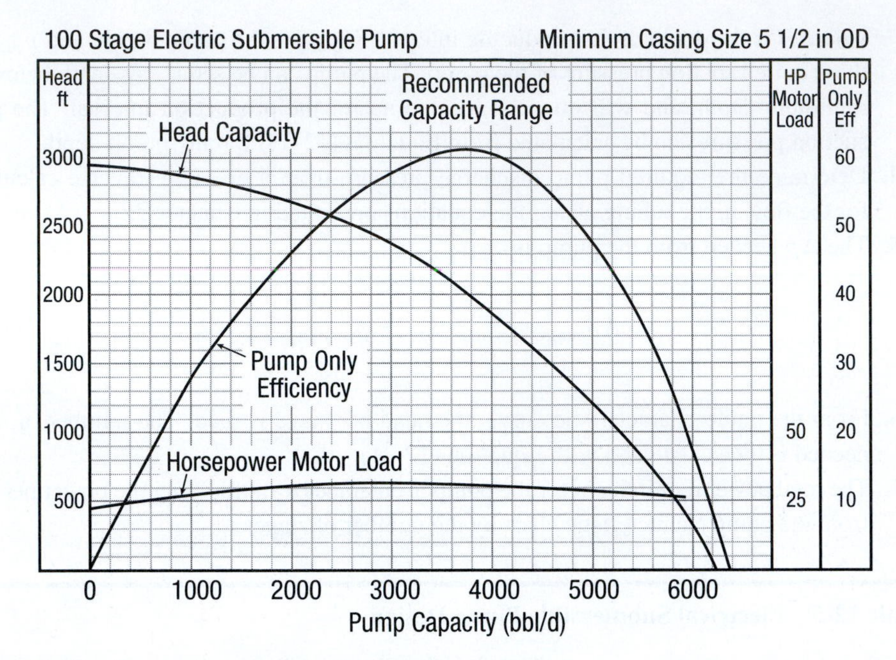

Figure 12-13 Pump characteristic chart for Example 12-5.

The pump can be set at any depth below this point. We will assume that the pump is set at 9800 ft; the pump suction pressure is then

$$p_{\text{suction}} = 2600 - \left(\frac{43}{144}\right)(10{,}000 - 9800) = 2540 \text{ psi.} \tag{12-36}$$

The pressure drop in the tubing must now be calculated to determine the pump discharge pressure. Following Example 12-1,

$$\Delta p_{\text{PE}} = \left(\frac{43}{144}\right)(9800) = 2926 \text{ psi.} \tag{12-37}$$

The Reynolds number is 130,600, the friction factor is 0.0054, and the mean velocity is 5.15 ft/sec, so

$$\Delta p_{\text{F}} = \frac{(2)(0.0054)(43)(5.15)^2(9800)}{(32.17)(2.992/12)(144)} = 78 \text{ psi.} \tag{12-38}$$

Therefore, the total Δp is 2926 + 78 = 3004 psi. The discharge pressure is

$$p_{\text{discharge}} = p_{\text{surf}} + \Delta p = 100 \text{ psig} + 14.7 \text{ psia} + 3004 \text{ psia} = 3119 \text{ psi} \tag{12-39}$$

and the pressure increase from the pump, from Equation (12-34), is

$$\Delta p_{\text{pump}} = 3119 - 2540 = 579 \text{ psi.} \tag{12-40}$$

The pressure increase from the pump can be expressed as feet of head using Equation (12-30):

$$h = \frac{579}{(43/144)} = 1939 \text{ ft.} \tag{12-41}$$

Reading from Figure 12-11, for a flow rate of 3870 bbl/d, the head for a 100-stage pump is about 2000 ft, or 20 ft/stage. The required number of stages is then

$$N_s = \frac{1939}{20} = 97 \text{ stages.} \tag{12-42}$$

A 100-stage pump would be selected. Finally, the horsepower required for the 100-stage pump at 3870 bbl/d from Figure 12-13 is 26 hp. The well in this example could produce at a much higher rate by using more pump stages to lower the liquid level and hence the bottomhole flowing pressure.

This example illustrates the optimal pump to produce this well at a fixed rate and fixed average reservoir pressure. As the reservoir pressure declines, the IPR curve will change, and hence the optimal pump design will gradually change. To design an ESP that will be optimal over a finite life span of the well involves a tradeoff between pumping requirements early in the life of the well and later in the life of the reservoir and is ultimately an economic question. Particularly in water-flooded reservoirs, where the water production rate may rise dramatically over the life span of the well, the pumping requirements may change drastically.

The pump characteristics given by ESP manufacturers are for pumping water and must be corrected if the fluid being pumped has a higher viscosity. High fluid viscosity decreases the efficiency of a centrifugal pump and can affect the head developed. Pump manufacturers provide correction charts to account for behavior with high-viscosity fluids.

12.4 Lifting Liquids in Gas Wells; Plunger Lift

Many gas wells produce liquid along with the gas, either condensate or water, or both. If the gas velocity in the tubing string is not sufficient to lift the liquids to the surface, liquid accumulates in the well, increasing the bottomhole pressure. In relatively low-rate gas wells, the liquid accumulation can severely restrict the gas production rate, or may in fact kill the well (prevent flow). Thus, many liquid-producing gas wells require some means of artificial lift to remove the liquid from the well. The same pumps used in oil-producing wells, such as rod pumps or electrical submersible pumps, can be used to pump liquid from a gas well. An alternative lifting method that is relatively inexpensive and requires no energy input is plunger lift. Plunger lift is a cyclical, intermittent lift method that uses the energy of the reservoir to lift the accumulated liquid from the tubing string.

The principle of plunger lift is illustrated in Figure 12-14 (Wikipedia, 2011). The plunger is a hollow tube that fits snugly in the tubing string. Associated with it is a ball that can seat in a standing valve at the bottom of the tubing string or in the plunger itself. The sketches on the left side of Figure 12-14 show the plunger and ball falling down the tubing string, until the ball seats on the standing valve. The well is typically shut-in during this period. The plunger continues to fall through the gas and then the standing liquid column in the tubing string until it reaches the ball in the standing valve. Once the ball has seated in the standing valve, the pressure below the valve builds up as fluid is produced into the well from the reservoir. Once the pressure in the casing below the plunger exceeds the pressure above the plunger, the plunger is displaced up the tubing, pushing the slug of accumulated liquid ahead of it. When the plunger reaches the top of the well, the ball is mechanically displaced from the plunger, the well is shut-in, and the ball and plunger fall down the tubing string, beginning a new lift cycle. The plunger lift cycle, including the automatic opening and closing of surface valves, is illustrated in Figure 12-15 (Lea, Nickens, and Wells, 2008). Figure 12-16 (Lea et al., 2008) shows a typical plunger lift well completion.

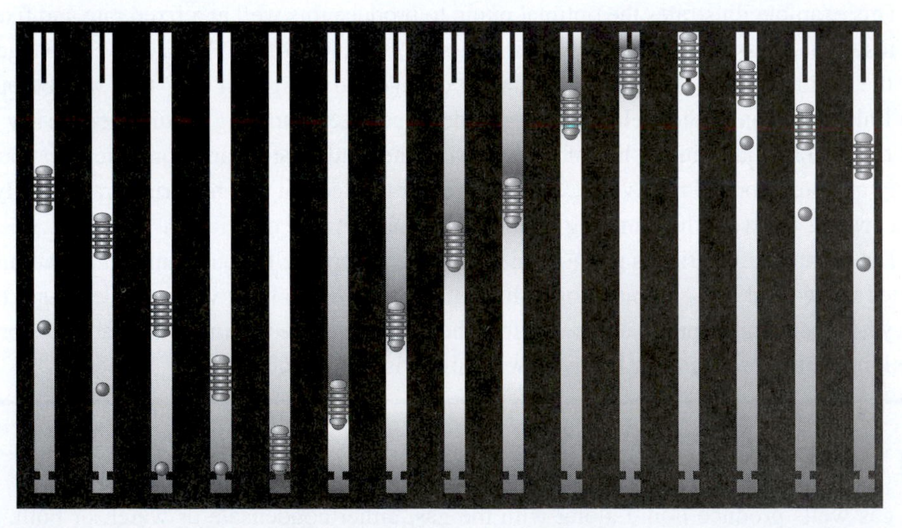

Figure 12-14 Plunger lift operation. (From Wikipedia, 2011.)

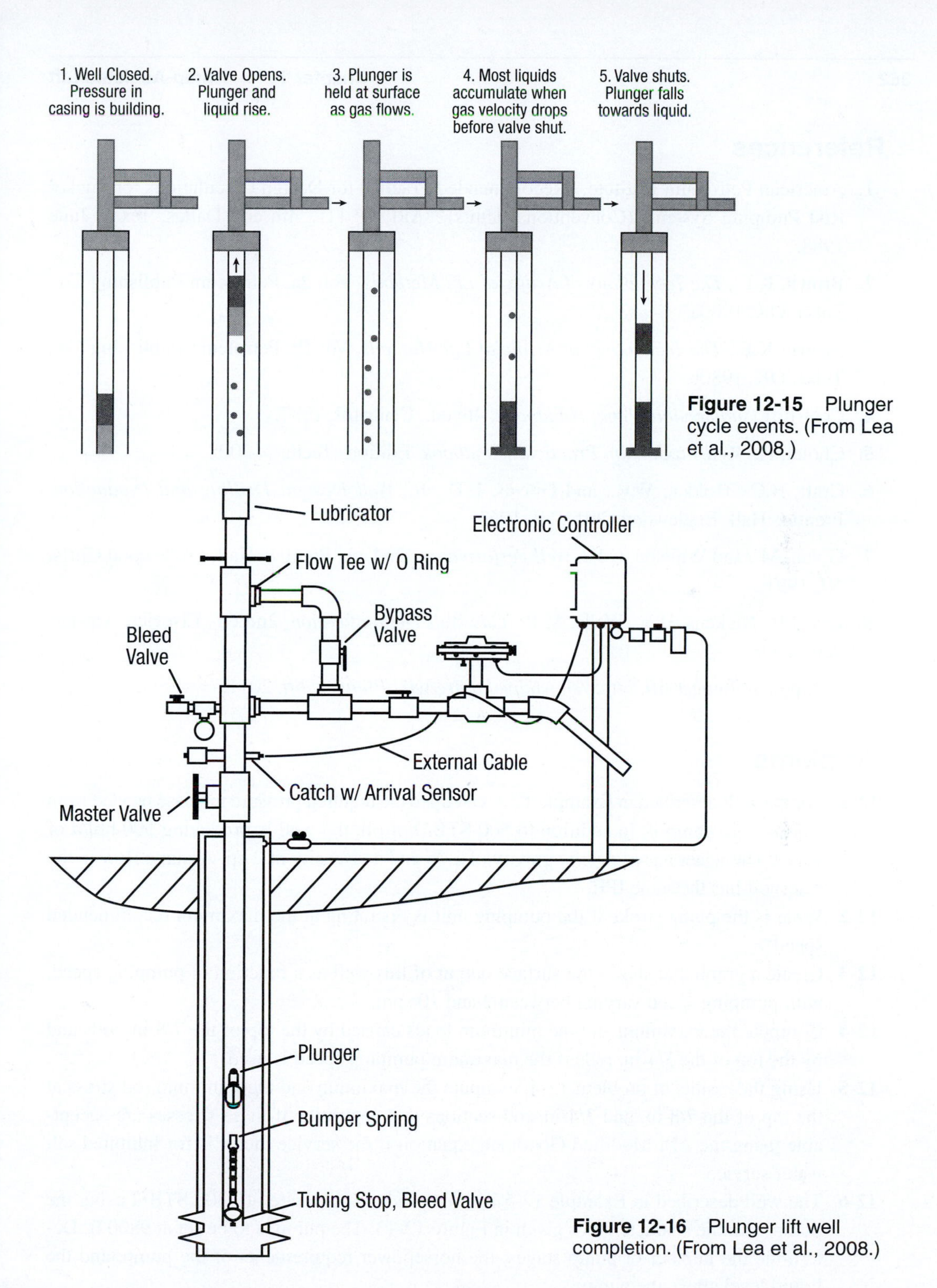
1. Well Closed. Pressure in casing is building.
2. Valve Opens. Plunger and liquid rise.
3. Plunger is held at surface as gas flows.
4. Most liquids accumulate when gas velocity drops before valve shut.
5. Valve shuts. Plunger falls towards liquid.

Figure 12-15 Plunger cycle events. (From Lea et al., 2008.)

Lubricator
Electronic Controller
Flow Tee w/ O Ring
Bypass Valve
Bleed Valve
External Cable
Catch w/ Arrival Sensor
Master Valve
Plunger
Bumper Spring
Tubing Stop, Bleed Valve

Figure 12-16 Plunger lift well completion. (From Lea et al., 2008.)

References

1. American Petroleum Institute, "Recommended Practice for Design Calculations for Sucker Rod Pumping Systems (Conventional Units)," API RP 11L, 4th ed., Dallas, Texas, June 1988.

2. Brown, K.E., *The Technology of Artificial Lift Methods,* Vol. 2a, Petroleum Publishing Co., Tulsa, OK, 1980a.

3. Brown, K.E., *The Technology of Artificial Lift Methods,* Vol 2b, Petroleum Publishing Co., Tulsa, OK, 1980b.

4. Centrilift, *Submersible Pump Handbook,* 4th ed., Centrilift, 1987.

5. Cholet, H., *Well Production Practical Handbook,* Editions Technip, 2008.

6. Craft, B.C., Holden, W.R., and Graves, E.D., Jr., *Well Design: Drilling and Production,* Prentice Hall, Englewood Cliffs, NJ, 1962.

7. Golan, M., and Whitson, C.H., *Well Performance,* 2nd ed., Prentice Hall, Englewood Cliffs, NJ, 1991.

8. Lea, J.F., Nickens, H.V., Wells, M.R., *Gas Well Deliquification,* 2nd ed., Elsevier, Amsterdam, 2008.

9. Wikipedia, Plunger lift, *http://en.wikipedia.org/wiki/Plunger_lift,* 2011.

Problems

12-1 For the well described in Example 12-1, calculate the required pressure increase needed from a downhole pump if, in addition to 500 STB/d of oil, the well is producing 500 bbl/d of water. The water has a specific gravity of 1.05 and a viscosity of 2 cp. Assume flow in the reservoir has the same IPR.

12-2 What is the pump stroke if the pumping unit is operating at the maximum recommended speed?

12-3 Create a graph that shows the surface output of this well as a function of pumping speed, with pumping speed varying between 2 and 10 spm.

12-4 Compute the maximum and the minimum loads carried by the top of the 7/8-in. rods and by the top of the 3/4-in. rods if the maximum pumping speed is used.

12-5 Using the results of problem 12-4, compute the maximum and the minimum rod stress at the top of the 7/8-in. and 3/4-in. rod sections and determine if these stresses are acceptable using the API Modified Goodman equation if the service factor is for inhibited salt water service.

12-6 The well described in Example 12-5 is to be produced at a rate of 1300 STB/d using the pump with the characteristics given in Figure 12-13. The pump is to be set at 9800 ft. Determine the number of pump stages, the horsepower requirements of the pump, and the liquid level above the pump.

Table 12-4 Data for Rod Pumping Problems 12-1–12-5

Pump Depth = 7,000 ft.
Working Liquid Level = 6,800 ft. from the surface
Surface Pumping Unit Stroke = 144 in.
Maximum Recommended Pumping Speed = 10 spm
Rod Combination: 7/8 in. on top with 3/4 in. below, both Class D *See Table 12-1 for percentages of each*
Pump Plunger Diameter = 1.50 in.
Pump Volumetric Efficiency = 80%
Tubing: 2 7/8-in. O.D. × 2.441-in. I.D. tubing is anchored. Required tubing pressure = 50 psig.
Liquid: 1.00 specific gravity, formation volume factor for the produced mixture = 1.05 bbl/STB

Chapter 13

Well Performance Evaluation

13.1 Introduction

The productive formation can be evaluated during drilling, before completing the well, after the well has been completed, and after the well has been worked over or stimulated. This chapter provides a brief overview of ways to quantify parameters essential to well performance evaluation and well completion or workover design. While much of this book is focused on using various parameters to model well performance or design well treatments, the objective of this chapter is to learn how those parameters can be determined in the first place.

The original static formation evaluation relies on well logs selected for their sensitivity to physical or electrical properties related to formation porosity, fluid saturation, fluid mobility, temperature, pressure, and rock mechanical properties. Core and formation fluid samples can also be collected from the formation during or after drilling. This chapter begins with a brief description of formation evaluation during drilling and from open-hole logs.

Once the well has been cased and completed, cased hole logs provide mechanisms to evaluate cement, saturation changes, or fluid movement behind the casing and to locate fluid entries along the well.

Transient well analysis includes traditional pressure transient analysis and production data analysis. While stabilized pressure and flow measurements can provide the well inflow performance relationship, quantification of permeability and skin generally requires analysis of transient data. The key to transient analysis is to understand how to recognize key streamline flow geometries, known as flow regimes, from their characteristic trends in the transient data. The flow regime overview provided in this chapter emphasizes the use of the log–log diagnostic plot for flow regime identification and estimation of parameters that are important to production engineering applications.

The final section of this chapter addresses diagnostic fracture injection tests as a means to determine permeability in unconventional oil and gas reservoirs with permeability that is so low that conventional methods fail.

13.2 Open-Hole Formation Evaluation

Open-hole wireline (electric) logs are acquired after drilling while the drilling mud is still in the borehole. When productive intervals are located above a casing point, the borehole segment must be logged before the casing is run.

Example 13-1 Open-Hole Log Determination of Porosity and Fluid Saturations

Figure 13-1 shows a sample of typical logs and their responses to different lithologies (limestone, shale, and sandstone) with different fluid saturations (gas, oil, and water). The first lithology track is drawn by a geologist. Study the log tracks carefully to answer the following questions: (a) Which log or logs are sensitive to lithology? In what way(s)? (b) Which log or logs are sensitive to fluid saturation? In what way(s)?

Solution

a. A careful look at the six log tracks reveals that the gamma ray log position clearly indicates the shaley intervals. Furthermore, the photoelectric (PE) log is highest in limestone, second highest in shale, still lower in dolomite, and lowest in sandstone. The other 4 logs show additional variations related to fluid saturation.

b. The resistivity log deflects to the right (higher resistivity) opposite fresh water, gas, and oil, and registers lowest values opposite shale intervals. At the base of the log, the rightward shift in resistivity shows a contrast between limestone and dolomite. The spontaneous potential (SP) log deflects to the right in shale intervals; shows similar behavior to the resistivity opposite gas, oil, and water; and fails to distinguish between limestone and dolomite. The compensated neutron log (CNL) and the density log show similar responses in response to changes in fluid saturation and opposite shale intervals.

Actual open-hole logs would include a depth track and typically a caliper log sensing the hole diameter. Log interpretation equations exist to enable estimation of lithology, mineralogy, porosity, and fluid saturations from a combined suite of several logs.

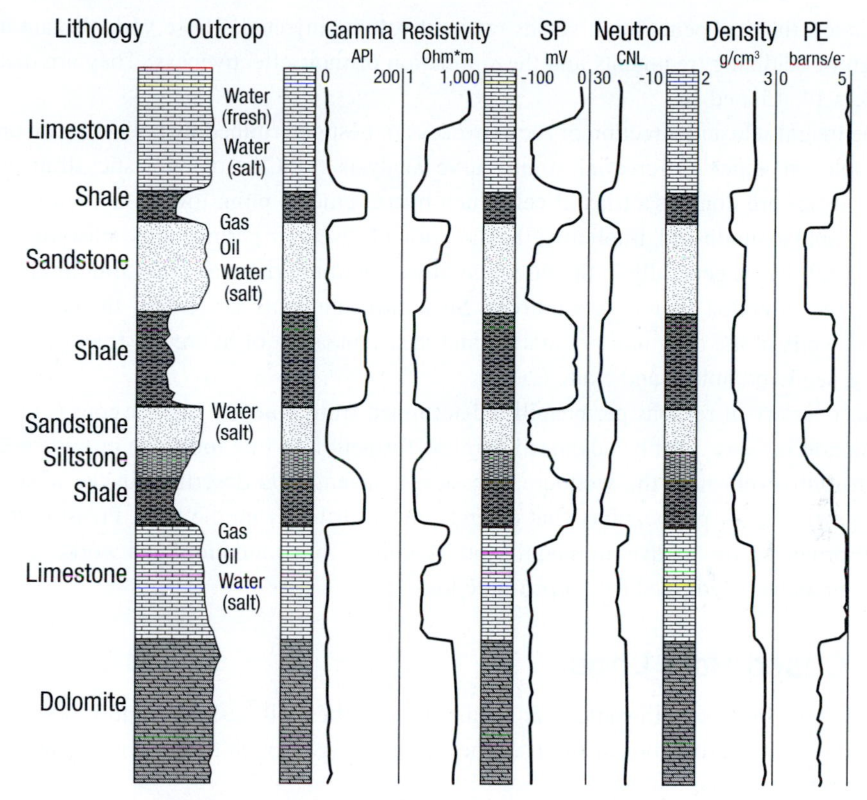

Figure 13-1 Open-hole log for Example 13-1. (From Evenick, 2008.)

Wireline formation test tools described in Section 13.3.1 enable determination of formation pressure and temperature and acquisition of a reservoir fluid sample as well and horizontal and vertical permeability at a particular depth. Well site equipment may enable on-site determination of fluid properties, and a pressurized fluid sample can be collected for more detailed laboratory analysis in a PVT laboratory.

Wireline tools can also be used to perform a microfrac test that can provide the fracture initiation pressure described in Chapter 17, Section 17.4. As well, sonic and dipole sonic logs can provide estimates for rock mechanics properties versus depth, including Young's modulus and Poisson's ratio. Lithology changes identified from the gamma ray log, especially from sandstone or carbonate to shale, help to estimate hydraulic fracture height.

There are many uses for core samples that are important to production engineering. For example, clay minerals can be analyzed that are needed for matrix acid design. A more advanced set of measurements involves the relationship between porosity and permeability and, in the case of carbonates, the acid penetration versus rate and volume injected. These variables are used in the design of acidizing treatments and the simulation of their effectiveness. They are discussed in Chapters 14, 15, and 16.

The magnitude and direction of rock stresses are best determined by collecting an oriented core sample on which differential strain curve analysis (DSCA) or anelastic strain recover (ASR) analyses are done in a triaxial cell. Such measurements point toward hydraulic fracture orientation and azimuth and, implicitly, the direction of reservoir permeability anisotropy, maximum permeability is generally in the direction of maximum horizontal stress, minimum permeability in the direction of minimum stress. Such measurements are useful in the drilling of horizontal wells in the optimum orientation and the delineation of hydraulic fractures. For further study, see Economides and Nolte (2004).

For a variety of reasons permeability determined from porosity logs or core is unreliable for production forecasts. While advanced wireline formation testing tools can provide estimates for permeability very near the wellbore, pressure transient tests described in the next section provide a measure for permeability that extends much further from the well. Pressure transient tests also provide a mechanism to quantify post completion, stimulation, and workover properties that cannot be determined from open-hole logs.

13.3 Cased Hole Logs

Cased hole logs include both static logs acquired while the well is shut in and production logs designed to measure properties of fluids as they are flowing through the wellbore. The latter are discussed in the next section.

Compared to open-hole logs, the formation signal in cased hole logs is attenuated by greater distance between the logging tool and the formation (standoff) that is occupied by wellbore fluid, casing, and cement. Anything between the logging tool and the formation impacts the signal detected by sensors in the tool and quantitative formation analysis must compensate for these effects. An uneven cement bond or pockets of gas or water in the cement can greatly confuse the formation analysis.

The first two subsections briefly review cased hole logs for cement and formation evaluation. Then the third subsection reviews key well evaluation strategies from production logs.

13.3.1 Cement Evaluation

Drilling and workover operations are often highly sensitive to the quality of the seal provided by the cement between the well casing and the formation. Figure 13-2 shows a sample of two different cement logging approaches. The cement evaluation tool (CET), shown by a combination of three log tracks on the left, combines gamma ray, caliper, fluid transit time, and relative bearing in the first track with minimum and maximum compressive strength and an angular average of acoustic impedance in the second track, and the circumferential bond image in the third track to the left of a track showing gas flags and reflection flags. The easiest track to follow in the CET is the circumferential bond image. White spaces occur in this image when the cement bond

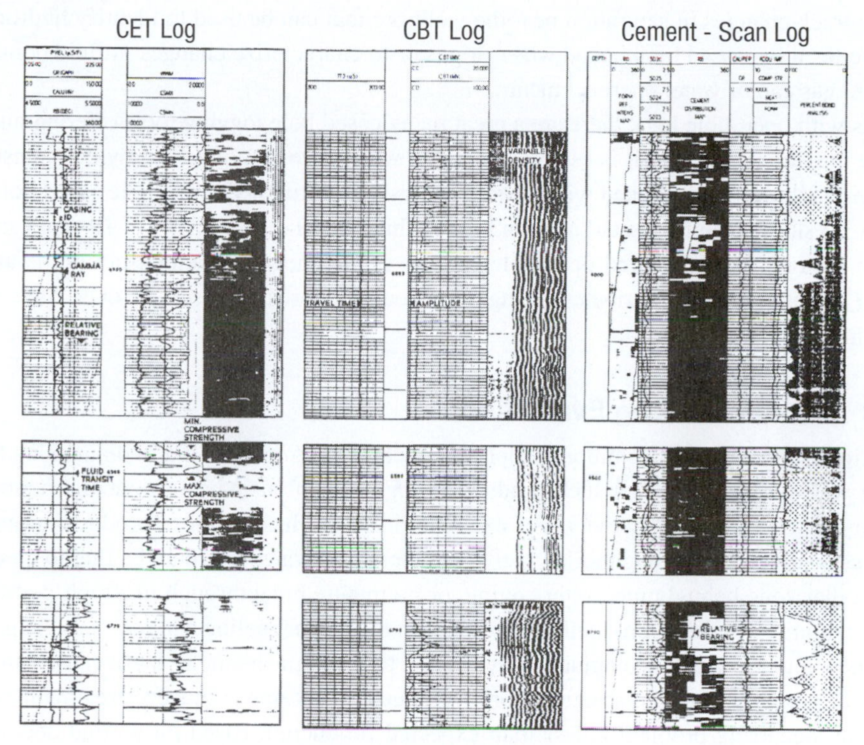

Figure 13-2 Example cement bond log. (From Schlumberger, 1990.)

is lacking and coincide with lower compressive strengths and higher incidence of gas and reflection flags.

The cement bond tool (CBT) shown by the three middle tracks, based on measurement of the time and amplitude of acoustic signals that travel through casing, cement, and formation rock in various annular paths along the borehole. The CBT includes an average sonic transit time log in the first track, average amplitude in the middle track, and variable density imaged in the right-most track. Again, the right-most track gives a visual impression where good cement bond occurs, but the variable density image is not as straightforward to interpret as the circumferential bond image in the newer CET log presentation.

The final cement-scan log provides a quantitative analysis of the log responses. In particular, the right-most track shows percentages of solid cement (black), liquid mud (white), gas, (dotted area), and gas cut cement (gray).

13.3.2 Cased Hole Formation Evaluation

Because cased hole logs can be run any time and many times after the well has been completed and on production or injection, they can provide a rich variety of information about many parameters important to production and reservoir engineering. For example, a series of cased hole logs can track changes in saturation near the wellbore that can be used to identify hydrocarbons that have been bypassed by injected water or gas or to characterize changes in fluid contact levels due to gas cap or water encroachment.

As with open-hole logs, there are a great many cased hole logging tools offering numerous possible measurements, and it is essential to know what various tools can provide to ensure that the logging job design is aligned with desired analysis objectives. The example cased hole log in Figure 13-3 shows a compensated neutron log run three times over a period of 10 years and compared to the originally acquired open-hole log. As a reference, the gamma ray log is also provided. The interpretation attributes changes in neutron porosity observed over time to sand production.

13.3.3 Production Log Evaluation

Production engineers most commonly apply production logging as an aid in diagnosing the case of poor well performance. As such production logs can often indicate remedial action to be taken to improve well productivity. For example, if a well has begun to produce an excessive water cut compared with neighboring wells, the increased water rate may be due to channeling from another zone behind pipe, water coning, or premature breakthrough in a high-permeability zone. By running a suite of production logs that can locate channeling and measure the profile of water entry into the well, the engineer may be able to distinguish among these causes and, more important, properly plan a corrective workover, such as a cement squeeze treatment. Another commonly occurring problem is less than expected production based on normal decline. This may be the result of damage in one of the layers in a co-mingled production. It becomes essential to identify the problematic zone and, perhaps, perform selective stimulation.

Production logging is not a well diagnosis panacea and should not be applied in a vacuum; rather, it should be used as a supplement to the information gained from the well flow rate and pressure history and other well tests. In this section we illustrate how production logging results can build upon other knowledge of well behavior to diagnose problem performance and assist in planning remediation.

The interpretation of production logs is not discussed here; rather, the results obtained from production logs (a well's flow profile, for example) serve as the starting point for illustrations of the application of production logging to well diagnosis. The engineer using production logging should always be mindful of the uncertainty that sometimes exists in these interpreted log results. For a thorough review of production logging practices and interpretation, the reader is referred to Hill (1990).

The section is organized according to the initial indicator of poor well performance or the objective of applying production logging. First, the use of production logs to diagnose low productivity is presented, followed by the evaluation of excessive gas and excessive water production. The application of production logging to well treatment planning and evaluation is then discussed. The chapter concludes with a discussion of injection well problem diagnosis.

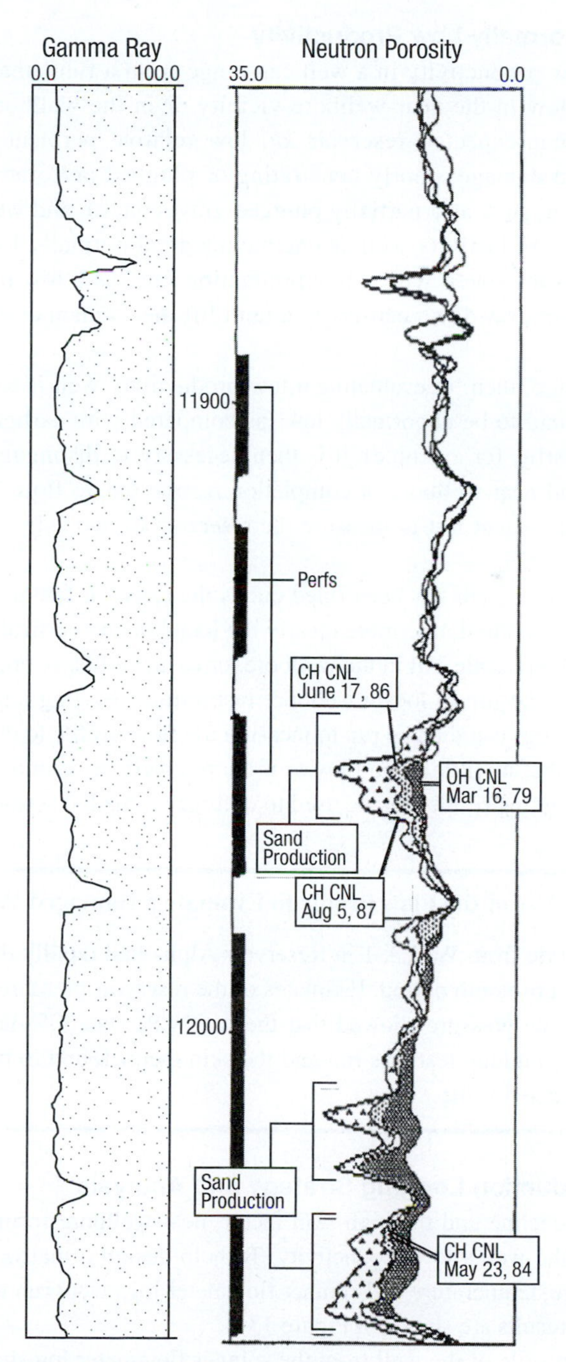

Figure 13-3 Example cased hole log. (From Dupree, 1989.)

13.3.3.1 Abnormally Low Productivity

The cause of low productivity in a well can range from a fundamental reservoir problem to restrictions to flow in the near-wellbore vicinity or in the wellbore itself. Potential causes include lower-than-expected reservoir *kh*, low relative permeability to the hydrocarbon phase, formation damage, poorly penetrating or plugged perforations (or other restrictions in the completion, such as a partially plugged gravel pack), and wellbore restrictions. Here, we define a low-productivity well as one having an abnormally low productivity index (J); this is distinct from a well with a low production rate, as a low rate of production may be due to insufficient drawdown caused by a fault lift mechanism or excessive pressure drop in the tubulars.

The first step, then, in evaluating a low-productivity well is to measure the productivity index. If it is found to be abnormally low (as compared with earlier in the well's life or with similar wells nearby, for example), it is then necessary to distinguish between low-formation flow capacity and near-wellbore or completion restrictions to flow. This is best accomplished with a pressure transient test to measure the reservoir *kh* and skin factor, as described later in this chapter.

If the reservoir itself has been ruled out as the cause of low productivity, production logging can now be used to define more clearly the location and vertical extent of productivity impairment. If wellbore scale, fill in the wellbore, or casing collapse are considered possibilities, a caliper log should be run to locate wellbore restrictions. Barring any obstructions in the wellbore, production logs can then be run to measure the flow profile to determine if parts of the formation are contributing little or no flow or if the productivity is uniformly low. In this instance, the production logging results can be used to optimize remedial action.

Example 13-2 Use of the Flow Profile to Evaluate a Damaged Well

The production rate from Well A-1 in Reservoir Alpha had rapidly declined to less than half of its initial rate in a 6-month period. Estimates of the reservoir pressure and a measurement of the flowing bottomhole pressure showed that the well's PI was 50% below those of surrounding wells. A pressure buildup test was run and the skin factor found to be 20, while the *kh* product was near the expected value.

13.3.3.2 Production Logging Strategy and Analysis

From the rapid decline and the high skin factor, near-wellbore formation damage is the likely culprit causing the well's low productivity. To help design a matrix acidizing treatment to remove the damage, temperature and spinner flowmeter logs were run to measure the flow profile. The interpreted results are shown in Figure 13-4.

The flow profile of the well from the spinner flowmeter log shows that zone A is producing less than 10% of the total flow, zone B is producing almost 70% of the total rate, and zone C

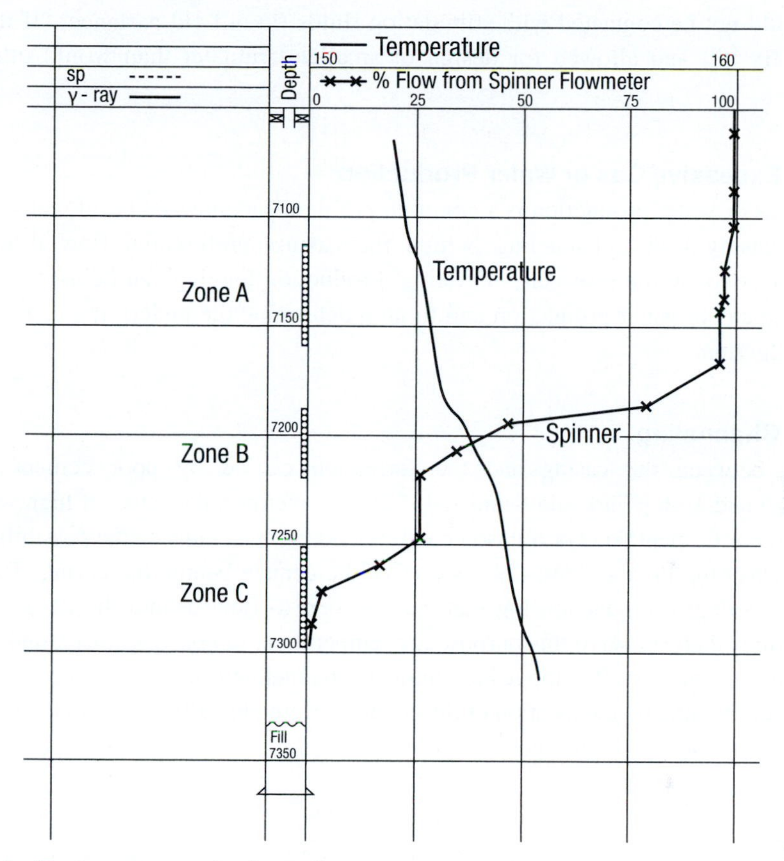

Figure 13-4 Temperature and spinner flowmeter-derived flow profile for Example 13-2.

is contributing about 25% of the production. The temperature log qualitatively confirms the interpretation of the spinner flowmeter log.

Apparently, zone A has been significantly damaged during the course of production, perhaps by fines migration near the wellbore. The production log shows that the needed treatment for this well is the selective stimulation of zone A, with perhaps a lesser amount of stimulation of zone C. Zone B does not need treatment; in fact, the flow profile shows that good diversion is needed during the stimulation treatment to minimize injection into zone B.

A matrix acidizing treatment to stimulate this well should begin with a diverter stage (ball sealers or particulate diverting agents) to prevent injection of acid into zone B. Subsequent treatment volumes and rates should be selected based on treating only zones A and C. In this example, the information from the production logs showed that the highly productive zone B should not be contacted with stimulation fluids (in oilfield parlance, "If it isn't broken, don't fix it") and allowed for design of smaller treatment than would otherwise be planned.

13.3.3.3 Excessive Gas or Water Production

Excessive gas or water production is a common problem encountered in oil wells and can be caused by casing leaks, channeling behind the casing, preferential flow through high-permeability zones in the reservoir, or coning. Production logging can be used to locate the source of the gas or water production and to help determine the underlying cause of the unwanted production.

13.3.3.4 Channeling

Channeling between the casing and the formation caused by poor cement conditions (Figures 13-5 and 13-6 [Clark and Schultz, 1956]) is sometimes the cause of high water or gas production rates. Cement bond or ultrasonic pulse-echo logs can indicate the possibility of channeling by measuring the mechanical properties of the cement behind the casing. To positively identify channeling, a production log that can respond to flow behind the casing is needed. Among the logs that can serve this purpose are temperature, radioactive tracer, and noise logs. The remedial treatment usually applied to eliminate channeling is a cement squeeze; to design a cement squeeze treatment, the location of the channel, preferably all the way to the source of the unwanted production, should be known.

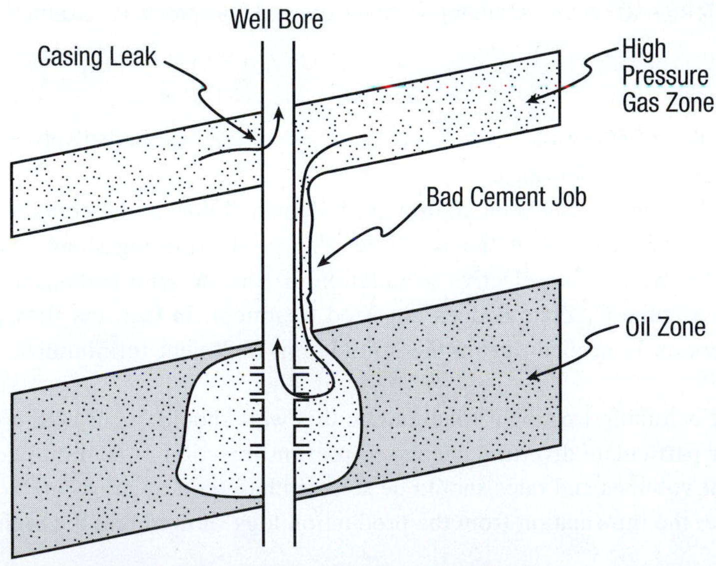

Figure 13-5 Gas channeling. (From Clark and Schultz, 1956.)

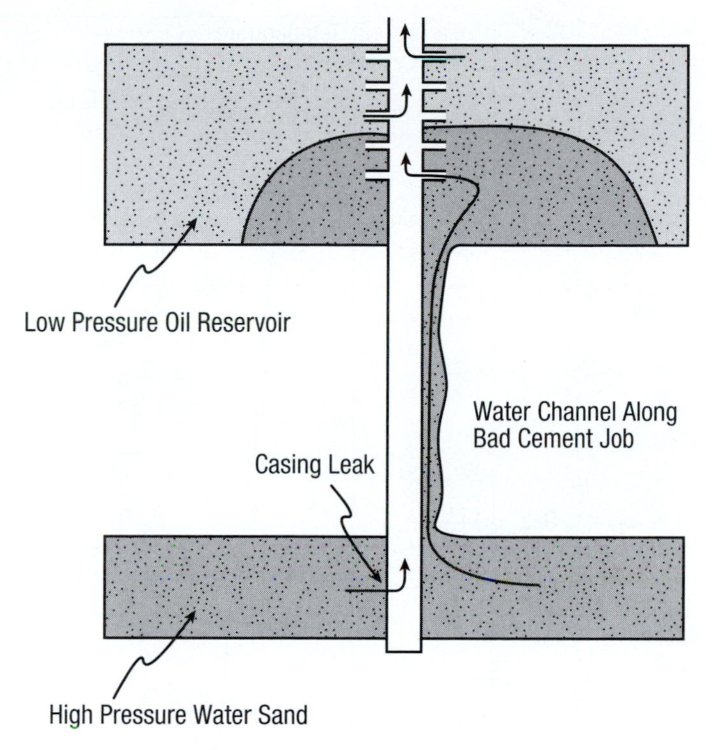

Figure 13-6 Water channeling. (From Clark and Schultz, 1956.)

Example 13-3 Locating a Gas Channel with Temperature and Noise Logs

The temperature and noise logs shown in Figure 13-7 were obtained in an oil well producing at an excessively high GOR. Both logs clearly indicate that gas is being produced from an upper gas sand and channeling down to the upper perforations in the oil zone. Both logs respond to gas expanding through restrictions; the temperature log exhibits cool anomalies caused by Joule-Thomson cooling at gas expansion locations, while the noise log measure increased noise amplitude at the same locations. Thus both logs respond to gas flow at the gas source, at a restriction in the channel behind the casing, and at the location of gas entry into the wellbore.

To eliminate the excessive gas production, a cement squeeze can be performed to block the flow in the channel. This can best be accomplished by perforating near the gas source and circulating cement through the channel (Nelson, 1990).

Note that measuring the flow profile in this well would not be particularly helpful in locating the cause of high gas production or in planning corrective measures. A flow profile would show gas production into the wellbore from the upper part of the oil zone. This could be due to channeling (as was actually the case) or high gas saturation in the upper part of the oil zone, as can occur with secondary gas cap development. Only by applying logs that clearly identify channeling can the appropriate workover be planned.

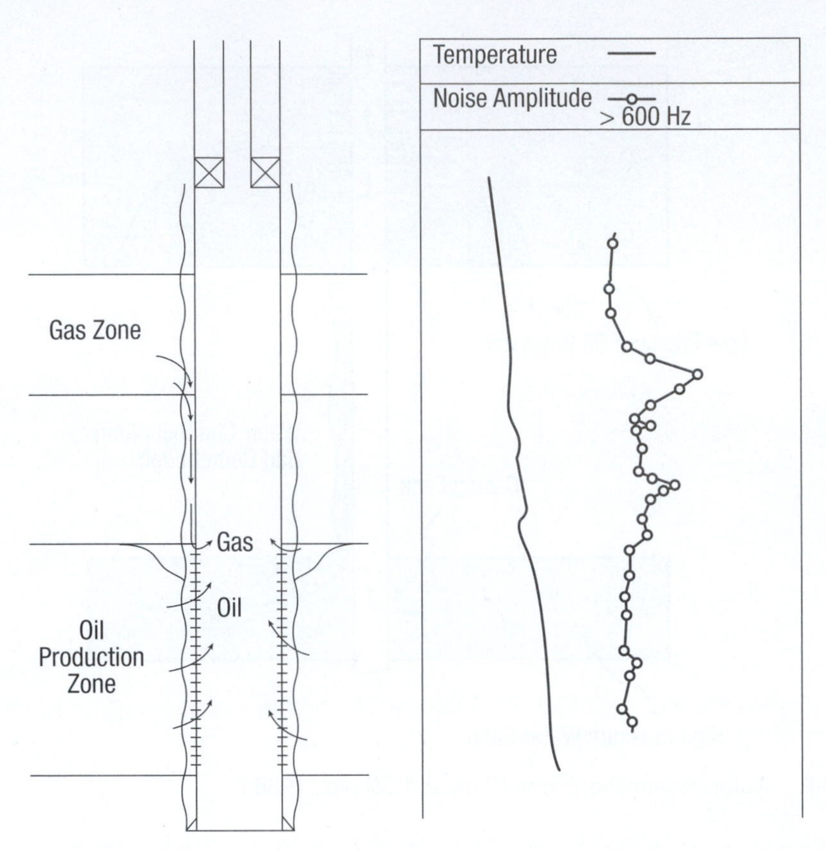

Figure 13-7 Temperature and noise logs identifying gas channeling for Example 13-3.

13.3.3.5 Preferential Gas or Water Flow Through High-Permeability Layers

Preferential flow of water or gas through high-permeability layers (often referred to as thief zones), as illustrated in Figures 13-8 and 13-9 (Clark and Schultz, 1956), is a common cause of high gas or water production in oil wells. Unwanted gas or water entries of this nature can sometimes be located with production logs.

Excessive water production may result from injected water in a waterflood or from water encroaching from an aquifer. An accurate flow profile of the production well can identify the location of the high-permeability zone or zones contributing to the high production rate. However, the location of the water entry is not generally sufficient information to identify the case of water production as being flow through a thief zone. Particularly if the water entry location is at the bottom of the completed interval, the water source may be channeling or coning from lower zones. Additional logs or tests are needed to distinguish among these possibilities (see Section 13.3.3). Because the measurement and interpretation of the logs in the multiphase flow in production wells are generally less accurate (and more expensive) than those in a single-phase flow, in waterflood operations, water

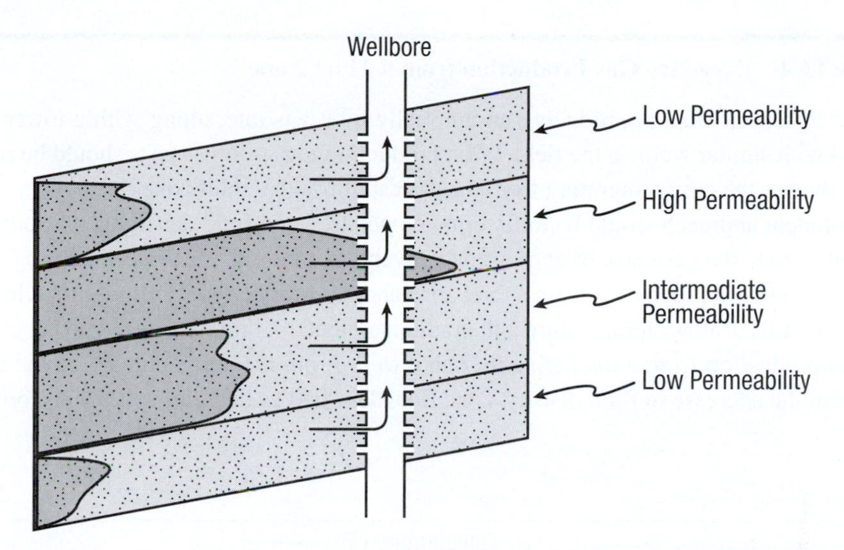

Figure 13-8 Early water breakthrough in high-permeability layers. (From Clark and Schultz, 1956.)

distribution in a reservoir is often monitored by measuring injection well profiles and assuming continuity of the reservoir layers between injectors and producers.

Excessive gas production can similarly result from flow of injected gas or from a gas cap. Again, a flow profile measured in a production well will identify the offending entry locations, or the high-permeability zones causing high gas production can be inferred from profiles measured in gas injection wells when gas is being injected into the reservoir. However, as with bottomwater production, if the gas production is from the upper part of the oil zone, it may be due to coning or channeling and further information beyond the flow profile is needed for complete diagnosis.

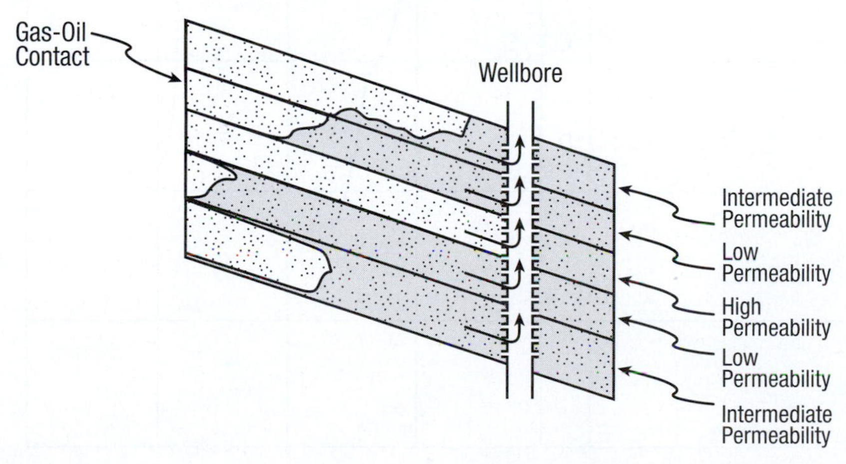

Figure 13-9 Early gas breakthrough in high-permeability layers. (From Clark and Schultz, 1956.)

Example 13-4 Excessive Gas Production from a Thief Zone

A well in Reservoir Beta is producing an unusually high gas rate, along with a lower oil rate, compared with similar wells in the field. What production logs or other tests should be run to determine whether the gas is migrating from the gas cap through a thief zone?

A prudent approach would be to first run temperature and fluid density logs. Both of these logs should locate the gas entry or entries qualitatively; in addition, the temperature log will help differentiate between production from a thief zone and gas production resulting from channeling.

Figure 13-10 shows temperature and gradiomanometer (fluid density) logs that clearly indicate gas production from a thief zone in such a well. From the cool anomaly on the temperature log and the decrease in fluid density, zone B is identified as the thief zone. Since oil is being

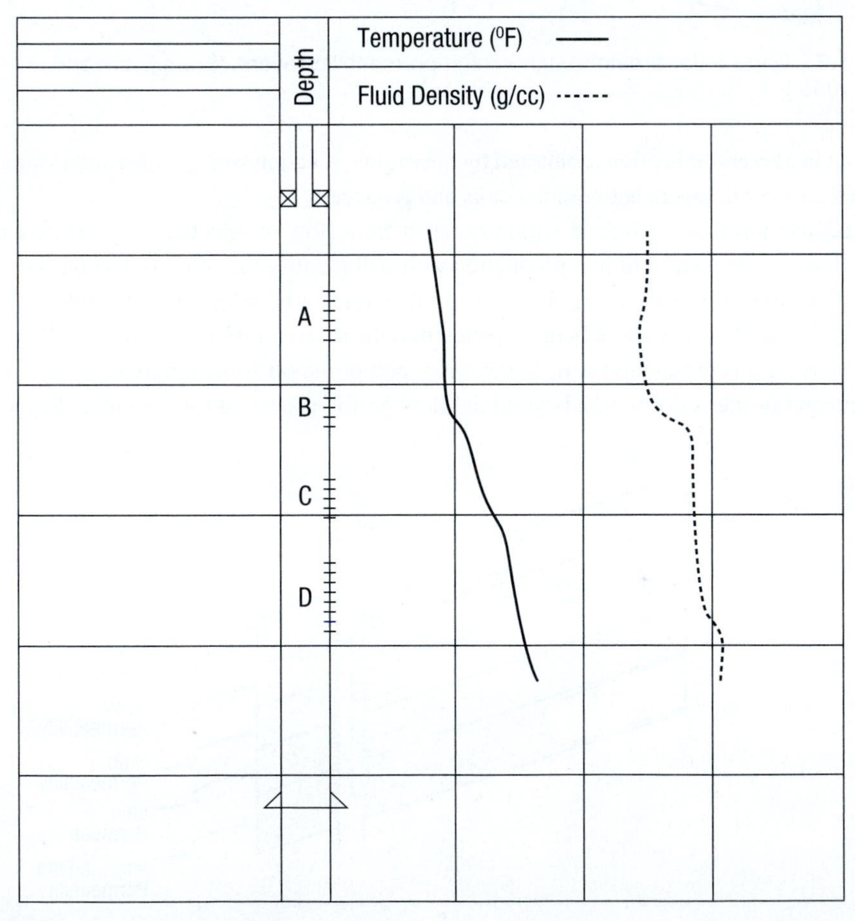

Figure 13-10 Temperature and fluid density logs locating gas entry for Example 13-4.

produced from zone A above this zone, as shown by the slight increase in fluid density across zone A, the gas production from zone B is not channeling or coning up to this level. This temperature log also gives no indication that channeling is occurring.

13.3.3.6 Gas or Water Coning

Gas or water coning, illustrated in Figures 13-11 and 13-12 (Clark and Schultz, 1956), is another possible source of excessive gas or water production. Gas coning results when a well is completed near a gas/oil contact and sufficient vertical permeability exists for gas to migrate downward to the wellbore as the pressure is drawn down around the well. Similarly, water can cone up from an underlying aquifer if vertical permeability is high enough. Discussions of the reservoir engineering aspects of coning are given by Frick and Taylor (1962) and Timmerman (1982).

This water could result from channeling from below the perforated interval, flow through a high-permeability zone located in the lower part of the interval, or coning. A log responding to flow outside the casing, such as a noise log, might be able to eliminate channeling as the water source (note that the temperature log will respond similarly to coning and channeling and thus will not distinguish between them). This technique requires that it is possible to log a sufficient distance below the lowest perforations. Distinguishing between coning and flow through a high-permeability layer will be difficult with production logs alone. The most conclusive test for coning would be to produce the well at several different flow rates or drawdowns, as coning is inherently a rate-sensitive phenomenon (Muskat, 1949).

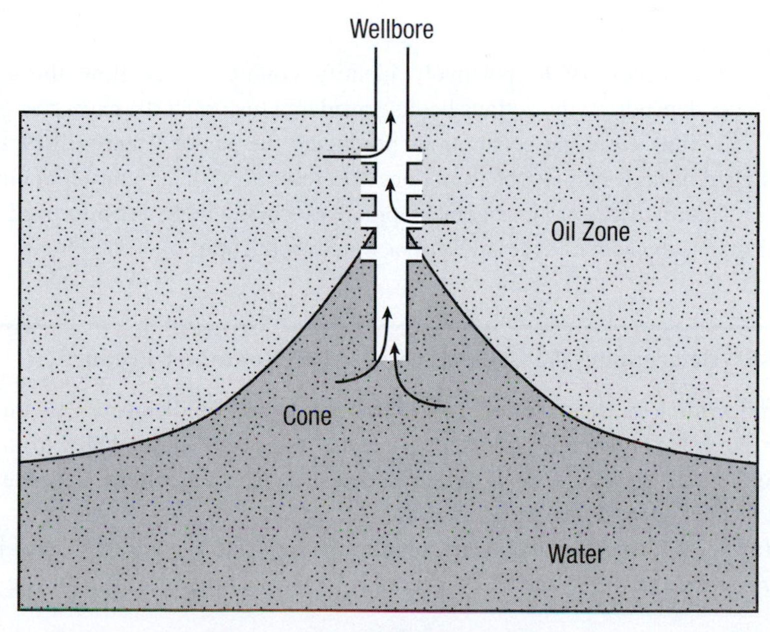

Figure 13-11 Water coning. (From Clark and Schultz, 1956.)

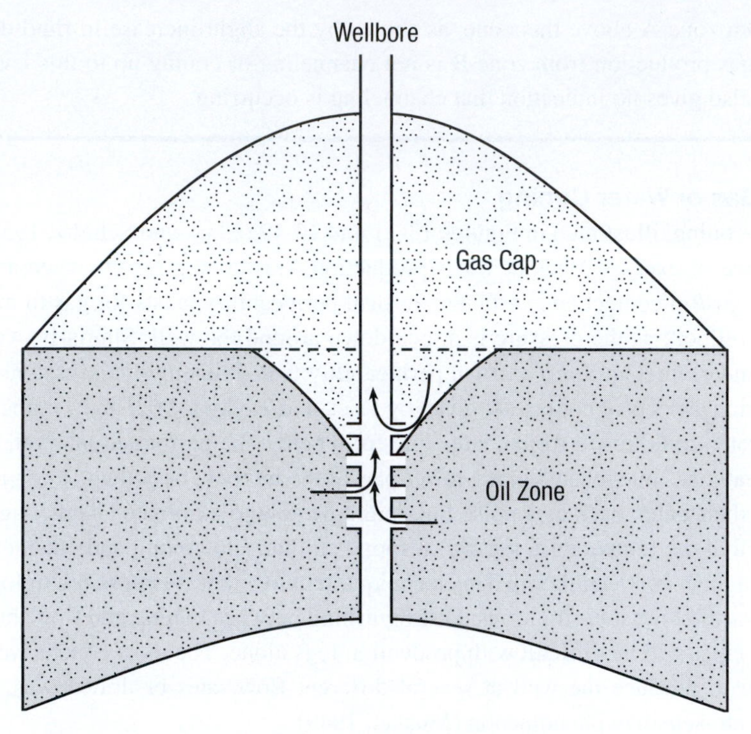

Figure 13-12 Gas coning. (From Clark and Schultz, 1956.)

Whether it is necessary to positively identify coning versus flow through a high-permeability layer depends on the actions being considered to correct the excessive gas or water production. If, in either case, the perforations producing the gas or water will be squeeze cemented or plugged back, it may not be important to determine with certainty if coning is occurring. However, future reservoir management practices could likely be improved by a clear knowledge of the mechanism of excessive gas or water production.

Example 13-5 Determining the Cause of Excessive Bottomwater Production

A suite of production logs (temperature, basket flowmeter, and fluid density) run on an oil well producing an excessive amount of water shows that the water is being produced from the bottom 20 ft of perforations. The field is being waterflooded in a five-spot pattern, with the nearest injection wells located approximately 800 ft away.

What further information (well tests, production logs, etc.) would be useful in planning corrective action for this well, and more generally, improved management of the waterflood operation?

The primary information needed is whether the excess water production is a problem with this particular well (channeling or coning) or is due to a surplus of injection water in the lower part of the reservoir. First, the production logs from the problem well should be reexamined for any evidence of channeling or coning. The temperature log, in particular, may indicate if flow is occurring from below the perforations. Next, the behavior of the surrounding injection wells should be investigated. Injection profiles should be run if they have not been obtained recently. If the pattern of water injection in one or more injection wells shows high water rates into the lowest zone, preferential flow through a high-permeability layer is the likely cause of the high water production. Finally, the total rate from the production well could be lowered—if a rate is found where water production ceased, coning has been identified as the cause of excessive water production.

13.3.3.7 Use of Production Logging for Well Treatment Evaluation

As has already been seen, production logging can provide useful information for the planning of well workovers, primarily by providing information about the distribution of flow of the various phases into the well. Production logs can in a similar way aid in evaluating the success (or failure) of well treatments. Among the workovers that can benefit from post-treatment production logging are remedial cementing, reperforating, acidizing, fracturing, and water shutoff or profile modification treatments.

The most straightforward application of production logging to well treatment evaluation is the before- and after-treatment measurement of the flow profile. For example, in a low-productivity well that is being reperforated, comparison of the flow profile before and after perforating should indicate the productivity of the reperforated intervals directly. Such evaluation would normally be applied only when the outcome of the workover is less than desired.

Thus, in evaluating well treatments, the use of production logs is similar to their application in treatment planning: They indicate what regions of the well were affected by the workover and to what extent. In addition to the measurement of the flow profile after the treatment, some production logs can be used for a more direct evaluation of the treatment itself. The most common examples of this type of application are the uses of temperature logs or radioactive tracer logs to measure fracture height near the wellbore.

Example 13-6 Measurement of Fracture Height

Well D-2 was one of the first wells in Field D to be hydraulically fractured. To aid in designing future fracture treatments in this field, temperature logs were run before and after fracturing, and the last 10,000 lb of proppant were radioactively tagged in order to measure the fracture height. From the temperature logs and post-fracturing γ-ray survey, determine the fracture height.

13.3.3.8 Temperature Log Interpretation of Fracture Height

Because the injected fracturing fluid is generally significantly cooler than the formation being fractured at typical injection rates, the fracture fluid temperature where it exits the wellbore will be near the surface temperature. As the treatment proceeds, the unfractured formation around the wellbore is cooled by radial heat conduction, as in any injection well, while cool fluid is placed in the fracture. When the well is shut in, the wellbore opposite the unfractured formation begins to return to the geothermal temperature by unsteady radial heat conduction, while in the fractured region the wellbore temperature is affected by linear heat conduction from the formation to the fracture. Because the radial heat transfer in the unfractured region is more rapid than the linear heat conduction in the fracture, the fractured region will warm more slowly, yielding a cool anomaly on a temperature log. Thus the fracture height can be identified by the location of a cool anomaly on a temperature log run after a brief (a few hours) shut-in period after fracturing.

However, numerous factors sometimes complicate this straightforward interpretation. In particular, warm anomalies often appear in the region that may have been fractured; these may be due to variations in the formation thermal diffusivity, frictional heating of the fracturing fluid as it is injected at a high rate through perforations or in the fracture, from the fluid movement within the fracture after shut-in (Dobkins, 1981), or deviation of the fracture plan from intersection with the wellbore. Warm anomalies caused by formation thermal property variations can be distinguished from those caused by fluid movement effects by running a prefracture shut-in temperature log after circulating cool fluid in the wellbore. The warm anomalies on the postfracture temperature log that correspond to warm anomalies on the prefracture log result from thermal property variations; these regions should not be included in the interpreted fractured zone. When a warm anomaly appears on a post-fracture temperature log and no corresponding anomaly exists on the prefracture log, the warm anomaly is apparently caused by fluid movement in the fracture after shut-in derivation of the fracture from intersection with the wellbore. The warm-anomaly region would then be included as part of the interpreted fractured zone.

The temperature logs run in Well D-2 after circulation of cool fluid before fracturing and after a short shut-in period after fracturing are shown in Figure 13-13. The vertical extent of the fracture is indicated by the region where the two logs diverge, showing the fracture being located in this case from about 10,100 to about 10,300 ft. Temperature anomalies farther up the well on the post-fracturing log are apparently due to variations in thermal diffusivity of the formation.

13.3.3.9 Measuring Fracture Height with Radioactively Tagged Proppant

A measure of propped fracture height can be obtained by radioactively tagging the final portion of the proppant, then running a postfracture γ-ray log to locate the tagged proppant. The fracture height interpreted in this manner can be misleading if the tagged proppant is not completely displaced from the wellbore or if it is displaced too far from the wellbore (the radiation from the tagged proppant can only be detected within about a foot from the wellbore). As with the temperature log, this method fails if the fracture plane is not coincident with the wellbore for the entire height of the fracture.

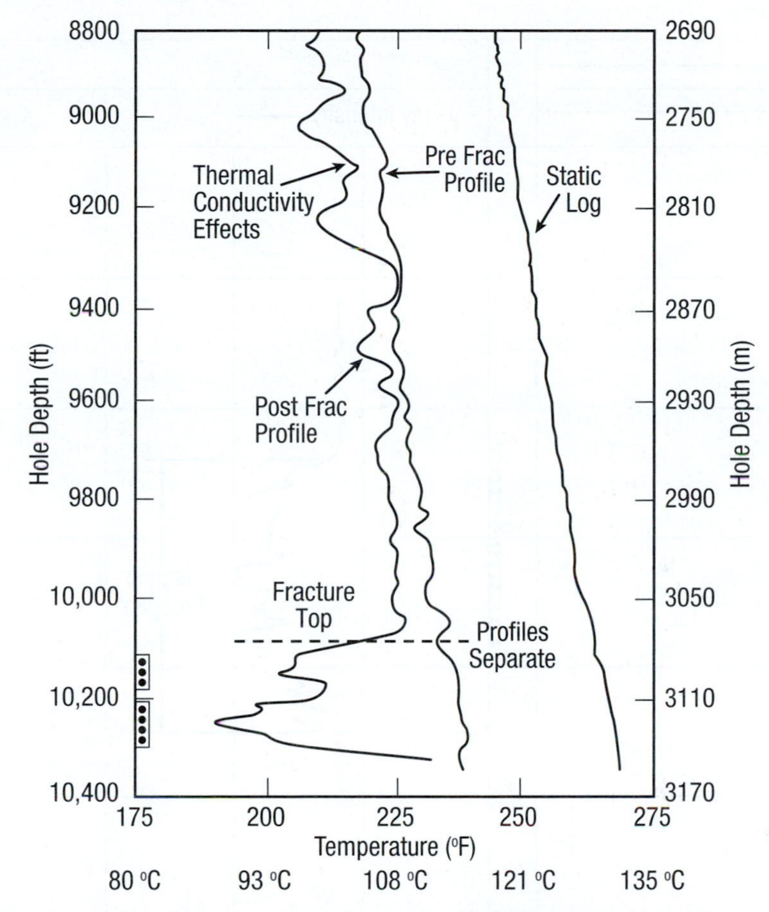

Figure 13-13 Pre- and post-fracture temperature profile for Example 13-6. (From Dobkins, 1981.)

Figure 13-14 is the postfracture gamma ray survey from Well D-2 after injection of 10,000 lb of tagged proppant. The log shows tagged proppant detected from about 10,130 ft to about 10,340 ft. Comparing with the temperature log results, the top of the propped fracture is about 30 ft below the top of the created fracture. Proppant was also detected extending about 40 ft below the bottom of the fracture located by the temperature log. However, there is a good chance that the proppant detected below 10,300 ft is residual proppant in the wellbore.

13.3.3.10 Injection Well Diagnosis

Production logs are used in injection wells to monitor reservoir behavior and to evaluate problems observed with the individual injection well or the reservoir. Among the problems that may arise are abnormally low or high injectivity, abnormal pressure or fluid level in the annulus, and low productivity or high water production in offset producers. Production logs are

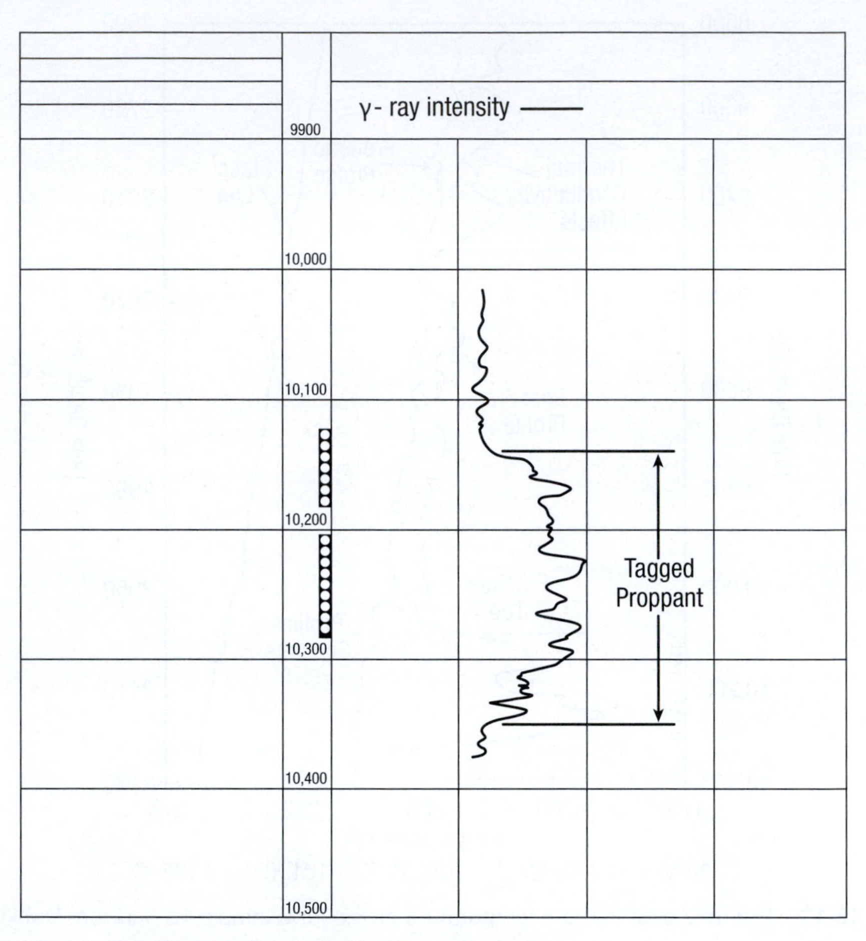

Figure 13-14 *γ-Ray* log after fracturing for Example 13-6.

used to evaluate such problems in injection wells in a manner similar to that already described for production wells, that is, by measuring the amount of flow to each reservoir interval, or by checking for interval isolation, by finding leaks in the well equipment.

The fundamental information sought with a production log in an injection well is the flow profile, the amount of fluid being injected into each interval. Flow profiles are measured in injection wells with temperature, radioactive tracer, and spinning flowmeter logs. A temperature log will yield a qualitative indication of the formation injection intervals, while the spinner flowmeter or radioactive tracer logs more precisely define the distribution of flow exiting the wellbore.

The flow profile shows where fluids leave the wellbore, but there is no guarantee that fluids are entering the formation at these same locations because they may move through channels behind the casing and enter zones other than those intended. The capability of the well completion to isolate the injection zones from other formations is crucial to proper reservoir management and is thus an important property to be evaluated with production logs. To identify channeling positively, a production log that can respond to flow behind the casing is needed. Among the logs that will serve this purpose are temperature, radioactive tracer, and noise logs.

A change in rate and/or wellhead pressure often indicates a serious well or reservoir problem. Abnormally low injectivity or a marked drop in injection rate can result from formation damage around the wellbore, plugged perforations, restrictions in the casing or tubing, or scaling. An unusually high injection rate may be caused by leaks in the tubing, casing, or packer, by channeling to other zones, or by fracturing of the reservoir. Techniques that combine production logging with transient testing, such as the production log test, selective inflow performance, and the multilayer transient test provide the most complete information about abnormally high or low injectivity.

The cause of a rate change in a well is often easier to diagnose if production logs have been run periodically throughout the life of the well. For example, the flow profile in a water injection well may change gradually throughout the life of the well as the saturation distribution changes in the reservoir. Logs obtained occasionally through the life of the well would show this as being a natural progression in a waterflood. Without knowledge of this gradual change, a profile obtained several years after the start of injection might appear sufficiently different from the initial profile to lead to an erroneous conclusion that channeling had developed or that some other drastic change had taken place.

Example 13-7 Abnormally High Injection Rate

In a waterflood operation in Reservoir A, water is being distributed to several injection wells from a common injection system; that is, water is supplied to all the wells at approximately the same wellhead pressure. Routine measurement of the individual well injection rates showed that one well was receiving approximately 40% more water than its neighbors. The sum of the *kh* products for all of the injection wells were approximately the same, and they were all completed at approximately the same depth. What are the possible causes of the abnormally high injection rate in this well, and what production logs or other tests might be run to diagnose the problem and plan remedial action?

The most likely causes of the high injection rate are leaks in the tubing, casing, or packer, or channeling to another zone. Fracturing is not a likely cause because the similarly completed surrounding injectors have the same wellhead pressure, yet do not exhibit abnormally high rate. Another unlikely, but possible, cause is that all the surrounding injection wells are damaged to similar extents, while the high-rate well is relatively undamaged.

For this scenario, production logs that can detect leaks or channeling should be run on the high-rate injection well. A combination of a temperature and a noise log would be a good choice to attempt to locate the suspected leak or channel.

Example 13-8 Abnormally High Injection Rate

The temperature and noise logs shown in Figure 13-15 were obtained in the injection well described in Example 13-7. What is the cause of the abnormally high injection rate in this well, and what corrective actions might be taken?

In a water injection well, the lowest point of water injection should be indicated clearly on the temperature log as the depth where the temperature (on both the flowing and shut-in logs) increases sharply toward the geothermal temperature. If such a sharp break does not occur, water is moving downward past the lowest depth logged.

The temperature logs in Figure 13-12 show no sudden increase in the lower part of the well, indicating that injection water is moving downward at least to 9150 ft. Thus the

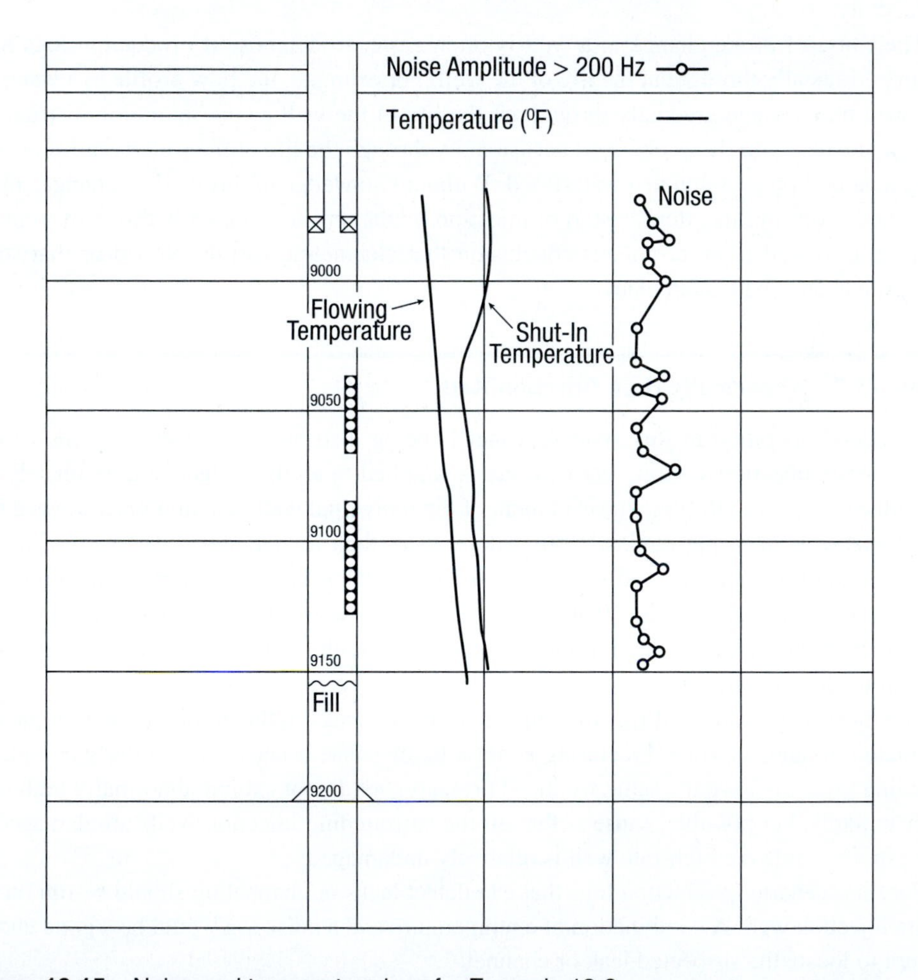

Figure 13-15 Noise and temperature logs for Example 13-8.

excessive water injection is exiting the wellbore through a casing leak below the lowest perforations or is channeling down from the lowest perforations. A log measuring the flow profile (a spinner flowmeter or radioactive tracer log) would distinguish between these two possibilities. The noise log is not very diagnostic in this well. The small increase in noise amplitude at about 9140 ft could be due to flow through a restriction in a channel or flow through a casing leak.

To eliminate the unwanted injection, the well could be plugged back to about 9120 ft. This would plug any casing leak below this depth and would probably eliminate flow into a channel from the lowest perforations.

13.4 Transient Well Analysis

This section takes a broad view of transient well analysis considering the analysis of pressure and flow rate data measured over an interval of time ranging from a few minutes for wireline formation tests to a few hours or days for pressure buildup tests to months or years for long-term production data. The first subsection briefly describes rate transient analysis primarily for estimation of the well expected ultimate recovery (EUR). This is followed by a brief description of wireline formation tests designed for measurement of formation pressure versus depth and for bottomhole fluid sampling.

Then the emphasis moves to the main thrust of this section: pressure and rate transient analysis based on geometric flow regimes that enable estimation of key parameters of interest to production engineering including the formation pore pressure, horizontal and vertical permeability values, damage skin, hydraulic fracture conductivity and half-length, and the well drainage pore volume. Flow regimes are used to analyze traditional pressure buildup tests as well as long-term production data. Absent from this treatment are type curves. Instead of using type curves we show that flow regimes are easily recognized as straight trends with characteristic slopes by processing data to produce the log–log diagnostic plot. While the literature is full of a great many type curves, there are only five flow regimes critical to production engineering: radial, spherical, linear, bilinear, and compression/expansion flow.

The final subsection in this chapter describes analysis of flow regimes found in fracture calibration tests that enable estimation of reservoir permeability along with the minimum stress, fluid leakoff coefficient, and fluid efficiency needed for hydraulic fracture treatment design. This will introduce one additional flow regime associated with fracture closure.

13.4.1 Rate Transient Analysis

In most of the United States operators are required to report gas, oil, and water production on a regular basis, and the data are publicly available. Rate transient analysis (RTA) methods look for empirical trends in the rate data. Because this analysis is typically done on production wells with declining rate, this is also called decline curve analysis.

Table 13-1 Arps Decline Curve Trend Functions

Curve Type	Exponential	Harmonic	Hyperbolic
Instantaneous producing rate at time t	$q(t) = q_i e^{-at}$	$q(t) = \dfrac{q_i}{1 + a_i t}$	$q(t) = \dfrac{q_i}{\left(1 + \dfrac{a_i t}{n}\right)^n}$
Instantaneous decline factor	$a = \text{constant}$	$a = a_i \dfrac{q_i}{q}$	$a = a_i \left(\dfrac{q_i}{q}\right)^{1/n}$
Straight line rate plot	$\ln q(t) = \ln q_i - at$	$\dfrac{1}{q(t)} = \dfrac{1}{q_i} + \dfrac{a_i}{q_i} t$	$\left(\dfrac{q_i}{q_i}\right)^{1/n} = 1 + \dfrac{a_i}{n} t$
Straight line cumulative production plot	$q(t) = q_i - aQ(t)$	$\ln q(t) = \ln q_i - \dfrac{a_i}{q_i} Q(t)$	No straight line

Traditional Arps (1945) decline curve analysis functions are summarized in Table 13-1. When the late-time rate decline behavior follows one of these functions, it is convenient to graph the data with plot axes that provide a straight line trend. The trend is then easily extrapolated to the time when the rate reaches the economic limit rate corresponding to the rate below which there is insufficient production to pay for continued operation of the well. The operator is losing money if the well continues to produce at rates below the economic limit rate.

The rate function can be integrated over time to determine the cumulative production up until the time the well reaches the economic limit rate. The resulting cumulative production is called the expected ultimate recovery, or the EUR.

Example 13-9 Expected Ultimate Recovery from Rate Decline Data

Eight years of production data for an oil well shown in Figure 13-16 (top) are graphed in Figure 13-16 (bottom) as ln q versus time, and a straight line trend is observed. Which type of decline behavior does this represent? What will be the EUR for an economic limit rate of 1 bpd?

Solution

The graph in Figure 13-16 (bottom) shows a straight line trend for a graph of ln q versus time with equation

$$q = y = 40.09e^{-0.000362t} \tag{13-1}$$

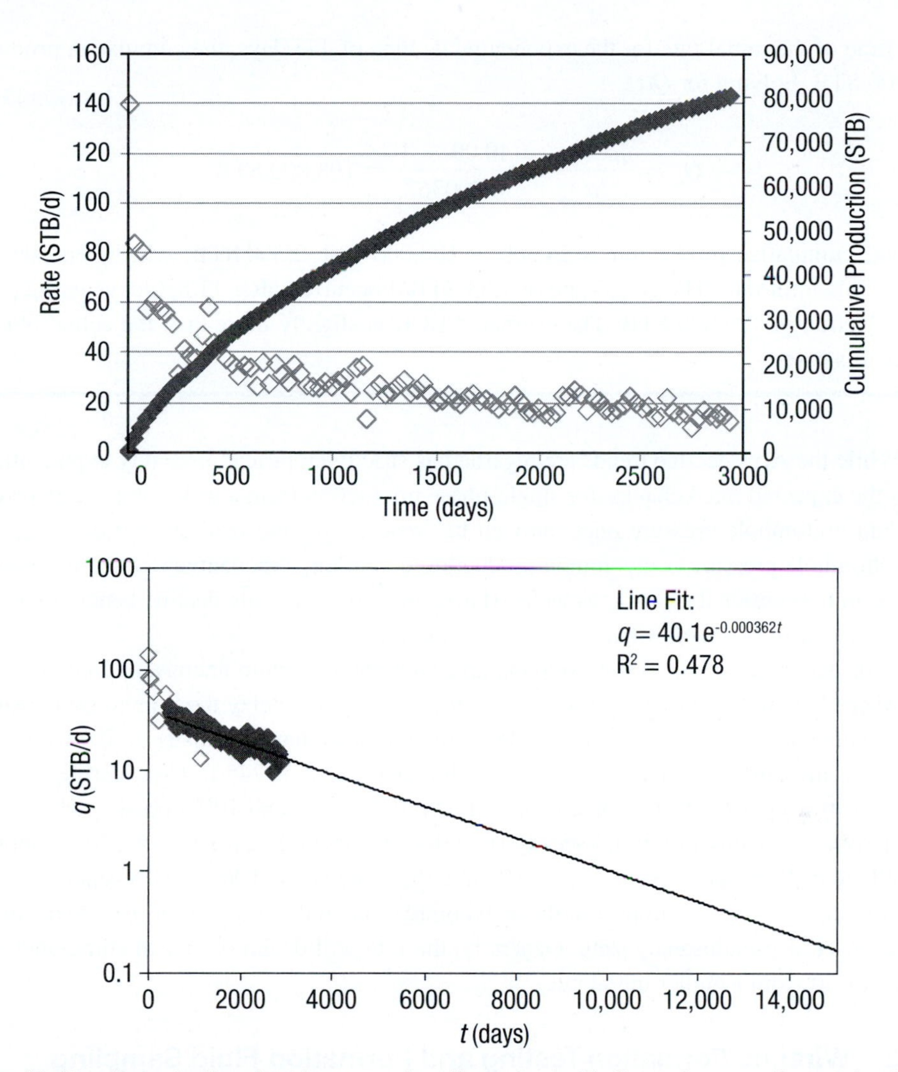

Figure 13-16 Rate decline data from an oil well graphed with Cartesian coordinates (top), and as ln *q* versus *t* (bottom).

with least-squares difference between data points and the trend line given by 0.777. From Table 13-1, this represents exponential rate decline with initial rate 40 bpd and with an instantaneous decline factor of $-0.0003624t$ per day or 0.000362 (365) = 13.2% per year. Extrapolation of the trend line until the abandonment rate, $q_a = 1$ STB/d corresponds to the time

$$t = \frac{\ln\left(\frac{q_i}{q_a}\right)}{a} = \frac{\ln\left(\frac{40.09}{1}\right)}{0.000362} = 10{,}200 \text{ days.} \tag{13-2}$$

At the time of the initial rate for the exponential decline of 443 days, the cumulative production is 24,300 STB. Solving for $Q(t)$,

$$Q_a = \frac{q_i - q_a}{a} = \frac{40.09 - 1}{0.000362} = 108{,}000 \text{ STB}. \tag{13-3}$$

The total cumulative production is 24,300 + 108,000 = 132,300 STB, which represents the EUR for this limit rate. The actual rate of 1.63 STB/d occurred after 11,830 days with cumulative production of 129,000 STB. The estimated EUR is slightly more than the actual observed value in this case.

While the Arps decline trends are regarded as strictly empirical, in reality exponential decline is the expected rate behavior for single-phase production from a well that is produced with a constant bottomhole pressure once the well has reached pseudosteady state flow. Since constant bottomhole pressure is a common production condition, exponential decline is frequently observed in conventional oil and gas wells. Harmonic and hyperbolic decline behaviors seem to be associated with flow below bubble or dew point pressure.

Arps decline behavior is less common and even uncommon in unconventional oil and gas wells where low mobility due usually to low permeability may delay the time to pseudosteady-state flow almost indefinitely. Equation 2-35 easily demonstrates why this is so. If we consider a well that drains a 40 acre square, the effective drainage radius is 745 ft. Considering a light oil with viscosity 1 cp, total formation compressibility 10^{-5} psi^{-1}, and 10% porosity, when permeability is 10 md, the onset of pseudosteady-state flow occurs at the time t_{pss} = 130 hr, about 5.5 days. When, instead, for permeability of 0.01 md, this time is 5500 days, or 15 years.

We must consider the impact of the well completion on the behavior of the rate decline before the onset of pseudosteady state. Generally, the rate will decline faster in stimulated wells, but it will start from a higher initial rate.

13.4.2 Wireline Formation Testing and Formation Fluid Sampling

The formation test (Moran and Finklea, 1962) is not a well test. As shown in Figure 13-17 on the left, the wireline formation test is typically performed in a mud-filled open hole. Hydraulically operated arms press against one side of the borehole, forcing the testing device against the opposing side. A doughnut-shaped rubber packer provides a seal, inside which a probe is forced into the formation. When exposed to low pressure, reservoir fluids flow into the tool. When a valve is shut, flow stops and a buildup occurs. Because the flow is only for a few seconds or minutes, the pressure builds up to the reservoir pressure in a short period of time. The test radius of investigation is at most only a few feet, but unlike in most well tests, the pressure disturbance expands spherically, not radially. This test is repeated at successive locations, providing a measure of reservoir pressure versus depth.

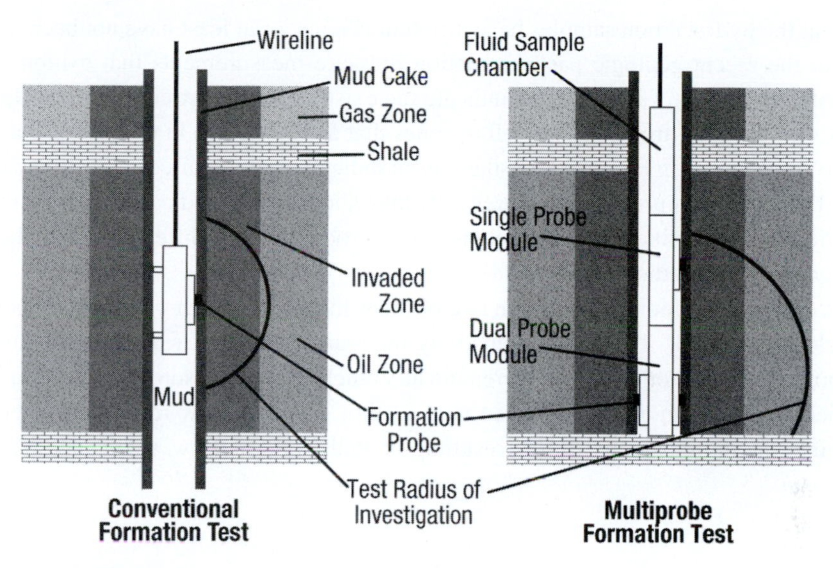

Figure 13-17 Formation test; multiprobe formation test.

Because the test is usually conducted on a newly drilled well, the pressure measurements represent the static pressure profile with depth for the area in hydraulic communication with the well horizontally.

Recent improvements in the wireline formation test tool include the addition of observation probes located opposite the active probe and offset at another depth, as shown in Figure 13-17 on the right. The observation probes permit reliable determination of vertical and horizontal permeability between the source and sink location and can also characterize partial barriers to vertical flow. In addition, the tool string offers a packer option with a larger flow chamber permitting flow for a longer period. This option, which is analogous to a mini drill stem test, extends the test radius of investigation, thus enabling determination of more representative permeability values. The device can also be used to inject fluids into the formation. One application for this option is determination of the formation minimum stress by creating a microfracture. Repetitive applications of the measurements provide the stress profile along the interval of interest as well as in layers above or below. Such measurements are very useful in predicting hydraulic fracture height migration and containment.

The formation test tool also includes a device that senses the phase of fluids withdrawn from the reservoir so that representative samples of reservoir fluids can be obtained, and multiple chambers are available that permit collection of pressurized downhole fluid samples at more than one depth in a logging trip. Formation fluid properties can sometimes be determined at the wellsite, and samples can also be sent to a PVT laboratory for more rigorous analysis.

Fluid samples and formation pressure measurements also help to locate barriers to flow. Gross changes in fluid composition from one depth to another or from one well to another may suggest that the hydrocarbon samples have different origins, or at least have not been in communication in the recent geologic past. Formation pressure measurements that exhibit an abrupt change with depth, as in Figure 13-18, indicate the existence of a barrier to vertical flow. When pressures vary abruptly from well to well in zones that are correlated in the geological model of the reservoir, also seen in Figure 13-18, there is an indication of impaired or obstructed flow between wells. Formation pressure measurements may show clearer indications of vertical and/or areal barriers in wells drilled after the field is put on production and differential depletion effects have become evident in the pressure data.

Differential depletion occurs when two or more intervals of high permeability contrast are commingled in the well. The high conductivity interval is produced with less pressure drop in the wellbore than in the other(s), which remain at a much higher pressure. Whenever a well with differential depletion among commingled units is shut in, fluids flow from the higher-pressure intervals into the more depleted zones, resulting in wellbore crossflow, also called backflow. In

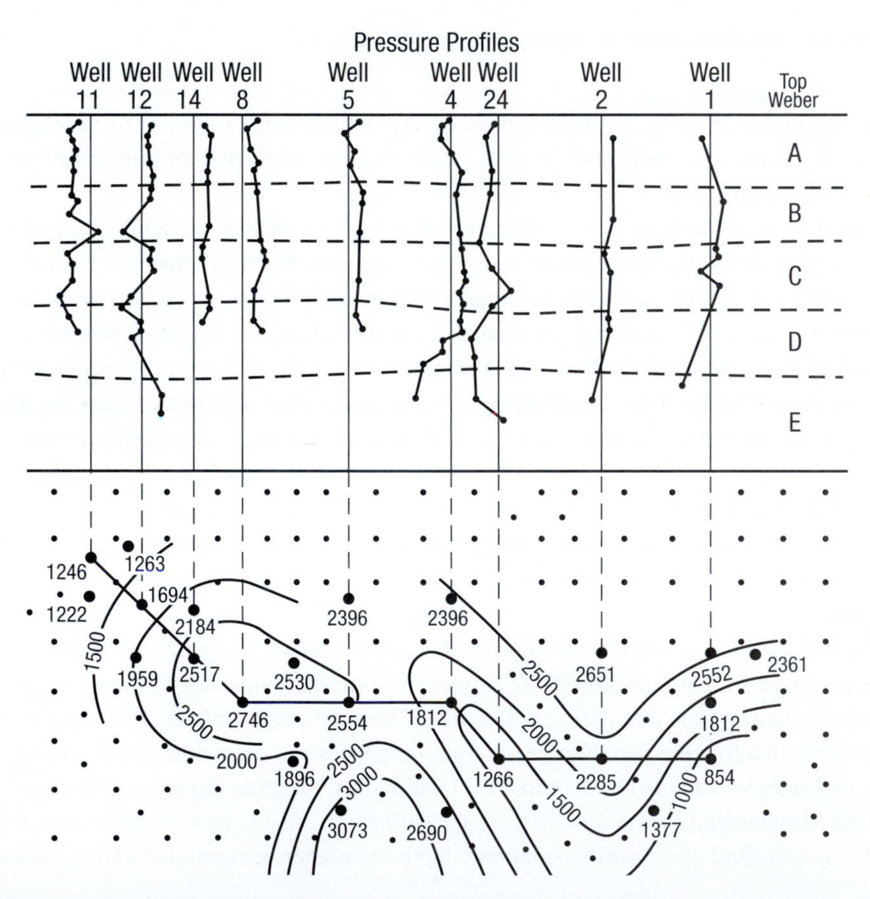

Figure 13-18 Formation test measurements and reservoir description.

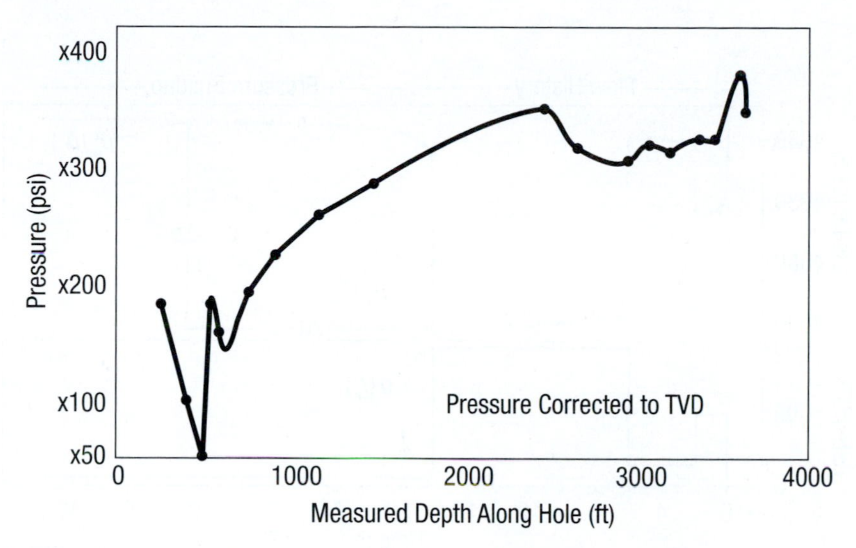

Figure 13-19 Formation test measurements along a horizontal borehole.

such a case, to enable production of the lower permeability reserves it may be necessary eventually to plug off the depleted zone(s).

Formation test pressures measured along a horizontal borehole, as in Figure 13-19, often show significant pressure variation laterally in the formation. These measurements are often an eye opener, but in addition, they signal cause for caution in planning a pressure buildup test too early in the life of a horizontal well. Since such a test would be dominated by backflow in the horizontal wellbore as reservoir fluids seek pressure equilibrium, a drawdown test would have a better chance for success. Otherwise, an even better plan would be to put the well on production for some time before testing.

Formation pressure and fluid sampling modules are also now available for operation during drilling (Villareal et al., 2010). While these measurements can save considerable time, their real value is shown when hole stability problems or high hole deviation make subsequent wireline logging risky or impossible. In that case measurements while drilling provide the only means to determine the formation and fluid properties.

Permeability and pressure determination combined with downhole fluid sampling have enabled wireline testing to achieve functions that were once only performed by drill stem tests. Increasingly strict restrictions on flaring of produced natural gas during exploration well testing have further elevated the importance and the value of wireline formation tests.

13.4.3 Well Rate and Pressure Transient Analysis

Rate transient analysis enables estimation of the well EUR by extrapolation of rate versus time data. Similarly, wireline formation tests enable measurement of the formation pressure by extrapolation of pressure versus time data. In contrast, the focus of this section is on well analysis involving both rate and pressure measured versus time.

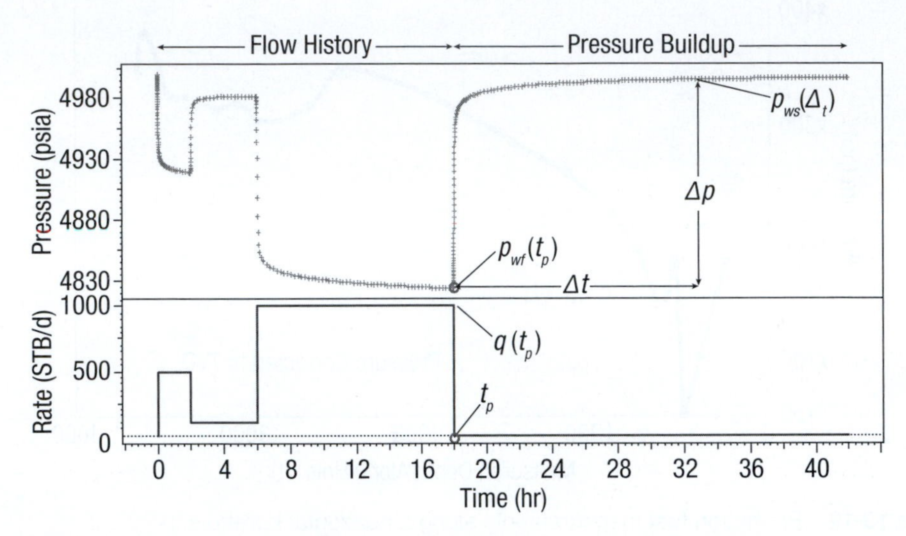

Figure 13-20 Graph of pressure buildup data in Cartesian coordinates.

The most common pressure transient testing strategy is a pressure buildup test. When a production well is shut in, the downhole pressure builds up, as seen in Figure 13-20. Pressure buildup data can be acquired any time flow from the well is stopped, whether for the purpose of a test or for other operational reasons, provided there is a pressure guage available to sense time-dependent pressure changes and either transmit the pressure measurements in real time or record them in memory for later retrieval. Many wells are equipped with permanent downhole pressure gauges that provide a continuous pressure record for very long periods, including several years. Otherwise, when a pressure transient test is planned, the pressure gauge can be temporarily lowered on wireline or slick line.

Chapters 2–4 described transient, steady, and pseudosteady-state production modes and illustrated the sensitivity of the well inflow performance relationship to permeability and skin. In a stabilized flow test, the stabilized bottomhole flowing pressure is measured for a series of surface flow rates. While common, stabilized flow tests are subject to error because a well must be in the steady-state production mode to achieve a truly constant flowing pressure under constant rate production. A well in the transient or pseudosteady-state production mode may reach an apparent pressure stabilization under which the pressure change with time is small. Furthermore, neither permeability nor skin can be quantified from the productivity index determined from a stabilized flow test unless the other is known from some other measurement.

In Chapter 2 we saw that during transient mode production the pressure is a linear function of the natural logarithm of time

$$p_{wf} = p_i - \frac{q\mu}{4\pi kh}\left(\ln\frac{4kt}{\gamma\phi\mu c_t r_w^2} + 2s\right) = p_i - \frac{\alpha_p q\mu}{2kh}\left(\ln\frac{\alpha_t kt}{\phi\mu c_t r_w^2} 0.80907 + 2s\right) \quad \textbf{(2-23)}$$

with coefficients α_p and α_t depending on the units system [Equation (2-5) has been slightly modified to include the skin, s]. For oilfield units, $\alpha_p = 141.2$, and $\alpha_t = 0.00026375$. This equation shows that when pressure is measured while the well is flowed at a constant rate, a graph of the flowing pressure, p_{wf}, versus the natural logarithm of time will have a slope $-\alpha_p q\mu/2kh$. Since all of the other parameters in the slope are known, the value of the slope can be used to quantify permeability. Furthermore, once permeability is known, the skin, s, can also be determined.

Because small rate fluctuations while flowing cause fluctuations in pressure that can obscure interpretation, most pressure transient data are interpreted when the well is not flowing as either a pressure buildup in a producing well or a pressure falloff in an injection well.

Transient well analysis is most straightforward when the data are presented on a log–log diagnostic plot. This plot effectively reveals trends that at a particular time may be directly related to the streamline flow geometry in the reservoir. As time increases each flow regime revealed in the transient data is further away from the well. Flow regime trends provide a mechanism to compute parameters of interest to production engineers.

The log–log diagnostic plot for the buildup data in Figure 13-20 is shown in Figure 13-21. The pressure change in Figure 13-21, $\Delta p_{ws}(\Delta t)$, is computed for each elapsed time $\Delta t = t - t_p$, as

$$\Delta p_{ws}(\Delta t) = p(t) - p_{wf}(t_p) \tag{13-4}$$

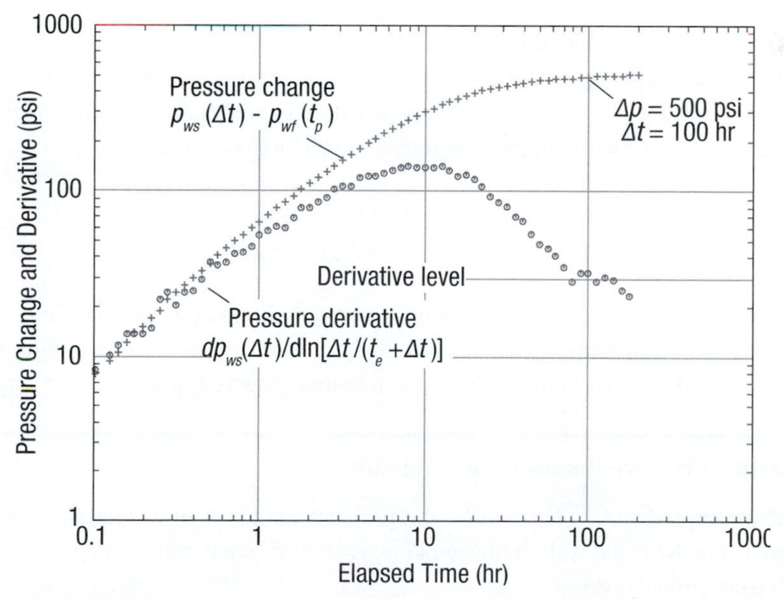

Figure 13-21 Log–log diagnostic plot of buildup data for Example 13-11.

where t_p is the time when the well is shut in. The lower curve shown in Figure 13-21 is the derivative of the pressure change computed as

$$\Delta p' = \frac{dp}{d \ln \tau} = \frac{dp}{d\tau}\tau \quad \textbf{(13-5)}$$

where the superposition time, τ, is defined as

$$\ln \tau = \ln\left(\frac{\Delta t}{t_e + \Delta t}\right) \quad \textbf{(13-6)}$$

with the material balance time, t_e, computed as

$$t_e = \frac{24Q}{q} \quad \textbf{(13-7)}$$

In the log–log diagnostic plot, the upper curve is pressure change, and the lower curve is the pressure change derivative. For gas wells, the diagnostic plot should be constructed as the change in the real gas pseudopotential and its derivative versus elapsed time, where q_g is the gas flow rate in MSCF/d before the well is shut in, T_R is the reservoir temperature, and the real gas pseudopotential

$$m(p) = 2\int_0^p \frac{pdp}{\mu Z} \quad \textbf{(4-71)}$$

and has units of squared pressure/cp.

Because the data are acquired while the well is shut in, pressure buildup data typically last from a few hours to a few days. For a vertical well without a hydraulic fracture, the pressure signal penetrates into the formation during the transient production mode as

$$r_i = \sqrt{\frac{kt}{948\phi\mu c_t}} \quad \textbf{(13-8)}$$

where r_i is the pressure radius of investigation for permeability, k, porosity, ϕ, fluid viscosity, μ, and total formation compressibility, c_t, with time, t, measured in hours. The longer the buildup continues, the more distant from the well are the features observed in the transient signal.

Example 13-10 Pressure Radius of Investigation

For the fluid in Appendix A, what will be the pressure radius of investigation after 2 minutes, 24 hours, and 1 month? If the well drainage radius is 1000 ft, what will be the time of the start of pseudosteady-state production?

Repeat for the gas in Appendix C.

Solution

Using Equation 13-8, the pressure radius of investigation after 2 minutes (0.0333 hr) is

$$r_i = \sqrt{\frac{kt}{948\phi\mu c_t}} = \sqrt{\frac{(8.2)(0.0333)}{948(0.19)(1.72)(12.9\times10^{-6})}} = 8 \text{ ft.} \tag{13-9}$$

After 24 hours, $r_i = 223$ ft, and after 1 month (720 hr) $r_i = 1{,}223$ ft, provided this is less than the effective well drainage radius, r_e.

If the well drainage radius is 1000 ft, the drainage area is $A = 3{,}141{,}593$ ft^2 from Equation 2-41,

$$t_{pss} = \frac{\phi\mu c_t A}{0.0002637k} t_{DA} = \frac{(0.19)(1.72)(12.9\times10^{-6})(3141593)}{0.0002637(8.2)}(0.1)$$
$$= 605 \text{ hr} = 25 \text{ days} \tag{13-10}$$

Once the pressure disturbance reaches the drainage area boundary nearest the well, Equation 13-1 no longer applies.

For the gas in Appendix B, evaluating viscosity and compressibility at the initial pressure gives $\mu = 0.024$ cp and $c_t = 0.000162$ psi^{-1}. In that case, the pressure radius of investigation is 24 ft after 2 min and 635 ft after 1 day. For the 1000-ft drainage radius, the onset of pseudosteady state occurs before 5 days on production.

Example 13-10 illustrates the importance of time to transient well analysis. The succession of later and later features on a log–log diagnostic corresponds to flow geometries further and further from the well. When the well is hydraulically fractured or horizontal, the pressure penetrates from the extended wellbore, which itself may extend several thousand feet. For this reason, a pressure buildup test may be dominated by transients related to the well, and not the reservoir, for a very long period of time, especially when the permeability is low. Under such conditions a pressure buildup test cannot provide estimates for permeability, skin, and formation pressure.

As mentioned earlier, production rate data are always recorded and frequently public, at least in the United States. While a well is on production, both rate and pressure are likely to be varying over time, and considered along, neither rate nor pressure data can be readily interpreted beyond the simple empirical trends shown at the beginning of this section. When pressure is also recorded on a regular basis, it is possible to analyze long-term production in a much more rigorous way.

Two common production data analysis methods that can be used to estimate well and reservoir parameters are rate-normalized pressure (reciprocal productivity index) and pressure-normalized rate (productivity index). While less commonly used, rate-normalized pressure

(RNP) is actually the most straightforward to use for well and reservoir characterization. RNP is computed as

$$\mathrm{RNP} = \frac{p_i - p(t)}{q(t)} q_{\mathrm{ref}} \tag{13-11}$$

and graphed versus the material balance time, t_e, computed as in Equation 13-7. The rate used for q_{ref} to keep the RNP in pressure units is a matter of convention.

RNP is interpreted as a virtual drawdown at constant rate. As for pressure transient data, the RNP derivative with respect to the logarithm of time provides the same flow regime trends as would be seen for pressure drawdown under constant rate production. The RNP derivative is best computed with respect to time (not material balance time). If the result is then graphed versus material balance time, less apparent noise appears in the derivative, and important trends are more visible. The integral of RNP (IRNP) is computed as

$$\mathrm{IRNP} = \frac{q_{\mathrm{ref}}}{t_e} \int \frac{p_i - p(\tau)}{q(\tau)} d\tau. \tag{13-12}$$

The IRNP and its derivative usually look smoother than the RNP, but sometimes they can be severely distorted, thereby obscuring diagnosis of the data trends. In such cases the RNP and its derivative may work better. The IRNP and derivative for the data in Figure 13-22 are shown in Figure 13-23.

The advantage of production data is that it is collected for a much longer time than pressure buildup data, which is only available whenever the well is shut in. Figure 13-22 shows rate and pressure data recorded for more than 34 years for the same well as was featured in Example 13-9. The well production rate was recorded on a monthly basis. No pressure buildup data are available for this well.

While the occasional well may be equipped with a permanent downhole pressure gauge and its own continuous flow meter, typically production data are recorded no more frequently than on a daily basis, and often no more than once a month. When the well completion includes a permanent downhole pressure gauge, this provides a continuous record of the well pressure transients. Especially for dry gas wells, even a permanent surface pressure gauge may provide meaningful long-term production data.

Unfortunately when production data are recorded for several wells in a common gathering system, the rate from each individual well may be measured infrequently, and estimates via a rate allocation technique introduce uncertainties. Nonetheless, when well pressure and rate data are recorded continuously, the data may contain interpretable trends, and the more frequently the data are recorded the better. The next subsection shows how to recognize interpretable trends in both buildup and long-term production data.

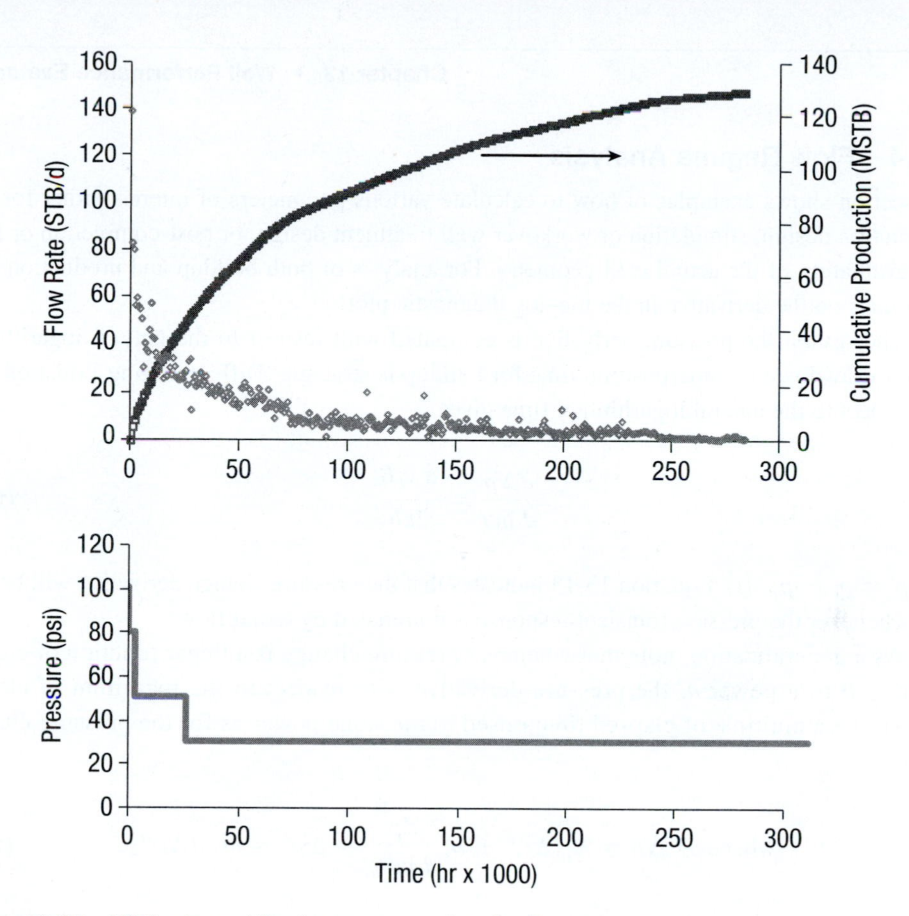

Figure 13-22 Well rate and flowing pressure data for Example 13-14.

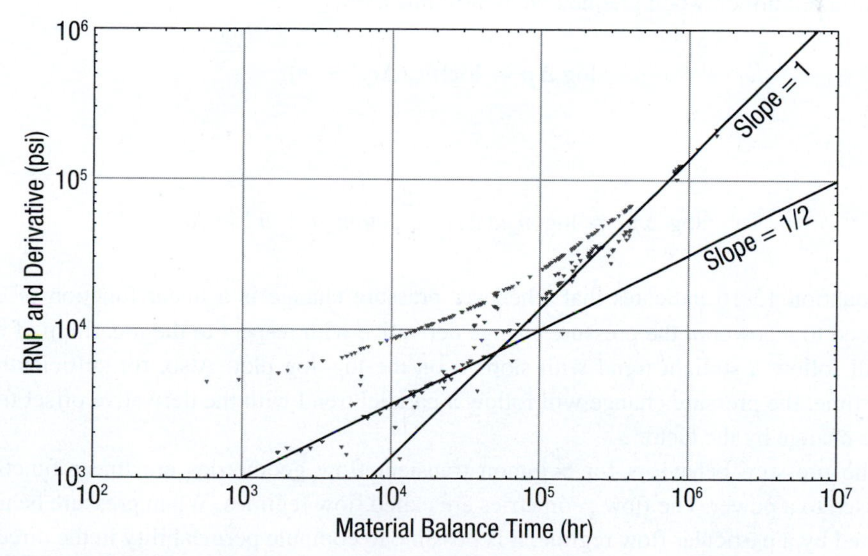

Figure 13-23 IRNP and derivative for the data in Figure 13-22.

13.4.4 Flow Regime Analysis

This section shows examples of how to calculate various parameters of interest either for well construction design, stimulation or workover well treatment design, or post-completion or treatment evaluation of the actual well geometry. For analysis of both buildup and production data, the focus is on the derivative in the log–log diagnostic plot.

The reason the pressure derivative is computed with respect to the natural logarithm of time for drawdown or superposition time for buildup is strategic. Differentiating Equation 2-23 with respect to the natural logarithm of time gives

$$\frac{d\Delta p}{d \ln t} = \frac{\alpha_p q B}{2kh} \tag{13-13}$$

for $\Delta p = p_i - p_{wf}(t)$. Equation 13-13 indicates that the pressure change derivative will be constant whenever the pressure transient response is dominated by radial flow.

As a generalization, note that whenever pressure change is a linear function of elapsed time raised to a power a, the pressure derivative with respect to the logarithm of elapsed time will be a multiple of elapsed time raised to the same power as for the pressure change. That is,

$$\text{whenever } \Delta p = m_a(\Delta t)^a + b, \frac{dp}{d \ln(t)} = \Delta p' = m_a a(\Delta t)^a, \tag{13-14}$$

using $\Delta p'$ as a shorthand notation for the derivative of pressure with respect to the natural logarithm of time. Further, when graphed on logarithmic axes,

$$\log \Delta p = \log[m_a(\Delta t)^a + b], \tag{13-15}$$

and

$$\log \Delta p' = \log m_a a(\Delta t)^a = \log m_a a + a \log \Delta t. \tag{13-16}$$

Equation 13-16 indicates that whenever pressure change is a linear function of elapsed time raised to a power a, the pressure change derivative with respect to the logarithm of elapsed time will follow a straight trend with slope a on the log–log plot. Also, for sufficiently large elapsed time, the pressure change will follow a parallel trend with the derivative offset from the pressure change by the factor a.

The pressure behaviors for common transient flow geometries are linear functions of time raised to a power. The flow geometries are called flow regimes. When pressure behavior is dominated by a particular flow regime, it is possible to compute permeability in the direction of flow and dimensions of flow geometry. Table 13-2 summarizes the equations for flow regimes described in this subsection. A succession of flow regimes can be identified on the log–log diagnostic plot as a series of straight trends in the pressure derivative. Because the pressure signal travels in time according to Equation 13-8, as time increases, the flow geometry is increasingly distant from the well. In Table 13-2, the symbols for units conversion coefficients are α with a subscript. Table 13-3 provides the α values for four commonly encountered units systems.

13.4.4.1 Determination of Permeability, Skin, and Pressure from the Radial Flow Regime

The radial flow regime is the most strategic of all the flow regimes because it enables estimation of the permeability parallel to the bedding plane. This flow regime could also be called cylindrical flow because it actually represents flow to a cylinder. Figure 13-24 shows diagrams of commonly encountered radial flow geometries in vertical wells. The partial radial flow shown in Figure 13-24a occurs in very early time and may be obscured by wellbore storage transients, described in the next subsection. The height of the partial radial flow corresponds to the length of the flowing interval. The height of the complete radial flow, shown in Figure 13-24b is the entire interval thickness. Chapter 17 describes flow to a well with a vertical fracture as pseudoradial flow shown in Figure 13-24c because flow near the well has a geometry governed by the fracture. Far enough away from the fractured well, pseudoradial flow is like flow to a well with a larger apparent radius. Hemi-radial flow in Figure 13-24d occurs if a well is near a planar sealing boundary like a fault. Far enough away from the well, the hemi-radial flow is like flow to the well plus its image across the fault.

Permeability is actually directionally dependent, and the permeability value in the direction perpendicular to the bedding plane is often significantly less than the permeability parallel to the bedding plane. The term *reservoir permeability* is the permeability value in the direction parallel to the bedding plane, and this is often called *horizontal permeability* even though the bedding plane may be dipping. Likewise, the permeability value for the direction perpendicular to the bedding plane is called *vertical permeability.* When there is significant variation in permeability for different directions, this is called *permeability anisotropy.* We have seen already in Chapter 5 the importance of permeability anisotropy to horizontal well productivity.

On the log–log diagnostic plot, radial flow appears as a level (constant) derivative trend. The reservoir permeability in md can be estimated from the log–log diagnostic plot from the constant value, m', of the pressure derivative:

$$k = \frac{\alpha_p q B \mu}{2.303 m' h} = \frac{70.6 q B \mu}{m' h} \text{ (oil)}$$

$$k = \frac{10.1 \alpha_p q_g T_R}{2.303 m' h} = \frac{711 q_g T_R}{m' h} \text{ (gas)} \tag{13-17}$$

Table 13-2 Flow Regime Equations

Flow Regime	Equation[a]	Parameters to Compute
Infinite-acting radial flow (radial flow)[b]	$\Delta p = mt_{\text{sup}} + p_{1hr}$ $\Delta p = p_i - p_{wf}(\Delta t)$ for flowing well $\Delta p = p_{ws}(\Delta t) - p_{wf}(t_p)$ for shut-in well $m = \dfrac{1.151\alpha_p qB\mu}{kh}$ $p_{1hr} = m\left(\log\dfrac{\alpha_t k}{\phi\mu c_t r_w^2} + 0.351 + 0.87s\right)$	$k = \dfrac{1.151\alpha_p qB\mu}{mh}$ $s = 1.151\left(\dfrac{\Delta p_{1hr}}{m} - \log\dfrac{\alpha_t k}{\phi\mu c_t r_w^2} - 0.351\right)$ $\Delta p_{1hr} = p_i - p_{1hr}$ for following well $\Delta p_{1hr} = p_{wf} - p_{1hr}$ for shut-in well
Wellbore storage (fluid expansion/ compression)[b]	$\Delta p = m_c\Delta t,\ m_c = \dfrac{qB}{\alpha_c C}$	$C = \dfrac{qB}{\alpha_c m_c}$
Partial penetration (spherical flow)[d]	$\Delta p = \dfrac{\alpha_p qB\mu}{2k_{sph}r_w} - \dfrac{m_{pp}}{\sqrt{\Delta t}}$ $m_{pp} = \dfrac{\alpha_p qB\mu}{2k_{sph}}\left(\dfrac{\phi\mu c_t}{\pi\alpha_t k_{sph}}\right)^{1/2}$	$k_{sph} = \left(\dfrac{\alpha_{pp} qB\mu}{m_{pp}}\sqrt{\phi\mu c_t}\right)^{2/3}$
Finite-conductivity vertical fracture (bilinear flow)[c]	$\Delta p = m_{bf}\ \sqrt[4]{\Delta t}$ $m_{bf} = \dfrac{2.45\alpha_p qB\mu}{kh\sqrt{k_f w/kx_f}}\left(\dfrac{\alpha_t k}{\phi\mu c_t x_f^2}\right)^{1/4}$	$k_f w\sqrt{k} = \left(\dfrac{\alpha_{bf} qB\mu}{m_{bf}h}\right)^2\left(\dfrac{1}{\phi\mu c_t}\right)^{1/2}$

Flow Regime	Equation[a]	Parameters to Compute
Infinite-conductivity vertical fracture (linear flow)[c]	$\Delta p = m_{lf}\sqrt{\Delta t} + \left(\frac{\pi}{3}\right)\left(\frac{m}{1.151}\right)\frac{kx_f}{k_f w}$ $m_{lf} = \frac{\alpha_p q B \mu}{kh}\left(\frac{\pi\alpha_t k}{\phi\mu c_t x_f^2}\right)^{1/2}$	$x_f\sqrt{k} = \left(\frac{\alpha_{lf} q B}{m_{lf} h}\right)\left(\frac{\mu}{\phi c_t}\right)^{1/2}$ $\frac{k_f w}{k x_f} = \left(\frac{\pi}{3}\right)\left(\frac{m}{1.151}\right)\left(\frac{1}{\Delta p_{\text{int}}}\right)$
Pseudosteady-state (fluid expansion/compression)	$\Delta p = m_{pss}\Delta t + p_{\text{int}}$	$\phi hA = \frac{\alpha_{pss} q B}{c_t m_{pss}\Delta t}$

[a] For a simple drawdown test, the generalized superposition time function, t_{sup}, simplifies to log Δt, and for a buildup after a single drawdown flow period to the Horner time function given by $(t_p + \Delta t)/\Delta t$, where t_p is the flow time before shutting in the well.

[b] Earlougher (1977).

[c] Economides and Nolte (1989).

[d] Chatas (1966).

Table 13-3 Units Conversion Factors for Equations in Table 13-2

Quantity	Oilfield	SI	API	cgs
Production rate, q	STB/D	m^3/s	dm^3/s	cm^3/s
Formation thickness, h	ft	m	m	cm
Permeability, k	md	m^2	μm^2	darcy
Viscosity, μ	cp	Pa-s	kPa-s	cp
Pressure difference, Δp	psi	Pa	kPa	atm
Pressure, p				
Radius, r	ft	m	m	cm
Fracture half length, x_f				
Fracture width, w				
Effective horizontal well length, L_p				
Time, t	hr	s	hr	s
Porosity, ϕ, fraction	unitless	unitless	unitless	unitless
Total system compressibility, c_t	psi^{-1}	Pa^{-1}	kPa^{-1}	atm^{-1}
Wellbore storage, C	bbl/psi	$m^3/Pa\text{-}s$	$dm^3/kPa\text{-}s$	$cm^3/atm\text{-}s$
Skin, s	unitless	unitless	unitless	unitless
Horizontal well early pseudoradial skin s_{epr}				
Horizontal well damage skin, s_m				
Conversion Factor				
α_p	141.2	$1/(2\pi)$	$1/(2\pi)$	$1/(2\pi)$
α_t	0.0002637	1	3.610^{-6}	1
α_c	24	1	1/3600	1
α_{lf}	4.064	$1/(2\sqrt{\pi})$	$1.016 \cdot 10^{-6}$	$1/(2\sqrt{\pi})$
α_{bf}	44.1	0.3896	0.01697	0.3896
α_{pp}*	2453	0.00449	2.366	0.00449
α_f	0.000148	0.7493	$2.698 \cdot 10^{-6}$	0.7493
α_{pss}	0.234	1	$3.6 \cdot 10^{-6}$	1
α_{CA}	5.456	5.456	5.456	5.456

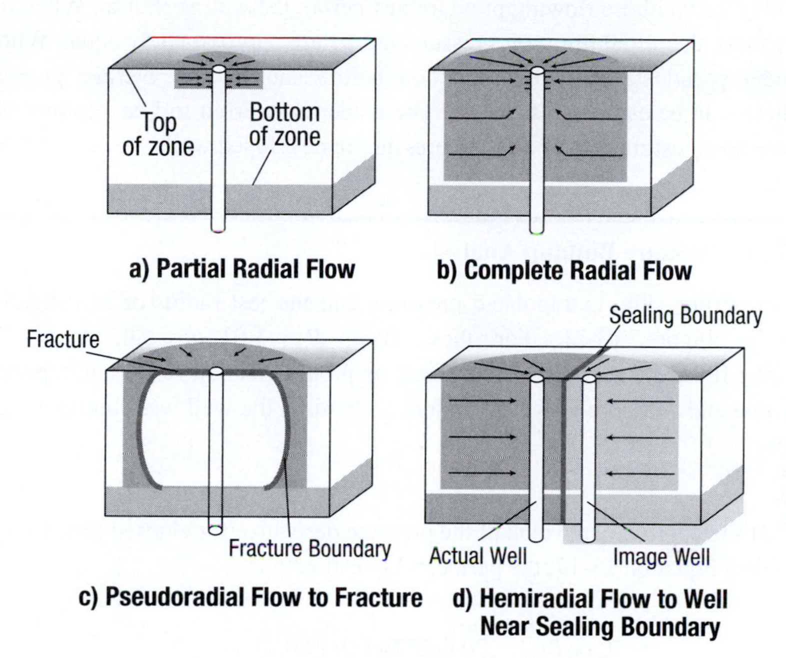

Figure 13-24 Radial flow geometries for vertical wells.

where q is the flow rate in STB/d before the well is shut in, B is the fluid formation volume factor in res bbl/STB, μ is the fluid viscosity in cp, and h is the net formation thickness in ft.

The skin factor can be estimated from the diagnostic plot using the following equation:

$$s = 1.151\left[\frac{\Delta p_{IARF}}{2.303m'} - \log\frac{kt_e}{1688\phi\mu c_t r_w^2} + \log\left(\frac{t_e + \Delta t}{\Delta t}\right)\right] \text{(oil)}$$

$$s = 1.151\left[\frac{\Delta m(p)_{IARF}}{2.303m'} - \log\frac{kt_e}{1688\phi\overline{\mu c_t} r_w^2} + \log\left(\frac{t_e + \Delta t}{\Delta t}\right)\right] \text{(gas)} \quad \textbf{(13-18)}$$

where Δp_{IARF} is pressure change at an elapsed time, Δt_{IARF}, when the derivative is constant, ϕ is the porosity, r_w is the well radius, and t_e is the material balance time when the well was shut in.

Finally, the reservoir pressure is estimated from the extrapolated pressure, p^*, given by

$$p^* = m'\ln\left(\frac{t_e + \Delta t}{\Delta t}\right) + \Delta p_{IARF} + p_{wf}(t_p) \quad \textbf{(13-19)}$$

where $p_{wf}(t_p)$ is the wellbore flowing pressure just before the well is shut in. When the time on production is short, the initial reservoir pressure and p^* are approximately equal. When the well has been under pseudosteady-state production before shut-in, the average pressure in the drainage volume can be estimated from p^*. The reader is referred to Lee, Rollins, and Spivey (2003) or more about estimation of average pressure in developed wells.

Example 13-11 Pressure Buildup Analysis

Estimate permeability, skin, extrapolated pressure, and the test radius of investigation for the data shown in Figure 13-21. For these data, $B = 1.01$, $\mu = 80.7$ cp, $h = 79$ ft, $\phi = 0.11$, $c_t = 1.56 \cdot 10^{-5}$, and $r_w = 0.3$ ft. The time on production before the buildup was 2542 hr, and the flow rate and wellbore flowing pressure at the time the well was shut in are 277 STB/d and 7893.5 psi.

Solution

In Figure 13-21 the approximate value of the pressure derivative for elapsed time from 80 to 150 hr is 30 psi. From Equation 13-17, the permeability estimate is

$$k = \frac{70.6 q B \mu}{m' h} = \frac{70.6(277)(1.01)(80.7)}{(30)(79)} = 673 \text{ md.} \tag{13-20}$$

At elapsed time $\Delta t = 100$ hr, the derivative is approximately constant, and the pressure change $\Delta p_{IARF} = 500$ psi. From Equation 13-7 the material balance time before the buildup is given by

$$t_e = t_p = 2542 \text{ hr.} \tag{13-21}$$

Using Equation 13-18, the skin is

$$\begin{aligned} s &= 1.151\left[\frac{\Delta p_{IARF}}{2.303m'} - \log\frac{k t_e}{1688\phi\mu c_t r_w^2} + \log\left(\frac{t_e + \Delta t}{\Delta t}\right)\right] \\ &= 1.151\left[\frac{500}{2.303(30)} - \log\frac{(673)(2542)}{1688(0.11)(80.7)(1.56 \cdot 10^{-5})(0.3)^2} \right. \\ &\quad \left. + \log\left(\frac{2542 + 100}{100}\right)\right] = 0.86 \end{aligned} \tag{13-22}$$

Using the same elapsed time and pressure change, the extrapolated pressure p^* is determined from Equation 13-19 as

$$p^* = m' \ln\left(\frac{t_e + \Delta t_{IARF}}{\Delta t_{IARF}}\right) + \Delta p_{IARF} + p_{wf}(t_p)$$

$$= (30)\ln\left(\frac{2542 + 100}{100}\right) + 500 + 7893.5 = 8492 \text{ psi.} \quad \textbf{(13-23)}$$

In this case the extrapolated pressure may be approximately equal to the initial reservoir pressure because the well has not been produced for a very long time.

Finally, the test radius investigation estimate with Equation 13-8 for the total duration of the buildup transient data of 200 hr is

$$r_i = \sqrt{\frac{kt}{948\phi\mu c_t}} = \sqrt{\frac{(673)(200)}{948(0.11)(80.7)(15.6 \cdot 10^{-5})}} \cong 1000 \text{ ft.} \quad \textbf{(13-24)}$$

13.4.4.2 Wellbore Storage

In Example 13-10, the pressure and derivative behavior seen in Figure 13-21 before $\Delta t = 80$ hr is characteristic of what is termed *wellbore storage,* so called because in early time most pressure transient tests show behavior dominated by expansion and/or redistribution of wellbore fluids. At the beginning of the wellbore storage transient, both pressure and derivative may be overlain in a straight trend with unit slope. The unit slope trend in the derivative corresponds to pressure change that is directly proportional to elapsed time, and both pressure change and the derivative follow the same trend. During this time, essentially no change occurs in the flow rate between the wellbore and the sandface. As the pressure change and derivative trends separate, the flow rate across the sandface changes from the value before the surface flow rate is changed to the rate after the change. For pressure buildup during this time the sandface rate changes from the flow rate before the buildup to zero. After the sandface rate is zero, the observed pressure change and derivative are dominated by pressure transient behavior in the reservoir.

The transition from the early wellbore storage trend with unit slope derivative to a reservoir-dominated transient looks like a hump in the derivative that ends when the derivative develops a recognizable reservoir flow unit trend. The duration of constant wellbore storage behavior is estimated as

$$t_{ewbs} = \alpha_w(60 + 3.5s)\frac{\mu C}{kh} \quad \textbf{(13-25)}$$

where the constant wellbore storage coefficient, C, is approximated as

$$C = V_w c_w \quad \text{for single phase wellbore fluid}$$

$$C = \frac{\alpha_l V_u}{\rho_l(g/g_c)} \quad \text{for changing liquid level} \quad \textbf{(13-26)}$$

where V_w is the wellbore volume, V_u is the wellbore volume per unit length, c_w is the wellbore fluid compressibility, ρ_l is wellbore liquid density, g is acceleration of gravity, g_c corrects for gravity units, and α_l is a units conversion factor. In oilfield units V_w is in bbl, V_u is in bbl/ft, c_w is in psi^{-1}, ρ_l is in lb$_m$/ft^2, $g_c = 32.17$, $\alpha_w = 3387$, and $\alpha_l = 144$. The duration of wellbore storage behavior is minimized when flow is controlled by a downhole shut-in valve located near the productive interval instead of being controlled at the surface by a valve at the wellhead.

Wellbore storage behavior can be complex and may exhibit changing wellbore storage due to redistribution of wellbore fluid phases. For oil wells, as gas bubbles migrate upward in the wellbore, the average compressibility of the wellbore fluids increases, causing an increasing effective wellbore storage coefficient. In gas wells, gas condensate droplets dropping in the wellbore cause a decrease in the effective wellbore storage coefficient. It is very important to be able to distinguish wellbore storage transients from reservoir behavior. More on wellbore storage can be found in Chapter 4 of Lee et al. (2003).

When the pressure buildup diagnostic plot begins with both pressure change and derivative aligned in a unit slope trend, the wellbore storage coefficient can be estimated as

$$C = \frac{qB}{24\frac{\Delta p}{\Delta t}} \tag{13-27}$$

where Δp and Δt are from a point on the unit slope trend.

Example 13-12 Wellbore Storage Estimation

Estimate the wellbore storage coefficient for the data shown in Example 13-11. What value for C is estimated for a well depth of 10,000 ft, tubing shoe at 9,900 ft, tubing I.D. 2.5 in., casing I.D. 3.5 in., and oil with compressibility 10^{-5} psi^{-1} as the wellbore fluid? Estimate the time at the end of wellbore storage.

Solution

Using the point $\Delta p = 8$ psi and $\Delta t = 0.1$ hr in Figure 13-21, the wellbore storage coefficient is estimated from Equation 13-27 as

$$C = \frac{qB}{24\frac{\Delta p}{\Delta t}} = \frac{(277)(1.01)}{24\left(\frac{8}{0.1}\right)} = 0.15 \tag{13-28}$$

Assuming single phase oil in the wellbore, the volume of the wellbore is

$$V_w = \frac{9{,}900\pi\left(\frac{2.5}{12}\right)^2 + 100\pi\left(\frac{3.5}{12}\right)^2}{5.615} = 245 \text{ bbl.} \tag{13-29}$$

From Equation 13-26, C is estimated for single phase wellbore fluid as

$$C = V_w c_w = 245(10^{-5}) = 0.00245, \tag{13-30}$$

which is clearly lower than computed from the pressure transient data. From Equation 13-26 for changing liquid level, the effective wellbore fluid density corresponding to the wellbore storage value computed from buildup data is 57 lb_m/ft^3, which is quite consistent to oil with viscosity 80.7 cp.

From Equation 13-25, the end of wellbore storage should occur at approximately

$$t_{ewbs} = \alpha_w(60 + 3.5s)\frac{\mu C}{kh} = 3387(60 + 3.5(0.87))\frac{(80.7)(0.15)}{(673)(79)} = 49 \text{ hr.} \tag{13-31}$$

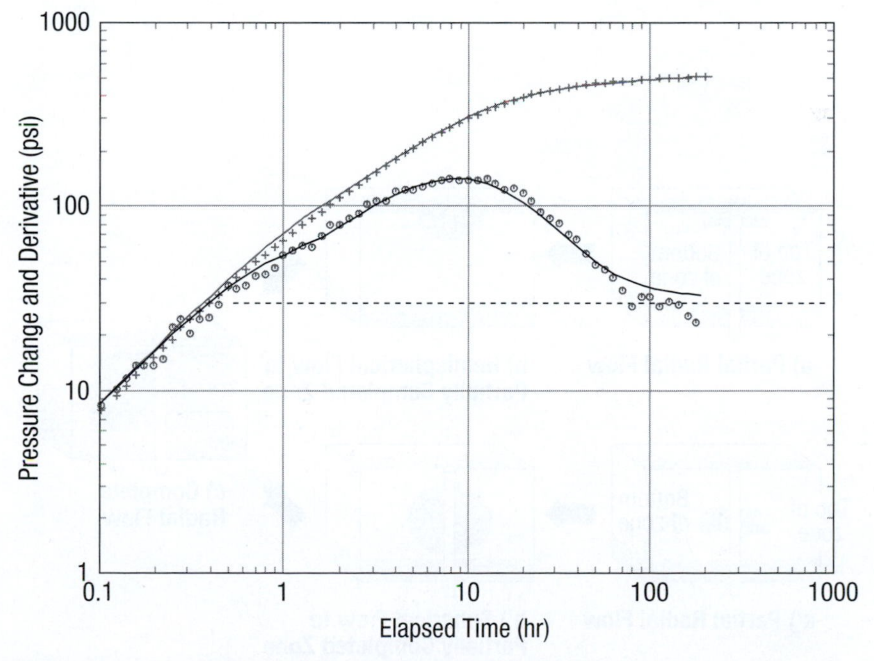

Figure 13-25 Model match for buildup data analyzed in Examples 13-11 and 13-12.

Figure 13-21 shows the radial flow starting after about 80 hours. A detailed match using a global model is shown in Figure 13-22. This match is achieved using variable wellbore storage. The value determined in Equation 13-28 corresponds to the initial wellbore storage. The initial wellbore storage in the model is 0.13, and the final value is 0.21, which corresponds to wellbore storage ending at about 70 hr. The need for variable wellbore storage to match the data explains why the changing liquid level form for Equation 13-26 provides a reasonable estimate for the wellbore storage coefficient. The permeability and skin for this match are 682 md and 0.86, very close to the values estimated from the flow regimes.

Matching pressure transient data with a global model accounting for all parameters provides more accurate parameter estimates than analysis by individual flow regimes. Typically the global model is found using a least-square regression calculation such as would be provided in commercial software for pressure transient test analysis. However, more often than not flow regime estimates are quite close and serve as good starting values for the least-square regression calculation.

13.4.4.3 Limited-entry Vertical Well

Figure 13-26 shows diagrams of successive flow regimes that may be observed in the buildup transient behavior of a limited-entry well. The partial and complete radial flow regimes were described previously in Figure 13-24a and Figure 13-24b, respectively. The centered partial radial flow shown in Figure 13-26a is conceptually the same as that shown in Figure 13-24a. The new

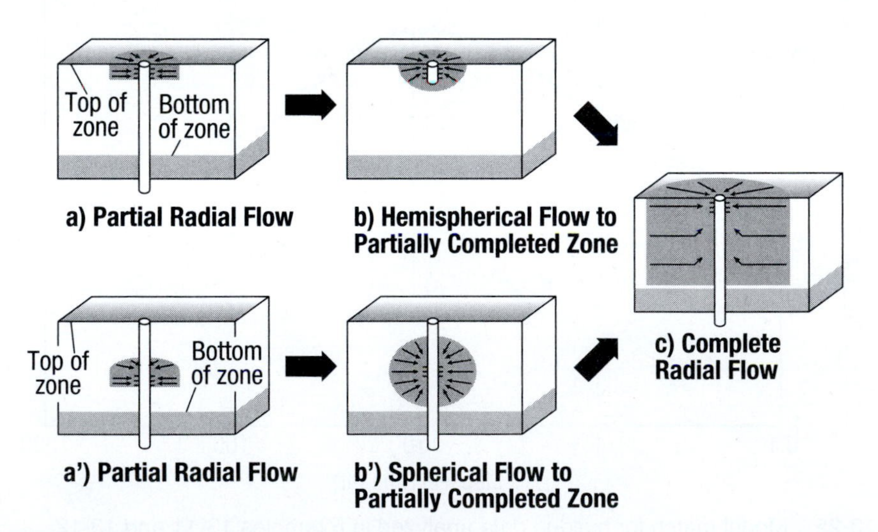

Figure 13-26 Flow regimes for a limited-entry well.

flow regimes are shown in Figure 13-26b as hemispherical flow and in Figure 13-26b as spherical flow. Both hemispherical and spherical flow regimes represent flow effectively to a point. These occur if the flowing interval is small compared to the total formation thickness.

This condition is often intentional, or it can occur unintentionally when severe formation or perforation damage prevents flow to the well along a portion of the productive interval. Limited entry is called partial completion when only a portion of the cased and cemented completion is perforated. This completion strategy is often used when oil is directly underlain by water; only the top portion of the well is perforated in an effort to try to delay or avoid water coning. If the oil is overlain by gas, the partial completion would be near the bottom of the productive interval, or if the oil is sandwiched between gas cap and aquifer, the perforations may be near the middle of the interval. The limited-entry geometry is called partial penetration when the borehole only penetrates a portion of the interval, as may happen in geothermal wells when the high reservoir temperature impairs the ability to drill through very hot rock.

Because the pressure change in radial flow is a function of the logarithm of time, the pressure derivative will be flat whenever this flow regime is dominant. In general,

$$\frac{kh_{eff}}{\mu} = \frac{\alpha_p qB}{2m'} \tag{13-32}$$

where h_{eff} is the effective length of the cylinder to which flow converges radially.

In contrast, streamlines for spherical and hemispherical flow converge approximately to a point. The equation for the pressure change during transient spherical flow is

$$\Delta p = \frac{\alpha_p qB\mu}{2k_{sph}}\left[\frac{1}{r_w} - \left(\frac{\phi\mu c_t}{\pi\alpha_t k_{sph}\Delta t}\right)^{1/2}\right] \text{(oil)}$$

$$\Delta m(p) = \frac{10.1\alpha_p qT_R}{2k_{sph}}\left[\frac{1}{r_w} - \left(\frac{\phi c_t}{\pi\alpha_t k_{sph}\,\mu\Delta t}\right)^{1/2}\right]. \tag{13-33}$$

Taking the derivative with respect to ln Δt gives

$$\Delta p' = -\frac{1}{2}\frac{\alpha_p qB\mu}{2k_{sph}}\left[-\left(\frac{\phi\mu c_t}{\pi\alpha_t k_{sph}\Delta t}\right)^{1/2}\right] = \frac{1}{2}\frac{\alpha_p qB\mu}{2k_{sph}}\left(\frac{\phi\mu c_t}{\pi\alpha_t k_{sph}\Delta t}\right)^{1/2} = \frac{1}{2}\frac{m_{pp}}{\sqrt{\Delta t}}. \tag{13-34}$$

Solving for the spherical permeability, k_{sph}, gives

$$k_{sph} = \sqrt[3]{k_r^2 k_z} = \sqrt[3]{k_x k_y k_z} = \left(\frac{\alpha_{pp} qB\mu}{m_{pp}}\sqrt{\phi\mu c_t}\right)^{2/3} \text{(oil)}$$

$$= \left(\frac{10.1\alpha_{pp} qT_R}{m_{pp}}\sqrt{\frac{\phi c_t}{\mu}}\right)^{2/3} \text{(gas)} \tag{13-35}$$

where k_x, k_y, and k_z are the principle permeability values in the permeability tensor. Generally, k_x and k_y are parallel to bedding, and k_z is perpendicular to bedding. Also, horizontal permeability, k_r, is the geometric mean of k_x and k_y given by $k_r = \sqrt{k_x k_y}$, and spherical flow, k_{sph}, is the geometric mean of k_x, k_y, and k_z. The conversion coefficient, α_{pp}, is 2453 for spherical flow and 867.3 for hemispherical flow (in oilfield units).

If pressure is graphed versus the reciprocal square root of time, the slope of the graph will be m_{pp}. However, m_{pp} can be determined directly from any point (Δt_{pp}, $\Delta p'_{pp}$) in the portion of the log–log diagnostic plot where the slope of the derivative trend is –1/2 by solving Equation 13-18 for m_{pp} as

$$m_{pp} = 2\sqrt{\Delta t_{pp}}\Delta p'_{pp}. \tag{13-36}$$

Pressure buildup data illustrating a limited-entry transient response are shown in Figure 13-27.

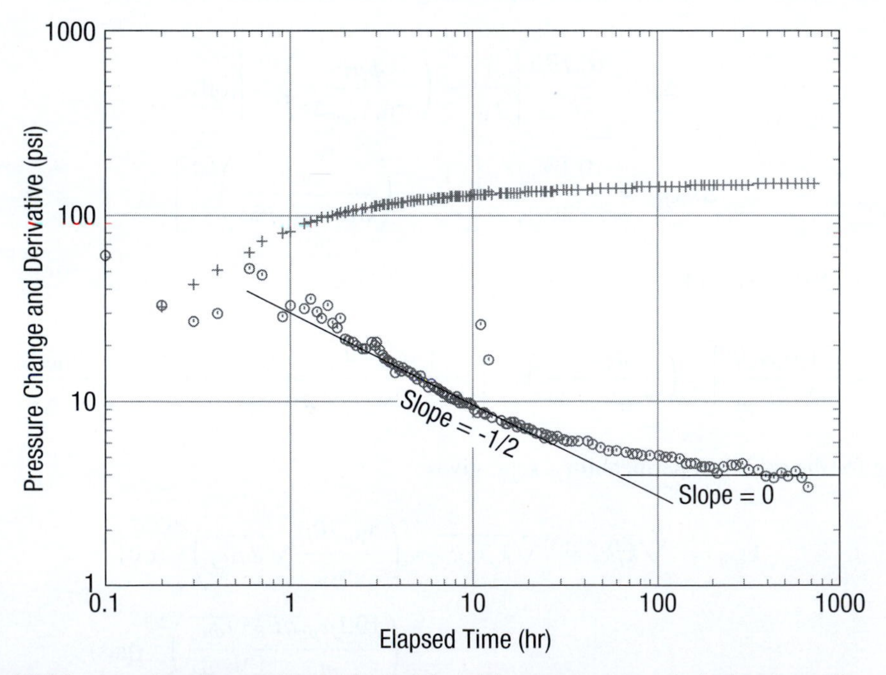

Figure 13-27 Log–log diagnostic plot for buildup data acquired in a limited-entry well.

Example 13-13 Buildup Analysis for a Limited-Entry Well

Pressure buildup data in Figure 13-25 are from a well producing oil with $B = 1.08$, $\mu = 2.7$ cp, and $c_t = 2.1 \cdot 10^{-5}$ psi^{-1}, from the top 50 ft of a formation with $h = 138$ ft and $\phi = 0.213$ and with $r_w = 0.29$ ft. Determine horizontal and vertical permeability and the total skin factor. From an estimate of the total limited-entry skin factor from Equation 6-20, determine the damage skin opposite the flowing interval. Cumulative production before the buildup is 77069 STB, and the flow rate and wellbore flowing pressure before shut-in are 921 STB/d and 1670.24 psi.

Solution

The constant value of the pressure derivative for elapsed time from 29 to 70 hr is 4 psi. From Equation 13-17, the bedding plane (horizontal) permeability estimate is

$$k = \frac{70.6 q B \mu}{m' h} = \frac{70.6(921)(1.08)(2.7)}{(4)(138)} = 343 \text{ md.} \tag{13-37}$$

At elapsed time $\Delta t = 70$ hr, where in Figure 13-22 the derivative is approximately constant, the pressure change $\Delta p_{IARF} = 150$ psi. From Equation 13-7 the material balance time before the buildup is given by

$$t_e = \frac{24Q}{q} = \frac{24(77069)}{(921)} = 2008 \text{ hr} \tag{13-38}$$

Using Equation 13-18, the total skin is

$$s = 1.151\left[\frac{\Delta p_{IARF}}{2.303m'} - \log\frac{k t_e}{1688\phi\mu c_t r_w^2} + \log\left(\frac{t_e + \Delta t_{IARF}}{\Delta t_{IARF}}\right)\right]$$

$$= 1.151\left[\frac{150}{2.303(4)} - \log\frac{(343)(2008)}{1688(0.213)(2.7)(2.1 \cdot 10^{-5})(0.29)^2}\right.$$

$$\left. + \log\left(\frac{2008 + 70}{70}\right)\right] = 10.5. \tag{13-39}$$

From $\Delta t = 1$ to 10 hr, the derivative trend is approximately straight with slope somewhat less than 1/2. With 50 ft flowing, the flow is not converging to a small enough interval to appear as rigorous spherical flow. Nonetheless, from Equation 13-36 for the time $\Delta t_{pp} = 3$ hr, $\Delta p'_{pp} = 6.4$ psi, the value for m_{pp} is determined as

$$m_{pp} = 2\sqrt{\Delta t_{pp}\Delta p'_{pp}} = 2\sqrt{3(6.4)} = 22. \tag{13-40}$$

From Equation 13-35

$$k_{sph} = \left(\frac{2453qB\mu}{m_{pp}}\sqrt{\phi\mu c_t}\right)^{2/3}$$

$$= \left(\frac{2453(921)(1.08)(2.7)}{22}\sqrt{(0.213)(2.7)(2.1\cdot 10^{-5})}\right)^{2/3} = 103 \text{ md.} \tag{13-41}$$

Since the horizontal permeability is known from Equation 13-37, vertical permeability is

$$k_z = \frac{k_{sph}^3}{k_r^2} = 9.3 \text{ md.} \tag{13-42}$$

From Equation 6-20, the total partial penetration skin is 10.6 (see Example 6-4) for $h_w = 50$ ft. The very small difference between this value and the total skin indicates that the damage skin opposite the interval open to flow is approximately 0. The smooth curves in Figure 13-28 show a global match for the data with very similar results to what are computed in this example from the radial and spherical flow regimes.

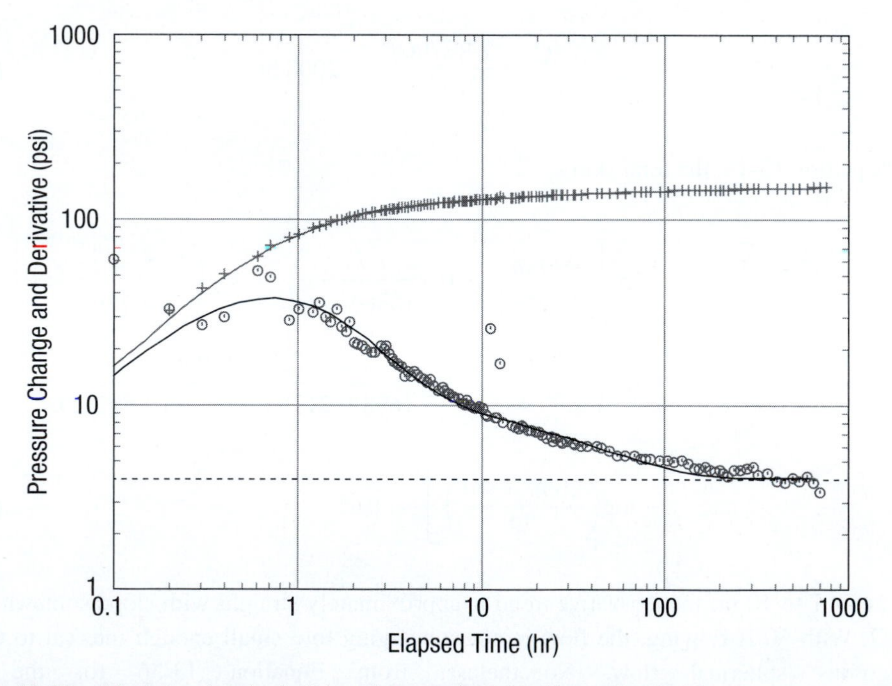

Figure 13-28 Model match for buildup data analyzed in Example 13-12.

The previous example illustrates fundamentally simple and profound ways to quantify values important to production engineers. More importantly, this example shows how to determine vertical permeability from a conventional buildup test. Because vertical permeability is critical to forecasting horizontal well performance, a test like that diagrammed in Figure 13-29

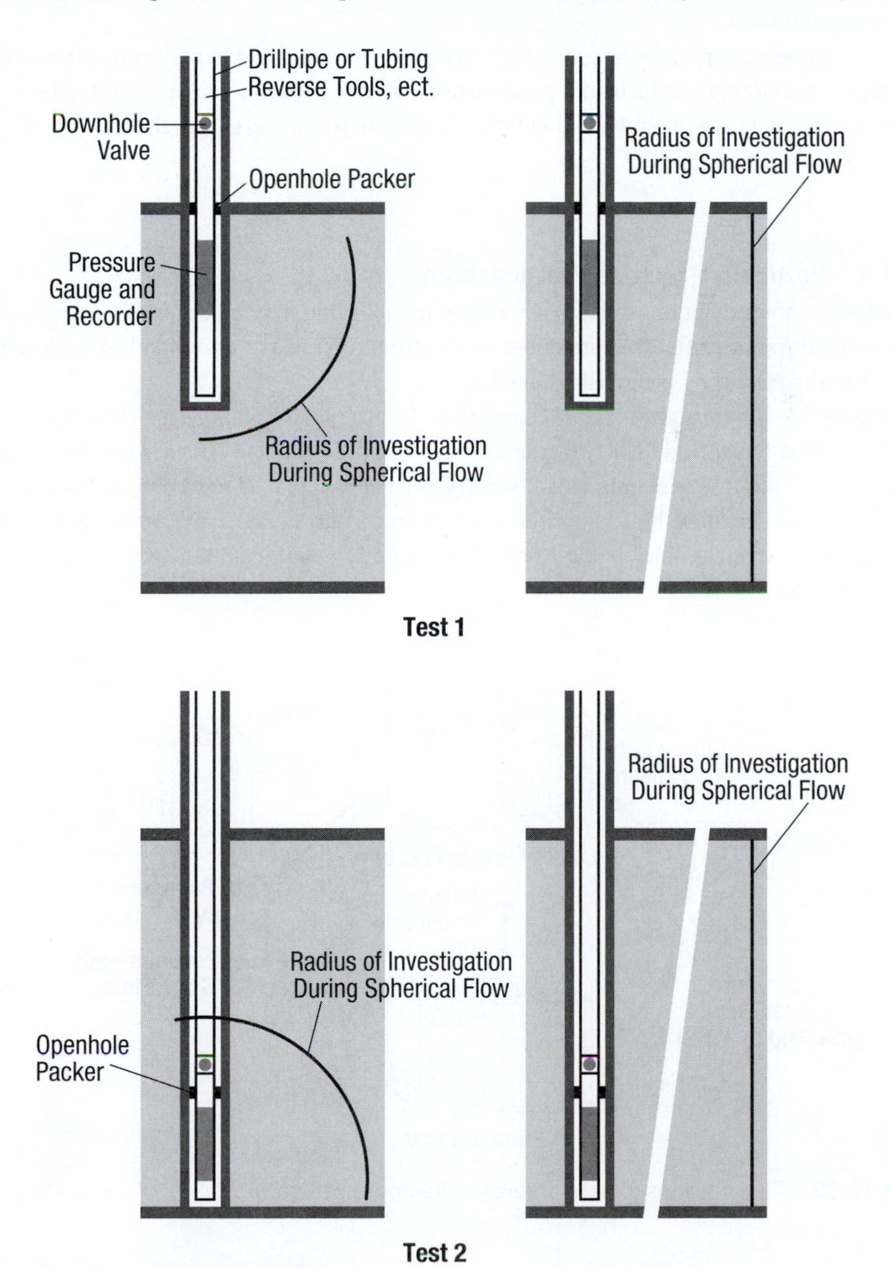

Figure 13-29 Limited-entry test design.

can be designed to determine horizontal and vertical permeability in a pilot hole drilled through the productive interval before drilling a horizontal well. Because spherical flow occurs very early in the pressure transient, downhole shut-in should be employed in the test execution. An open-hole log in the pilot hole will also provide formation porosity and thickness and the connate water saturation.

An alternative approach for estimating vertical permeability is the multiprobe formation test.

As more and more wells install permanent downhole pressure gauges, buildup tests are becoming increasingly common. Knowing how to estimate parameters from the log–log diagnostic plot is a skill worth mastering.

13.4.4.4 Vertically Fractured Vertical Well

Figure 13-30 shows diagrams of successive flow regimes that may be observed in a vertical well with a vertical fracture. The fracture could be a natural fracture, or more likely a hydraulic fracture. Hydraulic fracturing is described in Chapter 17.

Figure 13-30b illustrates pseudolinear flow to the fracture plane. This flow regime is also called formation linear flow. The flow regime is named pseudolinear because true linear flow occurs only when the flow streamlines are strictly normal to a planar source, and this only occurs when the vertical fracture half-length is equal to the width of the well drainage area. During pseudolinear flow, streamlines in the reservoir are strictly parallel, and the equation for transient pseudolinear flow is

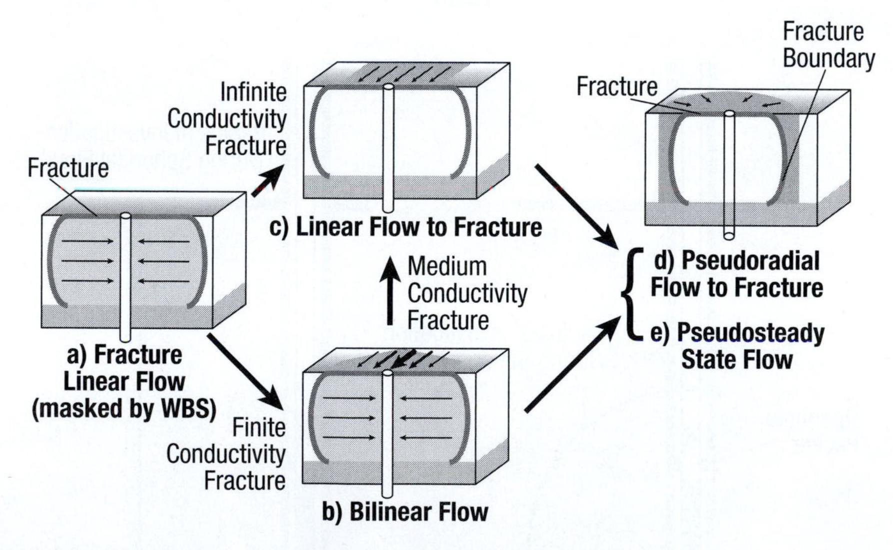

Figure 13-30 Flow regimes for a hydraulically fractured vertical well.

$$\Delta p = \frac{\alpha_p q B \mu}{kh}\left[\left(\frac{\pi \alpha_t k \Delta t}{\phi \mu c_t x_f^2}\right)^{1/2} + \left(\frac{\pi}{3}\right)\frac{k x_f}{k_f w}\right] \text{(oil)}$$

$$\Delta m(p) = \frac{10.1 \alpha_p q T_R}{kh}\left[\left(\frac{\pi \alpha_t k \Delta t}{\phi \mu c_t x_f^2}\right)^{1/2} + \left(\frac{\pi}{3}\right)\frac{k x_f}{k_f w}\right] \text{(gas).} \tag{13-43}$$

The second term in the brackets in Equation 13-43 is $\pi/3$ divided by the dimensionless fracture conductivity,

$$c_{fD} = \frac{k_f w}{k x_f} \tag{13-44}$$

where k_f is the permeability in the fracture, w is the fracture width, and x_f is the fracture half length. For effectively infinite conductivity fractures with $C_{fD} > 100$, this term can be neglected. In any case, the pressure derivative during linear flow is given by

$$\Delta p' = \frac{\alpha_p q B \mu}{2kh}\left(\frac{\pi \alpha_t k \Delta t}{\phi \mu c_t x_f^2}\right)^{1/2} = \frac{m_{lf}}{2}\sqrt{\Delta t} \tag{13-45}$$

and

$$x_f \sqrt{k} = \frac{\alpha_{lf} q B}{m_{lf} h}\left(\frac{\mu}{\phi c_t}\right)^{1/2} \text{(oil)}$$

$$x_f \sqrt{k} = \frac{10.1 \alpha_{lf} q T_R}{m_{lf} h}\left(\frac{1}{\phi \mu c_g}\right)^{1/2} \text{(gas)} \tag{13-46}$$

with α_{lf} equal to 4.064 in oilfield units.

Furthermore, under an assumption of negligible fracture skin, Equation 13-43 provides another analysis relationship, this time including the fracture conductivity, $k_f w$.

Figure 13-30a is a diagram of bilinear flow. This flow regime occurs for finite conductivity fractures in which the pressure gradient within the fracture plane is not insignificant compared to the pressure gradient between the reservoir and the fracture plane. During bilinear flow,

$$\Delta p = \frac{2.45 \alpha_p q B \mu}{kh\sqrt{k_f w / k x_f}}\left(\frac{\alpha_t k \Delta t}{\phi \mu c_t x_f^2}\right)^{1/4}, \tag{13-47}$$

the pressure derivative trend is

$$\Delta p' = \frac{2.45\alpha_p qB\mu}{4kh\sqrt{k_f w/kx_f}}\left(\frac{\alpha_t k\Delta t}{\phi\mu c_t x_f^2}\right)^{1/4} = \frac{m_{bf}}{4}\sqrt[4]{\Delta t}, \tag{13-48}$$

and

$$k_f w\sqrt{k} = \left(\frac{\alpha_{bf} qB}{m_{bf}h}\right)^2\left(\frac{\mu^3}{\phi c_t}\right)^{1/2} \text{(oil)}$$

$$k_f w\sqrt{k} = \left(\frac{10.1\alpha_{bf} qT_R}{m_{bf}h}\right)^2\left(\frac{1}{\phi\overline{\mu c}_g}\right)^{1/2} \text{(gas)} \tag{13-49}$$

with $\alpha_{bf} = 44.1$ for oilfield units. For moderate conductivity fractures, bilinear flow may appear before (but never after) linear flow.

After bilinear or pseudolinear flow, pseudoradial flow may occur, as diagrammed in Figure 13-30c. In this context, pseudoradial flow is cylindrical flow to an effective wellbore radius given by

$$r_w'' = r_w e^{-s_f} \tag{13-50}$$

where r_w'' is the effective radius accounting only for the geometric skin of the fracture. For an effectively infinite conductivity fracture, $r_w'' = x_f/2$. For an undamaged fracture without choke or fracture face skin, $r_w'' = r_w$.

When $x_f < r_e/10$, the time, t_{eplf}, at the end of pseudolinear flow to an effectively infinite conductivity fracture is given by

$$t_{eplf-d} = \frac{0.016\phi\mu c_t x_f^2}{\alpha_t k}. \tag{13-51}$$

Equation 13-51 applies when the derivative bends downward from the $^1/_2$-slope trend and should only be used on drawdown data, or virtual drawdown data like the RNP, because buildup data can be distorted in late time by superposition effects.

The time, t_{spr}, at the start of pseudoradial flow appears approximately at the time

$$t_{spr} = \frac{25\phi\mu c_t r_w''^2}{\alpha_t k}. \tag{13-52}$$

Pseudoradial flow appears only when the effective wellbore radius is less than 1/10 of the well effective drainage radius. Equation 13-17 applies for pseudoradial flow; and permeability, skin, and the reservoir pressure can be determined when this flow regime is present in the transient response. For hydraulically fractured wells, typically the skin is less than zero, and the effective wellbore radius is larger than the actual well radius. For effectively infinite conductivity fractures, the effective wellbore radius approaches half the fracture half length.

After sufficient time on production, the well will reach either steady-state or pseudosteady-state flow. For unconventional reservoirs with very low permeability, the time to reach pseudoradial, pseudo-steady, or steady-state flow may be very long. If the well spacing is not sufficiently small, the well may remain in the fractured dominated pseudolinear or bilinear flow regime for its entire life. For a fractured well in a very low permeability reservoir, pseudolinear, or bilinear flow can last for months or even decades.

In constant rate drawdown or RNP data if the derivative bends upward from the $^1/_2$-slope pseudolinear flow trend or from the $^1/_4$-slope bilinear flow trend, this signals that the pressure has penetrated the distance to the nearest drainage area boundary. This transition will occur approximately when

$$t_{eplf-u} = \frac{\phi \mu c_t x_{el}^2}{4\alpha_t k}. \tag{13-53}$$

In Equation 13-53, x_{el} is the distance in the direction perpendicular to the fracture face from the fracture to the nearest of two parallel rectangular drainage area boundaries between which the fracture is centered.

During pseudosteady-state flow, the RNP and IRNP follow the behavior of a drawdown during constant rate production given by

$$\Delta p = \frac{2\pi \alpha_p \alpha_t q B}{\phi c_t h A} t + p_{int} \tag{13-54}$$

with $q = q_{\text{norm}}$, and p_{int} is the intersection on a Cartesian graph of pressure change (RNP or IRNP) versus time. The derivative is given by

$$\Delta p' = \frac{\alpha_{pps} q B}{\phi c_t h A} t = m_{pss} t \text{ (oil)}$$

$$\Delta p' = \frac{10.1 \alpha_{pps} q T_R}{\phi \mu c_t h A} t = m_{pss} t \text{ (gas)}. \tag{13-55}$$

The coefficient $\alpha_{pps} = 0.234$ is in oilfield units.

Figure 13-22 shows production data for a hydraulically fractured oil well, and Figure 13-23 shows an IRNP plot of the production data.

Example 13-14 Production Data Analysis for a High-Conductivity Fractured Oil Well

The diagnostic plot of integral IRNP and its derivative in Figure 13-23 is from a hydraulically fractured oil well with $B = 1.16$, $\mu = 0.8$ cp, $c_t = 1.57 \cdot 10^{-5}$ psi^{-1}, $h = 120$ ft, and $\phi = 0.12$, and with $r_w = 0.32$ ft. The reference rate for the diagnostic plot is 0.001 m^3/s = 542 STB/d. What can be quantified from the flow regimes observed in these data?

Solution

Two flow regimes dominate the data. In late time, starting at 160,000 hr, the IRNP derivative trend is a unit slope indicating pseudosteady-state production. In early time until about 25,000 hr, the predominant IRNP derivative slope is $\frac{1}{2}$. There is no evidence of a leveling in the derivative that could represent pseudoradial flow.

The absence of pseudoradial flow means that permeability cannot be determined in the manner used for Examples 13-11 and 13-13. Instead, we can determine the drainage area from the pseudosteady-state flow regime and assume that the lack of pseudoradial flow is explained by a fully penetrating hydraulic fracture. Then permeability can be estimated from the pseudolinear trend.

Selecting the point I-RNP′ = 36,000 for $t = 280{,}000$ hr from the IRNP graph in Figure 13-21, rearranging Equation 13-55 gives

$$m_{pss} = \frac{\Delta p'}{\Delta t} = \frac{IRNP'}{t_e} = \frac{36{,}000}{280{,}000} = 0.129 \text{ psi/hr.} \tag{13-56}$$

and solving Equation 13-55 for A,

$$A = \frac{\alpha_{pps} q B}{\phi c_t h m_{pss}} = \frac{0.234(542)(1.16)}{(0.12)(1.57 \cdot 10^{-5})(120)(0.129)} = 5.0 \cdot 10^6 \text{ ft}^2. \tag{13-57}$$

Assuming the fracture is centered in a square drainage area, the length of each side of the square would be the square root of $5.0 \cdot 10^6$, or 2236 ft. Further, the half-length of a fully penetrating hydraulic fracture would be

$$x_f = 2236/2 = 1118 \text{ ft.} \tag{13-58}$$

Having found an estimate for the hydraulic fracture half-length, solving Equation 13-46 for k enables an estimate for the fracture half length. Selecting IRNP′ = 3100 psi at $t_e = 10{,}000$ hr, Equation 13-45 gives

$$m_{lf} = \frac{2IRNP'}{\sqrt{t}} = \frac{2(3100)}{\sqrt{10{,}000}} = 62, \tag{13-59}$$

and

$$k = \left(\frac{\alpha_{lf} q B}{m_{lf} h x_f}\right)^2 \left(\frac{\mu}{\phi c_t}\right) = \left(\frac{4.064(542)(1.16)}{(62)(120)(1118)}\right)\left(\frac{0.8}{0.12(1.57 \cdot 10^{-5})}\right) = 0.0401 \text{ md.} \quad \textbf{(13-60)}$$

The global model represented by the square drainage area with permeability and fracture half-length as computed above is shown in Figure 13-31a. A positive skin value was introduced to improve the match with observed cumulative production. While not significantly off, we expect a better match with IRNP and its derivative.

The explanation for the apparent skin could be finite fracture conductivity. Noting the IRNP value of 8722 for $t_e = 10{,}000$ hr, Equation 13-43 can be solved for fracture conductivity as follows:

$$k_f w = \frac{\dfrac{\alpha_p q B \mu}{h}\left(\dfrac{\pi}{3}\right) x_f}{IRNP - m_{lf}\sqrt{t_e}} = \frac{\dfrac{141.2(542)(1.16)(0.8)}{120}\left(\dfrac{\pi}{3}\right)(1118)}{8722 - 62(100)} = 275 \text{ md-ft.} \quad \textbf{(13-61)}$$

However, if the permeability is 0.0401 md, Equation 13-53 estimates that the distance to the nearest boundary parallel to the fracture is

$$x_{el} = \left(\frac{k t_{eplf-u}}{948 \phi \mu c_t}\right)^{1/2} = \left(\frac{.0401(25{,}000)}{948(0.12)(0.8)(1.57 \cdot 10^{-5})}\right)^{1/2} = 838 \text{ ft.} \quad \textbf{(13-62)}$$

If the fracture is centered between boundaries each at a distance of 838 ft, the width of the drainage area must be $2(838) = 1676$ ft, and its length must be $5.0 \cdot 10^6/(1676) = 2983$ ft. The global model for this case is also shown in Figure 13-31b. The second match uses a finite conductivity fracture and zero skin. This match is better with the IRNP, but the match with cumulative production is poor.

The final global model match shown in Figure 13-31c has permeability 0.03 md and fracture half-length 1250 ft and conductivity 275 md-ft (resulting in dimensionless conductivity of about 20) centered in an elongated drainage area with the longer side length, $x_e = 2720$ ft, 1.4 times the shorter side length, $y_e = 1840$ ft, and with the fracture aligned parallel to the longer side. This implies a very effective fracture treatment. The unified fracture design approach described in Chapter 17 can be used to show that this fracture provides very nearly maximum well productivity.

This example shows that analysis without pseudoradial flow is challenging. Also, without external knowledge of the likely well drainage area shape, the final global match is nonunique. However, the possible ranges of values for permeability, fracture half length, and drainage area dimensions that will provide a match for the data are not large. Table 13-4 summarizes the model parameters for the three global model matches.

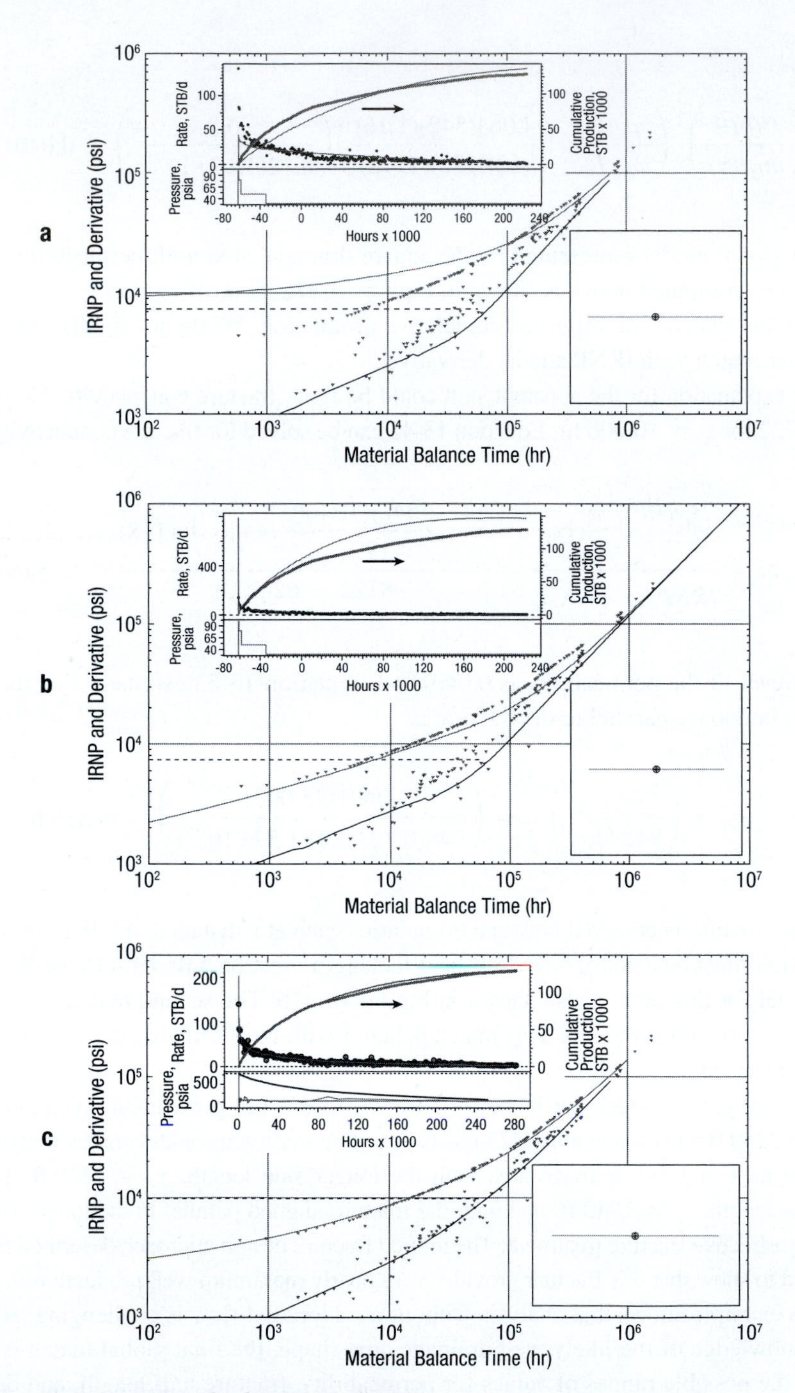

Figure 13-31 Model matches for fractured oil well.

Table 13-4 Parameters for Model Matches in Example 13-14

Parameter	(a)	(b)	(c)
Permeability, k (md)	0.0401	0.0401	0.03
Fracture half length, x_f (ft)	1120	1120	1250
Fracture conductivity, $k_f w$ (md-ft)	44,000	275	275
Fracture skin	0.6	0	0
Dimensionless conductivity, C_{fD}	2720	6.14	7.33
Drainage length (ft)	2240	2980	2720
Drainage width (ft)	2240	1676	1844
Oil in place, MMSTB	11.1	11.1	11.1

The previous example illustrates that a great deal can be learned from extended production data. No pressure buildup tests were reported on that well during the entire production history. Figure 13-32 shows production data for a hydraulically fractured gas well, this time including several pressure buildup flow periods. Figure 13-33 shows the IRNP and derivative for the production data with lines indicating possible flow regime trends. A log–log diagnostic plot for the longest buildup period is shown in Figure 13-34. The following example shows that for this well even a 324-hour buildup is dominated completely by the fracture behavior and does not provide a unique value for the reservoir permeability.

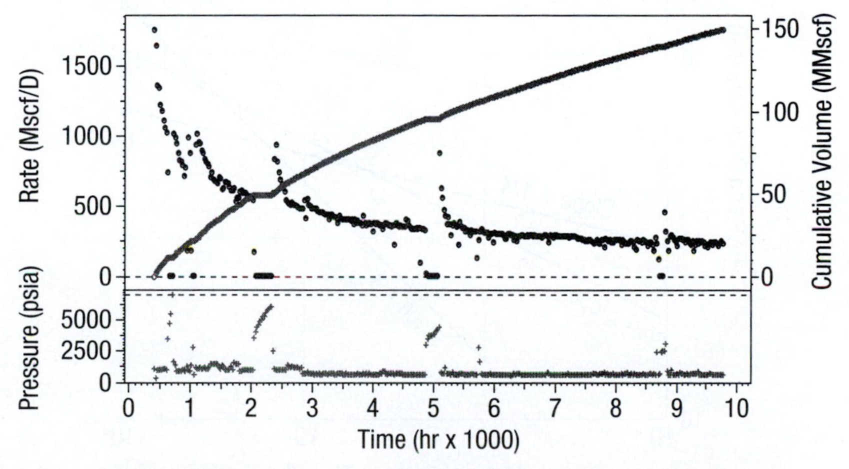

Figure 13-32 Production rate and pressure data for Example 13-15 and 13-16.

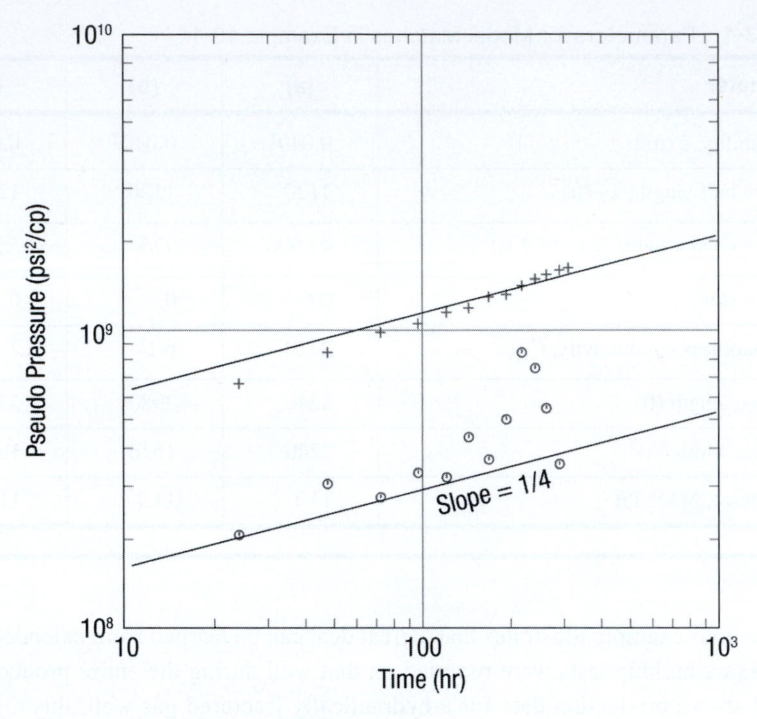

Figure 13-33 Pressure buildup data for Example 13-15.

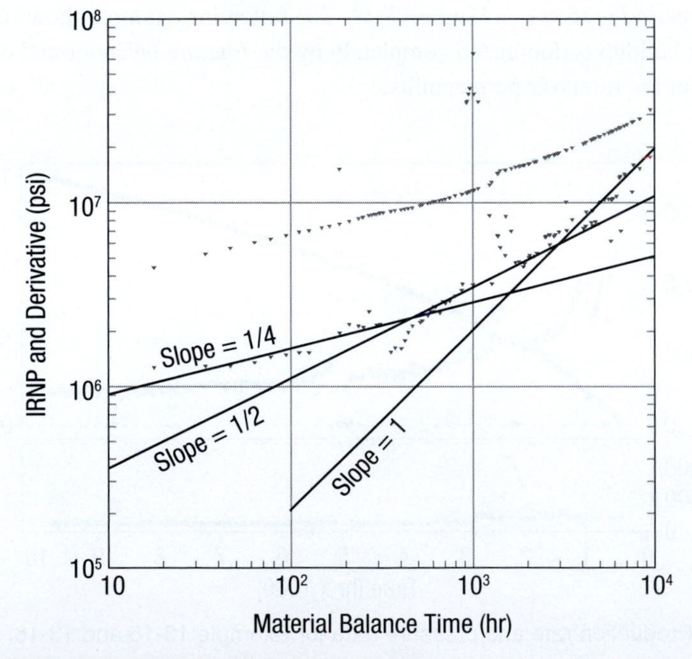

Figure 13-34 IRNP diagnostic plot for Example 13-16.

Example 13-15 Buildup Analysis for a Low-Conductivity Fractured Gas Well

The buildup test data in Figure 13-33 is from a hydraulically fractured gas well with initial reservoir pressure 9590 psi and reservoir temperature 300°F, h = 25 ft and $\phi = 0.09$, $S_w = 0.12$, and with $r_w = 0.333$. At these conditions $\mu = 0.036$ cp $c_g = 4.5 \cdot 10^{-5} \text{psi}^{-1}$, and $Z = 1.39$. Cumulative production before the buildup test was 40 BCF, and the flow rate and wellbore flowing pressure before shut-in were 548 MSCF/d and 975.6 psi. What can be quantified from these data?

Solution

As in Example 13-14, the data shown in Figure 13-33 show no evidence of pseudoradial flow. In fact, for the entire buildup, the pressure derivative has a straight trend with $^1/_4$ slope corresponding to bilinear flow. For the time $\Delta t_{bf} = 100$ hr, the pseudopressure change is $\Delta p_{bf} = 3.5 \cdot 10^8$ psi^2. From Equation 13-48, the value for m_{bf} is determined as

$$m_{bf} = \frac{4\Delta p'}{\sqrt[4]{\Delta t}} = \frac{4(3.5 \times 10^8)}{\sqrt[4]{100}} = 4.43 \times 10^8. \tag{13-63}$$

From Equation 13-49,

$$\begin{aligned} k_f w\sqrt{k} &= \left(\frac{10.1\alpha_{bf} q T_R}{m_{bf} h}\right)^2 \left(\frac{1}{\phi\overline{\mu c}_g}\right)^{1/2} \\ &= \left(\frac{10.1(44.1)(548)(760)}{(4.43 \times 10^8)(25)}\right)^2 \left(\frac{1}{(0.09)(0.036)(4.5 \times 10^{-5})}\right)^{1/2} \\ &= 0.736 \text{ md}^{3/2}\text{ - ft.} \end{aligned} \tag{13-64}$$

Without the appearance of pseudoradial flow, no individual parameters can be determined. If permeability were known from a pretreatment test, the fracture conductivity could be determined.

Because the maximum derivative value in the data is about $4 \cdot 10^8$ psi^2, the level, m', for pseudoradial flow cannot be less than this value. Therefore, the reservoir permeability cannot be more than

$$k < \frac{711 q T_R}{m' h} = \frac{711(548)(760)}{(4 \times 10^8)(25)} = 0.03 \text{ md.} \tag{13-65}$$

In turn,

$$k_f w > \frac{0.736}{\sqrt{0.03}} = 4.2 \text{ md-ft.} \tag{13-66}$$

In Example 13-15, only wellbore- and fracture-dominated behaviors appear even for a buildup lasting 324 hours (more than one week). For these wells, the use of long-term production data may provide greater information than any buildup test. The next example shows what is revealed in the long-term production data for the same well as in Example 13-15.

Example 13-16 Production Data Analysis for a Low-Conductivity Fractured Well

Long-term production (bottomhole pressure and surface flow rate) data for the well in Example 13-15 are graphed as IRNP in Figure 13-34. The reference rate for this analysis is 0.001 m^3/s = 3.05 MSCF/d. The data have been recorded while the well was on production for 1 year. What can be quantified from these data?

Solution

The IRNP and derivative mimic the $^1/_4$ slope behavior observed in the buildup data analyzed in Example 13-15 for the first 300 hr. Selecting IRNP′ $= 1.85 \cdot 10^6$ psi^2/cp for $t_e = 100$ hr, Equation 13-48 gives

$$m_{bf} = \frac{4IRNP'}{\sqrt[4]{t_e}} = \frac{4(1.85 \times 10^6)}{\sqrt[4]{100}} = 2.34 \times 10^6. \tag{13-67}$$

From Equation 13-49,

$$\begin{aligned} k_f w\sqrt{k} &= \left(\frac{10.1\alpha_{bf}qT_R}{m_{bf}h}\right)^2\left(\frac{1}{\phi\mu c_g}\right)^{1/2} \\ &= \left(\frac{10.1(44.1)(3.05)(760)}{(2.34 \times 10^6)(25)}\right)^2\left(\frac{1}{(0.09)(0.036)(4.5 \times 10^{-5})}\right)^{1/2} \\ &= 0.816 \text{ md}^{3/2}\text{-ft}. \end{aligned} \tag{13-68}$$

This result is similar to what was determined from the buildup analysis. After that the IRNP and derivative exhibit linear flow with a characteristic $^1/_2$ slope trend. The $^1/_2$ slope trend could represent pseudolinear flow to a moderate conductivity hydraulic fracture or parallel sealing boundaries. For a moderate conductivity hydraulic fracture and selecting IRNP′ $= 3.4 \cdot 10^6$ psi^2/cp for $t_e = 1000$ hr, Equation 13-45 gives

$$m_{lf} = \frac{2IRNP'}{\sqrt{t_e}} = \frac{2(3.4 \times 10^6)}{\sqrt{1000}} = 2.15 \cdot 10^5. \qquad \textbf{(13-69)}$$

In turn, from Equation 13-46

$$x_f\sqrt{k} = \frac{10.1\alpha_{lf}qT_R}{m_{lf}h}\left(\frac{1}{\phi\mu c_g}\right)^{1/2}$$

$$= \frac{10.1(4.034)(3.05)(760)}{2.15 \times 10^5(25)}\left(\frac{1}{0.09(0.036)(4.5 \times 10^{-5}}\right)^{1/2} = 46 \text{ md}^{1/2}\text{- ft.} \qquad \textbf{(13-70)}$$

Because the maximum derivative value in the data is about $2 \cdot 10^7$ psi^2/cp, the level, m', for pseudoradial flow cannot be less than this value. Therefore, the reservoir permeability cannot be more than

$$k < \frac{711qT_R}{m'h} = \frac{711(3.05)(760)}{(2 \times 10^7)(25)} = 0.0033 \text{ md.} \qquad \textbf{(13-71)}$$

In turn, as in Equation 13-64, $k_f w > 0.816/\sqrt{0.0033} = 14.2$ md-ft, and $x_f > 46/\sqrt{0.0033} = 801$ ft. Furthermore, if the unit slope line marked in Figure 13-34 represents pseudosteady-state flow, then from Equation 13-53 and using time 5000 hr for RNPI $= 7 \cdot 10^7$ psi^2/cp, the drainage area is at least

$$A = \frac{10.1\alpha_{pps}qT_R}{\phi\mu c_t h m_{pss}} = \frac{10.1(0.234)(3.05)(760)}{0.09(0.036)(4.5 \times 10^{-5})(25)(7 \times 10^7/5000)}$$

$$= 1.07 \times 10^5 \text{ ft}^2. \qquad \textbf{(13-72)}$$

The global model match shown in Figure 13-35 provides an excellent match for rate and cumulative production that does not include a drainage boundary. The parameters for this match are provided in Table 13-5.

Further work on this example is left as a problem at the end of the chapter.

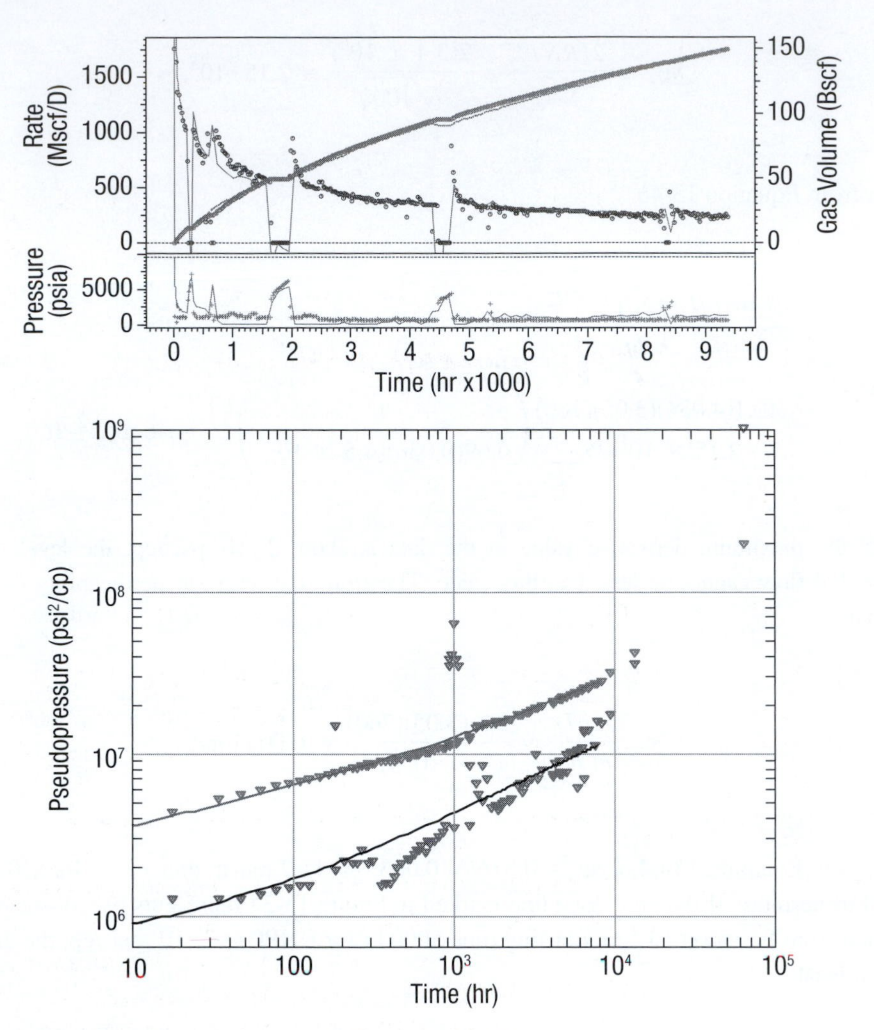

Figure 13-35 Global match for data in Example 13-16.

Table 13-5 Parameters for Model Match in Example 13-16

Parameter	
Permeability, k (md)	0.000974
Fracture half length, x_f (ft)	1300
Fracture conductivity, $k_f w$ (md-ft)	35
Fracture skin	0
Dimensionless conductivity, C_{fD}	28

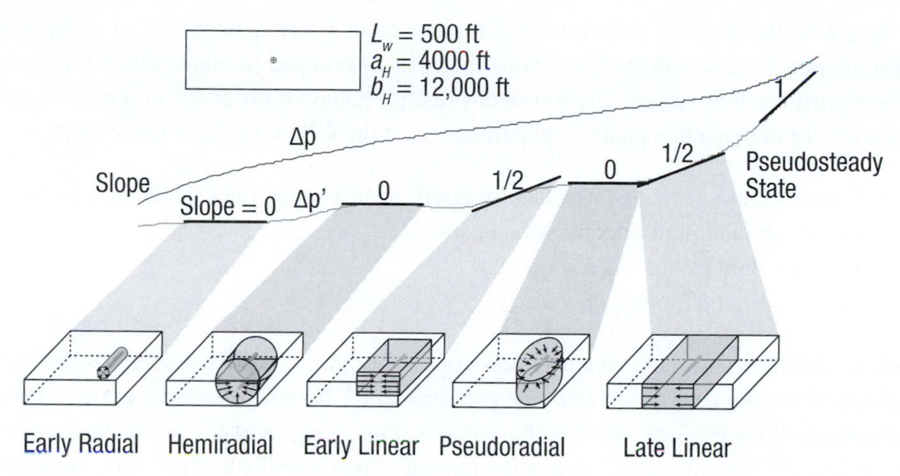

Figure 13-36 Flow regimes in horizontal wells.

13.4.4.5 Horizontal Well

Many flow regimes may appear in horizontal wells. Figure 13-36 shows diagrams of successive flow regimes that may be observed in a horizontal well. The number of unknown parameters for horizontal well analysis include horizontal permeability values k_x and k_y, vertical permeability k_z, the effective flowing length of the well, the well standoff (distance from water contact if an aquifer is below the well and/or from a gas cap if gas is above the well), formation thickness (because often the well is drilled without first a pilot hole through the entire formation thickness), and distances to drainage boundaries.

Like hydraulically fractured wells, horizontal wells can dominate the pressure transient behavior for too long a time for pressure buildup analysis, and production data may enable quantification of many more parameters of interest. As well, either a test diagrammed in Figure 13-29 or a multiprobe formation test can quantify vertical and horizontal geometric mean permeability values and the formation thickness, and would enable more reliable and complete analysis of production data or of a subsequent pressure buildup test in the horizontal well.

The Lee et al. (2003) textbook provides an entire chapter on horizontal well transient interpretation and provides flow regime equations that enable analysis analogous to what has been presented in this chapter. Because horizontal well interpretation is quite advanced, we leave it to the reader to consult this reference for more detail.

13.4.4.6 Fracture Calibration Injection Falloff

In low-permeability formations, it is often impossible to determine permeability from a conventional pretreatment buildup test because flow to the well cannot occur without hydraulic fracture stimulation. However, Chapter 17 demonstrates the importance of permeability to hydraulic fracture design. Furthermore, the examples in the previous section show how challenging it can be to determine permeability after the hydraulic fracture treatment and how long it can take to determine permeability from production data. The fracture calibration injection falloff test, commercially called a mini-frac or a diagnostic fracture injection test (DFIT), offers a way to estimate the formation permeability and pressure before the main hydraulic fracture treatment, even in very low-permeability formations.

Injecting the fracture fluid intended for the main fracture treatment at a pressure high enough to open a fracture in the formation creates an unpropped fracture. When injection stops, the unpropped fracture closes. The transient pressure behavior observed in the pressure falloff after the end of the injection enables quantification of the following three parameters:

- Formation closure stress, which corresponds to an average value along the lithological profile of the minimum principle in-situ stress
- Fracture fluid leakoff coefficient, C_L
- Fracture fluid efficiency, η

Chapter 17 indicates how these parameters are used to design the main hydraulic fracture treatment. These parameters are determined from the pressure falloff behavior observed up until the created fracture closes. If the pressure falloff is allowed to continue long enough after the fracture closure is observed, it is possible also to determine the formation permeability and pressure. Analysis of the falloff behavior observed after the fracture closes is called after-closure analysis (ACA).

Barree, Barree, and Craig (2009) summarized ACA approaches from various other authors. However, Mohamed, Nasralla, Sayed, Marongiu-Porcu, and Ehlig-Economides (2011) suggested using the previously described log–log diagnostic graph of pressure and derivative for this analysis, and Marongiu-Porcu, Ehlig-Economides, and Economides (2011) provided equations for estimating parameters directly from the diagnostic plot and developed a global model for the entire pressure falloff response.

Figure 13-37 shows a diagram of the usual sequence of events observed in a fracture calibration test. Injection of the same type of fluid to be used for the main treatment pressurizes the formation until the rock breakdown is achieved. After breakdown the fracture propagates until the pumps are shut down. At this point the pressure falls off. An instantaneous decrease in pressure to the value labeled as ISIP (instantaneous shut-in pressure) is due to friction losses. The instantaneous pressure drop is generally quite large if pressure is recorded at the wellhead, but when bottomhole pressure is recorded, the instantaneous pressure drop is mainly due to pressure losses in the near wellbore area.

To create the log–log diagnostic plot, the pressure change is calculated as $\Delta p = \text{ISIP} - p_{ws}(\Delta t)$. The pressure derivative is computed using the Horner time function to represent superposition and using material balance time for the time on injection. The log–log diagnostic plot for a fracture calibration test in a well from the Cotton Valley sandstone formation is shown in Figure 13-36.

The main flow regime before closure is dominated by the behavior described by the Nolte (1979) *g-function,* $g(\Delta t_D, \alpha)$. Valkó and Economides (1995) provided an analytical expression for the dimensionless *g-function,* $g(\Delta t, \alpha)$ for Nolte's well-known power law fracture surface growth assumption. While the actual analytical expression is quite complex, Nolte indicated an upper-bound limiting form for fluid efficiency $\approx 100\%$ as

$$g(\Delta t_D, \alpha = 1) = \frac{4}{3}[(1 + \Delta t_D)^{\frac{3}{2}} - \Delta t_D{}^{\frac{3}{2}}]. \tag{13-73}$$

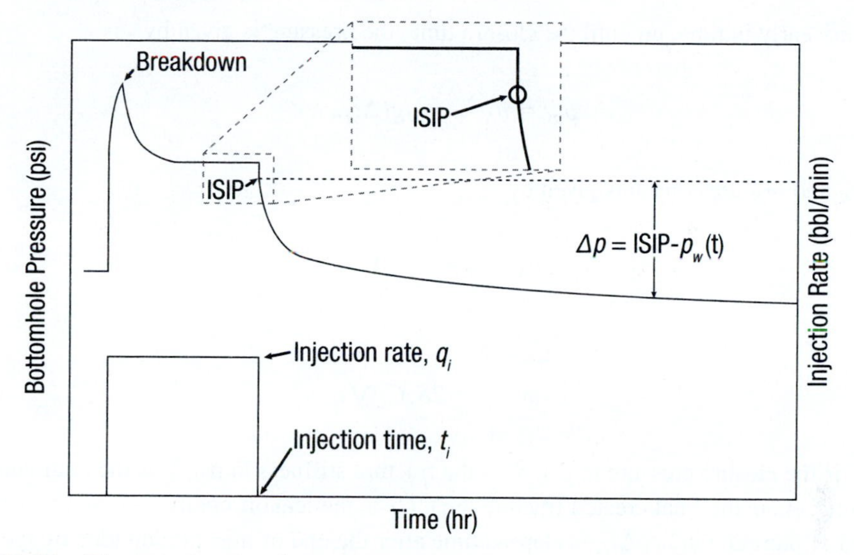

Figure 13-37 Diagram showing the sequence of events observed in a fracture calibration test.

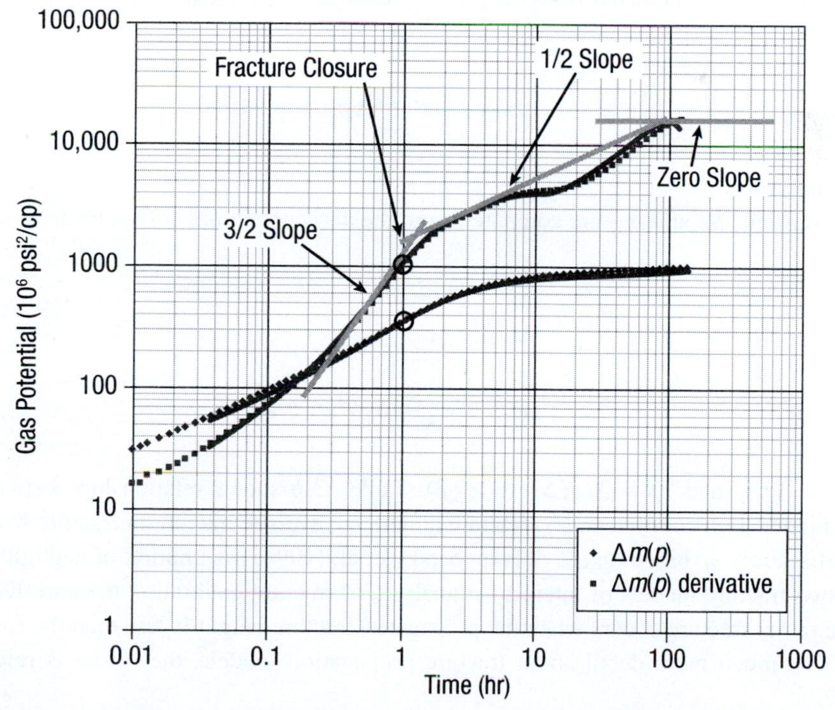

Figure 13-38 Log–log diagnostic plot for a fracture calibration test.

Except very early in time, up until the closure time, the pressure is given by

$$p_w = b_N + m_N g(\Delta t_D, \alpha). \tag{13-74}$$

Where b_N and m_N are constants given by

$$b_N = p_C + S_f V_i / A_e \tag{13-75}$$

and

$$m_N = -2 S_f C_L \sqrt{t_e} \tag{13-76}$$

where p_c is the closure pressure in psi, S_f is the fracture stiffness in psi, V_i is the total volume injected in ft^3, A_e is the final created fracture area, C_L is the leakoff coefficient, and t_e is the time at the end of pumping in hr; Δt_D is elapsed time after the end of injection divided by the time at the end of injection, t_e, and α is the fracture propagation model-dependent power law exponent. Therefore, the presence of a time exponent of 3/2 in Equation 13-73 reveals that the pressure derivative calculated with respect to the natural logarithm of the superposition time will exhibit a 3/2 slope. Therefore, before closure behavior is easily recognized on the log–log diagnostic plot, the closure pressure is calculated from the pressure change observed at the end of the 3/2 slope trend as

$$p_c(t_c) = ISIP - \Delta p(t_c) \tag{13-77}$$

for the closure time, t_c.

The constants b_N and m_N are computed from the derivative point at the closure time as

$$m_N = \frac{\Delta p'}{2 \Delta t_{D_c}^{5/2} \tau_c (1 - \tau_c^{1/2})} \tag{13-78}$$

$$b_N = p_c - m_N \frac{4}{3} \Delta t_{D_c}^{\frac{3}{2}} (\tau_c^{\frac{3}{2}} - 1) \tag{13-79}$$

with $\Delta t_{D_c} = \Delta t_c / t_e$ and $\tau_c = \Delta t_c / (\Delta t_c + t_{e,adj})$. Table 13-6 shows relationships between m_N and b_N and the leakoff coefficient, C_L, fracture height, h_f, average maximum fracture width, $\overline{w}_e$, and fluid efficiency, η, based on the Shlyapobersky et al. (1998) assumption of negligible spurt loss. The two fracture models of interest for analysis of fracture calibration tests are KGD and radial. The radial fracture occurs when the propagated fracture height is less than the formation thickness. For much more detail about fracture propagation models, the reader is referred to Chapter 17.

Table 13-6 Fracture Calibration Analysis Models

Fracture Model	Radial	PKN	KGD
α	8/9	4/5	2/3
Fracture height	$R_f = \sqrt[3]{\dfrac{3E'V_i}{8(b_N - p_c)}}$	$h_f = \sqrt{\dfrac{2E'V_i}{\pi x_f(b_N - p_c)}}$	$h_f = \dfrac{E'V_i}{\pi x_f^2(b_N - p_c)}$
Leakoff coefficient	$C_L = 3\pi \dfrac{8R_f}{4\sqrt{t_e E'}}(-m_N)$	$C_L = \dfrac{\pi h_f}{4\sqrt{t_e E'}}(-m_N)$	$C_L = \dfrac{\pi h_f}{2\sqrt{t_e E'}}(-m_N)$
Fracture width	$\overline{w}_e = \dfrac{V_i}{R_f^2 \dfrac{\pi}{2}} - 2.754 C_L \sqrt{t_e}$	$\overline{w}_e = \dfrac{V_i}{x_f h_f} - 2.830 C_L \sqrt{t_e}$	$\overline{w}_e = \dfrac{V_i}{x_f h_f} - 2.956 C_L \sqrt{t_e}$
Fluid efficiency	$\eta_e = \dfrac{\overline{w}_e R_f^2 \dfrac{\pi}{2}}{V_i}$	$\eta_e = \dfrac{\overline{w}_e x_f h_f}{V_i}$	$\eta_e = \dfrac{\overline{w}_e x_f h_f}{V_i}$

After closure, the pressure continues to fall off, but the remaining falloff behavior is increasingly dominated by pressure diffusion transients responding to fracture fluid leakoff through the propagated fracture face. Since during propagation the pressure gradient inside the fracture is negligible, the propagated fracture has effectively infinite conductivity. Therefore, the falloff behavior is dominated by only the pseudolinear and pseudoradial flow regimes. In some cases the pseudolinear flow regime may be masked by before-closure behavior. If the falloff data are collected long enough, the final pseudoradial flow regime can be used to estimate formation permeability and pressure. Then, if the pseudolinear flow regime is visible as a half-slope trend in the derivative, the fracture area opposite permeable formation rock, A_f, can be estimated from a rearrangement of Equation 13-46:

$$A_f = \frac{\alpha_{lf} qB}{m_{lf}} \left(\frac{\mu}{k\phi c_t} \right)^{1/2} \quad \text{(oil)}$$

$$A_f = \frac{10.1 \alpha_{lf} q T_R}{m_{lf}} \left(\frac{1}{k\phi\mu c_g} \right)^{1/2} \quad \text{(gas).} \tag{13-80}$$

In turn, the fracture half length is estimated as

$$\text{If } A_f < h^2,\ x_f = \sqrt{A_f} = \frac{h_f}{2},\ \text{else } x_f = \frac{A_f}{2h}. \tag{13-81}$$

Equation 13-81 considers two possibilities. When the fracture area opposite permeable formation rock is less than that of a square with sides equal to the formation thickness, the radial fracture propagation model is appropriate. When the fracture area opposite permeable formation rock is greater than h^2, $x_f > h_f$, the PKN fracture propagation model is appropriate.

Example 13-17 After-Closure Analysis from a Fracture Calibration Test

The pressure falloff behavior shown in Figure 13-38 are from a fracture calibration test in the Cotton Valley sandstone with $B_g = 0.0043$, $\mu_g = 0.0223$ cp, $c_t = 9.1 \cdot 10^{-5}$ psi^{-1}, $h = 15$ ft, $\phi = 0.065$ $r_w = 0.354$ ft, and with $T_R = 270°$F. This test was conducted as a "step-rate test," starting at 0.5 bpm and ending at 2.5 bpm, for a total volume of 36.25 bbls of 3% KCl brine injected in 19 minutes, corresponding to an adjusted injection time of 36.25/2.5 = 14.5 minutes (0.242 hr).

The final injection pressure is 7287 psi for m(p) = $2.42 \cdot 10^9$ psi^2/cp. Estimate formation permeability and pressure and the fracture half length.

Solution

In Figure 13-38, the derivative level at the end of the falloff is $m' = 1.8 \cdot 10^{10}$ psi^2/cp. Noting that the 2.5 bpm injection of brine produces a volumetric displacement at wellbore conditions equivalent to a rate of 4700 Mscf/d, from Equation 13-17

$$k = \frac{711 q_g T_R}{m'h} = \frac{711(4700)(730)}{1.8 \times 10^{10}(15)} = 0.0090 \text{ md} \tag{13-82}$$

and for the adjusted injection time $t_{e,adj} = 0.242$ hr, Equation 13-19 gives

$$\begin{aligned} m(p^*) &= m' \ln\left(\frac{t_{e,adj} + \Delta t}{\Delta t}\right) + \Delta m(p_{IARF}) + m(p_{wf}(t_p)) \\ &= -1.8 \times 10^{10} \ln\left(\frac{0.242 + 105}{105}\right) - 1.06 \times 10^9 + 2.42 \times 10^9 \\ &= 1.32 \times 10^9 \text{ psi}^2/\text{cp} \end{aligned} \tag{13-83}$$

$1.32 \cdot 10^9$ psi^2/cp corresponds to a pressure 4813 psi.

Having found an estimate for permeability, Equation 13-46 can be used to calculate an estimate for the fracture half length. Selecting $\Delta m(p)' = 3.25 \cdot 10^9$ psi^2/cp at $\Delta t = 4$ hr, Equation 13-45 gives

$$m_{lf} = \frac{2\Delta m(p')}{\sqrt{t}} = \frac{2(3.25 \cdot 10^9)}{\sqrt{4}} = 3.25 \cdot 10^9 \tag{13-84}$$

and Equation 13-80 gives

$$\begin{aligned} A_f = 2x_f h &= \frac{10.1\alpha_{lf} q T_R}{m_{lf}} \left(\frac{1}{\phi \mu c_g k} \right)^{1/2} \\ &= \frac{10.1(4.064)(4700)(730)}{(3.25 \cdot 10^9)} \left(\frac{1}{0.065(0.0223)(9.1 \cdot 10^{-5})(0.009)} \right)^{1/2} \\ &= 1257 \text{ ft}^2. \end{aligned} \tag{13-85}$$

In this case $\sqrt{A_f} = 35$ ft is greater than the formation thickness. Therefore, from Equation 13-81 the fracture half-length estimate is 42 ft.

While they are not estimated from standard pressure transient analysis, the closure stress, leakoff coefficient, and fracture fluid efficiency are critical inputs for main treatment hydraulic fracture design. The next example illustrates estimation of these parameters from the diagnostic plot in Figure 13-38.

Example 13-17 Before-Closure Analysis from a Fracture Calibration Test

Using again the pressure falloff behavior shown in Figure 13-38, for ISIP = 7,287 psi, estimate the closure stress, leakoff coefficient, and fracture fluid efficiency. The plane strain modulus is $E' = 6 \cdot 10^6$ psi.

Solution

In Figure 13-38, the closure time value of 1 hr is found at the time the derivative begins to depart from the 3/2 slope trend observed while the fracture is closing. At a time 1 hr after injection stops, the recorded pressure is 6371 psi and the pressure derivative is

$$\Delta p' = \frac{dp}{d\ln\tau} = \frac{dp}{dm(p)}RNP' = (2.211 \times 10^{-6})(1.35 \times 10^{9}) = 2985 \text{ psi/cycle.} \quad \textbf{(13-86)}$$

For an injection time of 14.5/60 = 0.242 hr, the dimensionless elapsed closure time $\Delta t_{Dc} = \Delta t_c/t_e = 1/0.242 = 4.138$, and the superposition time, $\tau_c = (\Delta t_c + t_e)/\Delta t_c = (14.5 + 60)/60 = 1.242$. The injected volume in one fracture wing in ft^3 is $V_i = 36.25(5.615)/2 = 102$ ft^3.

From Equation 13-78,

$$m_N = \frac{\Delta p'}{2\Delta t_{D_c}^{5/2}\tau_c(1 - \tau_c^{1/2})} = \frac{2{,}985}{2(4.138)^{5/2}1.242(1 - 1.242^{1/2})} = -301.5 \quad \textbf{(13-87)}$$

and from Equation 13-79,

$$b_N = p_c - m_N\frac{4}{3}\Delta t_{D_c}^{3/2}\left(\tau_c^{\frac{3}{2}} - 1\right)$$

$$= 6{,}371 + 301.5\left(\frac{4}{3}\right)(4.138^{3/2}(1.242^{3/2} - 1) = 7{,}671 \text{ psi.} \quad \textbf{(13-88)}$$

From the equation in Table 13-6 for the PKN model fracture height and using a plane strain modulus $E' = 6 \cdot 10^6$ psi,

$$h_f = \sqrt{\frac{2E'V_i}{\pi x_f(b_N - p_c)}} = \sqrt{\frac{2(6 \times 10^6)(102)}{\pi(42)(7{,}671 - 6371)}} = 84.5 \text{ ft} \quad \textbf{(13-89)}$$

leakoff coefficient,

$$C_L = \frac{\pi h_f}{4\sqrt{t_e}E'}(-m_N) = \frac{\pi(84.5)}{4\sqrt{14.5}(6 \times 10^6)}(301.5) = 0.00088 \text{ ft/min}^{1/2} \quad \textbf{(13-90)}$$

fracture width,

$$\overline{w}_e = \frac{V_i}{x_f h_f} - 2.830C_L\sqrt{t_e}$$

$$= \frac{102}{42(84.5)} - 2.830(0.00088)\sqrt{14.5} = 0.0138 \text{ ft} = 0.232 \text{ in} \quad \textbf{(13-91)}$$

and fracture fluid efficiency,

$$\eta_e = \frac{\overline{w}_e x_f h_f}{V_i} = \frac{(0.0138)(42)(84.5)}{102} = 67\% \quad \textbf{(13-92)}$$

Figure 13-39 shows a global match for these data based on the parameters estimated from Examples 13-17 and 13-18.

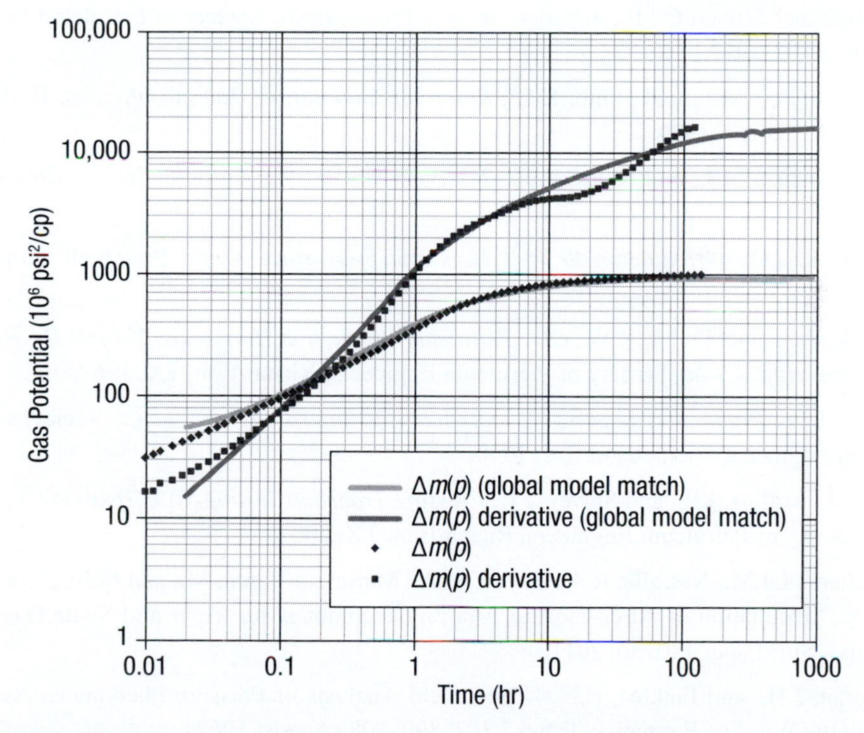

Figure 13-39 Global pressure match for Examples 13-16 and 13-17.

References

1. Arps, J.J., "Analysis of Decline Curves," *Trans. AIME,* **160:** 228–247 (1945).
2. Barree, R.D., Barree, V.L., and Craig, D.P., "Holistic Fracture Diagnostics: Consistent Interpretation of Prefrac Injection Test Using Multiple Analysis Methods," *SPE Production & Operations,* 396–406 (August 2009).
3. Chatas, A.T., "Unsteady Spherical Flow in Petroleum Reservoirs," *SPEJ,* 102–114 (June 1966).
4. Clark, N.J., and Schultz, W.P., "The Analysis of Problem Wells," *The Petroleum Engineer* 28: B30–B38 (September 1956).
5. Dobkins, T.A., "Improved Methods to Determine Hydraulic Fracture Height," *JPT,* 719–726 (April 1981).
6. Dupree, J.H., "Cased-Hole Nuclear Logging Interpretation, Prudhoe Bay, Alaska," *The Log Analyst,* 162–177 (May–June 1989).
7. Earlougher, Robert C., Jr., *Advances in Well Test Analysis,* Society of Petroleum Engineers, Dallas, TX, 1977.
8. Economides, M.J., and Nolte, K.G., *Reservoir Stimulation,* 2nd ed., Prentice Hall, Englewood Cliffs, NJ, 1989.
9. Economides, M.J., and Nolte, K.G., *Reservoir Stimulation,* third ed., Wiley, Hoboken, NJ, 2000.
10. Evenick, J.C., *Introduction to Well Logs and Subsurface Maps,* PennWell Corporation, 2008.
11. Frick, T.C., and Taylor, R.W., eds., *Petroleum Production Handbook, Volume II—Reservoir Engineering,* 43–46, Society of Petroleum Engineers, Richardson, TX, 1962.
12. Hill, A.D., *Production Logging: Theoretical and Interpretive Elements,* Society of Petroleum Engineers, Richardson, TX, 1990.
13. Lee, J., Rollins, J.B., and Spivey, J.P., *Pressure Transient Testing,* SPE Textbook Series Vol. 9, Society of Petroleum Engineers, Richardson, TX, 2003.
14. Mohamed, I.M., Nasralla, R.A., Sayed, M.A., Marongiu-Porcu, M., and Ehlig-Economides, C.A., "Evaluation of After-Closure Analysis Techniques for Tight and Shale Gas Formations," SPE Paper 140136, 2011.
15. Moran, J.H., and Finklea, E.E., "Theoretical Analysis of Pressure Phenomena Associated with the Wireline Formation Tester," *JPT,* 899–908 (August 1962).
16. Morangiu-Porcu, M., Ehlig-Economides, C.A., and Economides, M.J., "Global Model for Fracture Falloff Analysis," SPE Paper 144028, 2011.
17. Muskat, Morris, *Physical Principles of Oil Production,* McGraw-Hill, New York, 1949.

18. Nelson, E.B., ed., *Well Cementing,* Elsevier, Amsterdam, 1990.

19. Nolte, K.G., "Determination of Fracture Parameters from Fracturing Pressure Decline." SPE Paper 8341, 1979.

20. Schlumberger, "Schlumberger's Cement-Scan Log–A Major Advancement in Determining Cement Integrity," SMP Paper 5058, Houston, TX, 1990.

21. Shlyapobersky, J., Walhaug, W.W., Sheffield, R.E., Huckabee, P.T., "Field Determination of Fracturing Parameters for Overpressure Calibrated Design of Hydraulic Fracturing," SPE Paper 18195, 1998.

22. Timmerman, E.H., *Practical Reservoir Engineering, Volume II,* PennWell Books, Tulsa, OK, 1982.

23. Valkó, P.P., and Economides, M.J., *Hydraulic Fracture Mechanics,* John Wiley & Sons, Chichester, England, 1995.

24. Villareal, S., Pop, J., Bernard, F., Harms, K., Hoefel, H., Kamiya, A., Swinburne, P., and Ramshaw, S., "Characterization of Sampling While Drilling Operations," IADC/SPE Paper 128249, 2010.

Problems

13-1 A new zone (zone A) was perforated in a water injection well in Reservoir B. Prior to perforating the new zone, the injection rate into the well was 4500 bbl/d at 250 psi surface tubing pressure; after perforating, the rate was 4700 bbl/d at 250 psi. It had been expected that the new zone would take at least 2000 bbl/d of injection water at this pressure. The old zone (zone B) in the well is located 20 ft below the new zone and has been on injection for 8 years.

It is important to know why perforating the new zone did not increase the total injection rate by the desired amount. First, list the possible causes of the apparent low injectivity into zone A. Then, describe the production logs or other tests that you would recommend to diagnose this problem, giving a priority rank to the logs or tests recommended.

13-2 A production well in Reservoir A is producing an excessive amount of water (50% water cut). To locate the source of the excessive water production, the temperature, basket flowmeter, and fluid density logs shown (Figure 13-13) were obtained. For this well, $B_o = 1.3$, $B_w = 1.0$, and at bottomhole conditions, $\rho_o = 0.85$ g/cm^3 and $\rho_w = 1.05$ g/cm^3. Which zone appears to be producing most of the water? From these logs, can the cause of the high water production be determined? Explain your answers.

13-3 An oil well is producing an excessive gas rate because of gas coning down from a gas cap. The well produces no water. Assuming that there are two perforated zones, sketch the temperature, noise, and fluid density logs you would expect to obtain in this well.

13-4 An injection well in a waterflood has abnormally high injectivity. Described the production logs or other tests you would propose to diagnose the cause of the high injectivity.

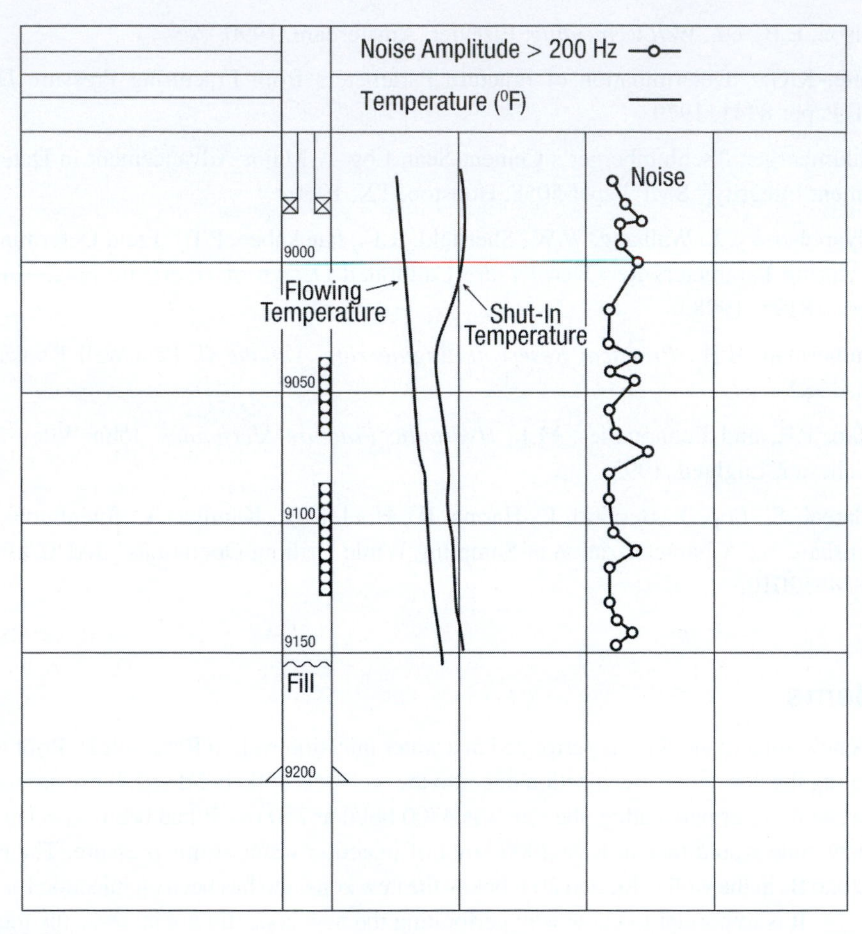

Figure 13-40

13-5 Estimate the extrapolated pressure, p^*, for the well in Example 13-13. Calculate I_{ani} for this reservoir. Is this formation a good candidate for a horizontal well? Why or why not?

13-6 The analysis in Example 13-13 estimates for horizontal and vertical permeability values of 343 md and 9.3 md, respectively, and skin 10.3, given that only the top 50 ft of the total formation thickness of 138 ft is perforated. Estimate the damage skin opposite the flowing interval using the Papazakos equation (Equation 6-20) for limited-entry skin.

13-7 Determine the productivity index for the final well characterization in Figure 13-31c. Then estimate the skin for this well. (*Hint:* Determine the average pressure by material balance; then use the data provided to find values for q and p_{wf} for a time when the well is exhibiting pseudosteady-state flow behavior.)

13-8 Figure 17-2 relates fracture half length, conductivity, and skin. Estimate the skin for the final well characterization in Figure 13-31 and compare with your result from the previous problem. Which skin characterization is correct? Why?

13-9 Calculate the minimum well drainage radius using Equation (13-72). Does this conflict with any of the other estimated parameter ranges? Does the model match in Figure 13-35 honor the unit slope trend marked in Figure 13-34? What does this tell you about the unit slope trend marked in Figure 13-34? Based on the model match shown in Table 13-15, what must be the minimum drainage radius for this well?

13-10 The diagnostic plot for a limited-entry test in a pilot hole drilled prior to drilling a horizontal well is shown below. The DST packers have opened flow 5 feet at the base of the formation. Reservoir, fluid, and well properties are the following: $\phi = 16\%$, $h = 85$ ft, $B_o = 1.3$ rb/STB, $\mu_o = 0.6$ cp, $c_t = 13 \cdot 10^{-6}$ psi^{-1}, $r_w = 0.33$ ft, $q = 50$ STB/d, $t_p = 6$ hr, $p_{wf}(t_p) = 3345.87$ psi. Estimate wellbore storage, total skin, horizontal and vertical permeability, and the extrapolated pressure. Estimate the damage skin opposite the flowing interval.

13-11 A post-treatment pressure buildup test in a hydraulically fractured oil well is shown in the graph below. Well, reservoir, and fluid properties follow: $t_p = 1000$ hr, $q = 1500$ bpd, $p_{wf}(t_p) = 4406$ psi, $\phi = 17\%$, $h = 100$ ft, $B_o = 1.2$ res bbl/STB, $\mu_o = 0.8$ cp, $c_t = 12 \cdot 10^{-6}$ psi^{-1}, $r_w = 0.328$ ft. Estimate the wellbore storage coefficient, horizontal permeability, fracture half length, and conductivity. Use Figure 17-2 to estimate the fracture skin. Calculate the well productivity for a 40-acre well spacing.

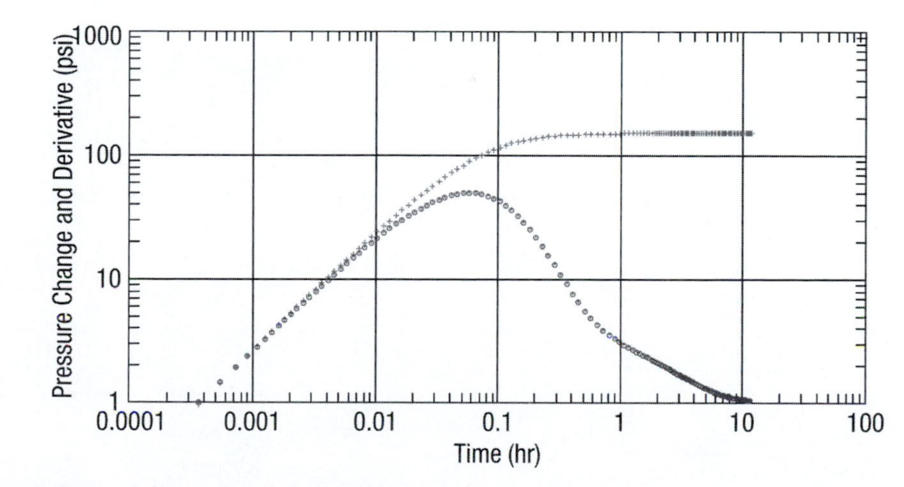

Figure 13-41 Diagnostic plot for problem 13-10.

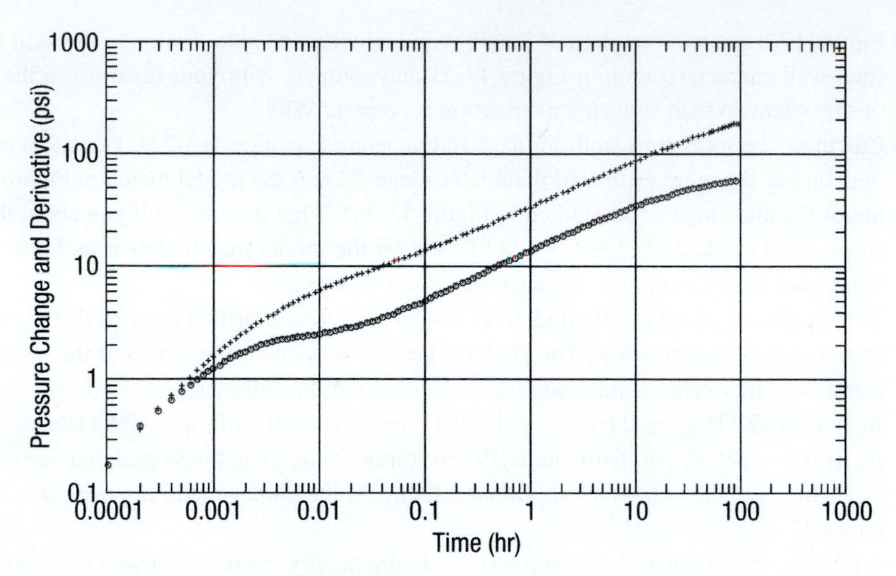

Figure 13-42 Buildup data for problem 13-11.

CHAPTER 14

Matrix Acidizing: Acid/Rock Interactions

14.1 Introduction

Matrix acidizing is a well stimulation technique in which an acid solution is injected into the formation in order to dissolve some of the minerals present, and hence recover or increase the permeability in the near-wellbore vicinity. The most common acids used are hydrochloric acid (HCl), used primarily to dissolve carbonate minerals, and mixtures of hydrofluoric and hydrochloric acids (HF/HCl), for attacking silicate minerals such as clays and feldspars. Because traditional drilling fluid (drilling "mud") was a suspension of bentonite clay, the mixture HF/HCl, effective in dissolving bentonite particles, has been known in the vernacular as mud acid. Other acids, particularly some of the weak organic acids, are used in special applications. Matrix acidizing is a near-wellbore treatment, with all of the acid reacting within about a foot of the wellbore in sandstone formations, and within a few to perhaps as much as 10–20 ft of the wellbore in carbonates.

An acidizing treatment is called a "matrix" treatment because the acid is injected at pressures below the parting pressure of the formation, so that fractures are not created. The objective is to greatly enhance or recover the permeability very near the wellbore, rather than affect a large portion of the reservoir.

Matrix acidizing should not be confused with acid fracturing. While both types of stimulation treatments can be applied to carbonate formations and both use acids, their purpose of application and, consequently, the candidate reservoirs are often very different. Acid fracturing, resulting from the injection of fluids at pressure above the fracturing pressure of the formation, is intended to create a path of high conductivity by dissolving the walls of the fracture in a nonuniform way. Therefore, for the petroleum production engineer, acid fracturing has more in common with proppant fracturing than with matrix acidizing. Acid fracturing is sometimes used to overcome formation damage in relatively high-permeability formations. However, carbonate reservoirs of relatively low permeability may also be candidates for acid fracturing. For such reservoirs, a comparison with propped fracturing must be done, taking into account the expected post-treatment production rate versus costs. Acid fracturing is discussed in Chapter 16.

Matrix acidizing can significantly enhance the productivity of a well when near-wellbore formation damage is present and, conversely, is of lesser benefit in an undamaged well. In sandstone reservoirs, matrix acidizing should generally be applied only when a well has a high skin effect that cannot be attributed to partial penetration, perforation efficiency, or other mechanical aspects of the completion. The wormholing process that occurs in carbonates makes it possible to obtain stimulation to sufficient radial penetrations that treating undamaged wells can be of great benefit. In carbonates, acid can also preferentially flow along natural fractures or in vug networks, making the potential depth of acid stimulation even greater. Thus, in carbonate reservoirs, matrix acidizing is often applied without the presence of formation damage.

Example 14-1 The Potential Benefits of Acidizing in Damaged and Undamaged Wells

Assume that the well described in Appendix A has a damaged region extending 1 ft beyond the wellbore ($r_w = 0.328$ ft). The well is on a 40-acre spacing ($r_e = 745$ ft). Calculate the ratio of the productivity index after removing this damage with acidizing to the productivity index of the damaged well for a permeability in the damaged region ranging from 5% to 100% of the undamaged reservoir permeability. Next, assume that the well is originally undamaged and acid is used to increase the permeability in a 1-ft region around the wellbore up to 20 times the original reservoir permeability. This 1-ft acid penetration distance is typical of what occurs in sandstone acidizing. Calculate the ratio of the stimulated productivity index to the productivity index of the undamaged well. In both cases, assume steady-state flow and no mechanical skin effects. Repeat the calculation, but now assume that the acid has stimulated the permeability to a radial distance of 20 ft beyond the wellbore, as can occur in carbonate formations.

Solution

The productivity of a well in an undersaturated reservoir is given by Equation (2-13). Taking the ratio of the stimulated productivity index to the damaged productivity index, noting that $s = 0$ when the damage has been removed, yields

$$\frac{J_s}{J_d} = \frac{\ln(r_e/r_w) + s}{\ln(r_e/r_w)}. \tag{14-1}$$

The skin effect is related to the permeability and radius of the damaged region by Hawkins' formula [Equation (5-4)]. Substituting for s in Equation (14-1) gives

$$\frac{J_s}{J_d} = 1 + \left(\frac{1}{X_d} - 1\right)\frac{\ln(r_s/r_w)}{\ln(r_e/r_w)} \tag{14-2}$$

where X_d is the ratio of the damaged permeability to that of the undamaged reservoir k_s/k in Equation (5-4). Applying Equation (14-2) for X_d ranging from 0.05 to 1, the results shown in Figure 14-1 are obtained. For the case of severe damage, where the permeability of the damaged region is 5% of the original permeability, the skin effect of the damaged well is 26 and removal of the damage with acidizing increases the productivity index of the well by a factor of 4.5. With a damaged permeability of 20% of the original (a typical amount of damage from drilling), the original skin effect is 5.6 and complete removal of the damage increases the productivity index by about 70%.

For the case of stimulating an undamaged well, the ratio of the stimulated well productivity index to the original is

$$\frac{J_s}{J_o} = \frac{1}{1 + [(1/X_s) - 1][\ln(r_s/r_w)/\ln(r_e/r_w)]} \tag{14-3}$$

where X_s is the ratio of the stimulated permeability to the original permeability. Figure 14-2 is obtained by applying Equation (14-3) for X_s ranging from 1 to 20. For this undamaged well, increasing the permeability by a factor of 20 in a 1-ft radius around the well increases the productivity index by only 21%. In fact, if the permeability in this 1-ft region were infinite (no resistance to flow), the productivity increase would be only 22%. However, if the acid stimulates the formation to a distance of 20 feet beyond the well, as may occur in carbonates, the productivity of the well is doubled.

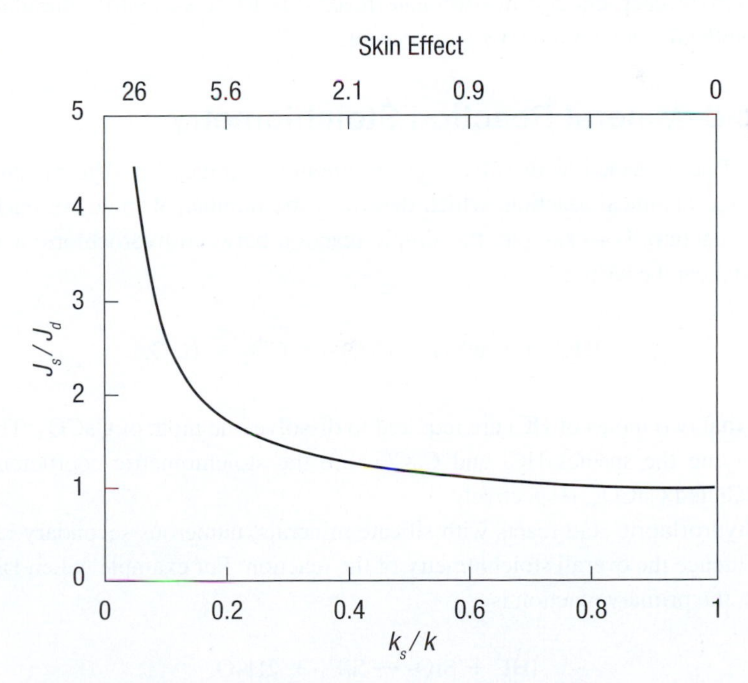

Figure 14-1 Potential productivity improvement from removing damage with acidizing.

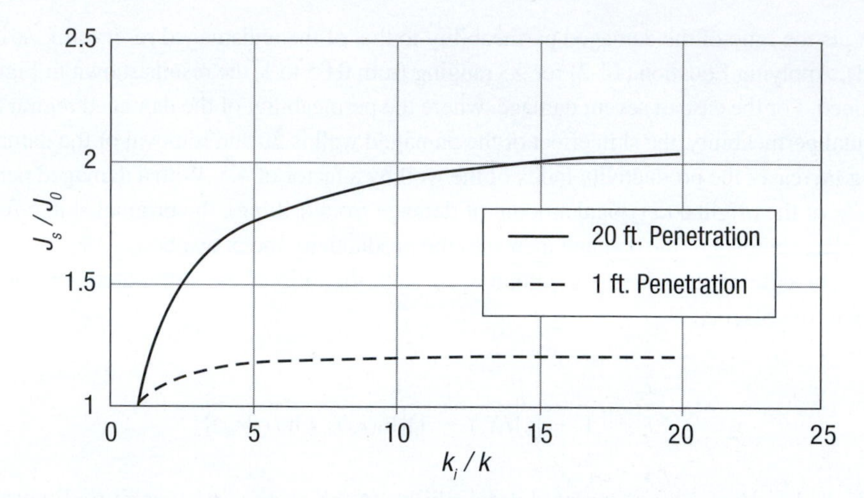

Figure 14-2 Potential productivity improvement from acidizing an undamaged well.

This example illustrates that matrix acidizing is primarily beneficial in removing damage when the depth of live acid penetration is shallow. For lower-permeability sandstone reservoirs, significant simulation of undamaged wells is generally possible only with hydraulic fracturing. Acid can penetrate deep enough in carbonate reservoirs to make matrix stimulation of these reservoirs beneficial, even with no damage present.

14.2 Acid–Mineral Reaction Stoichiometry

The amount of acid needed to dissolve a given amount of mineral is determined by the stoichiometry of the chemical reaction, which describes the number of moles of each species involved in the reaction. For example, the simple reaction between hydrochloric acid (HCl) and calcite ($CaCO_3$) can be written as

$$2HCl + CaCO_3 \rightarrow CaCl_2 + CO_2 + H_2O, \quad \textbf{(14-4)}$$

which shows that two moles of HCl are required to dissolve one mole of $CaCO_3$. The numbers 2 and 1 multiplying the species HCl and $CaCO_3$ are the stoichiometric coefficients, ν_{HCl} and ν_{CaCO_3}, for HCl and $CaCO_3$, respectively.

When hydrofluoric acid reacts with silicate minerals, numerous secondary reactions may occur that influence the overall stoichiometry of the reaction. For example, when HF reacts with quartz (SiO_2), the primary reaction is

$$4HF + SiO_2 \leftrightarrow SiF_4 + 2H_2O, \quad \textbf{(14-5)}$$

producing silicon tetrafluoride (SiF_4) and water. The stoichiometry of this reaction shows that 4 moles of HF are needed to consume one mole of SiO_2. However, the SiF_4 produced may also react with the HF to form fluosilicic acid (H_2SiF_6), according to

$$SiF_4 + 2HF \leftrightarrow H_2SiF_6. \tag{14-6}$$

If this secondary reaction goes to completion, six moles of HF, rather than four, will be consumed to dissolve one mole of quartz. A complication is that the fluosilicates may exist in various forms (Bryant, 1991), so that the total amount of HF required to dissolve a given amount of quartz depends on the solution concentration.

The most common reactions involved in acidizing are summarized in Table 14-1. For the reactions between HF and silicate minerals, only the primary reactions are listed; secondary reactions often occur that change the stoichiometry. For the reactions between HF and feldspars, for example, the primary reactions predict that 14 moles of HF are needed to consume one mole of feldspar. However, Schechter (1992) suggests that about 20 moles of HF are consumed for every mole of feldspar under typical acidizing conditions.

Secondary and tertiary reactions of HF/HCl solutions with alumino-silicate minerals (primarily clays and feldspars) have been shown to play important roles in sandstone acidizing (Gdanski, 1997a, 1997b, 1998; Gdanski and Shuchart, 1996; Ziauddin et al., 2005; Hartman, Lecerf, Frenier, Ziauddin, and Fogler, 2006). These reactions change the overall stoichiometry of the HF reactions and result in the precipitation of reaction products, primarily amorphous silica.

Typical secondary reactions are illustrated by the reactions of kaolinite with hydrofluoric acid. The primary reaction of HF with kaolinite is (Hartman et al., 2006)

$$32HF + Al_4Si_4O_{10}(OH)_8 \longrightarrow 4AlF_2^+ + 4SiF_6^{-2} + 18H_2O + 4H^+. \tag{14-7}$$

This reaction creates a new reactive species, fluosilicic acid (shown dissociated), which in the presence of sufficient clays or feldspars also reacts in the secondary reaction,

$$\begin{aligned} 2SiF_6^{-2} + 16H_2O + Al_4Si_4O_{10}(OH)_8 \longrightarrow\ & 6HF + 3AlF_2^+ + 10(OH)^- + Al^{+3} \\ & + 6SiO_2 \cdot 2H_2O. \end{aligned} \tag{14-8}$$

The secondary reaction produces amorphous silica, which will precipitate, and produce some HF, which can in turn react with other minerals. If only the primary reaction occurs, 32 moles of HF are consumed in dissolving one mole of kaolinite. However, if the secondary reaction proceeds to completion, the net consumption of HF per mole of kaolinite is only 20/3 because the fluosilicic acid produced in the primary reaction would consume two more moles of kaolinite and also produce 12 moles of HF.

Table 14-1 Primary Chemical Reactions in Acidizing

HCl
Calcite: $2HCl + CaCO_3 \rightarrow CaCl_2 + CO_2 + H_2O$
Dolomite: $4HCl + CaMg(CO_3)_2 \rightarrow CaCl_2 + MgCl_2 + 2CO_2 + 2H_2O$
Siderite: $2HCl + FeCO_3 \rightarrow FeCl_2 + CO_2 + H_2O$
HF/HCl
Quartz: $4HF + SiO_2 \leftrightarrow SiF_4$ (silicon tetrafluoride) $+ 2H_2O$ $SiF_4 + 2HF \leftrightarrow H_2SiF_6$ (fluosilicic acid)
Albite (sodium feldspar): $NaAlSi_3O_8 + 14HF + 2H^+ \leftrightarrow Na^+ + AlF_2^+ + 3SiF_4 + 8H_2O$
Orthoclase (potassium feldspar): $KAlSi_3O_8 + 14HF + 2H^+ \leftrightarrow K^+ + AlF_2^+ + 3SiF_4 + 8H_2O$
Kaolinite: $Al_4Si_4O_{10}(OH)_8 + 24HF + 4H^+ \leftrightarrow 4AlF_2^+ + 18H_2O$
Montmorillonite: $Al_4Si_8O_{20}(OH)_4 + 40HF + 4H^+ \leftrightarrow 4AlF_2^+ + 8SiF_4 + 24H_2O$

The tertiary reactions in sandstone acidizing involve the conversion of the AlF_2^+ to other aluminum fluoride species. The effects of these reactions on the overall reaction process is described either with kinetic expressions describing the reaction rates (Gdanski, 1998), or with equilibrium relationships (Hartman et al., 2006).

A more convenient way to express reaction stoichiometry is with the dissolving power, introduced by Williams, Gidley, and Schechter (1979). The dissolving power expresses the amount of mineral that can be consumed by a given amount of acid on a mass or volume basis. First, the gravimetric dissolving power, β, the mass of mineral consumed by a given mass of acid, is defined as

$$\beta = \frac{\nu_{\text{mineral}} \text{MW}_{\text{mineral}}}{\nu_{\text{acid}} \text{MW}_{\text{acid}}}. \tag{14-9}$$

Thus, for the reaction between 100% HCl and $CaCO_3$,

$$\beta_{100} = \frac{(1)(100.1)}{(2)(36.5)} = 1.37 \frac{\text{lb}_m \text{CaCO}_3}{\text{lb}_m \text{HCl}} \tag{14-10}$$

where the subscript 100 denotes 100% HCl. The dissolving power of any other concentration of acid is the β_{100} times the weight fraction of acid in the acid solution. For the commonly used 15 wt% HCl, $\beta_{15} = 0.15(\beta_{100}) = 0.21\ \text{lb}_m\text{CaCO}_3/\text{lb}_m$ HCl. The stoichiometric coefficients for common acidizing reactions are found from the reaction equations in Table 14-1, while the molecular weights of the most common species are listed in Table 14-2.

Table 14-2 Molecular Weights of Species in Acidizing

Species	Atomic or Molecular Weight (mass/mole)
Elements	
Hydrogen, H	1
Carbon, C	12
Oxygen, O	16
Fluorine, F	19
Sodium, Na	23
Magnesium, Mg	24.3
Aluminum, Al	27
Silicon, Si	28.1
Chlorine, Cl	35.5
Potassium, K	39.1
Calcium, Ca	40.1
Iron, Fe	55.8
Molecules	
Hydrochloric acid, HCl	36.5
Hydrofluoric acid, HF	20
Calcite, $CaCO_3$	100.1
Dolomite, $CaMg(CO_3)_2$	184.4
Siderite, $FeCO_3$	115.8
Quartz, SiO_2	60.1
Albite (sodium feldspar), $NaAlSi_3O_8$	262.3
Orthoclase (potassium feldspar), $KAlSi_3O_8$	278.4
Kaolinite, $Al_4Si_4O_{10}(OH)_8$	516.4
Montmorillonite, $Al_4Si_8O_{20}(OH)_4$	720.8

The volumetric dissolving power, χ, similarly defined as the volume of mineral dissolved by a given volume of acid solution, is related to the gravimetric dissolving power by

$$\chi = \beta \frac{\rho_{\text{acid solution}}}{\rho_{\text{mineral}}}. \tag{14-11}$$

A 15 wt% HCl solution has a specific gravity of about 1.07, and $CaCO_3$ has a density of 169 lb_m/ft^3. For the reaction of these species, the volumetric dissolving power is

$$\begin{aligned} \chi_{15} &= 0.21\left(\frac{lb_m CaCO_3}{lb_m 15\%HCl}\right)\left(\frac{(1.07)(62.4)(lb_m 15\%HCl)(ft^3 15\%HCl)}{169(lb_m CaCO_3)/(ft^3 CaCO_3)}\right) \\ &= 0.082\frac{ft^3 CaCO_3}{ft^3 15\%HCl}. \end{aligned} \tag{14-12}$$

The dissolving powers of various acids with limestone and dolomite and for HF with quartz and albite are given in Tables 14-3 and 14-4 (Schechter, 1992). These dissolving powers assume complete reaction between the acid and the dissolving mineral species. However, the reactions of weak acids with carbonate minerals is often limited by the equilibrium between the reactants and the reaction products. For example, for the reaction of formic acid with calcite, the equilibrium composition of the solution is governed by

$$K_d = \frac{[H^+][HCOO^-]}{[HCOOH]} \tag{14-13}$$

Table 14-3 Dissolving Power of Various Acids

			χ			
Formulation	**Acid**	β_{100}	**5%**	**10%**	**15%**	**30%**
Limestone:	Hydrochloric (HCl)	1.37	0.026	0.053	0.082	0.175
$CaCO_3$	Formic (HCOOH)	1.09	0.020	0.041	0.062	0.129
$\rho CaCO_3 = 2.71 g/cm^3$	Acetic (CH_3COOH)	0.83	0.016	0.031	0.047	0.096
Dolomite:	Hydrochloric	1.27	0.023	0.046	0.071	0.152
$CaMg(CO_3)_2$	Formic	1.00	0.018	0.036	0.054	0.112
$\rho CaMg(CO_3)_2 = 2.87 g/cm^3$	Acetic	0.77	0.014	0.027	0.041	0.083

From Schechter (1992).

Table 14-4 Dissolving Power for Hydrofluoric Acid[a]

	Quartz (SiO_2)		Albite ($NaAlSi_3O_8$)	
Acid Concentration (wt%)	β	χ	β	χ
2	0.015	0.006	0.019	0.008
3	0.023	0.010	0.028	0.011
4	0.030	0.018	0.037	0.015
6	0.045	0.019	0.056	0.023
8	0.060	0.025	0.075	0.030

[a]β, Mass of rock dissolved/mass of acid reacted; X, Volume of rock dissolved/volume of acid reacted
From Schechter, 1992.

where the quantities in brackets are the activities of the various species in solution and K_d is the dissociation constant. Because of the buffering action of the CO_2 created when weak acids dissolve carbonates, the equilibrium hydrogen ion concentration used in Equation 14-13 results in a significant amount of the weak acid never dissociating. This equilibrium limitation decreases the dissolving power of weak acids. Table 14-5 from Buijse, de Boer, Breukel, and Burgos (2004) shows the percentages of formic or acetic acid that is spent when reactions with calcite are complete for several common acid formulations.

Table 14-5 Spending of Acid Components in Acid Mixtures and Equivalent HCl Concentration

	Acid Component Spent (%)			
Acid Mix	HCl	Acetic	Formic	Equivalent HCl (%)
10% acetic	–	54	–	3.4
10% formic	–	–	85	6.8
13/9% acetic/formic	–	31	82	8.5
7/11% HCl/formic	100	–	78	14.1
15/10% HCl/acetic	100	24	–	16.5

Example 14-2 Calculating the HCl Preflush Volume

In sandstone acidizing treatments, a preflush of HCl is usually injected ahead of the HF/HCl mixture to dissolve the carbonate minerals and establish a low-pH environment. A sandstone with a porosity of 0.2 containing 10% (volume) calcite ($CaCO_3$) is to be acidized. If the HCl preflush is to remove all carbonates to a distance of 1 ft from the wellbore before the HF/HCl stage enters the formation, what minimum preflush volume (gallons of acid solution per foot of formation thickness) is required if the preflush is 15% HCl solution? The wellbore radius is 0.328 ft.

Solution

The minimum volume is determined by assuming that the HCl–carbonate reaction is very fast, so that the HCl front is sharp (discussed in Chapter 15). The required preflush volume is then the volume of acid solution needed to dissolve all of the calcite to a radius of 1.328 ft plus the volume of acid solution that will be left in the pore space in this region. The volume of acid needed to consume the calcite is the volume of calcite present divided by the volumetric dissolving power:

$$\begin{aligned} V_{CaCO_3} &= \pi(r_{HCl}^2 - r_w^2)(1 - \phi)x_{CaCO_3} \\ &= \pi(1.328^2 - 0.328^2)(1 - 0.2)(0.1) = 0.42 \text{ ft}^3/\text{ft CaCO}_3 \end{aligned} \tag{14-14}$$

$$V_{HCl} = \frac{V_{CaCO_3}}{X_{15}} = \frac{0.42}{0.082} = 5.12 \text{ ft}^3 \text{ 15 wt\% HCl solution/ft} \tag{14-15}$$

The volume of pore space within 1 ft of the wellbore after removal of the carbonate is

$$\begin{aligned} V_p &= \pi(r_{HCl}^2 - r_w^2)[\phi + (x_{CaCO_3})(1 - \phi)] \\ &= \pi(1.328^2 - .328^2)[.2 + .1(1 - .2)] = 1.46 \text{ ft}^3/\text{ft}, \end{aligned} \tag{14-16}$$

so the total volume of HCl preflush is

$$V_{HCl} = V_{HCl,1} + V_p = [(5.12 + 1.46)\text{ft}^3/\text{ft}](7.48 \text{ gal/ft}^3) = 49 \text{ gal/ft}. \tag{14-17}$$

This is a typical volume of HCl preflush used in sandstone acidizing.

A simple equation for calculating the HCl preflush location for any injected volume, applicable either to radial flow as in this example or to flow from perforations, is presented in Chapter 15.

14.3 Acid–Mineral Reaction Kinetics

Acid–mineral reactions are termed "heterogeneous" reactions because they are reactions between species occurring at the interface between different phases, the aqueous phase acid and the solid mineral. The *kinetics* of a reaction is a description of the rate at which the chemical reaction takes place, once the reacting species have been brought into contact. The reaction between an acid and a mineral occurs when acid reaches the surface of the mineral by diffusion or convection from the bulk solution. The overall rate of acid consumption or mineral dissolution will depend on two distinct phenomena: the rate of transport of acid to the mineral surface by diffusion or convection, and the actual reaction rate on the mineral surface. Often, one of these processes will be much slower than the other. In this case, the fast process can be ignored, since it can be thought to occur in an insignificant amount of time compared with the slow process. For example, the reaction rate for the $HCl-CaCO_3$ reaction is extremely high, so the overall rate of this reaction is usually controlled by the rate of acid transport to the surface, the slower of the two processes. On the other hand, the surface reaction rates for many HF–mineral reactions are very slow compared with the acid transport rate, and the overall rate of acid consumption or mineral dissolution is reaction rate controlled. Surface reaction rates (reaction kinetics) are discussed in this section, while acid transport is detailed in Section 14.4.

A reaction rate is generally defined as the rate of appearance in the solution of the species of interest in units of moles per second. A surface reaction rate depends on the amount of surface exposed to reaction, so these reactions are expressed on a per-unit surface area basis. In general, the surface reaction rate of an aqueous species A reacting with mineral B is

$$R_A = r_A S_B \tag{14-18}$$

where R_A is the rate of appearance of A (moles/sec), r_A is the surface area–specific reaction rate of A (moles/sec-m^2), and S_B is the surface area of mineral B. When A is being consumed, the reaction rates, r_A and R_A, are negative.

The reaction rate, r_A, will generally depend on the concentrations of the reacting species. However, in the reaction between an aqueous species and a solid, the concentration of the solid can be ignored, since it will remain constant. For example, a grain of quartz has a fixed number of moles of quartz per unit volume of quartz, regardless of reactions that may be occurring on the surface of the grain. Incorporating the concentration dependence into the rate expression yields

$$-R_A = E_f C_A^{\alpha} S_B \tag{14-19}$$

where E_f is a reaction rate constant with units of moles A/[m^2-sec-(moles A/m^3)$^\alpha$], C_A is the concentration of species A at the reactive surface, and α is the order of the reaction, a measure of how strongly the reaction rate depends on the concentration of A. The reaction rate constant depends on temperature and sometimes on the concentration of chemical species other than A. Finally, Equation (14-19) is written in the conventional manner for a species that is being consumed from solution, by placing a minus sign ahead of R_A so that E_f is a positive number.

14.3.1 Laboratory Measurement of Reaction Kinetics

In order to measure the surface reaction rate of acid–mineral reactions, it is necessary to maintain a constant mineral surface area or measure its change during reaction, and to ensure that the rate of acid transport to the mineral surface is fast relative to the reaction rate. The two most common methods of obtaining these conditions are with a well-stirred slurry of mineral particles suspended in an acid solution (a stirred reactor) and with a rotating-disk apparatus (Fogler, Lund, and McCune, 1976; Taylor and Nasr-El-Din, 2009). In the rotating-disk apparatus, a disk of the mineral is placed in a large container holding the acid solution. The disk is rotated rapidly, so that the acid mass transfer rate is high relative to the surface reaction rate. A third, more indirect method is by matching the core flood response to acidizing with a model of the process of flow with reaction. This approach has been common with sandstones and is discussed in Chapter 15.

14.3.2 Reactions of HCl and Weak Acids with Carbonates

Hydrochloric acid is a strong acid, meaning that when HCl is dissolved in water, the molecules almost completely dissociate to form hydrogen ions, H^+, and chloride ions, Cl^-. The reaction between HCl and carbonate minerals is actually a reaction of the H^+ with the mineral. With weak acids, such as acetic or formic acid, the reaction is also between H^+ and the mineral, with the added complication that the acid is not completely dissociated, thus limiting the supply of H^+ available for reaction. Because H^+ is the reactive species, the kinetics for the HCl reaction can also be used for weak acids by considering the acid dissociation equilibrium.

Lund et al. (Lund, Fogler, and McCune, 1973; Lund, Fogler, McCune, and Ault, 1975) measured the kinetics of the HCl–calcite and HCl–dolomite reactions, respectively. Their results were summarized by Schechter (1992) as follows:

$$-r_{\text{HCl}} = E_f C_{\text{HCl}}^{\alpha} \tag{14-20}$$

$$E_f = E_f^0 \exp\left(-\frac{\Delta E}{RT}\right) \tag{14-21}$$

The constants α, E_f^0, and $\Delta E/R$ are given in Table 14-6. SI units are used in these expressions, so C_{HCl} has units of kg-mole/m^3 and T is in K.

The kinetics of a weak-acid carbonate mineral reaction can be obtained from the HCl reaction kinetics by (Schechter, 1992)

$$-r_{\text{weak acid}} = E_f K_d^{\alpha/2} C_{\text{weak acid}}^{\alpha/2} \tag{14-22}$$

Table 14-6 Constants in HCl–Mineral Reaction Kinetics Models

Mineral	α	$E_f^0 \left[\frac{\text{kg moles HCl}}{\text{m}^2\text{-s-(kg-moles HCl/m}^3\text{ acid solution)}^{\alpha}} \right]$	$\frac{\Delta E}{R}$ (K)
Calcite ($CaCO_3$)	0.63	7.55×10^3	7.314×10^7
Dolomite ($CA\ Mg(CO_3)_2$)	$\frac{6.32 \times 10^{-4}T}{1 - 1.92 \times 10^{-3}T}$	7.9×10^3	4.48×10^5

where K_d is the dissociation constant of the weak acid and E_f is the reaction rate constant for the HCl–mineral reaction. A more complex approach to the overall reaction rate of weak acids with carbonate minerals, including mass transfer effects, is given by Buijse et al. (2004).

14.3.3 Reaction of HF with Sandstone Minerals

Hydrofluoric acid reacts with virtually all of the many mineral constituents of sandstone. Reaction kinetics have been reported for the reactions of HF with quartz (Bergman, 1963; Hill, Lindsay, Silberberg, and Schechter, 1981), feldspars (Fogler, Lund, and McCune, 1975), and clays (Kline and Fogler, 1981). These kinetic expressions can all be represented by

$$-r_{\text{mineral}} = E_f[1 + K(C_{\text{HCl}})^{\beta}]C_{\text{HF}}^{\alpha} \tag{14-23}$$

$$E_f = E_f^o \exp\left(-\frac{\Delta E}{RT}\right) \tag{14-24}$$

and the constants α, β, E_f^0, and $\Delta E/R$ are given in Table 14-7.

These expressions show that the dependence on HF concentration is approximately first order ($\alpha = 1$). For the feldspar reactions, the reaction rate increases with increasing HCl concentration, even though HCl is not consumed in the reaction. Thus, HCl catalyzes the HF–feldspar reactions. Also, the reaction rates between clay minerals and HF are very similar in magnitude, except for the illite reaction, which is about two orders of magnitude slower than the others.

Table 14-7 Constants in HF–Mineral Reaction Kinetics Models

Mineral	α	β	$K[(\text{kg} - \text{mole HCl/m}^3)^{-\beta}]$	$E_f^0\left[\frac{\text{kg mole mineral}}{\text{m}^2\text{-sec-(kg-mole HF/m}^3\text{ acid)}^\alpha}\right]$	$\frac{\Delta E}{R}(K)$
Quartz, SiO_2^a	1.0	—	0	2.32×10^{-8}	1150
Orthoclase, K-Feldspar, $KAlSi_3O_8$	1.2	0.4	$5.66 \times 10^{-2} \exp(956/T)$	1.27×10^{-1}	4680
Albite, Na-Feldspar, $NaAlSi_3O_8$	1.0	1.0	$6.24 \times 10^{-2} \exp(554/T)$	9.50×10^{-3}	3930
Kaolinite, $Al_4Si_4O_{10}(OH)_8$	1.0	—	0	0.33	6540^b
Sodium montmorillonite, $Al_4Si_8O_{20}(OH)_4 - nH_2O$	1.0	—	0	0.88	6540^b
Illite, $K_{0-2}Al_4(Al,Si)_8O_{20}(OH)_4$	1.0	—	0	2.75×10^{-2}	6540^b
Muscovite, $KAl_2Si_3O_{10}(OH)_2$	1.0	—	0	0.49	6540^b

[a] Based on 6 moles of HF per mole of SiO_2.

[b] Approximate values reported by Schechter (1992).

Example 14-3 Dissolution Rate of a Rotating Disk of Feldspar

A 2-cm.-diameter disk of albite (Na-feldspar) is immersed in a 3 wt% HF, 12 wt%HCl solution at 50°C and rotated rapidly for 1 hr. The density of the acid solution is 1.075 g/cm^3 and the density of the feldspar is 2.63 g/cm^3. If the acid concentration remains approximately constant during the exposure period, what thickness of the disk will be dissolved and what mass of HF will be consumed?

Solution

The change in the number of moles of feldspar, M_f, is equal to the reaction rate, R_f, or

$$\frac{dM_f}{dt} = R_f = r_f S_f. \tag{14-25}$$

The specific reaction rate, r_f, is constant, since the acid concentration is approximately constant, and neglecting the reaction at the edge of the disk and any effects of surface roughness, the surface area of the disk is constant. Thus, Equation (14-25) is readily integrated to yield

$$\Delta M_f = r_f s_f \Delta t. \tag{14-26}$$

The surface area of the disk is πcm^2 or $\pi \times 10^{-4}$ m^2. The specific reaction rate is obtained from Equations (14-23) and (14-24), using the data in Table 14-7. First, the acid concentrations must be expressed in kg-moles/m^3 of solution:

$$C_{\mathrm{HF}} = \left(\frac{0.03 \text{ kg HF}}{\text{kg solution}}\right)\left(\frac{1075 \text{ kg solution}}{\text{m}^3 \text{ solution}}\right)\left(\frac{1 \text{ kg-mole HF}}{20 \text{ kg HF}}\right)$$
$$= 1.61 \text{ kg-mole HF/m}^3 \text{ solution} \tag{14-27}$$

$$C_{\mathrm{HCl}} = \left(\frac{0.12 \text{ kg HCl}}{\text{kg solution}}\right)\left(\frac{1075 \text{ kg solution}}{\text{m}^3 \text{ solution}}\right)\left(\frac{1 \text{ kg-mole HCl}}{36.5 \text{ kg HCl}}\right)$$
$$= 3.53 \text{ kg-mole HCl/m}^3 \text{ solution} \tag{14-28}$$

Then

$$K = 6.24 \times 10^{-2} \exp\left(\frac{554}{273 + 50}\right) = 0.347\left(\frac{\text{kg moles HCl}}{\text{m}^3 \text{ solution}}\right)^{-1} \tag{14-29}$$

$$E_f = 9.5 \times 10^{-3} \exp\left(-\frac{3930}{273 + 50}\right)$$

$$= 4.94 \times 10^{-8} \frac{\text{kg-mole feldspar}}{m^2 - \text{sec} - (\text{kg-mole HF/m}^3 \text{ solution})} \tag{14-30}$$

$$r_f = -4.94 \times 10^{-8}[1 + 0.347(3.53)](1.61)$$

$$= -1.77 \times 10^{-7} \text{ kg-mole feldspar/m}^2 \text{ sec} \tag{14-31}$$

and the number of moles of feldspar dissolved in 1 hr is

$$\Delta M_f = (-1.77 \times 10^{-7})(\pi \times 10^{-4})(3600) = 2 \times 10^{-7} \text{ kg-moles feldspar.} \tag{14-32}$$

The change in thickness of the disk is the volume dissolved divided by the surface area:

$$\Delta h = \frac{(-2 \times 10^{-7} \text{ kg-mole feldspar})(262 \text{ kg feldspar/kg-mole feldsapar})(\text{m}^3 \text{ feldspar/2630 kg feldspar})}{\pi \times 10^{-4}\, m^2}$$

$$= -6.3 \times 10^{-5}\, m = -0.063 \text{ mm.} \tag{14-33}$$

The minus sign indicates that the surface is receding.

The mass of acid consumed is related to the mass of feldspar dissolved through the stoichiometry. Assuming that 20 moles of HF is consumed per mole of feldspar, the amount of acid consumed is

$$\Delta W_{\text{HF}} = (2 \times 10^{-7} \text{ kg moles feldspar})\left(\frac{20 \text{ kg-moles HF}}{\text{kg-mole feldspar}}\right)\left(\frac{20 \text{ kg HF}}{\text{kg-mole HF}}\right)$$

$$= 8 \times 10^{-5} \text{ kg HF} = 0.08 \text{ g HF.} \tag{14-34}$$

Only 0.063 mm of the feldspar surface is dissolved in 1 hr. Since feldspar is one of the faster-reacting constituents of sandstone, this indicates that the reaction rates in sandstone acidizing are quite small. However, in this example, it was assumed that the feldspar surface was smooth, making the available surface for reaction small. In sandstones composed of small grains of minerals, the specific surface areas of the grains (the surface area per unit volume) are much larger, making the overall reaction rate much larger.

Example 14-4 The Relative Reaction Rates of Sandstone Minerals

A sandstone contains 85% quartz, 10% Na-feldspar, and 5% kaolinite by mass. The specific surface areas of the minerals are approximately 20 m^2/kg for the quartz and feldspar and 8000 m^2/kg for the clay (the reactive surface area of clays is less than the total surface area; Schechter, 1992). If this rock is contacted with a 3 wt% HF, 12 wt% HCl solution at 50 °C, what proportion of the HF will initially be consumed by each of the three minerals?

Solution

Per unit mass of rock, the specific surface area of each mineral is its specific surface area times the mass fraction of the mineral present in the sandstone. For example, the specific surface area of quartz per mass of sandstone, S_q^*, is (20 m^2/kg)(0.85) = 17 m^2/kg rock. Similarly, the specific surface areas of feldspar and kaolinite are 2 and 400 m^2/kg rock, respectively. The overall rate of reaction of HF with each mineral is

$$-R_{HF,\,q} = \gamma_q r_q S_q^* \tag{14-35}$$

$$-R_{HF,\,f} = \gamma_f r_f S_f^* \tag{14-36}$$

$$-R_{HF,\,q=k} = \gamma_k r_k S_k^* \tag{14-37}$$

where γ_i is the ratio of the stoichiometric coefficient of HF to that of mineral i ($\nu_{HF}/\nu_{\text{mineral } i}$). Obtaining the specific reaction rates with Equations 14-23 and 14-24, and the data in Table 14-7, and using for γ_i, 6 moles HF/mole quartz, 20 moles HF/mole feldspar, and 24 moles HF/mole kaolinite,

$$-R_{HF,\,q} = 9.4 \times 10^{-8} C_{HF} \tag{14-38}$$

$$-R_{HF,\,f} = 4.4 \times 10^{-6} C_{HF} \tag{14-39}$$

$$-R_{HF,\,k} = 5.1 \times 10^{-6} C_{HF} \tag{14-40}$$

The fraction of HF expended in a particular reaction is the overall reaction rate for that mineral divided by the sum of the reaction rates, showing that 1% of the HF is reacting with quartz, 46% is reacting with feldspar, and 53% is reacting with kaolinite. This is a typical result, illustrating that the reaction rates of HF with clays and feldspars are approximately two orders of magnitude higher than that between HF and quartz. Because the clay and feldspar reaction rates are relatively high and they generally comprise a small portion of the total rock mass, they will be consumed first in sandstone acidizing. The quartz reaction becomes important in regions where most of the clay and feldspar have already been dissolved.

14.3.4 Reactions of Fluosilicic Acid with Sandstone Minerals

As discussed in Section 14.2, fluosilicic acid, H_2SiF_6, is produced when HF dissolves silicate minerals, and the fluosilicic acid itself may then react with aluminosilicates. From models of core flood experiments, Bryant (1991) and da Motta et al. (1992) conclude that the reaction between fluosilicic acid and clays and feldspars is slow at room temperature, but that it is of the same order of magnitude as the HF reactions with these minerals at temperatures above 50 °C. Advanced models of the sandstone acidizing process include the reactions of fluosilicic acid with alumino-silicate minerals.

14.4 Acid Transport to the Mineral Surface

When the surface reaction rate is high, the overall rate of consumption of acid and dissolution of minerals will be controlled by the rate of acid transport to the mineral surfaces. This occurs most commonly with the reaction of HCl with carbonate minerals, but is possible with other reactions in acidizing, particularly at elevated temperatures. Acid is transported to the mineral surface by diffusion and by convection. The diffusional flux of acid is given by Fick's law,

$$J_y^A = -D_A \frac{\partial C_A}{\partial y} \tag{14-41}$$

where J_y^A is the acid flux [moles or mass/(cm^2-sec)], D_A is the acid diffusion coefficient (cm^2/sec), and y denotes the direction of diffusion. Equation (14-41) shows that acid diffusion to the reactive surface is driven by the concentration gradient between the bulk solution and the surface. The effective diffusion coefficient for HCl has been determined by Roberts and Guin (1975) and is shown in Figure 14-3.

In a porous medium composed of grains of rock and irregularly shaped, interconnected pore spaces, as reservoir rocks typically are, transport of acid to the mineral surfaces is likely due to both diffusion and convection. Modeling this process requires a hydrodynamic model of the flow through the pore structure, a model that is currently intractable except for a few idealized cases. Convective transport of acid to the reactive surface is more simply treated using the concept of fluid loss from a fracture or channel, as applied in hydraulic fracturing models. This application is presented in Chapter 16.

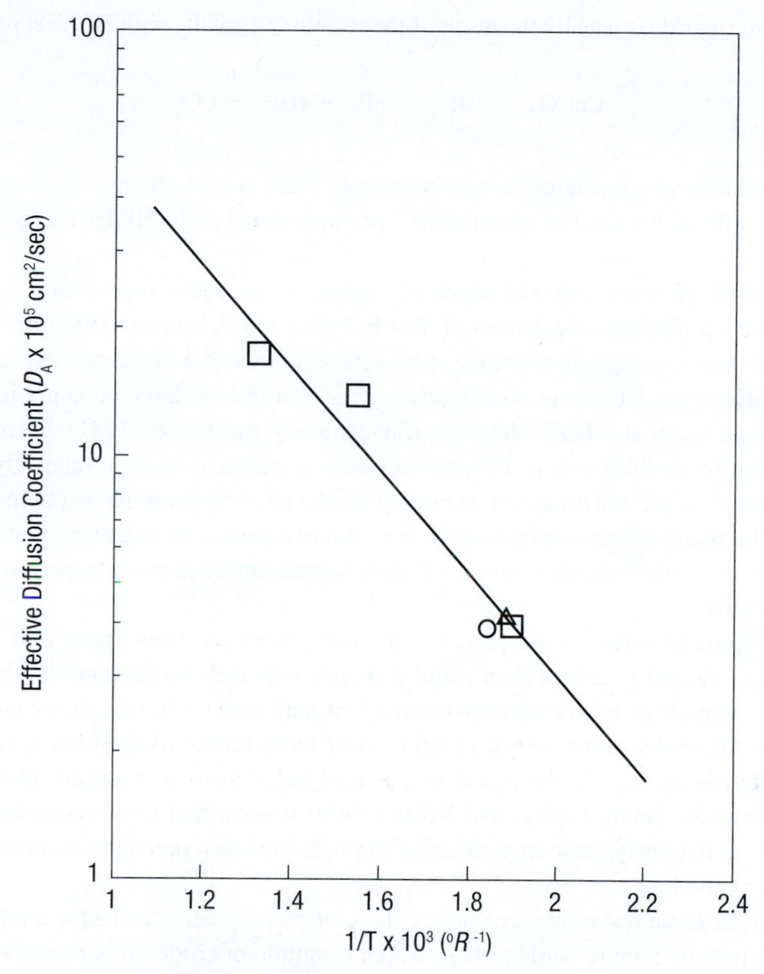

Figure 14-3 The effective diffusion coefficient of HCl. (From Roberts and Guin, 1975.)

14.5 Precipitation of Acid Reaction Products

A major concern in acidizing, particularly acidizing of sandstones, is damage caused by the precipitation of acid–mineral reaction products. In acidizing of sandstone with HF, the formation of some precipitates is probably unavoidable. However, the amount of damage they cause to the well productivity depends on the amount and location of the precipitates. These factors can be controlled somewhat with proper job design.

The most common damaging precipitates that may form in sandstone acidizing are calcium fluoride, CaF_2; colloidal silica, $Si(OH)_4$; ferric hydroxide, $Fe(OH)_3$; and asphaltene sludges.

Calcium fluoride is usually the result of the reaction of calcite with HF, according to

$$CaCO_3 + 2HF \leftrightarrow CaF_2 + H_2O + CO_2. \tag{14-42}$$

Calcium fluoride is very insoluble, so precipitation of CaF_2 is likely if any calcite is available to react with the HF. Inclusion of an adequate HCl preflush ahead of the HF/HCl stage prevents the formation of CaF_2.

Production of some colloidal silica precipitate is probably unavoidable in sandstone acidizing. The equilibrium calculations of Walsh, Lake, and Schechter (1982) show that there will virtually always be regions where the spent acid solution has the tendency to precipitate colloidal silica. However, laboratory core floods suggest that the precipitation is not instantaneous and in fact may occur at a fairly slow rate (Shaughnessy and Kunze, 1981). To minimize the damage caused by colloidal silica, it is probably advantageous to inject at relatively high rates, so that the potential precipitation zone is rapidly displaced away from the wellbore. Also, spent acid should be produced back immediately after the completion of injection, since shutting in the well for even a relatively short time may allow significant silica precipitation to occur in the near-well vicinity.

When ferric ions (Fe^{3+}) are present, they can precipitate from spent acid solutions as $Fe(OH)_3$ when the pH is greater than about 2. Ferric ions may result from the dissolution of iron-bearing minerals in an oxidative environment or may derive from the dissolution of rust in the tubing by the acid solution. When a high level of ferric ions is likely in the spent acid solution, sequestering agents can be added to the acid solution to prevent the precipitation of $Fe(OH)_3$. However, Smith, Crowe, and Nolan (1969) suggest that these sequestrates be used with caution, as they may cause more damage through their own precipitation than would have been caused by the iron.

Finally, in some reservoirs, contacting the crude oil by acid can cause the formation of asphaltenic sludges. Simple bottle tests in which a sample of crude oil is mixed with the acid can indicate whether the crude has a tendency for sludge formation when contacted by acid. When sludge formation is a problem, emulsions of acid in aromatic solvents or surface-active additives have been used to prevent asphaltene precipitation (Moore, Crowe, and Hendrickson, 1965).

Example 14-5 The Effect of a Precipitate Zone on Well Productivity

A well with a radius of 0.328 ft in a 745-ft drainage radius has been treated with acid and all damage removed. However, 1 ft^3/ft of reservoir thickness of spent acid will form a precipitate that reduces the permeability to 10% of the original permeability. By the design of the overflush, this spent acid may be left next to the wellbore or displaced farther into the formation. Determine how the productivity of the well is influenced by the location of this zone of precipitate. The rock porosity is 0.15.

Solution

When the precipitate zone is displaced away from the wellbore, there will be three zones around the well; the region between the well and the precipitate zone, where it can be assumed that the permeability is equal to the original reservoir permeability (the acid removed all damage); the precipitate zone, where the permeability is 10% of the original permeability; and the region beyond the precipitate zone, where the permeability is the original permeability of the reservoir. For steady-state radial flow with three regions in series, the productivity index is

$$J_p = \frac{h}{141.2B\mu\{[\ln(r_1/r_w)/k] + [\ln(r_2/r_1)/k_p] + [\ln(r_e/r_2)/k]\}} \tag{14-43}$$

where r_1 is the inner radius of the precipitate zone, r_2 is the outer radius of the precipitate zone, k_p is the permeability in the precipitate zone, and k is the original reservoir permeability. (*Note:* Equation 14-43 is simply derived by calculating a skin factor using Equation 6-7.) Dividing this equation by the equation describing the productivity index of the undamaged well and defining X_p as k_p/k yields

$$\frac{J_p}{J} = \frac{\ln(r_e/r_w)}{\ln(r_1/r_w) + (1/X_p)\ln(r_2/r_1) + \ln(r_e/r_2)}. \tag{14-44}$$

Finally, since the volume of spent acid creating the precipitate zone is fixed,

$$r_2 = \sqrt{r_1^2 + \frac{V_{\text{spent acid}}}{\pi\phi}}. \tag{14-45}$$

Using values of r_1 ranging from r_w to 2.5 ft., the results in Figure 14-4 are obtained. When the precipitate is surrounding the wellbore, the productivity index of the well is less than 40% of its potential; displacing the precipitate to beyond 2 ft from the wellbore restores the PI to more than 80% of the undamaged case.

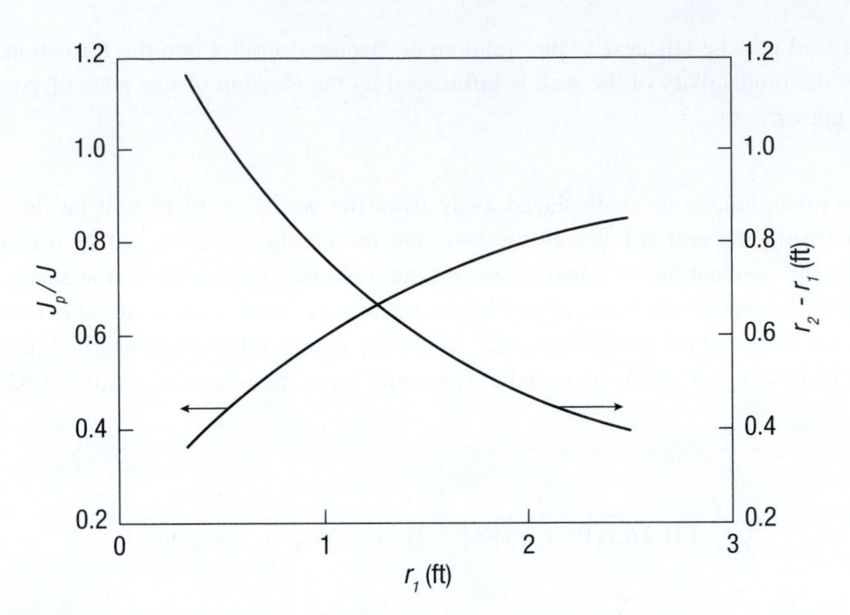

Figure 14-4 Effect of a zone of spent acid precipitate on well productivity.

References

1. Bergman, I., "Silica Powders of Respirable Sizes IV. The Long-Term Dissolution of Silica Powders in Dilute Hydrofluoric Acid: An Anisotropic Mechanism of Dissolution for the Courser Quartz Powders," *J. Appl. Chem.,* **3:** 356–361 (August 1963).

2. Bryant, S.L., "An Improved Model of Mud Acid/Sandstone Chemistry," SPE Paper 22855, 1991.

3. Buijse, M., de Boer, P., Breukel, B., and Burgos, G., "Organic Acids in Carbonate Acidizing," *SPE Production and Facilities,* 128–134 (August 2004).

4. da Motta, E.P., Plavnik, B., Schechter, R.S., and Hill, A.D., "The Relationship between Reservoir Mineralogy and Optimum Sandstone Acid Treatment," SPE Paper 23802, 1992.

5. Fogler, H.S., Lund, K., and McCune, C.C., "Acidization. Part 3. The Kinetics of the Dissolution of Sodium and Potassium Feldspar in HF/HCl Acid Mixtures," *Chem. Eng. Sci.,* **30** (11); 1425–1432, 1975.

6. Fogler, H.S., Lund, K., and McCune, C.C., "Predicting the Flow and Reaction of HCl/HF Mixtures in Porous Sandstone Cores," *SPEJ,* 248–260 (October 1976), *Trans. AIME,* **234.**

7. Furui, K., Zhu, D., and Hill, A.D., "A Rigorous Formation Damage Skin Factor and Reservoir Inflow Model for a Horizontal Well," *SPE Production and Facilities* 151–157, August 2003.

8. Gdanski, R., "Kinetics of the Primary Reaction of HF on Alumino-Silicates," SPE Paper 37459, 1997a.

9. Gdanski, R., "Kinetics of the Secondary Reaction of HF on Alumino-Silicates," SPE Paper 37214, 1997b.

10. Gdanski, R., "Kinetics of Tertiary Reactions of Hydrofluoric Acid on Alumino-Silicates," *SPE Production and Facilities,* 75–80 (May 1998).

11. Gdanski, R.D., and Shuchart, C.E., "Newly Discovered Equilibrium Controls HF Stoichiometry," *JPT,* 145–149 (February 1996).

12. Hartman, R.L., Lecerf, B., Frenier, W., Ziauddin, M., and Fogler, H.S., "Acid-Sensitive Aluminosilicates: Dissolution Kinetics and Fluid Selection for Matrix-Stimulation Treatments," *SPEPO,* 194–204 (May 2006).

13. Hill, A.D., Lindsay, D.M., Silberberg, I.H., and Schechter, R.S., "Theoretical and Experimental Studies of Sandstone Acidizing," *SPEJ,* **21:** 30–42 (February 1981).

14. Kline, W.E., and Fogler, H.S., "Dissolution Kinetics: The Nature of the Particle Attack of Layered Silicates in HF," *Chem. Eng. Sci.,* **36:** 871–884 (1981).

15. Lund, K., Fogler, H.S., and McCune, C.C., "Acidization I: The Dissolution of Dolomite in Hydrochloric Acid," *Chem. Eng. Sci.,* **28:** 691 (1973).

16. Lund, K., Fogler, H.S., McCune, C.C., and Ault, J.W., "Acidization II—The Dissolution of Calcite in Hydrochloric Acid," *Chem. Eng. Sci.,* **30:** 825 (1975).

17. Moore, E.W., Crowe, C.W., and Hendrickson, A.R., "Formation, Effect, and Prevention of Asphaltene Sludges during Stimulation Treatments," *JPT,* 1023–1028 (September 1965).

18. Roberts, L.D., and Guin, J.A., "A New Method for Predicting Acid Penetration Distance," *SPEJ,* 277–286 (August 1975).

19. Schechter, R.S., *Oil Well Stimulation,* Prentice Hall, Englewood Cliffs, NJ, 1992.

20. Shaughnessy, C.M., and Kunze, K.R., "Understanding Sandstone Acidizing Leads to Improved Field Practices," *JPT,* 1196–1202 (July 1981).

21. Smith, C.F., Crowe, C.W., and Nolan, T.J., III, "Secondary Deposition of Iron Compounds Following Acidizing Treatments, *JPT,* 1121–1129 (September 1969).

22. Taylor, K.C., and Nasr-El-Din, H.A., "Measurement of Acid Reaction Rates with the Rotating Disk Apparatus," *JCPT,* **48**(6): 66–70 (June 2009).

23. Walsh, M.P., Lake, L.W., and Schechter, R.S., "A Description of Chemical Precipitation Mechanisms and Their Role in Formation Damage during Stimulation by Hydrofluoric Acid," *JPT,* 2097–2112 (September 1982).

24. Williams, B.B., Gidley, J.L., and Schechter, R.S., *Acidizing Fundamentals,* Society of Petroleum Engineers, Richardson, TX, 1979.

25. Ziauddin, M., Gillard, M., Lecerf, B., Frenier, W., Archibald, I., and Healey, D., "Method for Characterizing Secondary and Tertiary Reactions Using Short Reservoir Cores," *SPE Production and Facilities,* 106–114 (May 2005).

Problems

14-1 A well with a wellbore radius of 0.328 ft and a drainage radius of 745 ft has a perforation skin effect of 3. In addition, a damaged zone where the permeability is 10% of the original permeability extends 9 in. from the wellbore. Calculate the productivity improvement that can be obtained by removing the damage and the skin effect before and after damage removal.

14-2 A vertical well in a reservoir with Appendix A properties, except that the permeabilities are 10 times higher ($k_H = 82$ md and $k_V = 9$ md), is producing 1000 bopd with a bottomhole flowing pressure of 4000 psi and a flowing tubing pressure of 100 psi. The drainage area of the well is 80 acres and the well can be assumed to be operating at steady state with the pressure at the drainage boundary equal to the initial pressure of 5651 psi. It is desired to double the production from the well, either by acidizing the well or by installing a gas-lift completion. The depth to the productive zone is 8000 ft. The field engineers have determined that acid treatments cost about $5000 per unit of skin factor reduction (e.g., to decrease the skin factor by 5 costs about $25,000). Installing gas-lift valves and a compressor for gas lifting the wells costs about $800/horsepower of compression requirement. For gas-lift design, the operating valve is assumed to be at the total depth of 8000 ft (i.e., $H = H_{inj} = 8000$ ft).

a. What is the initial skin factor of the well?

b. What is the cost of an acid treatment that would double the production (assuming that a sufficient amount of the positive skin factor is due to damage that is removable with acid)?

An analysis of the two-phase flow conditions in the tubing for this well has shown that the average pressure gradient in the tubing is decreased by 0.05 psi/ft for every 100 scf/bbl of gas that is added to the natural flow, up to a total GLR of 1000 scf/bbl. For gas-lift operation, the flowing tubing pressure is to be maintained at 100 psi. The injection gas has a gravity of 0.7, the average compressibility factor for the gas in the annulus is 0.9, the average temperature is 140°F, and gas can be supplied to the compressor at 100 psi.

a. What is the gas injection rate needed to double production with no stimulation?

b. What is the surface injection pressure for the gas if the pressure drop across the operating gas lift valve is 100 psi?

c. What is the gas compression requirement (HHP)?

d. Based only on initial cost, without considering operating costs, which method, acidization or artificial lift, appears to be the most economical way to double production for this well?

14-3 A 2000-ft long horizontal well is completed with a slotted liner with a completion skin factor of 1.5 if there were no damage (non-Darcy effects are negligible, and the slots are not plugged). The formation is 50 ft thick, has an undamaged permeability of 20 md, and has a damage zone extending 12 in. into the formation. The damaged zone permeability is 10% of the undamaged zone permeability. The vertical permeability is equal to the horizontal permeability in this homogeneous formation. Steady-state flow can be assumed for all calculations, with a reservoir boundary pressure of 3000 psi, and the well is fully penetrating in the x direction, so that the Furui et al. (2003) inflow model is appropriate for steady-state calculations. The wellbore radius is 0.25 ft and the distance to the drainage boundary in the y-direction is 2000 ft.

a. What is the initial total skin factor for the well?

b. If it is decided to acidize the well and acid injection is started at a constant rate of 5 bpm, what is the bottomhole injection pressure, assuming 1 cp viscosity for all fluids?

c. If the formation is sandstone with a porosity of 0.25 and 5 volume % of the solid material is $CaCO_3$, what volume of 5 wt% acetic acid preflush solution is required to dissolve all the carbonate to a distance of 6 in. beyond the wellbore?

d. Assume that at the end of acid injection, the following conditions have resulted: the permeability in the formerly damaged region is now 200 md; a precipitation zone has been created that extends from 9 in. from the wellbore to 12 in. from the wellbore, and in this region, the permeability is 50 md. Calculate the total skin factor at the end of the treatment.

e. What is the stimulation ratio, J_a/J_d, for this well and treatment?

14-4 Based on the primary reactions in Table 14-1, calculate the gravimetric and volumetric dissolving powers of 3 wt% HF reacting with: (a) orthoclase feldspar; (b) kaolinite; (c) montmorillonite.

14-5 Calculate the specific reaction rate of 3 wt% HF with orthoclase feldspar at 120°F.

14-6 Determine which overall reaction rate is higher at 100°F: the reaction between 3 wt% HF and illite with a specific surface area of 4000 m^2/kg or the reaction between 3 wt% HF and albite with a specific surface area of 20 m^2/kg.

14-7 A 3 wt% HF/12 wt% HCl treatment is injected without an HCl preflush into a sandstone reservoir containing 10% $CaCO_3$. If half of the HF is consumed in reaction with $CaCO_3$ to form CaF_2, what will be the net porosity change, considering both $CaCO_3$ dissolution and CaF_2 precipitation? Assume that all $CaCO_3$ is dissolved in the region contacted by acid. The density of CaF_2 is 2.5 g/cm^3.

CHAPTER 15

Sandstone Acidizing Design

15.1 Introduction

As described in Chapter 14, matrix acidizing is a stimulation technique that is beneficial in sandstone formations primarily when near-wellbore damage exists that can be removed with acid. Therefore, the first step in the planning of any matrix acidizing treatment for a sandstone formation should be a careful analysis of the cause(s) of impaired well performance. This analysis should begin with the measurement of the well skin effect. For highly deviated wells, there is a negative skin effect contribution from the extra exposure to the reservoir. A slightly positive or even zero skin effect that is obtained from a well test should be investigated. Potential recovery of a negative skin effect could greatly enhance well productivity. When a large positive skin effect is present, sources of mechanical skin (partial penetration, perforation skin, etc.) should be investigated, as described in Chapter 6. If mechanical effects do not explain the flow impairment, formation damage is indicated. The well history should then be studied to determine if the damage present is amenable to removal with acid; in general, damage due to drilling mud invasion or fines migration can be successfully treated with acid.

Once acid-treatable formation damage has been identified as the cause of poor well productivity, acidizing treatment design can begin. A typical acid treatment in sandstones consists of the injection of an HCl preflush, with 50 gal/ft of formation being a common preflush volume, followed by the injection of 50–200 gal/ft of HF/HCl mixture. A post-flush of diesel, brine, or HCl then displaces the HF/HCl from the tubing or wellbore. Once the treatment is completed, the spent acid should be immediately produced back to minimize damage by the precipitation of reaction products.

A sandstone acidizing treatment design begins with the selection of the type and concentration of acid to be used. The volume of preflush, HF/HCl mixture, and post-flush required, and the desired injection rate(s), are considered next. In virtually all acid treatments, the placement of the acid is an important issue—a strategy to ensure that sufficient volumes of acid contact all productive parts of the formation should be carefully planned. Proper execution of the treatment is critical to acidizing success, so the conduct of the treatment, including the mechanical arrangements for introducing the acid to the formation and the methods of treatment monitoring, should be planned in detail. Finally, numerous additives are incorporated with acid solutions for various purposes. The types and amounts of additives to be used in the treatment must be determined based on the completion, the formation, and the reservoir fluids. All of these design factors are considered in this chapter.

15.2 Acid Selection

The type and strength (concentration) of acid used in sandstones or carbonates is selected based primarily on field experience with particular formations. For years, the standard treatments consisted of 15 wt% HCl for carbonate formations and a 3 wt% HF, 12 wt% HCl mixture, preceded by a 15 wt% HCl preflush, for sandstones. In fact, the 3/12 HF/HCl mixture has been so common that it is referred to generically as "full-strength mud acid." In recent years, however, the trend has been toward the use of lower-strength HF solutions (Brannon, Netters, and Grimmer, 1987). The benefits of lower-concentration HF solutions are the reduction in damaging precipitates from the spent acid and lessened risk of unconsolidation of the formation around the wellbore.

Based on extensive field experience, McLeod (1984, 1989) presented guidelines for acid selection that were later modified by McLeod and Norman (2000). These later recommendations for sandstone reservoirs are shown in Table 15-1. These guidelines should not be taken as hard-and-fast rules, but rather as starting points in a treatment design.

For a more careful selection of optimal acid formulations, particularly when many wells may be treated in a particular formation, laboratory tests of the responses of cores to different acid strengths are worthwhile. Such tests usually consist of flowing the acid through a small core (typically 1–1½ in. in diameter by 1–3 in. long) while monitoring the permeability response of the core by measuring the pressure drop across the core. A plot of the permeability of the core as a function of acid throughput in pore volumes is termed an "acid response curve" and is a common means of comparing different acidizing conditions. Figure 15-1 (Smith and Hendrickson, 1965) shows acid response curves for three different HF concentrations in Berea sandstone. These curves show that a lower-strength HF solution yields less damage in the early stages of injection; a conservative treatment design would select this low concentration as the best choice of those tested.

The results of a test in a short core cannot be expected to predict the response in a well accurately, but instead should be used as a guideline for field treatments or to provide input to a comprehensive acidizing model. Experiments in log cores, such as those reported by Cheung and Van Arsdale (1992), more accurately reflect field conditions; however, such experiments are expensive and difficult to perform.

Table 15-1 Sandstone Acidizing Fluid Selection Guidelines

Preflush Fluids			
Mineralogy	**Permeability**		
	>100 md	20–100 md	<20 md
<10% silt and <10% clay	15% HCl	10% HCl	7.5% HCl
>10% silt and >10% clay	10% HCl	7.5% HCl	5% HCl
>10% silt and <10% clay	10% HCl	7.5% HCl	5% HCl
<10% silt and >10% clay	10% HCl	7.5% HCl	5% HCl
Main Acid Fluids			
Mineralogy	**Permeability**		
	>100 md	20–100 md	<20 md
<10% silt and <10% clay	12% HCl-3% HF	8% HCl-2% HF	6% HCl-1.5% HF
>10% silt and >10% clay	13.5% HCl-1.5% HF	9% HCl-1% HF	4.5% HCl-0.5% HF
>10% silt and <10% clay	12% HCl-2% HF	9% HCl-1.5% HF	6% HCl-1% HF
<10% silt and >10% clay	12% HCl-2% HF	9% HCl-1.5% HF	6% HCl-1% HF

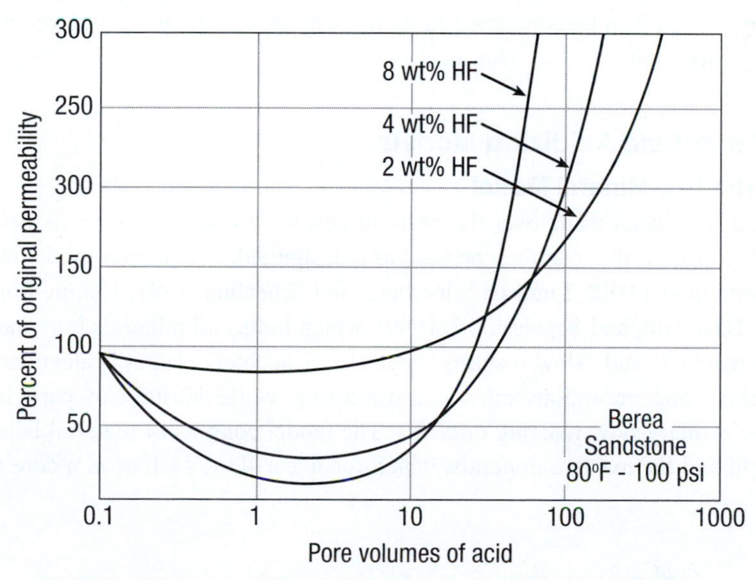

Figure 15-1 Acid response curves. (From Smith and Hendrickson, 1965.)

15.3 Acid Volume and Injection Rate

15.3.1 Competing Factors Influencing Treatment Design

The obvious goal of a sandstone acidizing treatment is to remove near-wellbore formation damage. A less obvious but equally important goal is to minimize the damage caused by the acidizing process itself. At times, these two goals may be at odds. For example, if a shallow damaged zone exists around perforations (e.g., 2 in. or less of damaged region), the damage can be removed with the least amount of acid by injection at a slow rate so that most of the acid is reacted within the 2-in. damaged region. However, such a slow rate may allow precipitates to form from the spent acid very near the perforations, reducing the overall effectiveness of the acid. Thus, what may be an optimal rate based only on the dissolution of minerals in the damaged zone may be less than optimum when the overall acidizing process is considered.

The selection of an optimal acid volume is similarly complicated by competing effects. First, the volume of acid that is needed depends strongly on the depth of the damaged zone, and this depth is seldom known with any accuracy. With an assumption being made about the depth of the damaged zone, an optimal volume for a particular location in the well may be selected based on a laboratory acid response curve or an acidizing model. Such an analysis can yield a minimum recommended acid volume, such as 25 gal per perforation or 50 gal/ft of formation thickness. However, the acid will not in general be distributed equally to all parts of the formation. To ensure that an adequate amount of acid contacts most of the formation, a larger amount of acid may be necessary, depending mainly on the techniques used for proper acid placement.

Thus, design of a sandstone acidizing treatment will be imprecise. The approach recommended here will be to select a target acid volume based on a model of the acidizing process. The treatment should then be conducted in such a way that acidizing can be optimized "on the fly" for a particular well.

15.3.2 Sandstone Acidizing Models

15.3.2.1 The Two-Mineral Model

Several efforts have been made over the years to develop a comprehensive model of the sandstone acidizing process that can then be used as a design aid. A common model in use today is the two-mineral model (Hill, Lindsay, Silberberg, and Schechter, 1981; Hekim, Fogler, and McCune, 1982; Taha, Hill, and Sepehrnoori, 1989), which lumps all minerals into one or two categories: fast-reacting and slow-reacting species. Schechter (1992) categorizes feldspars, authogenic clays, and amorphous silica as fast-reacting, while detrital clay particles and quartz grains are the primary slow-reacting minerals. The model consists of material balances applied to the HF acid and the reactive minerals, which for linear flow, such as in a core flood, can be written as

$$\frac{\partial(\phi C_{\mathrm{HF}})}{\partial t} + u\frac{\partial C_{\mathrm{HF}}}{\partial x} = -(S_F^* V_F E_{f,F} + S_S^* V_S E_{f,S})(1 - \phi)C_{\mathrm{HF}} \qquad \text{(15-1)}$$

$$\frac{\partial}{\partial t}[(1-\phi)V_F] = \frac{-MW_{\mathrm{HF}}S_F^* V_F \beta_F E_{f,F} C_{\mathrm{HF}}}{\rho_F} \tag{15-2}$$

$$\frac{\partial}{\partial t}[(1-\phi)V_S] = \frac{-MW_{\mathrm{HF}}S_S^* V_S \beta_S E_{f,S} C_{\mathrm{HF}}}{\rho_S} \tag{15-3}$$

In these equations, C_{HF} is the concentration of HF in solution, u is the acid flux, x is distance, S_F^* and S_S^* are the specific surface areas, V_F and V_S are the volume fractions, $E_{f,F}$ and $E_{f,S}$ are the reaction rate constants (based on the rate of consumption of HF), MW_F and MW_S are the molecular weights, β_F and β_S are the dissolving powers of 100% HF, and ρ_F and ρ_S are the densities of the fast- and slow-reacting minerals, respectively. When made dimensionless, assuming that porosity remains constant, these equations become:

$$\frac{\partial \psi}{\partial \theta} + \frac{\partial \psi}{\partial \varepsilon} + (N_{\mathrm{Da},F}\Lambda_F + N_{\mathrm{Da},S}\Lambda_S)\psi = 0 \tag{15-4}$$

$$\frac{\partial \Lambda_F}{\partial \theta} = -N_{\mathrm{Da},F}N_{\mathrm{Ac},F}\psi\Lambda_F \tag{15-5}$$

$$\frac{\partial \Lambda_S}{\partial \theta} = -N_{\mathrm{Da},S}N_{\mathrm{Ac},S}\psi\Lambda_S \tag{15-6}$$

where the dimensionless variables are defined as

$$\psi = \frac{C_{\mathrm{HF}}}{C_{\mathrm{HF}}^0} \tag{15-7}$$

$$\Lambda_F = \frac{V_F}{V_F^0} \tag{15-8}$$

$$\Lambda_S = \frac{V_S}{V_S^0} \tag{15-9}$$

$$\varepsilon = \frac{x}{L} \tag{15-10}$$

$$\theta = \frac{ut}{\phi L} \tag{15-11}$$

so that ψ is the dimensionless HF concentration, Λ is dimensionless mineral composition, ε is dimensionless distance, and θ is dimensionless time (pore volumes). For a core flood, L is the core length. In Equations (15-4) through (15-6), two dimensionless groups appear for each mineral, N_{Da}, the Damkohler number, and N_{Ac}, the acid capacity number. These two groups describe the kinetics and the stoichiometry of the HF–mineral reactions. The Damkohler number is the ratio of the rate of acid consumption to the rate of acid convection, which for the fast-reacting mineral is

$$N_{\mathrm{Da},F} = \frac{(1 - \phi_0)V_F^0 E_{f,F} S_F^* L}{u}. \quad \text{(15-12)}$$

The acid capacity number is the ratio of the amount of mineral dissolved by the acid occupying a unit volume of rock pore space to the amount of mineral present in the unit volume of rock, which for the fast-reacting mineral is

$$N_{\mathrm{Ac},F} = \frac{\phi_0 \beta_F C_{\mathrm{HF}}^0 \rho_{\mathrm{acid}}}{(1 - \phi_0)V_F^0 \rho_F}. \quad \text{(15-13)}$$

The Damkohler and acid capacity numbers for the slow-reacting minerals are similarly defined. In this expression for the acid capacity number, the acid concentration is in weight fraction, not moles/volume.

As acid is injected into a sandstone, a reaction front is established by the reaction between the HF and the fast-reacting minerals. The shape of this front depends on $N_{\mathrm{Da},F}$. For low Damkohler numbers, the convection rate is high relative to the reaction rate and the front will be diffuse. With a high Damkohler number, the reaction front is relatively sharp because the reaction rate is high compared with the convection rate. Figure 15-2 (da Motta, Plavnik, and Schechter, 1992a), shows typical concentration profiles for high and low values of $N_{\mathrm{Da},F}$.

Equations (15-4) through (15-6) can only be solved numerically in their general form. However, analytical solutions are possible for certain simplified situations. Schechter (1992) presented an approximate solution that is valid for relatively high Damkohler numbers ($N_{\mathrm{Da},F} > 10$) and is useful for design purposes. This solution approximates the HF-fast-reacting mineral front as a sharp front, so that behind the front all of the fast-reacting minerals have been removed. Conversely, ahead of the front, no dissolution has occurred. The reaction between slow-reacting minerals and HF behind the front serves to diminish the HF concentration reaching the front. The location of the front is given by

$$\theta = \frac{\exp(N_{\mathrm{Da},S}\varepsilon_f) - 1}{N_{\mathrm{Ac},F} N_{\mathrm{Da},S}} + \varepsilon_f, \quad \text{(15-14)}$$

which relates dimensionless time (or, equivalently, acid volume) to the dimensionless position of the front, ε_f, defined as the position of the front divided by the core length for linear flow. The dimensionless acid concentration behind the front is

$$\psi = \exp(-N_{\mathrm{Da},S}\varepsilon). \quad \text{(15-15)}$$

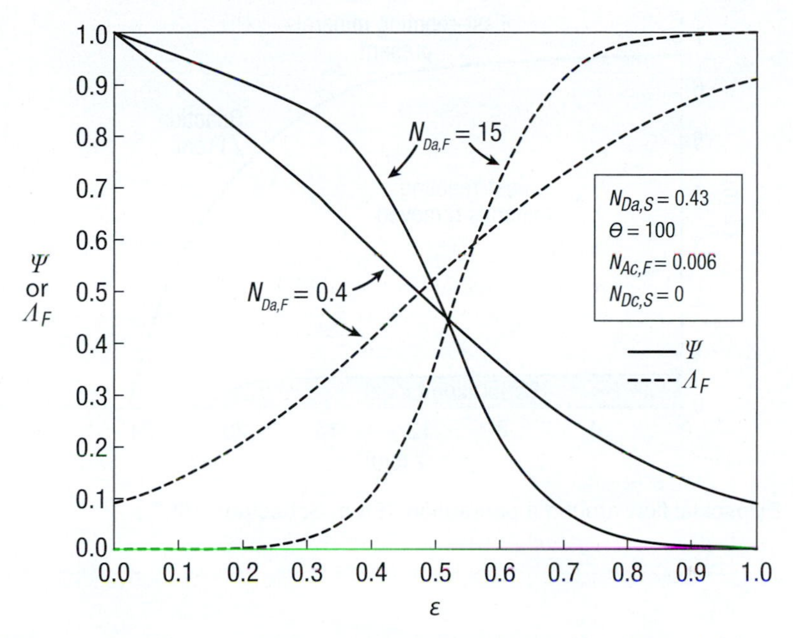

Figure 15-2 Acid and fast-reacting mineral concentration profiles. (From da Motta et al., 1992a.)

A particularly convenient feature of this approximation is that it is applicable to linear, radial, and ellipsoidal flow fields with appropriate definition of dimensionless variables and groups. Radial flow represents the flow of acid from an open-hole completion and may also be a reasonable approximation to the flow from a perforated well with sufficient perforation density. The ellipsoidal flow geometry approximates the flow around a perforation and is illustrated in Figure 15-3. The proper dimensionless variables and groups for these three flow fields are given in Table 15-2. For the perforation geometry, the position of the front, ε_f, depends on the position along the perforation. In Table 15-2, expressions are given for the position of the acid extending directly from the tip of the perforation and for the acid penetration along the wellbore wall. These two positions should be sufficient for design purposes; the reader is referred to Schechter (1992) for methods for calculating the complete acid penetration profile for this geometry.

It is interesting to note that the Damkohler number for the slow-reacting mineral and the acid capacity number for the fast-reacting mineral are the only dimensionless groups that appear in this solution. $N_{\mathrm{Da},S}$ regulates how much live HF reaches the front; if the slow mineral reacts fast relative to the convection rate, little acid will be available to propagate the fast-mineral front. The acid capacity number for the slow-reacting mineral is not important because the supply of slow-reacting mineral is almost constant behind the front. $N_{\mathrm{Ac},F}$ affects the frontal propagation rate directly: The more fast-reacting mineral is present, the slower the front will move. $N_{\mathrm{Da},F}$ does not appear because the front was assumed to be sharp, implying that $N_{\mathrm{Da},F}$ is infinite. This solution can be used to estimate the volume of acid needed to remove the fast-reacting minerals from a given region around a wellbore or perforation.

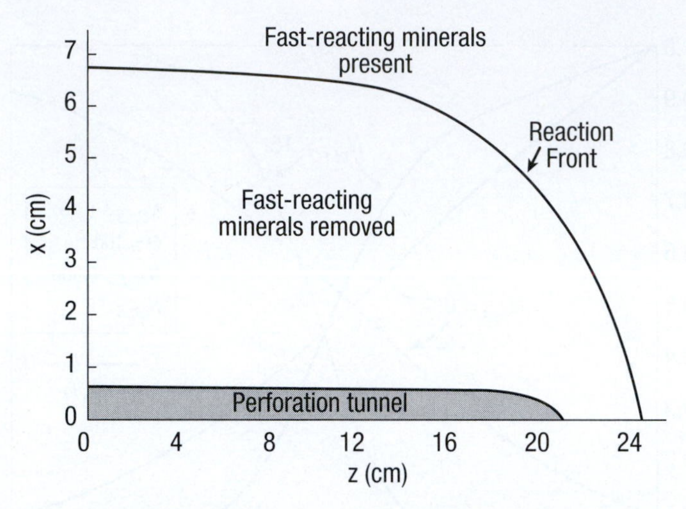

Figure 15-3 Ellipsoidal flow around a perforation. (From Schechter, 1992.)

Table 15-2 Dimensionless Groups in Sandstone Acidizing Model

Flow Geometry	ϵ	θ	$N_{Da,S}$
Linear	$\dfrac{x}{L}$	$\dfrac{ut}{\phi L}$	$\dfrac{(1-\phi)V_S^0 E_{f,S} S_S^* L}{u}$
Radial	$\dfrac{r^2}{r_w^2} - 1$	$\dfrac{q_i t}{\pi r_w^2 h \phi}$	$\dfrac{(1-\phi)V_S^0 E_{f,S} S_S^* \pi r_w^2 h}{q_i}$
Ellipsoidal	Penetration from the tip of the perforation $\dfrac{1}{3}\bar{z}^3 - \bar{z} + \dfrac{2}{3};\ \bar{z} = \dfrac{z}{l_{perf}}$	$\dfrac{q_{perf} t}{2\pi l_{perf}^3 \phi}$	$\dfrac{2\pi(1-\phi) l_{perf}^3 S_S^* V_S^0 E_{f,S}}{q_{perf}}$
	Penetration adjacent to the wellbore $\dfrac{1}{3}\left(\bar{x} + \dfrac{1}{\bar{x} + \sqrt{\bar{x}^2+1}}\right)^3 - \dfrac{1}{3}$ $\bar{x} = \dfrac{x}{l_{perf}}$		

Note: Ψ, Λ, and $N_{Ac,F}$ are the same for all geometries.

The dimensionless groups, $N_{Ac,F}$ and $N_{Da,S}$, can be calculated with Equation (15-13) and Table 15-2 based on the rock mineralogy or can be obtained from experiments, as will be illustrated in the following example.

Example 15-1 Determining $N_{Ac,F}$ and $N_{Da,S}$ from Laboratory Data

The effluent acid concentration measured in a core flood of a 0.87-in. diameter by 1.57-in.-long Devonian sandstone core with 1.5 wt% HF, 13.5 wt% HCl by da Motta et al. (1992a) is shown in Figure 15-4. The acid flux was 0.346 cm/min (0.0115 ft/min). Determine $N_{Da,S}$ and $N_{Ac,F}$ from the data.

Solution

In a typical acidizing core flood, the effluent acid concentration will start at a low value, then gradually increase as the fast-reacting minerals are removed from the core. The sharper the fast-reacting mineral front, the more rapid will be this rise in effluent concentration. When virtually all the fast-reacting minerals have been dissolved, the acid concentration will level off to a plateau value that reflects the consumption of HF by the remaining quartz and other slow-reacting minerals in the core. The effluent concentration is the concentration at $\varepsilon = 1$ as a function of time. Therefore, from Equation (15-15), after all fast-reacting minerals have been dissolved,

$$N_{Da,S} = -\ln(\psi_e). \tag{15-16}$$

From Figure 15-4, the dimensionless effluent acid concentration, ψ_e, gradually increases from 0.61 to 0.65 during the last 50 pore-volumes of the flood (the slight increase is due either to the removal of small amounts of fast-reacting mineral or decreasing surface area of the slow-reacting mineral). Using the average value of 0.63 in Equation (15-16), $N_{Da,S} = 0.46$.

The acid capacity number for the fast-reacting minerals can be estimated from the breakthrough time of the reaction front, using Equation (15-15). The location of the front can be approximated as that position where the acid concentration is half the value supplied to the front. Using the value of 0.63 as the dimensionless concentration supplied to the front, the front emerges from the core when the effluent concentration is 0.315. This occurs at about 180 pore volumes. Solving for $N_{Ac,F}$ from Equation (15-15), when $\varepsilon_f = 1$ gives

$$N_{Ac,F} = \frac{\exp N_{Da,S} - 1}{(\theta_{bt} - 1)N_{Da,S}}, \tag{15-17}$$

so

$$N_{Ac,F} = \frac{\exp(0.46) - 1}{(180 - 1)(0.46)} = 0.0071. \tag{15-18}$$

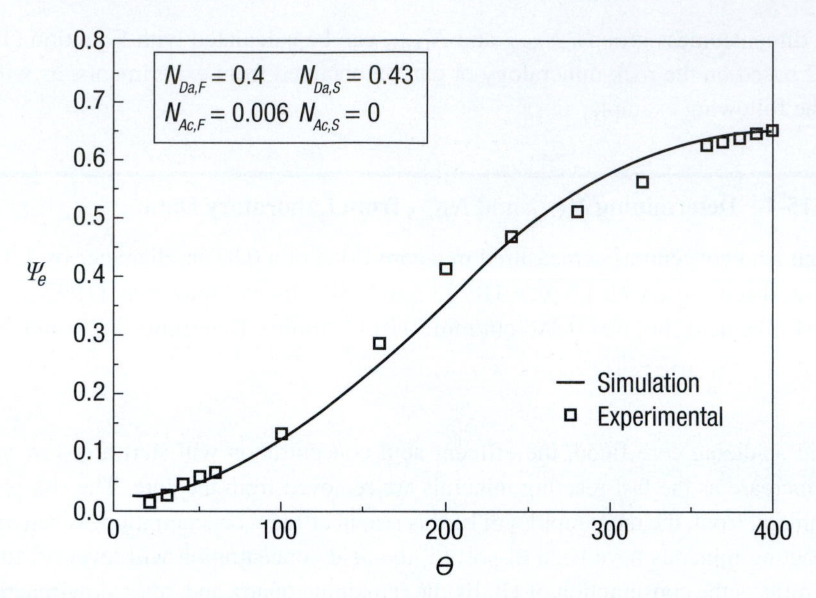

Figure 15-4 Effluent acid concentration from a core flood. (From da Motta et al., 1992a.)

These results compare well with the values of $N_{\mathrm{Da},S} = 0.43$ and $N_{\mathrm{Ac},F} = 0.006$ determined by da Motta et al. (1992a) with a more complex numerical model.

Example 15-2 Acid Volume Design for Radial Flow

Using the acid capacity and Damkohler numbers and other laboratory data from Example 15-1, determine the acid volume (gal/ft) needed to remove all fast-reacting minerals to distances of 3 in. and 6 in. from a wellbore radius of 0.328 ft, assuming that the acid flows radially into the formation, such as would occur in an open-hole completion. The acid injection rate is 0.1 bbl/min-ft of thickness, and the porosity is 0.2.

Solution

From Table 15-2, the radial flow Damkohler number is related to that for linear flow by

$$(N_{\mathrm{Da},S})_{\mathrm{radial}} = (N_{\mathrm{Da},S})_{\mathrm{linear}}\left(\frac{\pi r_w^2 h}{q_i}\right)_{\mathrm{well}}\left(\frac{u}{L}\right)_{\mathrm{core}} \tag{15-19}$$

so

$$(N_{\mathrm{Da},S})_{\mathrm{radial}} = (0.46)\left[\frac{(\pi)(0.328\ \mathrm{ft})^2}{(0.1\ \mathrm{bbl/min\text{-}ft})(5.615\ \mathrm{ft^3/bbl})}\right]\left(\frac{0.0114\ \mathrm{ft/min}}{(1.57/12)\ \mathrm{ft}}\right)$$

$$= 0.024. \tag{15-20}$$

The dimensionless position of the front for radial flow is related to the radial penetration, as shown in Table 15-2:

$$\varepsilon_f = \left(\frac{0.328\ \text{ft} + (3/12)\ \text{ft}}{0.328\ \text{ft}}\right)^2 - 1 = 2.1 \tag{15-21}$$

Using Equation (15-14),

$$\theta = \frac{\exp[(0.024)(2.1)] - 1}{(0.007)(0.024)} + 2.1 = 310. \tag{15-22}$$

From the definition of θ for radial flow in Table 15-2 and recognizing that $q_i t/h$ is simply the volume injected per unit thickness,

$$\frac{q_i t}{h} = \theta \pi r_w^2 \phi_o \tag{15-23}$$

$$\frac{q_i t}{h} = (310)(\pi)(0.328\ \text{ft})^2(0.2)(7.48\ \text{gal/ft}^3) = 160\ \text{gal/ft} \tag{15-24}$$

The dimensionless front position for a penetration of 6 in. from the wellbore is 5.37. Repeating the calculations yields an acid volume of 420 gal/ft, an unreasonably large volume. This particular formation has a high concentration of fast-reacting minerals (low $N_{\text{Ac},F}$) and a relatively high slow-mineral reaction rate (high $N_{\text{Da},S}$), making it difficult to propagate HF far into the formation. A higher acid concentration or higher injection rate would be needed for deeper penetration with the same volume of acid.

Example 15-3 Acid Volume Design for Ellipsoidal Flow from a Perforation

Calculate the acid volume needed (gal/ft) to remove all fast-reacting minerals a distance of 3 in. from the tip of the perforations for a well completed with 4 shots/ft, based on the laboratory data in Example 15-1. The perforation length is 6 in., the porosity is 0.2, and the injection rate is 0.1 bbl/min-ft of thickness. Also calculate the position of the acid front immediately adjacent to the wellbore for this volume of acid injection.

Solution

From Table 15-2, the ellipsoidal flow Damkohler number is related to that for linear flow by

$$(N_{\text{Da},S})_{\text{ellipsoidal}} = (N_{\text{Da},S})_{\text{linear}}\left(\frac{2\pi l_{\text{perf}}^3}{q_{\text{perf}}}\right)\left(\frac{u}{L}\right). \tag{15-25}$$

Since there are 4 shots/ft, q_{perf} is $q_i/4$. Thus,

$$(N_{Da,S})_{ellipsoidal} = (0.46)\left(\frac{(2\pi)(0.5\text{ ft})^3}{(0.025\text{ bbl/min})(5.615\text{ ft}^3/\text{ bbl})}\right)\left(\frac{0.0114\text{ ft/min}}{(1.57/12)\text{ ft}}\right) = 0.224. \quad \textbf{(15-26)}$$

Using the expression from Table 15-2 relating the dimensionless front position to the acid penetration from the tip of the perforation, and noting that z is measured from the wellbore (the start of the perforation),

$$\bar{z} = \frac{z}{l_{perf}} = \frac{6\text{ in.} + 3\text{ in.}}{6\text{ in.}} = 1.5 \quad \textbf{(15-27)}$$

$$\varepsilon_f = \frac{1}{3}(1.5)^3 - 1.5 + \frac{2}{3} = 0.292 \quad \textbf{(15-28)}$$

Then

$$\theta = \frac{\exp[(0.224)(0.292)] - 1}{(0.007)(0.224)} + 0.292 = 43.4 \quad \textbf{(15-29)}$$

The volume injected per perforation is $q_{perf}\,t$, so, from Table 15-2,

$$q_{perf}t = 2\pi l_{perf}^3\phi\theta \quad \textbf{(15-30)}$$

$$q_{perf}t = (2\pi)(0.5\text{ ft})^3(0.2)(43.4)(7.48\text{ gal/ft}^3) = 51\text{ gal/perf} \quad \textbf{(15-31)}$$

Since there are 4 shots/ft, the total acid volume per foot is 204 gal/ft.

The dimensionless acid penetration distance, ε_f, is the same at any location around the perforation. The acid penetration at the wellbore is found from the second expression in Table 15-2, using the value of 0.292 for ε_f and solving for x by trial and error. The result is a penetration distance of 4.3 in. around the base of the perforation adjacent to the wellbore.

Example 15-4 Acid Volume Needed to Remove Drilling Mud Damage

A very common application of acidizing in sandstones is to remove the damage caused by the invasion of drilling mud into the formation. When damage is caused by the invasion of HF-soluble solids like bentonite clay, an acidizing treatment is a very effective method of increasing well productivity if the damage is not too deep. The amount of acid needed depends strongly on the depth of the bentonite damage.

Assume that the Damkohler and acid capacity numbers from Example 15-3 describe a sandstone formation that has been damaged by bentonite mud. The bentonite mud has reduced the porosity in the damaged zone to 0.1 and the permeability in the damaged zone is low enough to create a skin factor of 20. Assume the same perforated completion as described in Example 15-3. Calculate the volume of acid needed to remove the damage and create a zero skin factor for damage depths up to 6 in. (0.5 ft).

Solution

Repeating the calculations shown in Example 15-3 (except changing the porosity to 0.1) for depths of damage varying up to 6 inches, Figure 15-5 is created. As the damage depth increases, the acid volume required to remove the damage increases dramatically. The 350 gal/ft of acid required to remove the damage 6 in. beyond the end of a perforation is more than twice as much is typically pumped. Successful acidizing treatments in sandstones are usually treating very shallow, but severe, formation damage.

15.3.2.2 The Two-Acid, Three-Mineral Model

The two-mineral model just presented greatly simplifies the complex chemistry of sandstone acidizing to allow for analytical calculations of the movement of reaction fronts in sandstone acidizing. However, as was reviewed in Chapter 14, there are numerous secondary and tertiary reactions in sandstone acidizing that alter the process. The simplest extension of the two-mineral model is to include the reaction of fluosilicic acid with the lumped fast-reacting minerals, which also requires explicit consideration of the precipitation of amorphous silica, as described by

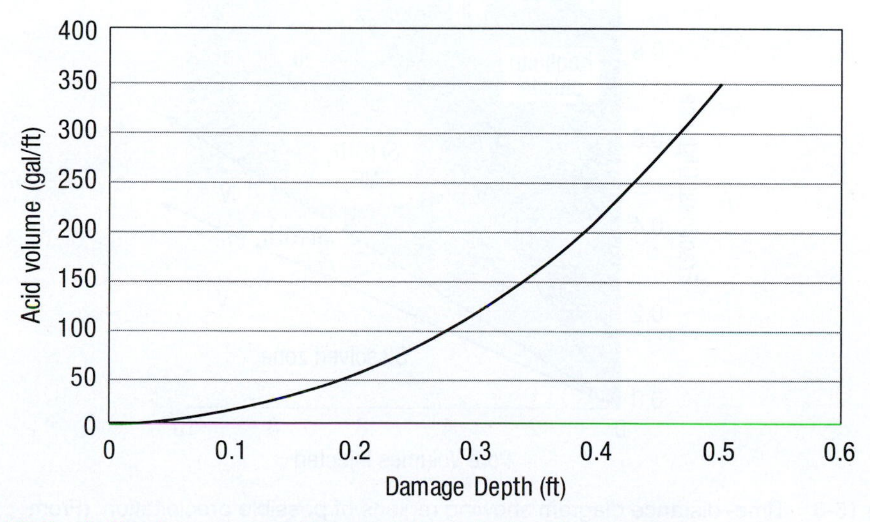

Figure 15-5 Acid volume required to penetrate a shallow damage zone.

Bryant (1991) and da Motta, Plavnik, Schechter, and Hill (1992b). This adds the following reaction to the two-mineral model:

$$H_2SiF_6 + \text{fast-reacting mineral} \rightarrow \nu Si(OH)_4 + \text{Al fluorides}. \quad (15\text{-}32)$$

The practical implications of this reaction being significant are that less HF will be needed to consume the fast-reacting minerals with a given volume of acid because the fluosilicic acid will also be reacting with these minerals, and the reaction product, $Si(OH)_4$ (silica gel), will precipitate. This reaction allows live HF to penetrate further into the formation; the price is the risk of the formation of possible damaging precipitates.

The two-acid, three-mineral model suggests that using the two-mineral model for predicting required acid volumes will be a conservative approach, particularly for high-temperature applications.

15.3.2.3 Precipitation Models

The two-acid, three-mineral model considers the precipitation of silica gel in its description of the acidizing process. However, there are other reaction products that may precipitate. The tendency for reaction product precipitation has been studied with comprehensive geochemical models that consider a large number of possible reactions that may take place in sandstone acidizing.

The most common type of geochemical model used to study sandstone acidizing is the local equilibrium model, such as that described by Walsh, Lake, and Schechter (1982). This type of model assumes that all reactions are in local equilibrium (i.e., all reaction rates are infinitely fast). A typical result from this model is shown in Figure 15-6, a time–distance diagram for the

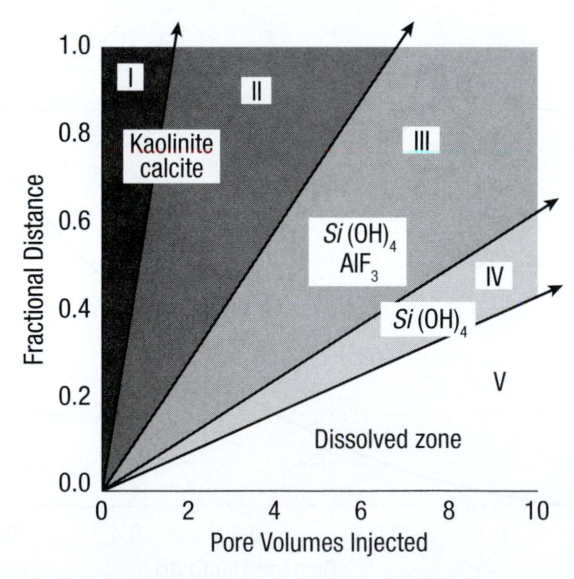

Figure 15-6 Time–distance diagram showing regions of possible precipitation. (From Schechter, 1992.)

injection of 4 wt% HF/11 wt% HCl into a formation containing calcite, kaolinite, and quartz. This plot shows regions where amorphous silica and aluminum fluoride tend to precipitate. A vertical line on this plot represents the mineral species present as a function of distance if all reactions are in local equilibrium.

Sevougian, Lake, and Schechter (1995) presented a geochemical model that includes kinetics for both dissolution and precipitation reactions. This model shows that precipitation damage will be lessened if either the dissolution or the precipitation reactions are not instantaneous. For example, Figure 15-7 shows predicted regions and concentrations of precipitates for four different precipitation reaction rates (minerals AC and DB are precipitates). As the reaction rate decreased, the amount of damaging precipitate formed decreased. Other comprehensive geochemical models that incorporate numerous solid and aqueous phase species are described in the literature (Ziauddin, Kotlar, Vikane, Frenier, and Poitrenaud, 2002a; Ziauddin et al., 2002b).

15.3.2.4 Permeability Models

To predict the response of a formation to acidizing, it is necessary to predict the change in permeability as acid dissolves some of the formation minerals and as other minerals precipitate. The permeability change as a result of acidizing is an extremely complicated process, since it is affected by several different, sometimes competing phenomena in the porous media. The permeability increases as pores and pore throats are enlarged by mineral dissolution. At the same time,

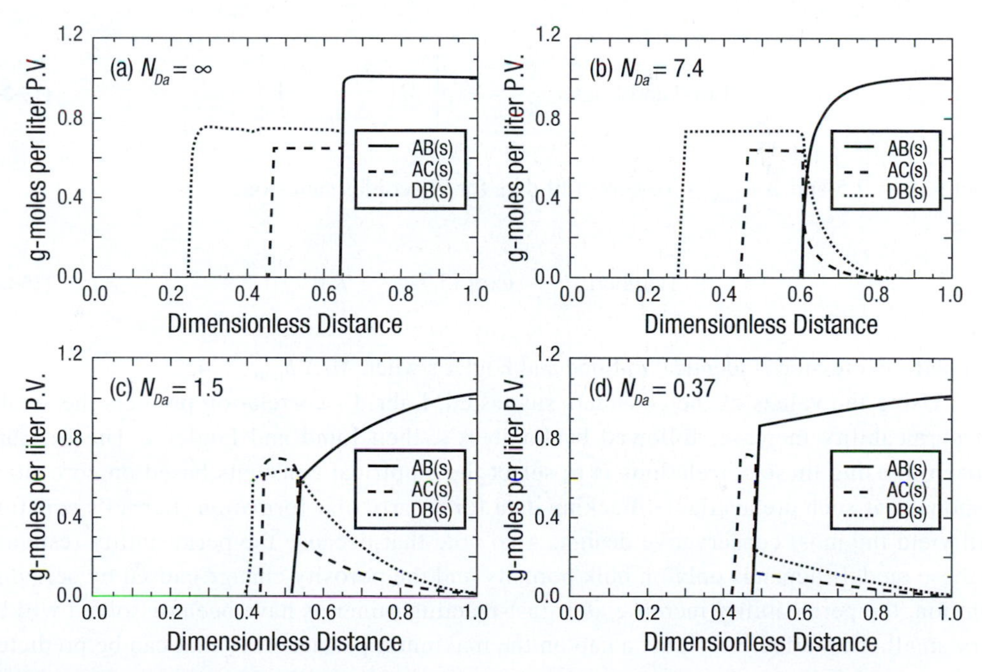

Figure 15-7 Effect of precipitation kinetics on regions of precipitate formation. (From Sevougian et al., 1995.)

small particles are released as cementing material is dissolved, and some of these particles will lodge (perhaps temporarily) in pore throats, reducing the permeability. Any precipitates formed will also tend to decrease the permeability. The formation of CO_2 as carbonate minerals are dissolved may also cause a temporary reduction in the relative permeability to liquids. The result of these competing effects is that the permeability in core floods usually decreases initially; with continued acid injection, the permeability eventually increases to values considerably higher than the original permeability.

The complex nature of the permeability response has made its theoretical prediction for real sandstones impractical, although some success has been achieved for more ideal systems such as sintered disks (Guin, Schechter, and Silberberg, 1971). As a result, empirical correlations relating the permeability increase to the porosity change during acidizing are used. The most common correlations used are those of Labrid (1975), Lund and Fogler (1976), and Lambert (1981). These correlations are:

$$\text{Labrid: } \frac{k_i}{k} = M\left(\frac{\phi_i}{\phi}\right)^n \tag{15-33}$$

where k and ϕ are the initial permeability and porosity and k_i and ϕ_i are the permeability and porosity after acidizing. M and n are empirical constants, reported to be 1 and 3 for Fontainbleau sandstone.

$$\text{Lund and Fogler: } \frac{k_i}{k} = \exp\left[M\left(\frac{\phi_i - \phi}{\Delta\phi_{\max}}\right)\right] \tag{15-34}$$

where $M = 7.5$ and $\Delta\phi_{\max} = 0.08$ best fit data for Phacoides sandstone.

$$\text{Lambert: } \frac{k_i}{k} = \exp[45.7(\phi_i - \phi)] \tag{15-35}$$

Lambert's expression is identical to Lund and Fogler's when $M/\Delta\phi_{\max} = 45.7$.

Using the values of the constants suggested, Labrid's correlation predicts the smallest permeability increase, followed by Lambert's, then Lund and Fogler's. The best approach in using these correlations is to select the empirical constants based on core flood responses, if such are available. Lacking data for a particular formation, Labrid's equation will yield the most conservative design. Also note that because the permeability response in these models depends only on bulk porosity and the porosity change caused by acid dissolution, the permeability increase after fast-reacting minerals have been dissolved will be very small. This effectively puts a cap on the maximum permeability that can be predicted by these models.

Example 15-5 Predicting the Permeability Response to Acidizing

A sandstone with an initial porosity of 0.2 and initial permeability of 20 md contains a total of 10 vol% of carbonate and fast-reacting minerals in the damaged region. Calculate the permeability after removing all of these minerals using each of the permeability correlations.

Solution

The change in porosity due to acidizing is the initial fraction of the bulk volume that is solid $(1 - \phi)$ multiplied by the fraction of solid dissolved,

$$\Delta\phi = (1 - 0.2)(0.1) = 0.08, \tag{15-36}$$

so the porosity after acidizing is 0.28. The predicted permeabilities after acidizing are as follows.

$$\text{Labrid: } k_i = (20 \text{ md})\left(\frac{0.28}{0.2}\right)^3 = 55 \text{ md} \tag{15-37}$$

$$\text{Lund and Fogler: } k_i = (20 \text{ md}) \text{ EXP}\left[\frac{(7.5)(0.08)}{0.08}\right] = 3.6 \times 10^4 \text{ md} \tag{15-38}$$

$$\text{Lambert: } k_i = (20 \text{ md}) \exp[(45.7)(0.08)] = 770 \text{ md} \tag{15-39}$$

The predicted permeabilities with the three correlations differ by orders of magnitude, illustrating the great uncertainty involved in trying to predict the permeability response to acidizing in sandstones. Each of these correlations is based on experiments with particular sandstones and, clearly, the manner in which the pore structures of these formations responded to acid differed markedly.

Table 15-3 Example 15-5 Results

q_i (bpm)	N_{Re}	f	Δp_F (psi)	$p_{ti,max}$ (psi)
0.5	50,800	0.0061	48	2370
1.0	102,000	0.0056	175	2500
2.0	203,000	0.0053	660	2980
4.0	406,000	0.0051	2550	4870

15.3.3 Monitoring the Acidizing Process, the Optimal Rate Schedule

Selecting an optimal injection rate for sandstone acidizing is a difficult process because of uncertainties about the damaged zone and the competing effects of mineral dissolution and reaction product precipitation. Based on mineral dissolution alone, and given the depth of the damaged zone, an optimal injection rate can be derived using the two-mineral model, with the optimal rate being the rate that maximizes the spending of the HF in the damaged zone. Such an approach has been presented by da Motta et al. (1992a). However, the depth of the damage is seldom known with any certainty, and the detrimental effects of reaction product precipitates may be lessened by using rates higher than the optimum based on mineral dissolution alone. In fact, results with the two-acid, three-mineral model have shown that acidizing efficiency is relatively insensitive to injection rate, and that the highest rate possible yields the best results. Figure 15-8 (da Motta, 1993) shows that with shallow damage, injection rate has little effect on the skin effect after 100 gal/ft of acid injection; with deeper damage, the higher the injection rate, the lower the skin effect achieved by acidizing.

Field experience has also been contradictory regarding the optimal injection rate for sandstone acidizing. McLeod (1984) recommends relatively low injection rates, based on the observation that acid contact times with the formation of 2–4 hours appear to give good results. On the other hand, Paccaloni, Tambini, and Galoppini (1988) and Paccaloni and Tambini (1993) reported high success rates in numerous field treatments using the highest injection rates possible. It is likely that the success of Paccaloni's "maximum Δp, maximum rate" procedure is due to

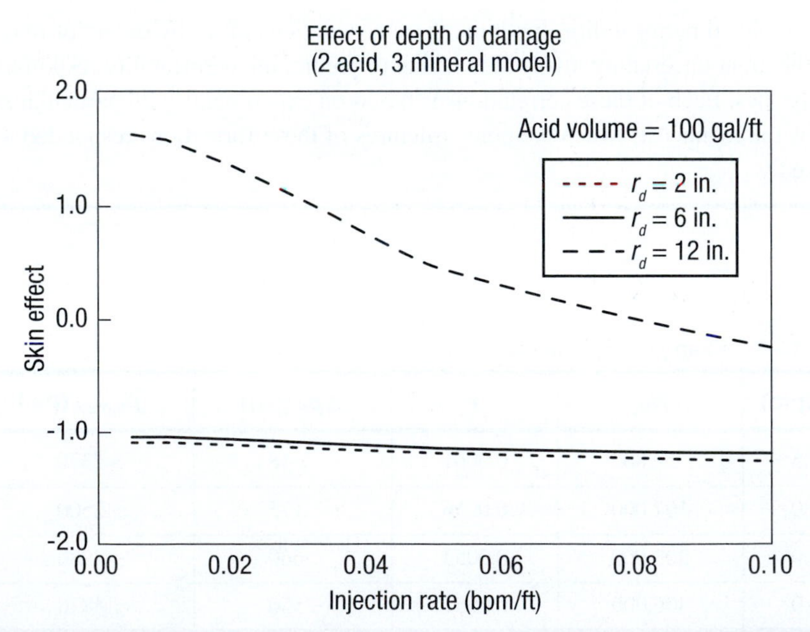

Figure 15-8 Effect of injection rate on skin evolution. (From da Motta, 1993; courtesy of E. P. da Motta.)

improved acid coverage (discussed in the next section) and the minimization of precipitation damage. Unless experience in a particular area suggests otherwise, Paccaloni's maximum Δp, maximum rate procedure is recommended and is described here.

There is always an upper bound on injection rate, since, to ensure a matrix treatment, the injection rate must be kept below the rate that will cause hydraulic fracture formation. The bottomhole pressure that will initiate a fracture (the breakdown pressure) is simply the fracture gradient, FG, times depth, or

$$p_{bd} = \text{FG}(H). \tag{15-40}$$

(See Chapter 18 for a thorough discussion of the breakdown pressure.) Assuming pseudosteady-state flow, the breakdown pressure is related to the injection rate by Equation (2-34), which for injection is

$$p_{bd} - \overline{p} = \frac{141.2 q_{i,\text{max}} \mu}{kh} \left(\ln \frac{0.472 r_e}{r_w} + s \right). \tag{15-41}$$

Most commonly, only the surface rate and tubing pressure are monitored during an acid job, so it is necessary to relate the surface pressure to the bottomhole pressure by

$$p_{ti,\text{max}} = p_{wf} - \Delta p_{\text{PE}} + \Delta p_{\text{F}} \tag{15-42}$$

where the potential energy and frictional pressure drops are both positive quantities. If the fluid is Newtonian, these can be computed using the methods described in Chapter 7. When foams or other non-Newtonian fluids are used, the bottomhole pressure should be monitored continuously (e.g., by monitoring annulus pressure), or pumping should occasionally be stopped for a brief period to determine the frictional pressure drop. Before a matrix acidizing treatment is begun, the relationship between the maximum rate and tubing pressure should be calculated so that the breakdown pressure is never exceeded. This should be done for a wide range of injection rates, since the injection rate that the formation can accommodate without fracturing will increase as the skin effect is reduced by the acid.

Example 15-6 Maximum Injection Rate and Pressure

A well in Reservoir A, at a depth of 9822 ft, is to be acidized with an acid solution having a specific gravity of 1.07 and a viscosity of 0.7 cp down 2-in. I.D. coiled tubing with a relative roughness of 0.001. The formation fracture gradient is 0.7 psi/ft. Plot the maximum tubing injection pressure versus acid injection rate. If the skin effect in the well is initially 10, what is the maximum allowable rate at the start of the treatment? Assume that $r_e = 1000$ ft and $\overline{p} = 4500$ psi.

Solution

The breakdown pressure from Equation (15-40) is

$$p_{bd} = (0.7 \text{ psi/ft})(9822 \text{ ft}) = 6875 \text{ psi} \tag{15-43}$$

The maximum tubing injection pressure is given by Equation (15-42). Using Equation (7-22), the potential energy pressure drop, which is independent of flow rate, is

$$\Delta p_{\mathrm{PE}} = (0.433\ \mathrm{psi/ft})(1.07)(9822\ \mathrm{ft}) = 4551\ \mathrm{psi} \tag{15-44}$$

The frictional pressure drop is given by Equation (7-31), which, for q_i in bpm and all other quantities in the usual oilfield units, is

$$\Delta p_F = \frac{1.525 \rho q_i^2 f L}{D^5}. \tag{15-45}$$

The friction factor is obtained from the Reynolds number and the relative roughness using Equation (7-35) or Figure 7-7. The results for flow rates ranging from 0.5 to 10 bpm are given in Table 15-3 and plotted in Figure 15-9. As injection rate is increased, the surface tubing pressure to maintain a constant bottomhole pressure increases because of the increased frictional pressure drop.

The maximum injection rate at the start of the treatment can be obtained from Equation (15-41). Solving for $q_{i,\max}$ yields

$$q_{i,\max} = \frac{(6875 - 4500)(8.2)(53)}{(141.2)(1.7)\{\ln[(0.472)(1000)/0.328] + 10\}}$$
$$= 250\ \mathrm{bbl/d} = 0.17\ \mathrm{bpm}. \tag{15-46}$$

Thus, the initial injection rate should be less than 0.2 bpm and the surface tubing pressure should be less than 2330 psi. The viscosity of the reservoir oil was used in this calculation since almost

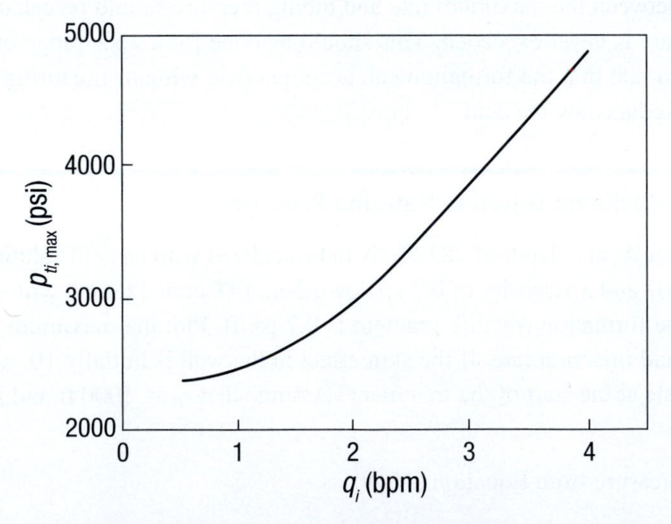

Figure 15-9 Maximum tubing injection pressure (Example 15-6).

all of the pressure drop in the reservoir occurs where oil is flowing, even though an aqueous fluid is being injected.

As the skin effect decreases during acidizing, a higher injection rate is possible without fracturing the formation. To ensure that fracturing does not occur, it is usually recommended that the surface tubing pressure be maintained about 10% below the pressure that would initiate fracturing. For this well, a prudent design would begin with an injection rate of 0.1 bpm, which could be increased while keeping the surface pressure below 2000 psi. If the rate became higher than about 1 bpm, the tubing pressure could increase above 2000 psi, but should be held below 10% of the pressure shown in Figure 15-9 for any particular rate.

15.3.3.1 Paccaloni's Maximum Δp, Maximum Rate Procedure

Paccaloni's maximum Δp, maximum rate procedure has two major components: (1) The acid should be injected at the maximum rate and hence, maximum p_{wf} possible without fracturing the formation; and (2) the treatment should be monitored in real time to ensure that the maximum rate objective is being met and to determine when sufficient acid has been injected into the formation.

To monitor the progress of a stimulation treatment, Paccaloni calculates a steady-state skin effect, according to

$$s = \frac{0.00708kh\Delta p}{\mu q_i} - \ln\frac{r_b}{r_w} \tag{15-47}$$

where all of the variables are in standard oilfield units. In Equation (15-47), r_b is the radius affected by acid injection and Paccaloni suggests that a value of 4 ft be used. Equation (15-47) relates q_i, p_{wf} (through Δp), and s. To monitor a stimulation treatment, Paccaloni constructs a graph of p_{ti} versus q_i, with s as a parameter, where p_{ti} is related to p_{wf} by Equation (15-42). As the treatment proceeds, the wellhead pressure is plotted against the injection rate and the evolving skin effect can be read from the graph. The fracture pressure versus injection rate is also plotted on this graph, so that the rate can be held at the highest possible value without fracturing the formation. Figure 15-10 shows such a design chart for a matrix stimulation treatment.

A drawback to Paccaloni's method is that it will tend to overestimate the skin effect because it neglects transient flow effects, as pointed out by Prouvost and Economides (1988) and confirmed by Paccaloni and Tambini (1993). This error can be severe when abrupt rate changes occur; however, in most treatments, it is not a serious problem because the error will be relatively constant and the evolution in skin effect is more important that its absolute value. Prouvost and Economides presented a procedure to calculate more accurately the changing effect during matrix acidizing. This technique is more applicable to a careful post-treatment analysis, when the injection rate schedule from the treatment is known.

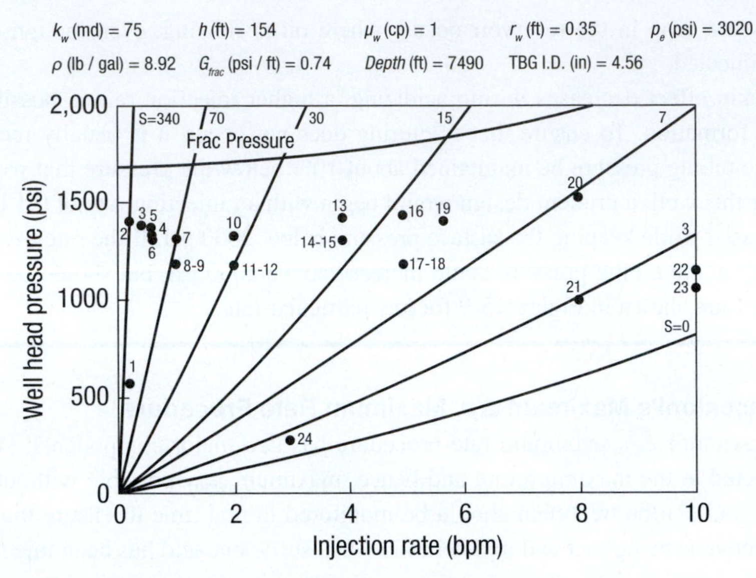

Figure 15-10 Paccaloni design chart for monitoring acidizing. (From Paccaloni and Tambini, 1993.)

15.3.3.2 Calculating Skin Factor Evolution During Acidizing

An alternative real-time monitoring procedure that accounts for transient effects and explicitly handles rate changes was presented by Zhu and Hill (1998). This method begins with the transient flow equation for injection:

$$p_{wi} - p_i = \frac{162.6qB\mu}{kh}\left(\log(t) + \log\left(\frac{k}{\phi\mu c_t r_w^2}\right) - 3.23 + 0.87s\right) \quad \textbf{(15-48)}$$

where p_{wi} is the bottomhole injection pressure, and all other variables are the same as defined for Equation 2-26.

Rearranging Equation 15-48,

$$\frac{p_{wi} - p_i}{q} = m \log(t) + b \quad \textbf{(15-49)}$$

where

$$m = \frac{162.6B\mu}{kh} \quad \textbf{(15-50)}$$

and

$$b = m\left(\log\left(\frac{k}{\phi\mu c_t r_w^2}\right) - 3.23 + 0.87s\right) \tag{15-51}$$

In Equation 15-49, the left-hand term is called *inverse injectivity,* which contains the data measured during the treatment, and the initial reservoir pressure. Equation 15-49 states that there exists a linear relationship between inverse injectivity and the logarithm of time, with a slope of m and an intercept of b defined by Equations 15-50 and 15-51. For a varying injection rate, the superposition principle leads to, from the above equations,

$$\frac{p_i - p_{wf}}{q} = m\Delta t_{\text{sup}} + b \tag{15-52}$$

where the superposition time function, Δt_{sup}, in Equation 15-52 is

$$\Delta t_{\text{sup}} = \sum_{j=1}^{N} \frac{q_j - q_{j-1}}{q_N} \log(t_N - t_{j-1}). \tag{15-53}$$

The skin evolution method uses the above equations to calculate skin factor as a function of time. At any time, the superposition time function can be calculated from the flow rate history at the current time from Equation 15-53, and then the parameter, b (the intercept of the plot of inverse injectivity versus superposition time function) can be obtained from Equation 15-52. Once b is known, the skin at the current time can be computed from Equation 15-51. Notice that the parameter m (the slope of the plot of inverse injectivity versus superposition time function) does not change as injection proceeds, and it can be obtained before treatment from evaluation of the reservoir.

Example 15-7 Skin Factor Evolution during Acidizing

For the data given in Table 15-4, determine the skin factor as a function of time for this acid treatment. Additional data about the well are:

$p_i = 2250$ psi $\quad B_o = 1.2 \quad \phi = 0.2 \quad c_t = 9 \times 10^{-5}$ psi^{-1} $\quad h = 253$ ft

$k = 80$ md $\quad$ initial skin factor, $s = 40 \quad r_w = 0.51$ ft $\quad \mu = 2.4$ cp

Solution

From the given data, first we calculate

$$m = \frac{162.6B_o\mu}{kh} = \frac{(162.6)(1.2)(2.4\text{ cp})}{(80\text{ md})(253\text{ ft})} = 0.023\frac{\text{psi}}{\text{bbl}}. \tag{15-54}$$

Table 15-4 Acid Treatment Data

Fluid at perf	Time (hr)	q (bpm)	p_{wf} (psi)
Water	0.15	2	4500
Water	0.30	2	4489
Water	0.45	2	4400
Aromatic solvent	0.60	2	4390
Aromatic solvent	0.75	2.5	4350
Aromatic solvent	0.90	2.4	4270
Aromatic solvent	1.05	2.7	4250
15% HCl	1.20	2.5	4200
15% HCl	1.35	3.2	3970
3%HF/12%HCl	1.50	3.7	3800
3%HF/12%HCl	1.65	4.1	3750
3%HF/12%HCl	1.80	4.4	3670
3%HF/12%HCl	2.10	3.9	3400
3%HF/12%HCl	2.25	5.6	3730
3%HF/12%HCl	2.40	5.6	3750
3%HF/12%HCl	2.55	5.6	3650
3%HF/12%HCl	2.70	5.7	3600
3%HF/12%HCl	2.85	5.8	3400

Then Δt_{sup} is calculated using Equation 15-53 and b at each time, where

$$b = \frac{p_{wf} - p_i}{q} - m\Delta t_{\text{sup}}. \tag{15-55}$$

Then the skin, s, is calculated from

$$s = \frac{1}{0.87}\left[\frac{b}{m} - \log\left(\frac{k}{\phi\mu c_t r_w^2}\right) + 3.23\right]. \tag{15-56}$$

Table 15-5 Skin Evolution Example Results

Input				Calculated		
Fluid at Perfs.	**Time, t (hr)**	**q_i (bpm)**	**p_{wf} (psi)**	**Δt_{sup}**	**b**	**s**
Water	0.15	2	4500	−0.824	0.799	**38**
Water	0.30	2	4480	−0.532	0.786	**37**
Water	0.45	2	4400	−0.347	0.754	**36**
Aromatic solvent	0.60	2	4390	−0.222	0.748	**35**
Aromatic solvent	0.75	2.5	4350	−0.265	0.589	**27**
Aromatic solvent	0.90	2.4	4270	−0.113	0.587	**27**
Aromatic solvent	1.05	2.7	4250	−0.121	0.517	**23**
15% HCl	1.20	2.5	4200	0.036	0.541	**24**
15% HCl	1.35	3.2	3970	−0.111	0.376	**15**
3% HF / 12% HCl	1.50	3.7	3800	−0.117	0.293	**11**
3% HF / 12% HCl	1.65	4.1	3750	−0.092	0.256	**9**
3% HF / 12% HCl	1.80	4.4	3670	−0.051	0.225	**7**
3% HF / 12% HCl	2.10	3.9	3400	−0.041	0.206	**6**
3% HF / 12% HCl	2.25	5.6	3730	−0.186	0.187	**5**
3% HF / 12% HCl	2.40	5.6	3750	0.207	0.182	**5**
3% HF / 12% HCl	2.55	5.6	3650	0.292	0.167	**4**
3% HF / 12% HCl	2.70	5.7	3600	0.348	0.157	**3**
3% HF / 12% HCl	2.85	5.8	3400	0.404	0.129	**2**

The calculation results are detailed in Table 15-5 and the skin factor versus time is plotted in Figure 15-11. Overall, the matrix stimulation treatment was successful as it reduced the skin factor from about 40 to 2. The sandstone contains a considerable amount of carbonates as the 15% HCl preflush resulted in skin being reduced from 24 to 11. The actual main acid (3%HF/12%HCl) reduced skin from 11 to 2, which is close to complete removal of damage.

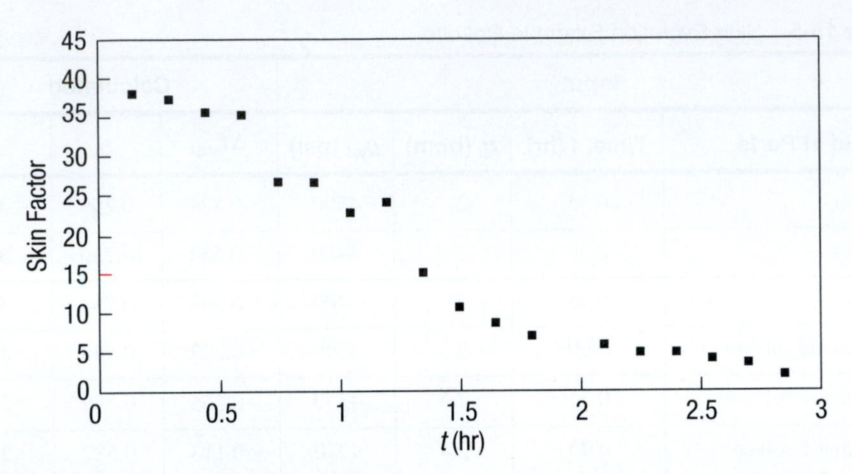

Figure 15-11 Skin evolution (Example 15-7).

15.3.3.3 Acid Injection into a Gas Reservoir

For treatments in gas wells, the single-phase liquid flow assumption made in the previous derivation is not valid anymore, and we need a transient flow equation with a proper format (a linear relationship between inverse injectivity, $\Delta p/q$, and a time function) for gas flow in the reservoir. When the reservoir pressure is high enough, gas viscosity, μ, and compressibility, Z, are directly proportional to pressure, and the pressure transient flow equation for a constant flow rate can be simplified to

$$p_{wf} = p_i - 50300\frac{z_i\mu_{gi}}{2p_i}\frac{p_{sc}}{T_{sc}}\frac{qT}{kh}\left(\log\left(\frac{kt}{\phi\mu c r_w^2}\right) + 0.87s - 3.23\right). \tag{15-57}$$

Equation 15-57 can be written in the exact format of Equation 15-49, but now the definition of m is different:

$$m = 50300\frac{z_i\mu_{gi}}{2p_i}\frac{p_{sc}}{T_{sc}}\frac{T}{kh}. \tag{15-58}$$

The same approach can be applied to obtain the skin factor for gas wells as for oil wells. The equation for skin calculation is

$$s = \frac{1}{0.868}\left[\frac{b}{m} - \log\left(\frac{k}{\phi\mu c_t r_w^2}\right) + 3.23\right]. \tag{15-59}$$

When injecting a fluid with a different viscosity than the reservoir fluid viscosity, the viscosity contrast affects the pressure response to injection. This effect on the injection pressure of a bank of fluid of one viscosity displacing a fluid of another can be accounted for with an apparent viscous skin factor, s_{vis}, given by

$$s_{vis} = \left(\frac{\mu_{acid}}{\mu_g} - 1\right) \ln\frac{r_{acid}}{r_w}. \tag{15-60}$$

This equation is exactly analogous to the Hawkins' formula, which gives the skin factor for a region of altered permeability around a well. Assuming the acid penetrates uniformly into a formation of thickness h, and neglecting any residual saturation of the displaced fluid behind the acid front, the radial position of the acid is related to the injected acid volume at any time by

$$r_{acid} = \sqrt{r_w^2 + \frac{V_{acid}}{\pi\phi h}}. \tag{15-61}$$

Combining Equations 15-60 and 15-61, the apparent viscous skin is given by

$$s_{vis} = \frac{1}{2}\left(\frac{\mu_{acid}}{\mu_g} - 1\right) \ln\left(1 + \frac{V_{acid}}{\pi\phi h r_w^2}\right) \tag{15-62}$$

and the damage skin factor during acid injection is obtained by subtracting the viscous skin factor from the total:

$$s_d = s_{app} - s_{vis}. \tag{15-63}$$

Because the viscosity of the injected acid is typically many times larger than the viscosity of the gas being displaced, the apparent viscous skin factor can be quite large. For example, using Equation 15-62 for an acid with a viscosity of 1 cp displacing a gas with a viscosity of 0.02 cp, after the injection of 100 gal/ft of acid into a formation with a porosity of 0.2 and thickness of 100 ft from a 0.25-ft radius wellbore, the viscous skin factor is 32. The viscous skin factor is zero at the start of injection, increases rapidly during the early stages of injection, then more slowly later in time, according to the logarithmic relationship in Equation 15-62.

The manner in which viscous skin effects can mask the well response to an acid treatment in a gas well is illustrated in Figure 15-12 (Zhu, Hill, and da Motta, 1998). In this figure, the upper curve is the total apparent skin factor calculated from the rate/pressure record during a treatment. This total skin increases rapidly because of the viscous effect. The middle curve is the viscous skin factor calculated with Equation 15-62. When this is subtracted from the apparent skin curve, the true evolution of the damage skin is revealed (the lower curve).

On the other hand, when acidizing oil wells, because the viscosity of the oil is often similar to the viscosity of the acid or larger, the viscous skin factor is small, and will be negative

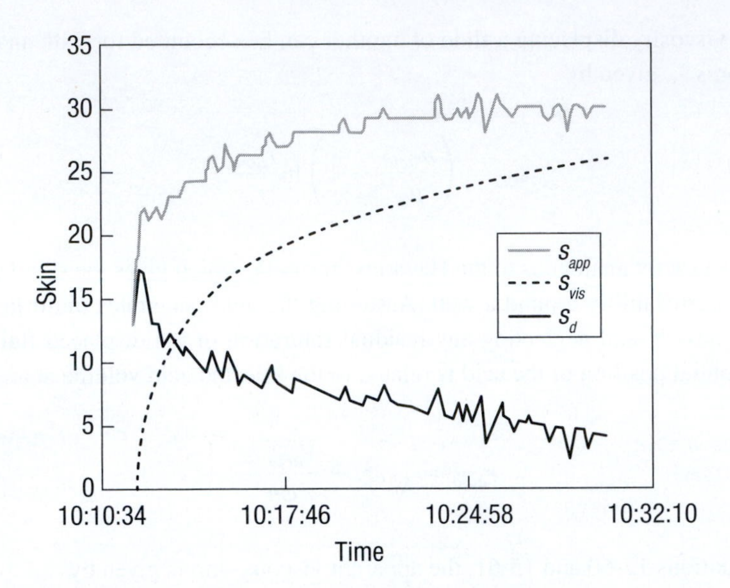

Figure 15-12 Accounting for viscous skin factor in a gas well treatment. (From Zhu et al., 1998.)

when oil viscosity is greater than acid viscosity. For 100 gal/ft of injection of acid with a viscosity half that of the oil, the apparent viscous skin factor is -1.4, a minor correction to the skin factor obtained from assuming the acid has the same viscosity as the displaced fluid.

15.4 Fluid Placement and Diversion

A critical factor to the success of a matrix acidizing treatment is proper placement of the acid so that all of the productive intervals are contacted by sufficient volumes of acid. If there are significant variations in reservoir permeability, the acid will tend to flow primarily into the highest-permeability zones, leaving lower-permeability zones virtually untreated. Even in relatively homogeneous formations, the damage may not be distributed uniformly; without the use of techniques to improve the acid placement, much of the damage may be left untreated. Thus the distribution of the acid into the formation is a very important consideration in matrix acidizing, and a treatment design should include plans for acid placement. When it is determined that sufficient acid coverage cannot be obtained by allowing the acid to choose its own path, the acid should be diverted by mechanical means, with ball sealers, with participate diverting agents, or with gels or foams.

15.4.1 Mechanical Acid Placement

The surest method for obtaining uniform acid placement is to isolate individual zones mechanically and treat all zones successively. Mechanical isolation can be accomplished with an opposed cup packer (perforation wash tool), the combination of a squeeze packer and a retrievable bridge plug, or inflatable straddle packers. These methods are described by McLeod (1984). Mechanical placement techniques require the removal of tubulars from the well, adding significant cost to a treatment. However, this cost may often be justified by the improved placement, particularly in horizontal wells.

15.4.2 Ball Sealers

Ball sealers are rubber-coated balls that are designed to seat in the perforations in the casing, thereby diverting injected fluid to other perforations. Ball sealers are added to the treating fluid in stages so that after a number of perforations have received acid, they are blocked, diverting acid to other perforations. Most commonly, ball sealers are more dense than the treating fluid, so that after treatment, the ball sealers will fall into the rathole. However, Erbstoesser (1980) showed that ball sealers that are slightly buoyant in the carrying fluid seat more efficiently than dense ball sealers. Buoyant ball sealers will be produced back to the surface after completion of the treatment, so a ball trap must be added to the flow line. A general guideline for ball-sealer use is to use twice as many dense ball sealers as perforations; with buoyant ball sealers, a 50% excess of ball sealers is recommended. The seating efficiency of ball sealers increases as injection rate increases and ball sealers are generally not recommended for low rate (<1 bpm) treatments. Ball sealers are not effective in deviated or horizontal wells unless the perforations are along the high side and/or the low side of the wellbore. Dense balls are more likely to seat on low-side perforations, and buoyant balls on the high-side perforations.

15.4.3 Particulate Diverting Agents

A common method of improving placement in matrix acidizing treatments is through the use of particulate diverting agents. Particulate diverting agents are fine particles that form a relatively low-permeability filter cake on the formation face. The pressure drop through this filter cake increases the flow resistance, diverting the acid to other parts of the formation where less diverting agent has been deposited. Diverting agents are either added to the treating fluid continuously, or in batches between stages of acid.

A particulate diverting agent must form a low-permeability filter cake and must be easily removed after treatment. To form a low-permeability filter cake, small particles and a wide range of particle sizes are needed. To ensure cleanup, materials that are soluble in oil, gas, and/or water are chosen. Commonly used particulate diverting agents and their recommended concentrations are given in Table 15-6 (McLeod, 1984).

The presence of diverting agents complicates the diagnosis of acid treatments because the diverting agent is continuously adding a pressure drop at the wellbore. Simultaneously, the acid is removing damage, decreasing the flow resistance near the wellbore. These two effects can be properly accounted for only with a comprehensive acidizing model, such as that presented by Taha et al. (1989). However, approximate models of the overall process can be helpful in designing diversion and interpreting the effect of diversion on the real-time rate/pressure behavior during an acid treatment.

Table 15-6 Summary of Particulate Diverters with Recommended Concentrations

Diverting Agent	Concentration
Oil-soluble resin or polymer	0.5 to 5 gal/1000 gal
Benzoic acid	1 lb_m/ft of perforations
Rock salt	0.5 to 2 lb_m/ft (do not use with HF acid)
Unibeads (wax beads)	1 to 2 lb_m/ft
Naphthalene flakes or mothballs	0.25 to 1 lb_m/ft (do not use in water-injection wells)

Hill and Galloway (1984) presented a model of oil-soluble resin diverting agent performance that has been improved and extended by Doerler and Prouvost (1987), Taha and Sepehrnoori (1989), and Schechter (1992). These models all assume that the diverting agent forms an incompressible filter cake, and thus are applicable to any diverting agent that exhibits this behavior.

The pressure drop across an incompressible filter cake can be expressed by Darcy's law as

$$\Delta p_{\text{cake}} = \frac{\mu u l}{g_c k_{\text{cake}}} \tag{15-64}$$

where u is the flux through the filter cake (q/A), l is the thickness of the filter cake, μ is the viscosity of the fluid carrying the diverting agent, and k_{cake} is the cake permeability. The thickness of the filter cake increases continuously as diverting agent is deposited and is related directly to the cumulative volume of diverting agent solution that has been injected. Incorporating this relationship into Equation (15-64) yields

$$\Delta p_{\text{cake}} = \frac{\mu u C_{\text{da}} V}{g_c(1 - \phi_{\text{cake}})k_{\text{cake}} A} \tag{15-65}$$

where C_{da} is the concentration of diverting agent particles in the carrying solution (volume of particles/volume of solution), V is the cumulative volume of solution injected, A is the cross-sectional area of the filter cake, and ϕ_{cake} is the porosity of the filter cake. This expression contains the intrinsic cake properties, ϕ_{cake} and k_{cake}, which are constant for an incompressible filter cake. However, these properties are difficult to measure independently. Instead, the pressure drop through the filter cake can be expressed in terms of the specific cake resistance, α, defined as

$$\alpha = \frac{1}{\rho_{\text{da}}(1 - \phi_{\text{cake}})k_{\text{cake}}} \tag{15-66}$$

where ρ_{da} is the density of the diverting agent particles, so that

$$\Delta p_{cake} = \frac{\alpha \mu u C_{da} \rho_{da} V}{A g_c}. \quad \textbf{(15-67)}$$

Taha et al. (1989) showed that α can be readily obtained from a variety of common laboratory tests. For example, if a diverting agent solution is injected into a core at constant rate and the pressure drop across the filter cake measure, Equation (15-67) shows that α can be obtained from the slope of a Δp_{cake} versus cumulative volume curve. Taha et al. reported values of α for oil-soluble resin-diverting agents ranging from about 10^{13} to about 10^{15} ft/lb$_m$.

To calculate the effect of diverting agents during an acid treatment, the pressure drop through the filter cake can be expressed as a skin effect (Doerler and Prouvost, 1987):

$$s_{cake} = \frac{2\pi kh}{q\mu} \Delta p_{cake}. \quad \textbf{(15-68)}$$

Substituting for Δp_{cake} with Equation (15-67) and reconciling all units gives

$$s_{cake} = \frac{2.26 \times 10^{-16} \alpha C_{da} \rho_{da} k \overline{V}}{r_w^2} \quad \textbf{(15-69)}$$

where α is in ft/lb$_m$, C_{da} is ft^3 of diverting agent particles per cubic foot of solution, ρ_{da} is in lb$_m$/ft^3, k is the permeability of the formation in md, $\overline{V}$ is the specific cumulative volume injected (gal/ft), and r_w is the wellbore radius (ft). This expression for s_{cake} can now be incorporated into an equation describing the flow into a particular reservoir zone j. For example, assuming steady-state flow in the reservoir yields

$$\left(\frac{q_i}{h}\right)_j = \frac{(4.92 \times 10^{-6})(p_{wf} - p_e)k_j}{\mu[\ln(r_e/r_w) + s_j + s_{cake,j}]} \quad \textbf{(15-70)}$$

where $(q_i/h)_j$ is in bpm/ft and all other quantities are in standard oilfield units. Recognizing that $(q_i/h)_j$ is the change in specific cumulative volume with time and incorporating the expression for $s_{cake,j}$, Equation (15-70) becomes

$$\frac{d\overline{V}_j}{dt} = \frac{(2.066 \times 10^{-4})(p_{wf} - p_e)k_j}{\mu[\ln(r_e/r_w) + s_j + c_{1,j}\overline{V}_j]} \quad \textbf{(15-71)}$$

where

$$c_{1,j} = \frac{2.26 \times 10^{-16} \alpha c_{da} \rho_{da} k_j}{r_w^2}. \quad \textbf{(15-72)}$$

Equation (15-71) is an ordinary differential equation in $\overline{V}$ and t. If $p_{wf} - p_e$ remains constant, as in the maximum Δp, maximum rate method, Equations (15-70) and (15-71) can be solved directly to give the injection rate and cumulative volume injected into each layer as a function of time if the changing damage skin effect, s_j, is not considered. Alternatively, if the total injection rate is held constant, a system of j coupled ordinary differential equations results. Equations of this type and solutions for a few cases, neglecting the changing damage skin effect, were presented by Doerler and Prouvost (1987).

To model the effects of diverting agents during acidizing more correctly, Equations (15-70) and (15-71) can be coupled with a comprehensive model of the acidizing process, as presented by Taha et al. (1989). However, such an approach results in a complex numerical model. A simpler model can be obtained by assuming a relationship for the evolution of the damage skin effect and incorporating this relationship into Equations (15-70) and (15-71).

As acidizing proceeds, the skin effect will decrease until the damage has been almost entirely removed. After this point has been reached, the skin effect will change very slowly and can be approximated as being zero. The volume of acid needed to reduce the skin effect to zero can be estimated based on an acidizing model, from laboratory results, or from field experience. The simplest approach is to assume that the skin effect will decrease linearly with cumulative acid injected until it reaches zero, and that it then remains constant. Thus,

$$s_j = s_{0,j} - c_{2,j}\overline{V}_j \qquad \overline{V}_j < \overline{V}_c \tag{15-73}$$

$$s_j = 0 \qquad \overline{V}_j > \overline{V}_c \tag{15-74}$$

where $s_{0,j}$ is the initial skin effect in layer j, $c_{2,j}$ is the rate of change of the skin effect as acid is injected, and $\overline{V}_{c,j}$ is the volume of acid needed to reduce the skin effect to zero. For example, if the initial skin effect is 10 and 50 gal/ft of acid are needed to reduce the skin effect to zero, $s_{0,j} = 10$, $c_2 = 0.2$, and $\overline{V}_{c,j} = 50$. Bringing Equation (15-73) to Equations (15-70) and (15-71) gives

$$\left(\frac{q_i}{h}\right)_j = \frac{(4.92 \times 10^{-6})(p_{wf} - p_e)k_j}{\mu[\ln(r_e/r_w) + s_{0,j} + (c_1 + c_2)_j\overline{V}_j]} \tag{15-75}$$

$$\frac{d\overline{V}_j}{dt} = \frac{(2.66 \times 10^{-4})(p_{wf} - p_e)k_j}{\mu[\ln(r_e/r_w) + s_{0,j} + (c_1 + c_2)_j\overline{V}_j]}. \tag{15-76}$$

For convenience, these can be written as

$$\left(\frac{q_i}{h}\right)_j = \frac{a_{3,j}}{42(a_{1,j} + a_{2,j}\overline{V}_j)} \tag{15-77}$$

$$\frac{d\overline{V}_j}{dt} = \frac{a_{3,j}}{a_{1,j} + a_{2,j}\overline{V}_j} \tag{15-78}$$

where

$$a_{1,j} = \ln\left(\frac{r_e}{r_w}\right) + s_{0,j} \tag{15-79}$$

$$a_{2,j} = (c_1 - c_2)_j \tag{15-80}$$

$$a_{3,j} = \frac{2.066 \times 10^{-4}(p_{wf} - p_e)k_j}{\mu}. \tag{15-81}$$

When it is integrated, Equation (15-78) yields

$$t_j = \frac{a_{1,j}\overline{V}_j + (a_{2,j}/2)\overline{V}_j^2}{a_{3,j}} \tag{15-82}$$

$$\overline{V}_j = \frac{-a_{1,j} + \sqrt{a_{1,j}^2 + 2a_{2,j}a_{3,j}t}}{a_{2,j}}. \tag{15-83}$$

Equations (15-75) through (15-83) apply to any layer for which $\overline{V}_j$ is less than $\overline{V}_{c,j}$. The injection time, $t_{c,j}$, at which $\overline{V}_j$ is equal to $\overline{V}_{c,j}$ is obtained from Equation (15-82) by setting $\overline{V}_j$ equal to $\overline{V}_{c,j}$. For any layer in which $\overline{V}_j$ is greater than $\overline{V}_{c,j}$, $s_j = 0$ and the following equations hold:

$$a_{1,j} = \ln\left(\frac{r_e}{r_w}\right) \tag{15-84}$$

$$a_{2,j} = c_{1,j} \tag{15-85}$$

$$a_{4,j} = -\left[a_{3,j}(t - t_{c,j}) + a_{1,j}\overline{V}_{c,j} + \frac{a_{2,j}}{2}\overline{V}_{c,j}^2\right] \tag{15-86}$$

$$t_j = t_{c,j} + \frac{a_{1,j}(\overline{V}_j - \overline{V}_{c,j}) + (a_{2,j}/2)(\overline{V}_j^2 - \overline{V}_{c,j}^2)}{a_{3,j}} \tag{15-87}$$

$$\overline{V}_j = \frac{-a_{1,j} + \sqrt{(a_{1,j})^2 - 2a_{2,j}a_{4,j}}}{a_{2,j}}. \tag{15-88}$$

The constant $a_{3,j}$ is independent of the injected volume, and Equation (15-77) applies both before and after $\overline{V}_j$ is equal to $\overline{V}_{c,j}$ as long as the appropriate constants are used.

At any time during injection, the total injection rate and total injected volume are

$$\left(\frac{q_i}{h}\right)_t = \frac{\sum_{j=1}^{J}(q_i/h)_j h_j}{\sum_{j=1}^{J} h_j} \tag{15-89}$$

$$\overline{V}_t = \frac{\sum_{j=1}^{J}\overline{V}_j h_j}{\sum_{j=1}^{J} h_j}. \tag{15-90}$$

The flow rate and cumulative volume injected for each layer as functions of injection time or total cumulative volume can be obtained from Equations (15-77) through (15-90) as follows. The highest-permeability layer, denoted as layer 1, serves as a convenient reference layer because the cumulative volume of acid will be highest for this layer at any given time. First, the injection rate and injection time as functions of the cumulative volume injected into the highest permeability layer are calculated with Equations (15-77) and (15-85) for $\overline{V}_1$ less than $\overline{V}_{c,1}$, and with Equations (15-77) and (15-87) for $\overline{V}_1$ greater than $\overline{V}_{c,1}$. Next, the cumulative volume injected into any other layer j at the same time as when a known cumulative volume has been injected into the highest permeability layer is calculated with Equation (15-83) or (15-88), depending on whether $\overline{V}_j$ is less than or greater than $\overline{V}_{c,j}$. The injection rate for that layer is then calculated with Equation (15-77). Finally, the total injection rate and cumulative volume injected for the time of interest are calculated by summing the contributions from all the layers, according to Equations (15-89) and (15-90).

Example 15-8 Flow Distribution during Acidizing with Diverting Agents

Two reservoir zones of equal thickness and having permeabilities of 100 and 10 md are to be acidized with an acid solution containing an oil-soluble resin diverting agent. The diverting agent is continuously added to the acid solution so that the concentration of diverting agent is 0.1 vol%. The diverting agent has a specific gravity of 1.2 and a specific cake resistance of 10^{13} ft/lb$_m$, while the acid solution has a viscosity of 0.7 cp at downhole conditions.

The damage skin effect in both layers is initially 10, and 50 gal/ft of acid will remove all of the damage. The well radius is 0.328 ft, the drainage radius of the reservoir is 1980 ft, and the injection rate will be adjusted throughout the treatment to maintain $p_{wi} - p_e = 2000$ psi.

Calculate the injection rate and cumulative volume injected for each layer, and the total injection rate, versus time up to a total injected volume of at least 100 gal/ft.

Solution

First, the relevant constants for Equations (15-77) through (15-90) are calculated. From the given acidizing response, $\overline{V}_{c,j} = 50$ gal/ft and $c_{2,j} = 0.2$ for both layers. Labeling the high-permeability layer as layer 1, from Equation (15-60),

$$C_{1,1} = \frac{(2.26 \times 10^{-16})(10^{13})(0.001)(1.2)(62.4)(100)}{(0.328)^2} = 0.157 \quad \textbf{(15-91)}$$

and similarly, $c_{1,2} = 0.0157$. The constants $a_{i,j}$ are then calculated with Equations (15-79) through (15-81) for $\overline{V}_j$ less than $\overline{V}_{c,j}$, and with Equations (15-84) through (15-86) for $\overline{V}_j$ greater than $\overline{V}_{c,j}$. The results of these calculations are given in Table 15-7. The constants $a_{4,j}$ were not needed in this example because the volume injected into the lower-permeability layer never exceeds 50 gal/ft.

Next the injection rate into layer 1 and the time that a given volume has been injected are calculated using Equations (15-77) and (15-82) for volumes up to 50 gal/ft. For example, at 50 gal/ft (which is $\overline{V}_{c,1}$),

$$\left(\frac{q_i}{h}\right)_1 = \frac{59.0}{42(18.7 - (0.043)(50))} = 0.085 \text{ bpm/ft} \quad \textbf{(15-92)}$$

and

$$t_{c,1} = \frac{(18.7)(50) + (-0.043/2)(50^2)}{59.0} = 14.9 \text{ min.} \quad \textbf{(15-93)}$$

For larger injection volumes , the rate is again calculated with Equation (15-77), but with the constants for $\overline{V}_j$ greater than $\overline{V}_{c,j}$, and the time corresponding to each volume is calculated with Equation (15-87).

Table 15-7 Constants for Example 15-7

	$\overline{V}_j$ Less than $\overline{V}_{c,j}$			$\overline{V}_j$ Greater than $\overline{V}_{c,j}$		
Layer *j*	$a_{1,j}$	$a_{2,j}$	$a_{3,j}$	$a_{1,j}$	$a_{2,j}$	$a_{3,j}$
1	18.7	−0.043	59.0	8.7	0.157	59.0
2	18.7	−0.1843	5.90	8.7	0.0157	5.90

Now, for all injection times calculated previously, the flow rate and cumulative volume injected for layer 2 are calculated with Equations (15-77) and (15-83). Finally, the total injection rate and cumulative volume are calculated for each time with Equations (15-89) and (15-90). The results are given in Table 15-8.

Table 15-8 Example 15-7 Results

Time (min)	q_1 (bpm/ft)	q_2 (bpm/ft)	q_t (bpm/ft)	$\bar{V}_1$ (gal/ft)	$\bar{V}_2$ (gal/ft)	$\bar{V}_t$ (gal/ft)
0.00	0.075	0.0075	0.041	0.0	0.0	0.0
3.13	0.077	0.0076	0.042	10.0	1.0	5.5
6.20	0.079	0.0077	0.043	20.0	2.0	11.0
9.19	0.081	0.0077	0.044	30.0	2.9	16.5
12.10	0.083	0.0078	0.045	40.0	3.9	21.9
14.94	0.085	0.0079	0.046	50.0	4.8	27.4
18.15	0.077	0.0080	0.043	60.0	5.9	32.9
21.40	0.071	0.0081	0.040	70.0	7.0	38.5
24.71	0.066	0.0082	0.037	80.0	8.1	44.1
28.06	0.061	0.0083	0.035	90.0	9.3	49.6
31.47	0.058	0.0084	0.033	100.0	10.5	55.2
34.93	0.054	0.0085	0.031	110.0	11.7	60.8
38.45	0.051	0.0086	0.030	120.0	13.0	66.5
42.01	0.048	0.0087	0.028	130.0	14.3	72.1
45.63	0.046	0.0089	0.027	140.0	15.6	77.8
49.30	0.044	0.0090	0.026	150.0	17.0	83.5
53.03	0.042	0.0092	0.025	160.0	18.4	89.2
56.80	0.040	0.0093	0.025	170.0	19.9	94.9
60.63	0.038	0.0095	0.024	180.0	21.4	100.7
64.51	0.036	0.0097	0.023	190.0	22.9	106.5
68.44	0.035	0.0099	0.022	200.0	24.6	112.3

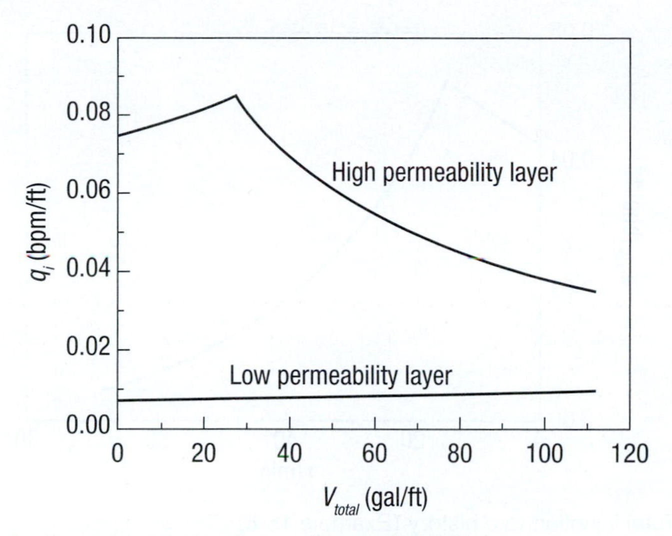

Figure 15-13 Injection rates into two layers (Example 15-8).

The injection rate and cumulative volume into each layer are shown in Figures 15-13 and 15-14. The rate into layer 1 increases until 50 gal/ft have been injected, after which it declines. This behavior is due to the damage skin effect decreasing faster than the diverting agent skin increases until all damage is removed from the layer. After this time, the injection rate declines as diverting agent continues to build up on the sandface. The diverting agent was not very effective in this treatment, as the high-permeability layer received considerably more acid than necessary while the low-permeability layer did not receive enough acid to remove all damage. Either a higher diverting agent concentration or a diverter with a higher specific cake resistance would result in better acid placement.

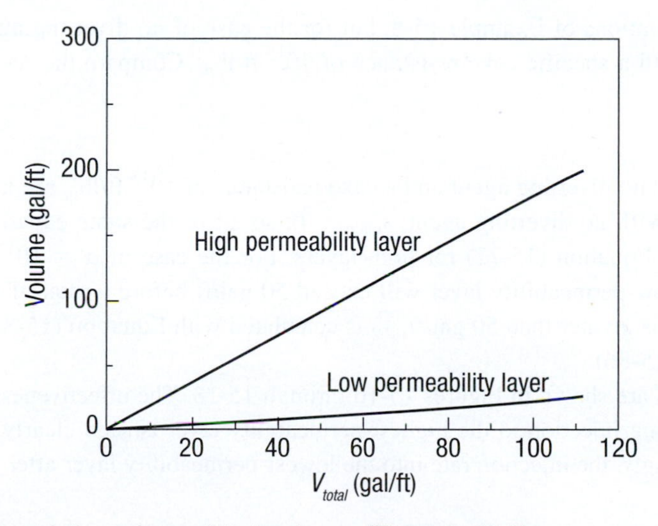

Figure 15-14 Cumulative volumes into two layers (Example 15-8).

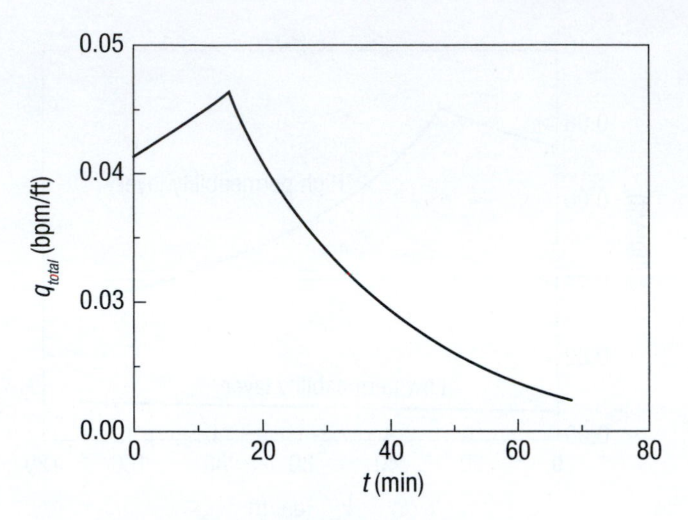

Figure 15-15 Total injection rate history (Example 15-8).

The total injection rate is shown in Figure 15-15. The total rate increases as the high-permeability layer responds to acid, then declines as the diverting agent builds resistance to flow faster than the acid stimulates injectivity.

Example 15-9 The Effect of the Specific Cake Resistance on Acid Placement

Repeat the calculations of Example 15-8, but for the case of no diverting agent and for a diverting agent with a specific cake resistance of 10^{15} ft/lb$_m$. Compare the results for the three cases.

Solution

The solutions for no diverting agent and a cake resistance of 10^{15} ft/lb$_m$ are obtained just as in Example 15-8. With no diverting agent, $s_{cake} = 0$, so all of the same equations can be used, with $c_{1,j} = 0$ in Equation (15-72) for both layers. For the case of $\alpha = 10^{15}$, the volume injected into the low-permeability layer will exceed 50 gal/ft before a total of 100 gal/ft are injected. When $\overline{V}_2$ is greater than 50 gal/ft, $\overline{V}_2$ is calculated with Equation (15-88), obtaining $a_{4,2}$ from Equation (15-86).

The results are shown in Figures 15-16 through 15-18. The effectiveness of the diverting agent in decreasing injection to the highest-permeability layer is seen clearly in Figure 15-16. Perhaps surprisingly, the injection rate into the lowest-permeability layer after 100 gal/ft of acid

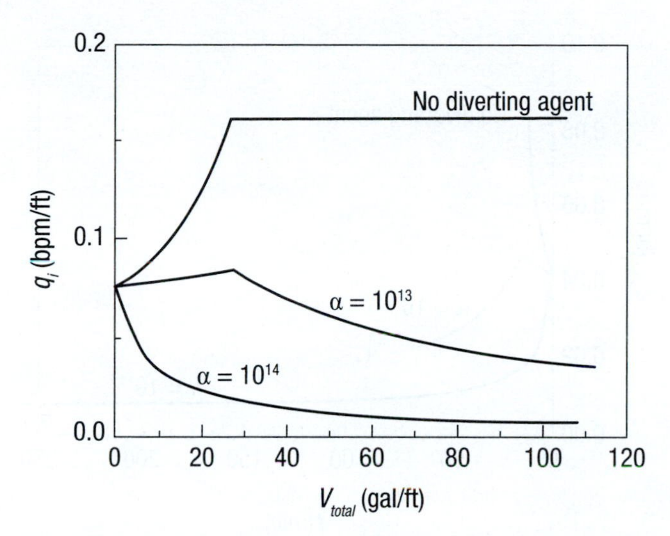

Figure 15-16 Injection rate into the high-permeability layer (Example 15-9).

injection is highest when no diverting agent is used (Figure 15-17), but paradoxically, the smallest cumulative amount of acid is injected into the low-permeability layer in this case. This is because the total injection rate is much higher with no diverting agent (recall that the bottomhole pressure is being held constant). With a high total injection rate, the time of injection is small, resulting in a small cumulative volume placed in the low-permeability layer.

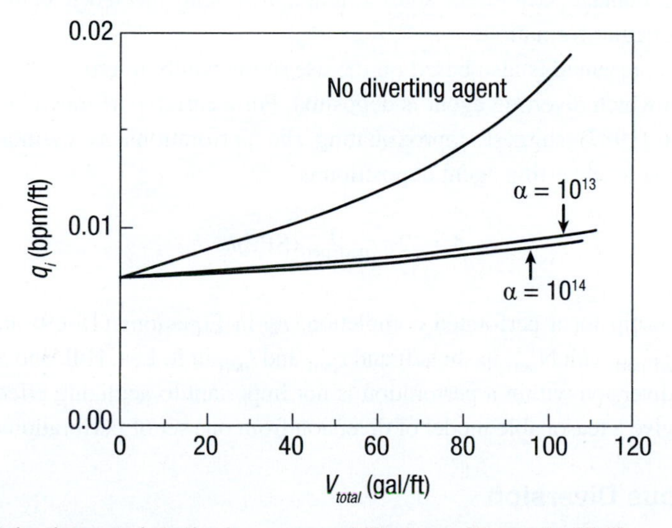

Figure 15-17 Injection rate into the low-permeability layer (Example 15-9).

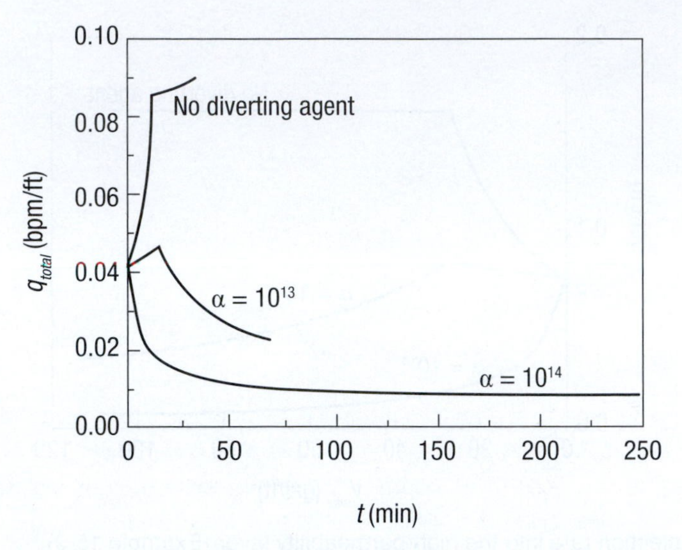

Figure 15-18 Total injection rate history (Example 15-9).

The method presented here is based on particular assumptions about the effect of acid on damage removal; for example, it assumes that the damage skin effect decreases linearly with the cumulative acid injected. However, the same approach taken here can be applied with other relationships between damage skin effect and the amount of acid injected if better information is available for a particular formation.

The method presented is also based on the use of the wellbore cross-sectional area to represent the area on which diverting agent is deposited. For a cased, perforated completion, Doerler and Prouvost (1987) suggest approximating the perforations as cylinders, so that the cross-sectional area for diverting agent deposition is

$$A = 2\pi r_{\text{perf}} l_{\text{perf}} (\text{SPF}) h. \quad \text{(15-94)}$$

Using this relationship for a perforated completion, r_w in Equations (15-69) and (15-72) is replaced by $N_{\text{perf}} r_{\text{perf}} l_{\text{perf}}$ with N_{perf} in shots/ft and r_{perf} and l_{perf} in ft. Lea, Hill, and Schechter (1991) have shown that diversion within a perforation is not important to acidizing effectiveness, so this approach should give a reasonable model of diversion from one set of perforations to another.

15.4.4 Viscous Diversion

Gels, foamed acid, and viscoelastic surfactant solutions are being used increasingly as a means of improving acid placement. With these fluids, the mechanism of diversion is viscous diversion, the increase in flow resistance in higher-permeability regions due to the presence of a bank of viscous fluid. This can be modeled in a similar manner to that just shown for particulate diverting agents by defining a viscous skin effect as

$$s_{\text{vis}} = \left(\frac{\mu_{\text{gel}}}{\mu} - 1\right) \ln \frac{r_{\text{gel}}}{r_w} \tag{15-95}$$

where μ_{gel} is the viscosity of the gelled acid, μ is the viscosity of the formation fluid being displaced, and r_{gel} is the radius penetrated by the gel. Equation (15-95) is exactly analogous to Hawkins' formula, with viscosity instead of permeability being altered in the region near the well. The radius of gel penetration is related to the injected volume as done previously for acid solutions containing diverting agents.

Diversion by foams is a more complex phenomenon than viscous diversion because the amount of flow resistance developed can depend on the local permeability of the formation and the foam propagation rate can differ from the bulk transport. The simplest approach to modeling foam diversion is to assume the permeability has been altered by the presence of the foam to an effective permeability, k_f. The resistance caused by the foam is then calculated using Hawkins' formula as (Hill and Rossen, 1994)

$$s_{\text{vis}} = \left(\frac{k}{k_f} - 1\right) \ln \frac{r_f}{r_w} \tag{15-96}$$

where k_f is the effective permeability in the foam invaded region and r_f is the radius of the foam bank. More complex models of foam diversion are presented by Cheng, Kam, Delshad, and Rossen (2002) and Rossen and Wang (1999).

15.5 Preflush and Postflush Design

15.5.1 The HCl Preflush

Before the HF/HCl mixture is injected in a sandstone acidizing treatment, a preflush of hydrochloric acid, most commonly a 15 wt% solution, is usually injected. The primary purpose of the preflush is to prevent the precipitation of species such as calcium fluoride when the HF solution contacts the formation. The HCl preflush accomplishes this by dissolving the carbonate minerals, displacing ions such as calcium and magnesium from the near-wellbore vicinity, and lowering the pH around the wellbore.

In situations where injectivity cannot be established with HCl, the HF/HCl mixture can be injected without a preflush. In this instance, the objective is not to obtain an optimal stimulation treatment, but to remove enough damage to establish communication with the formation for testing purposes. Paccaloni and Tambini (1990) reported several successful well responses without an HCl preflush in wells with very high skin effects that did not respond to conventional treatments using a preflush. It is likely that these wells had perforations clogged with drilling mud or other debris that was not HCl soluble.

When the HCl solution enters the formation, it dissolves carbonate minerals rapidly but does not react to a great extent with other minerals. HCl does leach aluminum from clay minerals while not dissolving the entire clay molecule structure (Hartman, Lecerf, Frenier, Ziauddin, and Fogler, 2006). As seen in Chapter 14, the reaction rates between HCl and carbonate minerals are high, so that the movement of the HCl into the formation can be approximated as a shock front. Behind the front, all carbonate minerals have been dissolved and the HCl concentration is equal to the injected concentration. Ahead of the front, no reactions have occurred and a bank of spent HCl solution gradually builds up. Using the same dimensionless variables as defined for the movement of the HF front in Table 15-2, the position of the HCl front is

$$\varepsilon_{\mathrm{HCl}} = \frac{\theta}{1 + [(1-\phi)/\phi]V^0_{\mathrm{CO_3}} + (1/N_{\mathrm{Ac,HCl}})} \tag{15-97}$$

where ϕ is the initial porosity in the region contacted by HCl, $V^0_{\mathrm{CO_3}}$ is the initial volume fraction of the rock that is carbonate mineral, and the acid capacity number for HCl is defined similarly to that for HF, as

$$N_{\mathrm{Ac,HCl}} = \frac{\phi \beta_{\mathrm{HCl}} C^0_{\mathrm{HCl}} \rho_{\mathrm{acid}}}{(1-\phi)V^0_{\mathrm{CO_3}} \rho_{\mathrm{CO_3}}}. \tag{15-98}$$

Equations (15-97) and (15-98) can be used to calculate the amount of HCl needed to remove all carbonates from a given region for radial flow around a wellbore or ellipsoidal flow around a perforation. The preflush volume is selected by assuming a distance to which all carbonates should be removed by the preflush and then calculating the acid volume required. A common procedure is to design the preflush to remove carbonates to a distance of 1 ft, as live HF is not likely to penetrate this far.

Example 15-10 Preflush Volume Design for a Perforated Completion

Calculate the volume of 15 wt% HCl preflush needed to dissolve all carbonates to a distance of 1 ft beyond the tip of a 6-in.-long, 0.25-in.-diameter perforation if there are 4 shots/ft. The density of the acid solution is 1.07 g/cm^3. The formation contains 5 vol% $CaCO_3$ and no other HCl-soluble minerals and has an initial porosity of 0.2.

Solution

The definition of the dimensionless penetration distance from the tip of a perforation is given in Table 15-2. For penetration of 1 ft from the tip of a 6-in. perforation, $\bar{z}$ = 1.5 ft/0.5 ft = 3. Then

$$\varepsilon_{\mathrm{HCl}} = \frac{1}{3}(3)^3 - 3 + \frac{2}{3} = 6\frac{2}{3}. \tag{15-99}$$

From Table 13-3, $\beta_{100,\mathrm{HCl}} = 1.37$ and $\rho_{\mathrm{CaCO_3}} = 2.71\ \mathrm{g/cm^3}$. Using Equation (15-98), the acid capacity number is

$$N_{\mathrm{Ac,HCl}} = \frac{(0.2)(1.37)(0.15)(1.07)}{(1-0.2)(0.05)(2.71)} 0.406. \tag{15-100}$$

Solving Equation (15-97) for the dimensionless acid volume gives

$$\theta = \varepsilon_{\mathrm{HCl}}\left(1 + \frac{1-\phi}{\phi}V^0_{\mathrm{CO_3}} + \frac{1}{N_{\mathrm{Ac,HCl}}}\right) \tag{15-101}$$

and substituting the known values yields

$$\theta = 6.67\left[1 + \frac{(1-0.2)}{0.2}(0.05) + \frac{1}{0.406}\right] = 24.4 \tag{15-102}$$

From Table 15-2, the volume of acid per perforation is

$$V_{\mathrm{HCl}} = q_{i,\mathrm{perf}}t = (2\pi)(0.5)^3(24.4)(0.2) = 3.8\ \mathrm{ft^3/perf} = 29\ \mathrm{gal/perf}. \tag{15-103}$$

With 4 shots/perf, about 120 gal/ft of HCl preflush are needed.

Designing the HCl preflush to remove all carbonates from 1 ft of formation before the injection of HF/HCl will be a conservative design because the HCl in the mud acid will continue to dissolve carbonates ahead of the live HF. Hill, Sepehrnoori, and Wu (1991) showed that preflushes as small as 25 gal/ft ensured that live HF would not contact regions of high pH, even in heterogeneous formations. However, Gdanksi and Peavy (1986) pointed out that some HCl will be consumed by reactions with clays, so an excess above that needed for carbonate dissolution may be appropriate.

15.5.2 The Postflush

After a sufficient volume of HF/HCl solution has been injected, it is displaced from the tubing and the wellbore with a post-flush. A variety of fluids have been used for the post-flush, including diesel, ammonium chloride (NH_4Cl) solutions, and HCl. The post-flush displaces the spent acid farther from the wellbore so that any precipitates that may form will be less damaging. As a minimum, the post-flush volume should be the volume of the tubing plus twice the volume of the wellbore below the tubing. Because of gravity segregation effects (Hong and Millhone, 1977), at least this much post-flush is needed to displace all acid from the wellbore.

15.6 Acid Additives

Besides diverting agents, numerous other chemicals are frequently added to acid solutions, the most common of these being corrosion inhibitors, iron-sequestering compounds, surfactants, and mutual solvents. Acid additives should be tested carefully to ensure compatibility with other chemicals in the acid solution and with the formation fluids; only those that provide a clear benefit should be included in the treating fluids.

Corrosion inhibitors are needed in virtually all acid treatments to prevent damage to the tubulars and casing during acidizing. Corrosion of steel by HCl can be very severe without inhibition, particularly at high temperatures. The corrosion inhibitors used are organic compounds containing polar groups that are attracted to the metal surface. Corrosion inhibitors are usually proprietary formulations, so it is the responsibility of the service company supplying the acid to recommend the type and concentration of corrosion inhibitor to be used in a particular acid treatment.

An iron-sequestering compound, usually EDTA, is sometimes added to acid solutions when it is thought that ferric ions (Fe^{3+}) are present in the near-wellbore region in order to prevent the precipitation of $Fe(OH)_3$ in the spent acid solution. This situation should be rare, and the sequestering agents themselves are potentially damaging to the formation (see Chapter 14). In general, these materials should be used only when there is a clear indication of $Fe(OH)_3$ precipitation during acidizing.

Surfactants are added to acid solutions to prevent the formation of emulsions, to speed cleanup of spent acid, and to prevent sludge formation. Like sequestering agents, surfactants may have deleterious effects on the formation and should be used only after careful testing with the formation fluids and core samples.

An additive that has shown clear benefits in some sandstone acidizing applications is a mutual solvent, usually ethylene glycol monobutyl ether (EGMBE) (Gidley, 1971). A mutual solvent is added to the post-flush—it improves productivity apparently by removing corrosion inhibitor that has absorbed on formation surfaces (Crowe and Minor, 1985) and by restoring water-wet conditions.

15.7 Acidizing Treatment Operations

Three guiding principles should be followed when conducting an acidizing treatment: (1) All solutions to be injected should be tested to ensure that they conform to the design formulations; (2) all necessary steps should be taken to minimize the damage caused by the acidizing process itself; and (3) the acidizing process should be monitored by measuring the rate and pressure (surface and/or bottomhole). McLeod (1984), Brannon et al. (1987), Paccaloni and Tambini (1988), and numerous others have clearly shown the importance of these guidelines.

As mentioned previously, before an acidizing treatment in a sandstone formation is undertaken, there should be a clear indication that low well productivity is due, at least in part, to acid-soluble formation damage. This can be determined through a pretreatment pressure transient test to measure the skin effect, an analysis of other possible sources of skin, as shown in Chapter 6, and an assessment of the source of formation damage. A prestimulation production log is also helpful in some cases in planning the treatment and as a baseline for post-treatment analysis.

Before conducting the treatment, samples of the acid solutions should be taken, and at a minimum, the HCl concentrations should be checked on site with simple titrations. Some companies have developed acid quality control kits (Watkins and Roberts, 1983) that can be used to check acid concentrations rapidly.

A very important step in an acidizing treatment is cleaning of all surface tanks, surface flow lines, and the tubing used to inject the acid. Hydrochloric acid will partially dissolve and loosen rust, pipe dope, and other contaminants from pipe walls, so any source of such material should be cleaned before acid is introduced to the formation. Surface equipment can be cleaned prior to being brought to the location or on site with the acid solution itself. If the production tubing is to be used for acid injection, an HCl solution or other cleaning fluids should be circulated down the tubing and back to the surface before injection into the formation begins. This process is called *tubing pickling*. Guidelines for the required volumes needed for tubing pickling with HCl were given by Al-Mutairi, Hill, and Nasr-El-Din (2007). Alternatively, clean coiled tubing can be employed for acid injection.

During the treatment, the injection rate and surface or bottomhole pressure should be monitored. The real-time monitoring discussed in this chapter can then be applied to optimize the treatment. Occasional samples of the injected solutions should also be taken; if any problems occur in the course of the treatment, samples of the injected solutions may be an important diagnostic aid.

When acid injection has been completed, the well should be flowed back immediately to minimize damage by reaction product precipitation. A post-treatment pressure transient test run shortly after a stabilized rate has been obtained is the most positive means of assessing the effect of the acid treatment. A production log will also be helpful in diagnosing the treatment outcome in some wells.

References

1. Al-Mutairi, S.H., Hill, A.D., and Nasr-El-Din, H.A., "Pickling Well Tubulars Using Coiled Tubing: Mathematical Modeling and Field Application," *SPE Production and Operations,* **22**(3): 326–334 (August 2007).
2. Brannon, D.H., Netters, C.K., and Grimmer, P.J., "Matrix Acidizing Design and Quality-Control Techniques Prove Successful in Main Pass Area Sandstone," *JPT,* 931–942 (August 1987).
3. Bryant, S.L., "An Improved Model of Mud Acid/Sandstone Chemistry," SPE Paper 22855, 1991.
4. Cheng, L., Kam, S.I., Delshad, M., and Rossen, W.R., "Simulation of Dynamic Foam-Acid Diversion Processes," *SPE Journal* (September 2002).
5. Cheung, S.K., and Van Arsdale, H., "Matrix Acid Stimulation Studies Yield New Results Using a Multi-Tap, Long-Core Permeameter," *JPT,* 98–102 (January 1992).

6. Crowe, C.W., and Minor, S.S., "Effect of Corrosion Inhibitors on Matrix Stimulation Results," *JPT,* 1853–1860 (October 1985).

7. da Motta, E.P., "Matrix Acidizing of Horizontal Wells," Ph.D. dissertation, University of Texas at Austin, TX, 1993.

8. da Motta, E.P., Plavnik, B., and Schechter, R.S., "Optimizing Sandstone Acidizing," *SPERE,* 159–153 (February 1992a).

9. da Motta, E.P., Plavnik, B., Schechter, R.S., and Hill, A.D., "The Relationship between Reservoir Mineralogy and Optimum Sandstone Acid Treatment," SPE Paper 23802, 1992b.

10. Doerler, N., and Prouvost, L.P., "Diverting Agents: Laboratory Study and Modeling of Resultant Zone Injectivities," SPE Paper 16250, 1987.

11. Erbstoesser, S.R., "Improved Ball Sealer Diversion," *JPT,* 1903–1910 (November 1980).

12. Gdanski, R.D., and Peavy, M.A., "Well Return Analysis Causes Re-evaluation of HCl Theories," SPE Paper 15825, 1986.

13. Gidley, J.L., "Stimulation of Sandstone Formations with the Acid-Mutual Solvent Method," *JPT,* 551–558 (May 1971).

14. Guin, J.A., Schechter, R.S., and Silberberg, I.H., "Chemically Induced Changes in Porous Media," *Ind. Eng. Chem. Fund.,* **10**(1): 50–54 (February 1971).

15. Hartman, R.L., Lecerf, B., Frenier, W., Ziauddin, M., and Fogler, H.S., "Acid-Sensitive Aluminosilicates: Dissolution Kinetics and Fluid Selection for Matrix-Stimulation Treatments," *SPEPO,* 194–204 (May 2006).

16. Hekim, Y., Fogler, H.S., and McCune, C.C., "The Radial Movement of Permeability Fronts and Multiple Reaction Zones in Porous Media," *SPEJ,* 99–107 (February 1982).

17. Hill, A.D., and Galloway, P.J., "Laboratory and Theoretical Modeling of Diverting Agent Behavior," *JPT,* 1157–1163 (July 1984).

18. Hill, A.D., Lindsay, D.M., Silberberg, I.H., and Schechter, R.S., "Theoretical and Experimental Studies of Sandstone Acidizing," *SPEJ,* 30–42 (February 1981).

19. Hill, A.D. and Rossen, W.R., "Fluid Placement and Diversion in Matrix Acidizing," SPE Paper 27982 presented at the Tulsa/SPE Centennial Petroleum Engineering Symposium, Tulsa, OK, August 29–31, 1994.

20. Hill, A.D., Sepehrnoori, K., and Wu, P.Y., "Design of the HCl Preflush in Sandstone Acidizing," SPE Paper 21720, 1991.

21. Hong, K.C., and Millhone, R.S., "Injection Profile Effects Caused by Gravity Segregation in the Wellbore," *JPT,* 1657–1663 (December 1977).

22. Labrid, J.C., "Thermodynamic and Kinetic Aspects of Argillaceous Sandstone Acidizing," *SPEJ,* 117–128 (April 1975).

23. Lambert, M.E., "A Statistical Study of Reservoir Heterogeneity," MS thesis, University of Texas at Austin, TX, 1981.

24. Lea, C.M., Hill, A.D., and Sepehrnoori, K., "The Effect of Fluid Diversion on the Acid Stimulation of a Perforation," SPE Paper 22852, 1991.

25. Lund, K., and Fogler, H.S., "Acidization V. The Prediction of the Movement of Acid and Permeability Fronts in Sandstone," *Chem. Eng. Sci.,* **31**(5): 381–392 (1976).

26. McLeod, H.O., Jr., "Matrix Acidizing," *JPT,* **36:** 2055–2069 (1984).

27. McLeod, H.O., "Significant Factors for Successful Matrix Acidizing," Paper presented at the SPE Centennial Symposium at New Mexico Tech, Socorro, New Mexico, 1989.

28. McLeod, H.O., Jr. and Norman, W.D., "Sandstone Acidizing," Ch. 18 in Reservoir Stimulation, 3rd edition, Economides, M.J., and Nolte, K.G., eds., John Wiley and Sons, Chichester, UK, 2000.

29. Paccaloni, G., and Tambini, M., "Advances in Matrix Stimulation Technology," *JPT,* 256–263 (March, 1993).

30. Paccaloni, G., Tambini, M., and Galoppini, M., "Key Factors for Enhanced Results of Matrix Stimulation Treatments," SPE Paper 17154, 1988.

31. Prouvost, L.P., and Economides, M.J., "Applications of Real-Time Matrix Acidizing Evaluation Method," SPE Paper 17156, 1988.

32. Rossen, W.R., and Wang, M.W., "Modeling Foams for Acid Diversion," *SPE Journal,* **4**(2): 92–100 (June 1999).

33. Schechter, R.S., *Oil Well Stimulation,* Prentice Hall, Englewood Cliffs, NJ, 1992.

34. Sevougian, S.D., Lake, L.W., and Schechter, R.S., "KGEOFLOW: A New Reactive Transport Simulator for Sandstone Matrix Acidizing," SPE Production and Facilities, 13–19 (February 1995).

35. Smith, C.F., and Hendrickson, A.R., "Hydrofluoric Acid Stimulation of Sandstone Reservoirs," *JPT,* 215–222 (February 1965); *Trans. AIME,* **234.**

36. Taha, R., Hill, A.D., and Sepehrnoori, K.,"Sandstone Acidizing Design Using a Generalized Model," *SPE Production Engineering*, 4, No. 1, pp. 49–55, February 1989.

37. Walsh, M.P., Lake, L.W., and Schechter, R.S., "A Description of Chemical Precipitation Mechanisms and Their Role in Formation Damage during Stimulation by Hydrofluoric Acid," *JPT,* 2097–2112 (September 1982).

38. Watkins, D.R., and Roberts, G.E., "On-Site Acidizing Fluid Analysis Shows HCl and HF Contents Often Vary Substantially from Specified Amounts," *JPT,* 865–871 (May 1983).

39. Ziauddin, M., Kotlar, H.K., Vikane, O., Frenier, W., and Poitrenaud, H., "The Use of a Virtual Chemistry Laboratory for the Design of Matrix-Stimulation Treatments in the Heidrun Field," SPE Paper 78314, 2002a.

40. Ziauddin, M., et al., "Evaluation of Kaolinite Clay Dissolution by Various Mud Acid Systems (Regular, Organic and Retarded)," Paper presented at the 5th International Conference and Exhibition on Chemistry in Industry, Manama, Bahrain, 2002b.

41. Zhu, D., and Hill, A.D., "Field Results Demonstrate Enhanced Matrix Acidizing Through Real-Time Monitoring," *SPE Production and Facilities* (November 1998).

42. Zhu, D., Hill, A.D., and da Motta, E.P., "On-site Evaluation of Acidizing Treatment of a Gas Reservoir," Paper SPE 39421 presented at the SPE International Symposium on Formation Damage Control, Lafayette, LA, February 18–19, 1998.

Problems

15-1 Select the acid formulation or formulations to be used in the following formations:
a. $k = 200$ md, $\phi = 0.2$, 5% carbonate, 5% feldspar, 10% kaolinite, 80% quartz
b. $k = 5$ md, $\phi = 0.15$, 10% carbonate, 5% feldspar, 5% kaolinite, 80% quartz
c. $k = 30$ md, $\phi = 0.25$, 20% carbonate, 5% chlorite, 75% quartz

15-2 In a core flood, $N_{\text{Ac},F} = 0.024$ and $N_{\text{Da},S} = 0.6$. How many pore volumes of acid must be injected for the fast-reacting front to break through at the end of the core?

15-3 In a core flood in a 12-in.-long core, the dimensionless acid concentration at the fast-reacting mineral front is 0.7 when the front has moved 3 in. into the core. What is the dimensionless acid concentration at the front when the front has progressed 6 in.? 9 in.?

15-4 In a core flood in a 6-in.-long core, the fast-reacting mineral front breaks through at the end of the core after 50 pore volumes have been injected. The dimensionless acid concentration after breakthrough is 0.8. Calculate $N_{\text{Ac},F}$ and $N_{\text{Da},S}$.

15-5 In a core flood of a 1-in.-diameter by 6-in.-long core, the acid flux is 0.04 ft/min. From this experiment, it was found that $N_{\text{Da},S} = 0.9$ and $N_{\text{Ac},F} = 0.02$. For this formation and acid concentration and an injection rate of 0.1 bpm/ft, calculate and plot the acid volume as a function of the distance from the wellbore that the fast-reacting minerals have all been removed, out to a distance of 6 in. from the wellbore. Assume radial flow.

15-6 Repeat Problem 15-5, but for penetration of the front from the tip of a perforation.

15-7 For a particular sandstone, after all carbonates have been dissolved, the permeability is 10 md and the porosity is 0.15. The remaining minerals are 5 vol% clay, which reacts rapidly with HF. Calculate the permeability of this sandstone after all clays have been removed using the Labrid, Lund and Fogler, and Lambert correlations.

15-8 Design a matrix acidizing treatment for the well described below. The design should give recommended acid types, concentrations, and volumes for the preflush and the main acid stages, a recommended injection procedure (rate and pressure), and a predicted stimulation ratio (J_s/J_d).

Reservoir fluid data: $\gamma_o = 30°$API, $\gamma_g = 0.68$, $p_i = 4000$ psi, $p_b = 4000$ psi
$T = 160°$F, $\phi = 0.2$, $k_h = 20$ md, $k_v = 2$ md, $h = 100$ ft
and $\mu_o = 1$ cp.

Well completion: $r_w = 0.35$ ft, $r_e = 1490$ ft.

The well is vertical, the top formation depth is 8000 ft.

Perforations: 40 ft perforated (measured along the well) starting at 8000 ft vertical depth, 4 spf, 90° phasing, 6 in long perfs.

Formation damage: $k_s = 2$ md, damage depth extends 0.5 ft from the wellbore.

Formation: 5% kaolinite, 5% albite, 5% $CaCO_3$, and 85% quartz.

15-9 For the data given below for a matrix acidizing treatment, calculate the skin factor evolution. What do you conclude about this treatment?

Table 15-9 Acid Treatment Data (Problem 15-9)

Reservoir and Well Data		
$p_i = 4200$ psi	$B_o = 1.0$	$\phi = 0.25$
$c_t = 5 \times 10^5$ psi^{-1}	$h = 56.5$ ft	$k = 150$ md
$s = 45$	$r_w = 3.14$ in	
$\mu = 0.8$ cp		

Time (min)	**Rate (bpm)**	**p_{wf} (psi)**
1	2.92	5138
3	3.5	5809
5	2.9	5920
7	3.5	6341
9	3.9	6040
11	4.12	5736
13	4.12	5696
15	4.19	5658
17	4.21	5605
19	4.37	5540
21	4.6	5321
23	4.49	5440
25	4.51	5406
27	4.6	5307
29	4.1	5931
31	4.1	5927
33	4.18	5832
35	4.07	5931
37	4.34	5611
39	4.04	5594
41	4.04	5554

Time (min)	Rate(bpm)	p_{wf} (psi)
43	4.18	5418
45	4.23	5340
47	4.3	5259
49	4.29	5257
51	4.39	5115
53	4.42	5050
55	4.75	4924

15-10 An acid solution containing an oil-soluble resin diverting agent is injected into a two-layer reservoir. The diversity agent concentration in the acid is 0.1 vol%, the diverting agent particles have a density of 1.2 g/cm^3, the specific cake resistance is 5×10^{13} ft/lb$_m$, and the viscosity of the acid solution at downhole conditions is 0.6 cp. For either layer, 50 gal/ft of acid will remove all damage. The wellbore radius is 0.328 ft, the drainage radius is 1590 ft, and the acid is injected at a pressure difference, $p_{wf} - p_e$, of 3000 psi. Layer 1 has a permeability of 75 md and an initial skin effect of 20, while layer 2 has a permeability of 25 md and an initial skin effect of 10. Plot $(q/h)_j$ and $\overline{V}_j$ for each layer as functions of the total cumulative volume per foot injected.

15-11 Repeat Problem 15-10, but for the case of no diverting agent in the acid.

15-12 Derive an equation for $(q/h)_j$ [analogous to Equation (15-63)] for viscous diversion with a gel. Leave the expression in terms of μ_{gel}, μ, $\overline{V}_j$, and other constants.

15-13 Calculate the HCl preflush volume needed to remove all carbonates to a distance 6 in. from the wellbore in radial flow. The acid is a 10 wt% solution with a specific gravity of 1.05. The formation contains 7 vol% $CaCO_3$ and has a porosity of 0.19.

CHAPTER 16

Carbonate Acidizing Design

16.1 Introduction

The acidizing process in carbonate formations is fundamentally different from that in sandstones. In a clastic formation, the surface reaction rates are slow and a relatively uniform acid front moves through the porous medium. In carbonates, surface reaction rates are very high, so mass transfer often limits the overall reaction rate, leading to highly nonuniform dissolution patterns. Often, a few large channels, called wormholes, are created, such as shown in Figure 16-1, caused by the nonuniform dissolution of limestone by HCl in a large block experiment (McDuff, Jackson, Schuchart, and Postl, 2010). The structure of these wormhole patterns depends on many factors, including (but not limited to) flow geometry, injection rate, reaction kinetics, and mass transfer rates. In Figure 16-2, also from McDuff et al. (2010), CT images of wormholes created in corefloods illustrate how the wormholes change from large, conical-shaped tubes at relatively low injection rates, to much narrower wormholes with few branches at an intermediate rate, to a highly branched structure at a high injection rate. Experiments such as these show that for any particular carbonate rock and acid system, there are optimal conditions for acid injection that will result in the longest wormholes that can be created with a given volume of acid.

Since wormholes are much larger than the pores in nonvuggy carbonates, the pressure drop through the region penetrated by wormholes is small and can often be neglected. Thus, in matrix acidizing, knowledge of the depth of penetration of wormholes allows a prediction of the effect of acidizing on the skin effect. The objective of a carbonate acidizing treatment is to create wormholes penetrating deep enough into the formation to bypass any damaged region and, often, to create a significantly negative skin factor (Figure 16-3). Extrapolation of laboratory test results and direct indications from field treatment responses show that wormholes commonly penetrate

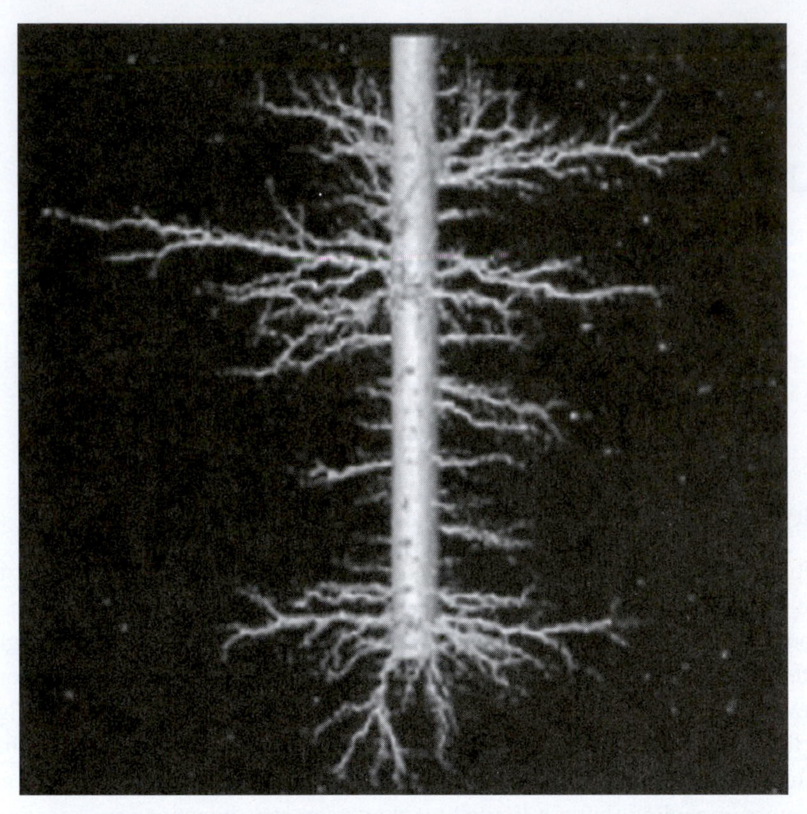

Figure 16-1 Wormholes created in a large block experiment. (From McDuff et al., 2010.)

distances on the order of 20 feet beyond the wellbore, resulting in skin factors of about −4. Figure 16-4 (Furui et al., 2010) shows the post-acid stimulation skin factors for over 400 wells, with the average being about −4. Thus, unlike sandstone acidizing where the objective is to overcome formation damage effects, matrix stimulation of carbonates creates sufficient productivity enhancement to make it an attractive procedure even in the absence of formation damage. For this reason, and because of its simplicity and relatively low cost, most wells in carbonate formations are acid stimulated.

Acid fracturing is a stimulation method in which strong acid solutions, usually viscosified by adding polymers or creating emulsions, are injected at pressures above the fracturing pressure. The acid injected into the hydraulic fracture etches the faces of the fracture. If this etching creates uneven surfaces and channels that remain open under the closure stress that occurs when pumping is stopped, this process can create a conductive fracture without the use of proppants. Depending primarily on rock characteristics, in some formations, acid fracturing is a cost-effective way to create a conductive hydraulic fracture.

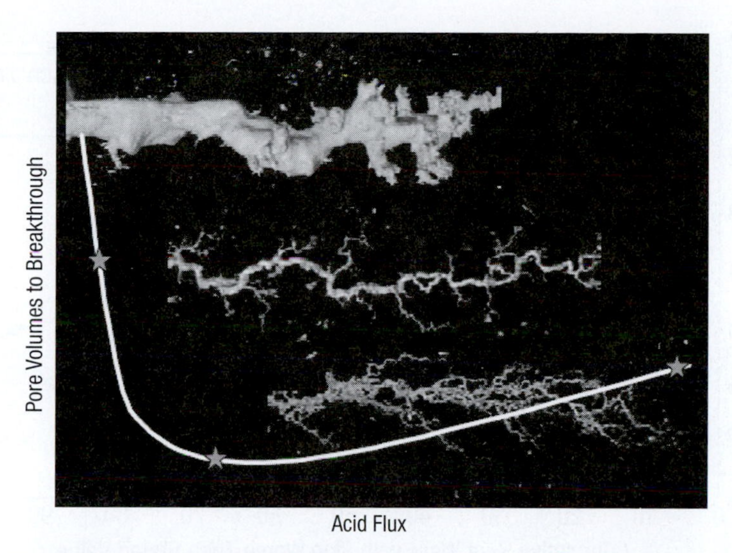

Figure 16-2 Wormhole morphologies at different injection rates (From McDuff et al., 2010.)

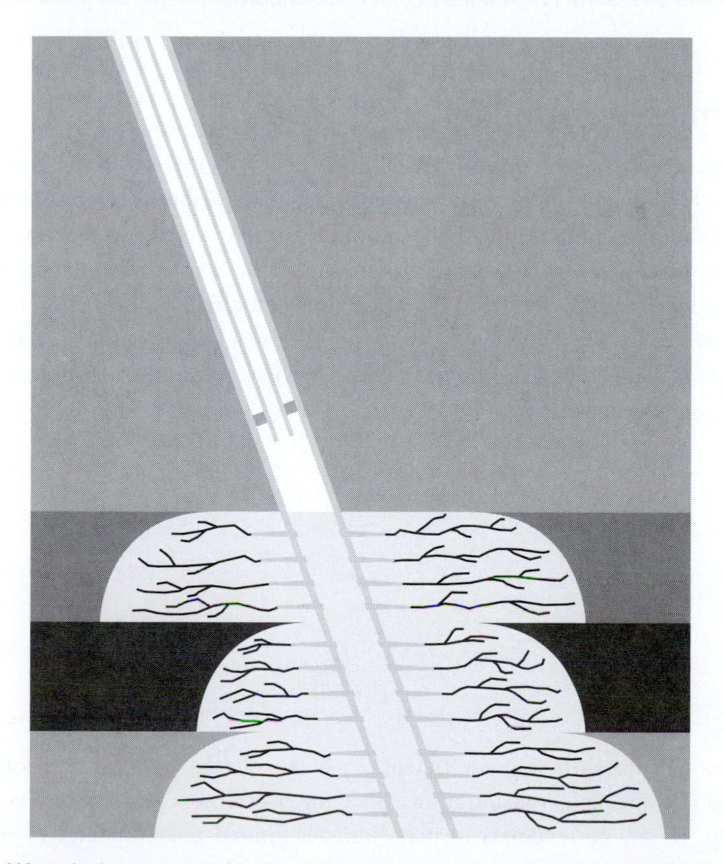

Figure 16-3 Wormholes created from acid injection into a perforated completion.

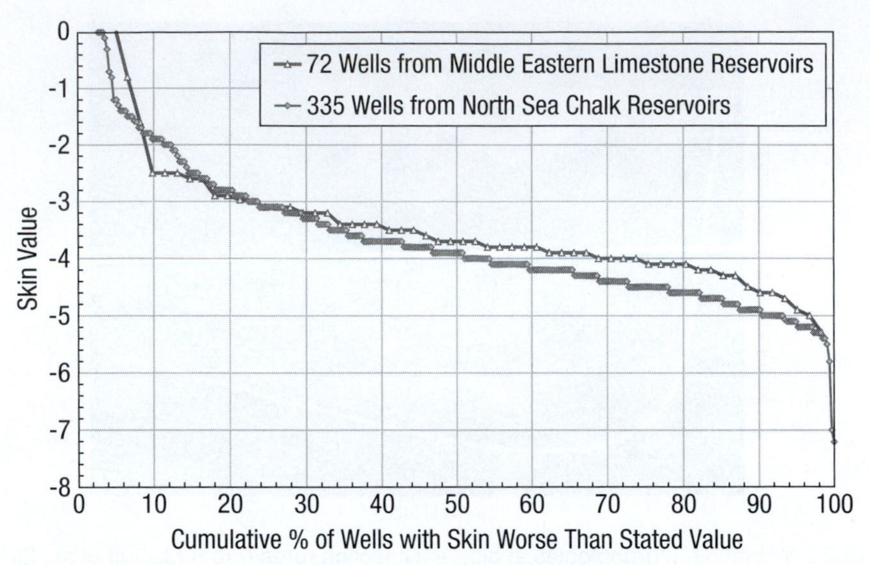

Figure 16-4 Field post-stimulation buildup test data for carbonate matrix acidizing. (From Furui et al., 2010.)

16.2 Wormhole Formation and Growth

Wormholes form in a dissolution process when the large pores grow at a rate substantially higher than the rate at which smaller pores grow, so that a large pore receives an increasingly larger proportion of the dissolving fluid, eventually becoming a wormhole. This occurs when the reactions are mass transfer limited or mixed kinetics prevail, that is, the mass transfer and surface reaction rates are similar in magnitude. For flow with reaction in a circular pore, the relative effects of mass transfer and surface reaction rates can be expressed by a kinetic parameter, *P*, the inverse of the Theile modulus, defined as the ratio of the diffusive flux to the flux of molecules consumed by surface reaction (Daccord, 1989),

$$P = \frac{u_d}{u_s} \tag{16-1}$$

or

$$P = \frac{D}{E_f r C^{n-1}} \tag{16-2}$$

where D is the molecular diffusion coefficient, E_f is the surface reaction rate constant, r is the pore radius, and C is the acid concentration. Since the overall reaction kinetics is controlled by the slowest step, $P \rightarrow 0$ corresponds to mass transfer-limited reactions and $P \rightarrow \infty$ indicates surface reaction rate-limited reactions. When P is near 1, the kinetics are mixed, and both surface reaction rate and mass transfer rate are important.

The natural tendency for wormholes to form when reaction is mass transfer limited has been demonstrated theoretically by Schechter and Gidley (1969), using a model of pore growth and collision. In this model, the change in the cross-sectional area of a pore can be expressed as

$$\frac{dA}{dt} = \psi A^{1-n} \tag{16-3}$$

where A is the pore cross-sectional area, t is time, and ψ is a pore growth function that depends on time. Examination of Equation (16-3) shows that if $n > 0$, smaller pores grow faster than larger pores and wormholes cannot form; when $n < 0$, larger pores grow faster than smaller pores, and wormholes will develop. From an analysis of flow with diffusion and surface reaction in single pores, Schechter and Gidley found that $n = 1/2$ when surface reaction rate controls the overall reaction rate, and $n = -1$ when diffusion controls the overall reaction rate.

Models of flow with reaction in the collections of cylindrical pores can predict the tendency for wormhole formation, but do not give a complete picture of the wormholing process because they do not include the effects of fluid loss from the pores. As acid moves through a large pore or channel, acid moves to the reactive surface by molecular diffusion, but also by convective transport as acid flows to the small pores connected with the large pore. As the large pore grows, the fluid loss flux is a larger and larger portion of the flow of acid to the wormhole wall and is ultimately the limiting factor for wormhole growth. The fluid loss through the walls of the wormhole leads to the branches seen in Figures 16-1 and 16-2.

Thus, whether or not wormholes will form, and the structure of the wormholes that do form, depend on the relative rates of surface reaction, diffusion, and fluid loss, all of which depend on the overall convection rate of the acid. For a given rock, acid system, and temperature, a progression of dissolution patterns occur as injection rate is increased. The typical dissolution patterns created when HCl is injected into limestone are illustrated by the radiographs of wormholes created in corefloods shown in Figure 16-5 (Fredd and Fogler, 1998). At a very low injection rate, the inlet face of the rock is slowly consumed as acid diffuses to the surface, yielding compact or face dissolution of the inlet rock surface. This mode of acid attack obviously requires the most acid to propagate the dissolution front, and should be avoided in a field application if at all possible. With increasing flow rate as in the second radiograph, a large-diameter wormhole referred to as a conical wormhole forms and propagates into the porous medium. This mode of wormholing is also inefficient, requiring large volumes of acid to propagate the wormhole any significant distance into the matrix. As injection rate is increased further, the wormholes formed become narrower and narrower, as in the middle two radiographs. This is the dominant wormhole structure that is most desirable in carbonate acidizing because it results in the deepest penetration of wormholes into the formation with the least amount of acid. When the injection rate is increased even more, the wormhole structure becomes more and more branched (last two radiographs), and hence more and more acid is consumed to propagate the wormhole

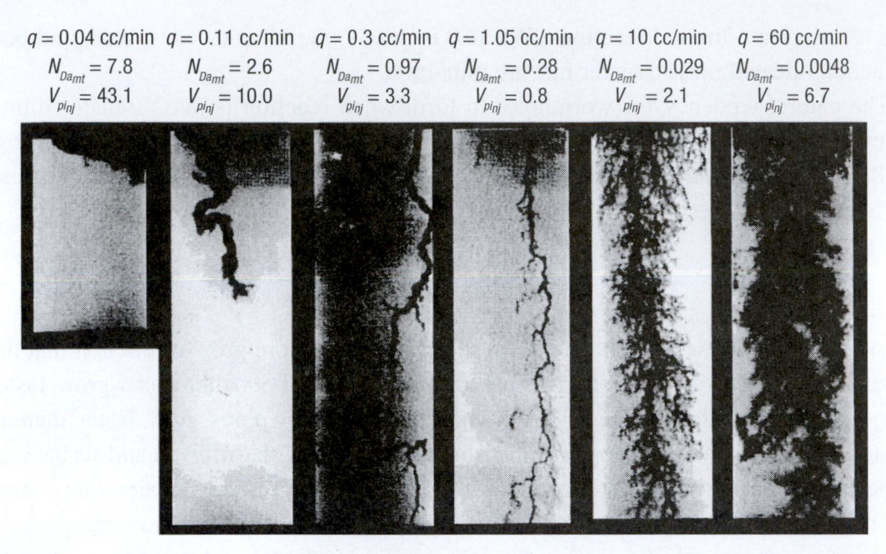

Figure 16-5 Wormhole structures created by different injection rates. (From Fredd and Fogler, 1998.)

system into the formation. Eventually, at sufficiently high rates, the dissolution pattern becomes uniform like that in sandstone, in which the entire pore system is being gradually enlarged. This uniform dissolution pattern may never occur in many carbonate acid systems because the rock would fracture before such high flow rates could be achieved in the matrix. Thus, the key to most efficient acid stimulation is to design an acid system that can be injected into the well at flow rates that would lead to dominant wormholing.

The progression presented here depends on the relative magnitudes of the diffusion and surface reaction rates and also on the flow geometry because of the role of fluid loss from the wormholes. Thus, predictions based on linear flow, such as standard core floods, may not be valid for other flow geometries, such as radial flow around a well or flow from a perforation.

Numerous studies have shown that for a given rock/acid system, and temperature, a distinct minimum in the amount of acid required to propagate a wormhole through the core exists as a function of flow rate (Wang, Hill, and Schechter, 1993; Fredd and Fogler, 1998; Hoefner and Fogler, 1989). Figure 16-6 (Buijse and Glasbergen, 2005) shows the typical wormholing behavior that occurs when HCl is injected into limestone. When the flow rate is below the optimum, the volume of acid needed to propagate a wormhole a certain distance decreases rapidly as the injection rate increases, but when the wormhole propagation is above the optimal, the volume of acid required increases only gradually with increasing injection rate. This means that it is better to inject at a rate above the optimum than at too low a rate.

The HCl–dolomite reaction rate is significantly lower than that for HCl and limestone. With a lower reaction rate, substantially more acid is needed to propagate a wormhole a given distance, as shown in Figure 16-7.

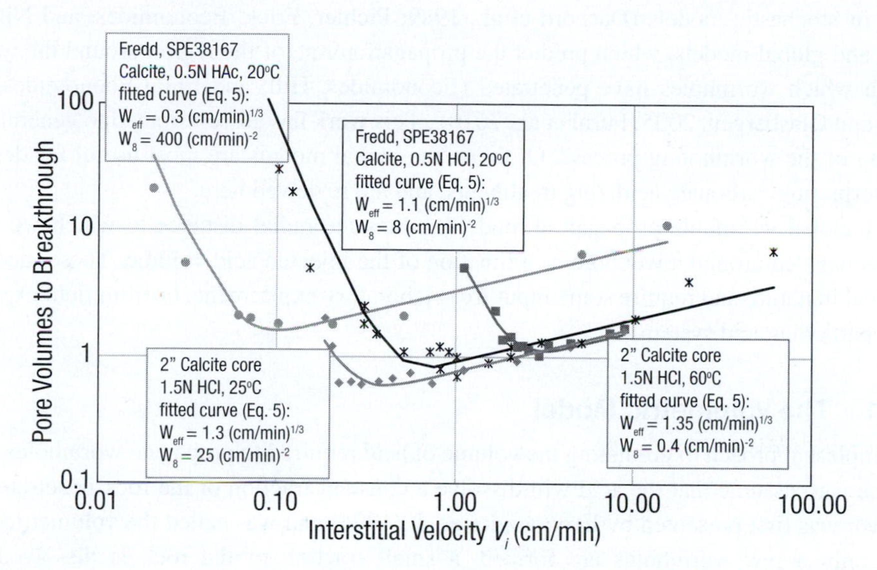

Figure 16-6 Laboratory characterization of wormhole propagation efficiency. (From Buijse and Glasbergen, 2005.)

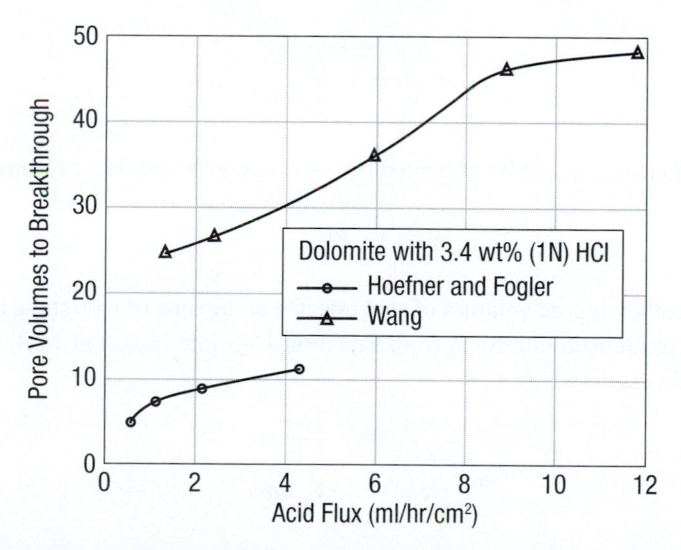

Figure 16-7 Acid volumes needed to propagate wormholes through linear San Andres dolomite cores. (From Wang et al., 1993.)

16.3 Wormhole Propagation Models

Many models of the wormholing process have been developed including mechanistic models of a single wormhole or a collection of wormholes (Hung, Hill, and Sepehrnoori, 1989; Schechter, 1992), network models (Hoefner and Fogler, 1988; Daccord, Touboul, and Lenormand, 1989), fractal or stochastic models (Daccord et al., 1989; Pichler, Frick, Economides, and Nittmann, 1992), and global models, which predict the propagation rate of the region around the wellbore through which wormholes have penetrated (Economides, Hill, and Ehlig-Economides, 1994; Buijse and Glasbergen, 2005; Furui et al., 2010) . This work has all contributed to general understanding of the wormholing process. Of these, the global models are most useful for designing and interpreting carbonate acidizing treatments, and are reviewed here.

A global wormhole propagation model predicts the radial distance to which wormholes have propagated around a wellbore as a function of the injected acid volume. These models are empirical in nature and require some input from laboratory experiments or from field experience with a particular acid system.

16.3.1 The Volumetric Model

The simplest approach to predicting the volume of acid required to propagate wormholes a given distance is to assume that the acid will dissolve a constant fraction of the rock penetrated. This approach was first presented by Economides et al. (1994) and was called the volumetric model. When only a few wormholes are formed, a small fraction of the rock is dissolved; more branched wormhole structures dissolve larger fractions of the matrix. Defining η as the fraction of the rock dissolved in the region penetrated by acid, for radial flow it can be shown that

$$r_{\text{wh}} = \sqrt{r_w^2 + \frac{N_{\text{Ac}}V}{\eta\pi\phi h}}. \tag{16-4}$$

The wormholing efficiency, η, can be estimated from linear core flood data as being

$$\eta = N_{\text{Ac}}PV_{\text{bt}} \tag{16-5}$$

where PV_{bt} is the number of pore volumes of acid injected at the time of wormhole breakthrough at the end of the core. Substituting for η from Equation 16-5 into Equation 16-4, the radius of wormhole propagation is

$$r_{wh} = \sqrt{r_w^2 + \frac{V}{PV_{bt}\pi\phi h}}. \tag{16-6}$$

Equation 16-6 shows that the only parameter needed concerning wormhole propagation for this model is the pore volumes to breakthrough, which is obtainable from coreflood experiments. This model does not consider the dependence of pore volumes to breakthrough on acid flux, a dependence illustrated in Figure 16-6. A reasonable average of the PV_{bt} over the range of fluxes that will occur in the wormhole region must be used for this model to correctly predict wormhole propagation distances. If the fluxes of interest are higher than the optimal flux, such an average is easy to estimate because the pore volumes to breakthrough changes slowly above the optimum. If fluxes are below the optimum, more accurate models should be used. The volumetric model's primary utility is as a means to easily estimate the distance to which wormholes propagate.

Example 16-1 Volumetric Model of Wormhole Propagation

Figure 16-8 (Furui et al., 2010) shows that for a 15 weight% HCl solution injected into 1-in. diameter chalk cores at 150°F, the average pore volumes to breakthrough for interstitial velocities ranging from about 1 cm/min to about 3 cm/min, is approximately 0.7. From the curve fit through this data, the optimal pore volumes to breakthrough is 0.5 at an optimal interstitial flux of 1.5 cm/min. Using the volumetric model, calculate the radius of the region penetrated by wormholes as a function of injected acid volume, up to a volume of 100 gal/ft.

Recent research has shown that the value of pore volumes to breakthrough that should be used for well calculations is much lower than those measured in the laboratory with small diameter cores. Repeat the calculations assuming an average pore volume to breakthrough of 0.05.

Finally, calculate the injection rate (bpm) required to create an interstitial flux at the wellbore equal to the optimal flux of 1.5 cm/min for a reservoir thickness of 100 ft. For all calculations, assume that the porosity is 0.3 and the wellbore radius is 0.328 ft.

Solution

For incremental values of injected acid volume per unit thickness of reservoir (V/h), the radius to which wormholes propagate is calculated with Equation 16-6. For 10 gal/ft of acid injection,

$$r_{wh} = \sqrt{(0.328 \text{ ft})^2 + \frac{(10 \text{ gal/ft})(1 \text{ ft}^3/7.48 \text{ gal})}{\pi(0.7)(0.3)}} = 1.46 \text{ ft}. \tag{16-7}$$

Calculating the radius of the wormhole-penetrated region for acid volumes up to 100 gal/ft, Figure 16-9 is generated.

For much more efficient wormhole propagation, as indicated by an average pore volume to breakthrough of 0.05, the wormhole region is significantly larger, as shown in Figure 16-10.

The interstitial velocity is simply the volumetric rate divided by the flow cross-sectional area multiplied by the porosity. Thus,

$$v_i = \frac{q}{2\pi r_w h\phi} \tag{16-8}$$

So the injection rate to create a desired interstitial flux at the wellbore radius is

$$q_i = 2\pi v_i r_w h\phi. \tag{16-9}$$

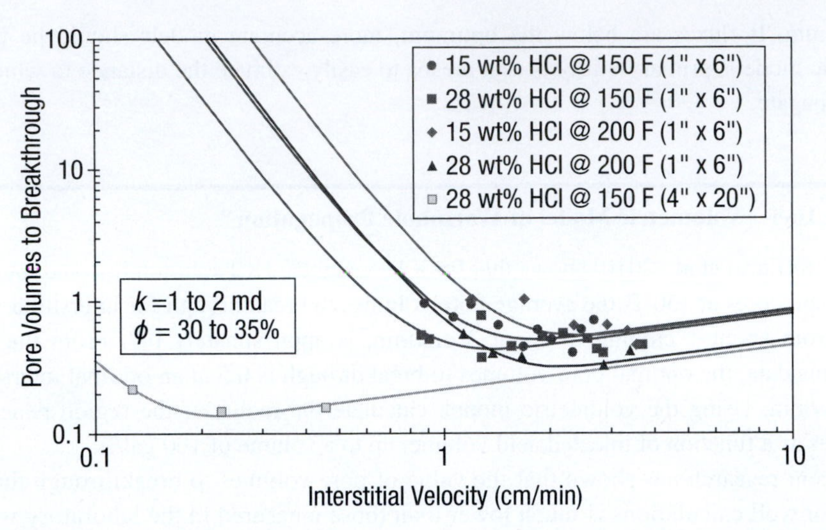

Figure 16-8 Wormhole propagation efficiency with strong acids in chalk core samples. (From Furui et al., 2010.)

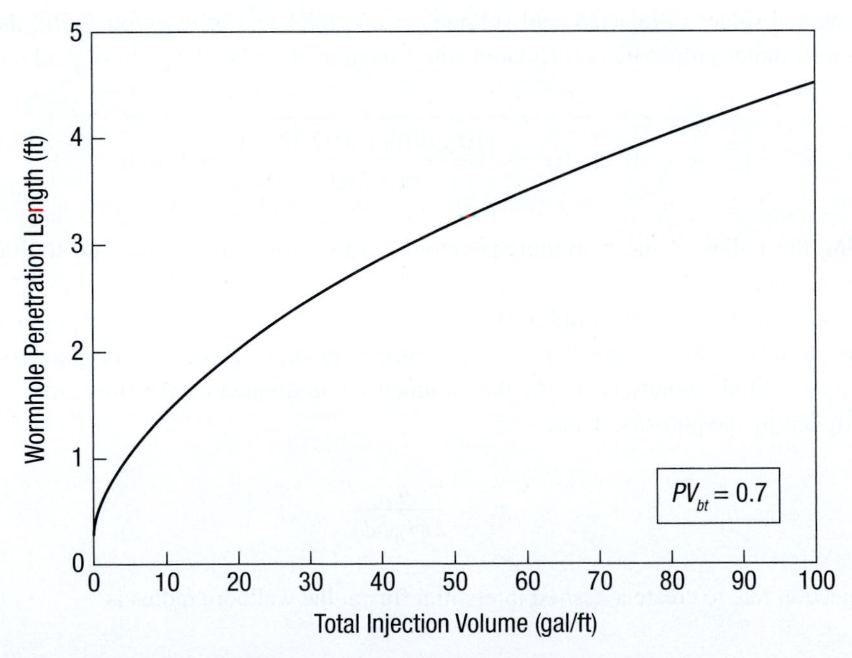

Figure 16-9 Wormhole propagation distance for $PV_{bt} = 0.7$ (Example 16-1).

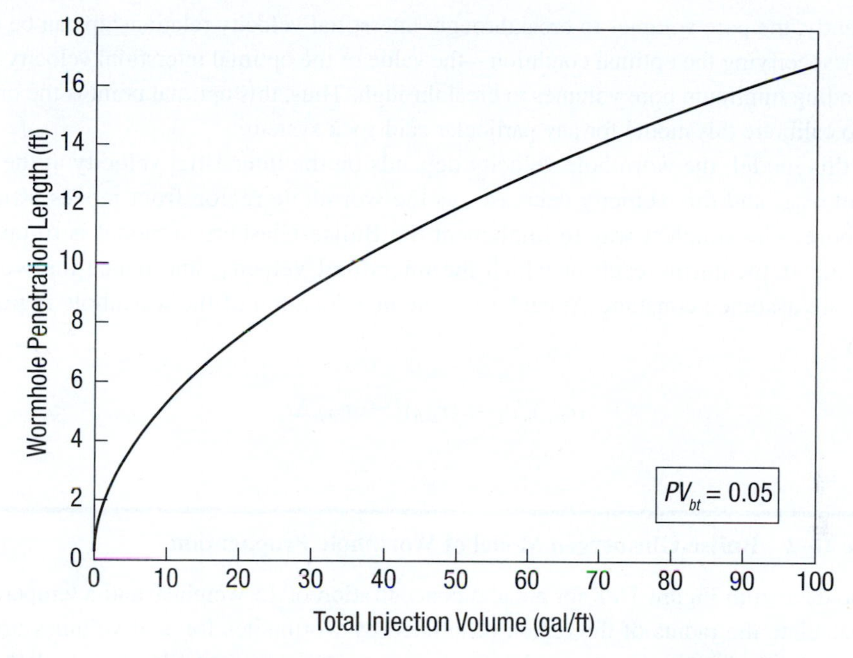

Figure 16-10 Wormhole propagation distance for $PV_{bt} = 0.05$ (Example 16-1).

To create an interstitial flux of 1.5 cm/min into a 100-ft-thick reservoir, the injection rate is

$$q_i = 2\pi(1.5 \text{ cm/min})\left(\frac{1 \text{ ft}}{30.48 \text{ cm}}\right)(0.328 \text{ ft})(100 \text{ ft})(0.3)\left(\frac{1 \text{ bbl}}{5.615 \text{ ft}^3}\right) = 0.54 \text{ bpm}. \quad \textbf{(16-10)}$$

This injection rate should be readily achievable in reservoirs with permeabilities greater than 1 md and having no severe formation damage.

16.3.2 The Buijse-Glasbergen Model

Buijse and Glasbergen (2005) presented an empirical model of wormhole propagation based on the characteristic dependence of the pore volumes to breakthrough in acid corefloods on the interstitial velocity. They recognized that the wormhole propagation velocity, which is inversely related to the pore volumes to breakthrough, has a consistent functional dependence on velocity for different rocks and different acid systems. Based on this supposition, they derived a function that captures this dependence. Their model can be expressed as

$$v_{wh} = \frac{dr_w}{dt} = \left(\frac{v_i}{PV_{bt,opt}}\right) \times \left(\frac{v_i}{v_{i,opt}}\right)^{-\gamma} \times \left\{1 - \exp\left[-4\left(\frac{v_i}{v_{i,opt}}\right)^2\right]\right\}^2. \quad \textbf{(16-11)}$$

Conveniently, the pore volumes to breakthrough–interstitial velocity relationship can be defined by simply specifying the optimal condition—the value of the optimal interstitial velocity and the corresponding minimum pore volumes to breakthrough. Thus, this optimal point is the only data needed to calibrate this model for any particular acid-rock system.

In this model, the wormhole velocity depends on the interstitial velocity at the wormhole front, r_{wh}, and this velocity decreases as the wormhole region front moves away from the wellbore. The simplest way to implement the Buijse-Glasbergen model is to take a series of time steps, during each of which the interstitial velocity, and hence the wormhole velocity, are assumed constant. At each step, the new location of the wormhole region front is simply

$$(r_{wh})_{n+1} = (r_{wh})_n + v_{wh}\Delta t. \tag{16-12}$$

Example 16-2 Buijse-Glasbergen Model of Wormhole Propagation

Using the data from Figure 16-8 for an acid concentration of 15 weight% and a temperature of 150°F, calculate the radius of the region penetrated by wormholes for acid volumes up to 100 gal/ft using the Buijse-Glasbergen model. Assume an injection rate of 1 bpm into a 100-ft-thick reservoir and all other data is the same as in Example 16-1.

Solution

From Figure 16-8 for an acid concentration of 15 weight% and temperature of 150°F, $PV_{bt,opt} = 0.5$ and $v_{i,opt} = 1.5$ cm/min.

In the Buijse-Glasbergen model, the wormhole velocity depends on the interstitial velocity at the radial distance to which wormholes have propagated (r_{wh}). To solve this problem, a series of time steps are taken, during each of which the interstitial velocity is assumed constant.

When $t = 0$, $r^0_{wh} = r_w = 0.328$ ft, the initial interstitial velocity at r^0_{wh} is calculated by Equation (16-8),

$$v^0_i = \frac{1 \text{ bpm}}{2\pi(0.328 \text{ ft})(100 \text{ ft})(0.3)}\left(\frac{5.615 \text{ ft}^3}{1 \text{ bbl}}\right)\left(\frac{30.48 \text{ cm}}{1 \text{ ft}}\right) = 2.77 \text{ cm/min}. \tag{16-13}$$

In a similar way, at a given time $t = n\Delta t$, the interstitial velocity v^n_i at r^n_{wh} is

$$v^n_i = \frac{1 \text{ bpm}}{2\pi(r^n_{wh} \text{ ft})(100 \text{ ft})(0.3)}\left(\frac{5.615 \text{ ft}^3}{1 \text{ bbl}}\right)\left(\frac{30.48 \text{ cm}}{1 \text{ ft}}\right) = \frac{0.91}{r^n_{wh}} \text{ cm/min}. \tag{16-14}$$

Then, the velocity of wormhole propagation at $t = n\Delta t$ is calculated with Equation (16-11),

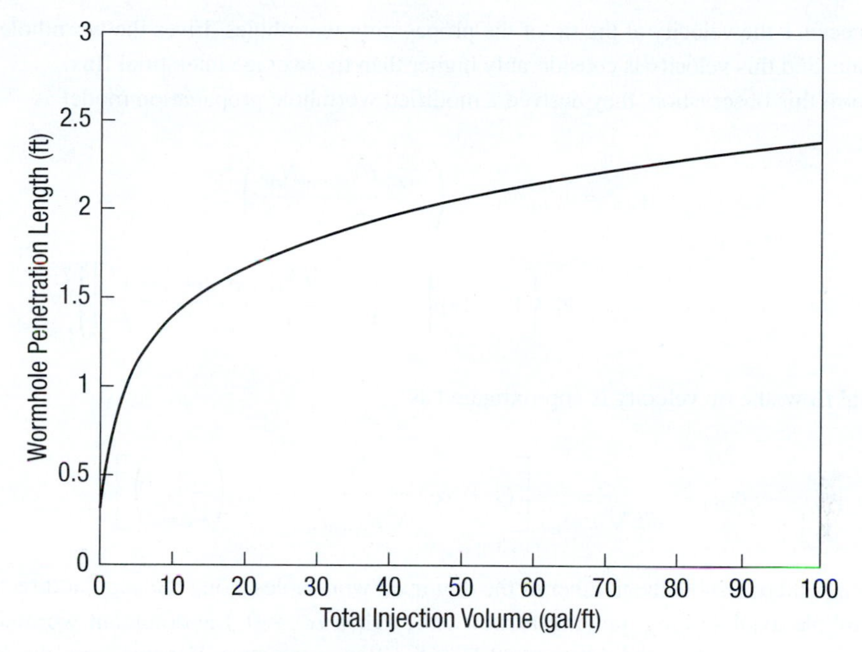

Figure 16-11 Buijse and Glasbergen model results (Example 16-2).

$$v_{wh}^{n} = \frac{(0.91/r_{wh}^{n}\ \text{cm/min})}{0.5}\left(\frac{0.91/r_{wh}^{n}\ \text{cm/min}}{1.5\ \text{cm/min}}\right)^{-\frac{1}{3}}$$

$$\left\{1 - \exp\left[-4\left(\frac{0.91/r_{wh}^{n}\ \text{cm/min}}{1.5\ \text{cm/min}}\right)^{2}\right]\right\}^{2}\left(\frac{1\ \text{ft}}{30.48\ \text{cm}}\right)$$

$$= \frac{1.82}{r_{wh}^{n}}\left(\frac{0.607}{r_{wh}^{n}}\right)^{-\frac{1}{3}}\left\{1 - \exp\left[\frac{-1.472}{(r_{wh}^{n})^{2}}\right]\right\}^{2}(\text{ft/min}) \qquad \textbf{(16-15)}$$

Subsequently, the location of wormhole front at $t = (n + 1)\Delta t$ is

$$r_{wh}^{n+1} = r_{wh}^{n} + v_{wh}^{n}\Delta t. \qquad \textbf{(16-16)}$$

Calculating the radius of the region penetrated by wormholes for acid volumes up to 100 gal/ft using the above algorithm, Figure 16-11 is generated.

16.3.3 The Furui et al. Model

From comparisons of the wormhole lengths predicted by the Buijse-Glasbergen model using optimal rate and pore volumes to breakthrough data from small diameter corefloods with field responses, Furui, Burton, Burkhead, Abdelmalek, Hill, Zhu, and Nozaki (2010) concluded that using the average interstitial flux, *v*, in this model underestimated wormhole propagation velocity.

This is because the velocity at the tip of the propagating wormholes drives the wormhole propagation rate, and this velocity is considerably higher than the average interstitial flux.

From this observation, they derived a modified wormhole propagation model as

$$v_{wh} = v_{i,tip} N_{Ac} \times \left(\frac{v_{i,tip} PV_{bt,opt} N_{Ac}}{v_{i,opt}} \right)^{-\gamma} \times \left\{ 1 - \exp\left[-4 \left(\frac{v_{i,tip} PV_{bt,opt} N_{Ac} L_{core}}{v_{i,opt} r_{wh}} \right)^2 \right] \right\}^2. \tag{16-17}$$

For radial flow, the tip velocity is approximated as

$$v_{i,tip} = \frac{q}{\phi h \sqrt{\pi m_{wh}}} \left[(1 - \alpha_z) \frac{1}{\sqrt{d_{e,wh} r_{wh}}} + \alpha_z \left(\frac{1}{d_{e,wh}} \right) \right] \tag{16-18}$$

where m_{wh} and α_z denote the number of the dominant wormholes along the angular direction and the wormhole axial spacing, ranging from 0 to 1. When $\alpha_z = 0$, the dominant wormholes are closely spaced along the axial direction (2-D radial flow geometry). For this case, the injection velocity at the wormhole tip declines proportionally to $1/r_{wh}^{0.5}$. This extreme case represents formations where the permeability in the longitudinal direction is significantly lower than that in the radial direction. When $\alpha_z = 1$, the dominant wormholes are sparsely spaced along the axial direction with very little effect of low-length/stalled wormholes. In this case, the injection velocity at the wormhole tip does not decline as r_{wh} increases. For typical acid stimulation design, it is recommended to use $\alpha_z = 0.25 - 0.5$ for vertical wells ($k_a < k_r$) and $\alpha_z = 0.5 - 0.75$ for horizontal wells ($k_a \approx k_r$). Also it is important to note that substituting $m_{wh} = 4\pi$, $d_{e,wh} = r_{wh}$, and $\alpha_z = 0$ into Equation 16-18 reduces to the conventional radial flow equation

$$v_{i,tip} = \frac{q}{2\pi \phi h r_{wh}}. \tag{16-19}$$

In Equation 16-18, $d_{e,wh}$, the diameter of the wormhole cluster is an empirical parameter accounting for the presence of branches around the main wormholes. This effective diameter can be estimated as

$$d_{e,wh} = d_{core} N_{AC} PV_{bt,opt}. \tag{16-20}$$

Example 16-3 Furui Model of Wormhole Propagation

Repeat Example 16-2, but using the Furui model instead of the Buijse-Glasbergen model. Compare the results with those obtained with the Buijse-Glasbergen model.

Solution

Reading from Figure 16-8 for an acid concentration of 15 wt% and temperature of 120°F, $PV_{bt,opt} = 0.5$, $v_{i,opt} = 1.5$ cm/min, and $L_{core} = 6$ in. The density of 15 wt% HCl and the chalk are 1.07 g/cm^3 and 2.71 g/cm^3, respectively.

The acid-dissolving power β_{15} for 15 wt% HCl solution is 0.21 g $CaCO_3$/g 15% HCl (Chapter 14).

From Equation (15-13), the acid capacity number is

$$N_{AC} = \frac{(0.3)(0.21)(1.07 \text{ g/cm}^3)}{(1 - 0.3)(2.71 \text{ g/cm}^3)} = 0.035. \tag{16-21}$$

Similar to Example 16-2, a series of time steps are taken. At a certain time $t = n\Delta t$, the interstitial velocity at the wormhole tip $v_{i,tip}^n$ is calculated by Equation (16-18),

$$v_{i,tip}^n = \frac{1 \text{ bpm}}{(0.3)(100 \text{ ft})(\sqrt{\pi m_{wh}})}\left(\frac{5.615 \text{ ft}^3}{1 \text{ bbl}}\right)\left[(1 - \alpha_z)\frac{1}{\sqrt{d_{e,wh} r_{wh}^n}} + \alpha_z\left(\frac{1}{d_{e,wh}}\right)\right] \left(\frac{30.48 \text{ cm}}{1 \text{ ft}}\right)(\text{cm/min}) \tag{16-22}$$

where

$$d_{e,wh} = d_{core} N_{AC} PV_{bt,opt} = \frac{1 \text{ in}}{12 \text{ in/ft}}(0.035)(0.5) = 0.00145 \text{ ft}. \tag{16-23}$$

For an isotropic formation,

$$\alpha_z = 0.75 \times \left(\frac{1}{I_{ani}}\right)^{0.7} = 0.75. \tag{16-24}$$

Then the velocity of wormhole propagation at $t = n\Delta t$ is calculated with Equation (16-17),

$$v_{wh}^n = v_{i,tip}^n(0.035)\left(\frac{v_{i,tip}^n(0.5)(0.035)}{1.5 \text{ cm/min}}\right)^{-\gamma} \cdot \left\{1 - \exp\left[-4\left(\frac{v_{i,tip}^n(0.5)(0.035)(6 \text{ in})(2.54 \text{ in/cm})}{(1.5 \text{ cm/min}) r_{wh}^n}\right)^2\right]\right\}^2 \left(\frac{1 \text{ ft}}{30.48 \text{ cm}}\right)(\text{ft/min}). \tag{16-25}$$

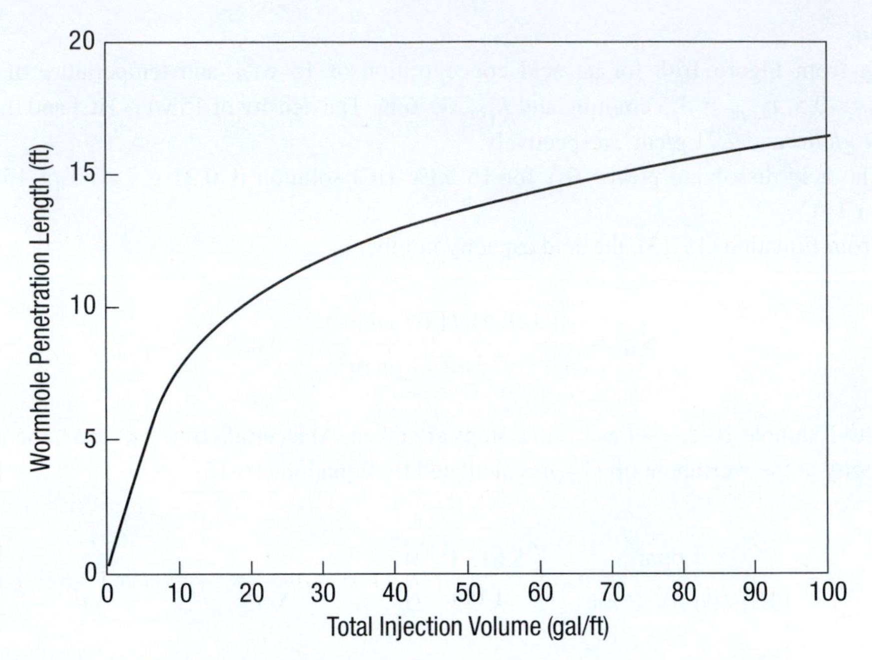

Figure 16-12 Furui et al. model results (Example 16-3).

So, the location of the wormhole front at $t = (n + 1)\Delta t$ is

$$r_{wh}^{n+1} = r_{wh}^{n} + v_{wh}^{n}\Delta t. \quad \textbf{(16-26)}$$

Substituting $m_{wh} = 6$, $\gamma = 1/3$ into the above equations, the wormhole penetration length versus total injection volumes up to 100 gal/ft is calculated (Figure 16-12). The wormhole penetration length after 100 gal/ft of acid injection is over six times greater than that predicted by the Buijse-Glasbergen model.

The models just presented show that the pore volumes to breakthrough for a given rock/acid system is a critical parameter in predicting how deep into the formation wormholes can propagate. For highly heterogeneous carbonates, particularly naturally fractured or vuggy limestones, this parameter is much lower than that of more homogeneous rocks. Figure 16-13 (Izgec, Keys, Zhu, and Hill, 2008) shows measured pore volume to breakthrough values as low as 0.04 from lab tests using 4-in.-diameter vuggy limestone cores.

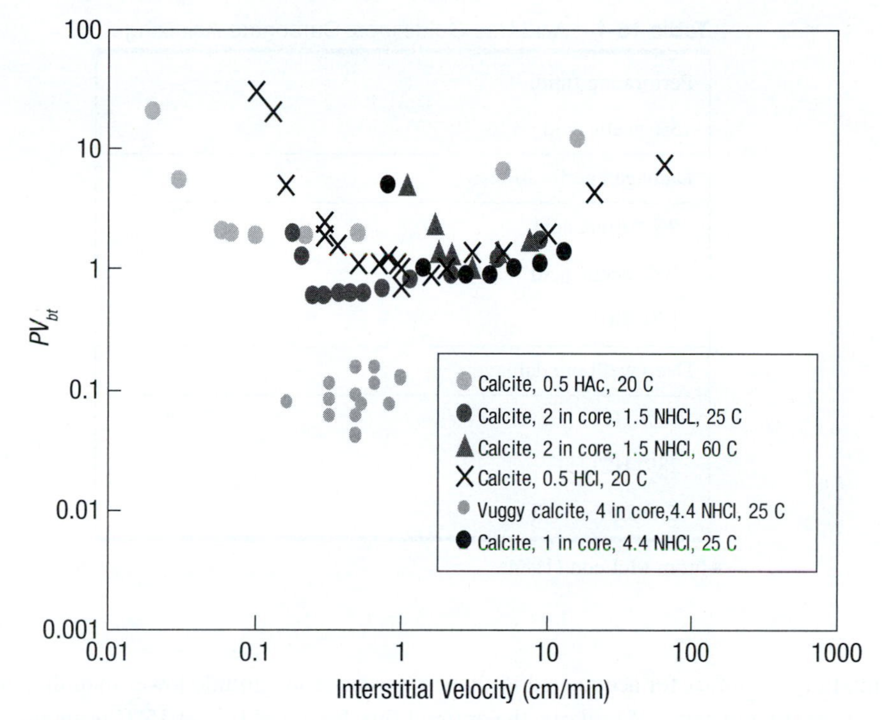

Figure 16-13 Wormhole propagation efficiency in limestones. (From Izgec, Zhu, and Hill, 2010.)

16.4 Matrix Acidizing Design for Carbonates

16.4.1 Acid Type and Concentration

Hydrochloric acid is by far the most common acid used in carbonate matrix acidizing. Table 16-1 shows the acids suggested by McLeod (1984) for various acid treatments of carbonate formations. Weak acids are suggested for perforation cleanup and perforating fluid, but otherwise, strong solutions of HCl are recommended. For a matrix treatment, HCl should be used unless corrosion considerations require a weaker acid (this would occur only in deep, hot wells). All models of wormhole propagation predict deeper penetration for higher acid concentrations, so a high concentration of HCl is preferable. Also, in carbonates, there are usually no precipitation reactions to limit the acid concentrations used, as is the case in sandstones. However, if sulfate-rich water, such as seawater, is used to mix the acid solution, or is used in conjunction with acid injection as a spacer, preflush, or post-flush fluid, precipitation of $CaSO_4$ may occur and cause formation damage (He, Mohamed, and Nasr-El-Din, 2011).

Besides simple rules of thumb, the acid type and concentration can be selected based on the optimal injection conditions for wormhole propagation. For example, Figure 16-6 shows that

Table 16-1 Acid Use Guidelines: Carbonate Acidizing

Perforating fluid:
5% acetic acid
Damaged perforations:
9% formic acid
10% acetic acid
15% HCl
Deep wellbore damage:
15% HCl
28% HCl
Emulsified HCl

From McLeod (1984).

the optimal injection flux for acetic acid is about an order of magnitude lower than that of 1.5 N HCl at the same temperature. Similarly, the optimal flux for 1.5 N HCl at 25°C is about an order of magnitude lower than the optimal flux of 1.5 N HCl at 60°C. If the well injectivity does not allow a flux that is near the optimal or above for a strong acid solution, it may be advantageous to use a weaker acid, such as acetic acid, because the optimal flux for the weaker acid is lower.

Applying a theoretical model of the reaction versus convection conditions that lead to optimal wormholing, Huang, Hill, and Schechter (2000) developed design charts that can be used to guide acid system selection, based on the expected treating temperature. Figure 16-14 is such a chart for limestone to be treated with hydrochloric, formic, or acetic acid. If the treating temperature is about 85°C (358°K; $1/T = 0.0028$), the optimal flux for formic or acetic acid is near the maximum flux possible for an open-hole well with certain assumed injectivity conditions and fracturing pressure. The optimal flux for 15% HCl, however, is much higher and would not be obtainable while maintaining matrix injection conditions. In this well, the weaker acids would propagate wormholes more efficiently than the stronger HCl solution. Other models for predicting the optimal injection rate are reviewed by Glasbergen, Kalia, and Talbot (2009).

16.4.2 Acid Volume and Injection Rate

The volume of acid to be injected in carbonate acidizing is based on the desired level of stimulation to be achieved. The more acid injected, the deeper into the formation wormholes propagate, and the lower the created skin factor. However, the penetration rate of wormholes slows as they grow longer, so at some acid volume, the growth rate is no longer sufficient to justify continued injection. The wormhole length as a function of volume of acid injected can be predicted using a

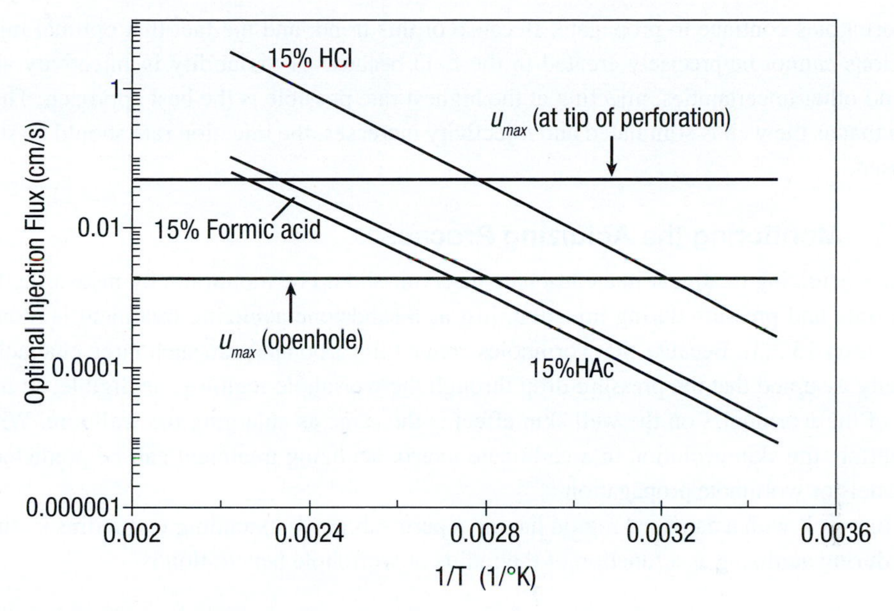

Figure 16-14 Acid selection design chart. (From Huang et al., 2000.)

wormhole propagation model, as illustrated in Examples 16-1 through 16-3. Coupling these predictions with a skin factor model (Section 16.3.3) allows for a simple economic analysis of the optimal acid volume. However, as with sandstones, such an approach can be calibrated by evaluating the skin factor evolution in actual treatments.

The injection rate schedule for a carbonate acidizing treatment is a very important design variable because of the sensitivity of the wormholing process to the flux of acid into the unstimulated rock ahead of the dissolution front. However, once an acid system has been selected, the best rate schedule is to simply inject at the highest rate possible without fracturing the formation, as has been suggested previously by Paccaloni and Tambini (1993) and Economides et al. (1994), and again justified by Glasbergen et al. (2009).

This preference for the highest possible injection rate is explained by the nature in which the wormholing process depends on rate. At injection rates below the optimum, wormholing becomes very inefficient, while at rates above the optimum, the efficiency decreases slowly with increased rate. For example, in Figure 16-6, consider the response curve for 0.5 N HCl at 20°C. For this acid stimulating limestone, the optimal interstitial flux is a little less than 1 cm/min and the minimum pore volumes to breakthough at this flux is about 0.9. If the acid is injected at a flux 10 times lower than the optimum, the pore volumes to breakthrough is over 100—in theory, more than 100 times as much acid would be needed to stimulate a given distance. This is because at this low a rate, the dissolution process is approaching face dissolution conditions. If, on the other hand, the injection rate creates a flux that is 10 times higher than the optimum, only about twice as much acid is needed. The higher injection rate causes more wormhole branching, but wormholes continue to propagate. Because of this trend, and the fact that optimal injection conditions cannot be precisely created in the field because of variability in injectivity along a well and other uncertainties, injecting at the highest rate possible is the best approach. This also means that as the well is stimulated and injectivity increases, the injection rate should be steadily increased.

16.4.3 Monitoring the Acidizing Process

A matrix acidizing treatment in a carbonate reservoir should be monitored by measuring the injection rate and pressure during injection, just as a sandstone acidizing treatment is monitored (see Section 15.3.3). Because the wormholes created in carbonates are such large channels, it is generally assumed that the pressure drop through the wormhole region is negligible, so that the effect of the wormholes on the well skin effect is the same as enlarging the wellbore. With this assumption, the skin evolution in a carbonate matrix acidizing treatment can be predicted with the models of wormhole propagation.

In a well with a damaged region having a permeability k, extending to a radius r_s, the skin effect during acidizing as a function of the radius of wormhole penetration is

$$s = \frac{k}{k_s}\ln\frac{r_s}{r_{wh}} - \ln\frac{r_s}{r_w}. \tag{16-27}$$

Equation (16-27) applies until the radius of wormhole penetration exceeds the radius of damage. If the well is originally undamaged or the wormhole radius is greater than the original damage radius, the skin effect during acidizing is obtained from Hawkins' formula assuming that k_s is infinite [or from Equation (16-27), setting $k_s = \infty$ and $r_s = r_{wh}$], which gives

$$s = -\ln\frac{r_{wh}}{r_w}. \tag{16-28}$$

Using Equations (16-27) and (16-28), if the injection rate is held constant throughout the treatment, the skin effect predicted by the volumetric model during injection is (with a damaged zone)

$$s = -\frac{k}{2k_s}\ln\left[\left(\frac{r_w}{r_s}\right)^2 + \frac{V}{PV_{bt}\pi r_s^2\phi h}\right] - \ln\frac{r_s}{r_w} \tag{16-29}$$

and, with no damage or the wormholes penetrating beyond the damaged region,

$$s = -\frac{1}{2}\ln\left(1 + \frac{V}{PV_{bt}\pi r_w^2\phi h}\right). \tag{16-30}$$

When using the Buijse-Glasbergen or the Furui et al. wormhole propagation models, the radius of the wormhole region, r_{wh}, is calculated for discrete values of injection time. For each r_{wh} calculated with these models, the skin factor is calculated with Equation 16-27 or Equation 16-28.

Example 16-4 Skin Evolution during Carbonate Matrix Acidizing

Using the wormhole propagation results from Examples 16-1 through 16-3 (Figures 16-9, 16-11, and 16-12), calculate and plot the skin factor as a function of injected volume. Assume that the formation permeability is 2 md and that there is a damaged zone with a permeability of 0.2 md extending one foot into the formation.

Solution

In the volumetric model, solving Equation (16-6) for *V/h* yields

$$\frac{V}{h} = \pi\phi(r_{wh}^2 - r_w^2)PV_{bt}. \tag{16-31}$$

For Example 16-1, the injection volume needed for wormholes to penetrate a distance of 1.328 ft into the formation can be calculated as

$$\frac{V}{h} = \pi(0.3)[(1.328 \text{ ft})^2 - (0.328 \text{ ft})^2](0.7) = 1.09 \text{ ft}^3/\text{ft} = 8.17 \text{ gal/ft}. \tag{16-32}$$

Thus, up to this cumulative volume of injection, Equation (16-29) applies,

$$s = -\frac{2}{(2)(0.2)} \ln\left[\left(\frac{0.328 \text{ ft}}{1.328 \text{ ft}}\right)^2 + \frac{(V/h)(\text{gal/ft})(1 \text{ ft}^3/7.48 \text{ gal})}{(\pi)(1.328 \text{ ft})^2(0.3)(0.7)}\right] - \ln\left(\frac{1.328 \text{ ft}}{0.328 \text{ ft}}\right) \tag{16-33}$$

or

$$s = -5 \ln\left[0.061 + 0.1149\left(\frac{V}{h}\right)\right] - 1.398. \tag{16-34}$$

When *V/h* is greater than 8.17 gal/ft, from Equation (16-30),

$$s = -\frac{1}{2} \ln\left[1 + \frac{(V/h)(1/7.48)}{(\pi)(0.328)^2(0.3)(0.7)}\right] \tag{16-35}$$

or

$$s = -\frac{1}{2} \ln\left[1 + 1.884\left(\frac{V}{h}\right)\right] \tag{16-36}$$

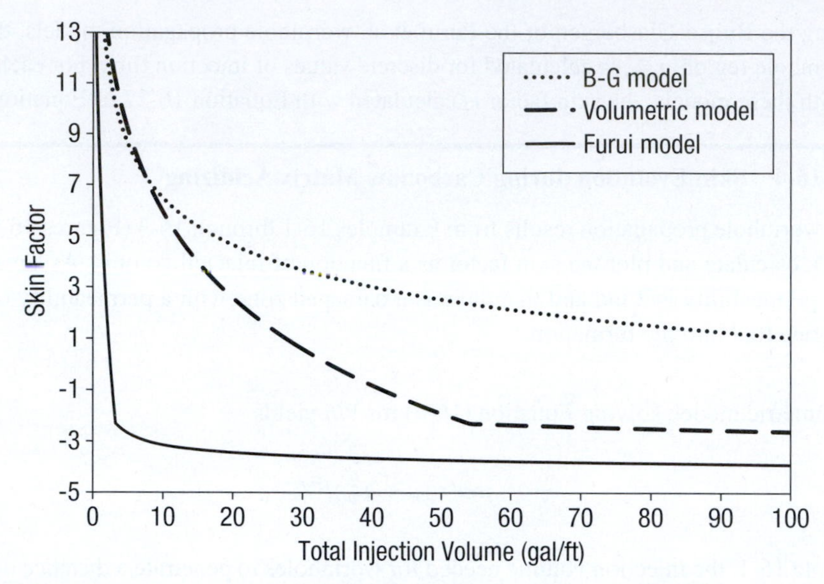

Figure 16-15 Skin factor evolution predicted by three wormhole models (Example 16-4).

When using the Buijse-Glasbergen model or the Furui model, a series of time steps are taken, and the radius of wormhole propagation, r_{wh}^n, is calculated at each discrete injection time. When the wormhole radius r_{wh}^n is shorter than the damage radius, Equation (16-27) applies; whereas for wormhole radius above the damage radius, Equation (16-28) should be used.

Calculating the skin factor versus total injection volume up to 100 gal/ft using the wormhole propagation results from Example 16-1 for the volumetric model, from Example 16-2 for the Buijse-Glasbergen model, and from Example 16-3 for the Furui model, Figure 16-15 is generated. These results are for an injection rate of 1 bpm. This damaged well is effectively stimulated with less than 10 gal/ft of acid, as this volume is sufficient to propagate wormholes through the damaged zone. Continued acid injection is predicted to gradually lower the skin factor as wormholes propagate deeper into the formation. A more effective deep stimulation could be obtained by injecting at a higher rate.

16.4.4 Fluid Diversion in Carbonates

Adequate placement of acid into all zones to be stimulated is as important in acidizing carbonates as it is in sandstones, and the same fluid placement techniques are applicable. However, because of the formation of wormholes, particulates should be used with caution because of potential difficulty in removing them. Viscous diversion is commonly used in carbonate acidizing, with viscosified acid systems including foamed acid, emulsified acid, and polymer-viscosified acid. A diverting material unique to carbonate acidizing is viscoelastic surfactant (VES) added to the acid solution. Except for the VES acid systems, the effect of the other viscosified acid systems on acid placement can be calculated using a viscous skin factor, as described for

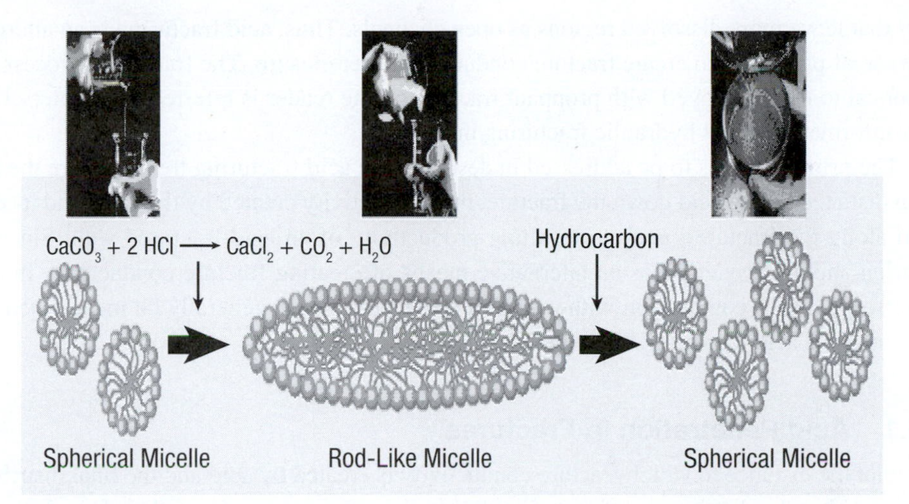

Figure 16-16 Mechanism of viscosity enhancement by viscoelastic surfactants. (From Nasr-El-Din et al., 2008.)

sandstone acidizing in Section 15.4.4. Nozaki and Hill (2010) described how viscous diversion can be incorporated into a wormhole propagation model to predict acid placement in carbonate acidizing.

Viscoelastic surfactant acid systems are unique diverting materials in that these fluids have low viscosity until the acid spends, then the viscosity increases dramatically in the spent acid in which the calcium concentration and pH have both increased greatly. The viscosity increase is caused by the formation of long, rod-like micelles as the chemical composition of the acid solution changes, as illustrated in Figure 16-16 (Nasr-El-Din, Al-Ghamdi, Al-Qahtani, and Samuel, 2008). When hydrocarbons contact the spent VES acid, the long micelles are broken, and the viscosity again becomes low, allowing for spent acid recovery. The viscous fluid is created in the region of spent acid in the matrix ahead of and surrounding the propagating wormholes (Tardy, Lecerf, and Christiani, 2007; Al-Ghamdi, Mahmoud, Hill, and Nasr-El-Din, 2010), not in the wormholes themselves. Because of the complex interaction between the wormholing process and the viscosification of the spent acid around the wormholes, the effect of VES diversion systems cannot be predicted by assuming the acid behaves as a bank of viscous fluid.

16.5 Acid Fracturing

Acid fracturing is a stimulation technique in which acid is injected at pressures above the parting pressure of the formation, so that a hydraulic fracture is created. Usually, a viscous pad fluid is injected ahead of the acid to initiate the fracture, then plain acid, gelled acid, foamed acid, or an emulsion containing acid is injected. Fracture conductivity is created by the acid differentially etching the walls of the fracture; that is, the acid reacts nonuniformly with the fracture walls so that, after closure, the fracture props itself open, with the relatively undissolved regions acting as pillars that leave more dissolved regions as open channels. Thus, acid fracturing is an alternative to the use of proppants to create fracture conductivity after closure. The fracturing process itself is identical to that employed with proppant fracturing. The reader is referred to Chapters 17 and 18 for information about hydraulic fracturing in general.

The primary issues to be addressed in designing an acid fracturing treatment are the penetration distance of live acid down the fracture, the conductivity created by the acid (and its distribution along the fracture), and the resulting productivity of an acid-fractured well. Since acid fracturing should be viewed as an alternative means of creating fracture conductivity in a carbonate formation, a comparison with proppant fracturing should generally be made when planning a possible acid fracturing treatment.

16.5.1 Acid Penetration in Fractures

To predict the distance to which fracture conductivity is created by acid and the final distribution of conductivity, the distribution of rock dissolution along the fracture is needed. This, in turn, requires a prediction of acid concentration along the fracture. The acid distribution along the fracture is obtained from an acid balance equation and appropriate boundary conditions. For linear flow down a fracture, with fluid leakoff and acid diffusion to the fracture walls, these are

$$\frac{\partial C}{\partial t} + \frac{\partial (u_x C)}{\partial x} - \frac{\partial (u_y C)}{\partial y} - \frac{\partial}{\partial y}\left(D_{\text{eff}} \frac{\partial C}{\partial y}\right) = 0 \tag{16-37}$$

$$C(x, y, t = 0) = 0 \tag{16-38}$$

$$C(x = 0, y, t) = C_i(t) \tag{16-39}$$

$$Cu_y - C_L q_L - D_{\text{eff}} \frac{\partial C}{\partial y} = E_f C^n (1 - \phi) \tag{16-40}$$

where C is the acid concentration, u_x is the flux along the fracture, u_y is the transverse flux due to fluid loss, D_{eff} is an effective diffusion coefficient, C_i is the injected acid concentration, E_f is the reaction rate constant, n is the order of the reaction, and ϕ is porosity. To solve Equation (16-37), continuity and transport equations are needed as well as a model of the fluid loss behavior. Complex numerical solutions of these equations that consider further complications such as the temperature distribution along the fracture, viscous fingering of low-viscosity acid through a viscous pad, the effect of the acid on leakoff behavior, and various fracture geometries have been presented by Ben-Naceur and Economides (1988), Lo and Dean (1989), and Settari (1991).

By assuming steady-state, laminar flow of a Newtonian fluid between parallel plates with constant fluid loss flux along the fracture, Nierode and Williams (1972) presented a solution for the acid balance equation for flow in a fracture, adapted from the solution presented by Terrill (1965) for heat transfer between parallel plates. The solution for the concentration profile is presented in Figure 16-17 as a function of the leakoff Peclet number, defined as

$$N_{\mathrm{Pe}} = \frac{\overline{u}_y w}{2D_{\mathrm{eff}}} \tag{16-41}$$

where $\overline{u}_y$ is average leakoff flux and w is the fracture width. The effective diffusion coefficient, D_{eff}, is generally higher than the molecular diffusion coefficient because of additional mixing caused by density gradients. Effective diffusion coefficients reported by Roberts and Guin (1974) are shown in Figure 16-18, based on the Reynolds number in the fracture $(wu_x\rho/\mu)$.

From Figure 16-17, it is apparent that at low Peclet numbers, the acid concentration becomes very low before reaching the end of the fracture; at high Peclet numbers, high acid concentrations reach the end of the fracture. At low Peclet numbers, diffusion controls acid propagation, while at high Peclet numbers, fluid loss is the controlling factor. To complete the prediction of acid penetration, a model of fracture propagation is necessary, as the actual distance of acid propagation depends on the fracture length.

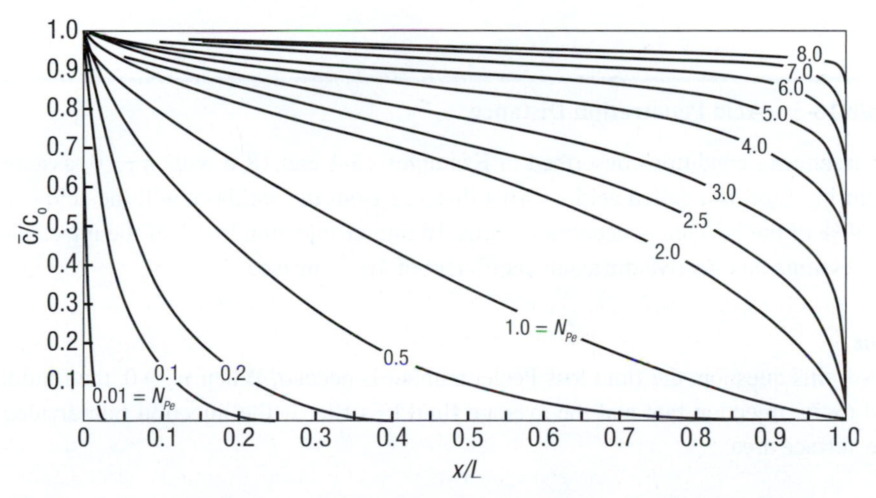

Figure 16-17 Acid concentration profiles along a fracture. (From Schechter, 1992.)

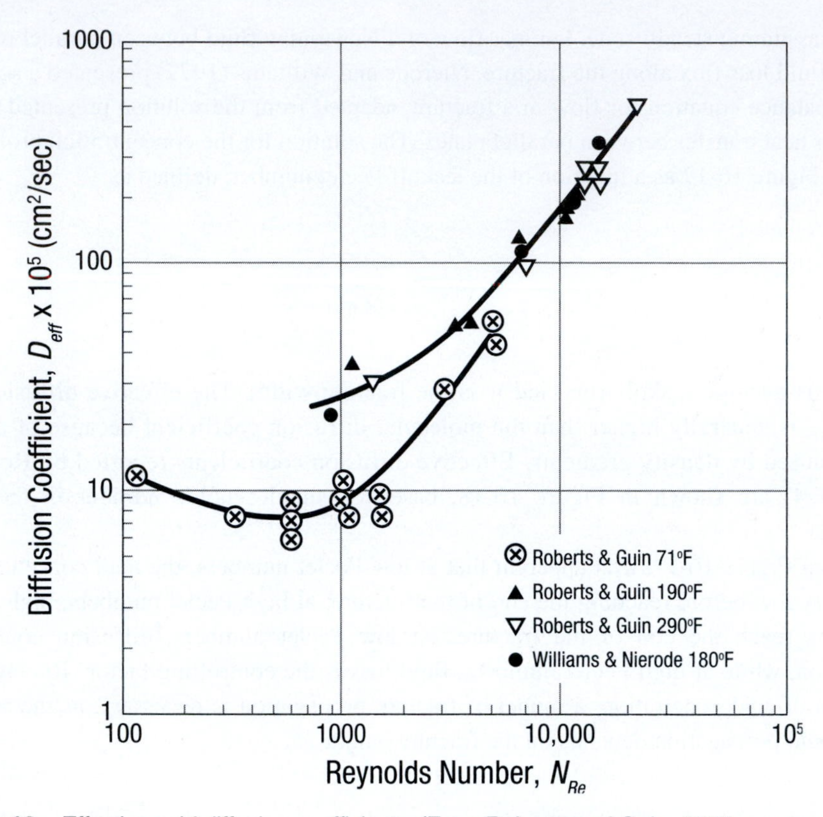

Figure 16-18 Effective acid diffusion coefficients. (From Roberts and Guin, 1974; courtesy of SPE.)

Example 16-5 Acid Penetration Distance

For the fracturing conditions described in Examples 18-4 and 18-6 with $\eta \rightarrow 0$, assuming that the fracturing fluid is a gelled acid, at what distance from the wellbore will the acid concentration be 50% of the injected concentration after 16 min of injection? 10% of the injected concentration? Assume an effective diffusion coefficient of 10^{-4} cm^2/sec.

Solution

To answer this question, the fluid loss Peclet number is needed. When $\eta \rightarrow 0$, the fluid loss rate is equal to the injection rate and the average fluid loss flux is the injection rate divided by the fracture surface area:

$$\bar{u}_y = \frac{q_i}{4hx_f}. \tag{16-42}$$

From Example 18-6, after 16 min of injection, $x_f = 286$ ft. Thus,

$$\overline{u}_y = \frac{(40 \text{ bpm})(5.615 \text{ ft}^3/\text{ bbl})}{(4)(100 \text{ ft})(286 \text{ ft})} = 1.96 \times 10^{-3} \text{ ft/min.} \tag{16-43}$$

Following Example 18-4, for a fracture length of 286 ft, the average fracture width is 0.14 in. (0.01167 ft). When the units are converted to ft^2/min, the effective diffusion coefficient is 6.46×10^{-6}. From Equation (16-41), the fluid loss Peclet number is

$$N_{\text{Pe}} = \frac{(0.01167 \text{ ft})(1.96 \times 10^{-3} \text{ ft/min})}{(2)(6.46 \times 10^{-6} \text{ ft}^2/\text{min})} = 1.8. \tag{16-44}$$

Reading from Figure 16-17 for a Peclet number of about 2, the acid concentration will be 50% of the injected value at a dimensionless distance of 0.69, or at (0.69)(286) = 197 ft. Similarly, a concentration of 10% of the injected concentration penetrates to a dimensionless distance of 0.98, or an actual distance of 280 ft.

16.5.2 Acid Fracture Conductivity

The conductivity ($k_f w$) of an acid fracture is difficult to predict because it depends inherently on a stochastic process: If the walls of the fracture are not etched heterogeneously, very little fracture conductivity will result after closure. Thus the approach taken to predict acid fracture conductivity has been an empirical one. First, based on the acid distribution in the fracture, the amount of rock dissolved as a function of position along the fracture is calculated. Then an empirical correlation is used to calculate fracture conductivity based on the amount of rock dissolved. Finally, since the conductivity usually varies significantly along the fracture, some averaging procedure is used to obtain an average conductivity for the entire fracture. The methods for predicting fracture conductivity of an acid fracture should not be expected to be very accurate. By measuring the effective conductivity of acid fractures in the field using pressure transient testing, these techniques can be "field calibrated."

The amount of rock dissolved in an acid fracture is represented by a parameter called the ideal width, w_i, defined as the fracture width created by acid dissolution before fracture closure. If all the acid injected into a fracture dissolves rock on the fracture face (i.e., no live acid penetrates into the matrix or forms wormholes in the fracture walls), the average ideal width is simply the total volume of rock dissolved divided by the fracture area, or

$$\overline{w}_i = \frac{XV}{2(1 - \phi)h_f x_f} \tag{16-45}$$

where X is the volumetric dissolving power of the acid, V is the total volume of acid injected, h_f is the fracture height, and x_f is the fracture half length. For Peclet numbers greater than about 5, the acid concentration is almost equal to the injected concentration along most of the fracture length, and the actual ideal width is approximately equal to the mean ideal width. For lower Peclet numbers, more acid is spent near the wellbore than farther along the fracture, and the concentration distribution must be considered to calculate the ideal width distribution. Schechter (1992) presented one such solution (Figure 16-19) based on the concentration profiles of Figure 16-17.

From the ideal fracture width, the conductivity of an acid fracture has been most commonly obtained from the correlation of Nierode and Kruk (1973). This correlation is based on extensive laboratory measurements of acid fracture conductivity and correlates conductivity with the ideal width, the closure stress, σ_c, and the rock embedment strength, S_{rock}. Nierode and Kruk's correlation is

$$k_f w = C_1 e^{-C_2 \sigma_c} \quad \textbf{(16-46)}$$

where

$$C_1 = 1.47 \times 10^7 w_i^{2.47} \quad \textbf{(16-47)}$$

and

$$C_2 = (13.9 - 1.3 \ln S_{rock}) \times 10^{-3} \quad \textbf{(16-48)}$$

for $S_{rock} < 20{,}000$ psi, and

$$C_2 = (3.8 - 0.28 \ln S_{rock}) \times 10^{-3} \quad \textbf{(16-49)}$$

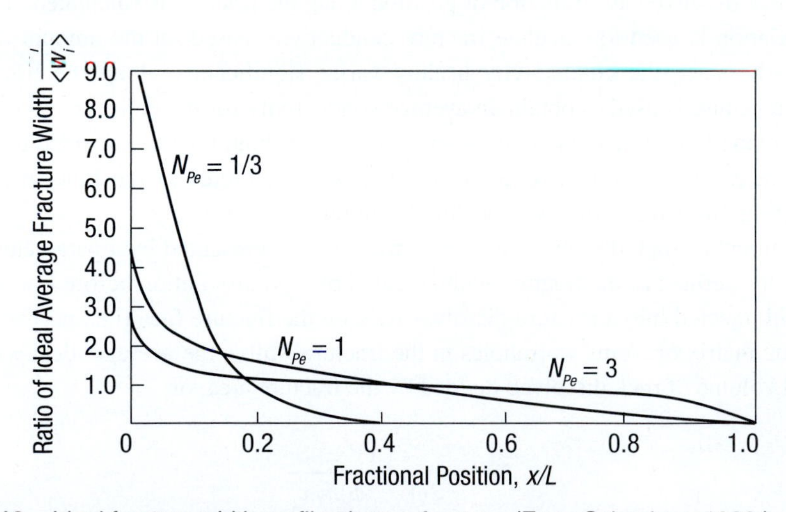

Figure 16-19 Ideal fracture width profile along a fracture. (From Schechter, 1992.)

for $S_{rock} > 20{,}000$ psi. [*Note:* In the original publication, there was a typographical error in the equation presented here as Equation (16-48). The original publication has the constant 19.9 in this equation; the value of 13.9 given here is the correct one.]

For the constants given in these equations, $k_f w$ is in md-ft, w_i is in in., and σ_c and S_{rock} are in psi. The rock embedment strength is the force required to push a metal sphere a certain distance into the surface of a rock sample. Laboratory-measured values of embedment strength and acid fracture conductivity for several common carbonate formations from Nierode and Kruk (1973) are given in Table 16-2.

Comparisons with field-measured acid fracture conductivities have shown that the Nierode-Kruk correlation sometimes overpredicts conductivity and sometimes underpredicts conductivity, and the errors can be large (Settari, Sullivan, and Hansen, 2001; Bale, Smith, and Klein, 2010). This is not surprising because it was based on laboratory measurements using very small core plugs, which cannot be expected to scale up to a field-scale fracture conductivity. Furthermore, the role of heterogeneity in formation properties, which is necessary for effective acid fracture conductivity to be created, is not considered in the Nierode-Kruk correlation.

A new correlation for acid fracture conductivity that explicitly depends on statistical variations in formation properties was recently presented by Deng, Mou, Hill, and Zhu (2011). For a formation in which variations in permeability, and thus the distribution of acid leakoff, cause differential etching, the correlation is as follows.

First, the conductivity at zero closure stress for permeability distribution dominant cases (Mou, Zhu, and Hill 2011) is:

$$(wk_f)_0 = 4.48 \times 10^9 \overline{w}^3 \left[1 + \left(a_1 erf(a_2(\lambda_{D,x} - a_3)) - a_4 erf(a_5(\lambda_{D,z} - a_6))\right)\sqrt{(e^{\sigma_D} - 1)}\right]$$

$$a_1 = 1.82, a_2 = 3.25, a_3 = 0.12, a_4 = 1.31, a_5 = 6.71, a_6 = 0.03 \qquad \textbf{(16-50)}$$

and $\overline{w}$ is average fracture width in inches. In this correlation, conductivity depends on statistical parameters describing the variation of the formation permeability. These are $\lambda_{D,x}$, the dimensionless correlation length in the horizontal direction (along the fracture); $\lambda_{D,z}$, the dimensionless correlation length in the vertical direction; and σ_D, the dimensionless standard deviation of permeability. The average width is the fracture width at zero closure stress. In practice, ideal fracture width, w_i, which is defined as dissolved rock volume divided by fracture surface area, is easier to obtain than average fracture width. For the high-leakoff coefficient (>0.004 ft/(min)$^{0.5}$),

$$\overline{w} = 0.56 erf(0.8\sigma_D) w_i^{0.83}. \qquad \textbf{(16-51)}$$

For the medium leakoff coefficient (~ 0.001 ft/(min)$^{0.5}$) with uniform mineralogy distributions,

$$\overline{w} = 0.2 erf(0.78\sigma_D) w_i^{0.81}. \qquad \textbf{(16-52)}$$

Table 16-2 Effective Fracture Conductivity for Well-Known Reservoirs as a Function of Rock Embedment Strength and Closure Stress

			Conductivity (md-in.) versus Closure Stress (psi)				
Reservoir	**Maximum Conductivity**	S_{rock}	**0**	**1000**	**3000**	**5000**	**7000**
San Andres dolomite	2.7×10^6	76,600	1.1×10^4	5.3×10^3	1.2×10^3	2.7×10^2	6.0×10^0
San Andres dolomite	5.1×10^8	63,800	1.2×10^6	7.5×10^5	3.0×10^5	1.2×10^5	4.7×10^4
San Andres dolomite	1.9×10^7	62,700	2.1×10^5	9.4×10^4	1.9×10^4	3.7×10^3	7.2×10^2
Canyon limestone	1.3×10^8	88,100	1.3×10^6	7.6×10^5	3.1×10^5	4.8×10^4	6.8×10^3
Canyon limestone	4.6×10^7	30,700	8.0×10^5	3.9×10^5	9.4×10^4	2.3×10^4	5.4×10^3
Canyon limestone	2.7×10^8	46,400	1.6×10^6	6.8×10^5	1.3×20^5	2.3×10^4	4.4×10^3
Cisco limestone	1.2×10^5	67,100	2.5×10^3	1.3×10^3	3.4×10^2	8.8×10^1	2.3×10^1
Cisco limestone	3.0×10^5	14,800	7.0×10^3	3.4×10^3	8.0×10^2	1.9×10^2	4.4×10^1
Cisco limestone	2.0×10^6	25,300	1.4×10^5	6.2×10^4	1.3×10^4	2.7×10^3	5.7×10^2
Capps limestone	3.2×10^5	13,000	9.7×10^3	4.2×10^3	7.6×10^2	1.4×10^2	2.5×10^1
Capps limestone	2.9×10^5	30,100	1.8×10^4	6.8×10^3	9.4×10^2	1.3×10^2	1.8×10^1
Indiana limestone	4.5×10^6	22,700	4.6×10^5	1.5×10^5	1.5×10^4	1.5×10^3	1.5×10^2
Indiana limestone	2.8×10^7	21,500	7.9×10^5	3.0×10^5	4.3×10^4	6.3×10^3	9.0×10^2
Indiana limestone	3.1×10^8	14,300	7.4×10^6	2.0×10^6	1.4×10^5	1.0×10^4	7.0×10^2
Austin chalk	3.9×10^6	11,100	5.6×10^4	1.6×10^3	1.3×10^0	–	–
Austin chalk	2.4×10^6	5,600	3.9×10^4	1.2×10^3	1.2×10^0	–	–
Austin chalk	4.8×10^5	13,200	1.0×10^4	1.7×10^3	4.9×10^1	1.4×10^0	–

Clearfork dolomite	3.6×10^4	35,000	3.4×10^3	1.7×10^3	4.1×10^2	1.0×10^2	2.4×10^1
Clearfork dolomite	3.3×10^4	11,800	9.3×10^3	1.6×10^3	4.5×10^1	1.3×10^0	–
Greyburg dolomite	8.3×10^6	14,400	2.5×10^5	4.0×10^4	1.0×10^3	2.5×10^1	–
Greyburg dolomite	3.9×10^6	12,200	2.1×10^5	7.9×10^4	1.0×10^4	1.5×10^3	2.0×10^2
Greyburg dolomite	3.2×10^6	16,600	8.0×10^4	1.5×10^4	4.8×10^2	1.6×10^1	–
San Andres dolomite	1.0×10^6	46,500	8.3×10^4	4.0×10^4	9.5×10^3	2.2×10^3	5.2×10^2
San Andres dolomite	2.4×10^6	76,500	1.9×10^4	6.8×10^3	8.5×10^2	1.0×10^2	1.3×10^1
San Andres dolomite	3.4×10^6	17,300	9.4×10^3	2.8×10^3	2.5×10^2	2.3×10^1	–

From Nierode and Kruk (1973).

Then, the correlation for overall fracture conductivity is:

$$wk_f = \alpha \exp[-\beta \sigma_c] \tag{16-53}$$

where

$$\alpha = \left(wk_f\right)_0 \left[0.22\left(\lambda_{D,x}\sigma_D\right)^{2.8} + 0.01\left(\left(1 - \lambda_{D,z}\right)\sigma_D\right)^{0.4}\right]^{0.52} \tag{16-54}$$

and

$$\beta = \left[14.9 - 3.78\ln(\sigma_D) - 6.81\ln(E)\right] \times 10^{-4}. \tag{16-55}$$

In Equations 16-53 through 16-55, σ_c is closure stress in psi, σ_D is normalized standard deviation, and E is Young's modulus in Mpsi (million psi). In addition, Young's modulus is required to be greater than 1 Mpsi. In general, soft rock with Young's modulus less than 2 Mpsi is not a good candidate for acid fracturing.

Often, the vertical correlation length of permeability distribution is low because the sedimentary carbonates are laminated. When the dimensionless vertical correlation length is low enough, for example, $\lambda_{D,z} < 0.02$, Equations 16-54 and 16-55 can be simplified to:

$$\alpha = 0.12(wk_f)_0(\lambda_{D,x}\sigma_D)^{0.1} \tag{16-56}$$

$$\beta = [15.6 - 4.5\ln(\sigma_D) - 7.8\ln(E)] \times 10^{-4} \tag{16-57}$$

For a discussion of how to obtain the statistical parameters needed in this correlation and to see an example of its application, the reader is referred to Beatty, Hill, Zhu, and Sullivan (2011).

Once the conductivity variation along the fracture is obtained, an average conductivity for the entire fracture can be calculated in order to estimate the productivity of the acid-fractured well. If the variation in the conductivity is not too great, Bennett (1982) has shown that a simple average of the conductivity predicts the well productivity after early production. This average is

$$\overline{k_f}\overline{w} = \frac{1}{x_f}\int_0^{x_f} k_f w dx. \tag{16-58}$$

This average should be adequate when the fluid loss Peclet number is greater than 3. For lower values of Peclet numbers, this average will overestimate the well productivity because it is too strongly influenced by the high conductivity near the well. It is possible in some acid fractures that the conductivity is a maximum at some distance from the wellbore (caused, for example, by increasing diffusion or reaction rates as the acid warms up; Elbel, 1989). For this type of distribution, a harmonic mean better approximates the behavior of the fractured well (Ben-Naceur and Economides, 1989):

$$\overline{k_f w} = \frac{x_f}{\int_0^{x_f} dx/k_f x}. \tag{16-59}$$

Example 16-6 Average Acid Fracture Conductivity

The fracturing treatment described in Example 16-5 is conducted using 15 wt% HCl in a 15% porosity limestone formation. The formation has an embedment strength of 60,000 psi, and the closure stress is 4000 psi. Acid is injected for a total of 15 min. Calculate the average fracture conductivity after fracturing using the Neirode-Kruk correlation.

Solution

First, the average ideal width is calculated with Equation (16-45). For a total of 600 bbl of injection [(40 bpm)(16 min)] into a 100-ft-high, 286-ft-long fracture, the average ideal width is

$$w_i = \frac{(0.082)(600 \text{ bbl})(5.615 \text{ ft}^3/\text{ bbl})}{(2)(1 - 0.15)(100 \text{ ft})(286 \text{ ft})}\left(\frac{12 \text{ in.}}{1 \text{ ft}}\right) = 0.068 \text{ in.} \tag{16-60}$$

The fluid loss Peclet number was found to be about 2 in Example 16-5. Interpolating between the curves for $N_{\text{Pe}} = 1$ and $N_{\text{Pe}} = 3$ in Figure 16-19, values for $\overline{w}_i/w_i$ are read. These are shown in Table 16-3.

Using the Neirode-Kruk correlation, C_2 is given by Equation (16-49),

$$C_2 = [3.8 - 0.28 \ln(60{,}000)] \times 10^{-3} = 7.194 \times 10^{-4} \tag{16-61}$$

and combining Equations (16-46) and (16-47),

$$k_f w = (1.47 \times 10^7) w_i^{2.47} e^{-(7.194 \times 10^{-4})(4000)} = (8.27 \times 10^5) w_i^{2.47}. \tag{16-62}$$

Table 16-3 Example 16-6 Results

x/x_f	$w_i/\overline{w_i}$	w_i (in.)	$k_f w$ (md-ft)
0.05	2.6	0.183	12,500
0.15	1.55	0.109	3500
0.25	1.3	0.092	2300
0.35	1.15	0.081	1700
0.45	1.0	0.070	1200
0.55	0.85	0.060	800
0.65	0.75	0.053	600
0.75	0.65	0.046	400
0.85	0.45	0.032	200
0.95	0.15	0.011	10

$k_f\overline{w} = 2300$ md-ft

The fracture conductivity for 10 equal increments along the fracture are shown in Table 16-3. Using the simple average, the average conductivity is 2300 md-ft. This value is probably high because of the high conductivity near the well. In contrast, using the harmonic average gives a value of 95 md-ft for the average conductivity.

16.5.3 Productivity of an Acid-Fractured Well

The productivity of an acid-fractured well can be predicted in the same manner as for a well with a propped fracture (see Chapter 17 for details on propped fracture performance) if an average fracture conductivity adequately describes the well flow behavior. One complicating factor is that the conductivity of an acid fracture depends strongly on the closure stress, and the closure stress increases as the bottomhole flowing pressure decreases during the life of the well. Since p_{wf} for a given flow rate depends on fracture conductivity and the fracture conductivity depends on the closure stress, and hence p_{wf}, an iterative procedure is required to determine the productivity of an acid-fractured well. Ben-Naceur and Economides (1989) presented a series of performance type curves for acid-fractured wells producing at a constant p_{wf} of 500 psi using the Nierode-Kruk conductivity correlation. An example of these curves is reproduced in Figure 16-20. In this graph, the dimensionless cumulative production, Q_D, and the dimensionless time, t_{Dxf}, are defined as

$$Q_D = \frac{3.73 \times 10^{-2} N_p B_o}{\phi h c_t x_f^2 (p_i - p_{wf})} \text{ for oil} \tag{16-63}$$

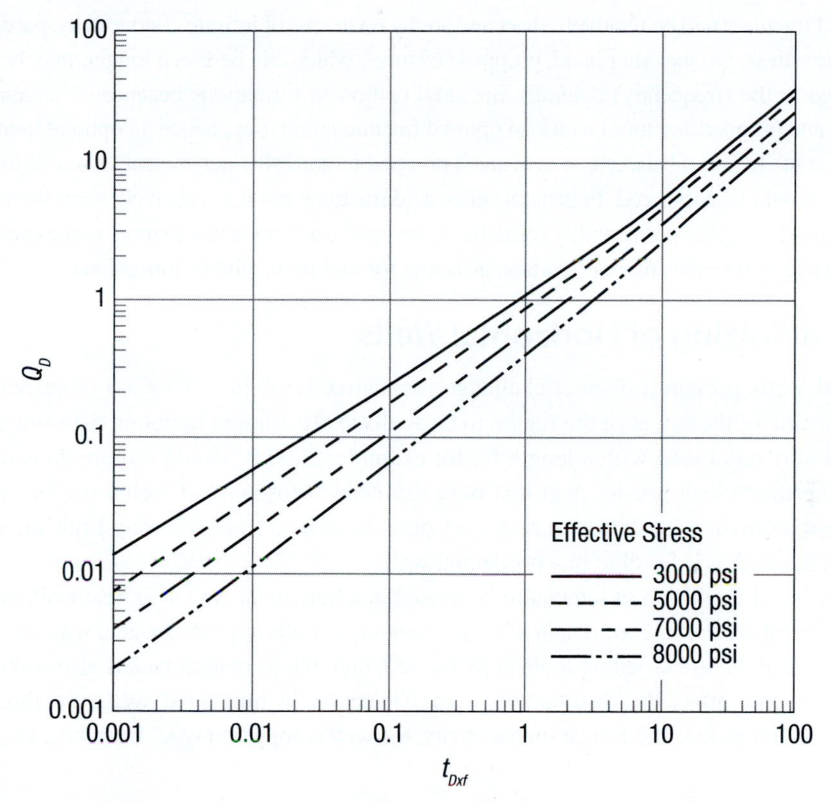

Figure 16-20 Performance curves for a 30,000-psi embedment strength formation as a function of closure stress. (From Ben-Naceur and Economides, 1988.)

$$Q_D = \frac{0.376 G_p Z T}{\phi h c_t x_f^2 (p_i^2 - p_{wf}^2)} \text{ for gas} \quad \textbf{(16-64)}$$

$$t_{Dxf} = \frac{0.000264 k t}{\phi \mu c_t x_f^2} \quad \textbf{(16-65)}$$

where N_p and G_p are the cumulative productions of oil (STB) and gas (MCSF), respectively, and all other variables and units are defined as in Chapter 17.

16.5.4 Comparison of Propped and Acid Fracture Performance

The decision to place a propped instead of an acid fracture in a carbonate formation should be taken based on the expected post-treatment performance and the costs of the treatments. The fact that acid fracturing is intended specifically for carbonate formations should not preclude the use of a propped fracture if it can be placed and has economic advantages.

Acid fractures will be relatively short and are by no means of infinite conductivity, particularly at high closure stress. On the other hand, propped fractures, which can be much longer, may be impossible to place in the (frequently) naturally fractured carbonate formations because of screenouts. For both acid- and propped-fractured wells, an optimal fracture length (and hence an optimal treatment design) will exist. To select between an acid and a propped fracture, the net present values of the optimal treatments should be compared. In general, since acid fractures result in relatively short fractures, they will be favored in higher-permeability formations; propped fractures become more and more favorable as fracture length becomes more important, as is true for low-permeability formations.

16.6 Acidizing of Horizontal Wells

Horizontal wells present a distinct challenge for matrix acidizing, if for no other reason than simply because of the extent of the region to be acidized. To attempt to obtain the same acid coverage in a horizontal well with a length of, for example, 2000 ft, would require an acid volume and treating time much greater than that typically needed for vertical wells. Besides the much greater cost, corrosion inhibition during long periods of acid injection may limit the extent of matrix acidizing that is feasible in a horizontal well.

Because of the length of interval to be treated in a horizontal well acid treatment, acid placement is a critical issue. It is common in acidizing horizontal wells to place the acid with coiled tubing or drill pipe, and the use of jetting tools to jet the acid onto the formation face is also common. The many factors that affect the distribution of acid injection in horizontal wells are illustrated in Figure 16-21 and include the formation injectivity, the well completion type, wormhole propagation,

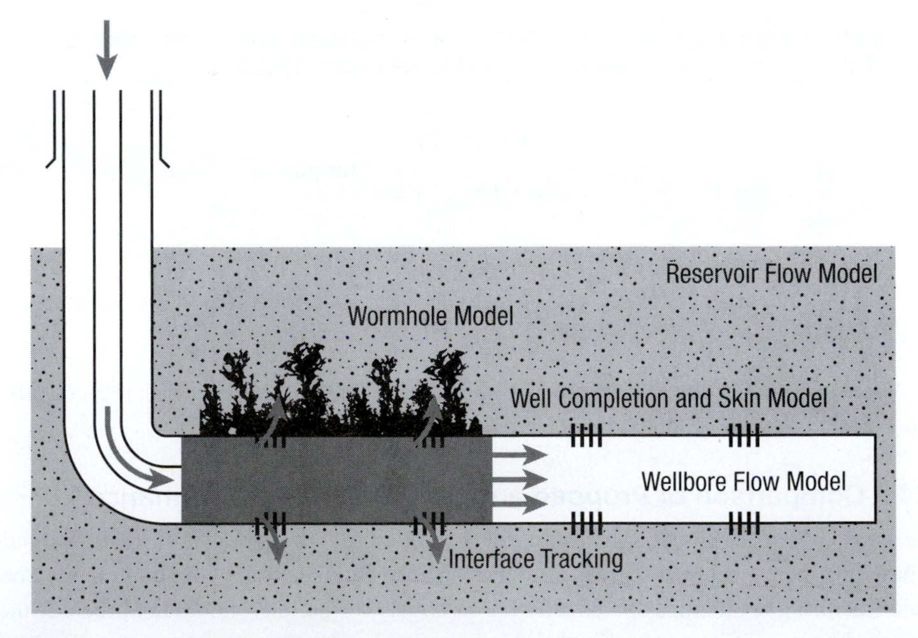

Figure 16-21 Acid injection into a horizontal well.

viscous effects as the treating fluids displace resident fluids in the formation, and wellbore flow conditions. Descriptions of such models of horizontal well acidizing in carbonates and comparisons with field rate-pressure data are given by Mishra, Zhu, and Hill (2007), Sasongko, Zhu, and Hill (2011), and Furui et al. (2011). Finally, it is now possible to diagnose the distribution of acid in matrix treatments of horizontal wells using distributed temperature measurements (DTS), as described by Glasbergen, Gualteri, van Domelen, and Sierra (2009).

References

1. Al-Ghamdi, A.H., Mahmoud, M.A., Hill, A.D. and Nasr-El-Din, H.A., "When Do Surfactant-Based Acids Work as Diverting Agents?", SPE Paper 128074, presented at the 2010 SPE International Symposium and Exhibition on Formation Damage Control, Lafayette, LA, February 10–12, 2010.
2. Bale, A., Smith, M.B., and Klein, H.H., "Stimulation of Carbonates Combining Acid Fracturing with Proppant (CAPF): A Revolutionary Approach for Enhancement of Sustained Fracture Conductivity and Effective Fracture Half-Length," SPE Paper 134307, presented at the SPE ATCE, Florence, Italy, September 19–22, 2010.
3. Beatty, C.V., Hill, A.D., Zhu, D., and Sullivan, R.B., "Characterization of Small Scale Heterogeneity to Predict Acid Fracture Performance," SPE Paper 140336, presented at the SPE Hydraulic Fracturing Technology Conference, The Woodlands, TX, January 24–26, 2011.
4. Ben-Naceur, K., and Economides, M.J., "The Effectiveness of Acid Fractures and Their Production Behavior," SPE Paper 18536, 1988.
5. Ben-Naceur, K., and Economides, M.J., "Acid Fracture Propagation and Production" in *Reservoir Stimulation,* M.J. Economides, and K.G. Nolte, eds., Prentice Hall, Englewood Cliffs, NJ, Chap. 18, 1989.
6. Bennett, C.O., "Analysis of Fractured Wells," Ph.D. thesis, University of Tulsa, Tulsa, OK, 1982.
7. Buijse, M.A. and Glasbergen, G., "A Semiempirical Model to Calculate Wormhole Growth in Carbonate Acidizing," SPE Paper 96892, presented at the SPE Annual Technical Conference and Exhibition, Dallas, TX, October 9–12, 2005.
8. Daccord, G., "Acidizing Physics," in *Reservoir Stimulation,* M.J. Economides and K.G. Nolte, eds., Prentice Hall, Englewood Cliffs, NJ, Chap. 13, 1989.
9. Daccord, G., Touboul, E., and Lenormand, R., "Carbonate Acidizing: Toward a Quantitative Model of the Wormholing Phenomenon," *SPEPE,* 63–68 (February 1989).
10. Deng, J., Mou, J., Hill, A.D., and Zhu, D., "A New Correlation of Acid Fracture Conductivity Subject to Closure Stress," *SPE Production and Operations*, 158–169 (May 2012).
11. Economides, M.J., Hill, A.D., and Ehlig-Economides, C.E., *Petroleum Production Systems,* Prentice-Hall, Upper Saddle River, NJ, 1994.

12. Elbel, J., "Field Evaluation of Acid Fracturing Treatments Using Geometry Simulation, Buildup, and Production Data," SPE Paper 19773, 1989.

13. Fredd, C.N., and Fogler, H.S., "Alternative Stimulation Fluids and Their Impact on Carbonate Acidizing," *SPEJ,* 34–41 (March 1998).

14. Furui, K., Burton, R.C., Burkhead, D.W., Abdelmalek, N.A., Hill, A.D., Zhu, D., and Nozaki, M. "A Comprehensive Model of High-Rate Matrix Acid Stimulation for Long Horizontal Well in Carbonate Reservoirs: Part 1—Scaling Up Core-Level Acid Wormholing to Field Treatments," *SPEJ,* **17**(1): 271–279 (March 2012).

15. Glasbergen, G., Kalia, N., and Talbot, M., "The Optimum Injection Rate for Wormhole Propagation: Myth or Reality?", SPE Paper 121464, presented at the SPE European Formation Damage Conference, Scheveningen, The Netherlands, May 27–29, 2009.

16. Glasbergen, G., Gualteri, D., van Domelen, M., and Sierra, J., "Real-Time Distribution Determination in Matrix Treatments Using DTS," *SPEPO,* 135–146 (February 2009).

17. He., J, Mohamed, I.M., and Nasr-El-Din, H., "Mixing Hydrochloric Acid and Seawater for Matrix Acidizing: Is It a Good Practice?", SPE Paper 143855, presented at the SPE European Formation Damage Conference, Noordwijk, The Netherlands, June 7–10, 2011.

18. Hoefner, M.L., and Fogler, H.S., "Pore Evolution and Channel Formation during Flow and Reaction in Porous Media," *AICHE J.,* **34:** 45–54 (January 1988).

19. Hoefner, M.L., and Fogler, H.S., "Fluid Velocity and Reaction-Rate Effects during Carbonate Acidizing: Application of Network Model," *SPEPE,* 56–62 (February 1989).

20. Huang, T., Hill, A.D., and Schechter, R.S., "Reaction Rate and Fluid Loss: The Keys to Wormhole Initiation and Propagation in Carbonate Acidizing," *SPE Journal,* **5**(3), 287–292 (September 2000).

21. Hung, K.M., Hill, A.D., and Sepehrnoori, K., "A Mechanistic Model of Wormhole Growth in the Carbonate Matrix Acidizing and Acid Fracturing," *JPT,* **41**(1): 59–66 (January 1989).

22. Izgec, O., Keys, R., Zhu, D., Hill, A.D., "An Integrated Theoretical and Experimental Study on the Effects of Multiscale Heterogeneities in Matrix Acidizing of Carbonates," SPE Paper 115143, presented at the SPE Annual Technical Conference and Exhibition, Denver, CO, September 21–24, 2008.

23. Izgec, O., Zhu, D., Hill, A. D., Numerical and experimental investigation of acid wormholing during acidization of vuggy carbonate rocks, *J. Pet. Science and Engr.,* **74:** 51–66 (2010).

24. Lo, K.K., and Dean, R.H., "Modeling of Acid Fracturing," *SPEPE,* 194–200 (May 1989).

25. McDuff, D., Jackson, S., Schuchart, C., and Postl, D., "Understanding Wormholes in Carbonates: Unprecedented Experimental Scale and 3D Visualization," *JPT,* **62**(10): 78–81 (2010).

26. McLeod, H.O., Jr., "Matrix Acidizing," *JPT,* **36:** 2055–2069 (1984).

27. Mishra, V., Zhu, D., Hill, A.D., "An Acid-Placement Model for Long Horizontal Wells in Carbonate Reservoirs," SPE Paper 107780, presented at the European Formation Damage Conference, Scheveningen, The Netherlands, May 30–June 1, 2007.

28. Mou, J., Zhu, D., and Hill, A.D., "New Correlations of Acid-Fracture Conductivity at Low Closure Stress Based on the Spatial Distributions of Formation Properties," *SPE Production and Operations,* **26**(21): 195–202 (May 2011).

29. Nasr-El-Din, H.A., Al-Ghamdi, A.H., Al-Qahtani A.A., and Samuel, M.M., "Impact of Acid Additives on the Rheological Properties of Viscoelastic Surfactants," *SPEJ,* **13**(1): 35–47 (2008).

30. Nierode, D.E., and Kruk, K.F., "An Evaluation of Acid Fluid Loss Additives, Retarded Acids, and Acidized Fracture Conductivity," SPE Paper 4549, 1973.

31. Nierode, D.E., and Williams, B.B., "Characteristics of Acid Reactions in Limestone Formations," *SPEJ,* 306–314, 1972.

32. Nozaki, M., and Hill, A. D., "A Placement Model for Matrix Acidizing of Vertically Extensive, Heterogeneous Gas Reservoirs," *SPE Production and Operations,* **25**(2): 388–397 (August 2010).

33. Paccaloni, G., and Tambini, M., "Advances in Matrix Stimulation Technology," *JPT,* 256–263 (March 1993).

34. Pichler, T., Frick, T.P., Economides, M.J., and Nittmann, J., "Stochastic Modeling of Wormhole Growth with Biased Randomness," SPE Paper 25004, 1992.

35. Roberts, L.D., and Guin, J.A., "The Effect of Surface Kinetics in Fracture Acidizing," *SPEJ,* 385–396 (August 1974); *Trans. AIME,* 257.

36. Sasongko, H., Zhu, D., and Hill, A.D., "Simulation of Acid Jetting Treatments in Long Horizontal Wells," SPE Paper 144200, presented at the SPE 2011 European Formation Damage Conference, Noordwijk, The Netherlands, June 7–10, 2011.

37. Schechter, R.S., *Oil Well Stimulation,* Prentice Hall, Englewood Cliffs, NJ, 1992.

38. Schechter, R.S., and Gidley, J.L., "The Change in Pore Size Distribution from Surface Reactions in Porous Media," *AICHE J.,* **16:** 339–350 (1969).

39. Settari, A., "Modelling of Acid Fracturing Treatment," SPE Paper 21870, 1991.

40. Settari, A., Sullivan, R.B., and Hansen, C., "A New Two-Dimensional Model for Acid-Fracturing Design," *SPEPF,* 200–209 (November 2001).

41. Tardy, P.M.J., Lecerf, B., and Christianti, Y., "An Experimentally Validated Wormhole Model for Sel-Diverting and Conventional Acids in Carbonate Rocks Under Radial Flow Conditions," SPE Paper 107854, 2007.

42. Terrill, R.M., "Heat Transfer in Laminar Flow between Parallel Porous Plates," *Int. J. Heat Mass Transfer,* **8:** 1491–1497 (1965).

43. Wang, Y., "The Optimum Injection Rate for Wormhole Propagation in Carbonate Acidizing," Ph.D. dissertation, University of Texas at Austin, TX, 1993.

44. Wang, Y., Hill, A. D., and Schechter, R. S., "The Optimum Injection Rate for Matrix Acidizing of Carbonate Formations," SPE Paper 26578, presented at the 68th Society of Petroleum Engineers Annual Technical Conference and Exhibition, Houston, TX, October 3–6, 1993.

Problems

16-1 Calculate the radius around the well to which wormholes penetrate using the volumetric model for acid volumes up to 100 gal/ft for the injection of 28% HCl at 0.05 bpm/ft into a chalk formation. Use the coreflood results from Figure 16-8. The treating temperature is 150°F, the formation porosity is 0.3, and the well radius is 0.25 ft.

16-2 Repeat problem 16-1 using the Buijse-Glasbergen model.

16-3 Repeat problem 16-1 using Furui et al.'s model.

16-4 Repeat problems 16-1 through 16-3, but for a 10% porosity dolomite formation for which the optimal pore volumes to breakthrough (PV_{bt}) is 10 at an optimal interstitial velocity of 0.02 cm/min. For the volumetric model, assume the average PV_{bt} is two times the optimal.

16-5 For the conditions of problem 16-3, vary the injection rate to find the constant rate injection conditions that yield the deepest wormhole penetration for a total injection volume of 100 gal/ft. Would varying injection rate during a treatment of this formation be helpful to the treatment? If so, should the rate be increased or decreased during the treatment?

16-6 Using the results from problem 16-1, calculate the skin factor evolution assuming that before acid injection, there was a damaged region extending 1 ft beyond the wellbore having a permeability of 10% of the undamaged formation permeability.

16-7 Using the results from problem 16-2, calculate the skin factor evolution assuming that before acid injection, there was a damaged region extending 1 ft beyond the wellbore having a permeability of 10% of the undamaged formation permeability.

16-8 Using the results from problem 16-3, calculate the skin factor evolution assuming that before acid injection, there was a damaged region extending 1 ft beyond the wellbore having a permeability of 10% of the undamaged formation permeability.

16-9 An acid fracture in a 10% porosity dolomite formation is 100-ft high and 400-ft long after the injection of 400 bbl of gelled 15% HCl solution. If the fracturing efficiency $\eta \rightarrow 0$, at what distance from the wellbore will the acid concentration be half of the injected concentration? $D_{\text{eff}} = 10^{-4}$ cm^2/sec.

16-10 For the fracturing treatment described in problem 16-8, assume that the rock is dissolved uniformly along the entire fracture. Using the Nierode-Kruk correlation, what is the resulting average fracture conductivity if $S_{\text{rock}} = 40{,}000$ psi?

Chapter 17

Hydraulic Fracturing for Well Stimulation

17.1 Introduction

When fluid is pumped into the reservoir at a sufficiently high pressure, the reservoir rock will crack open. Continued pumping of the fluid at this pressure will propagate the crack into the reservoir. A slurry can be pumped into the crack to create a slab-shaped zone of high permeability known as a hydraulic fracture. Figure 17-1 illustrates this concept.

Since the first edition of this textbook, hydraulic fracturing has been established as the premier production enhancement procedure in the petroleum industry. By 2012, hydraulic fracturing had emerged as a $20 billion annual activity, second only to drilling budgets in magnitude. Fracturing has continued to overwhelmingly dominate low-permeability reservoirs and has been instrumental in monetizing shale gas in North America, arguably one of the most important new activities of the petroleum industry.

The hydraulic fracture design is highly dependent on the reservoir permeability. As seen in Figure 17-2, in moderate permeability reservoirs (up to 50 md for oil and 1 md for gas), the fracture accelerates production without impacting the well reserves. In such reservoirs the rationale for fracturing frequently relies on net present value economics. However, Figure 17-3 shows that in low-permeability reservoirs (less than 1 md for oil and 0.01 md for gas), the hydraulic fracture greatly contributes both to well productivity and to the well reserves. More importantly, in such reservoirs the well would not produce an economic rate without the hydraulic fracture.

Very low- and very high-permeability reservoirs pose additional challenges for hydraulic fracturing. In very low permeability, such as tight gas and shale gas, the well drainage area is confined essentially to the stimulated reservoir volume created by multiple transverse hydraulic fractures in

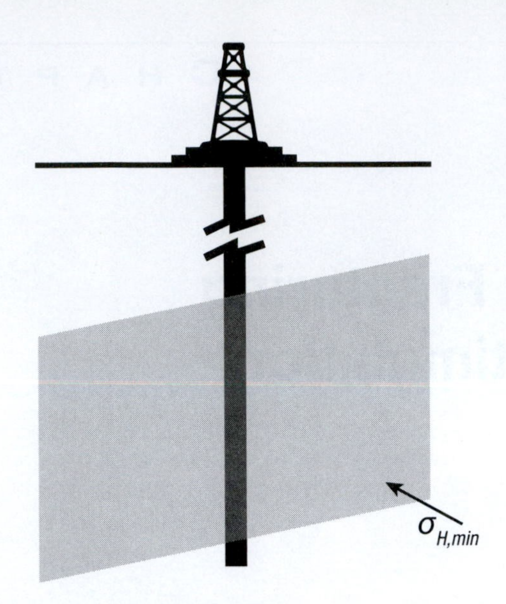

Figure 17-1 Diagram of vertical hydraulic fracture.

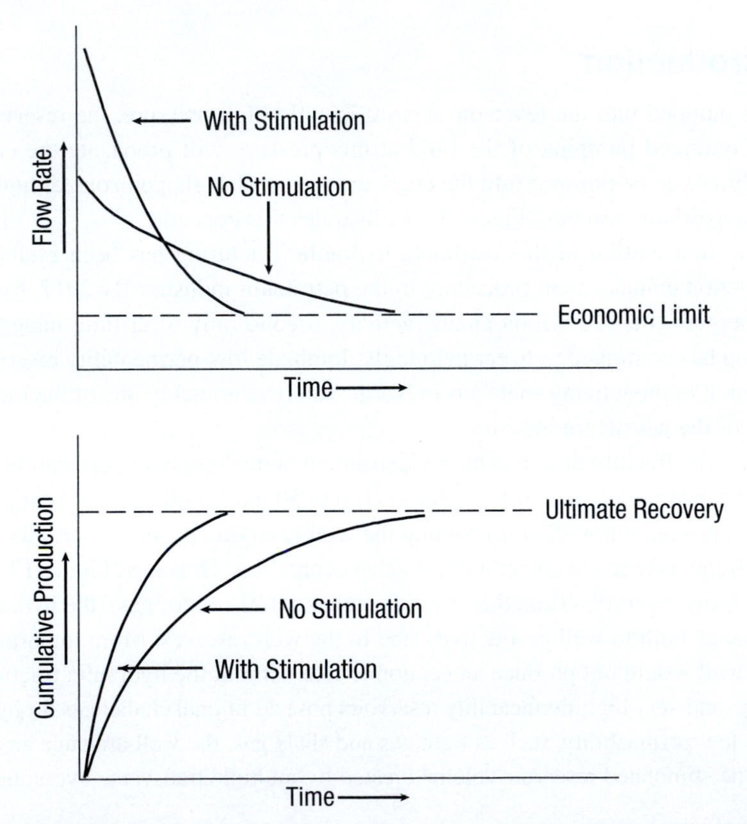

Figure 17-2 The concept of production and recovery acceleration, high-permeability reservoirs. (After Holditch, 2006.)

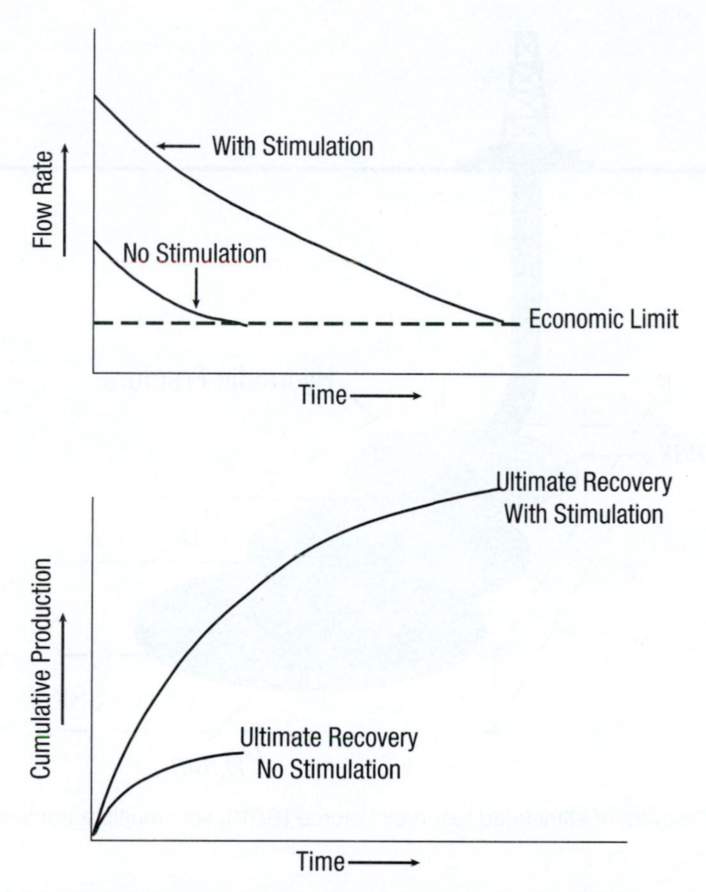

Figure 17-3 Production and reserves enhancement from hydraulic fracturing, low-permeability reservoirs. (After Holditch, 2006.)

horizontal wells, as diagrammed in Figure 17-4. In very high permeability reservoirs, hydraulic fractures have a dual purpose—to stimulate the well and to provide sand control—and the latter justifies postponing much of the discussion of this strategy until Chapter 19.

What fracturing accomplishes is to alter the way the fluids enter the wellbore, changing from near-well radial flow to linear or bilinear flow. The latter, described in Chapter 13, implies that the flow is from the reservoir into the fracture and then along the fracture into the well. It should be noted here that a hydraulic fracture bypasses the near-wellbore damage zone. Thus, the pretreatment skin effect has little or no impact on the post-fracture equivalent skin effect value.

What length, width, and fracture permeability are desirable in a hydraulic fracture? While this issue will be dealt with extensively in later sections of this chapter, a conceptual analogy is instructive.

Suppose that a reservoir is depicted as a countryside and the fracture is a road connecting two far-off points. Where there is an excellent and dense system of wide roads (high-permeability-reservoir analog), relative improvement of the traffic flow would require even

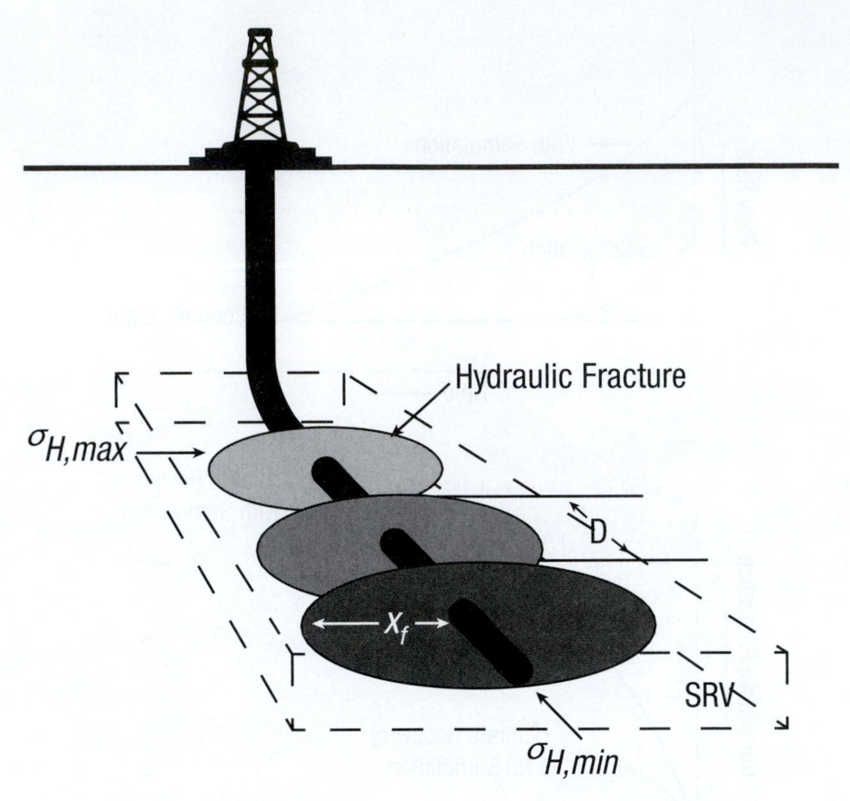

Figure 17-4 Diagram of stimulated reservoir volume (SRV), with multiple transverse fractures.

wider (more lanes) and faster freeways. This translates to high-permeability and wide fractures. Therefore, in moderate- to high-permeability reservoirs, a large fracture width is necessary, while the fracture length is of secondary importance.

Conversely, in a countryside with a sparse and poor road system (low-permeability analog), a long road connecting as much of the countryside as possible is indicated. Such a long road could bring about a major improvement in traffic flow. In fact, the poorer the existing road system, the lower the relative requirements would be from the new road. By analogy, the length of the fracture in a lower-permeability reservoir is the priority; fracture permeability and width are secondary. Thus, in designing a hydraulic fracture treatment, fracture length, width, and proppant pack permeability must be suited to the formation properties, especially formation permeability.

17.2 Length, Conductivity, and Equivalent Skin Effect

Every hydraulic fracture can be characterized by its length, conductivity, and related equivalent skin effect. In some cases there may be a distinction between the *created* and *conductive* fracture lengths. The created fracture length relates to the length of the crack propagated during the hydraulic fracture treatment execution. The conductive fracture length refers to the length of the fracture remaining open when the well is on production, thereby effectively providing an enhanced path for fluid production to the well. In fracture design and well productivity calculations, the *conductive* length is assumed to consist of two equal half-lengths, x_f, extending in opposite directions from the well. In the following definitions and in the remainder of this chapter and the following chapter, the term *fracture length,* denoted by x_f, will always refer to the conductive fracture half length.

The fracture conductivity is given by the product of the fracture permeability, k_f, and the fracture width, w. In 1961, Prats provided pressure profiles in a fractured reservoir as functions of the fracture half length and the relative capacity, a, which he defined as

$$a = \frac{\pi k x_f}{2 k_f w} \tag{17-1}$$

where k is the reservoir permeability.

Small values of a imply large fracture permeability-width product or small reservoir permeability–fracture length product, thus a high *conductivity* fracture. In subsequent work, Agarawal, Gardner, Kleinsteiber, and Fussell (1979) and Cinco-Ley and Samaniego (1981) introduced the fracture dimensionless conductivity, C_{fD}, which is exactly

$$C_{fD} = \frac{k_f w}{k x_f} \tag{17-2}$$

and is related to Prats's a by

$$C_{fD} = \frac{\pi}{2a}. \tag{17-3}$$

Prats (1961) also introduced the concept of dimensionless effective wellbore radius in a hydraulically fractured well,

$$r'_{wD} = \frac{r'_w}{x_f} \tag{17-4}$$

where r'_w is

$$r'_w = r_w e^{-s_f}. \tag{17-5}$$

The equivalent skin effect, s_f, resulting from a hydraulic fracture of a certain length and conductivity can be added to the well inflow equations in the usual manner. From Equation 2-34 it is evident that the pseudosteady-state flow equation for a hydraulically fractured oil well would be

$$q = \frac{kh(\overline{p} - p_{wf})}{141.2B\mu[\ln(0.472r_e/r_w) + s_f]} = \frac{kh(\overline{p} - p_{wf})}{141.2B\mu[\ln(r_e/r_w) - 0.75 + s_f]}. \tag{17-6}$$

Similarly, this skin effect can be added to equations for steady-state and transient conditions once pseudoradial flow has been established in the reservoir. As for horizontal wells, pseudoradial flow to a fractured vertical well applies when flow streamlines to the well at sufficient distance are mainly radial. The equivalent skin from a fracture and as well the effective wellbore radius in Equation 17-5 only apply after pseudoradial flow is established. When the fracture half length approaches the effective radius of the well drainage area, the concepts of equivalent skin and effective well radius do not apply. As mentioned in Chapter 13, radial and pseudoradial flow are also called infinite-acting radial flow and occur only as long as the radius of the pressure disturbance due to production is less than the distance to the nearest boundary of the well drainage area.

Prats (1961) correlated the relative capacity parameter, *a*, and the dimensionless effective wellbore radius (Figure 17-5). From Figure 17-5, for small values of *a*, or high conductivity fractures, the r'_{wD} is equal to 0.5, leading to

$$r'_w = \frac{x_f}{2}. \tag{17-7}$$

Such high conductivity fractures are said to have effectively infinite conductivity, and Equation (17-7) implies that in this case the reservoir drains to a well with an effective wellbore radius equal to half of the fracture half length.

A useful alternative graph of Figure 17-5 is Figure 17-6 (Cinco-Ley and Samaniego, 1981), relating C_{fD} and s_f directly. The relationships presented in Figures 17-5 and 17-6 can be calculated simply by (Meyer and Jacot, 2005)

$$r'_w = \frac{x_f}{\dfrac{\pi}{C_{fD}} + 2}. \tag{17-8}$$

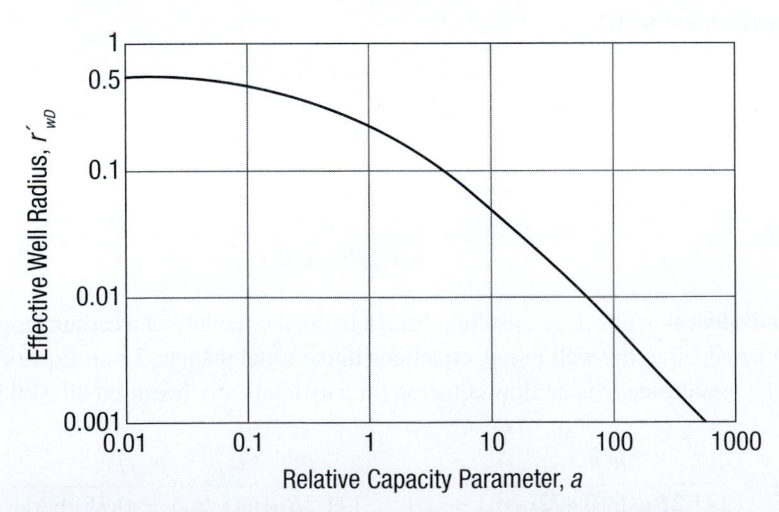

Figure 17-5 The concept of effective wellbore radius. (From Prats, 1961.)

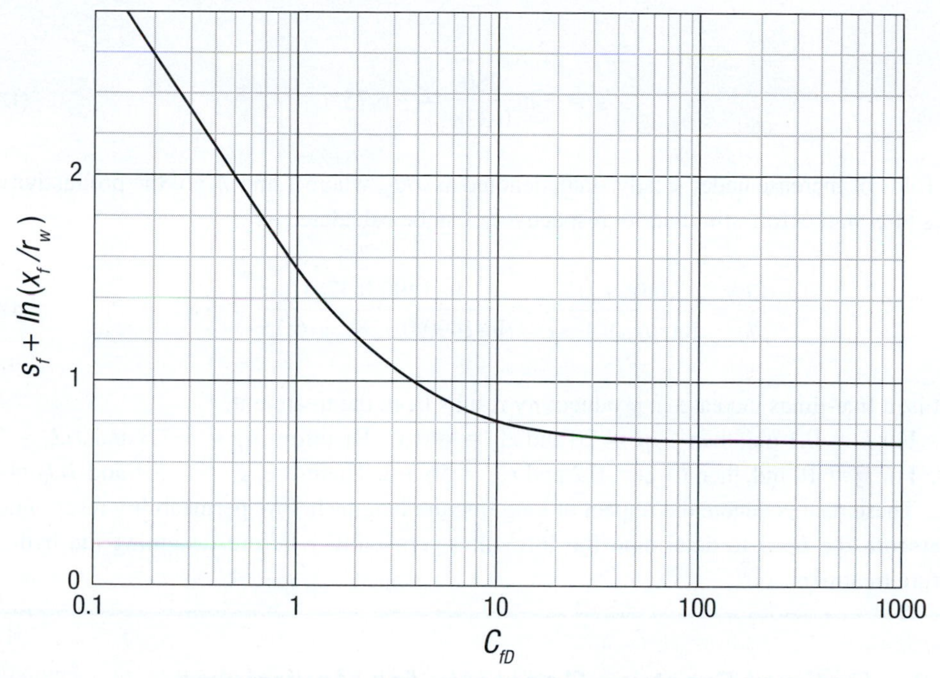

Figure 17-6 Equivalent fracture skin, length, and conductivity. (From Cinco-Ley and Samaniego, 1981.)

Example 17-1 Calculation of the Equivalent Skin Effect from Hydraulic Fractures

Assuming that $k_f w = 2000$ md-ft, $k = 1$ md, $x_f = 1000$ ft, and $r_w = 0.328$ ft, calculate the equivalent skin effect and the folds of well productivity increase (at steady-state flow) for a reservoir with drainage radius, $r_e = 1490$ ft (160 acres) and no pretreatment damage.

What would be the folds of increase for the same fracture lengths and $k_f w$ if $k = 0.1$ md and $k = 10$ md?

Solution

From Equation (17-2),

$$C_{fD} = \frac{k_f w}{k x_f} = \frac{2000}{(1)(1000)} = 2 \quad \textbf{(17-9)}$$

and therefore, from Equation 17-8,

$$r'_w = \frac{x_f}{\dfrac{\pi}{C_{fD}} + 2} = \frac{1000}{\dfrac{\pi}{2} + 2} = 280 \text{ ft} \quad \textbf{(17-10)}$$

and

$$s = -\ln\frac{280}{0.328} = -6.75. \tag{17-11}$$

The folds of increase under steady state, denoted as J/J_0, where J and J_0 are the productivity indexes after and before stimulation, respectively, can be calculated as

$$\frac{J}{J_0} = \frac{\ln(r_e/r_w)}{\ln(r_e/r_w) + s_f} = \frac{\ln(1490/0.328)}{\ln(1490/0.328) - 6.75} = 5.0. \tag{17-12}$$

That is, a five-times increase in productivity results from the treatment.

For $k = 0.1$ md, then $C_{fD} = 20$ and $r'_w = 464$ ft. Therefore, $s_f = -7.3$ and $J/J_0 = 7.5$.

For $k = 10$ md, then $C_{fD} = 0.2$ and $r'_w = 56.5$ ft. Therefore, $s_f = -5.1$ and $J/J_0 = 2.5$.

These results denote the impact of long fracture lengths in low-permeability reservoirs and underscore the need to determine the formation permeability before designing the hydraulic fracture treatment.

17.3 Optimal Fracture Geometry for Maximizing the Fractured Well Productivity

Before a discussion of the mechanics of fracture execution, it is useful to consider what fracture permeability, width, and half length will optimize the well productivity. Equations 2-9 (for steady state) and 2-34 (for pseudosteady state) relate the dimensionless productivity index, J_D, with the well skin effect for a vertical well under radial flow. Figure 17-6 was constructed for pseudoradial (and infinite-acting radial) flow and its original use was to corroborate well testing results for fractured wells. When pseudoradial flow emerges after fracture bilinear and/or linear flow, the skin determined from a pressure buildup test will be the equivalent skin effect indicated in Figure 17-6.

Starting with Equation 2-34 for the pseudosteady-state productivity index, a rather profound conclusion can be extracted. First, $\ln x_f$ is added and subtracted in the denominator of Equation 2-34, changing the skin to the fracture-equivalent skin, and with simple algebra:

$$J_{Dpss} = \frac{1}{\ln\frac{r_e}{x_f} - 0.75 + \ln\frac{x_f}{r_w} + s_f}. \tag{17-13}$$

The second step is to recognize that for a given propped fracture volume, V_f, the fracture dimensions are related to the fracture volume simply by

$$V_f = 2x_f w h_f \tag{17-14}$$

where h_f is the fracture height. When the fracture height is greater than the reservoir height, *h,* a distinction must be made between the productive fracture height, *h,* equal to the reservoir height, and the propped fracture height, h_f.

Next, combining Equations 17-2 and 17-14 and eliminating the fracture width, an expression for the fracture half length is obtained:

$$x_f = \left(\frac{k_f V_f}{2C_{fD} k h_f}\right)^{0.5}. \tag{17-15}$$

Finally, substituting for x_f in the first term in Equation 17-13 with the one in Equation 17-15 and applying simple algebra yields,

$$J_{Dpss} = \frac{1}{\ln r_e - 0.75 - 0.5 \ln \dfrac{k_f V_f}{2kh_f} + 0.5 \ln C_{fD} + \ln \dfrac{x_f}{r_w} + s_f} \tag{17-16}$$

To maximize the post-fracture productivity index the denominator in Equation 17-16 must be minimized. For a given reservoir drainage (r_e), permeability, *k,* and height, *h* (here net height, *h,* and fracture height, h_f, are assumed equal), and a given fracture volume of a specific proppant injected with permeability, k_f, the first three terms in the denominator of Equation 17-16 are constant. This means that the remaining three terms should be minimized.

Conveniently, Figure 17-6, a graph of $\ln(x_f/r_w) + s_f$ versus C_{fD} also yields a graph of the three terms and is shown in Figure 17-7. A minimum appears at exactly $C_{fD} = 1.6$. This means that fractures whose conductivity is more or less than this value will perform less. Thus, there exists only one optimum conductivity that maximizes the productivity index, and this conductivity corresponds to a particular fracture width and length. Narrower fracture width would mean that the fracture becomes a bottleneck; shorter fracture length would mean that the reservoir becomes the bottleneck.

Figure 17-7 applies whenever the fracture half length is sufficiently small compared to the well drainage area that pseudoradial flow can occur. Because a given volume of proppant with a specified permeability was assumed, thereby fixing the proppant cost, this represents not just *maximum* productivity, but *optimum* productivity as well for the assumed proppant expenditure.

17.3.1 Unified Fracture Design

The unified fracture design methodology provided by Economides, Oligney, and Valkó (2002), expands the above approach to include fracture and well drainage area dimensions that will not reach pseudoradial flow before the onset of pseudosteady state. Key to this approach is the idea that for a given proppant volume and well drainage area and shape there is a fracture half length, width, and conductivity that maximize the well productivity.

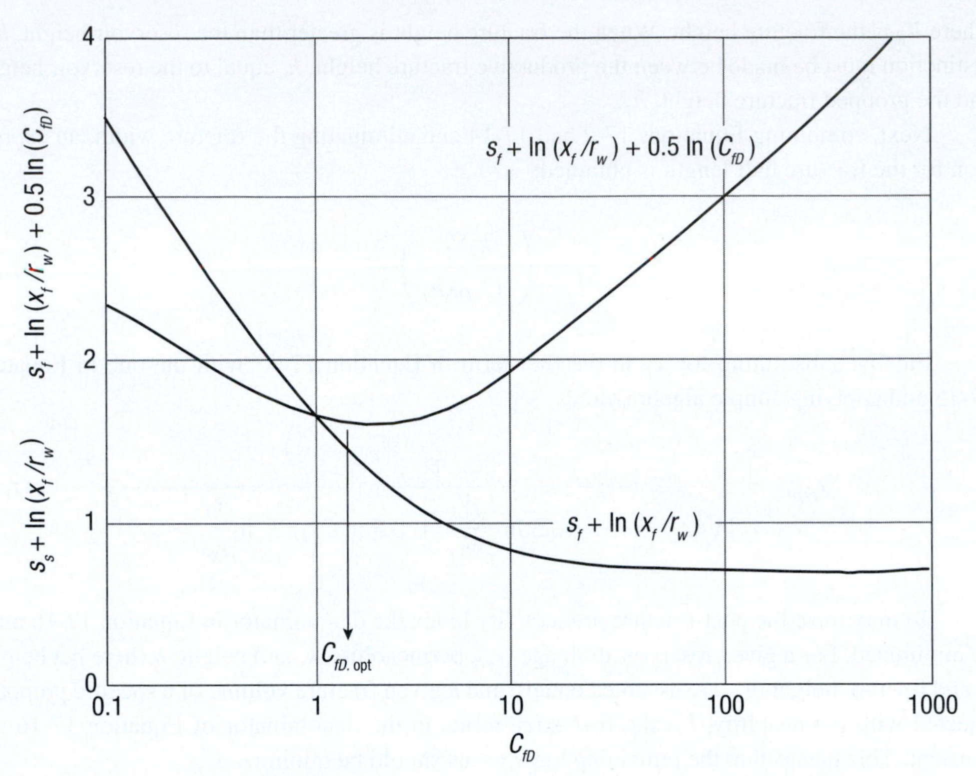

Figure 17-7 Optimization of fracture conductivity from Cinco-Ley and Samaniego (1981) equivalent skin concept under pseudoradial flow: Maximum productivity index at $C_{fD} = 1.6$.

For a given proppant volume, square well drainage area, and values for both proppant and reservoir permeability, the dimensionless *proppant number,* N_p, is defined as

$$N_p = I_x^2 C_{fD} = \frac{4x_f k_f w}{k x_e^2} = \frac{4k_f x_f w h}{k x_e^2 h} = \frac{2k_f V_f}{k V_r} \quad \textbf{(17-17)}$$

where x_e is the length of a square drainage area in which the well is centered, I_x is the penetration ratio ($I_x = 2x_f/x_e$), C_{fD} is the dimensionless fracture conductivity, V_r is the reservoir drainage volume, and V_f is the volume of the propped fracture *in the pay* equal to the total volume injected times the ratio of the net pay height to the fracture height. k_f is the proppant pack permeability and k is the reservoir permeability.

Figures 17-8 (for $N_p < 0.1$) and 17-9 (for $N_p > 0.1$) are graphs of the dimensionless fracture conductivity versus the dimensionless PI with the proppant number as the parameter.

At "low" proppant numbers (Figure 17-8), the optimal conductivity is $C_{fD} = 1.6$. It should not be surprising that this value is identical to the one obtained from Figure 17-7 since small proppant

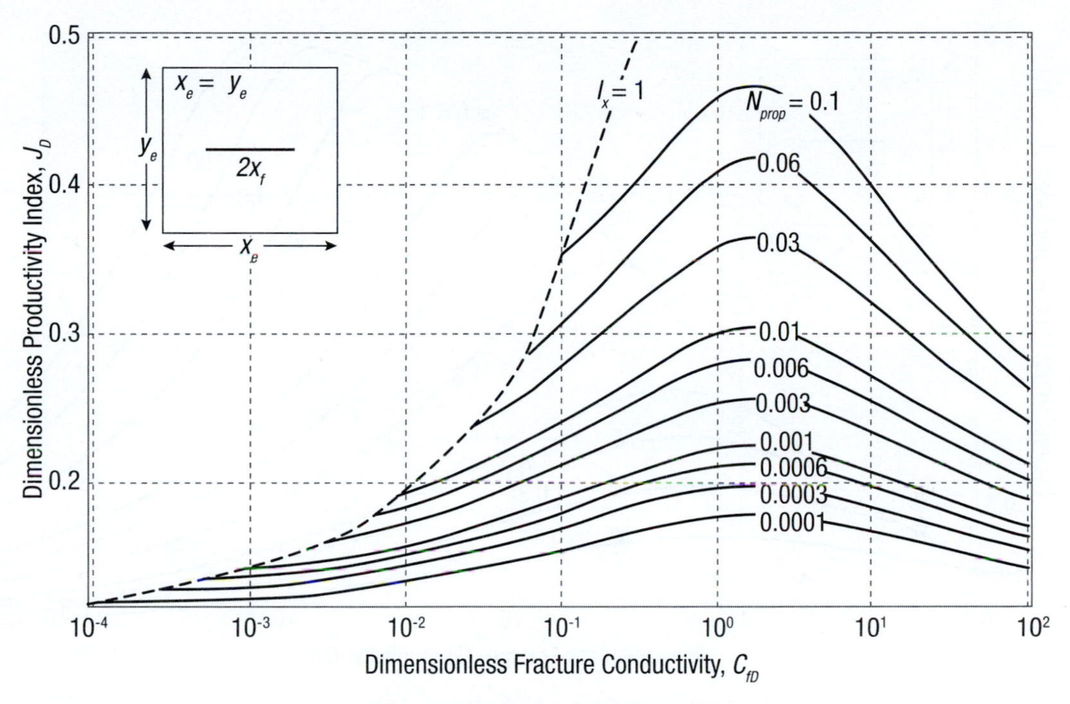

Figure 17-8 Maximum J_D at optimum C_{fD} for $N_p < 0.1$.

numbers would mean relatively small treatments or large drainages and thus fracture performance would reach pseudoradial flow before the onset of pseudosteady-state flow.

However, as the proppant number increases (Figure 17-9), the fracture penetration approaches the drainage boundaries ($I_x = 1$ is a fully penetrating fracture) and the boundaries normal to the fracture plane force the flow to remain linear or bilinear, both of which are far more efficient than radial, until eventually pseudosteady-state flow develops.

For square drainage area, the absolute maximum for the pseudosteady-state dimensionless productivity index, J_D, is $6/\pi = 1.909$. When the proppant number increases, due to increased propped volume or lower reservoir permeability, the optimum dimensionless fracture conductivity also increases.

The maximum achievable dimensionless pseudosteady-state productivity index as a function of the proppant number is given by

$$J_{D\,\max\,pss}(N_p) = \begin{cases} \dfrac{1}{0.990 - 0.5\ln N_p} & \text{if } N_p \leq 0.1 \\ \dfrac{6}{\pi} - \exp\left[\dfrac{0.423 - 0.311N_p - 0.089(N_p)^2}{1 + 0.667N_p + 0.015(N_p)^2}\right] & \text{if } N_p > 0.1 \\ \dfrac{6}{\pi} & \text{if } N_p \geq \sim 100 \end{cases} \tag{17-18}$$

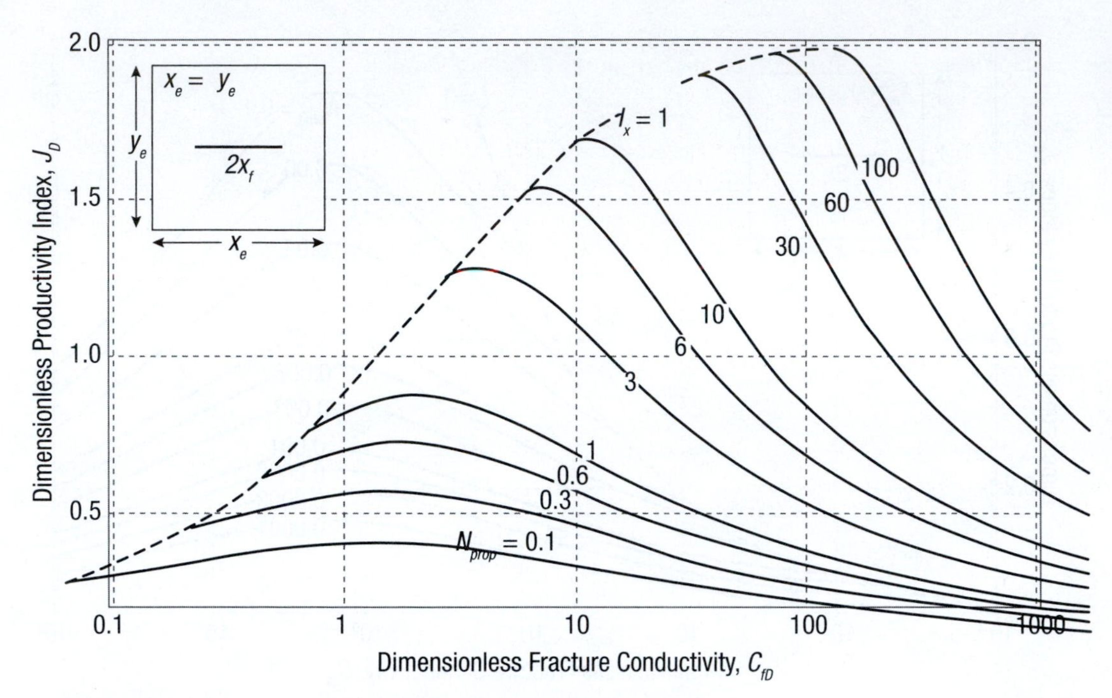

Figure 17-9 Maximum J_D at optimum C_{fD} for $N_p > 0.1$.

Similarly, the optimal dimensionless fracture conductivity for the entire range of proppant numbers is given as

$$C_{fDopt}(N_p) = \begin{cases} 1.6 & \text{if } N_p < 0.1 \\ 1.6 + \exp\left[\dfrac{-0.583 + 1.48 \ln N_p}{1 + 0.142 \ln N_p}\right] & \text{if } 0.1 \leq N_p \leq 10 \\ N_p & \text{if } N_p > 10 \end{cases} \quad \textbf{(17-19)}$$

Once the optimal dimensionless fracture conductivity is known, the optimal fracture length and width can be readily determined:

$$x_{fopt} = \left(\frac{k_f V_f}{2C_{fDopt} kh}\right)^{0.5}$$

$$w_{opt} = \left(\frac{C_{fDopt} k V_f}{2k_f h}\right)^{0.5}. \quad \textbf{(17-20)}$$

Example 17-2 Calculation of the Optimum Fracture Dimensions and Maximum Pseudosteady-State PI

Calculate the optimum fracture dimensions and the maximum pseudosteady-state productivity index from fracturing in STB/d/psi in an oil well with the following treatment and reservoir data. Repeat the calculation for a reservoir whose permeability is 15 times smaller. If the J_D of the unfractured well is 0.12, what are the folds of increase from fracturing?

Drainage area (~100 acre square) = 4×10^6 ft^2, $k = 15$ md, $B_o = 1.1$ resbbl/STB, $\mu = 1$ cp, mass of proppant = 150,000 lb, proppant specific gravity = 2.65, proppant porosity = 0.38, $k_f = 60{,}000$ md (20/40 mesh sand), $h = 50$ ft, and $h_f = 100$ ft.

Solution

The propped volume in the pay (V_f) and reservoir volume (V_r) are

$$V_f = \frac{\frac{h}{h_f}(M_p)}{\rho_p(62.4)(1-\phi_p)} = \frac{\frac{50}{100}(150{,}000)}{(2.65)(62.4)(1-0.38)} = 732 \text{ ft}^3 \tag{17-21}$$

$$V_r = Ah = 4 \times 10^6 \times 50 = 20 \times 10^7 \text{ ft}^3. \tag{17-22}$$

The proppant number can then be calculated using Equation 17-17,

$$N_p = \frac{2k_f}{k}\frac{V_f}{V_{res}} = \frac{2(60{,}000)}{(15)}\frac{732}{(20 \times 10^7)} = 0.0293. \tag{17-23}$$

For $N_p < 0.1$, the maximum possible fracture conductivity is 1.6 (from Equation 17-19) and from Equation 17-18 the maximum dimensionless pseudosteady-state productivity index is

$$J_{D\,\max\,pss}(N_p) = \frac{1}{0.990 - 0.5 \ln N_p} = \frac{1}{0.990 - 0.5 \ln(0.0293)} = 0.36. \tag{17-24}$$

From Equation 17-20,

$$x_{fopt} = \left(\frac{k_f V_f}{2C_{fDopt}kh}\right)^{0.5} = \left(\frac{(60{,}000)(732)}{2(1.6)(15)(50)}\right)^{0.5} = 135 \text{ ft} \tag{17-25}$$

and

$$w_{opt} = \left(\frac{C_{fDopt}kV_f}{2k_f h}\right)^{0.5} = \left(\frac{(1.6)(15)(732)}{2(60{,}000)(50)}\right)^{0.5} = 0.054 \text{ ft} = 0.65 \text{ in.} \tag{17-26}$$

Thus, the maximum pseudosteady-state productivity index would be

$$J = \frac{kh}{141.2B\mu} J_{D\,\max\,pss} = \frac{15(50)}{141.2(1.1)(1)}(0.36) = 1.75\ \text{STB/d/psi}. \tag{17-27}$$

The folds of increase from fracturing are 0.36/0.12 = 3.

The proppant number for $k = 1$ md is 0.44. Either from Figure 17-7 or Equations 17-18 and 17-19, $C_{fDopt} = 1.73$, and $J_D = 0.68$ from Equation 17-20. Thus, $x_{fopt} = 504$ ft and $w_{opt} = 0.17$ in., showing how much longer and narrower the fracture should be. Finally, even though in this case J is only 0.22 STB/d/psi, the folds of increase from fracturing are 0.68/0.12 = 5.7.

Productivity optimization in nonsquare drainage areas, appropriate for the partitioning of drainage in infield drilling applications and the execution of multiple hydraulic fractures intersecting a horizontal well transversely was presented by Daal and Economides (2006). The approach starts with an initial square drainage with dimensions x_e and y_e and a hydraulically fractured well positioned in the center. The x-direction is assumed to be the fracture propagation direction.

Then other fractured wells are added to the same drainage area, by placing them in rectangular areas obtained by slicing the square with respect to the y direction. Thus, the aspect ratio between the dimensions of the drainage will become increasingly different than 1, depending on the number of fractured wells. The following generalizes the proppant number for rectangular reservoirs:

$$N_p = \frac{2k_f V_f}{kV_r} = \frac{4k_f w h x_f}{k x_e y_e h} \times \frac{x_f x_e}{x_f x_e} = I_x^2 C_{fD} \frac{x_e}{y_e}, \tag{17-28}$$

which is the value for the square drainage area times the rectangular aspect ratio.

For proppant numbers less than 0.1, the optimum value of C_{fD} is still 1.6, thus the correlations for the calculation of the J_D and the C_{fD} presented in Section 17-3 are still suitable, with the condition that an equivalent proppant number, N_{pe}, is used:

$$N_{pe} = N_p \frac{C_A}{30.88} \tag{17-29}$$

where C_A is the classic Dietz shape factor, already introduced in Section 2.5.

For $0.1 < N_p < 100$, the maximum productivity index for the given proppant mass is given by

$$J_{D,\max} = \frac{1}{-0.63 - 0.5\ln(N_p) + F_{opt}} \tag{17-30}$$

and F_{opt} is given by

$$F_{opt} = \begin{cases} \dfrac{9.33y_{eD}^2 + 3.9y_{eD} + 4.7}{10y_{eD}} & \text{If } N_p < 0.1 \text{ and } 0.25 \geq y_{eD} \geq 0.1 \\ \dfrac{a + bu_{opt} + cu_{opt}^2 + du_{opt}^3}{a' + b'u_{opt} + c'u_{opt}^2} & \text{If } N_p \geq 0.1 \end{cases} \quad \textbf{(17-31)}$$

with

$$u_{opt} = \ln(C_{fD,opt}), \quad \textbf{(17-32)}$$

$$C_{fD,opt} = \frac{100y_{eD} - C_{fD,0.1}}{100} \times (N_p - 0.1) + C_{fD,0.1}, \quad \textbf{(17-33)}$$

$$y_{eD} = \frac{y_e}{x_e} \quad \textbf{(17-34)}$$

and

$$C_{fD,0.1} = \begin{cases} 1.6 & \text{If } \quad 1 \geq y_{eD} > 0.25 \\ 4.5y_{eD} + 0.25 & \text{If } \quad 0.1 \leq y_{eD} \leq 0.25 \end{cases} \quad \textbf{(17-35)}$$

The constants for the F-function for $N_p \geq 0.1$ are presented in Tables 17-1 and 17-2. As with the square drainage area, there is no benefit in $N_p > 100$.

Table 17-1 *F*-Function Constants

$y_e/x_e =$	1	0.7	0.5	0.25	0.2	0.1
a	17.2	17.4	21.4	38.3	35	30.6
b	54.5	55.5	54.3	46	59	89.6
c	52.5	53.3	56.3	71.1	70	70.2
d	16.9	16.9	16.9	15.84	16.3	17.8

Table 17-2 *F*-Function Prime Constants

For all Shapes	
a′	10
b′	35
c′	33

17.4 Fractured Well Behavior in Conventional Low-Permeability Reservoirs

For hydraulically fractured wells in low-permeability reservoirs, the time required to reach the pseudosteady-state flow condition described in the previous sections can be long, sometimes even longer than the productive life of the well. The productivity of such wells is predicted using transient flow models or numerically generated type curves that describe the transient flow period. Commonly used models are based on idealizations of the flow regime as being bilinear flow, linear flow in the formation, or pseudoradial flow, as described in Chapter 13. These models are very useful in pressure transient test analysis, but are less helpful for production forecasting because of gaps in the periods of applicability of each model and inconsistent transitions. The fractured well transient flow models all use the following dimensionless variables:

$$\text{Dimensionless time} \quad t_{Dxf} = \frac{0.0002637kt}{\phi \mu c_t x_f^2} \tag{17-36}$$

$$\text{Dimensionless oil rate} \quad \frac{1}{q_D} = \frac{kh[p_i - p_{wf}]}{141.2qB\mu} \tag{17-37}$$

$$\text{Dimensionless gas rate} \quad \frac{1}{q_D} = \frac{kh[m(p_i) - m(p_{wf})]}{1424q_gT}. \tag{17-38}$$

Low-permeability formations are commonly produced with the bottomhole pressure held approximately constant, so that is the well boundary condition for the well production models presented here.

17.4.1 Infinite Fracture Conductivity Performance

If the fracture conductivity is sufficiently greater than the reservoir's capacity to deliver fluid to the fracture, the pressure drop in the fracture is negligible, and the fracture is treated as having infinite conductivity. This condition is generally considered applicable when $C_{fD} > 300$, and for practical purposes is valid for $C_{fD} > 50$ (Poe and Economides, 2000). For an infinitely conductive fracture, with the fractured well in the middle of a rectangular bounded reservoir, a general equation for the dimensionless production rate was presented by Wattenbarger, El-Banbi, Villegas, and Maggard (1998):

$$\frac{1}{q_D} = \frac{\frac{\pi}{4}\left(\frac{y_e}{x_f}\right)}{\sum_{n=1}^{\infty} \exp\left[-\frac{(2n-1)^2\pi^2}{4} t_{Dye}\right]} \tag{17-39}$$

where y_e is the distance from the fracture to the drainage boundary in the direction perpendicular to the fracture, and the dimensionless time, t_{Dye} is

$$t_{Dye} = \frac{0.0002637kt}{\phi\mu c_t y_e^2} = \left(\frac{x_f}{y_e}\right)^2 t_{Dxf}. \tag{17-40}$$

Fortunately, Equation 17-39 can be replaced by two equations: one for early (transient) times and one for later (depletion) times. If $t_{Dye} < 0.25$, then we can use the early (transient) equation

$$\frac{1}{q_D} = \frac{\pi}{2}\sqrt{\pi t_{Dxf}} \tag{17-41}$$

Note that this equation differs from the formation linear flow equation for constant rate production used in pressure transient analysis by a factor of $\pi/2$ (Wattenbarger et al., 1998).

For $t_{Dye} > 1.25$, we use the exponential depletion equation

$$\frac{1}{q_D} = \frac{\pi}{4}\left(\frac{y_e}{x_f}\right)\exp\left[\frac{\pi^2}{4}t_{Dye}\right]. \tag{17-42}$$

For $0.25 > t_{Dye} > 1.25$ Equation 17-39 must be used.

Example 17-3 Long-Term Production Behavior for a Hydraulically Fractured Well

Consider an Appendix C gas well, but having a reservoir permeability of 0.02 md and with a hydraulic fracture with a half length, x_f, of 1000 feet. The reservoir boundaries in the horizontal direction perpendicular to the fracture are 2000 feet away. Calculate the expected production rate from the fractured well for the first 3 years of production assuming drawdown is kept constant at 1000 psi. Repeat the calculations for a fracture half length of 500 feet. Assume infinite fracture conductivity behavior.

Solution

From Appendix C, the properties needed are $\phi = 0.14$, $h = 78$ ft, $\mu = 0.0249$ cp, $T = 180°F$ (640°R), and $p_i = 4613$ psi. Also, for $c_t = 1.25 \times 10^{-5}$ psi^{-1}, and p_{wf} is 3613 psi, the real gas pseudopressures are $m(p_i) = 1.29 \times 10^9$ psi^2/cp and $m(p_{wf}) = 8.6 \times 10^8$ psi^2/cp. A convenient procedure to follow to generate a production forecast is:

1. Select a series of times for which flow rate is calculated, for example, every month after the start of production.

2. Calculate the dimensionless time, t_{Dye}, for each calculation time and decide whether $t_{Dye} < 0.25$ or $t_{Dye} > 0.25$.
3. Calculate the inverse dimensionless production rate, $1/q_D$, with Equation 17-41 or 17-42, depending on the value of t_{Dye}.
4. Calculate the actual flow rate by solving Equation 17-38 for q_g.

This procedure is illustrated for a time of 6 months after the well is put on production.

First, the dimensionless time, t_{Dxf}, is

$$t_{Dx_f} = \frac{0.0002637kt}{\phi\mu c_t x_f^2} = \frac{0.0002637(0.02)(6)(30)(24)}{(0.14)(0.0249)(1.25 \times 10^{-5})(1000)^2} = 0.523 \tag{17-43}$$

and

$$t_{Dye} = \left(\frac{x_f}{y_e}\right)^2 t_{Dx_f} = \left(\frac{1000}{2000}\right)^2 (0.523) = 0.131. \tag{17-44}$$

So, since $t_{Dye} < 0.25$, we use Equation 17-41. The inverse dimensionless rate is then

$$\frac{1}{q_D} = \frac{\pi}{2}\sqrt{\pi\, t_{Dxf}} = \frac{\pi}{2}\sqrt{\pi\,(0.523)} = 2.017. \tag{17-45}$$

Rearranging Equation 17-38 to solve for the gas flow rate yields

$$q_g = \frac{kh[m(p_i) - m(p_{wf})]}{1424T\left(\dfrac{1}{q_D}\right)} \tag{17-46}$$

and substituting,

$$q_g = \frac{(0.02)(78)(1.29 \times 10^9 - 8.6 \times 10^8)}{1424(640)(2.017)} = 366 \text{ Mscf/d}. \tag{17-47}$$

Performing similar calculations for each month for 3 years of production gives the rate forecast shown in Figure 17-10, noting that Equation 17-42 must be used for times when $t_{Dye} > 1.25$. Repeating this procedure for a fracture half length of 500 ft yields the lower curve in Figure 17-10. Substantially less gas would be recovered from this low-permeability gas reservoir with the shorter fracture.

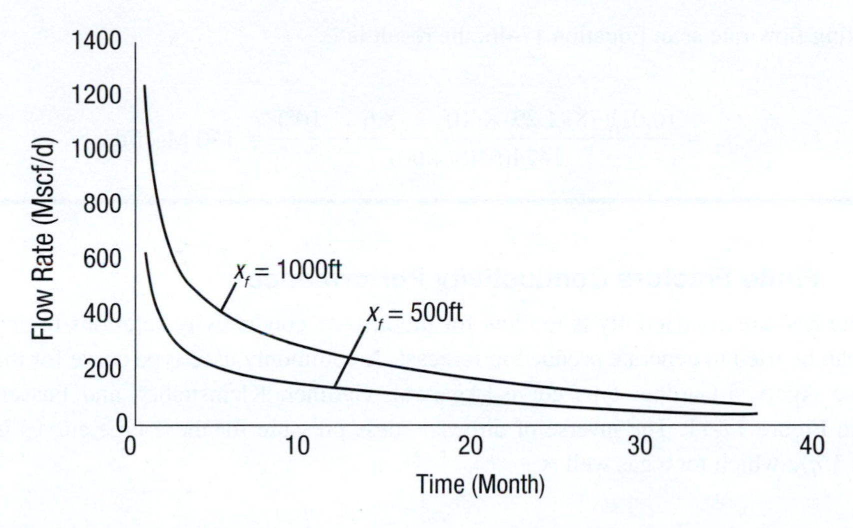

Figure 17-10 Gas well production forecast for infinite conductivity fracture (Example 17-3).

Example 17-4 Formation Linear Flow Approximation

For the well described in Example 17-3, calculate the time for the end of early (transient) linear flow. What is the production rate at 20 months?

Solution

The end of early (transient) flow occurs when $t_{Dye} = 0.25$. So from Equation 17-40, that occurs, in months, at

$$t_{Dye} = \frac{0.0002637kt}{\phi\mu c_t y_e^2} = \frac{0.0002637(0.02)(t_{\text{months}})(30)(24)}{(0.14)(0.0249)(1.25 \times 10^{-5})(2000)^2} = 0.25. \quad \textbf{(17-48)}$$

Solving this for t yields $t_{\text{months}} = 11.47$ months at the end of early (transient) flow. So Equation 17-42 must be used for times greater than 11.47 months. This is called *boundary dominated flow* or *exponential decline.*

For this example, $t = 20$ months, so

$$t_{Dye} = \frac{0.0002637kt}{\phi\mu c_t y_e^2} = \frac{0.0002637(0.02)(20)(30)(24)}{(0.14)(0.0249)(1.25 \times 10^{-5})(2000)^2} = 0.436. \quad \textbf{(17-49)}$$

We now use Equation 17-42 to solve for $1/q_D$, as follows:

$$\frac{1}{q_D} = \frac{\pi}{4}\left(\frac{y_e}{x_f}\right)\exp(\pi^2 t_{Dye}) = \frac{\pi}{4}\left(\frac{2000}{1000}\right)\exp(\pi^2\, 0.436) = 4.60 \quad \textbf{(17-50)}$$

Calculating flow rate as in Equation 17-46, the result is

$$q_g = \frac{(0.02)(78)(1.29 \times 10^9 - 8.6 \times 10^8)}{1424(640)(4.60)} = 160 \text{ Mscf/d.} \tag{17-51}$$

17.4.2 Finite Fracture Conductivity Performance

When the fracture conductivity is too low for the infinite conductivity solutions to apply, type curves can be used to generate production forecast. A commonly used type curve for this analysis is the Agarwal-Gardner type curve [Agarwal, Gardner, Kleinsteiber, and Fussell, 1999] shown in Figure 17-11. The inverse of dimensionless pressure for these type curves is the inverse of $1/q_D$, which for a gas well is

$$\frac{1}{p_D} = \frac{1424 q_g T}{kh[m(p_i) - m(p_{wf})]}. \tag{17-52}$$

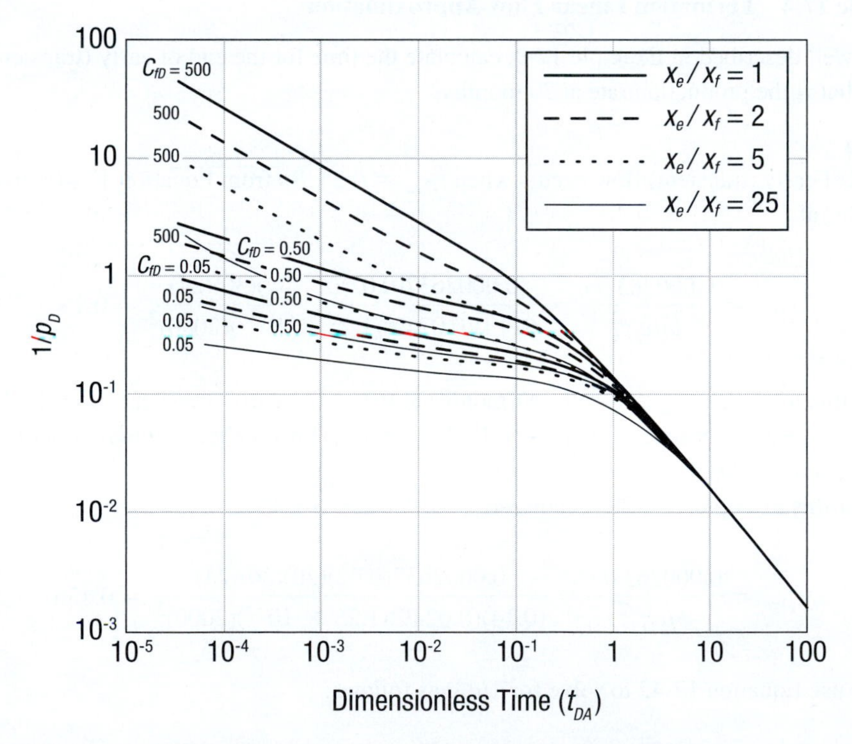

Figure 17-11 Agarwal-Gardner type curves. (From Agarwal et al., 1999.)

To use these type curves for gas production, the dimensionless time, t_{DA}, is calculated based on an equivalent time, t_a:

$$t_{DA} = \frac{0.0002637 k t_a}{\phi \mu c_t A} \tag{17-53}$$

where A is the drainage area of the fractured well and

$$t_a = \frac{1}{q(t)} (\mu c_g)_i \int_0^t \frac{q(t')dt'}{\mu(\overline{p}) c_g(\overline{p})} = \frac{1}{q(t)} (\mu c_g)_i \frac{Z_i G_i}{2 p_i} [m(p_i) - m(\overline{p})]. \tag{17-54}$$

The reader is referred to the Agarwal et al. paper or to Palacio and Blasingame (1993) for more information on how to compute the equivalent time to use these type curves.

17.5 The Effect of Non-Darcy Flow on Fractured Well Performance

In many hydraulically fractured gas wells, the additional pressure drop caused by non-Darcy flow in the fracture itself causes lower productivity than would occur with Darcy flow. Beginning with the laboratory propped fracture conductivity measurements of Cooke (1993), non-Darcy effects in the fracture have been recognized to be important. Non-Darcy effects increase with velocity and, therefore, are highest in higher-permeability formations, but can be significant even in tight gas wells (Miskimins, Lopez-Hernandez, and Barree, 2005).

As first shown by Holditch and Morse (1976), non-Darcy flow in the fracture can be accounted for by adjusting the fracture permeability to an effective, non-Darcy permeability, k_{nD},

$$\frac{k_{nD}}{k_f} = \frac{1}{1 + N_{Re,nD}} \tag{17-55}$$

where the non-Darcy flow Reynolds number is

$$N_{Re,nD} = \frac{\beta \rho v k_f}{\mu} \tag{17-56}$$

and β is the non-Darcy flow coefficient for the fracture, ρ is the fluid density, v is the velocity in the fracture, μ is the fluid viscosity, k_f is the fracture permeability. As with any dimensionless number, the parameters in this Reynolds number must be in consistent units. Non-Darcy flow effects are most important in the fracture near the wellbore where the velocity is the greatest. For high-rate gas wells, a simple, conservative approach, as suggested by Gidley (1991), is to use the velocity at the wellbore to estimate the effects of non-Darcy flow on well productivity.

Because the non-Darcy flow effect is rate-dependent, the flow rate must be known to determine productivity, and the problem is iterative. For conditions for which Figure 17-6 (Equation 17-8) can be used to calculate well productivity, the procedure is as follows:

1. For the Darcy flow C_{fD}, determine s_f and J based on Figure 17-6.
2. For an assumed bottomhole flowing pressure, calculate the velocity in the fracture at the wellbore.
3. Calculate the non-Darcy effective fracture permeability from Equation 17-55.
4. Calculate $C_{fD,nD}$:

$$C_{fD,nD} = \frac{k_{nD}}{k_f} C_{fD}. \tag{17-57}$$

5. Calculate a new productivity using the new value of $C_{fD,nD}$; if the new productivity is appreciably different, repeat the calculation procedure.

Non-Darcy flow effects in the fracture are most important when the fracture conductivity is low. For values of C_{fD} less than 0.2, Nolte and Economides (1991) showed that the effective wellbore radius depends only on fracture conductivity and reservoir permeability:

$$r'_w = \frac{0.28 k_f w}{k}. \tag{17-58}$$

Then, from Equation 17-5,

$$s_f = -\ln\left(\frac{0.28 k_f w}{k r_w}\right). \tag{17-59}$$

Equation 17-58 describes the linear portion of the curve at low C_{fD}. From the shape of this curve, these relationships appear valid up to a C_{fD} value of about 1.

Using this relationship, the effect of non-Darcy flow effects on well productivity can be determined as follows. First, for a gas well producing at pseudosteady-state and having a C_{fD} less than 1, the production rate without considering non-Darcy flow is

$$q_{go} = \frac{[m(\overline{p}) - m(p_{wf})]kh}{1424T} J_D \tag{17-60}$$

where

$$J_D = \frac{1}{ln\frac{r_e}{r_w} - 0.75 + s_f} = \frac{1}{ln\frac{r_e}{r_w} - 0.75 - \ln\left(\frac{0.28k_f w}{kr_w}\right)}. \tag{17-61}$$

We can write the ratio of the productivity index including non-Darcy flow in the fracture, J_{nD}, to the productivity index neglecting non-Darcy flow, J_o, as

$$\frac{J_{nD}}{J_o} = \frac{ln\frac{r_e}{r_w} - 0.75 - \ln\left(\frac{0.28k_f w}{kr_w}\right)}{ln\frac{r_e}{r_w} - 0.75 - \ln\left(\frac{0.28k_f w}{kr_w(1 + N_{Re,nD})}\right)}, \tag{17-62}$$

which can be rearranged to

$$\frac{J_{nD}}{J_o} = \frac{J_{Do}}{J_{Do} + \ln(1 + N_{Re,nD})}. \tag{17-63}$$

This equation still requires an iterative solution because the non-Darcy Reynolds number depends on velocity in the fracture, which in turn depends on the productivity index, the drawdown, and the bottomhole flowing pressure. However, this form is a convenient one to easily estimate the effect of non-Darcy flow in the fracture on gas well productivity.

Example 17-5 Effect of Non-Darcy Flow in the Fracture on Conventional Gas Well Productivity

Calculate the effect of non-Darcy flow in the fracture on the productivity of a gas well with the following properties: $k = 0.1$ md, $x_f = 500$ feet, $\overline{w} = 0.2$ inches, $k_f = 3000$ md, $\gamma_g = 0.71$,

$T = 120°F$, $\overline{p} = 4{,}000$ psi ($m(\overline{p}) = 1.005 \times 10^9$ psi^2/cp), $p_{wf} = 1{,}500$ psi, $m(p_{wf}) = 1.961 \times 10^8$ psi^2/cp), $\mu(p_{wf}) = 0.015$, $Z(p_{wf}) = 0.79$, $r_w = 0.328$ ft, and $r_e = 1490$ ft. Repeat the calculation for a bottomhole flowing pressure of 3500 psi, for which *m(p)* is 8×10^8 psi^2/cp.

Solution

For this well, C_{fD} is equal to 1, so it is at the upper limit of dimensionless conductivity for the applicability of Equation 17-59. First, we calculate the flow conditions neglecting non-Darcy effects:

With $C_{fD} = 1$, from Figure 17-6, $\ln(x_f/r_w) + s_f = 2$. Thus, s_f is -5.33. From Equation 17-61, the dimensionless productivity index in the absence of non-Darcy effects is

$$J_{Do} = \frac{1}{\ln \dfrac{1490}{0.328} - 0.75 - 5.33} = 0.43 \tag{17-64}$$

and

$$q_g = \frac{[m(\overline{p}) - m(p_{wf})]kh}{1424T} J_D$$

$$= \frac{(1.005 \cdot 10^9 - 1.961 \times 10^8)(0.1)(50)}{1424(580)}(0.427) = 2090 \text{ Mscf/d}. \tag{17-65}$$

To calculate the Reynolds number, *in-situ* volumetric properties are required at this point. From Equation 4-13,

$$B_g = 0.0283 \frac{(0.79)(580R)}{(1{,}500 \text{ psi})} = 0.00864 \text{ rescf/scf} \tag{17-66}$$

and

$$\rho_g = 1.22 \frac{\gamma_g}{B_g} = 1.22 \frac{0.71}{0.00864} = 100.25 \text{ kg/m}^3. \tag{17-67}$$

Linear velocity in the fracture at the wellbore is calculated as

$$v = \frac{B_g q_g}{2hw} = \frac{0.00864(2{,}090)(1{,}000)}{3{,}600(24)(2)(50)(0.008)} = 0.126 \text{ ft/s} = 0.038 \text{ m/s}. \tag{17-68}$$

The beta factor from the Frederick and Graves correlation (1994) for the given proppant permeability of 3000 md, a proppant pack porosity of 0.3, and no water saturation in the fracture is

$$\beta = 7.89 \times 10^{10} \frac{1}{(k_f)^{1.6}\phi^{0.404}} = 7.89 \times 10^{10} \frac{1}{(3{,}000)^{1.6}(0.3)^{0.404}} = 1.14 \times 10^{6}\ \mathrm{m}^{-1}. \quad \textbf{(17-69)}$$

The resulting Reynolds number is then calculated using SI units:

$$N_{\mathrm{Re},nD} = \frac{\beta k_{f,n} v \rho}{\mu} = \frac{1.14 \times 10^{6}(3{,}000)(9.869 \times 10^{-16})(0.126)(100.25)}{0.015 \times 10^{-3}} = 2.87 \quad \textbf{(17-70)}$$

The first estimate of the effect of non-Darcy flow is then calculated from Equation 17-63:

$$\frac{J_{nD}}{J_o} = \frac{0.43}{0.43 + \ln(1 + 2.87)} = 0.24 \quad \textbf{(17-71)}$$

We now use the estimate of the productivity index including non-Darcy flow in the fracture, J_{nD}, to calculate a new production rate, velocity, and Reynolds number, from which the estimate of the non-Darcy productivity index is updated. After a few iterations, we find that $J_{nD}/J_o = 0.53$. This means that non-Darcy flow in the fracture causes the productivity of the well to be only 53% of what it would be without non-Darcy effects.

For a lower drawdown of 500 psi ($p_{wf} = 3500$ psi), $J_{nD}/J_o = 0.86$, only a 14% reduction in productivity, illustrating the rate dependence of non-Darcy flow effects. It is clear that the well described would benefit from having a higher-conductivity fracture, even though the dimensionless conductivity is near the theoretical optimum in the absence of non-Darcy effects.

The results presented in Example 17-5 are pessimistic because we have used the conditions at the wellbore, where the velocity in the fracture is the greatest, to calculate the non-Darcy Reynolds number for the entire fracture. To calculate non-Darcy flow in the fracture considering the changing velocity all along the fracture requires a numerical solution coupling the flow in the reservoir to the fracture with the flow in the fracture itself. Using such an approach, Miskimins et al. (2005) predicted that non-Darcy flow effects in the fracture decrease the 10-year cumulative recovery from tight gas reservoirs with permeabilities of 0.01–0.1 md by 10–20% for typical fracture lengths and conductivities.

Non-Darcy effects become even more important in higher-permeability reservoirs, making achievement of high fracture conductivity critical. In addition, in higher-permeability fractured gas wells, non-Darcy effects in the reservoir and through the completion, particularly through perforation tunnels, also affect the well performance. Figure 17-12 from Lolon, Chipperfield, McVay, and Schubarth (2004) shows how the total pressure drop for a high-permeability gas well is distributed. The largest pressure drop is the non-Darcy pressure drop occurring in perforation

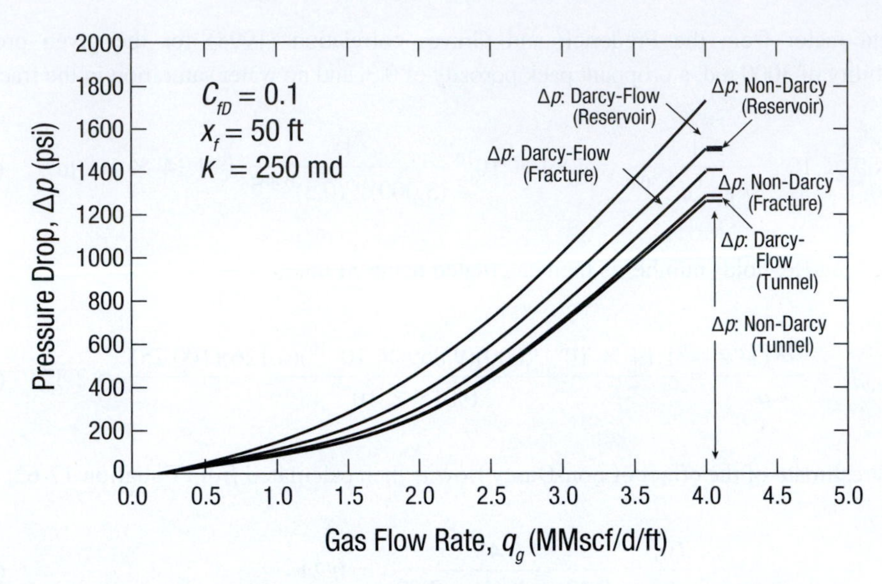

Figure 17-12 Pressure drop sources in a fractured well. (From Lolon et al., 2004.)

tunnels; the total pressure drop in the fracture itself is almost evenly divided between Darcy flow pressure drop and the non-Darcy contribution.

The critical parameter to predict non-Darcy flow in the fracture is the non-Darcy coefficient, β. Numerous correlations for β have been developed based on laboratory measurements of proppant pack conductivity, beginning with the work of Cooke (1973). Cooke's correlation is

$$\beta = 1 \times 10^8 \frac{b}{(k_f)^a} \tag{17-72}$$

where the empirical constants a and b are given in Table 17-3. As Cooke states in his paper, this correlation likely predicts β factors on the upper side of their possible range because of crushing and embedment that likely occurred in the experiments.

Table 17-3 Constants in Cooke β Correlation

Sand Size	*a*	*b*
8–12	1.24	3.32
10–20	1.34	2.63
20–40	1.54	2.65
40–60	1.60	1.10

Numerous other studies of the β factor for non-Darcy flow in proppant packs have been presented, including those of Geertsma (1974), Jin and Penny (1998), Frederick and Graves (1994), Barree and Conway (2004), and Vincent, Pearson, and Kullman (1999). Some of these studies proposed correlations that include the effects of a liquid phase being present in the proppant pack, such as that of Frederick and Graves:

$$\beta = \frac{159{,}160}{k_g^{0.5}(\phi(1 - S_w))^{5.5}}. \tag{17-73}$$

In this equation, k_g is effective gas permeability in md, ϕ is proppant pack porosity, S_w is water saturation, and β is in m^{-1}.

Note: Non-Darcy effects have an additional impact in production optimization. Hydraulic fractures should be redesigned to have a larger width to accommodate the reduction in the effective proppant pack permeability. Thus, for a given mass of proppant the geometry of the fracture and the relationship between length and width must be adjusted depending on the magnitude of the non-Darcy effects. The UFD approach does this in a coherent fashion.

17.6 Fractured Well Performance for Unconventional Tight Sand or Shale Reservoirs

An increasing amount of oil and, more commonly, gas is being produced from very low–permeability formations, referred to as unconventional resources. Such reservoirs include tight gas sand and shale resources, which share the common feature that economic recovery of hydrocarbons requires fracture stimulation. As described by the Holditch (2006) resource triangle (Figure 17-13), these unconventional resources are vast, but economic recovery of the hydrocarbon reserves contained in them is challenging and costly. Advanced hydraulic fracture stimulation technology is the key to economically accessing these reserves, with the creation of multiple hydraulic fractures from either vertical or horizontal wells often required.

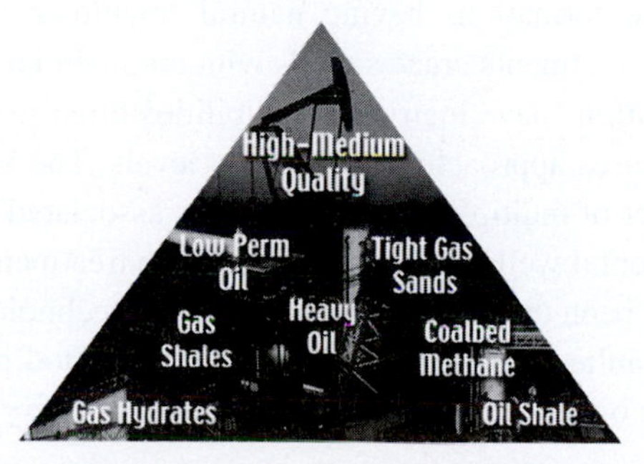

Figure 17-13 The oil and gas resource triangle. (From Holditch, 2006.)

17.6.1 Tight Gas Sands

Large hydraulic fracturing treatments have been applied to develop tight gas reservoirs, beginning in earnest in the 1970s. These reservoirs have permeabilities less than 0.1 md and as low as 0.001 md. To produce commercial quantities of gas, these reservoirs must be comprised of thick formations, or there must be multiple sand bodies that are penetrated by each well. A typical tight gas well is completed in very thick, stacked sands such as those of the Mesa Verde geological group in the Piceance Basin of the western United States. The braided stream deposits in this formation result in numerous isolated sand bodies distributed over as much as 5000 feet of gross vertical interval (Wolhart, Odegard, Warpinski, Waltman, and Machovoe, 2005). A typical stratigraphic column for tight gas wells in this area is shown in Figure 17-14 (Juranek, Seeburger, Tolman, Choi, Pirog, Jorgensen, and Jorgensen 2010), illustrating the need for multiple hydraulic fractures to stimulate the isolated lenticular sands of this formation.

The productivity of a vertical well having multiple fractures can be estimated by treating each fracture separately, assuming that they are not connected with each other or are not in pressure communication in the reservoir. However, if the pressure gradient in the wellbore differs from the pressure gradient in the reservoir, the drawdown for each fracture will be different, and in this case, the overall well productivity cannot be separated from the more general well deliverability problem. In this case, for N fractures spaced vertically along a well, the production rate from the well is

$$q = \sum_{n=1}^{N} J_n\{m(\overline{p}_n) - m(p_{wf,n})\} \quad \textbf{(17-74)}$$

and the bottomhole flowing pressures, $p_{wf,n}$, at each fracture location are not independent, but are related through the wellbore flow behavior.

17.6.2 Shale

In recent years, the development of effective hydraulic fracturing treatments to stimulate ultra-low permeability shale reservoirs has led to a dramatic increase in economically recoverable hydrocarbons. In brittle shale formations having natural fracture networks, the application of multiple hydraulic fracture treatments creates wells with commercial rates and reserves, in spite of the fact that these formations have matrix permeabilities often much less than a microdarcy ($<10^{-3}$ md) and in some cases approaching nanodarcy levels. The key to productivity in these formations is a large number of multiple fractures at times associated with complex fracture networks. Thousands of horizontal wells with multiple fracture treatments in the Barnett shale formation in north Texas have been the proving grounds for this technology.

When a rock is hydraulically fractured, the fractures created may range from simple bi-wing fractures extending in opposite directions from the well, to very complex fracture patterns,

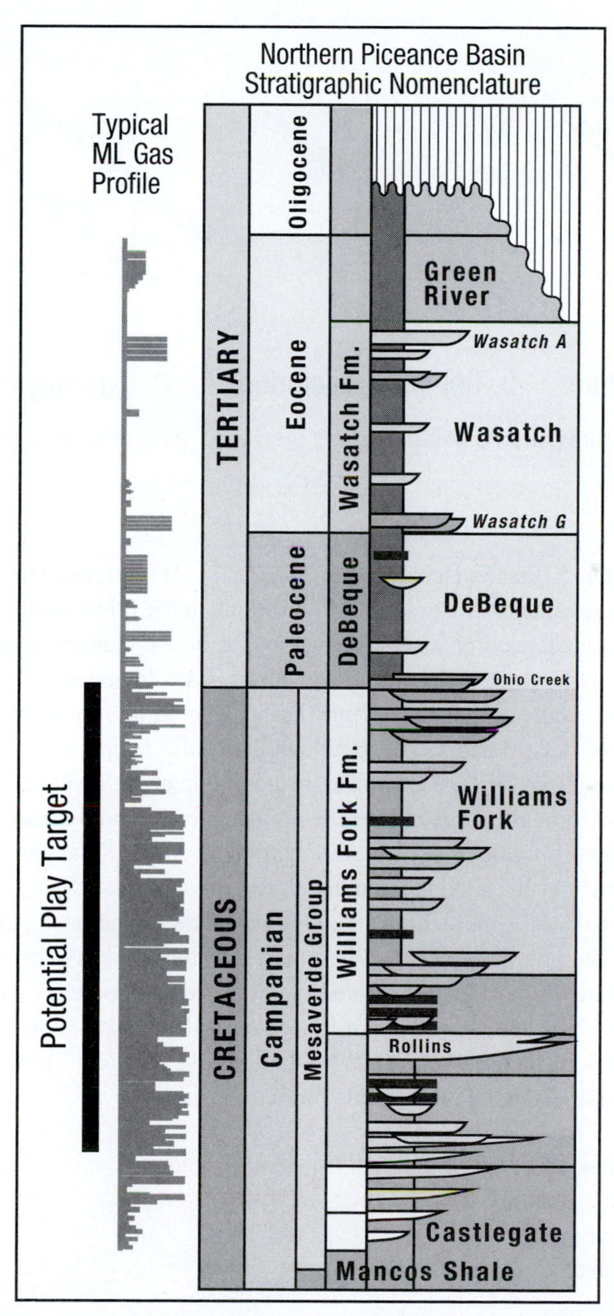

Figure 17-14 Piceance Basin stratigraphic column. (From Juranek et al., 2010.)

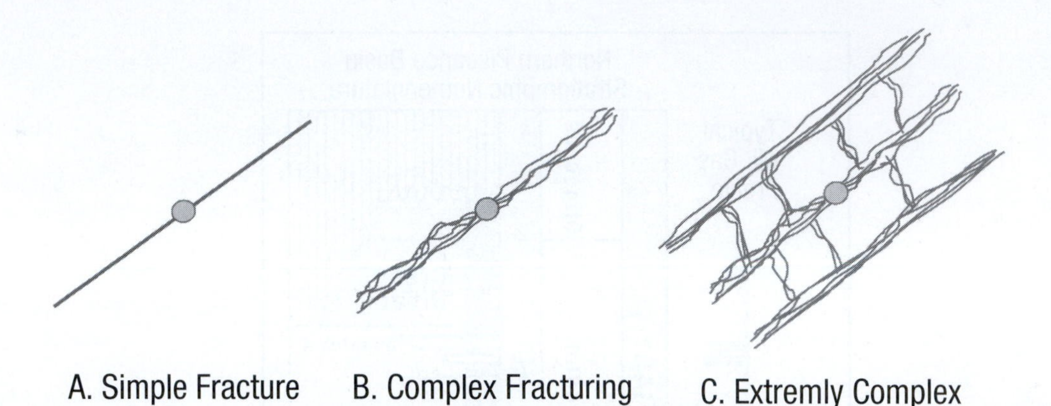

Figure 17-15 Fracture patterns created in tight reservoirs. (From Fisher et al., 2004.)

as shown in Figure 17-15 (Fisher, Heinze, Harris, Davidson, Wright, and Dunn, 2004). In brittle, naturally fractured shales near stress isotropic conditions in the horizontal plane, fracture treatments are designed to create complex fracturing, so that flow conductivity is created in a large volume of reservoir. The extent of the reservoir affected by fracturing is often monitored by recording the microseismic events occurring during fracturing. As shown in Figure 17-16 (Daniels, Waters, LeCalvez, Lassek, and Bentley, 2007), the region over which microseismic events are created during multistage fracturing of a horizontal well can extend a few thousand feet in all directions. Throughout the affected region, created hydraulic fractures and stimulated natural fracture systems provide flow conductivity to the wellbore. The part of the shale formation contacted by fractures is called the stimulated reservoir volume (SRV); the ultimate recovery from fractured shale wells increases as the SRV increases (Fisher et al., 2004).

Prediction of the productivity of a fractured shale formation is complicated by the extremely variable nature of the fractures created. Some current approaches typically assume that hydraulically created "primary" fractures are connected to the horizontal wellbore and that these intersect other, lower-conductivity fractures. Methods being applied to predict the productivity of multiply fractured horizontal wells in shale formations include:

- Analytical models of the production from each fracture
- Multiple sink solutions
- Decline curve analysis
- Reservoir simulators including fracture networks

The simplest approach is to treat each main fracture as draining a separate, isolated region, with no-flow boundaries separating the regions drained by each fracture, as depicted in Figure 17-17

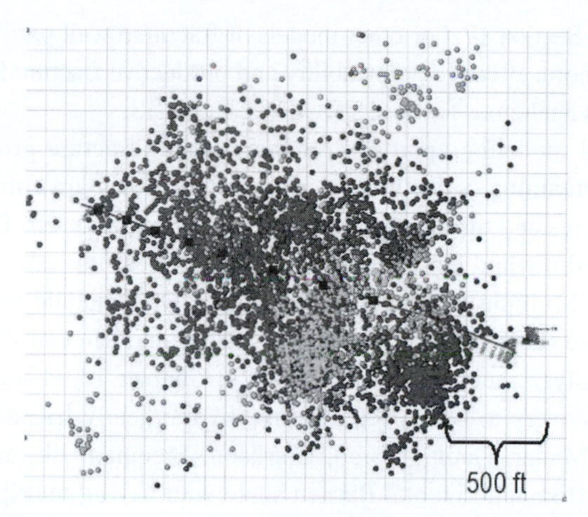

Figure 17-16 Microseismic map of complex fracture patterns. (From Daniels et al., 2007).

(Song, Economides, and Ehlig-Economides, 2011). Dividing the well drainage area in this manner, any fractured well model can be applied to the individual fractures and the contributions from all fractures summed to give the total production rate. The permeability used in such calculations should be an effective permeability that may reflect the contribution of the fracture network feeding the main fracture and/or true matrix permeability. The well production forecasting calculation

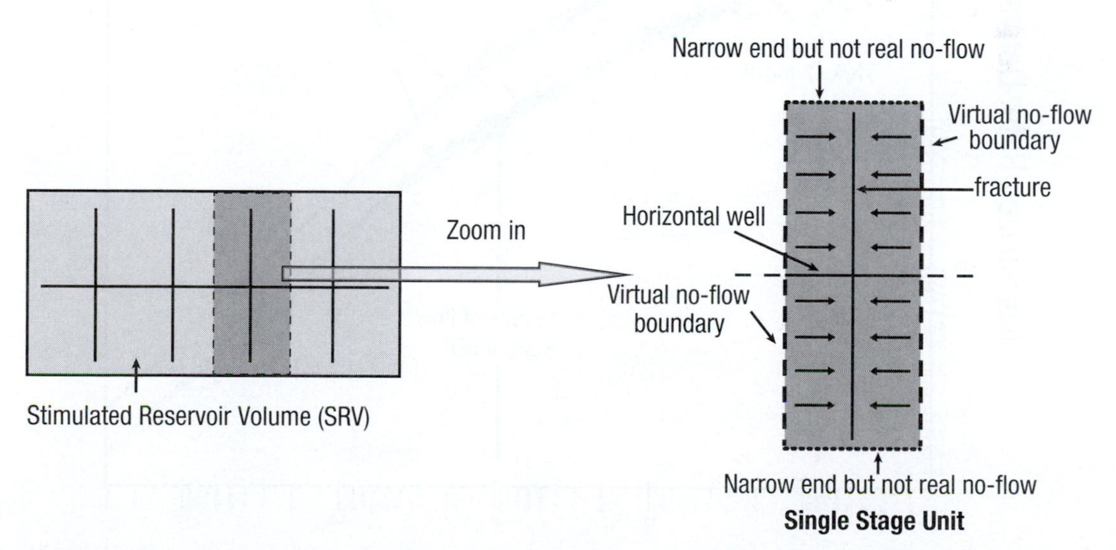

Figure 17-17 Drainage pattern from a horizontal well with multiple transverse fractures. (From Song et al., 2011.)

procedures presented in Section 17.4 or other models of fractured well performance can be applied in this manner. Furthermore, effective permeability and productive fracture length can be estimated by following procedures analogous to those shown in Chapter 13.

A semi-analytical modeling approach for the multiple fracture production problem is to treat the fractures as containing collections of pressure sinks that are communicating with each other through the fractures and through the low-permeability formation. This approach typically requires some numerical method be applied, such as numerical inversion of an analytical solution in Laplace space. Multiple fracture models of this type include those of Meyer, Bazan, Jacot, and Lattibeaudiere (2010), Valko and Amini (2007), and Lin and Zhu (2010).

Decline curve analysis, such as was described in Chapter 13, is commonly applied to determine the ultimate recovery to be expected from a hydraulically fractured shale well. For example, Figure 17-18 (Mattar, Gault, Morad, Clarkson, Freeman, Ilk, and Blasingame, 2008) shows production data from a horizontal Barnett shale well having multiple fractures plotted on an Agarwal-Gardner type curve. From the type curve match, an ultimate recovery of 5 Bcf of gas is predicted for this well.

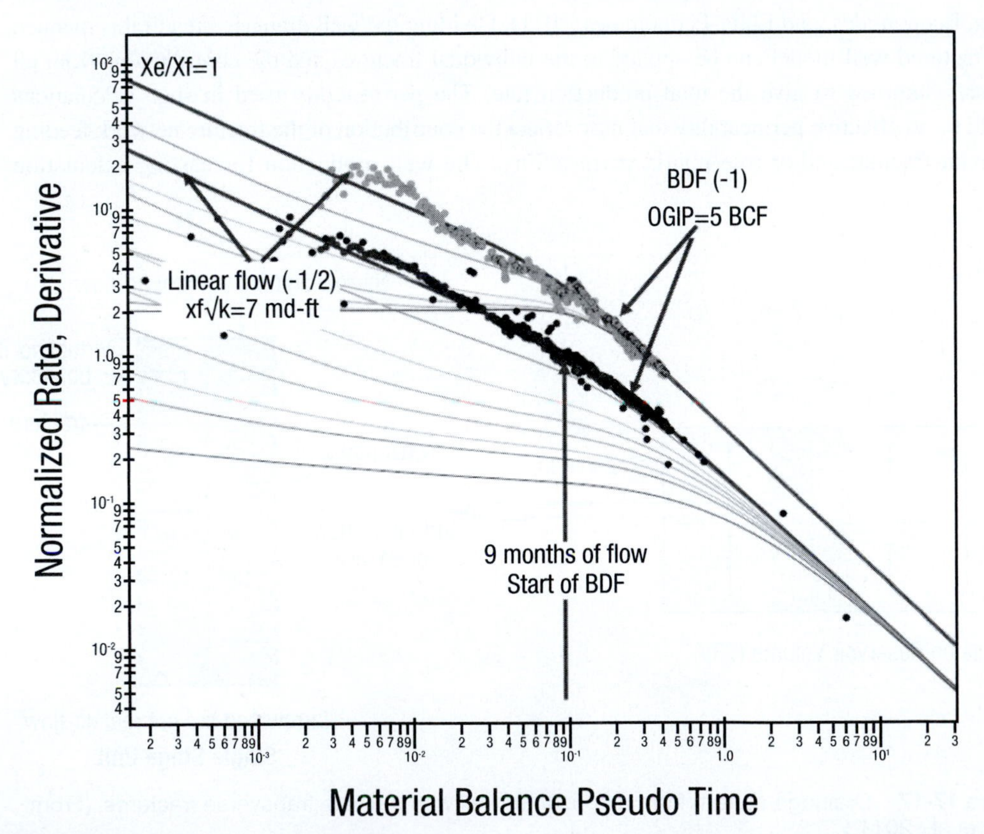

Figure 17-18 Type curve matching to forecast ultimate recovery from a shale gas well. (From Mattar et al., 2008.)

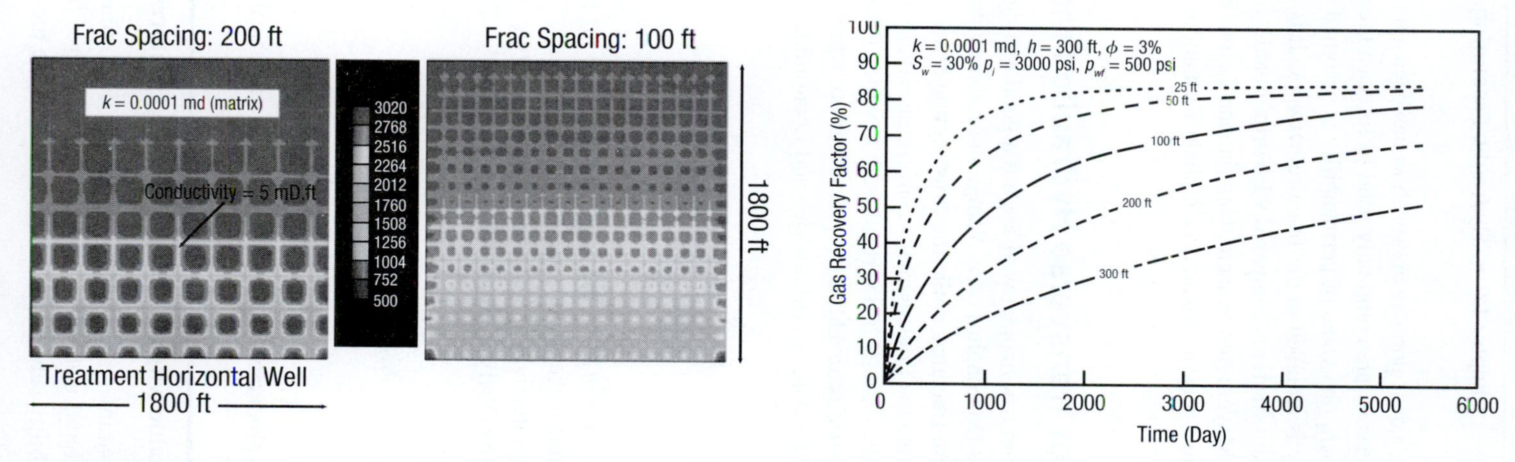

Figure 17-19 Reservoir simulator prediction of pressure field around a fractured shale well. (From Warpinski et al., 2009.)

Reservoir simulators using a dual-porosity approach as has been applied for many years to simulate naturally fractured reservoirs are commonly used to forecast the performance of fractured shale wells. In this approach, one value of permeability is assigned to the matrix blocks, and another, higher permeability is assigned to the fracture network. It is also possible to have different conductivities in the main fractures compared with orthogonal, stimulated natural fractures. Figure 17-19 (Warpinski, Mayerhofer, Vincent, Cipolla, and Lolon, 2009) shows the pressure field in the complex fracture network created in a shale formation, and the forecast gas recovery for different fracture spacing.

17.7 Choke Effect for Transverse Hydraulic Fractures

Multiple transverse fractures from a horizontal well are not a good strategy above 10 md in oil reservoirs and 1 md in gas wells (Marongiu-Porcu, Wang, and Economides, 2009; Economides and Martin, 2010). For transverse fractures, while the flow regime from the reservoir into each fracture is linear, inside the fracture the flow regime is converging radially to the horizontal well. (This is in contrast to the hydraulic fracture in a vertical well, in which flow inside the fracture to the well is linear.) The consequence is an additional pressure drop, which can be accounted for by a skin effect, denoted as "choke skin," s_c, (Mukherjee and Economides, 1991)

$$s_c = \frac{kh}{k_f w}\left[\ln\left(\frac{h}{2r_w}\right) - \frac{\pi}{2}\right] \tag{17-75}$$

where k is the formation permeability, h is the net pay thickness, k_f is the proppant-pack permeability, r_w is the well radius, and w is the fracture width.

Then, the dimensionless productivity index of a transverse fracture intersecting a horizontal well (J_{DTH}) can be obtained by (Wei and Economides, 2005)

$$J_{DTH} = \frac{1}{\dfrac{1}{J_{DV}} + s_c} \tag{17-76}$$

where J_{DV} is the dimensionless productivity index of the fractured vertical well.

Example 17-6 Optimum Productivity Index for Transverse Fractures in Horizontal Wells

Using the well and reservoir parameters from Example 17-2, determine the optimal fracture geometries for a 1000-ft horizontal well with two transverse fractures (one at the toe, one at the heel), for reservoir permeability values of 50 and 15 md. Include the choke skin effect and calculate the dimensionless productivity index for each transverse fracture. Each of the two transverse fractures should drain half of a 4×10^6 ft^2 area. As suggested for the 15 md case in Example 17-2, use proppant with 120,000 md permeability, and use half the total proppant mass (37,500 lb per fracture).

Solution

For this case, each fracture will use the same proppant volume used in the previous examples to drain half the reservoir volume, and the proppant number will be twice that determined in Example 17-2. We will double the proppant permeability and halve the proppant volume.

For each fracture, the propped volume in the pay (V_p) and reservoir volume (V_r) are

$$V_p = \frac{\frac{50}{100}(37{,}500)}{2.65(62.4)(1 - 0.38)} = 183 \text{ ft}^3 \tag{17-77}$$

$$V_r = Ah = 2 \times 10^6 \times 50 = 10^8 \text{ ft}^3. \tag{17-78}$$

The proppant number can then be calculated using Equation 17-17

$$N_p = \frac{2k_f}{k}\frac{V_p}{V_{res}} = \frac{2(120{,}000)}{(50)}\frac{183}{(10^8)} = 0.0088. \tag{17-79}$$

Equation 17-17 applies for $N_p < 0.1$:

$$N_p = N_{p,\,square}\frac{x_e}{y_e} = 0.0088\frac{1}{2} = 0.0044. \tag{17-80}$$

From Figure 2-3, the Dietz shape factor for a 2:1 rectangle is 21.8, and

$$N_{pe} = N_p\frac{C_A}{30.88} = (0.0044)\frac{21.8}{30.88} = 0.0031 \tag{17-81}$$

From Equation 17-18

$$J_{D\,\max\,pss}(N_p) = \frac{1}{0.990 - 0.5\ln N_p} = \frac{1}{0.990 - 0.5\ln(0.0031)} = 0.258. \tag{17-82}$$

From Equation 17-20,

$$x_{fopt} = \left(\frac{k_f V_f}{C_{fDopt} kh}\right)^{0.5} = \left(\frac{120{,}000)(183)}{(1.6)(50)(50)}\right)^{0.5} = 52 \text{ ft} \tag{17-83}$$

and

$$w_{opt} = \left(\frac{C_{fDopt} k V_f}{k_f h}\right)^{0.5} = \left(\frac{(1.6)(50)(183)}{(120{,}000)(50)}\right)^{0.5} = 0.035 \text{ ft} = 0.42 \text{ in.} \tag{17-84}$$

Equation (17-75) gives

$$s_c = \frac{kh}{k_f w}\left[\ln\left(\frac{h}{2r_w}\right) - \frac{\pi}{2}\right] = \frac{(50)(50)}{(120{,}000)(0.0035)}\left[\ln\left(\frac{50}{2(0.3)}\right) - \frac{\pi}{2}\right] = 1.7 \tag{17-85}$$

for $r_w = 0.3$ ft, and the dimensionless productivity of the transverse fracture will be

$$J_{DTH} = \frac{1}{\dfrac{1}{J_{DV}} + s_c} = \frac{1}{\dfrac{1}{0.258} + 1.7} = 0.179. \tag{17-86}$$

For permeability of 50 md the productivity of the transverse fracture is nearly 30% less than that of the same fracture from a vertical well. The choice to design the well with two transverse fractures would probably be considered only when the cost of a second vertical well is prohibitive. For example, there could be surface restrictions to drilling a well where the second fracture would be.

For the 15-md case, $s_c = 0.93$, and $J_{DTH} = 0.238$ compared to $J_D = 0.305$ for the vertical well. In this case the transverse fracture solution can be an attractive choice.

For gas wells, the choking effect in a transverse fracture is further penalized by non-Darcy flow, and the effective proppant permeability is reduced even more than for the case of the vertical fracture because of the radial flow convergence in the fracture.

References

1. Agarwal, R.G., Gardner, D.C., Kleinsteiber, S.W., and Fussell, D.D., "Analyzing Well Production Data Using Combined Type-Curve and Decline-Curve Analysis Concepts," *SPERE,* 2(5): 478–486 (October 1999).

2. Barree, R.D., and Conway, M.W., "Beyond Beta Factors: A Complete Model for Darcy, Forchheimer, and Trans-Forchheimer Flow in Porous Media," SPE Paper 89325, 2004.

3. Cinco-Ley, H., and Samaniego, F., "Transient Pressure Analysis for Fractured Wells," *JPT,* 1749–1766 (September 1981).

4. Cooke, C.E., Jr., "Conductivity of Proppants in Multiple Layers." *JPT,* 1101–1107 (October 1993).

5. Daal, J.A., and Economides, M.J., "Optimization of Hydraulically Fractured Wells in Irregularly Shaped Drainage Areas," SPE Paper 98047, 2006.

6. Daniels, J., Waters, G., LeCalvez, J., Lassek, J., and Bentley, D., "Contacting More of the Barnett Shale Through an Integration of Real-Time Microseismic Monitoring, Petrophysics and Hydraulic Fracture Design," SPE Paper 110562, 2007.

7. Economides, M.J., and Martin, A.N., "How to Decide Between Horizontal Transverse, Horizontal Longitudinal and Vertical Fractured Completion," SPE Paper 134424, 2010.

8. Economides, M.J., Oligney, R.E., and Valkó, P., *Unified Fracture Design,* Orsa Press, Houston, 2002.

9. Fisher, M.K., Heinze, J.R., Harris, C.D., Davidson, B.M., Wright, C.A., and Dunn, K.P., "Optimizing Horizontal Well Completion Techniques in the Barnett Shale Using Microseismic Fracture Mapping," SPE Paper 90051, 2004.

10. Frederick, D.C., and Graves, R.M., "New Correlations to Predict Non-Darcy Flow Coefficients at Immobile and Mobile Water Saturation," SPE Paper 28451, 1994.

11. Geertsma, J., "Estimating the Coefficient of Inertial Resistance in Fluid Flow through Porous Media," *SPEJ,* 445–450 (October 1974).

12. Gidley, J.L., "A Method for Correcting Dimensionless Fracture Conductivity for Non-Darcy Flow Effects," *SPEPE,* 391–394 (November 1991).

13. Holditch, S.A., "Tight Gas Sands," *JPT,* 84–86 (June 2006).

14. Holditch, S.A., and Morse, R.A., "The Effects of Non-Darcy Flow on the Behavior of Hydraulically Fractured Gas Wells," *JPT,* 1169–1179 (October 1976).

15. Jin, L., and Penny, G.S., "A Study on Two-Phase, Non-Darcy Gas Flow Through Proppant Packs," SPE Paper 49248, 1998.

16. Juranek, T.A., Seeburger, D., Tolman, R., Choi, N.H., Pirog, T.W., Jorgensen, D., Jorgensen, E., "Evolution of Mesaverde Stimulations in the Piceance Basin: A Case History of the Application of Lean Six Sigma Tools," SPE Paper 131731, 2010.

17. Lin, J., and Zhu, D., "Modeling Well Performance for Fractured Horizontal Gas Wells," SPE Paper 130794, 2010.

18. Lolon, E.P., Chipperfield, S.T., McVay, D.A., and Schubarth, S.K., "The Significance on Non-Darcy and Multiphase Flow Effects in High-Rate, Frac-Pack Gas Completions," SPE Paper 90530, 2004.

19. Marongiu-Porcu, M., Wang, X., and Economides, M.J., "Delineation of Application: Physical and Economic Optimization of Fractured Gas Wells," SPE Paper 120114, 2009.

20. Mattar, L., Gault, B., Morad, K., Clarkson, C.R., Freeman, C.M., Ilk, D., and Blasingame, T.A., "Production Analysis and Forecasting of Shale Gas Reservoirs: Case History-Based Approach," SPE Paper 119897, 2008.

21. Meyer, B.R., Bazan, L.W., Jacot, R.H., and Lattibeaudiere, M.G., "Optimization of Multiple Transverse Hydraulic Fractures in Horizontal Wellbores," SPE Paper 131732, 2010.

22. Meyer, B.R. and Jacot, R.H.,"Pseudosteady-State Analysis of Finite-Conductivity Vertical Fractures," SPE 95941, 2005.

23. Miskimins, J.L., Lopez-Hernandez, H.D., and Barree, R.D., "Non-Darcy Flow in Hydraulic Fractures: Does It Really Matter?" SPE Paper 96389, 2005.

24. Mukherjee, H., and Economides, M.J., "A Parametric Comparison of Horizontal and Vertical Well Performance," SPE Paper 18303, 1991.

25. Nolte, K.G., and Economides, M.J., "Fracture Design and Validation with Uncertainty and Model Limitations," *JPT,* 1147–1155 (September 1991).

26. Palacio, J.C., and Blasingame, T.A., "Decline-Curve Analysis Using Type Curves: Analysis of Gas Well Production Data," SPE Paper 25909, 1993.

27. Poe, B.D., Jr., and Economides, M.J., *Reservoir Stimulation,* 3rd ed., Wiley, New York, Chap. 12, 2000.

28. Prats, M., "Effect of Vertical Fractures on Reservoir Behavior—Incompressible Fluid Case," *SPEJ,* 105–118 (June 1961).

29. Song, B., Economides, M.J., and Ehlig-Economides, C., "Design of Multiple Transverse Fracture Horizontal Wells in Shale Gas Reservoirs," SPE Paper 140555, 2011.

30. Valko, P.P., and Amini, S., "The Method of Distributed Volumetric Sources for Calculating the Transient and Pseudosteady-State Productivity of Complex Well-Fracture Configurations," Paper presented at the SPE Hydraulic Fracturing Technology Conference, College Station, TX. Society of Petroleum Engineers, 2007.

31. Vincent, M.C., Pearson, C.M., and Kullman, J., "Non-Darcy and Multiphase Flow in Propped Fractures: Case Studies Illustrate the Dramatic Effect on Well Productivity," SPE Paper 54630, 1999.

32. Warpinski, N.R., Mayerhofer, M.J., Vincent, M.C., Cipolla, C.L., and Lolon, E.P., "Stimulating Unconventional Reservoirs: Maximizing Network Growth While Optimizing Fracture Conductivity," *JCPT,* **48**(10): 39–51 (October 2009).

33. Wattenbarger, R.A., El-Banbi, A.H., Villegas, M.E., and Maggard, J.B., "Production Analysis of Linear Flow into Fractured Tight Gas Wells," SPE Paper 39931, 1998.

34. Wei, Y., and Economides, M.J., "Transverse Hydraulic Fractures from a Horizontal Well," SPE Paper 94671, 2005.

35. Wolhart, S.L., Odegard, C.E., Warpinski, N.R., Waltman, C.K., and Machovoe, S.R., "Microseismic Fracture Mapping Optimizes Development of Low-Permeability Sands of the

Williams Fork Formation in the Piceance Basin," SPE Paper 95637, presented at the SPE Annual Technical Conference and Exhibition, Dallas, TX, October 9–12, 2005.

Problems

17-1 Calculate the fracture equivalent skin effect, s_f, for a fracture design resulting in length 500 ft and dimensioned conductivity, $k_f w$, equal to 1000 md-ft, but for a range of reservoir permeabilities from 0.001 to 100 md. If $r_e = 1500$ ft and $r_w = 0.328$ ft, calculate the corresponding productivity index increases.

17-2 Assume that the drainage is big enough to develop pseudoradial flow. For the range of permeabilities in problem 17-1 what should be the optimum fracture lengths if the $k_f w$ were kept constant and equal to 1000 md-ft? *Hint:* Use the results in Figure 17-7.

17-3 Calculate the fracture length and width that would provide the maximum productivity index after injecting 75,000 lb of proppant into a reservoir with $k = 3$ md, $h = 100$ ft, and a drainage area equal to 320 acres. Assume that the fracture height is 1.5 times the reservoir height. If $B\mu = 1$ cp-resbbl/STB, $r_w = 0.328$ ft, and $\Delta p = 1000$ psi, calculate the production before and after stimulation. First, use natural sand with $k_f = 60{,}000$ md. Repeat the calculation by increasing the mass of proppant sand to 150,000 and 500,000 lb. Also repeat the calculations using ceramic proppant ($k_f = 220{,}000$ md) and high-strength proppant ($k_f = 500{,}000$ md).

17-4 Develop a generalized graph of proppant number from 0.01 to 100 versus J_D for drainage aspect ratios of 0.1, 0.25, 0.5, 0.75 and 1.

17-5 Consider an oil well as in Appendix A, but with a reservoir permeability of 0.1 md and with a hydraulic fracture with a half length, x_f, of 700 feet. Assume an infinite conductivity fracture and forecast well performance for 3 years. The distance from the fracture to the drainage boundary perpendicular to the fracture is 2000 ft.

17-6 Perform a parametric study for a vertical gas well using UFD with the following data (you may not need everything. If anything is missing, make any assumptions you deem necessary):

- Reservoir permeability: $k = 0.01, 0.1, 1$ and 10 md, respectively
- Gas gravity: $\gamma = 0.7$
- Reservoir average pressure: $p = 4{,}500$ psi
- Flowing bottomhole pressure: $p_{wf} = 2{,}500$ psi
- Reservoir temperature: 180°F
- Wellbore radius: $r_w = 0.328$ ft
- Well drainage area: 320 acres
- Reservoir net height: $h = 100$ ft

Fracture treatment data:

- Proppant mass: 300,000 lbm
- Proppant specific gravity: 2.65
- Proppant porosity: 38%

- Proppant pack nominal permeability: $k_f = 220{,}000$ md
- Fracture height: $h_f = 150$ ft

Cooke constants:
$a = 1.54$
$b = 110{,}470$

Determine the dimensionless PI values, the optimum fracture geometries, and the non-Darcy effects reduction in the nominal proppant pack permeability.

17-7 Demonstrate the difference in the fracture geometry for the same permeabilities as in problem 17-6 if these were oil wells and not affected by non-Darcy effects. Comment on the geometry requirements for the gas wells.

17-8 Graph the folds of increase from fractured wells compared to nonfractured wells for both oil and gas reservoirs. Use the information and data in problem 17-6. Perform the study for the given permeability values. For the nonfractured oil well, use $J_D = 0.12$. The calculated J_D values for the fractured oil wells would suffice for the estimation of the folds of increase. For nonfractured gas wells, use Equations 4-53 and 4-54, allowing for reservoir turbulence. The reservoir porosity is equal to 0.2. For the fractured wells use the flow rate from the optimized well performance, the reduced proppant pack permeability, and the adjusted fracture geometry.

17-9 Two 2000-ft horizontal wells are intersected by transverse fractures. One is in a gas reservoir with $k = 0.01$ md and the other in $k = 10$ md, also gas. The mass of proppant for each treatment is 150,000 lb, and the proppant pack-permeability is 220,000 md. Other important variables are:

- Proppant specific gravity: 2.65
- Proppant pack porosity: 0.38
- Cooke constant *a:* 1.34
- Cooke constant *b:* 27539
- Well radius: $r_w = 0.328$ ft
- Net thickness: 50 ft
- Fracture height: 100 ft
- Reservoir temperature: 180°F
- Gas gravity: 0.596

Compare the cumulative production rate from all fractures with what a vertical well with fractures would have produced. Comment on your results.

Hint: Partition the drainage for one, two, three, and so on fractures. Assume the Reynolds number $N_{Re} = 20$; calculate the effective fracture permeability considering the effects of non-Darcy flow. Calculate the flow rate under optimal fracture geometry assuming that the flowing bottomhole pressure is 1500 psi and the average reservoir pressure is 3000 psi.

Calculate the Reynolds number under this flow rate and with the contact provided by the well and each fracture. An important issue is the reduced proppant pack permeability. Repeat the calculation iteratively until the calculated Reynolds number matches the assumed Reynolds number.

17-10 Forecast the production history for 3 years of production for a 4500-ft long well having ten equally spaced fractures, draining a region that is 5000-ft long, 2000-ft wide, and 200-ft thick. Use data given in Example 17-3, except use $k = 0.01$ md, $h = 200$ ft, and $\phi = 0.1$.

CHAPTER 18

The Design and Execution of Hydraulic Fracturing Treatments

18.1 Introduction

The previous chapter presented methods to determine what sort of hydraulic fracture would optimize well performance for a given reservoir, and methods to forecast well performance for assumed fracture conditions. In this chapter, we focus on the techniques needed to design and execute the hydraulic fracturing process to achieve the treatment goals. A fracturing treatment design begins with the desired fracture geometry and fracture conductivity. Then, appropriate fracture fluids and proppants are selected, and treatment volume, pumping schedule, and proppant loading schedule are designed to achieve the desired final fracture geometry.

The fracture treatment design process begins with a selection between a conventional, biwing fracture design and the planned creation of multiple fractures or, in some cases, the creation of multiple fractures coupled with activation of existing natural fractures. Except for very low–permeability reservoirs, biwing fracture design procedures are appropriate. For very low (microdarcy) and ultra-low (nanodarcy) reservoir permeabilities, the creation of high fracture density with multiple fractures is a critical part of the overall well completion and fracture treatment design. Finally, in low-permeability reservoirs having low contrast in horizontal stress and preexisting natural fracture networks, such as some shale formations, the goal of fracturing is to create complex fracture networks as a combination of new hydraulic fractures and stimulated natural fractures.

Models that can be applied to design conventional biwing fractures have been available for many years and are still useful for hydraulic fracture design. The created fracture geometry (height, width, length, and azimuth) can be approximated for given pumping conditions using analytical 2-D models of fracture propagation. These models combine elastic fracture mechanics, fluid transport in the fracture, fluid leakoff from the fracture, and material balances on the fluid and the proppant to calculate the created fracture geometry and the resulting proppant distribution. The classical Perkins-Kern-Nordgren (PKN); Kristianovich-Zheltov, Geertsma-deKlerk (KGD); and radial models are presented here.

The fracture geometry created is tailored by selecting the most appropriate fracture fluids and proppants. The physical properties of fracture fluids and proppants that must be known for fracture design are also reviewed here. Finally, modern methods for diagnosing the fracture created, often applied in real time, are presented.

18.2 The Fracturing of Reservoir Rock

Execution of a hydraulic fracture involves the injection of fluids at a pressure sufficiently high to cause tensile failure of the rock. At the fracture initiation pressure, often known as the "breakdown pressure," the rock opens. As additional fluids are injected, the opening is extended and the fracture propagates.

A properly executed hydraulic fracture results in a "path" connected to the well that has a much higher permeability than the surrounding formation. This path of large permeability (frequently five to six orders of magnitude larger than the reservoir permeability) is narrow but can be extremely long. Typical average widths of a hydraulic fracture for low-permeability reservoirs (e.g., 0.1 md or less) are of the order of 0.25 in. (or less), while the effective tip-to-tip length may be 3000 ft. For higher permeability reservoirs (e.g., 50 md), the fracture length may be as little as 25 ft but the width can be as much as 2 in.

A reservoir at depth is under a state of stress. This situation can be characterized by the stress vectors. In a geologically stable environment, three principal stresses can be identified. Their directions coincide with the directions where all shear stresses vanish. These are the vertical direction and two horizontal directions for minimum and maximum horizontal stresses.

A hydraulic fracture will be normal to the smallest of the three stresses because it will open and displace the rock against the least resistance. In the vast majority of reservoirs to be hydraulically fractured, the minimum horizontal stress is the smallest, leading to vertical hydraulic fractures (Hubbert and Willis, 1957). The direction (azimuth) of the hydraulic fracture plane will be perpendicular to the minimum horizontal stress direction.

The geometry of the fracture is affected by the state of stress and the rock properties. Fracture design for petroleum engineers must then take into account the natural state of the reservoir and rock and its influence on the fracture execution in attempting to design and execute the optimum stimulation treatment.

18.2.1 *In-Situ* Stresses

Formations at depth are subjected to a stress field that can be decomposed into its constituent vectors. The most readily understood stress is the vertical stress, which corresponds to the weight of the overburden. For a formation at depth *H*, the vertical stress, σ_v, is simply

$$\sigma_v = g \int_0^H \rho_f dH \tag{18-1}$$

where ρ_f is the density of the formations overlaying the target reservoir. This stress can be calculated from an integration of a density log. If an average formation density is used in lb/ft^3 and the depth is in ft, Equation (18-1) becomes

$$\sigma_v = \frac{\rho H}{144} \tag{18-2}$$

with σ_v in psi. For $\rho = 165$ lb/ft^3, which is the density of most sandstones, the vertical stress gradient is approximately 165/144 $\approx$ 1.1 psi/ft.

This stress is the *absolute stress,* and in the case of a porous medium, since the weight of the overburden will be carried by both the grains and the fluid within the pore space, the *effective stress,* σ'_v, is defined by

$$\sigma'_v = \sigma_v - \alpha p \tag{18-3}$$

where α is Biot's (1956) poroelastic constant, which for most hydrocarbon reservoirs is approximately equal to 0.7.

The vertical stress is *translated* horizontally through the Poisson relationship, which in its simplest expression has the form

$$\sigma'_H = \frac{\nu}{1-\nu}\sigma'_v \tag{18-4}$$

where σ'_H is the effective horizontal stress and ν is the Poisson ratio. This variable is a rock property. For sandstones it is approximately equal to 0.25, implying that the effective horizontal stress is approximately one-third the effective vertical stress.

The absolute horizontal stress, σ_H, would then be equal to the effective stress plus αp in the same manner as the relationship in Equation (18-3). The absolute horizontal stress decreases with fluid production.

The stress given by Equation (18-4) is not necessarily the one that would be in the reservoir because the Poisson translation implies free body expansion. Reservoirs are under tectonic stresses and also several geological processes may have intervened since the original deposition. Thus, while the absolute vertical stress is easy to measure and calculate, the horizontal stress is difficult to predict accurately. Fracture injection tests, known as "data fracs" or "minifracs," and their analysis are employed for such measurements.

Furthermore, the horizontal stress is not the same in all directions in the horizontal plane. Because of different tectonic components, there is a *minimum horizontal stress* and a *maximum horizontal stress.*

From the above, it is then obvious that three principal stresses can be identified in a formation, σ_v, $\sigma_{H,\min}$, and $\sigma_{H,\max}$. The fracture direction will be normal to the smallest of the three.

18.2.2 Breakdown Pressure

The magnitude of the breakdown pressure is characteristic of the values and the respective differences of the principal stresses, the tensile stress, and the reservoir pressure. An expression for the breakdown pressure has been given by Terzaghi (1923), and for a vertical well (i.e., coinciding with the direction of the vertical principal stress), this pressure, p_{bd}, is

$$p_{bd} = 3\sigma_{H,\min} - \sigma_{H,\max} + T_0 - p \tag{18-5}$$

where $\sigma_{H,\min}$ and $\sigma_{H,\max}$ are the minimum and maximum horizontal stresses, respectively, T_0 is the tensile stress of the rock, and p is the reservoir pressure.

For any other than the perfectly vertical direction, as in the case of deviated or horizontal wells, the breakdown pressure will be different than the one given by Equation (18-5), since there will be a nonvanishing shear stress component. This new breakdown pressure may be less, but usually it is more than the breakdown pressure for a vertical well (McLennan, Roegiers, and Economides, 1989).

Example 18-1 Calculation of Stresses versus Depth

Assume that a 75-ft sandstone formation is 10,000 ft deep. The rock density is 165 lb/ft^3, the poroelastic constant is 0.72, and the Poisson ratio is 0.25. Calculate and plot the absolute and effective vertical and minimum horizontal stresses.

The maximum horizontal stress is 2000 psi larger than the minimum horizontal stress. Plot the maximum stress also. Use hydrostatic reservoir pressure ($\rho_w = 62.4$ lb/ft^3).

Repeat the calculation, but plot the stress profile of the target interval and overlaying and underlaying shale layers, each 50 ft thick. Use first a Poisson ratio for the shales equal to 0.25 (as in the sandstone layer), and then repeat for $\nu = 0.27$ and 0.3.

Solution

From Equation (18-2) and $\rho = 165$ lb/ft^3, $\sigma_v = 1.15\,H$. Then from Equation (18-3),

$$\sigma'_v = 1.15H - \frac{(0.72)(62.4)(H)}{144} = 0.84H. \tag{18-6}$$

From Equation (18-4),

$$\sigma'_H = \frac{0.25}{1 - 0.25}(0.84H) = 0.28H \tag{18-7}$$

and therefore

$$\sigma_{H,\min} = 0.28H + \frac{(0.72)(62.4)(H)}{144} = 0.59H. \tag{18-8}$$

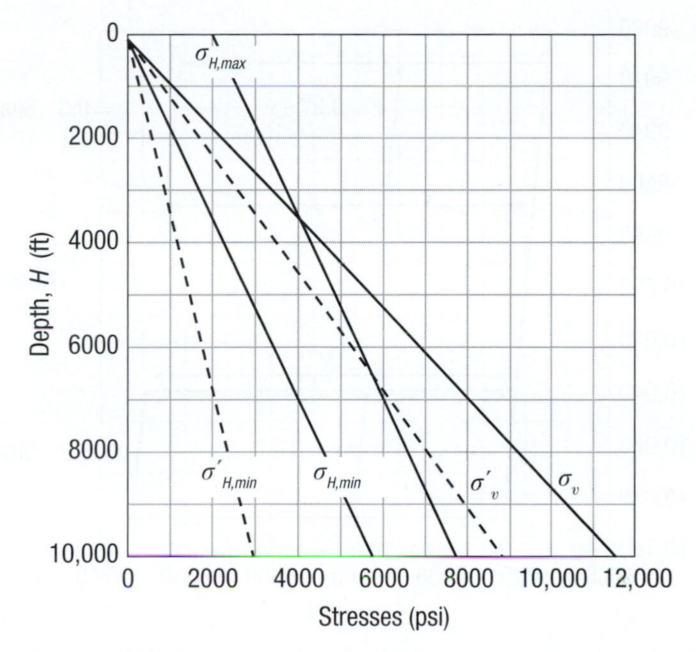

Figure 18-1 Absolute (σ_v, $\sigma_{H,max}$, $\sigma_{H,min}$) and effective (σ'_v, $\sigma'_{H,min}$) stresses for the formation of Example 18-1.

Figure 18-1 is a plot of all stresses: σ_v, σ'_v, $\sigma_{H,\min}$, $\sigma'_{H,\min}$, and $\sigma_{H,\max}$ ($=\sigma_{H,\min} + 2000$). At 10,000 ft these stresses are 11,500, 8,400, 5,900, 2,800, and 7,900 psi, respectively. Of the three principal absolute stresses, $\sigma_{H,\min}$ is the smallest, suggesting that at any depth (under the assumptions of this problem), a hydraulic fracture would be vertical and normal to the minimum horizontal stress direction.

If a depth of 10,000 ft is the midpoint of the 75-ft formation, the difference between the horizontal stresses at the top and bottom of the formation would be only 44 psi ($= 0.59\Delta H$; see Equation 18-8). When the two 50-ft shale layers are considered, the absolute vertical stresses at the top and bottom of the shale/sandstone and sandstone/shale sequence would be within ($1.15\Delta H = 1.15 \times 50$) 58 and ($1.15 \times 75 =$) 86 psi, respectively, and the absolute horizontal stresses would be within 30 and 44 psi (as per Equation 18-8), respectively (if $\nu = 0.25$ is used for all three layers). However, if $\nu = 0.27$ for the shales, while the horizontal stress in the sandstone remains the same, from Equation (18-4), and similarly to Equation (18-8), $\sigma_{H,\min} = 0.62\,H$, which at 10,000 ft results in an additional contrast of approximately 400 psi. For $\nu = 0.3$, this stress contrast is approximately 800 psi. A plot of the stress profiles in an expanded scale is shown in Figure 18-2.

The single most important reason for fracture height containment is this natural stress contrast resulting from the difference in the Poisson ratios. Without this difference, fractures would have largely uncontrolled height.

Fracture height migration and calculations are presented later in this chapter.

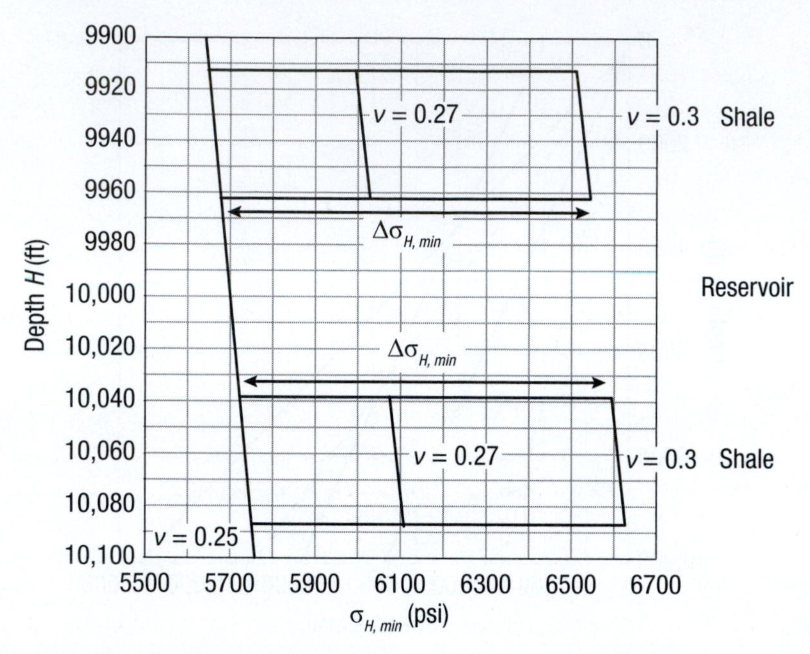

Figure 18-2 Idealized horizontal stress profiles and stress contrast between target, overlaying, and underlaying shales (Example 18-1).

Example 18-2 Calculation of Fracture Initiation Pressure

Estimate the fracture initiation pressure for a well in the reservoir described in the first paragraph of Example 18-1. The tensile stress is 1000 psi.

Solution

From Equation (18-5),

$$p_{\text{bd}} \approx 3(5900) - 7900 + 1000 - 4330 \approx 6500 \text{ psi}. \tag{18-9}$$

This breakdown pressure is bottomhole. The wellhead "treating" pressure, p_{tr}, would be

$$p_{\text{tr}} = p_{\text{bd}} - \Delta p_{\text{PE}} + \Delta p_F \tag{18-10}$$

where Δp_{PE} and Δp_F are the hydrostatic and frictional pressure drops, respectively.

18.2.3 Fracture Direction

Fracture direction is normal to the minimum resistance. This is represented by the absolute minimum stress.

In the previous section, Equations (18-1) and (18-4) and results of Example 18-1 (see Figure 18-1) suggest that the minimum horizontal stress is by definition smaller than the maximum horizontal stress and smaller than the vertical stress. Therefore, the conclusion would be that all hydraulic fractures should be vertical and normal to the minimum stress direction.

There are exceptions. In the previous section, the role of the overpressure on the magnitude of the horizontal stress was addressed. More important, though, is the magnitude of the vertical stress itself.

The Poisson relationship [Equation (18-4)] is valid during deposition, and the resulting horizontal stresses, contained within stiff boundaries, are "locked in place." The vertical stress, being directly proportional to the weight of the overburden, follows the geologic history (erosion, glaciation) of the top layers. Therefore, if ΔH is removed, the vertical stress at depth is $\rho g\,(H - \Delta H)$, where H is measured from the original ground surface. The slope of the vertical stress versus depth curve remains constant, but the curve is shifted to the left (of the curve in Figure 18-1).

While the intersection of the original vertical and minimum horizontal stresses was at the original ground surface, the removal of a portion of the overburden, coupled with largely constant values of the minimum horizontal stress, result in a new curve intersection, which marks a critical depth. Above this depth the original minimum horizontal stress is no longer the smallest of the three stresses. Instead, the vertical stress is the smallest and a hydraulic fracture would be horizontal, lifting the overburden.

Example 18-3 Calculation of Critical Depth for a Horizontal Fracture

Suppose that the formation described in the first paragraph of Example 18-1 lost 2000 ft of overburden over geologic history. Assuming that the horizontal stresses are the original and the reservoir pressure is hydrostatic, based on the original ground surface, calculate the critical depth for a horizontal fracture.

Solution

A graphical solution to the problem is shown in Figure 18-3. The new ground surface, 2000 ft below the original, is marked. A line parallel to the original vertical stress is drawn from the new zero depth, and it intersects the original minimum horizontal stress at $H = 4000$ ft from the original surface or 2000 ft from the current ground surface.

An algebraic solution is to set equal the stress values of the (new) vertical and minimum horizontal in their respective expressions and solve for H. If the original $\sigma_v = 1.15H$, after 2000 ft of overburden was removed, $\sigma_v = 1.15H - 2300$.

The horizontal stress is given by Equation (18-4) ($\sigma_{H,\min} = 0.59H$).

Setting the two stresses equal results in

$$0.59H = 1.15H - 2300. \tag{18-11}$$

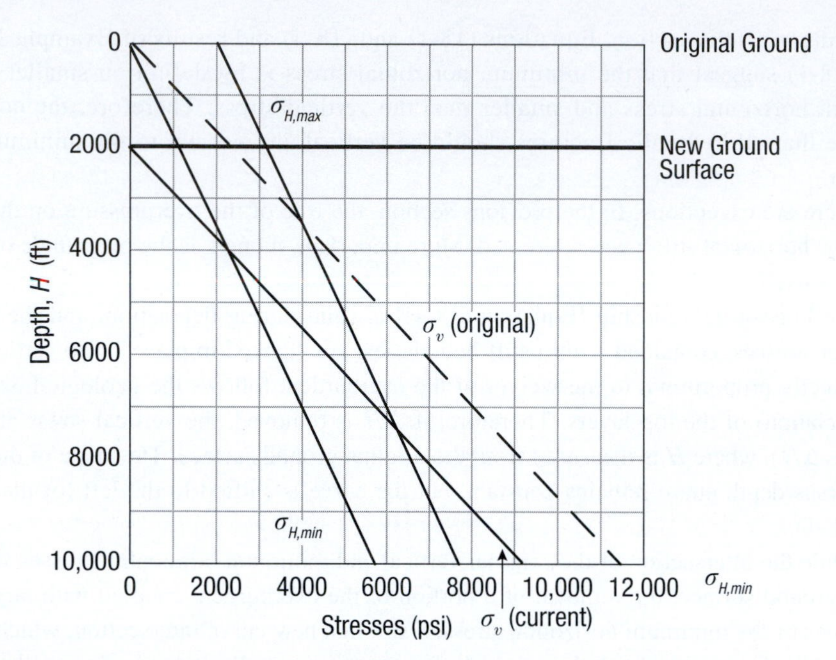

Figure 18-3 Critical depth for a horizontal fracture because of overburden removal (Example 18-3).

Therefore, at depths less than $H \approx 4100$ ft from the original ground surface or 2000 ft from the current surface, fractures will initiate horizontally.

Complex fracture geometries are likely to be encountered at depths greater than this level. Fractures may be *initiated* vertically but may turn horizontal (*T*-shape fractures) when (and if) the minimum stress plus the fracture net pressure exceed the overburden stress. A *T*-shape geometry is invariably undesirable, since the width of the horizontal branch is very small, resulting in substantial fluid loss during propped fracture propagation, slurry dehydration, and "screenouts." (A *screenout* is when there is not sufficient pressure to propagate the fracture because of extra resistance as the slurry cannot suspend the proppant. Other than the deliberate tip screenout [TSO], described in Section 18.3.4, all other involuntary screenouts are undesirable, resulting in the premature termination of a fracture treatment.)

Similar phenomena may occur during vertical fracture propagation if the values of the two horizontal stresses are nearly equal. The pressure in the fracture during execution may exceed the *seating* pressure of normal-to-the-trajectory natural fissures. Small widths will accept fluid but not proppant and increase the likelihood of a screenout.

The azimuthal direction of a vertical fracture is an important consideration when planning the locations of wells to be fractured, and when planning the direction to drill horizontal wells that will be fractured. A primary fracture propagates in the direction perpendicular to the minimum horizontal stress direction, and hence in the direction of the maximum horizontal stress.

The azimuthal direction of hydraulic fractures can sometimes be predicted from regional tectonics. For example, along a coastline in a sedimentary basin, normal faulting often occurs parallel to the coastline because the minimum horizontal stress is normal to the coastline. If the stress conditions that led to the normal faulting still exist, created hydraulic fractures should also be parallel to the coastline, and hence parallel to the normal faults in the region. Techniques to measure the azimuthal direction of hydraulic fractures are discussed in Section 18.7.

18.3 Fracture Geometry

Following fracture initiation, further fluid injection results in fracture propagation. Based on classical models, biwing fractures are created and the geometry of the created fracture can be approximated by taking into account the mechanical properties of the rock, the properties of the fracturing fluid, the conditions with which the fluid is injected (rate, pressure), and the stresses and stress distribution in the porous medium.

In describing fracture propagation, which is a particularly complex phenomenon, two sets of laws are required:

- Fundamental principles such as the laws of conservation of momentum, mass, and energy, and
- Criteria for propagation (i.e., what causes the tip of the fracture to advance). These include interactions of rock, fluid, and energy distribution. (See Ben-Naceur [1989] for an extensive treatment of the subject.)

Three general families of models are available: two-dimensional (2-D), pseudo-three-dimensional (p3-D) and, fully three-dimensional (3-D) models. The latter allow full three-dimensional fracture propagation with full two-dimensional fluid flow. The fracture is discretized, and within each block calculations are done based on the fundamental laws and criteria for propagation. The fracture is allowed to propagate laterally and vertically, and change plane of original direction, depending on the local stress distribution and rock properties. Such fully 3-D models require significant amounts of data to justify their use, are extremely calculation intensive, and are outside the scope of this textbook.

Two-dimensional models are closed-form analytical approximations assuming constant or some average fracture height. For petroleum engineering applications, three models have been used. For a fracture length much larger than the fracture height ($x_f \gg h_f$), the Perkins and Kern (1961) and Nordgren (1972), or PKN model, is an appropriate approximation. For $x_f \ll h_f$, the appropriate model has been presented by Khristianovic(h) and Zheltov (1955) and Geertsma and de Klerk (1969), frequently known as the KGD model. Finally, when the formation is thick enough or the fracture treatment is small enough that no vertical barriers to fracture growth are felt, the fracture created is approximately circular, and the radial or penny-shaped model is appropriate (Geertsma and de Klerk, 1969). In the modern fracturing context, the PKN model applies for deeply penetrating fractures appropriate in low-permeability reservoirs, and the KGD model relates better to short very high–conductivity fractures in high-permeability reservoirs. A limiting case, where $h_f = 2x_f$, is the radial or "penny-shape" model. The fracture height, h_f, used here is the dynamic value, that is, the fracture height at the time that the fracture length is equal to x_f.

For the purposes of this textbook the elegant 2-D models are used for approximate calculations of the fracture width and the fracture propagation pressure. Both Newtonian and non-Newtonian fracturing fluids are considered. For much greater exposure to models of fracture propagation, the reader is referred to Gidley, Holditch, Nierode, and Veatch (1989), Economides and Nolte (2000), and Economides and Martin (2007).

In formations that are naturally fractured and/or are composed of brittle rocks like shales, hydraulic fracturing with low viscosity fracture fluids (slickwater fracturing) in largely isotropic formations may create large networks of fractures, rather than single primary fractures emanating from the wellbore. This fracture pattern is called complex fracturing. Complex fracture geometries cannot be predicted with the simple models of fracture mechanics presented here.

18.3.1 Hydraulic Fracture Width with the PKN Model

The PKN model is depicted in Figure 18-4. It has an elliptical shape at the wellbore. The maximum width is at the centerline of this ellipse, with zero width at the top and bottom. For a *Newtonian* fluid, the maximum width, when the fracture half length is equal to x_f, is given (in coherent units and for a Newtonian fluid) by

$$w_{\max} = 3.27\left[\frac{q_i \mu x_f}{E'}\right]^{1/4} \tag{18-12}$$

where E' is the plane strain modulus and is related to Young's modulus, *E,* by

$$E' = \frac{E}{1 - \nu^2}. \tag{18-13}$$

In Equations (18-12) and (18-13), q_i is the injection rate, μ is the apparent fracturing fluid viscosity, and ν is the Poisson ratio. Equation (18-12) is particularly useful to understand the relationship among fracture width, treatment variables, and rock properties. The quarter-root relationship implies that to double the width, the viscosity of the fracturing fluid (or the injection rate) must be increased by a factor of 16.

On the other hand, rock properties have a large impact on the fracture width. The Young's modulus of common reservoir rocks may vary by almost two orders of magnitude, from 10^7 psi in tight and deep sandstones to 2×10^5 psi in diatomites, coals, and soft chalks. The difference in the fracture widths among these extremes even with unrestricted fracturing will be more than three times. The implication is that in stiff rocks, where the Young's modulus is large, for a given volume of fluid injected, the resulting fracture will be narrow but long. Conversely, in low-Young's-modulus formations, the same volume of fluid injected would result in wide but short

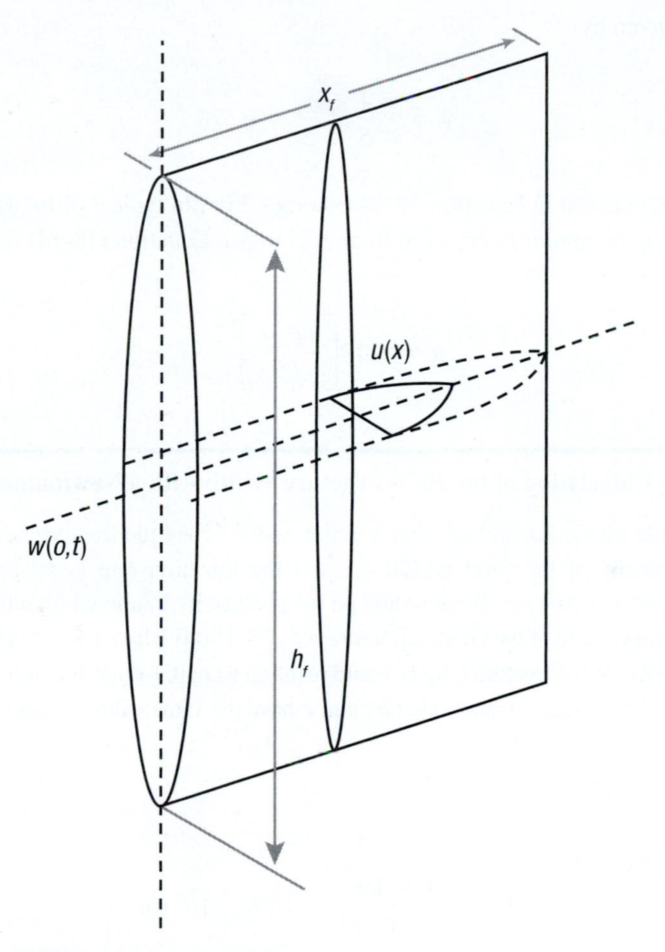

Figure 18-4 The PKN model geometry.

fractures. This is one of those phenomena where the natural state helps the success of fracture stimulation, since low-permeability reservoirs that require long fractures usually have large Young's modulus values.

The corollary is not always true. Low Young's moduli are not necessarily associated with higher-permeability formations, although in many cases this is true.

The elliptical geometry of the PKN model leads to an expression for the average width by the introduction of a geometric factor. Thus,

$$\overline{w} = 3.27\left[\frac{q_i \mu x_f}{E'}\right]^{1/4} \gamma. \tag{18-14}$$

The factor γ is given by

$$\gamma = \frac{\pi}{4}\frac{4}{5} = \frac{\pi}{5} = 0.628 \tag{18-15}$$

and therefore the constant in Equation 18-14 becomes 2.05. In typical oilfield units, where $\overline{w}$ is calculated in in., q_i in bpm, μ in cp, x_f in ft, and E' in psi, Equation (18-14) becomes

$$\overline{w} = 0.19\left[\frac{q_i \mu x_f}{E'}\right]^{1/4}. \tag{18-16}$$

Example 18-4 Calculation of the PKN Fracture Width with a Newtonian Fluid

What would be the maximum and average fracture widths when the fracture half length is 500 ft, the apparent viscosity of the fluid is 100 cp, and the injection rate is 40 bpm? Assume that $\nu = 0.25$ and $E = 4 \times 10^6$ psi. What would be the average hydraulic width when $x_f = 1000$ ft?

Calculate the volume of the created fracture if $h_f = 100$ ft when $x_f = 1000$ ft for the formation described above. What fracture length would result in a chalk formation with $E = 4 \times 10^5$ psi for the same fracture volume? Assume that h_f and ν have the same values as above.

Solution

From Equation (18-13),

$$E' = \frac{4 \times 10^6}{1 - 0.25^2} = 4.27 \times 10^6 \text{ psi} \tag{18-17}$$

and from Equation (18-16), when $x_f = 500$ ft,

$$\overline{w} = 0.19\left[\frac{(40)(100)(500)}{4.27 \times 10^6}\right]^{1/4} = 0.16 \text{ in.} \tag{18-18}$$

The maximum fracture width (at the wellbore), would then be (dividing by γ in Equation 18-13) 0.25 in.

When $x_f = 1000$ ft, the w_{max} and $\overline{w}$ are 0.30 and 0.19 in., respectively (simply the previous results multiplied by $2^{1/4}$).

The volume of the 1000-ft (half-length) fracture is

$$V = 2x_f h_f \overline{w} = (2)(1000)(100)\left(\frac{0.19}{12}\right) = 3170 \text{ ft}^3. \tag{18-19}$$

In the chalk formation, the product $x_f\overline{w}$ must also be 15.8 ft^2, since h_f has the same value. From Equation (18-72), $E' = 4.27 \times 10^5$ psi and rearrangement of Equation (18-16) results in

$$x_f = \left\{ \left[\frac{12(x_f\overline{w})}{0.19} \right]^4 \frac{E'}{q_i\mu} \right\}^{1/5} \tag{18-20}$$

and therefore $x_f = 638$ ft. The corresponding average width is 0.30 in.

18.3.2 Fracture Width with a Non-Newtonian Fluid

The expression for the maximum fracture width with a non-Newtonian fluid is (in oilfield units)

$$w_{max} = 12\left[\left(\frac{128}{3\pi}\right)(n'+1)\left(\frac{2n'+1}{n'}\right)^{n'}\left(\frac{0.9775}{144}\right)\left(\frac{5.61}{60}\right)^{n'}\right]^{1/(2n'+2)} \times \left(\frac{q_i^{n'}K'x_fh_f^{1-n'}}{E}\right)^{1/(2n'+2)} \tag{18-21}$$

where w_{max} is in inches. The average width can be calculated by multiplying the w_{max} by 0.628. The quantities n' and K' are the power-law rheological properties of the fracturing fluid. All other variables are as in Equation (18-14) for the Newtonian fluid.

Example 18-5 Fracture Width with a Non-Newtonian Fluid

Use the treatment and rock variables in Example 18-4 to calculate the fracture width at $x_f = 1000$ ft for a fracturing fluid with $K' = 8 \times 10^{-3}$ lb$_f$-sec$^{n'}$/ft^2 and $n' = 0.56$.

Solution

From Equation (18-21),

$$w_{max} = 12[(13.59)(1.56)(3.79)^{0.56}(0.0068)(0.0935)^{0.56}]^{0.32} \times \left[\frac{40^{0.56}(8 \times 10^{-3})(1000)(100)^{0.44}}{4 \times 10^6}\right]^{0.32} = 0.295 \text{ in.} \tag{18-22}$$

and therefore, multiplying by $\frac{\pi}{4}\gamma$, $\overline{w} = 0.18$ in.

18.3.3 Fracture Width with the KGD Model

The KGD model, depicted in Figure 18-5, is a 90° turn of the PKN model and is particularly applicable to approximate the geometry of fractures where $h_f \gg x_f$. Thus, it should not be used in cases where long fracture lengths are generated.

As can be seen from Figure 18-5, the shape of the KGD fracture implies equal width along the wellbore, in contrast to the elliptical shape (at the wellbore) of the PKN model. This width profile results in larger fracture volumes when using the KGD model instead of the PKN model for a given fracture length.

$$\overline{w} = 2.70\left[\frac{q_i \mu x_f^2}{E' h_f}\right]^{1/4}\left(\frac{\pi}{4}\right) \tag{18-23}$$

and in oilfield units, with $\overline{w}$ in in.,

$$\overline{w} = 0.34\left[\frac{q_i \mu x_f^2}{E' h_f}\right]^{1/4}\left(\frac{\pi}{4}\right) \tag{18-24}$$

Note: The value of γ is 1 for the KGD model.

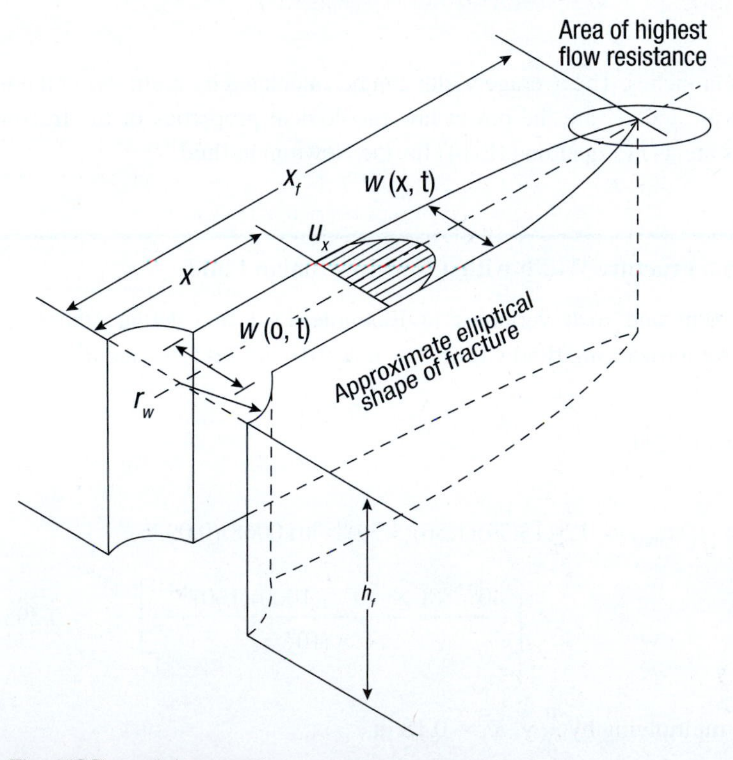

Figure 18-5 The KGD model geometry.

18.3.4 Fracture Width with the Radial Model

At early times during a fracturing treatment, the fracture is not confined by vertical barriers to fracture growth, and in a homogeneous formation, the fracture is approximately circular in shape when viewed from the side. When relatively small fracture treatments are applied in thick reservoirs, or in formations with little stress contract between layers to retard the vertical fracture growth, this circular geometry may persist throughout the entire fracturing process. Fractures of this nature are called radial or penny-shaped fractures. For this fracture geometry, the maximum fracture width is (Geertsma and DeKlerk, 1969)

$$w_{\max} = 2\left[\frac{q_i\mu(1-\nu)R}{G}\right]^{1/4}. \tag{18-25}$$

18.3.5 Tip Screenout (TSO) Treatments

The fracture geometry described by the PKN, KGD, and radial models is one that results from unrestricted fracturing. Starting in the early 1990's tip screenout (TSO) treatments became commonplace for medium- to higher-permeability reservoirs. The technique that has several variants, described in, among others, Economides and Martin (2007), is to arrest the fracture growth and then, subsequently, to continue injecting proppant in order to inflate the fracture width. As shown in Section 17.3, the productivity index is maximized for a given proppant volume by a fracture with a specified conductivity and half length. These can be achieved using the TSO technique to stop fracture propagation at the specified fracture half-length and inflate the fracture to the specified width after TSO.

One way to initiate TSO is to reduce the fracture width by injecting a low-viscosity fluid so that injection of the proppant stage will induce bridging at the fracture tip. Because leakoff could lead to fracture face damage, another method that does not rely on leakoff is to inject a special multidensity, multimeshed proppant stage that causes bridging through excessive pressure drop at the fracture tip. Even more sophisticated techniques involve the engineering of the slurry so that the fracture propagation self-stops because of excessive pressure drop along the created fracture. This subject is outside the scope of this textbook.

18.3.6 Creating Complex Fracture Geometries

In very low–permeability, largely stress-isotropic formations, particularly shales, satisfactory well performance depends on the creation of large networks of fractures, referred to as complex fracturing. To create complex fracture patterns, not only must fractures be created that propagate far away from the well, but in addition, secondary fractures perpendicular to the main fractures must be created, or opened in preexisting natural fracture systems. Warpinski, Mayerhofer, Vincent, Cipolla, and Lolon (2009) and King (2010) have reviewed the formation characteristics and fracturing practices that result in fracture complexity. These include:

1. *The presence of dense networks of natural fractures.* Complex fracture patterns are created by activating existing natural fracture systems with the propagating hydraulic fractures. Even when

the natural fractures are nonconductive as is typically the case in the Barnett shale, they are planes of weakness that are opened with sufficient net pressure in the fracture system. Conductivity can be created in these secondary fractures by shear slippage, even if no proppant is placed in them.

2. *Low horizontal stress contrast and high net pressures.* Secondary fractures orthogonal to the main flow direction are opened when the net pressure in the fracture exceeds the maximum horizontal stress. Obviously, the lower the contrast between the minimum and maximum horizontal stresses, the more likely it is that orthogonal fractures will open during the fracturing process. Very high pumping rates are commonly used to create high net pressures in the fractures. Microseismic monitoring and frac fluid production at offset wells have shown that orthogonal fractures can extend more than 1000 feet from the fractured wellbore.
3. *Low-viscosity fracture fluids.* Complex fracture patterns are most often created with "slickwater" fracturing fluids, water with friction-reducing polymers added to it. It is thought that low-viscosity fluid more easily flows into orthogonal, secondary fractures, creating more complexity than would be created with a viscous gel frac fluid.
4. *Rock brittleness.* Complex fracture networks are easier to create in brittle formations than in soft ones. To a first approximation, high Young's modulus is a measure of high brittleness in shales.
5. *Multiple stage fracturing with closely spaced stages.* Common practice in fracturing shale formations is to place multiple stages of fracture fluid and proppant from numerous perforated clusters along a horizontal well. The multiple fractures "interact and overlap," creating very complex fracture patterns. The high stress created by one fracture alters the stress conditions seen by subsequent fractures, reducing horizontal stress contract and thus promoting fracture complexity.

More details about how wells are completed and fracture treatments are pumped to create fracture complexity are presented in Section 18.8.

18.4 The Created Fracture Geometry and Net Pressure

18.4.1 Net Fracturing Pressure

The creation of a two-dimensional crack, with one dimension of largely infinite extent and the other of finite extent, *d,* has been described by Sneddon and Elliot (1946). The maximum width of the crack, which is proportional to this characteristic dimension, is also proportional to the net pressure ($p_f - \sigma_{\min}$) and inversely proportional to the plane strain modulus, E'.

Therefore the maximum width (at the wellbore) of a fracture during execution is given by

$$w_{\max} = \frac{2(p_f - \sigma_{\min})d}{E'}. \quad \textbf{(18-26)}$$

Equation 18-26 is particularly useful to understand the hydraulic fracture widths for different reservoirs and under different settings. For example, for the PKN model the characteristic dimension d is the fracture height h_f. During unrestricted fracturing a net pressure of 200 psi may be sufficient. Thus, for a tight reservoir with a stiff Young's modulus (e.g., $E = 5 \times 10^6$ psi) and a fracture height of 100 ft, Equation 18-26 would give a maximum fracture width equal to 0.1 in. But for a soft rock of a high-permeability reservoir, with Young's modulus 10 times smaller and under a TSO treatment with a net pressure equal to 1000 psi, the hydraulic width could approach 5 in. Very large fracture widths have been observed in high-permeability fracturing.

The average hydraulic width, $\overline{w}$, is

$$\overline{w} = \frac{\pi}{4}\gamma w_{\max}. \tag{18-27}$$

While for the PKN model the characteristic dimension d is the fracture height, h_f, for the KGD model, it is equal to the fracture length, tip to tip, $2x_f$. In Equation (18-27), the value of γ is 0.628 for the PKN model and 1 for the KGD model.

Nolte and Economides (1989) have shown that for a fracturing operation with efficiency $\eta(= V_f/V_i) \to 1$, the volume of the fracture, V_f, must be equal to the volume of fluid injected, V_i, and therefore

$$\overline{w}A_f = q_i t \tag{18-28}$$

where A_f is the fracture area and equal to $2x_f h_f$.

However, for $\eta \to 0$,

$$A_f = \frac{q_i\sqrt{t}}{\pi C_L r_p} \tag{18-29}$$

where C_L is the leakoff coefficient and r_p is the ratio of the permeable height to the fracture height. In a single-layer formation, the permeable height is the net reservoir thickness, h.

For $\eta \to 1$, Equation (18-28) leads to

$$x_f\overline{w} = \frac{q_i t}{2h_f}. \tag{18-30}$$

From Equation (18-21) (maximum width of the PKN model with non-Newtonian fluid flow) and the geometric factor for the average width, and using all multipliers of $x_f^{1/(2n'+2)}$ as a constant C_1 (without the factor 12 to convert width in in.), Equation (18-30) becomes

$$x_f(C_1 x_f^{1/(2n'+2)}) = \frac{5.615 q_i t}{2h_f} \tag{18-31}$$

where q_i is in bpm and t is in minutes. Therefore,

$$x_f = \frac{1}{C_1}\left(\frac{5.615 q_i t}{2h_f}\right)^{(2n'+2)/(2n'+3)}. \tag{18-32}$$

For $\eta \rightarrow 0$, x_f versus t can be obtained directly from Equation (18-29), and thus

$$x_f = \frac{5.615 q_i \sqrt{t}}{2\pi h C_L}. \tag{18-33}$$

The $h_f r_p$ product that would be in the denominator of Equation (18-33) becomes simply h.

From the Sneddon crack relationship [Equation (18-26) with $d = h_f$ for the PKN model], the net fracturing pressure is given by

$$\Delta p_f = p_f - \sigma_{\min} = \frac{C_1}{(\pi\gamma/4)} x_f^{1/(2n'+2)} \frac{E}{2(1-\nu^2)h_f}. \tag{18-34}$$

With Equation (18-32) for $\eta \rightarrow 1$ and using all multipliers of $t^{1/(2n'+3)}$ as C_2,

$$\Delta p_f = C_2 t^{1/(2n'+3)}. \tag{18-35}$$

Similarly, from Equations (18-33) and (18-34), for $\eta \rightarrow 0$,

$$\Delta p_f = C_3 t^{1/4(n'+1)} \tag{18-36}$$

where C_3 is the constant resulting from the combination of the equations.

Equations (18-32) and (18-33) for penetration and Equations (18-35) and (18-36) for net pressure represent the two extreme limits for $\eta \rightarrow 1$ and $\eta \rightarrow 0$, respectively.

The pressure expressions, in particular, are useful in monitoring the progress of a fracturing operation. Since η' is usually approximately equal to 0.5, the powers of time from Equations (18-35) and (18-36) should fall between 1/4 and 1/6. This observation was first made by Nolte and Smith (1981), who also presented analogous expressions for the KGD and Radial models. These are as follows:

KGD:

$$\Delta p_f \propto t^{-n'/(n'+2)} \qquad \text{for } \eta \rightarrow 1 \tag{18-37}$$

$$\Delta p_f \propto t^{-n'/2(n'+1)} \qquad \text{for } \eta \rightarrow 0 \tag{18-38}$$

Radial:

$$\Delta p_f \propto t^{-n'/(n'+2)} \qquad \text{for } \eta \rightarrow 1 \tag{18-39}$$

$$\Delta p_f \propto t^{-3n'/8(n'+1)} \qquad \text{for } \eta \rightarrow 0 \tag{18-40}$$

These results suggest that log–log plots during fracture execution can identify readily the morphology of the propagating fracture. In a plot of Δp_f versus time, positive slopes would suggest a normally extending, contained fracture, approximated by the PKN model. Negative slopes would imply either much larger fracture height than length growth, approximated by the KGD model, or radial fracture extension, approximated by the Radial model. This technique has been in wide use.

Finally, there are also approximate and easy-to-use expressions for the net fracturing pressure for the PKN and KGD models using a Newtonian fracturing fluid.

For the PKN model, in coherent units, this expression is

$$\Delta p_f = 1.37\left[\frac{E'^3 q_i \mu x_f}{h_f^4}\right]^{1/4}, \tag{18-41}$$

which in oilfield units becomes

$$\Delta p_f(\text{psi}) = 0.015\left[\frac{E'^3 q_i \mu x_f}{h_f^4}\right]^{1/4}. \tag{18-42}$$

For the KGD model, the analogous expression is

$$\Delta p_f(\text{psi}) = 0.030\left[\frac{G^3 q_i \mu}{(1-\nu)^3 h_f x_f^2}\right]^{1/4}. \tag{18-43}$$

Again, it can be concluded readily that for the PKN model [Equation (18-41)], as x_f grows, so does Δp_f. For the KGD model, as x_f grows, Δp_f declines [Equation (18-43)].

Example 18-6 Fracture Penetration and Net Pressure versus Time with the PKN Model

Calculate the fracture penetration and the corresponding net pressure for the two extreme cases (i.e., $\eta \rightarrow 1$ and $\eta \rightarrow 0$). Use $n' = 0.5$, $K' = 8 \times 10^{-3}$ $\text{lb}_f\text{-sec}^2/\text{ft}^2$, $q_i = 40$ bpm, $h = 100$ ft, $h_f = 150$ ft, $E = 2 \times 10^6$ psi, $\nu = 0.25$, and $C_L = 5 \times 10^{-3}$ $\text{ft}/\sqrt{\text{min}}$.

Plot the results of this calculation for the first 15 minutes.

Solution

From Equations (18-21) and (18-27) and the given variables, the constant C_1 in Equation (18-32) can be calculated:

$$C_1 = (0.628)(0.439)(6.8 \times 10^{-3}) = 1.87 \times 10^{-3}. \tag{18-44}$$

From Equation (18-32) for $\eta \rightarrow 1$,

$$x_f = 480t^{3/4}. \tag{18-45}$$

For $\eta \rightarrow 0$, from Equation (18-33),

$$x_f = 71.5t^{1/2}. \tag{18-46}$$

In both Equations (18-45) and (18-46), the time is in minutes.

Finally, Equation (18-35) for $\eta \rightarrow 1$ becomes

$$\Delta p_f = 171t^{1/4} \tag{18-47}$$

and for $\eta \rightarrow 0$, Equation (18-36) becomes

$$\Delta p_f = 87.9t^{1/6}. \tag{18-48}$$

Figure 18-6 is the plot of the results of this exercise. These net pressures increase with time, as should be expected from the PKN model. Of particular interest are the calculated fracture penetrations showing the enormous impact of the efficiency on fracture propagation. Thus, control of fluid leakoff should be of paramount importance for an effective hydraulic fracture treatment.

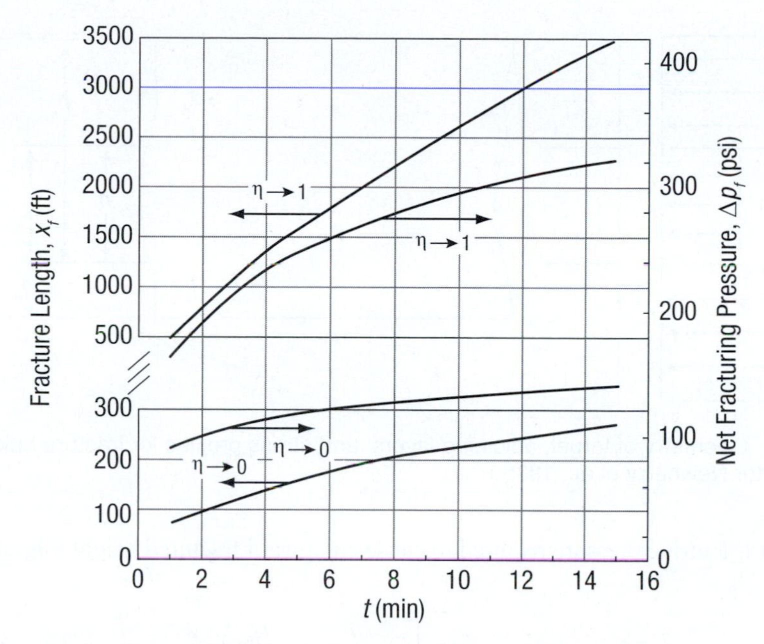

Figure 18-6 Penetration and net fracturing pressure using the PKN model for efficiency approaching 0 and 1 (Example 18-6).

18.4.2 Height Migration

An appropriate p3-D model would allow simultaneous lateral and vertical fracture height migration. An approximation for the fracture height at the wellbore (where it would have the maximum value) is presented below. The height prediction is based on log-derived mechanical properties and the wellbore net fracturing pressure.

In Section 18.2.2, the distribution of horizontal stresses along the vertical column was discussed. It was suggested that because different lithologies have different Poisson ratios, the vertical stress (weight of the overburden) is translated horizontally unevenly, resulting in stress contrasts between layers. Example 18-1 showed several hundred psi differences between the horizontal stress of a target sandstone and overlaying or underlaying shales.

A simple model by Simonson (1978) relates this stress contrast, net fracturing pressure, and fracture height migration at the wellbore. Additionally, the effects of the interlayer critical stress intensity factor, K_{IC} (fracture toughness), and gravity have been incorporated. The latter was done by Newberry, Nelson, and Ahmed (1985). Figure 18-7 is a schematic of the model. The horizontal stress value in the target layer of thickness, h, is σ. The overlaying layer has a stress σ_u, and the underlaying layer has a stress σ_d. Upward fracture migration h_u is measured from the bottom of the reservoir, and downward migration h_d is measured from the top of the reservoir.

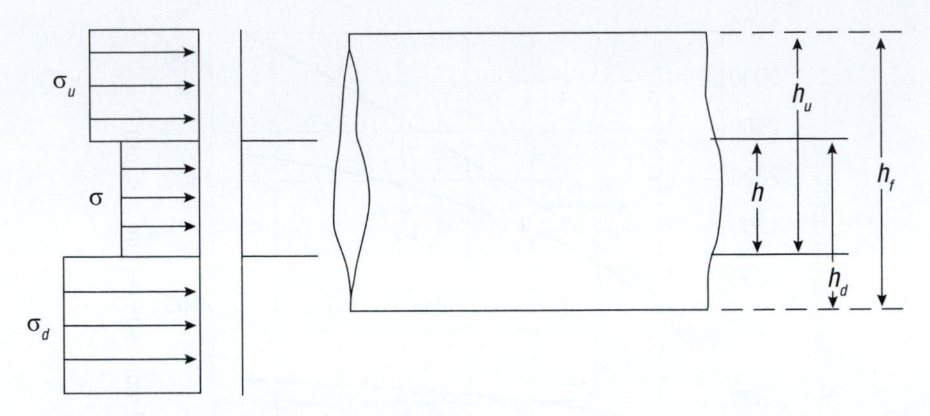

Figure 18-7 Geometry of target, adjoining layers, and stress profiles for fracture height migration. (After Newberry et al., 1985.)

The net fracturing pressure required to cause an upward fractured height migration h_u is

$$\Delta p_f = \frac{C_1}{\sqrt{h_u}}\left[K_{IC}\left(1 - \sqrt{\frac{h_u}{h}}\right) + C_2(\sigma_u - \sigma)\sqrt{h_u}\cos^{-1}\left(\frac{h}{h_u}\right)\right] + C_3\rho(h_u - 0.5h) \tag{18-49}$$

Similarly, the net fracturing pressure required for a downward height migration h_d is

$$\Delta p_f = \frac{C_1}{\sqrt{h_d}}\left[K_{IC}\left(1 - \sqrt{\frac{h_d}{h}}\right) + C_2(\sigma_d - \sigma)\sqrt{h_d}\cos^{-1}\left(\frac{h}{h_d}\right)\right] - C_3\rho(h_d - 0.5h) \tag{18-50}$$

In the expressions above, the contribution from the interlayer stress contrast (second term on the right-hand side) for almost all reservoirs is the largest. The first term, that of the critical stress intensity factor, contributes only a small amount. Finally, in upward migration, gravity effects are retarding, while in downward migration they are accelerating.

The constants C_1, C_2, and C_3 for oilfield units (σ, σ_u, σ_d in psi, h, h_u, h_d in ft, ρ in lb/ft^3, and K_{IC} in psi/$\sqrt{\text{in.}}$) are 0.163, 3.91, and 0.0069, respectively. The calculated net fracturing pressure would be in psi.

Finally, the inverse cosines must be evaluated in degrees.

Example 18-7 Estimation of Fracture Height Migration

A sandstone, 75 ft thick, has a minimum horizontal stress equal to 7100 psi. Overlaying and underlaying shales have stresses equal to 7700 and 8100 psi, respectively. $K_{IC} = 1000$ psi/$\sqrt{\text{in.}}$ and $\rho = 62.4$ lb/ft^3. Calculate the fracture height migration for net fracturing pressures equal to 200 and 500 psi.

If the overlaying layer is 100-ft thick, what would be the critical net fracturing pressure above which an undesirable breakthrough into another permeable formation may occur?

Solution

Figure 18-8 is a plot of $\Delta h_{u,d}$ ($= h_{u,d} - h$) versus Δp_f. An example calculation is presented below for the 100-ft migration above the target.

Since $\Delta h_u = 100$, then $h_u = 175$ ft and, from Equation (18-49),

$$\Delta p_f = \frac{(0.0217)}{\sqrt{175}}\left[(1000)\left(1 - \sqrt{\frac{175}{75}}\right) + (0.515)(7700 - 7100)\sqrt{175}\cos^{-1}\left(\frac{75}{175}\right)\right] + (0.0069)(55)[175 - (0.5)(75)]$$
$$= -6.5 + 431 + 59 = 484 \text{ psi.} \tag{18-51}$$

This is the critical net fracturing pressure that must not be exceeded during the treatment.

The three terms calculated above denote the relative contributions from the critical stress intensity factor, interlayer stress contrast, and gravity. For the type of stress contrast in this problem, the first term can be neglected and the gravity effects can be ignored for about a 10% error.

From Figure 18-8, the fracture height migrations above the reservoir for 200 and 500 psi net pressure are 9 and 110 ft, respectively. In view of the calculation for 100-ft migration, the second net pressure would result in fracture growth into another permeable formation, usually an undesirable occurrence.

Fracture migrations downward for 200 and 500 psi would be 6 and 35 ft, respectively.

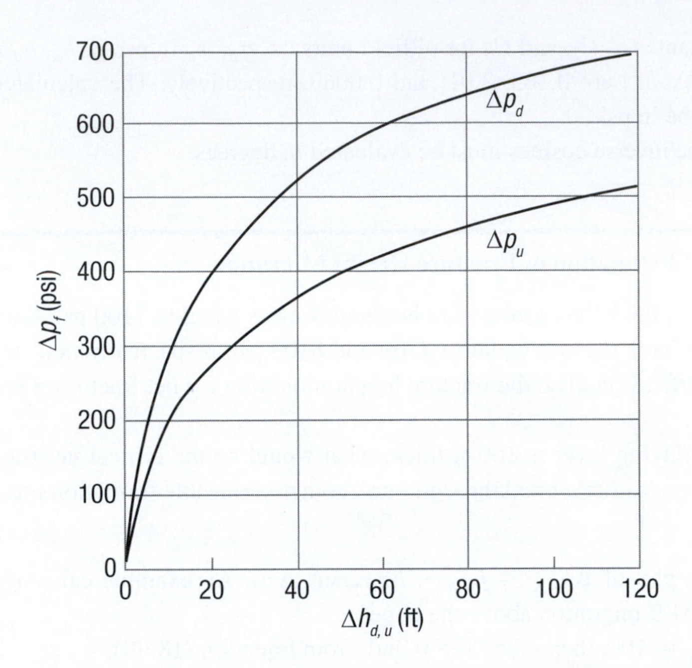

Figure 18-8 Pressures for upward and downward fracture height migration (Example 18-7).

18.4.3 Fluid Volume Requirements

A fracture execution consists of certain distinct fluid stages, each intended to perform a significant and specific task.

Pad is fracturing fluid that does not carry proppant. It is intended to initiate and propagate the fracture. During the fracture propagation, fluid *leakoff,* into the reservoir and normal to the created fracture area, is controlled through the buildup of a wall filtercake and, of course, also by the reservoir permeability. The volume of fluid leaking off is proportional to the square root of the residence time within the fracture. Therefore, pad, being the first fluid injected, acts as a sacrificial agent to the following proppant-carrying slurry.

After the pad injection, *proppant* is added to the fracturing fluid in increasing concentrations until at the end of the treatment the slurry concentration reaches a predetermined value. This value depends on the proppant-transporting abilities of the fluid and/or the capacity the reservoir and the created fracture can accommodate.

In general, excessive fluid leakoff may be caused by reservoir heterogeneities, such as natural fissures. Another problem may be encountered as a result of fracture height migration. Breaking through a thin layer that separates two permeable formations is likely to create a narrow opening (remember Sneddon's relationship; for a shale with larger horizontal stress, the *net* pressure will be smaller and hence the fracture width will be smaller). This narrow opening can allow fluid to escape leaving proppant behind. These phenomena may result in excessive slurry dehydration and a "screenout." As mentioned earlier, this term refers to an inability of the slurry to transport the proppant, and it leads to an excessive pressure increase that prevents further lateral fracture growth.

The created hydraulic fracture length differs from the propped length because proppant cannot be transported beyond the point where the fracture width is smaller than three proppant diameters.

An approximation of the relationship between total fluid volume requirements, V_i, and the volume that is pad, V_{pad}, based on the fluid efficiency, η, was given by Nolte (1986) and Meng and Brown (1987):

$$V_{\mathrm{pad}} \approx V_i\left(\frac{1-\eta}{1+\eta}\right). \tag{18-52}$$

An *over flush* is intended to displace the slurry from the well into the fracture. It should be less than well volume, because over-displacement would push the proppant away from the well and a "choked" fracture would result after the fracture pressure dissipates and the fracture closes. This should be a major concern of the stimulation treatment and should be avoided at all costs.

A material balance between total fluid injected, created fracture volume V_f, and fluid leakoff V_L can be written as

$$V_i = V_f + V_L. \tag{18-53}$$

Equation (18-53) can be expanded further by introducing constituent variables:

$$q_i t_i = A_f \overline{w} + K_L C_L (2A_f) r_p \sqrt{t_i} \tag{18-54}$$

where q_i is the injection rate, t_i is the injection time, A_f is the fracture area, C_L is the leakoff coefficient, and r_p is the ratio of the net to fracture height (h/h_f). The variable K_L is known as the *opening time distribution factor* and is related to the fluid efficiency by (Nolte, 1986)

$$K_L = \frac{1}{2}\left[\frac{8}{3}\eta + \pi(1-\eta)\right]. \tag{18-55}$$

The fracture area in the leakoff term is multiplied by 2 to accommodate both sides of the fracture. The fracture area, A_f, is simply equal to $2x_f h_f$.

For a given fracture length, the average hydraulic width, $\overline{w}$, can be calculated under the assumption of a fracture model [e.g., Equation (18-21) for the PKN model and a non-Newtonian fluid]. Knowledge of the fracture height, the leakoff coefficient, and the fluid efficiency would readily allow an inverse calculation using Equation (18-54). This is a quadratic equation, and it can provide the required time to propagate a fracture of certain length (and implied width) while undergoing the penalty of fluid leakoff. Of the two solutions for the square root of time, one will be positive and the other negative. Squaring the positive solution would result in the calculation of the total injection time, t_i, and the product $q_i t_i$ is equal to the required total fluid (pad and proppant slurry) volume.

Since the portion of the total fluid volume that is pad can be calculated from Equation (18-52), the onset of proppant addition can be obtained readily:

$$t_{\text{pad}} = \frac{V_{\text{pad}}}{q_i}. \quad \text{(18-56)}$$

The leakoff coefficient, C_L, in the material balance of Equation 18-54 can be obtained from a fracture calibration treatment as described by Nolte and Economides (1989).

Example 18-8 Total Fluid and Pad Volume Calculations

In Example 18-5, the average fracture width of a 1000-ft fracture was calculated to be equal to 0.18 in. Assuming that $r_p = 0.7$ (i.e., $h = 70$ ft), calculate the total volume and pad portion requirements. Use $C_L = 2 \times 10^{-3}$ ft/$\sqrt{\text{min}}$. Repeat the calculation for a small leakoff coefficient (2×10^{-4} ft/$\sqrt{\text{min}}$). The injection rate, q_i, is 40 bpm.

Solution

1. $C_L = 2 \times 10^{-3}$ ft/$\sqrt{\text{min}}$. The fracture area is $(2)(100)(1000) = 2 \times 10^5$ ft^2. Then, from Equation (18-54), with proper unit conversions, and assuming that $K_L = 1.5$,

$$(40)(5.615)t_i = 2 \times 10^5 \left(\frac{0.18}{12}\right) + (1.5)(2 \times 10^{-3})(4 \times 10^5)(0.7)\sqrt{t_i} \quad \text{(18-57)}$$

or

$$t_i - 3.74\sqrt{t_i} - 13.4 = 0, \quad \text{(18-58)}$$

which leads to $t_i = 36$ min.

The total volume required is then

$$V_i = (40)(42)(36) = 6 \times 10^4 \text{ gal}, \quad \text{(18-59)}$$

leading to an efficiency, η, of

$$\eta = \frac{V_f}{V_i} = \frac{(2 \times 10^5)(0.18/12)(7.48)}{6 \times 10^4} = 0.37 \quad (18\text{-}60)$$

where 7.48 is a conversion factor. From Equation (18-55), $K_L = 1.48$, justifying the assumption of $K_L = 1.5$. This calculation could be done by trial and error, although simple observation of Equation (18-55) suggests that K_L is bounded between 1.33 and 1.57 and the use of $K_L \simeq 1.5$ is generally appropriate. From Equation (18-52), the pad volume, V_{pad}, is

$$V_{pad} = (6 \times 10^4)\left(\frac{1 - 0.37}{1 + 0.37}\right) = 2.76 \times 10^4 \text{ gal}, \quad (18\text{-}61)$$

representing 46% of the total volume. At an injection rate of 40 bpm, it would require 17 min of pumping.

2. $C_L = 2 \times 10^{-4}$ ft/$\sqrt{\text{min}}$ The quadratic equation is

$$t_i - 0.374\sqrt{t_i} - 13.4 = 0 \quad (18\text{-}62)$$

and therefore $t_i = 15$ min.
The total volume requirements are

$$V_i = (40)(42)(15) = 2.5 \times 10^4 \text{ gal}. \quad (18\text{-}63)$$

The efficiency is

$$\eta = \frac{V_f}{V_i} = \frac{(2 \times 10^4)(0.18/12)(7.48)}{2.5 \times 10^4} = 0.9 \quad (18\text{-}64)$$

and the pad volume is

$$V_{pad} = (2.5 \times 10^4)\left(\frac{1 - 0.9}{1 + 0.9}\right) = 1.3 \times 10^3 \text{ gal}, \quad (18\text{-}65)$$

representing 5% of the total injected volume and requiring less than 1 min of pumping.

From these calculations it is evident how important the leakoff coefficient is on the portion of the total fluid that must be injected before proppant is added to the slurry. Unavoidably, this has a major impact on the total amount of proppant that can be injected and the resulting propped fracture width for a given fracture length.

The fracture volume balance equation (Equation 18-54) can be combined with a width equation for one of the 2-D fracture models (PKN, KGD, or Radial) to calculate fracture dimensions of length and width as functions of injected volume. Because the fracture width depends explicitly on fracture length, but not on injected volume, it is most convenient to assume a fracture length, then calculate the corresponding fracture width, and finally the volume of fluid pumped to create the described fracture geometry.

Example 18-9 Evolving Fracture Geometry

For the fracturing conditions described in Example 18-4, calculate the fracture length and width as functions of time up to a fracture length of 1000 ft. Assume the PKN model describes the fracture being created. In addition, assume that the leakoff coefficient is 0.001 ft/min$^{1/2}$.

Solution

The relevant formation properties and pumping conditions from Example 18-4 are $\nu = 0.25$, $E' = 4.27 \times 10^6$ psi, $h_f = 100$ ft, $q_i = 40$ bpm, and $\mu = 100$ cp. The calculation procedure is to first assume a value for fracture length, x_f. Then, the average fracture width is calculated with Equation 18-16. The length and fracture widths are then used in Equation 18-54 to calculate the pumping time required to create this geometry. In this calculation, K_L is first assumed to be 1.5. Then, knowing the total volume pumped (equal to $q_i t_i$) and the fracture volume, the efficiency is calculated and K_L checked with Equation 18-55. If K_L differs significantly from 1.5, the new value of K_L is used to calculate a new estimate of t_i. This procedure is continued iteratively until convergence is achieved.

First, assume a fracture length of 50 ft. From Equation 18-16,

$$\overline{w} = 0.19\left[\frac{(40)(100)(50)}{4.27 \times 10^6}\right]^{1/4} = 0.088 \text{ in.} \tag{18-66}$$

Then, using these values for fracture length and width, and assuming $K_L = 1.5$ and $r_p = 1$ in Equation 18-54 yields

$$(40)(5.615)t_i = 2(100)(50)\left(\frac{0.088}{12}\right) + (1.5)(1 \times 10^{-3})(4)(100)(50)\sqrt{t_i} \tag{18-67}$$

or

$$t_i - 0.1336\sqrt{t_i} - 0.326 = 0 \tag{18-68}$$

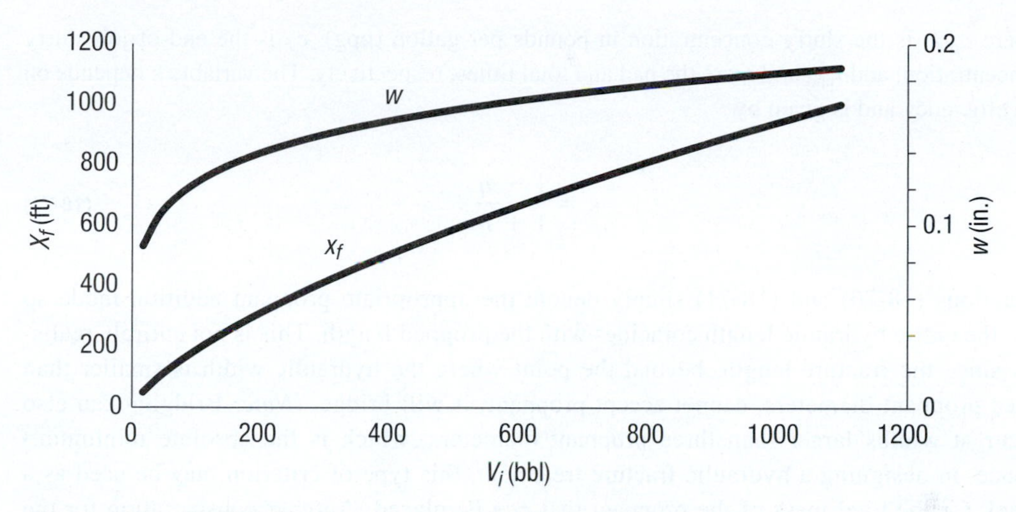

Figure 18-9 Predicted fracture geometry (Example 18-9).

So t_i is 0.41 minutes. Dividing the fracture volume by the injected volume at this time, the efficiency is 0.79 and Equation 18-55 gives

$$K_L = \frac{1}{2}\left[\frac{8}{3}(0.79) + \pi(1 - 0.79)\right] = 1.38. \quad \textbf{(18-69)}$$

Repeating the calculation of t_i using $K_L = 1.38$ yields an injection time of 0.406. Another check of K_L finds that 1.38 is the converged value.

This calculation procedure is repeated for different values of fracture length up to 1000 ft. The fracture length and width as functions of time are shown in Figure 18-9.

18.4.4 Proppant Schedule

Proppant addition, its starting point, and at what concentrations it is added versus time depend on the fluid efficiency. In the previous section the onset of proppant addition was determined after the pad volume was estimated [Equation (18-52)].

Nolte (1986) has shown that, based on a material balance, the continuous proppant addition, "ramped proppant schedule" versus time, should follow a relationship expressed by

$$c_p(t) = c_f\left(\frac{t - t_{\text{pad}}}{t_i - t_{\text{pad}}}\right)^{\varepsilon} \quad \textbf{(18-70)}$$

where $c_p(t)$ is the slurry concentration in pounds per gallon (ppg), c_f is the end-of-job slurry concentration, and t_{pad} and t_i are the pad and total times, respectively. The variable ε depends on the efficiency and is given by

$$\varepsilon = \frac{1 - \eta}{1 + \eta}. \tag{18-71}$$

Equations (18-70) and (18-71) simply denote the appropriate proppant addition mode so that the entire hydraulic length coincides with the propped length. This is not entirely realistic, since the fracture length, beyond the point where the hydraulic width is smaller than three proppant diameters, cannot accept proppant; it will bridge. (*Note:* Bridging can also occur at widths larger than three proppant diameters, which is the absolute minimum.) Hence, in designing a hydraulic fracture treatment, this type of criterion may be used as a check for the total mass of the proppant that *can* be placed. Another consideration for the end-of-job slurry concentration, c_f, is the proppant-transporting ability of the fracturing fluid. Certainly, in all cases the calculated average propped width cannot exceed the average hydraulic width.

Example 18-10 Determination of Proppant Schedule

Assume that the total injection time, t_i, is 245 min, and for efficiency $\eta = 0.4$, the pad injection time, t_{pad}, is 105 min. If the end-of-job slurry concentration, c_f, is 3 ppg, plot the continuous proppant addition schedule.

Solution

From Equation (18-71) and $\eta = 0.4$,

$$\varepsilon = \frac{1 - 0.4}{1 + 0.4} = 0.43 \tag{18-72}$$

and from Equation (18-128) and $c_f = 3$ ppg,

$$c_p(t) = 3\left(\frac{t - 105}{245 - 105}\right)^{0.43}. \tag{18-73}$$

For example, at $t = 150$ min, $c_p(t) = 1.84$ ppg. Of course, at $t = 105$ min, $c_p(t) = 0$, and at $t = 245$ min, $c_p(t) = 3$ ppg.

Figure 18-10 is a plot of the injection with the onset of proppant addition and proppant schedule.

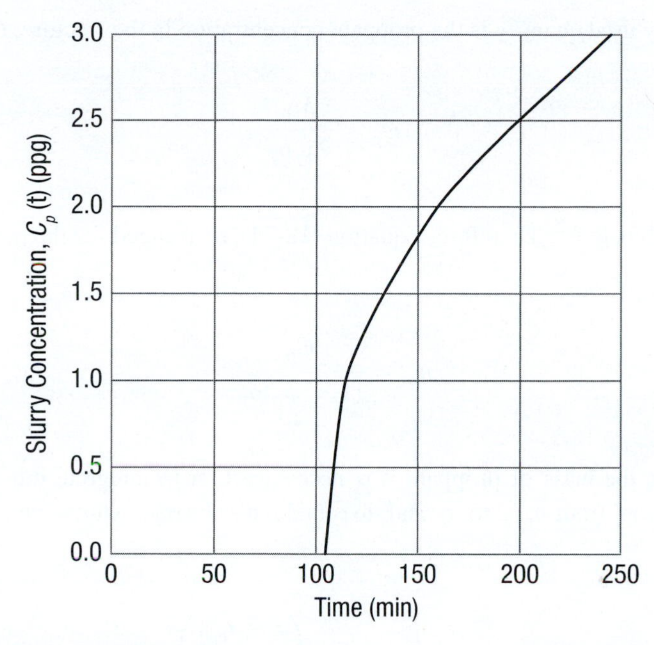

Figure 18-10 Onset of proppant slurry and continuous proppant addition (Example 18-10).

18.4.5 Propped Fracture Width

As was described in Sections 17.2 and 17.3, the propped width of the fracture, along with the fracture length, control post-treatment production. The dimensioned fracture conductivity is simply the product of the propped width and the proppant pack permeability. The dimensionless conductivity was given by Equation (17-2). The width in that expression is the propped width of the fracture.

As should be obvious from the last two sections, the relationship between hydraulic width and propped width is indirect; it depends greatly on the fluid efficiency and especially on the possible end-of-job proppant concentration.

Assuming that a mass of proppant, M_p, has been injected into a fracture of half length x_f and height h_f and the proppant is uniformly distributed, then

$$M_p = 2x_f h_f w_p (1 - \phi_p)\rho_p \tag{18-74}$$

where the product $2x_f h_f w_p (1 - \phi_p)$ represents the volume of the proppant pack and is characteristic of the proppant type and size. The density ρ_p is also a characteristic property of the proppant.

A frequently used quantity is the proppant concentration in the fracture, C_p, defined as

$$C_p = \frac{M_p}{2x_f h_f} \tag{18-75}$$

and the unit for C_p is lb/ft^2. Therefore, Equation (18-74), rearranged for the propped width, w_p, leads to

$$w_p = \frac{C_p}{(1 - \phi_p)\rho_p}. \tag{18-76}$$

To calculate the mass of proppant it is necessary first to integrate the ramped proppant schedule expression from t_{pad} to t_i and to obtain an average slurry concentration. From Equation (18-70),

$$\bar{c}_p = \frac{1}{t_i - t_{\text{pad}}} \int_{t_{\text{pad}}}^{t_i} c_f \left(\frac{t - t_{\text{pad}}}{t_i - t_{\text{pad}}} \right)^{\varepsilon} dt, \tag{18-77}$$

leading to

$$\bar{c}_p = \frac{c_f}{\varepsilon + 1}(1 - 0) = \frac{c_f}{\varepsilon + 1}. \tag{18-78}$$

The mass of the proppant would then be

$$M_p = \bar{c}_p(V_i - V_{\text{pad}}). \tag{18-79}$$

Equations (18-75) through (18-79) are sufficient to calculate the average propped width of a fracture.

Example 18-11 Calculation of the Propped Width

Suppose that 20/40 mesh sintered bauxite is injected ($\phi_p = 0.42$ and $\rho_p = 230$ lb/ft^3) into a fracture designed to have $x_f = 1000$ ft and $h_f = 150$ ft. If $c_f = 3$ ppg and $\varepsilon = 0.43$ (see Example 18-10), calculate the total mass of proppant, the propped width, and the proppant concentration in the fracture. The volume excluding the pad is $(4.12 \times 10^5) - (1.76 \times 10^5) = 2.36 \times 10^5$ gal.

Solution

The average slurry concentration can be calculated from Equation (18-78):

$$\bar{c}_p = \frac{c_f}{1 + \varepsilon} = \frac{3}{1.43} = 2.1 \text{ ppg.} \tag{18-80}$$

From Equation (18-79), the mass of proppant can be determined:

$$M_p = (2.1)(2.36 \times 10^5) = 4.9 \times 10^5 \text{ lb.} \tag{18-81}$$

The proppant concentration in the fracture, C_p, is [from (Equation 18-75)]

$$C_p = \frac{4.9 \times 10^5}{(2)(1000)(150)} = 1.63 \text{ lb/ft}^2. \tag{18-82}$$

Finally, from Equation (18-76),

$$w_p = \frac{1.63}{(1 - 0.42)(230)} = 0.012 \text{ ft} = 0.15 \text{ in.} \tag{18-83}$$

As discussed in Section 17.3, for each fracture job, there exists an optimal combination of fracture width and length that yields the highest productivity for the fractured well. In the course of fracture design, the fracture width and length should always attempt to match the optimal width and length as much as possible for a given job size (total volume of fluid and proppant). Realizing that design parameters are limited (injection time and rate, fluid and proppant selection, etc.), this optimal condition may not be achieved practically or economically, although it should always be the guideline for fracture design and execution.

The procedure to optimize a fracture design starts from calculating the optimal C_{fD} and maximum J_D from Equations 17-19 and 17-18 and then the optimal fracture length and width from Equation 17-20. Once the optimal geometry is determined, with the optimal fracture width for a given reservoir and proposed injection rate, the hydraulic fracture width can be obtained (Equation 18-16 for PKN model, for example). With the calculated width, injection time (Equation 18-54), and total volume and pad volume (Equation 18-52) now are available. Total proppant mass for a defined proppant is evaluated by Equation 18-79 and propped fracture width is from Equations 18-75 and 18-76. At this point, the propped width calculated should be compared with the width at the optimal condition and effort should be made to match the design fracture geometry with the optimal ones to achieve the maximum possible benefit of the fracture treatment. The injection rate and fracturing fluid rheological properties can also be adjusted to reach the fracture geometry as close to the optimum as possible.

Such designs are for unrestricted fracturing. With TSO treatments, appropriate for much higher permeability reservoirs, the appropriate propped width can be obtained by inflating the fracture, after the fracture propagation is arrested at the desired length.

Example 18-12 Design Fracture Treatment to Achieve the Optimal Dimensions

Calculate the propped fracture width with the PKN model with the following treatment and reservoir data, and compare the result with Example 17-2.

$k = 1$ md, $h = 50$ ft, $h_f = 100$ ft, $E = 1 \times 10^6$, $\nu = 0.2$, proppant specific gravity = 2.65, porosity of proppant = 0.38, k_f = 60,000 md (20/40 mesh sand), M_p = 150000 lb, apparent viscosity of fluid = 200 cp, $K_L = 1.5$, $C_L = 0.01$ ft/min$^{0.5}$, $c_f = 8$ ppg.

Solution

From Equation (18-13), E' is 1.04×10^6 psi. Assuming q_i is 50 BPM, from Equation (18-16), when x_f is 504 ft,

$$\overline{w} = 0.19\left[\frac{(50)(200)(504)}{1.04 \times 10^6}\right]^{1/4} = 0.28 \text{ in.} \tag{18-84}$$

The fracture area, A_f, is 100800 ft^2. Then, from Equation (18-54),

$$(50)(5.615)t_i = (100800)\left(\frac{0.28}{12}\right) + (1.5)(0.02)(2)(100800)\left(\frac{50}{100}\right)\sqrt{t_i}, \tag{18-85}$$

which leads to $t_i = 44$ min. The injection time gives a total injection volume of 9.3×10^4 gal, a fracture efficiency of 0.2 and ε of 0.67.

The average slurry concentration can be calculated from Equation (18-78):

$$\overline{c}_p = \frac{c_f}{1+\varepsilon} = \frac{8}{1+0.68} = 4.8 \text{ ppg.} \tag{18-86}$$

From Equations 18-79 and 18-75, the proppant mass and proppant concentration can be determined as $M_p = 1.46 \times 10^5$ lb, and $C_p = 1.5$ lb/ft^2.

Finally, from Equation (18-76),

$$w_p = \frac{1.5}{(1 - 0.38)(2.65)(62.4)} = 0.014 \text{ ft} = 0.17 \text{ in.} \tag{18-87}$$

The result matches the optimum fracture condition in Example 17-2 for a permeability of 1 md.

18.5 Fracturing Fluids

A wide variety of fluids have been used for fracturing, including water, aqueous solutions of polymers with or without crosslinkers, gelled oils, viscoelastic surfactant solutions, foams, and emulsions. The most common fluids used now are "slickwater"—water with a small amount of polymer added to reduce frictional pressure drop and solutions of the natural polymer guar, or its derivatives, particularly hydroxypropyl guar (HPG). Figure 18-11 (Brannon, 2007) shows the relative amounts of common fracture fluids pumped by a major service company, showing that slickwater and guar-based fluids comprised about 80% of all fluids used. The fracture fluid serves to create the fracture and to transport the proppant. High apparent viscosity is desirable to create greater fracture width, suspend the proppant as it is transported down the fracture, and

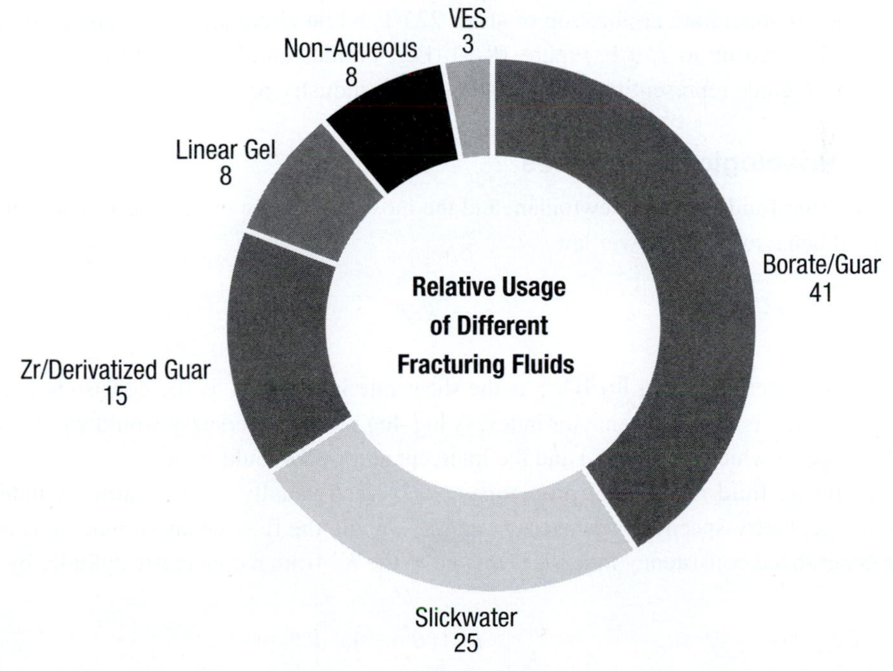

Figure 18-11 Fracture fluid usage. (From Brannon, 2007.)

minimize leakoff. Low-viscosity fluids, on the other hand, can be pumped at a higher rate because of lower frictional pressure drop and can more easily penetrate into narrow secondary fractures created in complex fracturing of ultra-low-permeability formations. Thus, the best fracturing fluid for a particular application depends on the desired fracture geometry.

Almost all desirable and undesirable properties of fracturing fluids are related to their apparent viscosity, which in turn is a function of the polymer load. Polymers such as guar, which is a naturally occurring material, or hydroxypropyl guar (HPG) have been used in aqueous solutions to provide substantial viscosity to the fracturing fluid. Polymer concentrations are often given in pounds of polymer per 1000 gallons of fluid (lb/1000 gal), and typical values range between 20 and 60, with perhaps 30 lb/1000 gal being the most common concentration in batch-mixed fracturing fluids that are currently in use.

Typical reservoir temperatures (175–200°F) lead to relatively low viscosities of the straight polymer solutions. For example, while a 40-lb/1000-gal HPG solution would exhibit an apparent viscosity of approximately 50 cp at room temperature at a shear rate of 170 sec^{-1}, the same solution at 175°F has an apparent viscosity of less than 20 cp. Thus, crosslinking agents, usually organometallic or transition metal compounds, are used to boost the apparent viscosity significantly. The most common crosslinking ions are borate, titanate, and zirconate. They form bonds with guar and HPG chains at various sites of the polymer, resulting in very high-molecular-weight compounds. The apparent viscosity at 170 sec^{-1} of 40-lb/1000-gal borate-crosslinked fluid is over 2000 cp at 100°F and about 250 cp at 200°F. Borate-crosslinked fluids have an upper temperature application of about 225°F, while zirconate- and titanate-crosslinked fluids may be used up to 350°F. Figure 18-12 (Economides and Martin, 2007) is a fracturing fluid selection guide representing current and evolving industry practices.

18.5.1 Rheological Properties

Most fracturing fluids are not Newtonian, and the most commonly used model to describe their rheological behavior is the power law

$$\tau = K\dot{\gamma}^n \tag{18-88}$$

where τ is the shear stress in lb_f/ft^2, $\dot{\gamma}$ is the shear rate in sec^{-1}, K is the consistency index in $\text{lb}_f\text{sec}^n/\text{ft}^2$, and n is the flow behavior index. A log–log plot of τ versus $\dot{\gamma}$ would yield a straight line, the slope of which would be n and the intercept at $\dot{\gamma} = 1$ would be K.

Fracturing fluid rheological properties are obtained usually in concentric cylinders that lead to the geometry-specific parameters η' and K'. While the flow behavior index n is equal to n', the generalized consistency index K is related to the K' from a concentric cylinder by

$$K = K'\left[\frac{B^{2/n'}(B^2 - 1)}{n'(B^{2/n'} - 1)B}\right]^{-n'} \tag{18-89}$$

where $B = r_{\text{cup}}/r_{\text{bob}}$ and r_{cup} is the inside cup radius and r_{bob} is the bob radius.

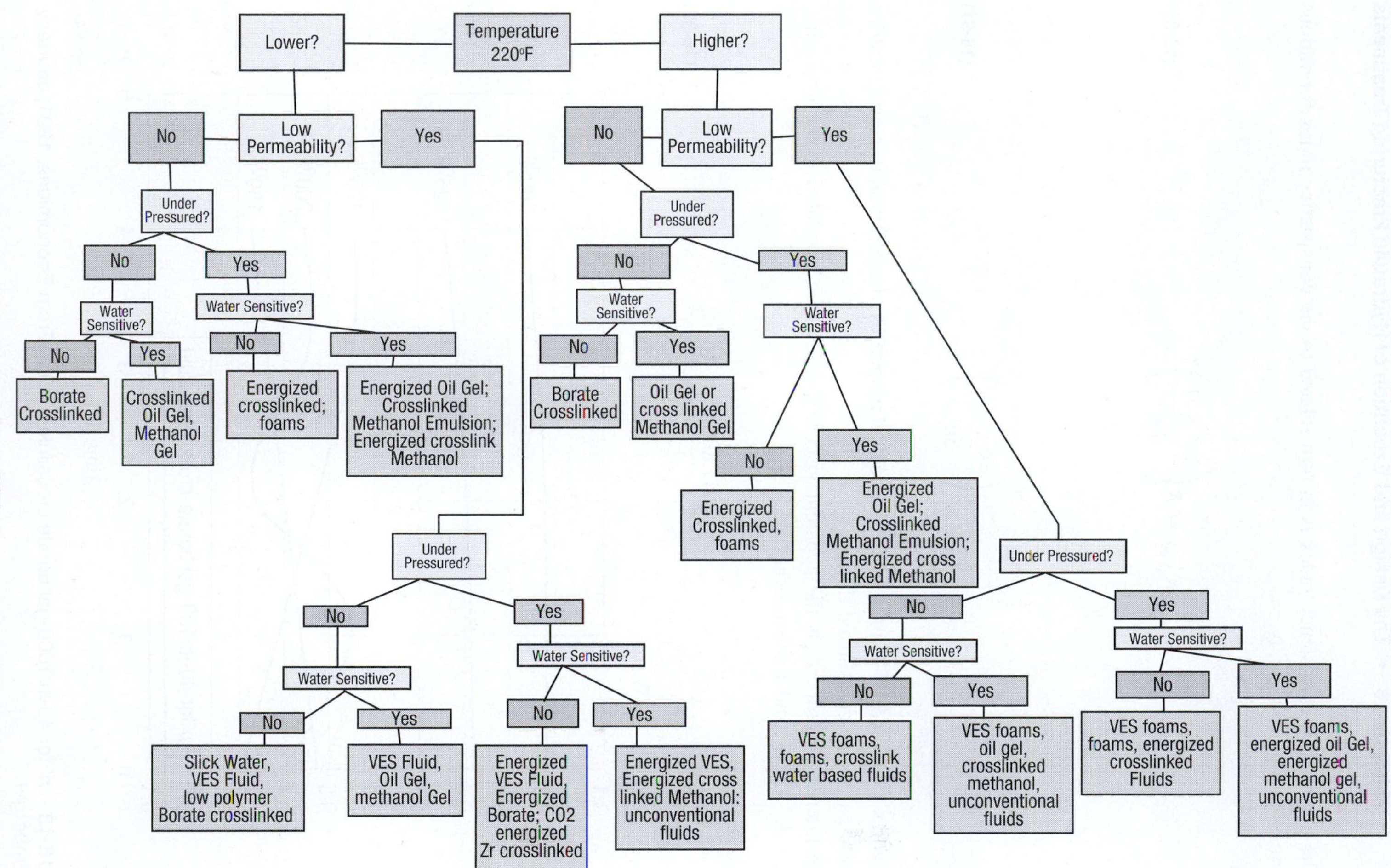

Figure 18-12 Fracturing fluid selection guide. (From Economides and Martin, 2007.)

The generalized consistency index is in turn related to the consistency index for various geometries. For a pipe, K'_{pipe}, is

$$K'_{\text{pipe}} = K\left(\frac{3n' + 1}{4n'}\right)^{n'} \quad \text{(18-90)}$$

and for a slot, K'_{slot}, is

$$K'_{\text{slot}} = K\left(\frac{2n' + 1}{3n'}\right)^{n'}. \quad \text{(18-91)}$$

Figures 18-13 and 18-14 present the rheological properties of a common fracturing fluid, a 40-lb/1000-gal borate-crosslinked fluid.

An interesting property is the apparent viscosity, μ_a, which is related to the geometry-dependent K', the n', and a given shear rate $\dot{\gamma}$:

$$\mu_a = \frac{47{,}800K'}{\dot{\gamma}^{1-n'}}. \quad \text{(18-92)}$$

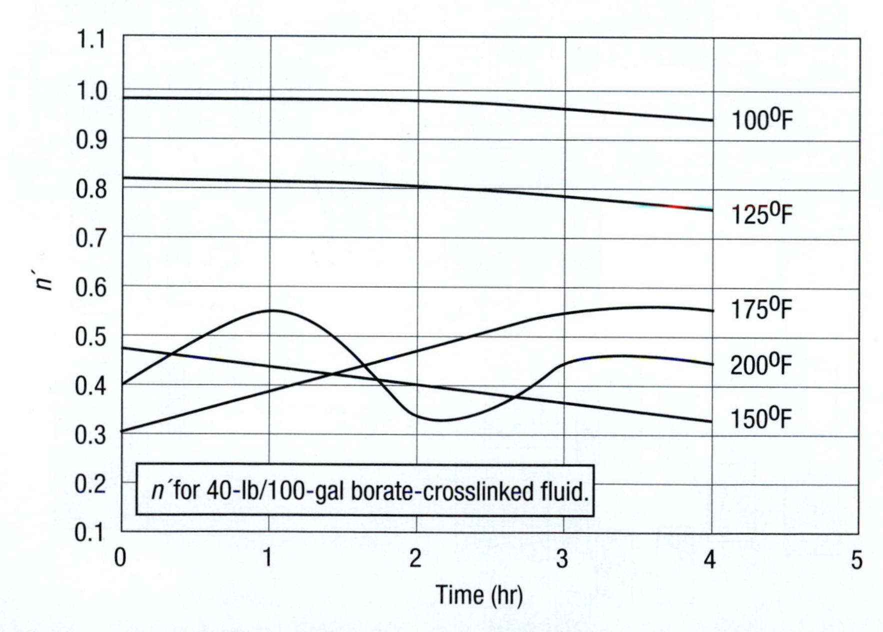

Figure 18-13 n' for 40-lb/1000-gal borate crosslinked fluid. (From Economides, 1991; courtesy of Schlumberger.)

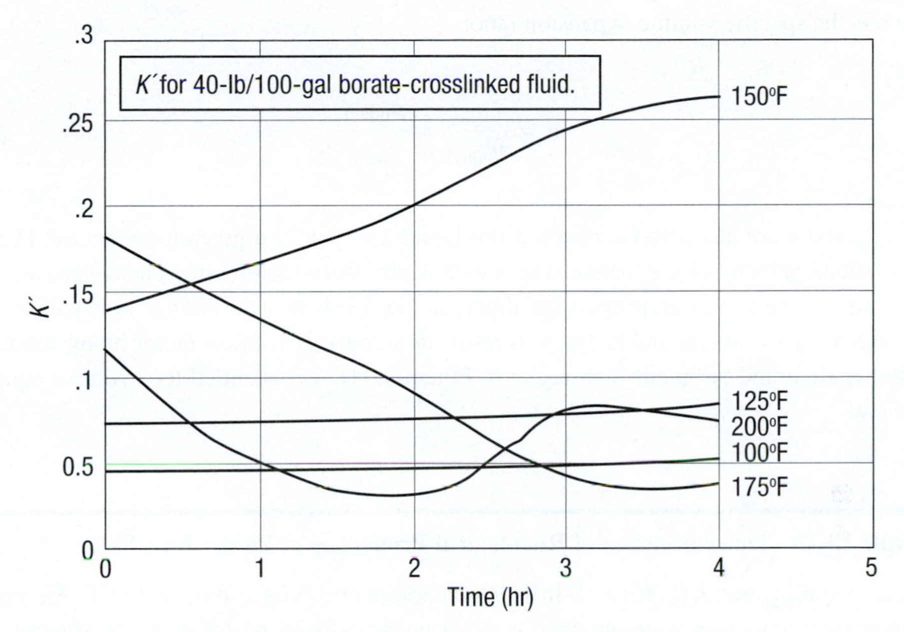

Figure 18-14 K' for 40-lb/1000-gal borate crosslinked fluid. (From Economides, 1991; courtesy of Schlumberger.)

In Equation (18-92), the viscosity is in cp.

Finally, the wall shear rate in a tube (well) for a power law fluid is

$$\dot{\gamma} = \left(\frac{3n' + 1}{4n'}\right)\frac{8u}{d} \tag{18-93}$$

where d is the diameter and u is the superficial velocity and is equal to q/A.

For a slot, which could approximate the geometry of a fracture, the wall shear rate is

$$\dot{\gamma} = \left(\frac{2n' + 1}{3n'}\right)\frac{6u}{w} \tag{18-94}$$

where w is the width of the slot.

For foam fluids, Valkó, Economides, Baumgartner, and McElfresh (1992) showed that the consistency index in Equation (18-88) can be expressed as

$$K = K_{\text{foam}}\varepsilon^{1-n} \tag{18-95}$$

where ε is the specific volume expansion ratio:

$$\varepsilon = \frac{\hat{v}_{\text{foam}}}{\hat{v}_{\text{liquid}}} = \frac{\rho_{\text{liquid}}}{\rho_{\text{foam}}} \tag{18-96}$$

and K_{foam} and n are characteristic for a given liquid–gas pair at a given temperature. In a well the superficial velocity of the foam changes with depth, since the pressure change causes a density variation. The important property of Equation (18-95) is that the change in K will compensate for density variations and hence will result in a constant friction factor along the tube in both the laminar and turbulent flow regimes. Equation (18-95) is called the "volume equalized power law."

Example 18-13 Determination of Rheological Properties of Power Law Fluids

Calculate the K'_{pipe} and K'_{slot} for a 40-lb/1000-gal borate-crosslinked fluid at 175°F. Assume that the rheological properties were obtained in a concentric cylinder where $B = 1.3$. If in a fracturing treatment the injection rate is 40 bpm, the fracture height is 250 ft, and the width is 0.35 in., calculate the apparent viscosity in the fracture.

Solution

From Figures 18-13 and 18-14 (at $T = 175°\text{F}$ and $t = 0$ hr), $n' = 0.3$ and $K' = 0.18\ \text{lb}_f$ $sec^{n'}/\text{ft}^2$. The generalized K is given by Equation (18-89):

$$K = (0.18)\left[\frac{1.3^{2/0.3}(1.3^2 - 1)}{0.3(1.3^{2/0.3} - 1)(1.3)}\right]^{-0.3} = 0.143\ \text{lb}_f\ \text{sec}^{n'}/\text{ft}^2. \tag{18-97}$$

Then, from Equation (18-90),

$$K'_{\text{pipe}} = (0.143)\left[\frac{(3)(0.3) + 1}{(4)(0.3)}\right]^{0.3} = 0.164\ \text{lb}_f - \text{sec}^{n}/\text{ft}^2 \tag{18-98}$$

and from Equation (18-91),

$$K'_{\text{slot}} = (0.143)\left[\frac{(2)(0.3) + 1}{(3)(0.3)}\right]^{0.3} = 0.17\ \text{lb}_f - \text{sec}^{n}/\text{ft}^2. \tag{18-99}$$

The shear rate in a fracture can be approximated by Equation (18-94). At first the superficial velocity (with proper unit conversions) is

$$u = \frac{(40/2)(5.615)}{(60)(250)(0.35/12)} = 0.26 \text{ ft/sec.} \tag{18-100}$$

The division of the flow rate by 2 implies that half of the flow is directed toward one wing of the fracture. Therefore,

$$\dot{\gamma} = \left[\frac{(2)(0.3) + 1}{(3)(0.3)}\right]\left[\frac{(6)(0.26)}{(0.35/12)}\right] = 95 \text{ sec}^{-1}. \tag{18-101}$$

The apparent viscosity is then [from Equation (18-92)]

$$\mu_a = \frac{(47{,}880)(0.17)}{95^{0.7}} = 336 \text{ cp.} \tag{18-102}$$

18.5.2 Frictional Pressure Drop during Pumping

Because of the high rate injection needed to create sufficient pressure to fracture the rock and to propagate hydraulic fractures deep into the reservoir, the frictional pressure drop caused by fracturing fluids is very important. To minimize friction during flow down the well tubulars, most fracturing fluids contain polymers that reduce the frictional pressure drop caused by the fluid. Fortuitously, the guar-based polymers used to create high apparent viscosity in fracture fluids also exhibit friction-reducing characteristics. These fracture fluids can be described as being thick (high viscosity), but also slippery. Slickwater fracture fluids contain low concentrations of polymers, most commonly polyacrylamide based, to reduce friction. The friction-reducing characteristics of fracturing fluids depends in a complex way on polymer concentration, temperature, salinity, and other factors, and is usually determined empirically. Typical levels of friction reduction are 50–60% compared with water having no friction reducer, and can be as high as 90%. The pipe flow calculation methods presented in Chapter 7 can be used to calculate the pressure drop in tubulars during fracture fluid injection by reducing the friction factor empirically.

To calculate the friction pressure drop for a power law fluid, first the Reynolds number must be determined:

$$N_{\text{Re}} = \frac{\rho u^{2-n'} D^{n'}}{K' 8^{n'-1}[(3n' + 1)/4n']^{n'}}. \tag{18-103}$$

Equation (18-103) is in consistent units. In oilfield units, it becomes

$$N_{\mathrm{Re}} = \frac{0.249\rho u^{2-n'} D^{n'}}{96^{n'} K'[(3n' + 1)/4n']^{n'}} \tag{18-104}$$

where ρ is the density in lb/ft^3, u is the velocity in ft/sec, D is the diameter in in., and K' is in lb$_f$-sec$^{n'}$/ft^2.

The velocity in oilfield units is

$$u = 17.17\frac{q_i}{D^2} \tag{18-105}$$

where q_i is the injection rate in bpm.

18.6 Proppants and Fracture Conductivity

The lasting conductivity of a hydraulic fracture is created by the proppant, solid particles pumped into the fracture as a slurry of solids mixed with the fracturing fluid. The most common material used as a proppant is sand, but several other materials including ceramic beads, resin-coated sand, and sintered bauxite are also commonly used (Figure 18-15; Brannon, 2007). The purpose of the proppant is to provide a conductive pathway from the reservoir to the wellbore by keeping the fracture from closing once the high pressure applied to create the fracture is relieved. When the fracture walls close on the proppant in the fracture, the proppant must be able to resist the resulting closure stress without being crushed to create effective conductivity. Thus, the strength of the proppant is a critical property that guides the selection of proppant type. The following characteristics of proppants are all important (Brannon, 2007):

Conductivity—The conductivity created by the proppant depends on the proppant mean size and size distribution, the proppant loading or concentration after closure (C_p), and the packing of the proppant.

Transportability—A desirable proppant can be transported far down a created hydraulic fracture. Because almost all proppants are considerably denser than the fracture fluids that carry them, proppant settling affects the ability to transport proppant down a fracture. Proppant transport depends on both properties of the proppant, including its density and particle size, and on transport conditions, such as velocity in the fracture and fluid apparent viscosity.

Particle strength—The proppant must be strong enough to not be crushed under the force of the closure stress applied. This limits the applicability of simple sand proppants to below about 6000 psi stress, leading to the use of higher strength proppants in deeper formations where stress is higher. It is also important to consider that the closure stress depends on the pressure in the fracture, and thus is likely to increase over the life of a fractured well as declining reservoir pressure results in lower wellbore pressure.

Figure 18-15 Proppants. (From Brannon, 2007.)

Inertness to reservoir fluids—Proppants must be chemically inert so that they are not dissolved away over the course of the well life.

18.6.1 Propped Fracture Conductivity

The fracture conductivity created by the proppant pack depends on many factors, including the proppant size and concentration, the presence of any damage to the proppant permeability, such as fine particles or residual fracture fluid gel, the strength of the proppant, and the closure stress.

Proppant conductivity is measured in the laboratory using a standard procedure (American Petroleum Institute, 2007) in which proppant at a loading of 2 lb/ft^2 is subjected to a range of loads in a standard conductivity cell (Figure 18-16). The results of such tests are often higher than the conductivity obtained in actual fractures, but are useful for comparing one proppant with another. Figure 18-17 (Predict K, 2007) shows how the conductivity of a variety of common proppants changes with increasing closure stress. At stresses above about 6000 psi, sand grains are crushed, and higher-strength proppants are required to maintain high conductivity.

The conductivity created by a proppant also depends on the size of the proppant, with larger proppant sizes creating higher-permeability proppant packs and hence, higher conductivities (Figure 18-18; Predict K, 2007). This increase is predicated on the proppant pack being undamaged, as finer particles added to larger size proppants can result in much lower conductivity. There is also a trade-off between conductivity and transportability as larger proppants settle

Figure 18-16 API Cell for the laboratory determination of proppant conductivity. (From Brannon, 2007.)

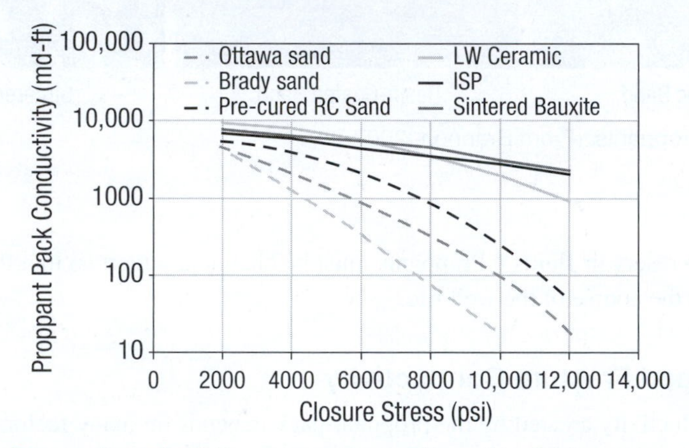

Figure 18-17 Proppant conductivity versus closure stress (depth) for various materials, all 20–40 mesh. (From Predict K, 2007.)

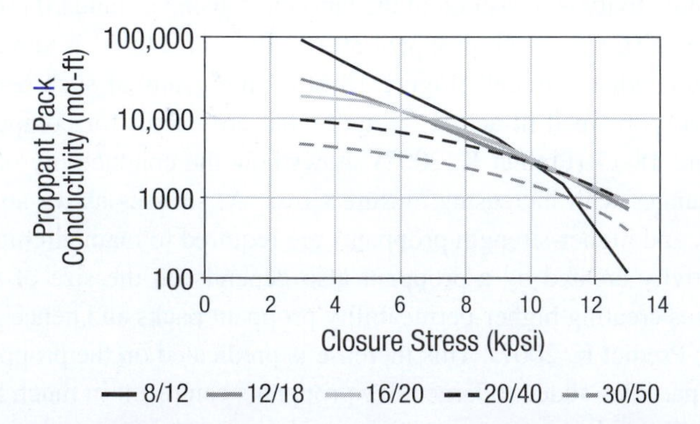

Figure 18-18 Proppant conductivity versus closure stress (depth) for a range of sizes of a ceramic proppant. (From Predict K, 2007.)

faster, and are thus more difficult to transport horizontally down the fracture, as described in the next section.

The conductivity of propped fractures is very susceptible to reductions caused by invasion of smaller particles into the proppant pack, crushing of some of the proppant particles, embedment of the proppant into the fracture surfaces, permeability reduction caused by unbroken polymer gels from the fracture fluid, two-phase flow effects, and non-Darcy flow effects (Cooke, 1973; Barree and Conway, 2004; Vincent, Pearson, and Kullman, 1999). For these reasons, the effective conductivity of fractures is often orders of magnitude less than that expected based on standard laboratory tests.

18.6.2 Proppant Transport

Ideally, the proppant is carried with the fracture fluid and does not settle or separate from the fracture fluid in any other way. However, because almost all proppants are significantly denser than the fracturing fluid transporting them, the proppant will settle, and more proppant is deposited in the lower part of the created fracture. The rate of proppant settling in a power law fluid can be estimated with a generalized form of Stokes' law [Brannon and Pearson, 2007]:

$$u_\infty = \left(\frac{(\rho_p - \rho_f) g d_p^{n'+1}}{3^{n'-1} 18 K'} \right)^{\frac{1}{n'}} \tag{18-106}$$

where u_∞ is the terminal settling velocity, ρ_p and ρ_f are the specific gravities of the proppant and the fluid, g is the acceleration of gravity, d_p is the particle diameter, and n' and K' are the usual flow behavior and consistency indices for a power law fluid. This equation is in consistent units.

Example 18-14 Proppant Settling Velocities

In the fracture design in Example 18-12, the proppant containing slurry would be pumped for about 14.5 minutes to create a 504-ft-long propped fracture. The mean diameter of the 20/40 mesh sand pumped is 0.024 in. and its density is 165 lb_m/ft^3. If K' is 0.07 $lb_f\text{-}sec^{n'}/ft^2$, n' is 0.4, and the fluid density is 65.5 lb_m/ft^3, typical of a borate-crosslinked gel as in Figures 18-13 and 18-14, how much will the proppant settle by the time it is transported to the end of the fracture? Repeat the calculation for a slickwater fracture fluid with a viscosity of 3 cp.

Solution

Applying Equation 18-106,

$$u_\infty = \left(\frac{(165 - 65.5)(1)\left(\frac{0.024}{12}\right)^{1.4}}{3^{(-0.6)} 18(0.07)} \right)^{\frac{1}{0.4}} = 0.0001 \text{ ft/s.} \tag{18-107}$$

Note that the g in this equation was replaced by g/g_c for English units. At such a low settling velocity, the proppant is predicted to settle only one inch as it moves 500 feet down the fracture. Such fluids, for which proppant settling is negligible, are called "perfect" fracture fluids.

The low-viscosity slickwater fracture fluid can be treated as a Newtonian fluid, for which Stokes' law for settling of spherical particles is

$$u_\infty = \frac{(\rho_p - \rho_f) g d_p^2}{18\mu_f}. \quad \text{(18-108)}$$

So, for the 3 cp fluid,

$$u_\infty = \frac{(165 - 65.5)(32.174)\left(\dfrac{0.024}{12}\right)^2}{18(3)(6.70 \times 10^{-4})} = 0.35 \text{ ft/s}. \quad \text{(18-109)}$$

In contrast to the gelled fracture fluid, in a low-viscosity, slickwater fluid, sand settles so rapidly that it is transported more as a moving dune in the lower part of the fracture than as a suspension.

18.7 Fracture Diagnostics

Fracture diagnostic methods that are aimed at determining the geometry of the fracture or fractures created, often applied in real time, have contributed greatly to improvements in the fracturing process. Beginning with the analysis of fracturing pressure to infer the type of fracture being created to the current real-time application of microseismic monitoring, fracture diagnostic methods are increasingly valuable aspects of hydraulic fracturing technology.

18.7.1 Fracturing Pressure Analysis

Nolte and Smith (1981) pioneered modern fracture diagnostic methods by describing how the characteristics of a log–log plot of net pressure versus time during pumping of a fracture treatment can be used to infer the mode of fracture propagation occurring. The hypothetical Nolte-Smith plot shown in Figure 18-19 illustrates this technique, which is based on 2-D fracture models, and Table 18-1 describes the modes of fracture propagation identified on this plot. A vertically confined fracture with length greater than its height corresponds to the PKN model, and according to Equations 18-35 and 18-36, the Nolte-Smith plot should show a straight line with a positive slope. If, on the other hand, the fracture is relatively unconfined vertically (KGD or radial fracture geometry), the plot will exhibit a straight line with a negative slope as

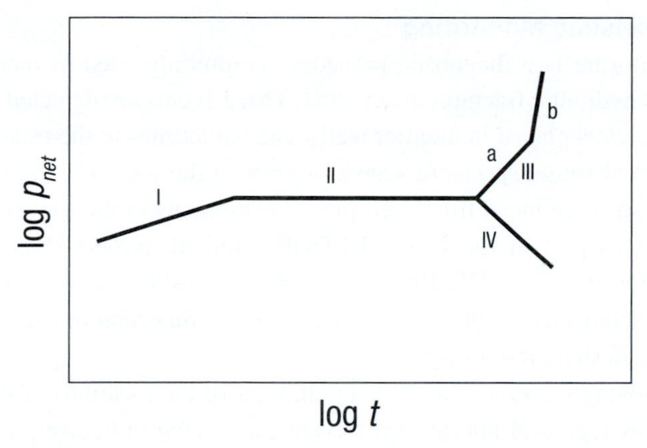

Figure 18-19 Nolte-Smith plot.

predicted by Equations 18-37 to 18-40. When fracture propagation is arrested when a screenout condition occurs, the Nolte-Smith plot has a positive unit slope. By monitoring the net pressure in real time, screenout conditions can be recognized quickly, and the treatment halted or injection conditions altered as necessary. Monitoring the net pressure in real time is a routine part of hydraulic fracturing operations.

18.7.2 Fracture Geometry Measurement

Diagnostic procedures have been developed to measure many aspects of the created fracture geometry, including height, length, and azimuth. In the case of complex fracturing, the region affected by the fracturing process can be detected. The primary techniques for monitoring fracture geometry are microseismic monitoring, tiltmeter monitoring, temperature profile measurements, and tracer methods.

Table 18-1 Nolte-Smith Analysis Pressure Response Modes (with reference to Figure 18-19)

Mode	Behavior
I	Propagation with PKN fracture geometry
II	Constant gradient $= 0$. Represents height growth in addition to length growth, increased fluid loss, or both. Can also be explained by a change in the relationship between p_{net} and w.
IIIa	Unit slope. This means that p_{net} is now directly proportional to time (and also to rate, as this is usually constant with respect to time). This behavior is usually associated with additional growth in w, such as during a tip screenout.
IIIb	Slope > 2. Screenout, usually a near-wellbore event with a very rapid rise in pressure.
IV	Negative slope. Rapid height growth. Potentially KGD or radial fracture geometry.

18.7.2.1 Microseismic Monitoring

Microseismic monitoring is a diagnostic procedure commonly used to map the microseismic events that occur as hydraulic fractures are created. These events are detected with arrays of geophones or accelerometers placed in monitor wells, and sometimes in the treated well itself. With assumed knowledge of sound speeds (a source of error in the method), the locations of the microseismic events can be deduced from multiple measurements of the same event. This technology was developed as part of the U.S. GRI/DOE multisite project (Sleefe, Warpinski, and Engler, 1995; Warpinski et al., 1995, 1996; Peterson et al., 1996) and is now routinely available to map hydraulic fractures. It has proven to be particularly important for mapping the large fractured region created in shale fracturing.

If a simple, biwing fracture is created, a plan view of the locations of microseismic events shows a long, narrow region of microseismic events, as shown in Figure 18-20 (Peterson et al., 1996). Although the width of the apparent fracture region is on the order of 100 ft, the width of the region actually fractured was determined to be about 5 ft by coring through the created fractures. The discrepancy is likely due to natural variations in sound speed of the rock between the microseismic sources and the receivers.

Microseismic mapping is extensively used with shale fracturing to measure the stimulated reservoir volume (SRV), the product of the area affected by complex fracturing and the height of the created fractures. This method is used to maximize the SRV. For example, Figure 18-21 (Cipolla, Lolon, and Dzubi, 2009) shows how the type of fracturing fluid used changes the SRV created in nearby wells in the same formation. Microseismic maps during multiple fracturing

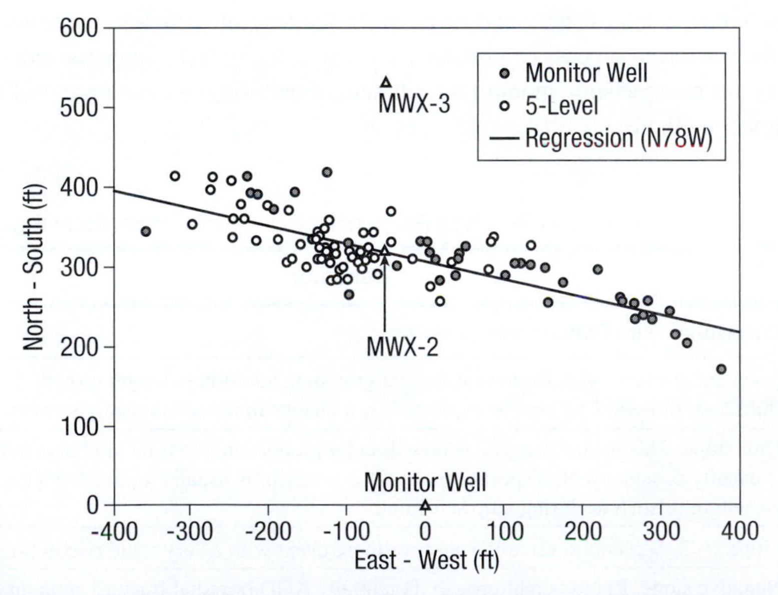

Figure 18-20 Plan view of microseismic events with a biwing fracture. (From Peterson et al., 1996.)

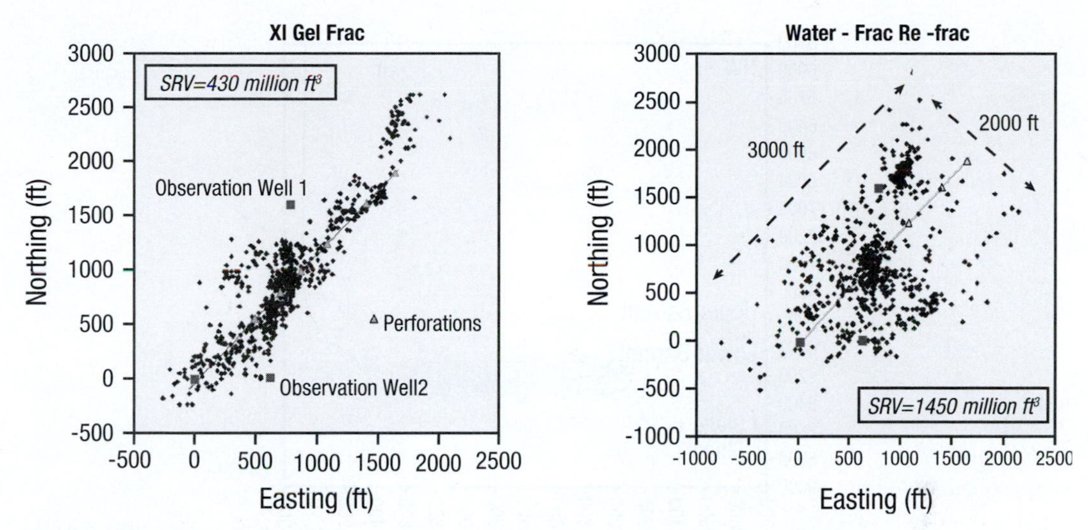

Figure 18-21 Microseismic map comparison of Gel frac versus Waterfrac. (From Cipolla et al., 2009.)

operations are also very diagnostic of the effectiveness of multi-stage fracturing, as illustrated in Figure 18-22 (Daniels, Waters, LeCalvez, Lassek, and Bentley, 2007), which maps the microseisms detected from a four-stage fracturing treatment. Finally, the vertical extent of created fractures can be detected if detectors are arrayed vertically, as illustrated by the cross section of microseismic events detected when a Barnett shale well was fractured (Figure 18-23; Fisher et al., 2004).

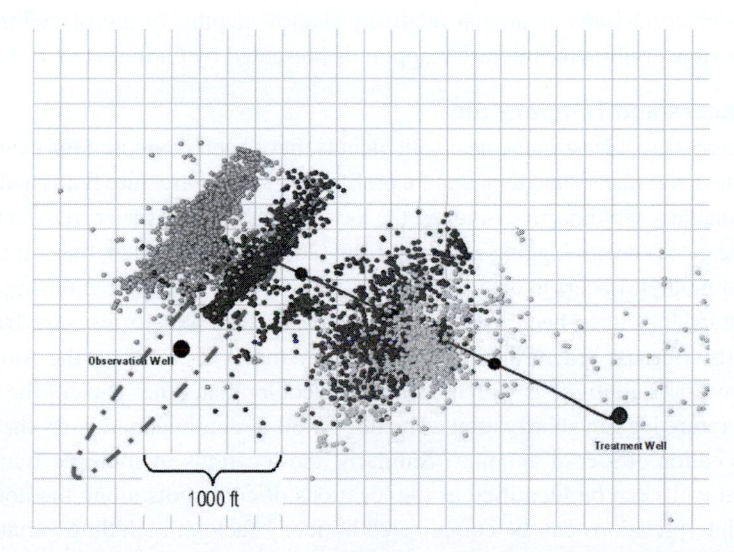

Figure 18-22 Microseismic map of multi-stage fracture treatment. (From Daniels et al., 2007.)

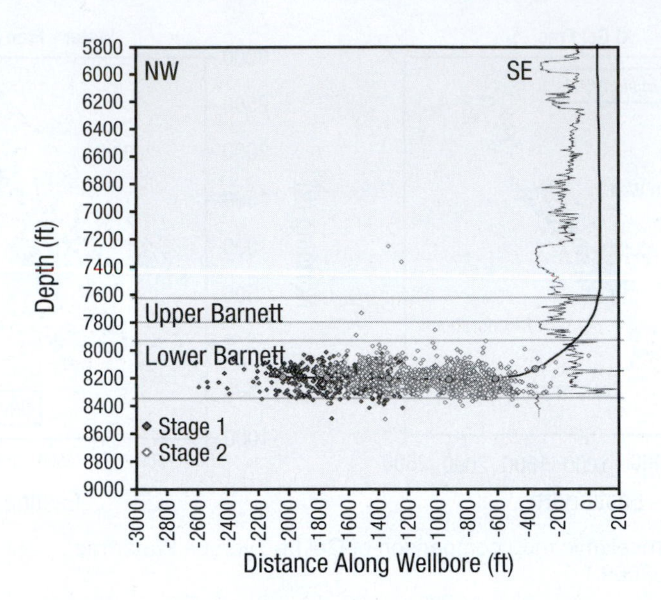

Figure 18-23 Mapping fracture height with microseismics. (From Fisher et al., 2004.)

18.7.2.2 Tiltmeter Mapping

Another means of detecting the geometry of hydraulic fractures is measuring very small earth movements with inclinometers or *tiltmeters.* As a fracture propagates, the formations around it are deformed, and these deformations can be detected with extremely sensitive measurements of the change in angle of a solid rod attached to the deforming surface. Tiltmeters are arrayed on the surface to detect fractures created at relatively shallow depths, or are placed in observation wellbores. A review of tiltmeter fracture mapping is presented by Gidley et al. (1989).

18.7.2.3 Tracers and Temperature

Fracture characteristics, at least in the near well vicinity, have been diagnosed for many years using chemical or radioactive tracers and temperature profiles. By tagging fracture fluids and/or proppants with chemical markers or radioactive isotopes, the location where these tracers remain near the wellbore can then be detected with logging methods. King (2010) reviewed these tracer methods, particularly the use of multiple tracers to diagnose where fractures were created in multistage fracturing.

Temperature logs have been used for many years to attempt to measure fracture height in vertical wells. Because fracturing fluid is typically much cooler than the formation being fractured, the wellbore contiguous with the fracture is cooler than other parts of the wellbore and hence, a temperature log run shortly after a fractured well is shut in can identify the fractured interval as the location of a cool anomaly. Similarly, the locations of multiple fractures created from horizontal wells can be identified as the locations of cool spots along the horizontal well. These simple interpretations can be complicated by many factors, including variations in wellbore trajectory (Davis, Zhu, and Hill, 1997).

18.8 Fracturing Horizontal Wells

The combination of horizontal wellbores and multiple hydraulic fractures provides a means to maximize wellbore contact with a reservoir, and is crucial to economic development of low-permeability reservoirs. Unlike vertical well fracturing, most fracturing jobs in horizontal wells are aimed at creating multiple fractures, either by single-stage pumping or multistage pumping with isolation provided by the completion. New challenges in horizontal well fracturing include, but are not limited to, orientation of horizontal wells for favorable stress fields to create the desired type of fractures, successfully propagating multiple fractures, less trips and effective isolation between fracturing stages, significantly larger volume of treatments, and creation of extremely complex fracture network systems in shale formations. This section discusses current practices in the field to create multiple fractures in horizontal wells.

18.8.1 Fracture Orientation in Horizontal Well Fracturing

Depending on the purpose of fracture stimulation, the fractures created from a horizontal well can be along the wellbore, referred to as longitudinal fractures, or perpendicular to the wellbore, referred to as transverse fractures. Figure 18-24 shows the two different types of fractures that may be created from a horizontal well. In general, the main reason to create longitudinal fractures is to break the vertical barriers when a horizontal well is placed in a relatively thick formation with impermeable laminations in the formation (low vertical permeability). Fractures in such a case will contribute the necessary vertical permeability for a horizontal well to produce. On the other hand,

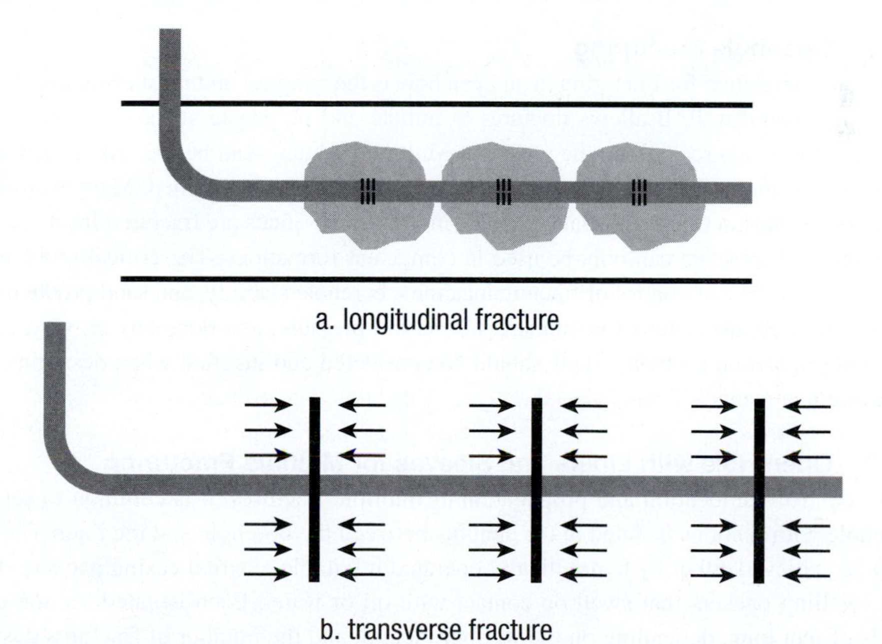

Figure 18-24 Longitudinal and transverse fracture geometries from horizontal wells.

creating transverse fractures is aimed at creating extensive contact with reservoirs that have too low a permeability to produce without stimulation. This type of fracturing has become a common practice in tight sand gas formation or shale oil and gas formations. In fact, multiple transverse fracturing is the only means by which unconventional shale formations can be economically producible.

Regardless of the applications of the type of fractures, one can only create certain types of fractures, longitudinal or transverse, based on the in-situ stress field of the formation that is going to be fractured. Fractures in a 3-D stress field tends to propagate in the direction of maximum stress (or perpendicular to the minimum stress direction). Once the purpose of fracturing is clarified, and the type of fractures desired is determined, it is important to orient the horizontal well correctly when drilling to ensure efficient propagation of fractures. Initially, a fracture propagates along the wellbore or perforations. To avoid fractures turning or twisting to align with the stress field once away from the wellbore, a horizontal well needs to oriented within 10–15° of the minimum stress direction to propagate longitudinal fractures (Yew, 1997). For any other well orientation relative to the stress field, transverse fractures are created.

18.8.2 Well Completions for Multiple Fracturing

There are three common ways to complete horizontal wells for multiple fracturing, either longitudinal or transverse: open-hole uncontrolled fracturing, open hole with liners and isolation for controlled multiple-stage fracturing, and cased hole with perforated completion for multiple fracturing. Open-hole uncontrolled fracturing can generate fractures in one or more stages; the other two completion types generate multiple fractures with multiple stages.

18.8.2.1 Openhole Fracturing

An openhole completion for fracturing in an open hole is the simplest and most economical way to fracture a horizontal well. It allows fractures to initiate and propagate at the locations of weak points in the formation rock along the wellbore. Multiple fractures can be created in such a treatment, but the locations of the fractures cannot be predetermined or controlled. Many wells in formations like the Austin Chalk and Bakken Shale in the United States are fractured from open-hole completions. This practice can only be used in competent formations. The critical problem with this method is the lack of control of fracture locations, borehole stability, and sand production. As production goes on, the status of in-situ stress, reservoir pressure, and flow may all vary, causing unpredicted production problems. This should be considered and justified when designing open-hole fracture treatments.

18.8.2.2 Open Hole with Liners and Sleeves for Multiple Fracturing

To better control the location and propagation of multiple fractures, it is common to set liners in open hole with sections isolated in the annulus between the borehole and the liner. The isolation can be achieved either by hydraulically operated inflatable external casing packers (ECPs) or with swelling packers that swell on contact with oil or water. Each isolated section can be hundreds of feet long, depending on the wellbore length and the number of fractures designed. After installing the liner and setting the packers, the well is ready for fracturing. There are several different methods of multistage fracturing with this type completion. Here we describe a procedure of using sleeves with isolation balls to sequentially place multiple fractures along a horizontal well, starting at the toe of the well.

Multiple sleeves (also commonly called fracture ports) are installed along a liner. Beginning with the sleeve at the toe of the well, each sleeve is progressively larger, so that balls of increasing sizes can be dropped sequentially to individually activate each sleeve. During the fracture execution, the isolation balls are dropped into the well one at a time. The balls have two functions: to open a fracture port and to isolate the downstream wellbore behind the port so that a fracture can be initiated at the port. The smallest ball is dropped first to the end of the horizontal wellbore to establish pressure integrity inside the liner. Once the isolation is established, the next ball that fits the first sleeve towards the toe is dropped to open the sleeve and the first stage fracture is ready to pump. When first-stage fracturing is completed, the wellbore is circulated clean for the next stage of fracturing. The next larger-sized ball is then pumped down to open the next sleeve for fracturing, and plug off the downstream section for isolation. The procedure repeats with the ball size getting larger and larger, and the last stage will have the largest sleeve size with the biggest ball. Currently this technique can create over 30 stages of fracturing. Figure 18-25 illustrates this technique.

When fracture stimulation is completed, the balls in the horizontal well should flow back by the well production. One of the problems with this procedure is that sometimes the production rate is not high enough to flow the balls back. When this happens, milling is required to rebuild the production flow path.

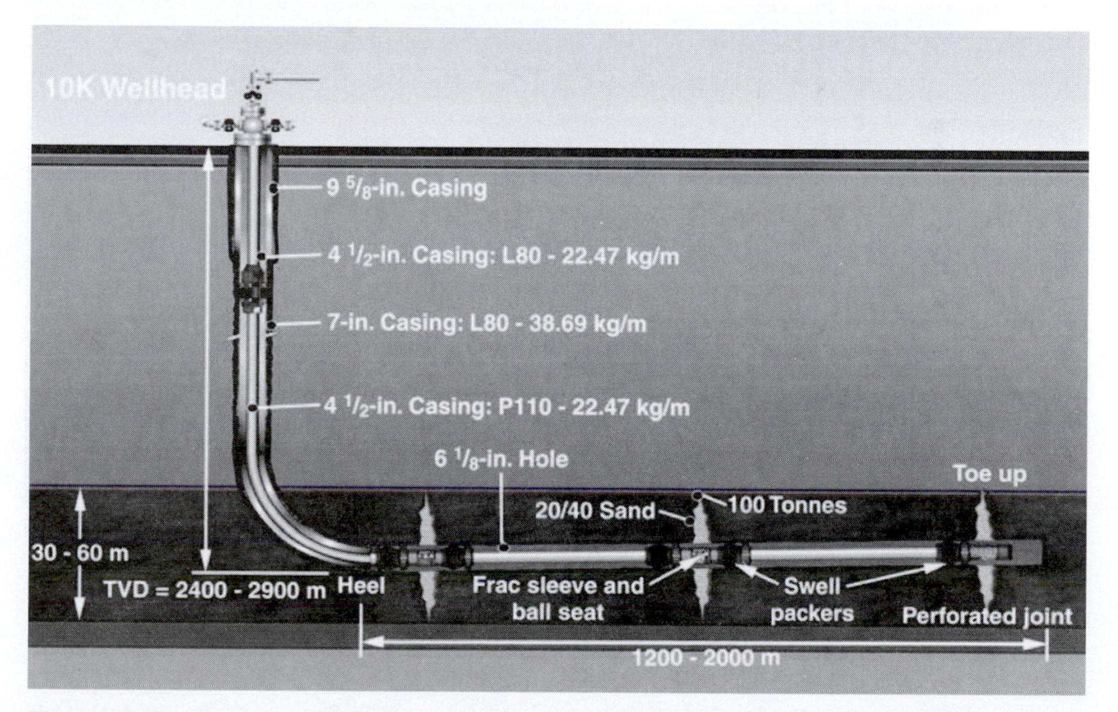

Figure 18-25 Multistage fracturing with sleeves. (From Thompson, Rispler, Hoch, and McDaniel, 2009.)

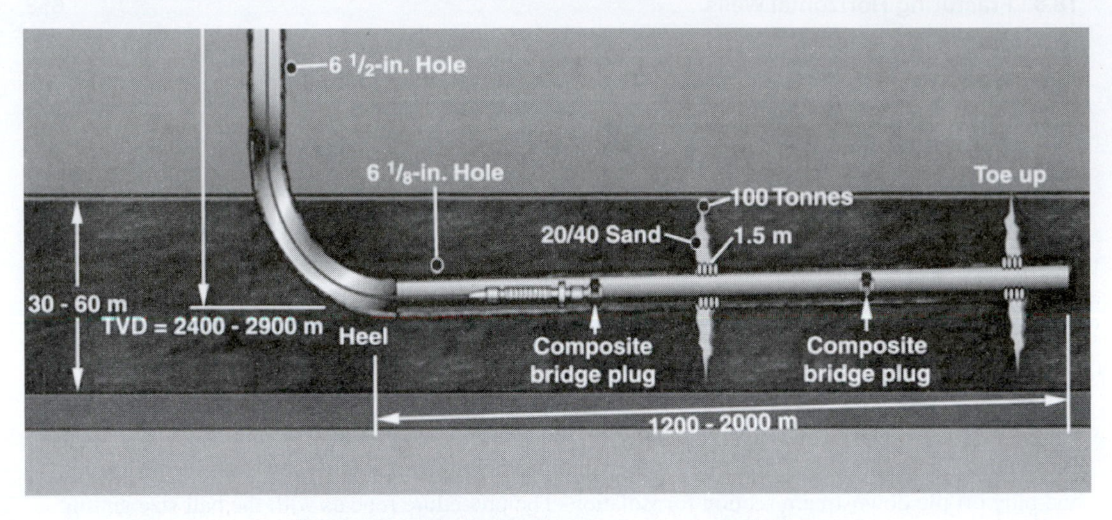

Pumpdown bridge plug perf gun completion

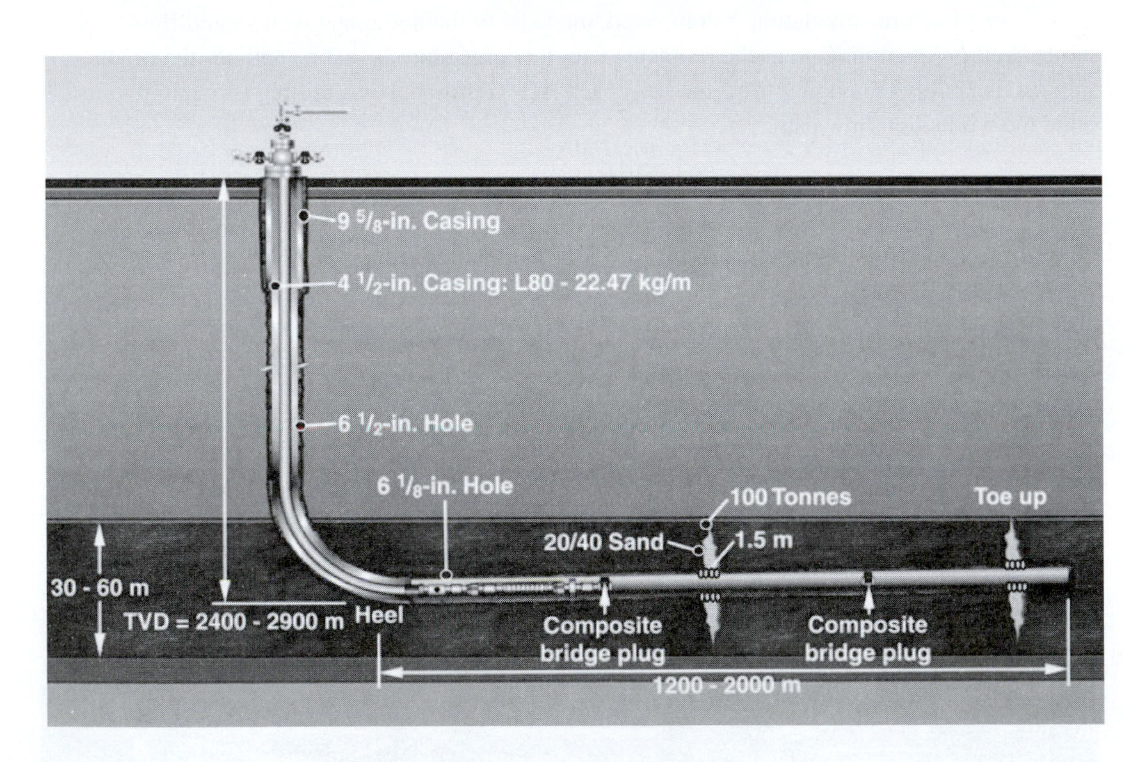

CT-deployed bridge plug perf gun completion

Figure 18-26 Completions for horizontal well multiple fracturing. (From Thompson et al., 2009.)

18.8.2.3 Perf and Plug Fracturing

In this approach, a perforation gun is deployed by a coiled tubing unit or a pump-down wireline tool (perf and plug tool) into a cased and cemented horizontal well. The perforation gun is designed to be able to perforate more than one interval. The first location toward the toe is perforated, then the perf and plug tool is pulled back to the next stage location and the fracture stage is pumped through the annulus. Then, the borehole is cleaned after fracturing, and a composite plug (made of gel, sand, gel and sand mixture, or other similar material) is pumped through the perf and plug tool to isolate the first fracture interval from the rest of the well. The well is then ready for the next stage of perforation and fracturing. Figure 18-26 shows a completion structure for perf and plug fracturing.

When perforating horizontal wells to create multiple transverse fractures, clusters of perforations are created spaced along the wellbore at the desired spacing of the main fractures. Each cluster has multiple perforations spaced over a few feet of wellbore. There are often several clusters treated at once during a fracture stage. For example, a typical wellbore configuration when fracturing a shale well has three to five perforation clusters spaced along an isolated 500–1000 feet of wellbore receiving a fracture treatment stage.

References

1. Agarwal, R.G., Carter, R.D., and Pollock, C.B., "Evaluation and Prediction of Performance of Low-Permeability Gas Wells Stimulated by Massive Hydraulic Fracturing," *JPT,* 362–372 (March 1979); *Trans. AIME,* **267.**
2. Barree, R.D., and Conway, M.W., "Beyond Beta Factors: A Complete Model for Darcy, Forchheimer, and Trans-Forchheimer Flow in Porous Media," SPE Paper 89325, 2004.
3. Ben-Naceur, K., "Modeling of Hydraulic Fractures," in *Reservoir Stimulation,* 2nd ed., M.J. Economides and K.G. Nolte, eds., Prentice Hall, Englewood Cliffs, NJ, 1989.
4. Biot, M.A., "General Solutions of the Equations of Elasticity and Consolidation for a Porous Material," *J. Appl. Mech.,* **23:** 91–96, 1956.
5. Brannon, H.D., "Fracturing Materials," keynote address, SPE Hydraulic Fracturing Conference, The Woodlands, TX, January 19–21, 2007.
6. Brannon, H.D., and Pearson, C.M., "Proppants and Fracture Conductivity," in *Modern Fracturing—Enhancing Natural Gas Production,* M.J. Economides and T. Martin, eds., Chap. 8, Energy Tribune Publishing, 2007.
7. Cipolla, C.L., Lolon, E.P., and Dzubi, B., "Evaluating Stimulation Effectiveness in Unconventional Gas Reservoirs," SPE Paper 124843, 2009.
8. Cooke, C.E. Jr., "Conductivity of Proppants in Multiple Layers." *JPT,* 1101–1107 (October 1993).
9. Daniels, J., Waters, G., LeCalvez, J., Lassek, J., and Bentley, D., "Contacting More of the Barnett Shale Through an Integration of Real-Time Microseismic Monitoring, Petrophysics and Hydraulic Fracture Design," SPE Paper 110562, 2007.
10. Davis, E.R., Zhu, D., and Hill, A.D., "Interpretation of Fracture Height from Temperature Logs—the Effect of Wellbore/Fracture Separation," *SPE Formation Evaluation* (June 1997).

11. Economides, M.J., *A Practical Companion to Reservoir Stimulation*, SES, Houston, Texas, 1991 and Elsevier, Amsterdam, 1991.

12. Economides, M.J., and Martin, T., *Modern Fracturing—Enhancing Natural Gas Production*, Gulf Publishing, Houston, TX, 2007.

13. Economides, M.J., and Nolte, K.G., *Reservoir Stimulation,* 3rd ed., Wiley, New York, 2000.

14. Fisher, M.K., Heinze, J.R., Harris, C.D., Davidson, B.M., Wright, C.A., and Dunn, K.P., "Optimizing Horizontal Well Completion Techniques in the Barnett Shale Using Microseismic Fracture Mapping," SPE Paper 90051, 2004.

15. Geertsma, J., and deKlerk, F., "A Rapid Method of Predicting Width and Extent of Hydraulically Induced Fractures," *JPT,* 1571–1581 (December 1969).

16. Gidley, J.L., Holditch, S.A., Nierode, D.E., and Veatch, Jr., R.W., *Recent Advances in Hydraulic Fracturing,* SPE monograph, Vol. 12, Society of Petroleum Engineers, Richardson, TX, 1989.

17. Hubbert, M.K., and Willis, D.G., "Mechanics of Hydraulic Fracturing," *Trans. AIME,* **210:** 153–168 (1957).

18. Khristianovic(h), S.A., and Zheltov, Y.P., "Formation of Vertical Fractures by Means of Highly Viscous Liquid," *Proc. Fourth World Petroleum Congress,* Sec. II, 579–586, 1955.

19. King, George E., "Thirty Years of Gas Shale Fracturing: What Have We Learned?", SPE Paper 133456, 2010.

20. McLennan, J.D., Roegiers, J.-C., and Economides, M.J., "Extended Reach and Horizontal Wells," in *Reservoir Stimulation,* 2nd ed., M.J. Economides and K.G. Nolte, eds., Prentice Hall, Englewood Cliffs, NJ, 1989.

21. Meng, H.-Z., and Brown, K.E. "Coupling of Production Forecasting, Fracture Geometry Requirements, and Treatment Scheduling in the Optimum Hydraulic Fracture Design," SPE Paper 16435, 1987.

22. Newberry, B.M., Nelson, R.F., and Ahmed, U., "Prediction of Vertical Hydraulic Fracture Migration Using Compressional and Shear Wave Slowness," SPE/DOE Paper 13895, 1985.

23. Nolte, K.G., "Determination of Proppant and Fluid Schedules from Fracturing Pressure Decline," *SPEPE,* 255–265 (July 1986).

24. Nolte, K.G., and Economides, M.J., "Fracturing, Diagnosis Using Pressure Analysis," in *Reservoir Stimulation,* 2nd ed., M.J. Economides and K.G. Nolte, eds., Prentice Hall, Englewood Cliffs, NJ, 1989.

25. Nolte, K.G., and Smith, M.B., "Interpretation of Fracturing Pressure," *JPT,* 1767–1775 (September 1981).

26. Nordgren, R.P., "Propagation of Vertical Hydraulic Fracture," *SPEJ,* 306–314 (August 1972).

27. Perkins, T.K., and Kern, L.R., "Widths of Hydraulic Fracture," *JPT,* 937–949 (September 1961).

28. Peterson, R.E., Wolhoart, S.L., Frohne, K.H., Branagan, P.T., Warpinski, N.R., and Wright, T.B., "Fracture Diagnostics Research at the GRI/DOE Multi-Site Project: Overview of the Concept and Results," SPE Paper 36449, 1996.

29. Predict K, v. 7.0, Proppant Conductivity Database, Stimlab, 2007.

30. Simonson, E.R., "Containment of Massive Hydraulic Fractures," *SPEJ,* 27–32 (February 1978).

31. Sleefe, G.E., Warpinski, N.R., and Engler, B.P., "The Use of Broadband Microseisms for Hydraulic Fracture Mapping," *SPE Formation Evaluation,* 233–239 (December 1995).

32. Sneddon, I.N., and Elliott, A.A., "The Opening of a Griffith Crack under Internal Pressure," *Quart. Appl. Math.,* **IV:** 262 (1946).

33. Terzaghi, K., "Die Berechnung der Durchlässigkeitsziffer des Tones aus dem Verlauf der Hydrodynamischen Spannungserscheinungen," *Sber. Akad. Wiss.,* Wien, **132:** 105 (1923).

34. Thompson, D., Rispler, K., Hoch, and McDaniel, B.W., "Operators Evaluate Various Stimulation Methods for Mutizone Stimulation of Horizontals in Northeast British Columbia," SPE Paper 119620, 2009.

35. Valkó, P., Economides, M.J., Baumgartner, S.A., and McElfresh, P.M., "The Rheological Properties of Carbon Dioxide and Nitrogen Foams," SPE Paper 23778, 1992.

36. Vincent, M.C., Pearson, C.M., and Kullman, J., "Non-Darcy and Multiphase Flow in Propped Fractures: Case Studies Illustrate the Dramatic Effect on Well Productivity," SPE 54630, 1999.

37. Warpinski, N.R., Engler, B.P., Young, C.J., Peterson, R.E., Branagan, P.T., and Fix, J.E., "Microseismic Mapping of Hydraulic Fractures Using Multi-Level Wireline Receivers," SPE Paper 30507, 1995.

38. Warpinski, N.R., Mayerhofer, M.J., Vincent, M.C., Cipolla, C.L., and Lolon, E.P., "Stimulating Unconventional Reservoirs: Maximizing Network Growth While Optimizing Fracture Conductivity," *JCPT,* **48**(10): 39–51 (October 2009).

39. Warpinski, N.R., Wright, T.B., Uhl, J.E., Engler, B.P., Young, C.J., and Peterson, R.E., "Microseismic Monitoring of the B-Sand Hydraulic Fracture Experiment at the DOE/GRI Multi-Site Project," SPE Paper 36450, 1996..

40. Yew, Ching H., *Mechanics of Hydraulic Fracturing,* Gulf Publishing Company, Houston, TX, 1997.

Problems

18-1 Assume that a formation is 6500 ft deep. What overpressure (above hydrostatic to water) is necessary to result in a horizontal fracture? The formation density, fluid density, and Poisson ratio are 165 lb/ft^3, 50 lb/ft^3, and 0.23, respectively.

18-2 In Terzaghi's relationship [Equation (18-5)], the breakdown pressure is in terms of absolute stresses and reservoir pressure. Assume that the reservoir described in Problem 18-1 is at hydrostatic pressure. Calculate its initial breakdown pressure and its value when the reservoir pressure declines by 500, 1000, and 2000 psi. The tensile stress is 500 psi. Assume that the maximum horizontal stress is 1500 psi large than the minimum horizontal stress.

18-3 The Terzaghi relationship [Equation (18-5)] is for a perfectly vertical well drilled along the vertical principal stress. If two horizontal wells, one along the minimum and a second along the maximum horizontal stresses, are drilled, what would their breakdown pressures be? Use the reservoir that is described in problems 18-1 and 18-2 and assume hydrostatic (to water) pressure.

18-4 Start with problem 18-1. Perform a parametric study for the critical depth where horizontal fractures will be created in which 500, 1000, 1500, and 2000 ft of overburden were removed over geologic time. For all cases use hydrostatic pressure but also repeat the calculation for 20% and 40% overpressure. Explain your results.

18-5 Plot the maximum and average fracture widths and fracture volume for fracture lengths from 100 to 2000 ft in increments of 100 ft. The apparent viscosity of the fluid is 100 cp, and the injection rate is 40 bpm. Assume that $v = 0.25$ but allow the Young's modulus, E, to range from 10^5 to 4×10^6 psi. Keep the height constant at 100 ft.

18-6 If $q_i = 40$ bpm, $h_f = 100$ ft, $E = 4 \times 10^6$ psi, and $K' = 0.2\ \text{lb}_f\ \text{sec}^{n'}/\text{ft}^2$, plot the average fracture width, using the non-Newtonian fluid expression for $n' = 0.3$, 0.4, 0.5, and 0.6. Assuming that $n' = 0.5$, what is the ratio of the average widths for $q_i = 40$ bpm and 20 bpm?

18-7 Graph and compare the average fracture widths obtained with the PKN and KGD models for $q_i = 40$ bpm, $\mu = 100$ cp, $v = 0.25$, $E = 4 \times 10^6$ psi, and $h_f = 100$ ft. Perform the calculation for x_f up to 1000 ft (although the KGD model should not be used for such a length).

18-8 In Equation (18-34), the constant C_1 is a function of the injection rate q_i and the consistency of the power law rheological model, K'. Assuming that all other variables are constant, develop an expression of the form $\Delta p_f = f(q_i, K', n')$. If $n' = 0.5$, show the corresponding changes in q_i (with K' constant) or K' (with q_i constant) to allow 10% reduction in Δp_f.

18-9 Develop a simple p-3-D fracture propagation model using Equations (16-49) and (16-50) for upward and downward fracture migration and Equations (16-21) (for width) and (16-26) relating this width with Δp_f. Note that $h_f = h_u + h_d - h$.

18-10 Range the injection rate from 10 to 50 bpm in increments of 10 bpm and use three values of the leakoff coefficient, $C_L = 8 \times 10^{-4}\ \text{ft}/\sqrt{\text{min}}$, $C_L = 4 \times 10^{-3}\ \text{ft}/\sqrt{\text{min}}$ and $C_L = 8 \times 10^{-3}\ \text{ft}/\sqrt{\text{min}}$. What would be the effects on the total fluid volume requirements? Use the non-Newtonian PKN width relationship and $n' = 0.56$, $K' = 8 \times 10^{-3}\ \text{lb}_f\ \text{sec}^{n'}/\text{ft}^2$, $E = 4 \times 10^6$ psi, $h_f = 100$ ft, $\eta = 0.3$ and $r_p = 0.7$. Graph V_i and the total time of injection, t_i, versus the injection rate with the leakoff coefficient as a parameter.

18-11 Repeat problem (18-10) for $q_i = 40$ bpm and $C_L = 8 \times 10^{-3}\ \text{ft}/\sqrt{\text{min}}$ but allow h_f to increase from 100 to 300 ft (still $h = 70$ ft). If the propped width w_p cannot exceed $0.8\overline{w}$, calculate and plot the allowable end-of-job proppant concentration c_f for increments of 50 ft in the h_f. Plot the proppant addition schedule for the same fracture heights.

18-12 Using the UFD concept and the following data, estimate the optimum fracture geometry and then calculate the total time of injection, pad time, fluid volume requirements, and proppant slurry concentrations as functions of time.

Proppant mass, lbm	300,000
Sp grav of proppant material (water = 1)	2.65
Porosity of proppant pack	0.38
Proppant pack permeability, md	200,000
Formation permeability, md	2
Permeable (leakoff) thickness, ft	70
Well radius, ft	0.30
Well drainage radius, ft	2,100
Fracture height, ft	140.0
Plane strain modulus, E′ (psi)	2.00E + 06
Slurry injection rate (two wings, liq + prop), bpm	20.0
Rheology, $K'(\text{lbf/ft}^2) \times s^{n'}$	0.0180
Rheology, n'	0.65
Leakoff coefficient in permeable layer, $\text{ft/min}^{0.5}$	0.00400
Maximum slurry concentration, ppg	12

18-13 A fracture design is intended to result in a fracture length of 600 ft and an average width of 0.5 in. The diameter of the 20/40 mesh ceramic that is pumped is 0.024 in. and its density is 200 lb_m/ft^3. The fluid density is 65.5 lb_m/ft^3, typical of a borate-crosslinked gel as in Figures 18-13 and 18-14. For the range of temperatures of those two figures and for a pumping time of 2 hours, sketch the trajectory of proppant settling (the height of the proppant pack as a function of position down the fracture). Pumping rate is 40 bpm and the fracture height is 120 ft.

18-14 Design a hydraulic fracture using the 2D PKN model using the following data:

$q_i = 40$ bpm, $h_f = 100$ ft, Poisson ratio $\nu = 0.25$, $E = 5 \times 10^6$ psi, $n' = 0.6$, $K' = 0.03$

Leakoff coefficient $C_L = 5 \times 10^{-4}\ \text{ft/min}^{1/2}$, $r_p = 1$, frac fluid volume pumped = 60,000 gal.

Proppant: $\rho_p = 165\ \text{lb}_m/\text{ft}^3$, $\phi_p = 0.42$, and $c_f = 3$ ppg, $k_f = 10$ Darcy

Reservoir: $\phi = 0.1$, $k = 0.1$ md, $r_e = 2980$ ft, $r_w = 0.25$ ft, $c_t = 1 \times 10^{-5}\ \text{psi}^{-1}$

Calculate:

1. Average fracture width at the end of pumping
2. Fracture length
3. Fracture efficiency
4. Pad volume, V_{pad} and pad injection time, t_{pad}
5. Proppant mass, M_p
6. Propped fracture width, w_p
7. Fracture conductivity
8. Dimensionless fracture conductivity

CHAPTER 19

Sand Management

19.1 Introduction

Although formation solids are not always sand, in the petroleum industry, the production of any solids from a well is generally called sand production. Often not anticipated, produced solids can accumulate in the well or in subsea or surface flow lines, destroy a downhole pump, or erode various well hardware including slotted liners or screens, gas-lift valves, the surface choke, or any bends in surface pipe.

The cost to remediate consequences of sand production can vary depending on the difficulty to remove accumulated sand or replace worn equipment or eroded hardware. In some cases these costs can be factored in as normal operating expenses, and sand production may be tolerated as part of normal operations. In other cases, such as offshore wells with subsea completions, remediation may be very expensive or even technically unfeasible.

At a low enough flow velocity, formation solids will remain in the formation. The critical sanding rate is the production rate at or below which sand production is avoided. The critical sanding rate may drop over time due to changes in the completion or the flow stream. For example, many wells do not produce formation fines until water breakthrough.

Sand management is a term applied to well completion designs that either prevent sand production by monitoring and controlling the well flow rate and/or drawdown, or permit sand production and plan for handling and disposing of produced sand.

Sand exclusion applies to well completions designed to prevent any sand production into the wellbore. One strategy to achieve sand exclusion is by filtering produced solids before they can be produced into the wellbore with a gravel pack. Another strategy is to reduce the flow velocity at the formation face by designing the flow geometry with more contact area between the completion and the formation, as with high-angle, horizontal, multilateral, and hydraulically fractured wells.

19.2 Sand Flow Modeling

In general, formation solids may become loose after a shear, volumetric, or tensile failure. Sand production occurs when the fluid flow velocity is sufficient to move loose fines, provided no barrier to the fines is encountered.

The purpose of sand flow modeling is to predict the conditions under which sand production is expected. Most of the sand flow models address formation failure mechanisms around perforation cavities. A motivation for much of this work may be to avoid the cost of a conventional gravel pack due to higher initial capital expenditure, maintenance such as a frequent need to acidize the well, and/or lost reserves due to a high gravel pack skin. Instead, the sand management approach is to perforate and complete the well without a gravel pack and then produce below the critical drawdown or below the critical sand flow rate indicated by the sand flow model.

If, for example, the production tubing limits the well flow rate to remain below the predicted critical sanding rate, then the well completion design need not require sand exclusion. As well, if drawdown is expected to stay well below critical levels because pressure maintenance either by natural gas cap or water drive or by gas or water injection is anticipated, again, there may be no need for sand exclusion. The latter example underscores the importance of reservoir management to the completion design.

Quantitative sand prediction studies enable a fit-for-purpose well design and may be worth the investment when compared to potential reductions in well completion and maintenance costs associated with gravel packing.

19.2.1 Factors Affecting Formation Sand Production

Arii, Morita, Ito, and Takano (2005) describe the five events to induce sand problems shown in Figure 19-1. Position 1 is located in the formation where the rock can fail due to either high effective stress or tensile failure. Failure due to high effective stress will occur when the difference between the *in situ* formation stress and the well pressure exceeds the compressive rock strength. Tensile failure occurs when the rock is under tension and may occur whenever the pressure gradient is high. The nature of tensile failure is typically a breakdown of the granular cementation. Position 2 inside the cavity illustrates that fluid flowing at a sufficiently high rate through the failed formation rock in position 1 can dislodge sand particles from the perforation cavity surface. In the horizontal perforation section at position 3, particles will flow when there is sufficient fluid flow velocity, but these particles may form sand bridges or arches at the mouth of the perforation tunnel, shown as position 4. In turn, sand arches can be destroyed under sufficiently high fluid flow velocity, or they may be eroded. Once into the wellbore, sand particles will either fall to collect at the bottom of the well, or at high enough fluid flow rate will be produced to the surface.

Economides and Nolte (2000) provide some basic fundamentals of rock mechanics. The Mohr circle shown in Figure 19-2 provides the information necessary to determine the two-dimensional stress state at any orientation in the sample. That is, for any plane in the sample at

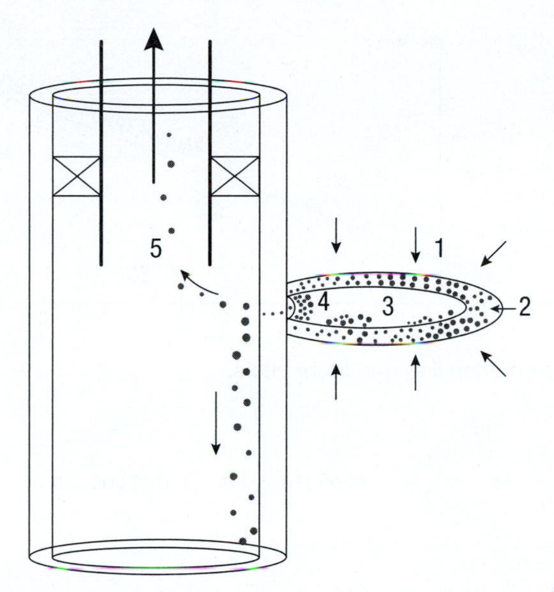

Figure 19-1 Five events to induce sand problems. (After Arii et al., 2005.)

an angle, θ, from the direction of the maximum principle stress, σ_1, the Mohr circle relates the shear stress, τ, to the corresponding normal stress, σ_n, at which the sample will fail, given also the minimum stress, σ_3. Because the case of interest is typically fluid-saturated rock, the stresses are effective stresses given by the difference between actual stress magnitude and the pore pressure. The following equations define the Mohr circle, and the point, $M = (\sigma_c, \tau_c)$, on the Mohr circle defining the normal stress, σ_c, and shear stress, τ_c, for the particular angle, θ_c, shown in the figure:

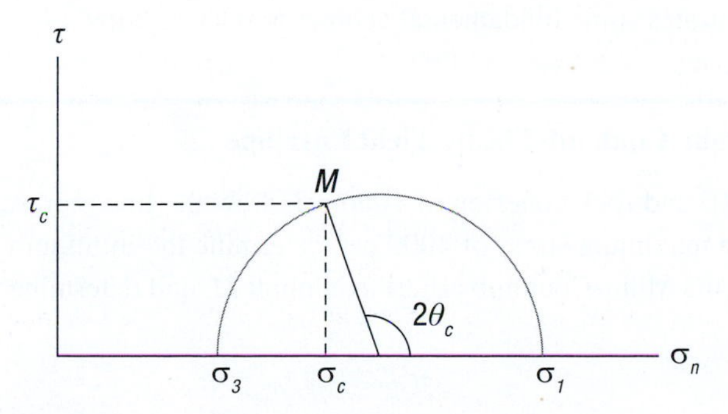

Figure 19-2 Mohr circle.

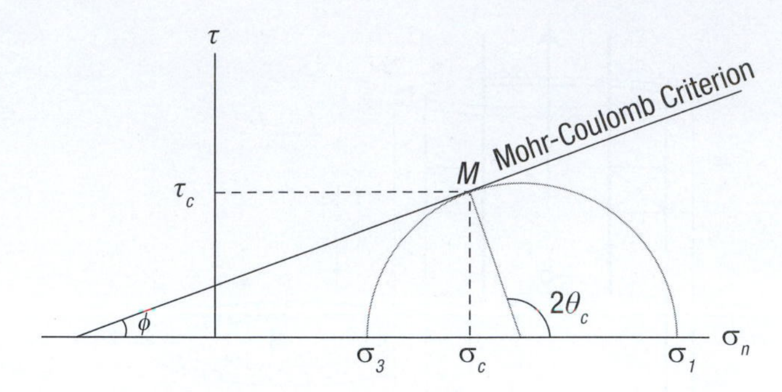

Figure 19-3 Coulomb criterion line with Mohr circle.

$$\sigma_c = \frac{1}{2}(\sigma_1 + \sigma_3) + \frac{1}{2}(\sigma_1 - \sigma_3)\cos 2\theta \tag{19-1}$$

and

$$\tau_c = \frac{1}{2}(\sigma_1 - \sigma_3)\sin 2\theta. \tag{19-2}$$

The Mohr-Coulomb criterion, shown in Figure 19-3, is a line passing through the point, M, and the rock cohesion, C_o, which is described by the equation

$$\tau = C_o + \sigma_n \tan\phi. \tag{19-3}$$

The friction angle, ϕ, is related to the angle θ by

$$2\theta = \phi + \frac{\pi}{2}. \tag{19-4}$$

Example 19-1 illustrates some fundamental geometric relationships.

Example 19-1 The Mohr-Coulomb Plastic Yield Envelope

For a friction angle of 20° and rock cohesion of 1000 psi, draw the line representing the Mohr-Coulomb criterion. For a maximum stress of 4000 psi, determine the minimum stress that gives a Mohr circle tangent to the Mohr-Coulomb criterion at point M, and determine the values of σ_c and τ_c at M.

Solution

For the given values for ϕ and C_o, from Equation 19-1 the intersection of the Mohr-Coulomb line with the effective stress axis is at

$$\sigma_n = -\frac{C_o}{\tan \phi} = -\frac{1000}{0.577} = -2747 \text{ psi.} \tag{19-5}$$

When $\sigma_n = 0$, $\tau = C_o$. The Mohr-Coulomb line must pass through these two points, as shown in Figure 19-4.

The solution now is iterative. First, assume a value for the minimum stress of 1000 psi. The Mohr circle can be drawn for the assumed value, as shown in Figure 19-4 by noting that the center of the circle is the point $\frac{1}{2}(\sigma_1 + \sigma_3)$, and its radius is given by $\frac{1}{2}(\sigma_1 + \sigma_3)$. The point M must have values for σ_n and τ given by Equations 19-1 and 19-2, where the angle θ is computed from the friction angle, ϕ using Equation 19-4:

$$\theta = \frac{\phi}{2} + \frac{\pi}{4} = 55°. \tag{19-6}$$

From Equation 19-4, the value for σ_n for $\theta = 55°$ is

$$\sigma_n = \frac{1}{2}(\sigma_1 + \sigma_3) + \frac{1}{2}(\sigma_1 - \sigma_3)\cos 2\theta = 2500 + 1500\cos(110°) = 1987 \text{ psi,} \tag{19-7}$$

and from Equation 19-3,

$$\tau = \frac{1}{2}(\sigma_1 - \sigma_3)\sin 2\theta = 1500\sin(110°) = 1410 \text{ psi.} \tag{19-8}$$

By iterating on the assumed value for the minimum stress, the correct value for σ_3 is 651 psi. Final values for σ_c and τ_c are 1692 psi and 1615 psi, respectively. The graph with the correct solution is provided in Figure 19-4.

In Location 1 of Figure 19-1 the stress state is that of a perfectly elastic material with stresses lying below the Mohr-Coulomb line. Example 19-2 examines how depletion may cause effective stresses that are unstable and result in flow conditions subject to failure and thereby the potential for sand production.

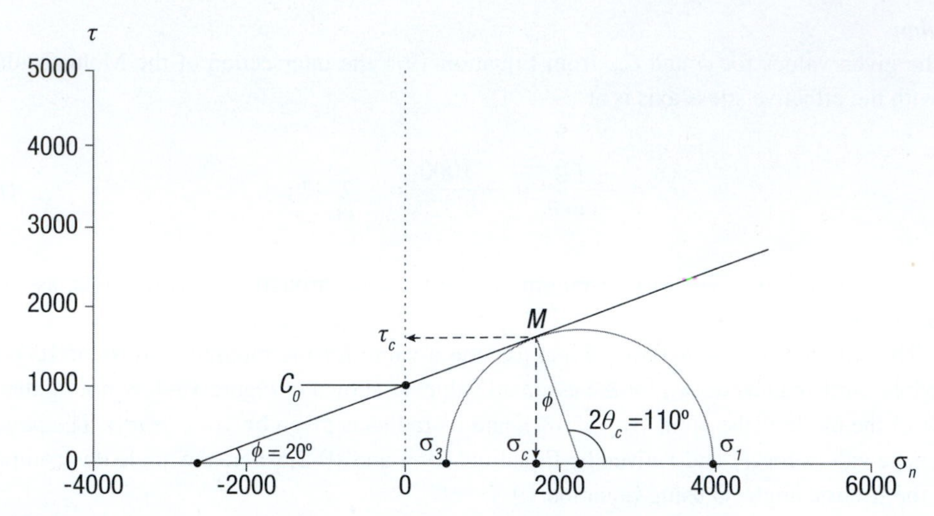

Figure 19-4 Coulomb criterion line with Mohr circle.

Example 19-2 Drawdown Failure Condition

Suppose a sandstone reservoir with formation rock characteristics like those in Example 19-1 is normally pressured at a depth of 10,000 ft. Assuming that the vertical stress gradient is 1.1 psi/ft and that the vertical stress, the friction angle, and the rock cohesion remain constant, draw the Mohr circle at initial reservoir pressure. At what drawdown will failure conditions commence in the formation?

Solution

A vertical stress gradient of 1.1 psi/ft indicates that the vertical stress at 10,000 ft is 11,000 psi (Chapter 18). A normally pressured reservoir has an initial reservoir pressure at hydrostatic pressure given by 0.433H = 4330 psi. From Equation 18-3, for this case the effective vertical stress is

$$\sigma_1' = \sigma_1 - \alpha p = 11{,}000 - (0.7)(4330) = 7969 \text{ psi} \tag{19-9}$$

where the Biot poroelastic constant, α, is assumed to be equal to 0.7. From Equation 18-4, the effective minimum stress for this case is

$$\sigma_3' = \frac{\nu}{1-\nu}\sigma_1' = \frac{0.25}{1-0.25}7969 = 2656 \text{ psi}, \tag{19-10}$$

assuming a Poisson ratio, ν, for a typical sandstone of 0.25. The Mohr circle for these values of effective maximum and minimum stress is shown in Figure 19-5 for 0 drawdown pressure. Also shown is the wellbore pressure. Because it falls inside the circle, no failure is expected.

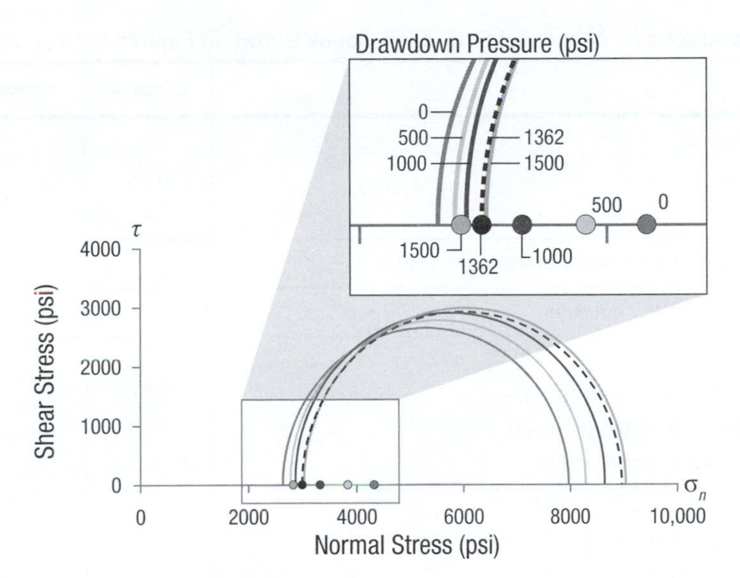

Figure 19-5 Mohr circles for Example 19-2.

For drawdown of 500 and 1000 psi, still no failure is expected, but for 1500 psi, the wellbore pressure falls left of the Mohr circle, and failure is expected. Iterating on the assumed drawdown finds that failure conditions occur at or above a drawdown of 1362 psi, which corresponds to a wellbore flowing pressure of 2968 psi.

For a very simplistic failure model, Example 19-2 indicates that sand production can be avoided by producing the well at a wellbore flowing pressure above 2968 psi. Actual models for predicting sand production are much more complex and require numerical simulation. Weingarten and Perkins (1995) derived an analytical model of perforation stability in gas wells based on flow into a hemispherical perforation. The model accounts for both drawdown, causing local failure around a perforation, and depletion, causing general shear failure in the reservoir. The model is intended to provide a quick assessment about the need for a gravel pack and indicates that the Mohr circle analysis is overly conservative.

Results from an example numerical model from Morita, Whitfill, Massie, and Knudsen (1989) are shown in Figure 19-6 for formation and fluid data shown in Table 19-1. These results are for a rock of intermediate strength. Critical flow and drawdown relationships are shown as solid curves for a slim cavity such as a newly formed perforation, and as dashed curves for an enlarged cavity after some borehole sand production.

Three types of failures that can lead to sand production are shown in Figure 19-6. First, the curves for shear failure occur for large drawdown, even when the flow rate is small. As flow rate increases, shear failure can occur for smaller drawdown. Comparing the initial perforation to the enlarged cavity (solid curve to dashed curve) reveals that shear failures become less likely after

Table 19-1 Conditions for Numerical Simulation Cases Shown in Figures 19-6 and 19-7

	Case A	Cases B & D
Cavity size		
Diameter, in.	0.65	3.6
Length, in.	7.5	7.5
Case A: Semiellipsoid with center at the outlet center		
Case B: Diameter of outlet holes, in.		0.85
Permeability		
Formation ($\beta = 0.9 \times 10^6$ ft^{-1}), md	222	222
Damaged ($\beta = 6 \times 10^6$ ft^{-1}), md	22.2	22.2
Thickness of damaged zone, in.	0.5	0.2
All Cases:		
Oil viscosity, cp		0.6
Oil density, g/cm^3		0.8
Perforation		
Density, shots/ft		4
Phasing, degrees		90
In-situ stress		
Overburden, psi/ft		0.8
K factor		0.7
Depth, ft		6,131
Initial pore pressure, psi		4,400

Cases A and B have intermediate formation strength.
Case D has weak formation strength with 150 psi unconfined compressive strength.

an initial failure enlarges the perforation cavity. The shear failures shown are due to compressional stresses like what were described in the introduction to this section and in Chapter 18. The maximum value for Δp_w provides the critical drawdown and Δp_w is the difference between the average reservoir pressure and the well flowing pressure.

The second type of failure shown in the figure is tensile failure, and there are two types of tensile failures shown: one during unloading and the other for a loaded path. Unloading occurs when the formation is subjected to a sudden drop in pressure, as may occur when the well first starts flowing or during underbalanced perforating. Unloading can also occur after the well has been shut in for a fairly long period.

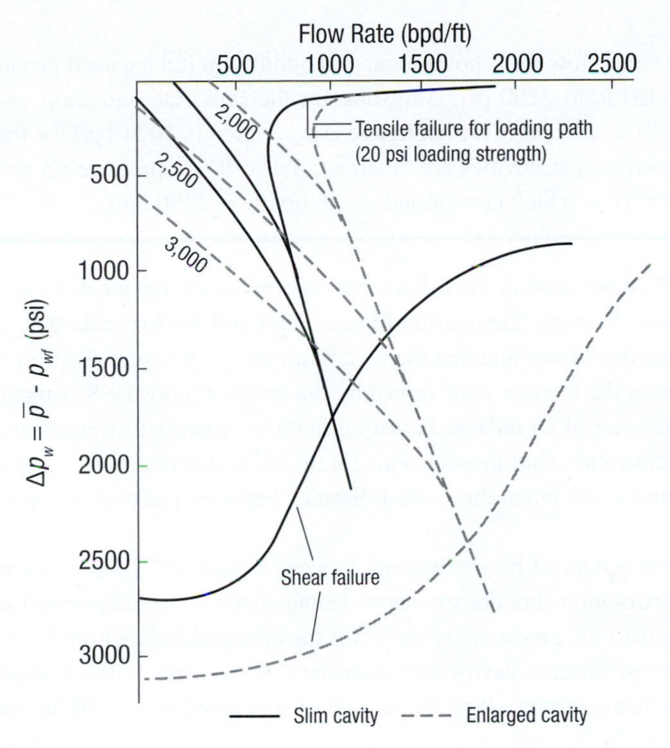

Figure 19-6 Simulated failure criteria for intermediate strength rock using input data for Cases A and B in Table 19-1. (After Morita et al., 1989.)

Morita et al. (1989) indicate that Figure 19-6 can be used for other permeability and viscosity values than those shown in Table 19-1 by multiplying the flow rate in the figure by (0.6/22.2) k/μ where k in md is the near wellbore (damaged zone) permeability and μ is the produced fluid viscosity in cp.

Example 19-3 Sand Production after Shut-in

For the reservoir in Example 19-2 with additional fluid and formation properties as in Appendix A, assume that the reservoir permeability is 82 md and the damage permeability is 8.2 md. Suppose the well drawdown was 2100 psi before being shut in. At what initial rate after shut-in will sand production occur based on the model in Figure 19-6 if the well is new and perforation cavities are still slim? If perforation cavities are enlarged from previous sand production?

Solution

Figure 19-6 shows that flow rate above about 150 bpd/ft will induce sand production due to tensile failure unloaded from 2100 psi. Adjusting for the flow rate and fluid viscosity gives 150 (0.6/22.2)(8.2/1.72) = 20 bpd/ft for negligible Δp_w, which is 1024 bpd for the 53-ft formation thickness. After perforation cavities are enlarged, the well can produce up to about 350 bpd/ft without sand production, which corresponds to 46 bpd/ft or 2390 bpd.

Figure 19-7 shows critical drawdown pressure behavior for weak rock, characterized by 150 psi compressive strength. Comparing Figures 19-6 and 19-7 reveals that weak rock fails at much lower critical drawdown and that the rock requires less pressure drop for shear failure than for tensile failure on the loading path shown by the arrow. Figure 19-8 compares critical drawdown pressure behavior of an enlarged cavity in both intermediate strength and weak rock as a function of the perforation shot density. This figure illustrates that high perforation density can result in cavity interaction when the closest distance between perforation cavities becomes less than 1–2 in.

Drawdown is enhanced by perforation damage. Figure 19-9 shows the productivity ratio as a function of perforation shot density for undamaged and damaged perforations. In this figure the top curves contrast the productivity ratios for the enlarged hole with diameter 3.6 in. and for convergent flow to perforation cavity with diameter 0.85 in., both without perforation or formation damage. The lower curves show the impact of a crushed zone with permeability 1/10 the

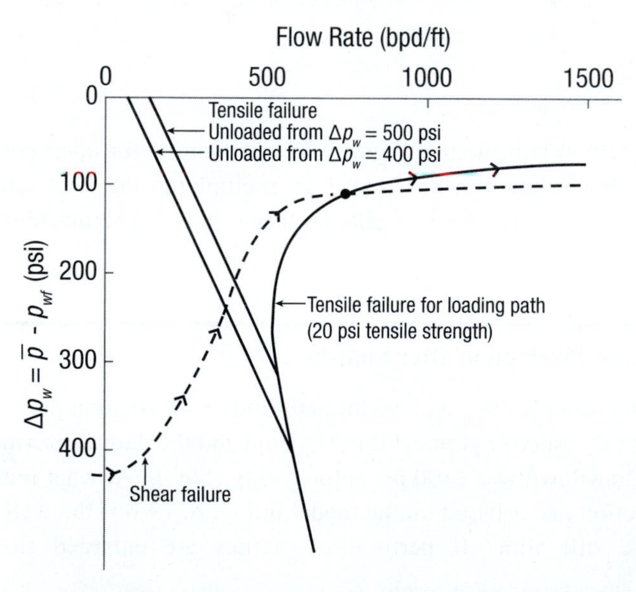

Figure 19-7 Simulated failure criteria for enlarged cavity in soft rock using input data for Case D in Table 19-1. (After Morita et al., 1989.)

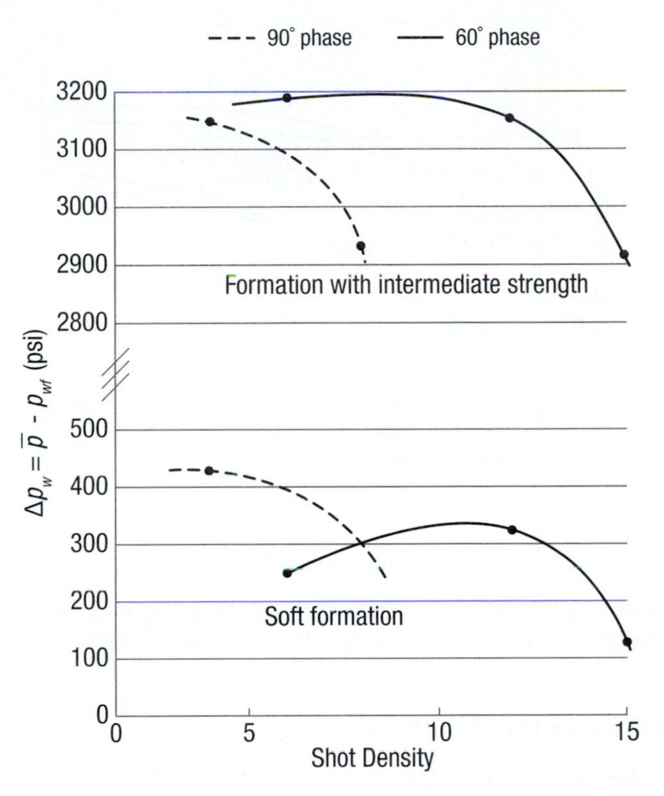

Figure 19-8 Simulated critical drawdown pressure for enlarged cavity with diameter 3.8 in. for perforation length of 7.5 in. and well diameter 12.25 in., with other data as in Table 19-1. (After Morita et al., 1989.)

formation permeability penetrating 0.5 in. from the perforation cavity and formation damage permeability ratio $k_s/k = 0.4$ and penetrating 5 in. For all cases the perforation length is 7.5 in., and the borehole diameter is 12.25 in. For the case simulated in this figure, even at 12 shots per foot, the drawdown will be 25% higher for the same production rate with the indicated perforation and wellbore damage.

Often sand production either begins or increases significantly after water breakthrough. Morita et al. (1989) suggest the reasons for this include (1) fines fixed by capillary pressure are released with free water saturation, (2) pressure gradient near the wellbore increases when oil and water flow together, (3) reservoir pressure may have depleted prior to the onset of waterflood pressure maintenance, (4) dissolution of cementing materials between sand grains by water, and (5) total fluid flow rate may be increased to maintain oil flow rate with increasing watercut. The last reason may lead to increased damage due to plugging caused by fines migration under higher flow velocity, in turn causing a pressure gradient in excess of the critical drawdown.

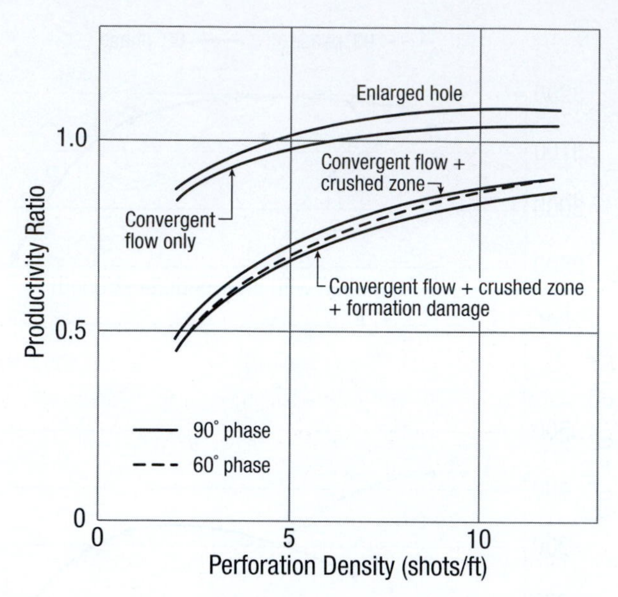

Figure 19-9 Productivity ratio versus perforation density showing sensitivity to crushed zone and formation damage and perforation phasing. (After Morita et al., 1989.)

19.2.2 Sand Flow in the Wellbore

When sand flows into the wellbore, it may flow to the surface, or it may settle in the rat hole depending on the wellbore flow velocity. Stokes' law can be used to determine the velocity sufficient to lift sand particles to the surface assuming spherical particles and Newtonian liquids in laminar flow. This velocity, known as the slip velocity, u_s, must be at least 50% more than the terminal settling velocity from Stokes' law and is given in m/s by Kelessidis, Maglione, and Mitchel (2011) for cuttings transport during drilling as

$$u_s = \frac{1}{18}\frac{d_p^2}{\mu}(\rho_s - \rho_f)g. \tag{19-11}$$

for particle diameter, d_p in m, particle and fluid densities ρ_s and ρ_f in kg/m^3, and viscosity, μ in Pa-s. The onset of turbulence can be related to a particle Reynolds number given by

$$N_{Re} = \frac{\rho_f u_s d_p}{\mu} \tag{19-12}$$

and Stokes' law is found to give acceptable accuracy for $N_{Re} < 0.1$. For $N_{Re} > 0.1$, an empirically determined factor, or drag coefficient, C_D, is used.

$$C_D = \frac{4}{3} g \frac{d_p}{u_s^2}\left(\frac{\rho_s - \rho_f}{\rho_f}\right). \tag{19-13}$$

For nonspherical particles, the drag coefficient can be adjusted using Figure 19-10. Use of this figure is illustrated in the following example.

Example 19-4 Sand Production to Surface

If sand particles with mean diameter 0.006 in., specific gravity 2.65, and sphericity of crushed silica are produced into the wellbore when the well in Example 19-3 is flowed at 1200 bpd, is sand produced at the surface? Assume the wellbore fluid density is 55 lbm/ft^3, and the casing diameter is 7 in.

Solution

In SI units the particle diameter d_p, the fluid density ρ_f, the fluid viscosity μ, and the solid density ρ_s are

$$d_p = \frac{0.006 \text{ in}}{(3.937)(10 \text{ in/m})} = 0.000152 \text{ m} \tag{19-14}$$

$$\rho_f = \frac{55 \text{ lbm/ft}^3}{0.062428(\text{lbm/ft}^3)/(\text{kg/m}^3)} = 881 \text{ kg/m}^3 \tag{19-15}$$

$$\mu = \frac{1.72 \text{ cp}}{10^3 \text{ cp/(Pa} \cdot \text{s)}} = 0.00172 \text{ Pa} \cdot \text{s} \tag{19-16}$$

$$\rho_s = \frac{2.65(62.4 \text{ lbm/ft}^3)}{0.062428(\text{lbm/ft}^3)/(\text{kg/m}^3)} = 2650 \text{ kg/m}^3 \tag{19-17}$$

As a first guess, assume Stokes' law applies, and use Equation 19-11:

$$u_s = \frac{1}{18}\frac{d_p^2}{\mu}(\rho_s - \rho_f)g = \frac{1}{18}\frac{(0.000152)^2}{0.0017}(2650 - 881)(9.81) = 0.0130 \text{ m/s}. \tag{19-18}$$

This slip velocity corresponds to a drag coefficient of

$$C_D = \frac{4}{3} g \frac{d_p}{u_s^2}\left(\frac{\rho_s - \rho_f}{\rho_f}\right) = \frac{4}{3}(9.81)\frac{0.000152}{0.0130^2}\left(\frac{2650 - 881}{881}\right) = 23.6 \tag{19-19}$$

and a Reynolds number of

$$N_{Re} = \frac{\rho_f u_s d_p}{\mu} = \frac{(881)(0.0130)(0.000152)}{0.00172} = 1.02. \quad \textbf{(19-20)}$$

Entering Figure 19-10 at the point ($C_D = 23.6$, $N_{Re} = 1.02$) and moving parallel to the slant lines to the curve for crushed silica yields an intersection point at ($C_D = 70$, $N_{Re} = 0.6$). Thus, the slip velocity is given by solving for u_s in Equation 19-13,

$$u_s = \sqrt{\frac{4}{3}\frac{g d_p}{C_D}\frac{(\rho_s - \rho_f)}{\rho_f}} = \sqrt{\frac{4}{3}\frac{9.81 * 0.000152}{70}\left(\frac{2650 - 881}{881}\right)}$$
$$= 0.00756 \text{ m/s} = 0.0248 \text{ ft/s} \quad \textbf{(19-21)}$$

Recomputing with the new value for u_s gives $C_D = 70$ and $N_{Re} = 0.6$, which corresponds to the same slip velocity.

At 1200 bpd, the wellbore flow velocity is

$$u = \frac{q}{A} = \frac{1200}{\pi\left(\frac{7}{24}\right)^2}\left(\frac{5.615}{86{,}400}\right) = 0.292 \text{ ft/s} = 0.0890 \text{ m/s}. \quad \textbf{(19-22)}$$

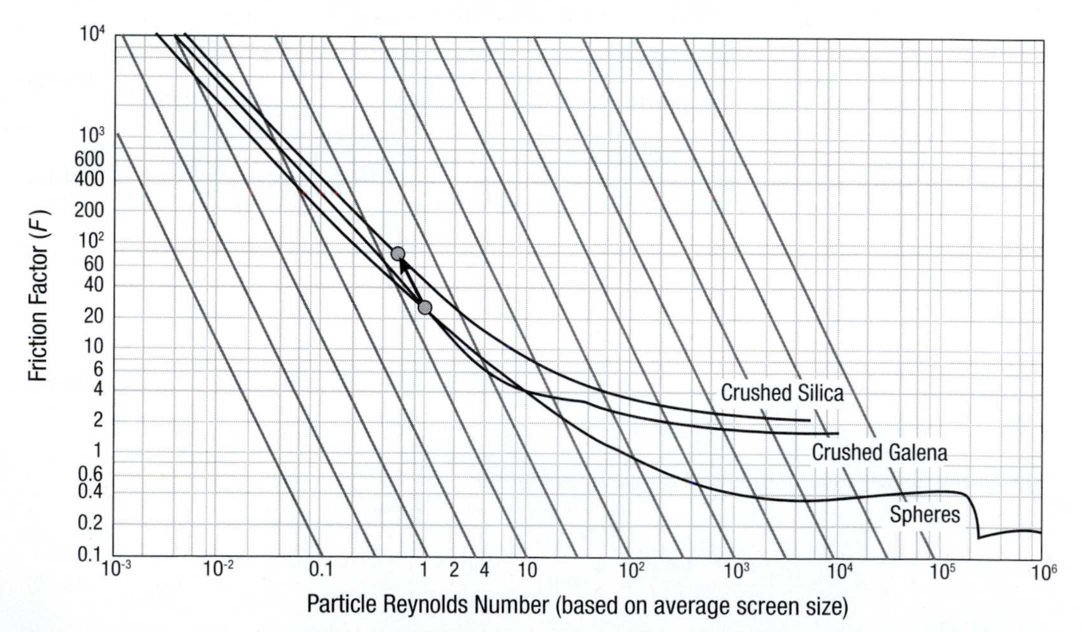

Figure 19-10 Friction factor (drag coefficient) for computing particle slip velocity. (From Mitchell and Miska, 2011.)

This is more than 50% above the slip velocity, and produced sand will flow to the surface. If, instead, the well is produced below 100 bpd, produced sand would fall and accumulate in the rat hole.

Example 19-5 Sand Accumulation in the Wellbore

If the sand is flowing at about 0.5% by volume, for the tubing shoe depth at 10,000 ft and production tubing ID 2 7/8 in., to what depth will sand settle in the rat hole at a depth of 10,200 ft when the well is shut in?

Solution

The volume of the wellbore is given by

$$V_w = 200\pi\left(\frac{7}{24}\right)^2 + 10{,}000\pi\left(\frac{2.875}{24}\right)^2 = 504 \text{ ft}^3. \tag{19-23}$$

If the sand settles with a porosity of 0.40, the sand volume will be approximately $504(0.005)/(1 - 0.4) = 4.2 \text{ ft}^3$, which corresponds to a depth of $4.2/[\pi(7/24)^2] = 15.7$ ft. Every time the well is shut in, additional sand will be deposited in the rat hole as long as sand production continues.

Example 19-6 Sand Erosion

A rate of 1 ft/s with 0.1% sand by volume is considered abrasive, removing 0.02 in. of metal thickness per month. After how much time will it go through a screen that is 0.25 in. thick? Assuming 200 perforations with $r_{perf} = 0.25$ in., what is the volumetric flow rate ($B_o = 1.2$ resbbl/STB) to give a velocity of 1 ft/s through each perforation and how much sand is produced each day? The density of the sand is 165 lb/ft^3.

Solution

For a 0.25-in. thick screen it will take about 12 months to be eroded at the given rate of erosion.

With $r_{perf} = 0.25$ in, the cross-sectional area of flow for each perforation is 1.36×10^{-3} ft^2, which translates to 0.27 ft^2 for all perforations. Furthermore, this means 0.27 ft^3/s or 4190 resbbl/d or 3500 STB/d (dividing by 1.2).

At 0.1% sand by volume, the sand production is 3.5 barrels or 3250 lb per day.

Frequently, sand production occurs briefly each time the well is brought back on production after it has been shut in. Morita et al. (1989) recommend to avoid increasing the drawdown beyond a critical cyclic value. In general, sand management without sand exclusion relies on avoiding flow velocity and drawdown sufficient to induce excessive sand production.

19.3 Sand Management

Any completion that does not provide for sand exclusion may be subject to sand production. Because sand exclusion completions are more expensive, and because gravel pack completions result in a positive skin from the beginning that grows in magnitude over time, the operator may prefer to avoid the need for sand exclusion, either by producing below the critical sanding rate or critical drawdown, or by deliberately producing sand in a well design that is not destroyed by sand production.

19.3.1 Sand Production Prevention

Example 19-7 Sand Production Prevention

For the reservoir in Example 19-2 with additional fluid and formation properties as in Appendix A except assuming a permeability of 82 md, if near-well damage results in a skin of 3, what initial flow rate is permissible? If fine particles collect in the perforation cavity and cause the skin to increase to 30, what flow rate is permissible? If the minimum bottomhole flowing pressure is 1323 psi, what percent of the oil in place can be recovered at the original rate if the skin is 3? If the skin is 30?

Solution

From Example 19-2, to avoid exceeding the maximum drawdown pressure drop of 1362 psi, the flowing bottomhole pressure must not drop below 2968 psi. For an initial pressure of 4330 psi, to prevent sand production the maximum allowable production rate is

$$q = \frac{kh\Delta p}{141.2B\mu\left[\ln\left(0.472\frac{r_e}{r_w} + s\right)\right]}$$

$$= \frac{(82)(53)(4330 - 2968)}{141.2(1.17)(1.72)\left[\ln\left(0.472\frac{1500}{0.328} + 3\right)\right]} = 1950 \text{ STB/d.} \quad \textbf{(19-24)}$$

With a skin of 30, the flow rate must not exceed 595 STB/d. For a minimum bottomhole flowing pressure of 1323 psi, to maintain the rate at 2090 STB/d, the average pressure in the well drainage area must be

$$\bar{p} = \Delta p_{\max} + p_{wf,\min} = 1362 + 1323 = 2685 \text{ psi.} \quad \textbf{(19-25)}$$

The oil recovery estimated from Equation 10.7 is

$$r = e^{c_t(p_i - \bar{p})} - 1 = e^{1.29\times10^{-5}(4330-2685)} - 1 = 2.1\%. \quad \textbf{(19-26)}$$

If the skin is 30, the recovery factor will be the same, but at the lower rate it will take much longer to produce the same volume of oil. After the bottomhole pressure drops to the minimum value, the well rate declines.

Example 19-7 illustrates that drawdown limits can be very restrictive. However, under pressure maintenance by waterflooding, according to this simple model, the well can be produced at 1950 STB/d with skin 3 or 595 STB/d with skin 30 at least until water breakthrough. If the breakthrough recovery factor is expected to be satisfactory, then the sand production prevention strategy may be an appropriate choice.

19.3.2 Cavity Completion

The cavity completion (cavity-like completion) risks some sand production in order to remove near wellbore damage and condition the completion. The result is a zone of higher porosity near the wellbore or even cavities. For some formations, modeling can predict approximately how much sand should be produced. Once the sand is produced, no further sand may be produced for a long time if drawdown is reduced somewhat. The strategy can be to produce sand before finalizing connection to the surface production facility. Eventual resumption of sand production may occur with water breakthrough, or by stressing the completion with repeated shut-ins over time. Also, shut-in and production resumption should avoid sudden rate changes. As a rule of thumb, this approach will work for a weakly cemented formation with unconfined compressive strength in the range of 20–50 psi in a new well, and proportionally higher compressive strength is needed for a depleted reservoir. Palmer and McClennan (2004) offer a comprehensive description of this completion strategy, which also provides considerable insight on the nature of sand management strategies in general. While the cavity completion strategy involves some risk, results often offset the initial costs to handle produced sand.

19.4 Sand Exclusion

Many reservoirs comprised of relatively young sediments are so poorly consolidated that sand will be produced along with the reservoir fluids unless the rate is restricted significantly. Sand production leads to numerous production problems, including erosion of downhole tubulars; erosion of valves, fittings, and surface flow lines; the wellbore filling up with sand; collapsed casing because of the lack of formation support; and clogging of surface processing equipment. Even if sand production can be tolerated, disposal of the produced sand is a problem, particularly at offshore fields. Thus, a means to eliminate sand production without greatly limiting production rates is desirable.

Originally, sand production was controlled by using gravel pack completions, slotted liner completions, or sand consolidation treatments, with gravel pack completions having been by far the most common approach. Reviews of conventional historical sand-control completion practices are given by Suman, Ellis, and Snyder (1983) and more recently by Penberthy and Shaughnessy (1992) and Ghalambor, Ali, and Norman (2009).

In the 1990s, operators noted that successful prepack gravel placement often posed a need to inject at pressures exceeding the formation fracture pressure. This led to development of the frac-pack approach designed to create a fracture extending past the radius of impaired permeability induced by drilling and completion fluid invasion. In a systematic study of completion trends in Gulf of Mexico fields, Tiner, Ely, and Schraufnagel (1996) showed that frac-pack completions had consistently lower skin compared to traditional gravel pack and high-rate water pack completions. More recent studies by Norman (2003) and Keck, Colbert, and Hardham (2005) have indicated that frac-pack completions also show much longer completion lifetime before failure.

A positive skin of 5–10 is considered excellent in traditional gravel pack completions, and much higher skins have been observed, leading to loss of as much as 50% of the well production capability. High-rate water pack techniques consistently reduce skins in gravel pack completions to a range of +2 to +5 by placing gravel outside the perforation tunnels between the cement and the formation during the prepack stage of the gravel packing operation. Modern unified hydraulic fracturing techniques employ tip screenout techniques, described in Chapter 17, to control fracture half length and width to design specifications.

In practice, high-rate water pack and frac-pack completions have achieved at best zero and even slightly negative total completion skins as low as −2. Reasons why hydraulic stimulation does not result in more negative skin even for typical fracture half lengths of more than 50 ft are provided in Section 19.4.2. However, in the high-rate water pack, all of the flow still passes through perforations subject to the same failure mechanisms described in the previous section. The gravel pack prevents sand production by trapping the particles, and over time the gravel pack skin will increase. While the gravel pack skin can be reduced with acid injection, this may not be an option in subsea completions common offshore, especially in deep water. In contrast, because the flow geometry in the frac-pack provides greater flow area and, hence, less pressure gradient at the formation face, the frac-pack skin is not expected to increase over time, reducing or eliminating a need for completion maintenance.

19.4.1 Gravel Pack Completion

In a gravel pack completion, sand that is larger than the average formation sand grain size is placed between the formation and a screen or slotted liner. The gravel pack sand (referred to as gravel, though it is actually sand in grain size), should retain most of the formation sand, but let very fine particles pass through it and be produced. The two most common types of gravel pack completions are an inside-casing gravel pack and an open-hole or underreamed-casing gravel pack (Figure 19-11). The underreamed-casing gravel pack provides better conductivity through the gravel, but is limited to single-zone completions. A successful gravel pack completion must retain the formation sand and offer the least possible resistance to flow through the gravel itself.

19.4.1.1 Gravel Pack Placement

For a successful gravel pack completion, gravel must be adjacent to the formation without having mixed with formation sand, and the annular space between the screen and the casing or formation

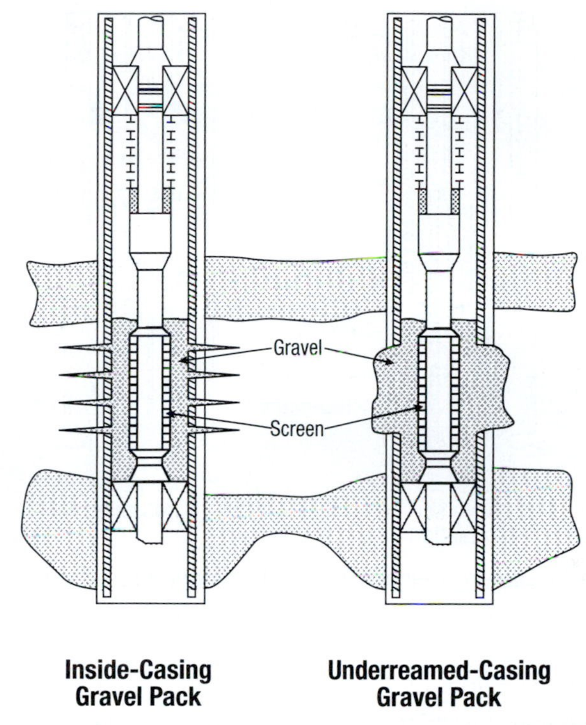

Figure 19-11 Common types of gravel pack completions.

must be completely filled with gravel. Special equipment and procedures have been developed over the years to accomplish good gravel placement.

Water or other low-viscosity fluids were first used as transporting fluids in gravel pack operations. Because these fluids could not suspend the sand, low sand concentrations and high velocities were needed. Now, viscosified fluids, most commonly solutions of hydroxyethylcellulose (HEC), are used so that high concentrations of sand can be transported without settling (Scheuerman, 1984). Just as with the fluids used in hydraulic fracturing, it is desirable that these solutions degrade to low viscosity with little residue, requiring the addition of breakers to the polymer solution.

In open-hole completions, the gravel-laden fluid can be pumped down the tubing casing annulus, after which the carrier fluid passes through the screen and flows back up the tubing. This is the reverse-circulation method depicted in Figure 19-12 on the left (Suman et al., 1983). A primary disadvantage of this method is the possibility of rust, pipe dope, or other debris being swept out of the annulus and mixed with the gravel, damaging the pack permeability. More commonly, a crossover method is used, in which the gravel-laden fluid is pumped down the tubing, crosses over to the screen-open-hole annulus, flows into a wash pipe inside the screen, leaving the gravel in the annulus, and then flows up the casing–tubing annulus to the surface (Figure 19-12 on the right). Notice that the open-hole section is usually underreamed through the productive interval to increase well productivity.

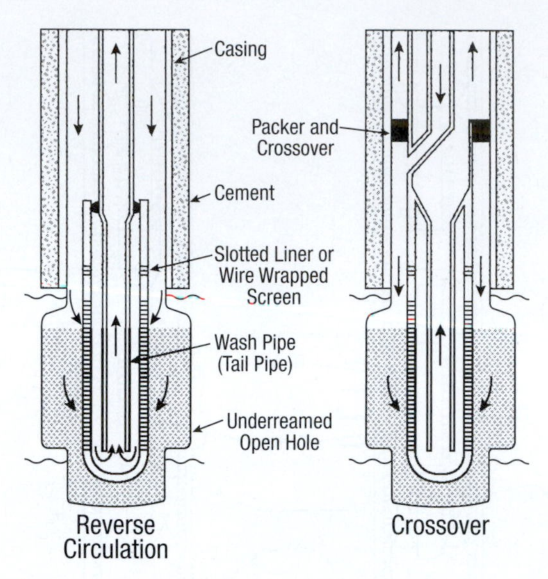

Figure 19-12 Gravel placement methods in open-hole or underreamed casing completions. (From Suman et al., 1983.)

For inside-casing gravel packing, washdown, reverse-circulation, and crossover methods are used (Figure 19-13) (Suman et al., 1983). In the washdown method, the gravel is placed opposite the productive interval before the screen is placed, and then the screen is washed down to its final position. The reverse-circulation and crossover methods are analogous to those used in open holes. A modern crossover method, described in detail by Schechter (1992), is shown in Figure 19-14. Gravel is first placed below the perforated interval by circulation through a section of screen called the telltale screen. When this has been covered, the pressure increases, signaling the beginning of the squeeze stage. During squeezing, the carrier fluid leaks off to the formation, placing gravel in the perforation tunnels. After squeezing, the washpipe is raised, and the carrier fluid circulates through the production screen, filling the casing production screen annulus with gravel. Gravel is also placed in a section of blank pipe above the screen to provide a supply of gravel as the gravel settles.

In deviated wells, gravel packing is greatly complicated by the fact that the gravel tends to settle to the low side of the hole, forming a dune in the casing–screen annulus. This problem is significant at deviations greater than 45° from vertical (Shryock, 1983). Gravel placement is improved in deviated wells by using a washpipe that is large relative to the screen (Gruesbeck, Salathiel, and Echols, 1979); this causes a higher velocity over the dune in the annulus between the screen and the casing by increasing the resistance to flow in the screen–washpipe annulus. Shryock's results also suggest that an intermediate-viscosity carrier fluid may be preferable to high-viscosity fluids in deviated wellbores.

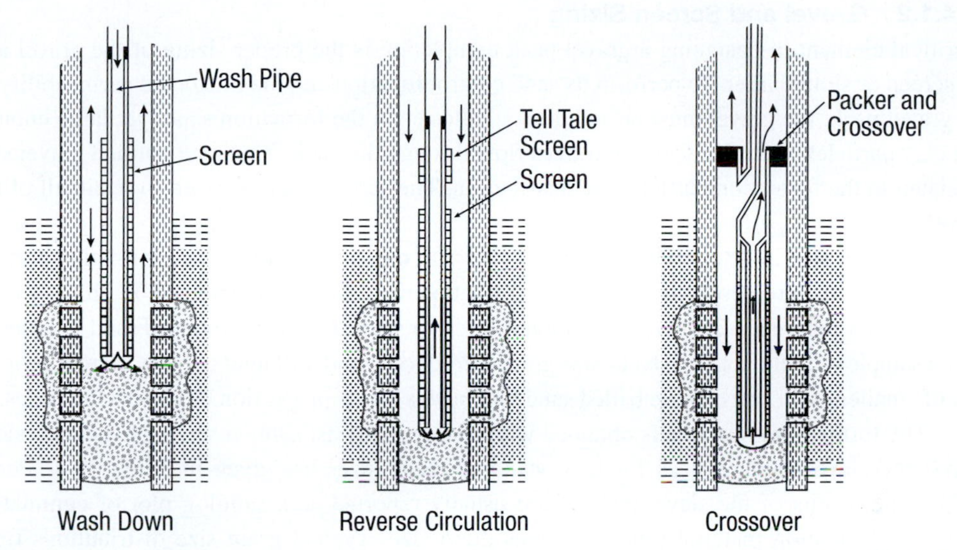

Figure 19-13 Gravel placement methods for inside-casing gravel packs. (After Suman et al., 1983.)

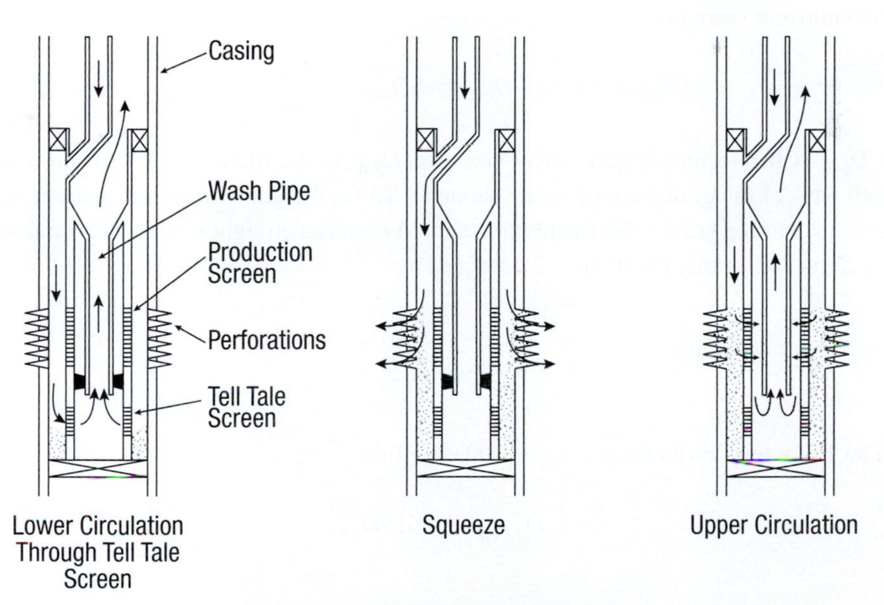

Figure 19-14 Steps in a crossover method of gravel pack placement. (After Suman et al., 1983.)

19.4.1.2 Gravel and Screen Sizing

A critical element in designing a gravel pack completion is the proper sizing of the gravel and the screen or slotted liner. To perform its sand control function and maximize the permeability of the gravel pack, the gravel must be small enough to retain the formation sand, yet large enough that clay particles and other formation fines flow through the pack. Thus, the optimal gravel size is related to the formation particle size distribution. The screen must be sized to retain all of the gravel.

The first step in determining the gravel size is to measure accurately the formation particle size distribution. A representative sample of formation material can be obtained, in order of preference, from rubber-sleeve cores, conventional cores, or sidewall cores. Produced sand samples or bailed samples should not be used to size gravel. Produced sand will tend to have a larger proportion of smaller grain sizes, while bailed sand will have a larger proportion of larger grain sizes.

The formation grain size is obtained with a sieve analysis, using a series of standard sieve strays; the sieve opening sizes for U.S. standard mesh sizes are given in Table 19-2 (Perry, 1963). The results of the sieve analysis are usually reported as a semilog plot of cumulative weight of formation material retained versus grain size. Typical grain size distributions from California and U.S. Gulf Coast unconsolidated sands are shown in Figure 19-15 (Suman et al., 1983).

Schwartz (1969) and Saucier (1974) presented similar (but slightly different) correlations for optimal gravel size based on the formation grain size distribution. Schwartz's correlation depends on the uniformity of the formation and the velocity through the screen but for most conditions (nonuniform sands) is

$$D_{g40} = 6D_{f40} \tag{19-27}$$

where D_{g40} is the recommended gravel size and D_{f40} is the diameter of formation sand for which 40 wt% of the grains are of larger diameter. To fix the gravel size distribution, Schwartz recommends that the gravel size distribution should plot as a straight line on the standard semilog plot, and a uniformity coefficient, U_c, defined as

$$U_c = \frac{D_{g40}}{D_{g90}} \tag{19-28}$$

should be 1.5 or less. From these requirements, we find

$$D_{g,\min} = 0.615D_{g40} \tag{19-29}$$

and

$$D_{g,\max} = 1.383D_{g40} \tag{19-30}$$

Table 19-2 Standard Sieve Sizes[a]

	Sieve Opening	
U.S. Standard Mesh Size	**(in.)**	**(mm)**
2 1/2	0.315	8.00
3	0.265	6.73
3 1/2	0.223	6.68
4	0.187	4.76
5	0.157	4.00
6	0.132	3.36
7	0.111	2.83
8	0.0937	2.38
10	0.0787	2.00
12	0.0661	1.68
14	0.0555	1.41
16	0.0469	1.19
18	0.0394	1.00
20	0.0331	0.840
25	0.0280	0.710
30	0.0232	0.589
35	0.0197	0.500
40	0.0165	0.420
45	0.0138	0.351
50	0.0117	0.297
60	0.0098	0.250
70	0.0083	0.210
80	0.0070	0.177
100	0.0059	0.149
120	0.0049	0.124

(*continued*)

Table 19-2 *Continued*

	Sieve Opening	
U.S. Standard Mesh Size	**(in.)**	**(mm)**
140	0.0041	0.104
170	0.0035	0.088
200	0.0029	0.074
230	0.0024	0.062
270	0.0021	0.053
325	0.0017	0.044
400	0.0015	0.037

[a]From Perry, (1963).

where $D_{g,\min}$ and $D_{g,\max}$ are the minimum and maximum sizes of gravel to be used, respectively. Equations (19-29) and (19-30) define the range of recommended gravel size.

Saucier recommends that the geometric mean gravel size be five or six times the mean formation grain size, or

$$D_{g50} = (5 \text{ or } 6)D_{f50}. \tag{19-31}$$

Saucier gave no recommendation about gravel size distribution. If we apply Schwartz's criteria, then

$$D_{g,\min} = 0.667D_{g40} \tag{19-32}$$

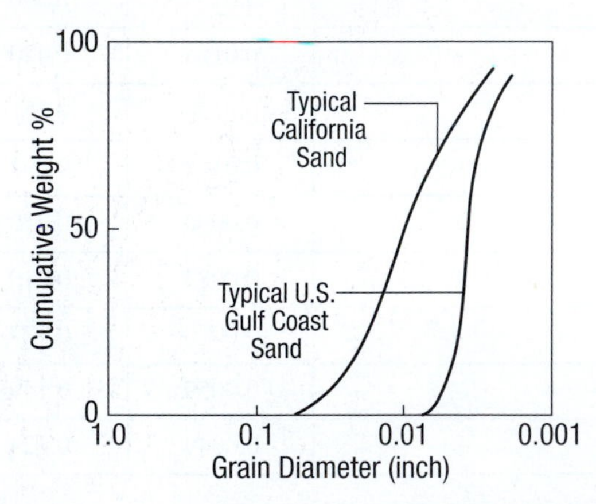

Figure 19-15 Grain size distributions for California and U.S. Gulf Coast sands. (From Suman et al., 1983.)

$$D_{g,\max} = 1.5D_{g50} \tag{19-33}$$

The screen openings should be small enough to retain all of the gravel, requiring that the screen openings be slightly less than the smallest gravel size.

Example 19-8 Selecting the Optimum Gravel and Screen Sizes

Using both the Schwartz and Saucier correlations, determine the optimal gravel and screen size for the California unconsolidated sand for which the grain size distribution was given in Figure 19-15.

Solution

Schwartz Correlation

The grain size distribution for the California unconsolidated sand is replotted in Figure 19-16. Reading from the plot for cumulative weight fraction of 40%, we find that $D_{f40} = 0.0135$ in. The 40% grain size for the gravel is then (6)(0.0135 in.) = 0.081 in. The 90% grain size is $D_{g40}/1.5 = 0.054$ in. The recommended grain size distribution is shown graphically as the dashed line on Figure 19-16; the intersections with the cumulative weight % = 100 and = 0 lines define the minimum and maximum gravel sizes, respectively calculated with Equations (19-29) and (19-30) to be 0.05 and 0.11 in.

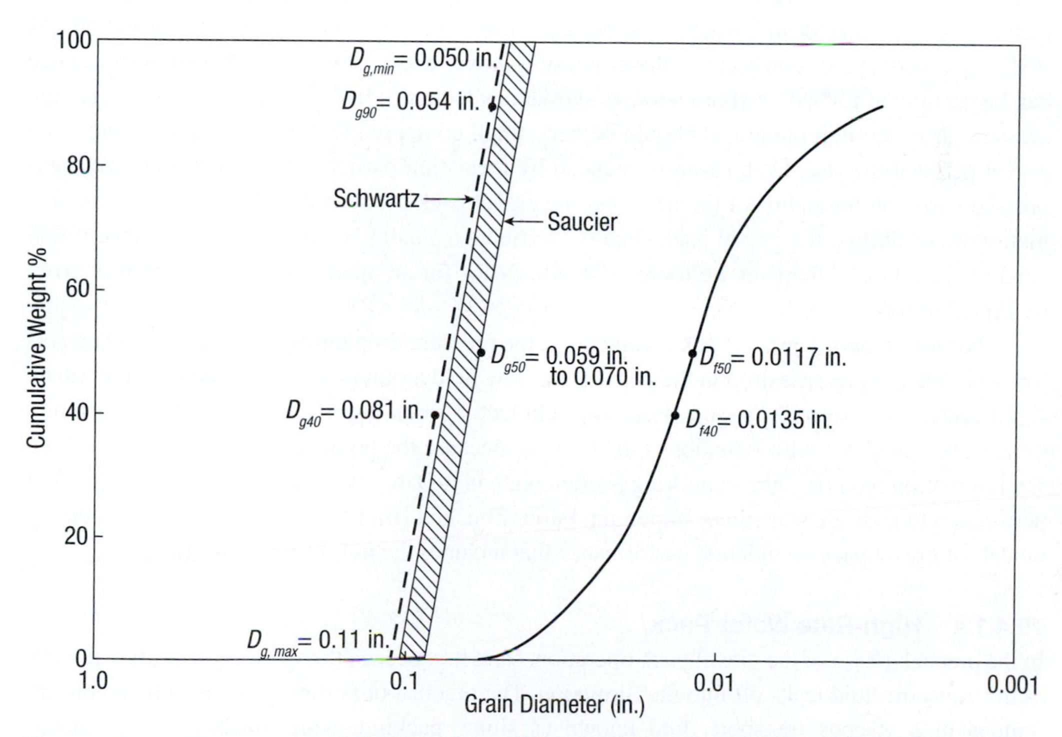

Figure 19-16 Gravel size distribution predicted by Saucier and Schwartz correlations.

From Table 19-2, the sieve sizes most closely corresponding to the maximum and minimum gravel sizes are 7 mesh and 16 mesh; an 8/16 mesh sand would probably be chosen, since 7 mesh is rarely used. The screen size should be less than 0.0469 in., so that all of the 16 mesh gravel is retained by the screen.

Saucier Correlation

From the formation grain size distribution, the mean grain size (D_{f50}) is found to be 0.0117 in. The recommended mean gravel size is (5 or 6) (0.0117 in.) = 0.059 or 0.070 in., and from Equations (19-32) and (19-33), the minimum gravel size is 0.039–0.047 in., while the maximum gravel size is 0.088–0.105 in. This range is shown as the shaded region in Figure 19-16. From Table 19-2, these grain sizes correspond to sieve sizes of 8 mesh and 16 or 18 mesh. An 8/16 mesh sand would again be chosen, with a screen size of less than 0.0469 in.

Once the gravel size has been chosen, it is important to check that the gravel used conforms closely to this size. The American Petroleum Institute (1986) recommends that a minimum of 96% of the gravel pack sand should pass the course-designated sieve and be retained on the fine-designated sieve. More detail about gravel pack sand quality control is given in API RP 58 (1986).

19.4.1.3 Productivity of Gravel-Packed Wells

The productivity of a gravel-packed well is affected by the pressure drop through the gravel pack, if this pressure drop is significant compared with the pressure drop in the formation. As with any other type of completion, the effect of a gravel pack completion on well performance can be accounted for with a skin factor, as shown in Chapter 6. In an open-hole gravel pack, the pressure drop through the gravel should be very small compared with the formation, unless the gravel permeability has been severely reduced by formation particles. If the productivity is expressed based on the radius of the liner and the gravel pack permeability is greater than the formation permeability, the gravel pack should contribute a small negative skin effect, since it will behave like a larger-diameter wellbore. The skin factor for an open-hole gravel pack was given by Equation 6-61.

For inside-casing gravel pack completions, the pressure drop through the gravel-packed perforations and the pressure drop in the converging flow field around the perforations can contribute significantly to the overall pressure drop. The skin factor for an inside-casing gravel pack completion can be calculated with Equations 6-62 to 6-70. Because the productivity of an unconsolidated sand formation requiring gravel packing is often high, non-Darcy flow effects in the gravel-packed perforation tunnels are sometimes important. Furui, Zhu, and Hill (2004) presented comprehensive models of gravel pack completion performance that includes the non-Darcy flow effects.

19.4.1.4 High-Rate Water Pack

In the prepack phase of the gravel pack operation, sand is squeezed though the perforation tunnels as the transport fluid leaks off into the formation. The practice of pumping gravel at high concentrations in a viscous transport fluid known as slurry packing, while enabling simultaneous prepacking and gravel packing phases and reducing both fluid-volume and pumping time, can lead to a nonuniform gravel pack, particularly in long, highly deviated completion intervals that are common in extended-reach offshore wells drilled from the platform location.

If a nonuniform gravel pack fails to pack the perforation tunnels, formation fines may plug the unpacked tunnels. Production logs in conventional gravel pack completions would indicate that only a few perforations are productive, usually at the bottom of the well where gravity assisted the gravel placement, but such evidence is rare because of the difficulty to log deviated offshore wells. The result is an unintended partial completion with a significant skin factor. Example 6-4 illustrates that positive skins on the order of +5 to +10 and more can result from the partial completion mechanism, which can explain high skin factors seen in buildup tests performed in gravel packed wells.

In a high-rate water pack, water is pumped at a high rate with low gravel concentration to ensure that all perforation tunnels are packed during the prepack stage and that the annulus between the casing and the gravel pack screen is fully packed during the gravel pack stage. This accounts for consistently lower skins than observed in conventional gravel packs, but skin due to formation permeability impairment from drilling and completion fluid invasion will remain. Furthermore, all gravel pack completions are subject to increasing skin over time as formation fines accumulate at the formation face or in the gravel pack itself. This mechanism is aggravated if most of the perforations become plugged with formation fines because the flow velocity in the formation is related to the well flow rate divided by the total flow area.

Example 19-9 Formation Flow Velocity

Suppose a well completed in a 100-ft thick zone produces oil at 5,000 STB/d with formation volume factor 1.1 rb/STB from an unconsolidated high-permeability formation with the gravel pack successfully packing only the lowest 20 ft of perforation cavities. What is the formation flow velocity into the productive interval? Will this exceed a critical flow velocity of 0.000095 ft/s measured in a core from the same formation?

Solution

Assuming the prepack stage provided uniform flow distribution along the 20-ft productive length, with a 0.328-ft wellbore radius the flow area is given by

$$A = 2\pi r_w h_p = 41 \text{ ft}^2. \tag{19-34}$$

In turn, the formation velocity at the completion face is given by

$$\frac{q}{A} = \frac{(5{,}000)(1.1)(5.615)}{(41)(86{,}400)} = 0.0087 \text{ ft/s}, \tag{19-35}$$

which exceeds the critical flow velocity for the formation. Assuming that the gravel pack grain distribution is effective in excluding fines production, this means that fines will accumulate at the completion face and the completion skin will increase over time. Periodic acid treatments may be required to reduce skin over the productive interval. However, acidizing will not extend the length of the productive interval limited by the plugged perforations. If the well were flowing uniformly along the entire length of 100 ft at the same rate, the formation flow velocity would be 0.0017 ft/s at the completion face. Because this would still result in formation flow that is above the critical flow velocity, the need for periodic acid treatments would remain.

19.4.1.5 Gravel Pack Evaluation

Unfocused γ-ray density logs can be used to locate the top of the gravel-packed section and to detect voids in gravel packs by measuring the density of the materials in the region of the well completion. The tools used are gamma-gamma density devices having a gamma-ray source, typically cesium-137, and a single γ-ray detector. In a gravel-packed well, the material through which the γ-rays travel should be constant except for the amount of gravel present—where the annulus is completely filled, the detector response is a minimum while void spaces yield higher count rates.

An unfocused γ-ray density log displays the regions of high γ-ray intensity in the packed zone when the gravel pack contains voids, as illustrated in Figure 19-17 (Neal and Carroll, 1985). Though a portion of the gravel-packed region produces a low count level, through a large region of the completion the γ-ray intensity is high, indicating an incomplete pack of gravel.

19.4.2 Frac-Pack Completion

In unconsolidated formations, pumping above formation parting pressure can result in gravel being packed between the cement and the formation in addition to packing the perforation tunnels. In the frac-pack literature this is called a halo. In the frac-pack completion, gravel is deliberately pumped above formation parting pressure using hydraulic fracture design techniques explained in Chapters 17 and 18. The objective is to create a high conductivity fracture plane with half length greater than the radius of near wellbore damage.

The properly designed and executed frac-pack completion both ensures sand exclusion and low completion skin. Assuming a vertical well, the geometric fracture skin, s_f, for the fracture is given by combining Equations 17-5 and 17-8, yielding

$$s_f = \ln\left[\frac{r_w\left(\dfrac{\pi}{C_{fD}} + 2\right)}{x_f}\right]. \tag{19-36}$$

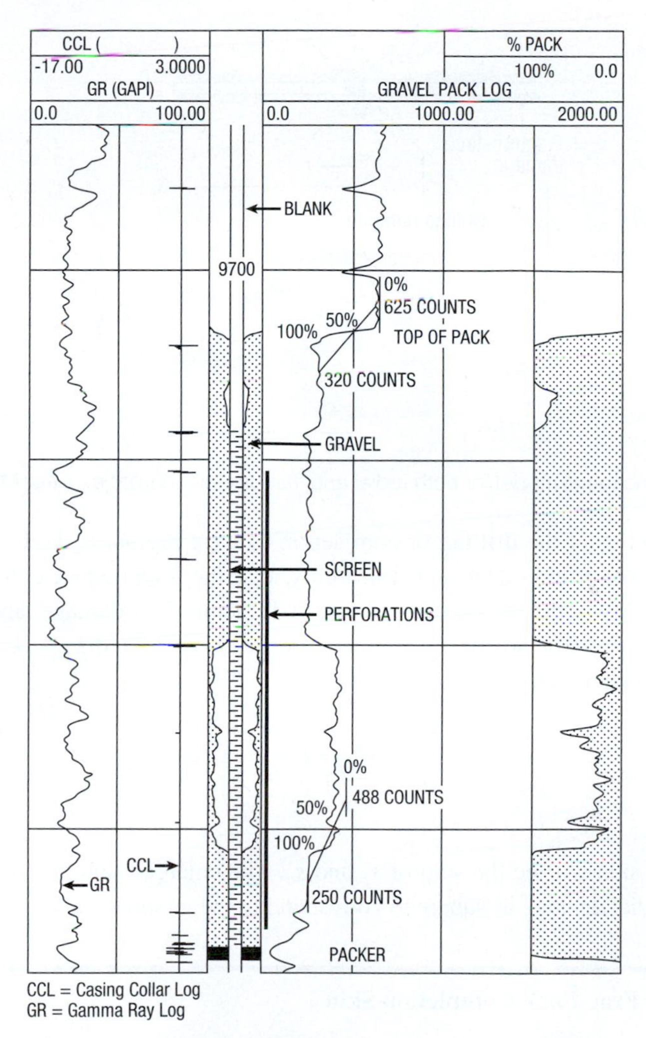

Figure 19-17 Gravel pack log showing response to voids in gravel pack. (From Neal and Carroll, 1985.)

However, often frac-pack completions have positive skin values between 5 and 10. Figure 19-18 illustrates components that may contribute to skin in a frac-pack completion. The following equation from Mathur, Ning, Marcinew, Ehlig-Economides, and Economides (1995) provides an estimate for damage skin in a frac-pack completion:

$$s_d = \frac{\pi k}{2}\left(\frac{b_1 k_s k_{sb}}{(b_1 - b_2)k_{sb} + b_2 k_s} + \frac{(x_f - b_1)k k_b}{(b_1 - b_2)k_b + b_2 k_s}\right)^{-1} - \frac{\pi b_1}{2x_f} \tag{19-37}$$

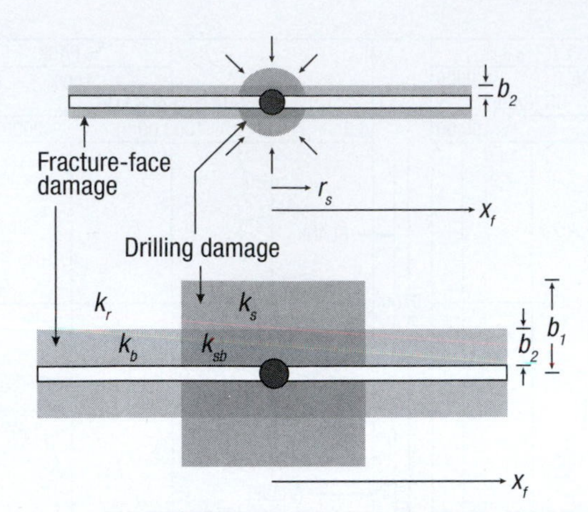

Figure 19-18 Conceptual model for both radial and fracture-face damage. (After Mathur et al., 1995.)

where b_1 is the radius of the drilling or completion damage region, b_2 is the depth of fracture face damage, k_s is the drilling damage permeability, k_b is the fracture face damage permeability, k_{sb} is the permeability in the region with drilling and fracture face damage, and k is undamaged formation permeability. For no radial wellbore damage, $k_1 = k_r$ and $k_3 = k_2$ and $b_1 = 0$, and only fracture face skin remains:

$$s_d = \frac{\pi b_2}{2x_f}\left(\frac{k_r}{k_2} - 1\right). \tag{19-38}$$

The total fracture skin will be the sum of s_f and s_d. The enlarged diagram in Figure 19-18 approximates the radial damage as square to correspond to the geometry used in Equation 19-37.

Example 19-10 Frac-Pack Completion Skin

For a frac-pack completion with $x_f = 25$ ft, and $C_{fD} = 1.6$, suppose the damage radius is 1 ft with $k_s/k_r = k_{sb}/k_r = 0.1$ for a well radius of 0.328 ft. Furthermore, suppose the fracture face skin is 2-in. thick with permeability impairment $k_b/k_r = 0.05$. What will be the total completion skin if the formation permeability is 100 md? What if the formation permeability is 1000 md?

Solution

First, applying Equation 19-36, the geometric skin, s_f, for the fracture is given by

$$s_f = \ln\left[\frac{r_w\left(\dfrac{\pi}{C_{fD}} + 2\right)}{x_f}\right] = \ln\left[\frac{0.328\left(\dfrac{\pi}{1.6} + 2\right)}{25}\right] = -2.96. \tag{19-39}$$

A damage radius of 1 ft corresponds to $b_1 = 1.328\pi^{0.5}/2 = 1.18$ ft. From Equation (19-37), for $k_r = 100$ md, $k_s = k_{sb} = 10$ md, and $k_b = 5$ md:

$$s_d = \frac{\pi(100)}{2}\left(\frac{(1.18)(10)(10)}{(1.18 - 0.0833)(10) + 0.0833(10)} + \frac{(25 - 1.18)(100)(5)}{(1.18 - 0.0833)(5) + 0.0833(10)}\right)^{-1} - \frac{\pi(1.18)}{2(25)} = 0.0087 \quad \text{(19-40)}$$

and the total skin is -2.95.

For $k_r = 1000$, $s_d = 0.0087$, and total skin is -2.95.

Example 19-10 illustrates that drilling damage has little impact as long as the fracture half length is longer than the radius of the altered permeability zone. Likewise, it may be that fracture face skin frequently contributes little to the damage skin. However, skins measured in pressure buildup tests in frac-pack wells are rarely negative and are frequently more than +5.

An explanation for additional skin may be pressure drop in the perforation tunnels. Wong, Fair, Bland, and Sherwood (2003) indicated that the difference between the total skin measured by a pressure buildup test (s_t) and the geometric skin assumed for the fracture (s_f), which he termed the mechanical skin (s_{mech}), is dominated by pressure drop in the perforation tunnels (Δp_{perf}), including both linear and turbulent flow, and with the magnitude of the latter increasing significantly if the number of perforations that are actually flowing is small. That is, given the gravel pack permeability, k_g, beta factor, β_g, and density, ρ, and the perforation tunnel length, l_{perf} (in in.), the mechanical skin is given by

$$\Delta p_{s_{\text{mech}}} \cong p_{\text{perf}} = \frac{1.138 \times 10^6 \,\mu l_{\text{perf}}}{k_g}(V_{\text{perf}}) + 1.799 \times 10^{-5}\,\rho\beta_g l_{\text{perf}}(V_{\text{perf}})^2. \quad \text{(19-41)}$$

where V_{perf} is the velocity in the perforation tunnels in ft/s. In turn, Wong et al. (2003) solves for the perforation velocity and then calculated the number of flowing perforations as

$$N = \frac{9.358 \times 10^{-3} qB}{V_{\text{perf}} A_p} \quad \text{(19-42)}$$

using the well production, q, and the perforation cross-sectional area, A_p, in square in. Finally, a perforation flow efficiency, SPF', is determined over the perforation length along the wellbore, h_p, as

$$SPF' = \frac{N}{h_p}. \quad \text{(19-43)}$$

Example 19-11 Number of Flowing Perforations

For the previous frac-pack completion with geometric skin −2.95, determine the perforation velocity assuming the same gravel properties as in Example 19-9 (k_g = 500,000 md, $\beta_g = 1.94 \times 10^4$, and l_{perf} = 2 in = 0.167 ft) and a light oil with viscosity 1 cp and density 50 lb_f/ft^3, if the measured total skin from a pressure buildup test is 7. For a downhole well production rate of 10,000 bpd and perforation radius 0.5 in., determine how many perforations are flowing, and determine the perforation flow efficiency for a perforation length of 120 ft with 12 SPF.

Solution

First, the mechanical skin is given by

$$S_{mech} = S_t - S_f = 7 - (-2.95) = 9.95. \quad (19\text{-}44)$$

Next, from Equation (19-41),

$$S_{mech} = 9.94\frac{1.138 \times 10^6\,\mu l_{perf}}{k_g}(V_{perf}) + 1.799 \times 10^{-5}\,\rho\beta_g l_{perf}(V_{perf})^2$$

$$\cong \frac{1.138 \times 10^6(1)(0.167)}{(500{,}000)}(V_{perf}) + 1.799 \times 10^{-5}(50)(1.94 \times 10^4)(0.167)(V_{perf})^2$$

$$= 0.38(V_{perf}) + 2.91(V_{perf})^2 \quad (19\text{-}45)$$

Solving for V_{perf} gives a velocity of 1.49 ft/s. For a perforation radius of 0.5 in., the perforation area is $\pi(0.5)^2 = 0.7854\ \text{in}^2$, and, from Equation 9-42, the number of perforations is

$$N = \frac{9.358 \cdot 10^{-3}(10{,}000)}{(1.49)(0.7854)} = 80. \quad (19\text{-}46)$$

Out of a total of (120)(12) = 1440 perforations, this represents a perforation flow efficiency of less than 6%.

Such startling results have led operators to presume that the gravel pack permeability is much less than the value supplied by the laboratory. Indeed, there is reason the proppant permeability may be somewhat less in the fracture due to some crushing during fracture closure, but the gravel in the perforation tunnels is not subjected to closure stress. If the gravel pack permeability is assumed to be only 50,000 md, the above example gives much more reasonable results for the perforation flow efficiency.

Table 19-3 β factor Correlations for Gravel Pack and Frac-Pack Media

Reference	Proppant Correlation	Units	
		β	k
Cooke (1973)	$\beta = \frac{3.41 \times 10^{12}}{k^{1.54}}$	1/ft	md
Ergun (1952)	$\beta = \frac{1.39 \times 10^{6}}{k^{0.5}\phi^{1.5}}$	1/ft	md
Penny and Jin (1995)	$\beta = \frac{5.19 \times 10^{11}}{k^{1.45}}$	1/ft	md
Purseli, Holditch, and Blakeley (1988)	$\beta = \frac{2.35 \times 10^{10}}{k^{1.12}}$	1/ft	md
Welling (oil) (1998)	$\beta = \frac{6.5 \times 10^{4}}{k_g^{0.996}}$	1/ft	md
Welling (gas) (1998)	$\beta = \frac{10^{7}}{k_g^{0.5}}$	1/ft	md
Martins, Milton-Taylor, and Leung (1990)	$\beta = \frac{8.32 \times 10^{9}}{k_g^{1.036}}$	1/ft	md

The previous calculation required a β factor value. Suitable correlations for the beta factors for gravel pack and frac-pack media are provided in Table 19-3.

19.4.3 High-Performance Fracturing

In remote or challenging drilling environments such as deepwater, the well is designed to deliver a very high rate in order to pay back the extremely high well costs. Sand exclusion is required because sand production cannot be tolerated in long subsea flow lines. In this case it should be a priority to ensure that the fracture area is sufficient to ensure that the formation flow velocity does not exceed the critical velocity when the well is flowed at the intended rate.

Example 19-12 Formation Flow Velocity to a High-Performance Fracture (HPF)

As in Example 19-9, suppose the well produces oil at 20,000 STB/d with formation volume factor 1.1 rb/STB from an unconsolidated high permeability formation of thickness 100 ft. What fracture half length will ensure that the formation flow velocity is less than the critical flow velocity of 0.000095 ft/s? What fracture conductivity, $k_f w$, will provide the maximum productivity for this fracture half length if the formation permeability is 100 md? 1000 md? What will be the resulting skin for each case?

Solution

For the formation flow velocity to be less than 0.000095 ft/s, the flow area must exceed the following:

$$A \geq \frac{q}{v_c} = \frac{(20{,}000)(1.1)(5.615)}{(0.000095)(86{,}400)} = 15{,}050 \text{ ft}^2 = 4hx_f. \quad \textbf{(19-47)}$$

Solving the above equation for x_f, the fracture half length must be at least 38 ft. From the principles of unified fracture design, productivity is maximized for a dimensionless conductivity $C_{fD} = 1.6$. If the formation permeability is 100 md, then

$$k_f w = (1.6)(38)(100) = 6080 \text{ md-ft} \quad \textbf{(19-48)}$$

for $k = 100$ md or 60,800 md-ft for $k = 1000$ md. If the proppant permeability is 200,000 md, the fracture width should be 0.36 in. for the 100 md reservoir or 3.6 in. for the 1000 md reservoir. In the latter case, it would be better to use a higher-permeability proppant.

From Equation 19-36,

$$s_f = \ln\left[\frac{r_w\left(\dfrac{\pi}{C_{fD}} + 2\right)}{x_f}\right] = \ln\left[\frac{0.328\left(\dfrac{\pi}{1.6} + 2\right)}{38}\right] = -3.4 \quad \textbf{(19-49)}$$

for either permeability value.

If designed according to the simple concepts in Example 19-12, the high-performance fracture should have a negative skin, and, in contrast to gravel packed wells, the skin should not increase over the life of the well. Also, unlike in gravel packed wells, the grain size distribution of the proppant is not important because no formation fines will flow at a velocity less than the critical velocity.

19.4.4 High-Performance Fractures in Deviated Production Wells

While some operators try to drill the last segment of a well that will be hydraulically fractured as close to vertical as possible, others prefer to let drillers maintain the extended reach angle used to reach the intended production location from the platform. In Chapters 17 and 18, it was assumed that a well to be fractured is either vertical or horizontal. Since, except for very shallow wells or highly overpressured formations, the maximum stress is the overburden stress, most hydraulic fractures propagate in a vertical plane. If the well trajectory is vertical, the hydraulic fracture will be aligned with the well.

In the case of a horizontal well, the fracture will align with the well as a longitudinal fracture if the well trajectory is in the direction normal to minimum stress, or will propagate a plane perpendicular to the well axis if the well is a transverse fracture if the trajectory is in the direction parallel to minimum stress. In the latter case, strategies such as cluster perforating were described in Chapter 18.

The well azimuth associated with extended reach may be planned in any arbitrary direction since the main purpose is to reach the intended production location in the shortest distance. Furthermore, the usual practice is to pump the fracture through perforations along the entire productive interval traversed by the well. The fracture results because the gravel is pumped at pressures exceeding formation parting pressure, and it may or may not employ rigorous hydraulic fracture design principles. There is a strong possibility that either multiple fracture planes will initiate (Yew, Schmidt, and Lee, 1989), or that the dominant fracture plane will turn to align normal to the minimum stress direction within a distance of 1–2 wellbore diameters, as illustrated in Figure 19-19. In that case, most of the perforations may lose connection to the fracture, resulting

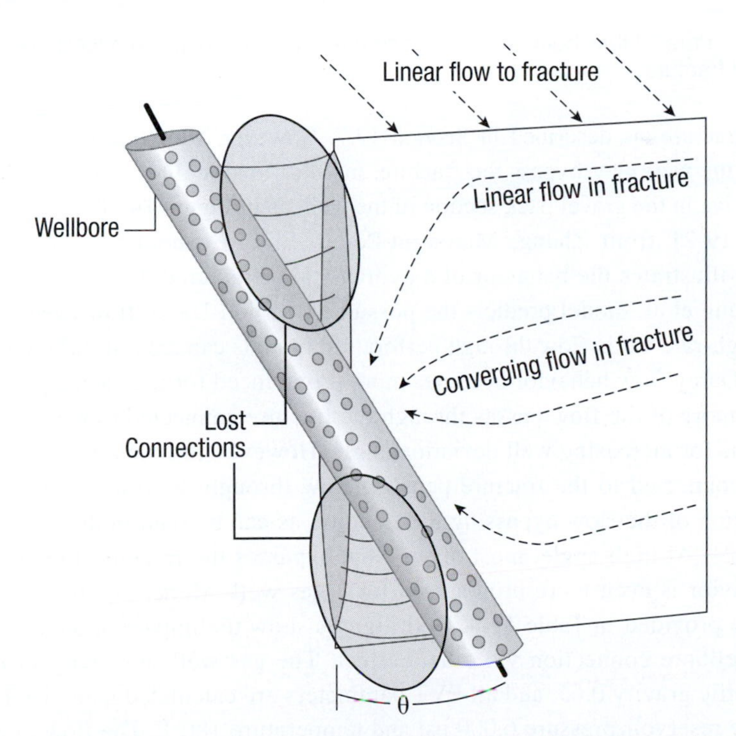

Figure 19-19 Diagram showing abandoned perforations for well trajectory not in fracture plane.

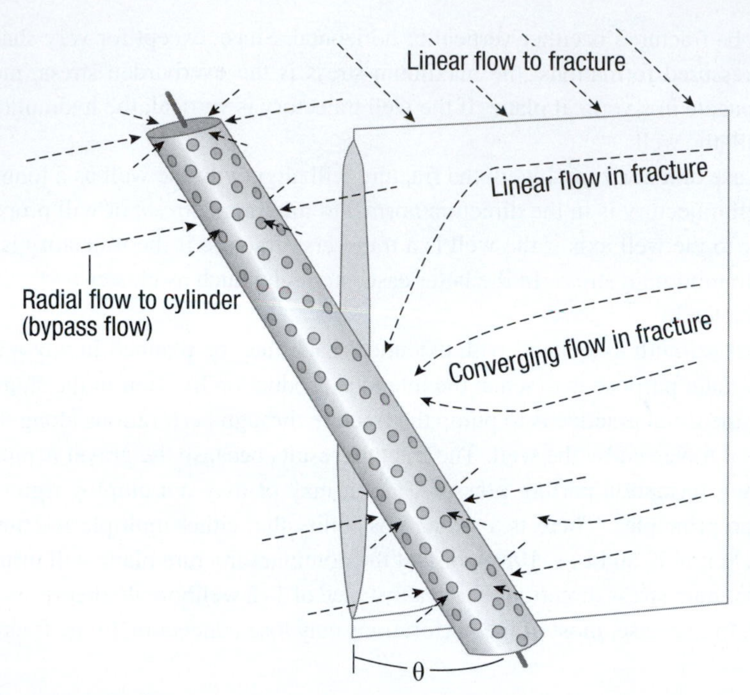

Figure 19-20 Parallel flow both to fracture and directly to gravel packed wellbore for wellbore not in plane of fracture.

in a choked fracture, as described in Section 17.7. However, when the fracture connection is poor, flow to the well may bypass the fracture, instead flowing directly to abandoned perforations that survive in the gravel pack section of the well, as in Figure 19-20.

Figure 19-21 from Zhang, Marongiu-Porcu, Ehlig-Economides, Tosic, and Economides (2010) illustrates the behavior of a hydraulically fractured deviated oil well. Interestingly, the Zhang et al. model predicts the possibility of non-Darcy flow even for an oil well because in high-rate wells flow through perforation tunnels can exhibit turbulent flow behavior. The non-Darcy flow behavior becomes more pronounced for increasing gravel pack skin, S_{gp}, because more of the flow passes through the fracture connected to a decreasing number of perforations for increasing well deviation angle. However, turbulent flow through perforation tunnels connected to the fracture penalize flow through the fracture, leading to an increasing fraction of the flow bypassing the fracture, as can be seen in the graph on the right in Figure 19-21. At high angle, most of the flow bypasses the fracture. Figure 19-22 shows that this behavior is even more pronounced for a gas well. Model inputs for Figures 19-21 and 19-22 are provided in Table 19-4. Both figures show the impact of an additional 1 ft of fracture to wellbore connection via a halo effect. The gas well sensitivity in Figure 19-22 uses gas-specific gravity 0.65, and all PVT parameters are calculated using PVT correlations under average reservoir pressure 6,000 psi and temperature 180°F. The flow rate used for all simulations is 100 MMscf/D.

Table 19-4 Reservoir and Fracture Properties for Figures 19-22 and 19-23

Wellbore radius	0.354 ft
Reservoir thickness	100 ft
Porosity	0.28
Reservoir horizontal permeability	202 md
Reservoir vertical permeability	30 md
Casing ID	0.5 ft
Perforation density and diameter	18 SPF/0.78 in.
Perforation phasing	140/20°
Fracture half length	27 ft
Fracture height	100 ft
Fracture width	1.13 inch
Fracture permeability	170,000 md
Perforation tunnel permeability	200,000 md

19.4.5 Perforating Strategy for High-Performance Fractures

To maximize productivity from high-performance fracturing, the connection between the fracture and the wellbore must be excellent. Some authors have advocated orienting perforations with 180° phasing in the direction normal to minimum stress to improve fracture to wellbore connectivity. Otherwise, flow from the fracture may have to pass through the halo to reach perforation tunnels, providing a flow restriction and increased skin. Also, big hole charges are preferred to deep penetrating perforations for frac-pack applications because the fracture does not follow the perforation cavity, and the larger the perforation cross-sectional area, the better.

In deviated wells the number of perforations connecting the fracture to the wellbore can be estimated geometrically, as in Zhang et al. (2010), and this number may be increased somewhat via connection through the halo. However, the actual number of flowing perforations may be much greater considering bypass flow directly to the gravel pack section of the well as in Figures 19-21 and 19-22. Nonetheless, flow bypassing the fracture will likely encounter near wellbore damage the fracture is designed to bypass. The net result is lower well productivity, in direct opposition to what high-performance fracturing completions are intended to provide.

The model by Zhang et al. (2010) presumes that one main fracture forms along the deviated wellbore. However, another possibility investigated by Yew et al. (1989) is formation of multiple *s*-shaped minifractures. Too many minifractures could result in failure to bypass near wellbore

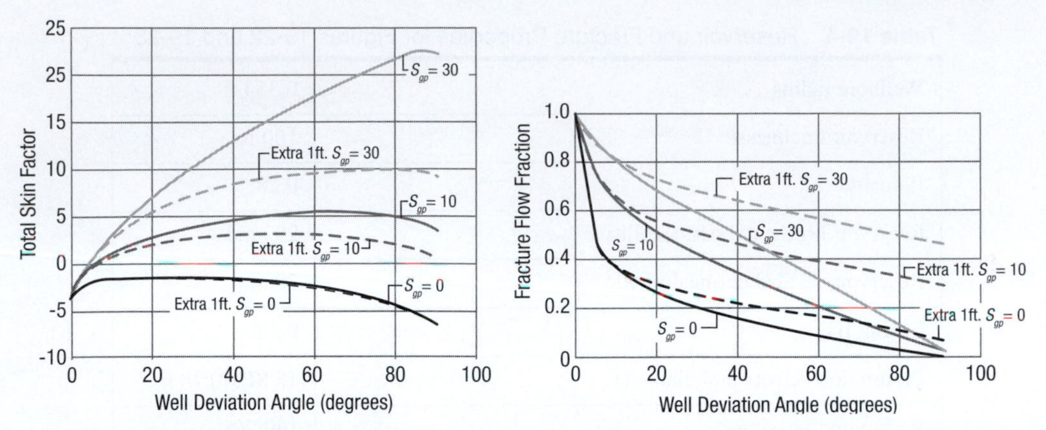

Figure 19-21 Total skin factor (left) and fracture flow fraction (right) versus well deviation angle for oil well. (From Zhang et al. 2010.)

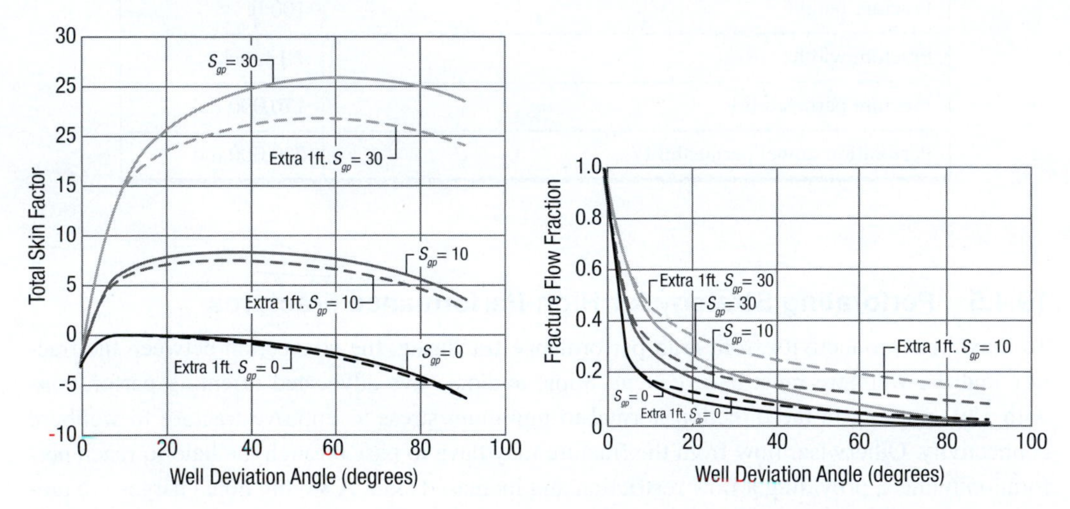

Figure 19-22 Total skin factor (left) and fracture flow fraction (right) versus well deviation angle for gas well. (From Zhang et al., 2010.)

damage and/or insufficient conductivity to compete with simple radial flow to the wellbore. Veeken, Davies, and Walters (1989) emphasized the importance of drilling vertically through the productive reservoir interval when the intent is to hydraulically fracture the well.

19.5 Completion Failure Avoidance

Common completion failure mechanisms include sand production, screen erosion, annular pack destabilization, screen collapse, and compaction. We have seen already that high-performance fracturing completions should avoid sand production altogether, while gravel pack completions may restrict sand production to only very fine particles. Chapter 10 from Penberthy and Shaughnessy (1992) shows gravel pack performance as success percent as a function of cumulative fluid production comparing open-hole and cased-hole gravel packs, prepacked and nonprepacked completions, wells with or without production tubing, perforated flow areas, and new (initial) gravel packs with old (worked-over) gravel packs. These studies showed success percent ranging from 20% to more than 95% for production of 1 million STB. However, these studies emphasized much lower well rates than are required from high-performance completions, especially for high-cost deepwater wells. For these wells, studies by Tiner et al. (1996), Norman (2003), and Keck et al. (2005) are more meaningful.

Screen erosion is a key reason for offshore completion failures, and high-rate completions in deepwater are frequently equipped with a permanent downhole pressure gauge for continuous surveillance. Each time the well is shut in for operational reasons, pressure buildup data are recorded by the permanent gauge, enabling relatively frequent assessment of the mechanical skin factor. Increasing skin may be a sign that flow velocity (flux) is concentrating to relatively few perforations. Screen erosion is typically limited to a short segment of the screen. High flow rate with negligible, but nonzero particulate over a short flow area results in sufficiently high velocity to erode the screen. Once eroded, the packed gravel may fall into the rat hole and plug off the well or flow toward the surface and erode the subsea tree or flow lines.

Nonuniform flow at a high rate can also lead to high pressure gradients that cause annular pack destabilization, screen collapse, or compaction.

This chapter has shown that sand exclusion is needed when the well rate without it is too low to be economical. We have also seen that the skin resulting from a high-rate water pack may be similar to what is achieved with a frac-pack. While reducing skin clearly improves well productivity, the overriding objective when designing completions for high-rate wells is prevention of loss of the well, not minimizing skin.

References

1. American Petroleum Institute, "Recommended Practices for Testing Sand Used in Gravel Packing Operations," API Recommended Practice 58 (RP 58), March 1986.
2. Arii, H., Morita, N., Ito, Y., and Takano, E., "Sand-Arch Strength Under Fluid Flow With and Without Capillary Pressure," SPE Paper 95812, presented at the 2005 SPE Annual Technical Conference and Exhibition in Dallas, TX, October, 9–12, 2005.
3. Cooke, C.E., Jr., "Conductivity of Fracture Proppants in Multiple Layers," *JPT,* 1101–1107 (September 1973).
4. Economides, M.J., and Nolte, K.G., *Reservoir Stimulation,* 3rd ed., Wiley, Hoboken, NJ, 2000.
5. Ergun, S., "Fluid Flow through Packed Columns." *Chemical Engineering Progress,* Vol. 48, No 2, 89–94, 1952.
6. Furui, K., Zhu, D., and Hill, A.D., "A New Skin Factor Model for Gravel-Packed Wells," SPE Paper 90433, presented at the SPE ATCE, September 26–29, 2004.

7. Ghalambor, A., Ali, S., and Norman, W.D., *Frac Packing Handbook,* Society of Petroleum Engineers, 2009.

8. Gruesbeck, C., Salathiel, W.M., and Echols, E.E., "Design of Gravel Packs in Deviated Wellbores," *JPT,* **31:** 109–115 (January 1979).

9. Keck, R.G., Colbert, J.R., and Hardham, W.D., "The Application of Flux-Based Sand-Control Guidelines to the Na Kika Deepwater Fields," SPE Paper 95294, presented at the Annual Technical Conference and Exhibition, Dallas, TX, October 9–12, 2005.

10. Kelessidis, V.C., Maglione, R., and Mitchel, R.F., *Drilling Hydraulics in Fundamentals of Drilling Engineering,* SPE Textbook Series No. 12, R.F. Mitchel and S.Z. Miska eds., Society of Petroleum Engineers, 2011.

11. Mathur, A.K., Ning, X., Marcinew, R.B., Ehlig-Economides, C.A., and Economides, M.J., "Hydraulic Fracture Stimulation of Highly Permeability Formations: The Effect of Critical Fracture Parameters on Oilwell Production and Pressure," SPE Paper 30652, presented at the Annual Technical Conference and Exhibition in Dallas, TX, October 1995.

12. Martins, J.P., Milton-Tayler, D., and Leung, H.K., "The Effects of Non-Darcy Flow in Propped Hydraulic Fractures," SPE 20709, presented at the SPE Annual Technical Conference and Exhibition in New Orleans, LA, September 23–26, 1990.

13. Mitchell, R.F., and Miska, S.Z., *Fundamentals of Drilling Engineering,* SPE Textbook Series No. 12, 2011.

14. Morita, N., Whitfill, D.L., Massie, I., and Knudsen, T.W., "Realistic Sand-Production Prediction: Numerical Approach," *SPEPE* (February 1989).

15. Neal, M.R., and Carroll, J.F., "A Quantitative Approach to Gravel Pack Evaluation," *JPT,* 1035–1040 (June 1985).

16. Norman, D., "The Frac-Pack Completion: Why has it Become the Standard Strategy for Sand Control," SPE Paper 101511, presented as an SPE Distinguished Lecture, 2003–2004.

17. Palmer, I., and McClennan, J., "Cavity Like Completions in Weak SandsPreferred Upstream Management Practices," DOE Report DE-FC26-02BC15275, April 2004.

18. Penberthy, W.L., and Shaughnessy, C.M., *Sand Control,* SPE Series on Special Topics Vol. 1, Henry L. Doherty Series, Society of Petroleum Engineers, 1992.

19. Penny, G.S., and Jin, L., "The Development of Laboratory Correlations Showing the Impact of Multiphase Flow, Fluid, and Proppant Selection Upon Gas Well Productivity," SPE 30494 presented at the 1995 SPE Technical Conference and Exhibition, Dallas, TX, October 22–25, 1995.

20. Perry, J.H., *Chemical Engineer's Handbook,* 4th ed., McGraw-Hill, New York, 1963.

21. Pursell, D.A., Holditch, S.A., and Blakeley, D.M., "Laboratory Investigation of Inertial Flow in High-Strength Fracture Proppants," paper SPE 18319 presented at the SPE Annual

Technical Conference and Exhibition of the Society of Petroleum Engineers, Houston, October 2–5, 1988.

22. Saucier, R.J., "Considerations in Gravel Pack Design," *JPT,* 205–212 (February 1974).

23. Schechter, R.S., *Oil Well Stimulation,* Prentice Hall, Englewood Cliffs, NJ, 1992.

24. Scheuerman, R.F., "New Look at Gravel Pack Carrier Fluid," SPE Paper 12476, presented at the Seventh Formation Damage Control Symposium, Bakersfield, CA, 1984.

25. Schwartz, D.H., "Successful Sand Control Design for High-Rate Oil and Water Wells," *JPT,* 1193–1198 (September 1969).

26. Shryock, S.G., "Gravel-Packing Studies in Full-Scale Deviated Model Wellbore," *JPT,* 603–609 (March 1983).

27. Suman, G.O., Jr., Ellis, R.C., and Snyder, R.E., *Sand Control Handbook,* 2nd ed., Gulf Publishing Co., Houston, TX, 1983.

28. Tiner, R.L., Ely, J.W, and Schraufnagel, R.S., "Frac Packs–State of the Art," SPE Paper 35456, presented at the 71st Annual Technical Conference and Exhibition, Denver, CO, 1996.

29. Veeken, C.A.M., Davies, D.R., and Walters, J.V., "Limited Communication between Hydraulic Fracture and (Deviated) Wellbore," SPE Paper 18982, 1989.

30. Weingarten, J.S., and Perkins, T.K., "Prediction of Sand Production in Gas Wells: Methods and Gulf of Mexico Case Studies," *JPT* (July 1995).

31. Welling, R.W.F., "Conventional High Rate Well Completions: Limitations of Frac-Pack, High Rate Water," SPE Paper 39475, February 1998.

32. Wong, G.K., Fair, P.S., Bland, K.F., and Sherwood, R.S., "Balancing Act: Gulf of Mexico Sand Control Completions, Peak Rate Versus Risk of Sand Control Failure," SPE Paper 84497, presented at the SPE Annual Technical Conference and Exhibition, Denver, CO, October 5–8, 2003.

33. Yew, C.H., Schmidt, J.H., and Li, Y., "On Fracture Design of Deviated Wells," SPE Paper 19722, presented at the SPE Annual Technical Conference and Exhibition, San Antonio, TX, October 8–11, 1989.

34. Zhang, L., and Dusseault, M.B., "Sand Production Simulation in Heavy Oil Reservoirs," SPE Paper 64747, presented at the International Oil and Gas Conference and Exhibition in Beijing, China, November 7–10, 2000.

35. Zhang, Y., Marongiu-Porcu, M., Ehlig-Economides, C.A., Tosic, S., and Economides, M.J., "Comprehensive Model for Flow Behavior of High-Performance Fracture Completions," *SPE Production and Operations,* 484–497, November 2010.

Problems

19-1 An oil well produces undersaturated oil (as in Appendix A) with the following well performance behavior:

$$q = 0.5(\overline{p} - p_{wf}) \text{ IPR}$$
$$q = 0.85(p_{wf} - p_{tf}) \text{ VFP.}$$

At what rate and flowing pressure does the well flow for average pressure of 3200 psi and tubing head pressure of 300 psi? If core measurements have determined that the formation critical flow rate for sand production is 0.0005 ft/sec, do you recommend sand management or sand exclusion?

19-2 For a formation thickness 50 ft and $r_w = 0.328$ ft, above what rate would sand production be a risk for formation critical flow rate values of 0.03 ft/sec, 0.0005 ft/sec, and 0.000095 ft/sec?

19-3 For the reservoir in Example 19-2 with additional fluid and formation properties as in Appendix A, assume that the reservoir permeability is 82 md and the damage permeability is 8.2 md. For drawdown of 2000 psi, use Figure 19-6 to determine at what rate shear failure will occur for the slim cavity case and for the enlarged cavity case.

19-4 Determine the optimal gravel size for the Gulf Coast formation sand grain size distribution given in Figure 19-15, using both the Schwartz and Saucier correlations.

19-5 The undersaturated oil described in Appendix A is being produced from a 50-ft-thick, 800 md unconsolidated sand reservoir with a drainage radius of 1490 ft. The well is completed with an inside-casing gravel pack with 20/40 mesh gravel. The casing is 7 in. (6 in. I.D.), and the drilled hole is 10 in. in diameter. Plot p_{wf} versus q for a range of p_{wf} from 2000 to 5000 psi, assuming 0.5-in.-diameter perforations and a shot density of (a) 1 shot/ft; (b) 2 shot/ft; (c) 4 shot/ft.

19-6 Assume that the grain distribution for a sand plot as a straight line on the usually semi-log of percent sand retained versus grain size. Show the relationship between the predicted mean particle sizes of the gravel from the Schwartz and Saucier correlations (i.e., compare D_{g50} from the two correlations).

19-7 Repeat the calculation for Example 19-10 using the same input parameters, except change x_f to 4 ft and damage radius to 5 ft.

19-8 Repeat the calculation for Example 19-11, assuming $k_g = 50{,}000$ md.

19-9 For the cases shown with dashed curves in Figure 19-21, graph dimensionless productivity versus well deviation angle for 80-acre well spacing. Repeat for the cases shown with dashed curves in Figure 19-22.

APPENDIX A

Well in an Undersaturated Oil Reservoir

$k_H = 8.2$ md

$k_v = 0.9$ md

$h = 53$ ft

$p_i = 5651$ psi

$p_b = 1697$ psi

$T = 220°F$

$c_o = 1.4 \times 10^{-5}$ psi^{-1}

$c_w = 3 \times 10^{-6}$ psi^{-1}

$c_f = 2.8 \times 10^{-6}$ psi^{-1}

$c_t = 1.29 \times 10^{-5}$ psi^{-1}

$R_s = 250$ SCF/STB

$\phi = 0.19$

$S_w = 0.34$

$\gamma_o = 28°API$

$\gamma_g = 0.71$

$r_w = 0.328$ ft (7 7/8 wellbore)

$p_{sep} = 100$ psi

$T_{sep} = 100°F$

At the Bubble Point

$$\mu = 1.03 \text{ cp}$$

$$B_o = 1.2 \text{ res bbl/STB}$$

At Initial Conditions

$$\mu = 1.72 \text{ cp}$$

$$B_o = 1.17 \text{ res bbl/STB}$$

APPENDIX B

Well in a Two-Phase Reservoir

$k_H = 13$ md

$k_v = 1$ md

$h = 115$ ft

$p_i = 4336$ psi

$p_b = 4336$ psi

$T = 220°F$

$c_o = 1.9 \times 10^{-5}$ psi^{-1}

$c_w = 3 \times 10^{-6}$ psi^{-1}

$c_f = 3.1 \times 10^{-6}$ psi^{-1}

$c_t = 1.25 \times 10^{-5}$ psi^{-1}

$R_s = 800$ SCF/STB

$\phi = 0.21$

$S_w = 0.3$

$\gamma_o = 32°API$

$\gamma_g = 0.71$

$r_w = 0.406$ ft

$p_{sep} = 100$ psi

$T_{sep} = 100°F$

At Bubble Point

$$\mu = 0.45 \text{ cp}$$

$$B_o = 1.46 \text{ res bbl/STB}$$

$$\rho = 38.9 \text{ lb/ft}^3$$

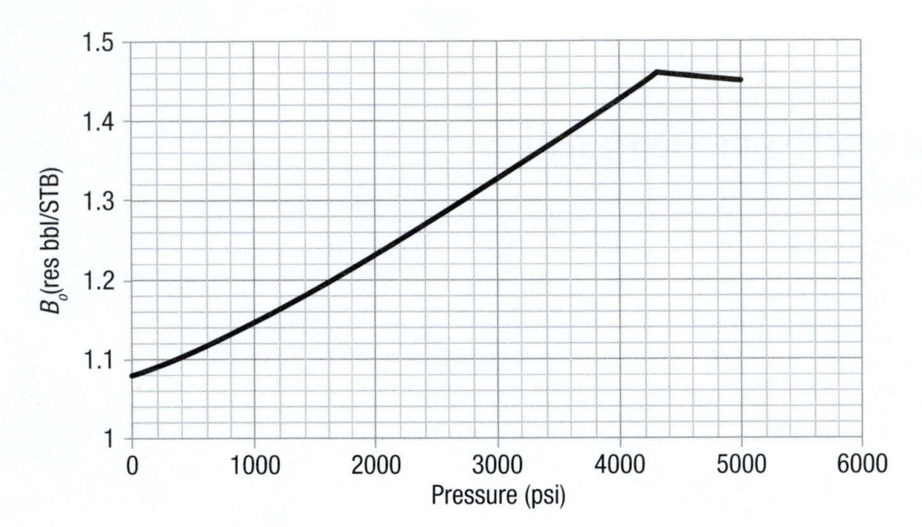

Figure B-1a

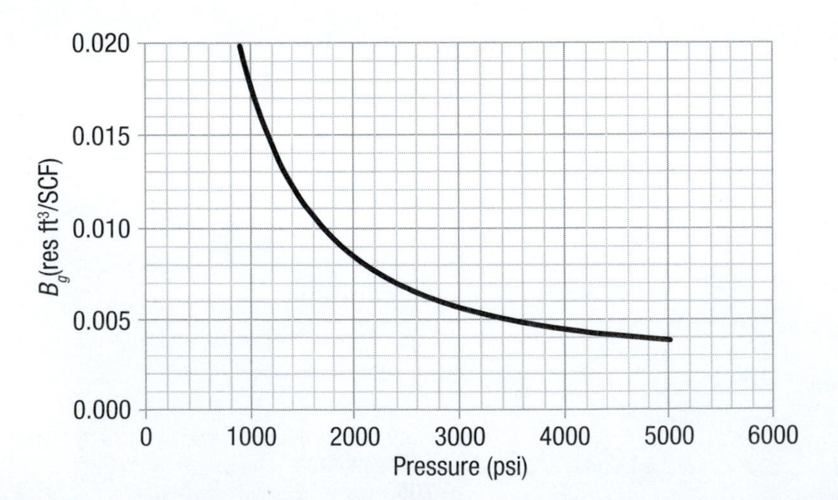

Figure B-1b

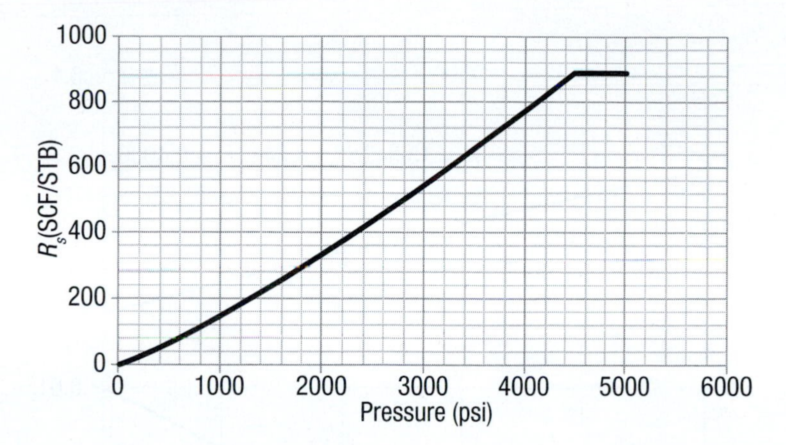

Figure B-1c

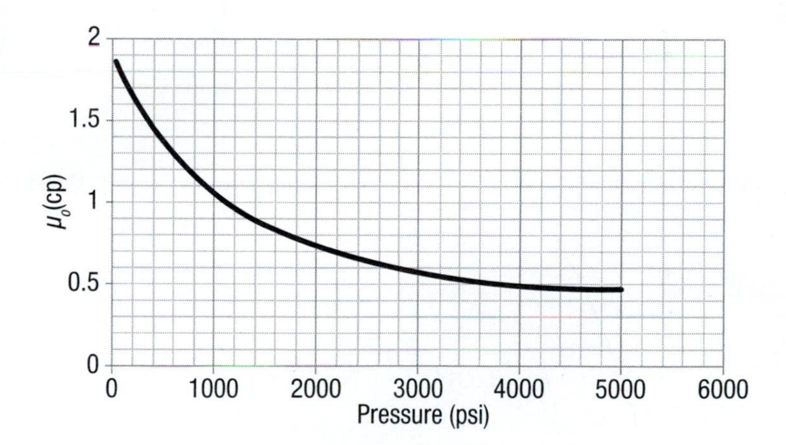

Figure B-1d

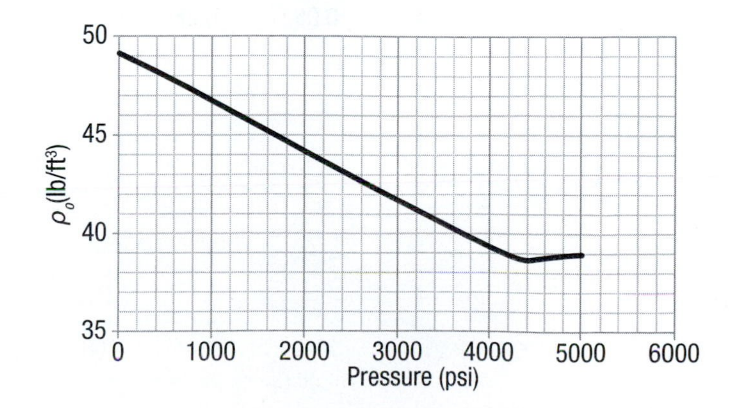

Figure B-1e

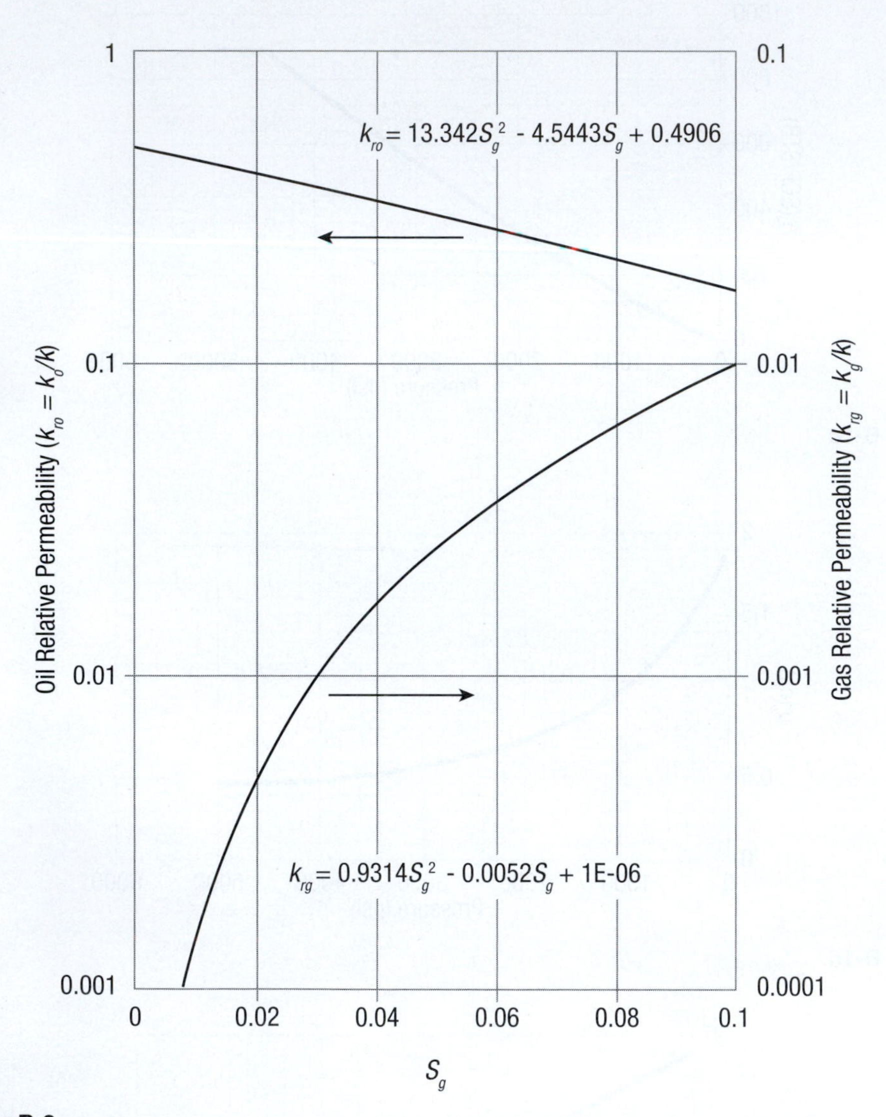

1
0.1
0.01
0.001
0.1
0.01
0.001
0.0001
0
0.02
0.04
0.06
0.08
0.1
$k_{ro} = 13.342S_g^2 - 4.5443S_g + 0.4906$
$k_{rg} = 0.9314S_g^2 - 0.0052S_g + 1E-06$
Oil Relative Permeability ($k_{ro} = k_o/k$)
Gas Relative Permeability ($k_{rg} = k_g/k$)
S_g

Figure B-2

Appendix C

Well in a Natural Gas Reservoir

Composition of gas and properties:

$C_1(0.875)$	$C_2(0.083)$	$C_3(0.021)$
$i-C_4(0.006)$	$n-C_4(0.002)$	$i-C_5(0.003)$
$n-C_5(0.008)$	$n-C_6(0.001)$	$C_7+(0.001)$
$T_{pc} = 375°\text{R}$	$p_{pc} = 671$ psi	$\gamma_g = 0.65$

Well and Reservoir Variables

$$T = 180°\text{F} = 640°\text{R}$$
$$p_i = 4613 \text{ psi}$$
$$Z_i = 0.96$$
$$\mu_i = 0.0249 \text{ cp}$$
$$B_{gi} = 0.00376 \text{ cft/Scf}$$
$$h = 78 \text{ ft}$$
$$S_w = 0.27$$
$$S_g = 0.73$$
$$\phi = 0.14$$
$$r_w = 0.328 \text{ ft (7 7/8-in. well)}$$
$$k = 0.17 \text{ md}$$

Index

Note: Page numbers with "f" indicate figures; those with "t" indicate tables.

A

D

G

H

I

J

K

L

M

N

O

P

R

S

T

U

X

Z

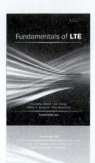

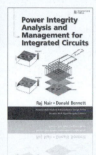